METEOROLOGY TODAY

AN INTRODUCTION TO WEATHER, CLIMATE, AND THE ENVIRONMENT

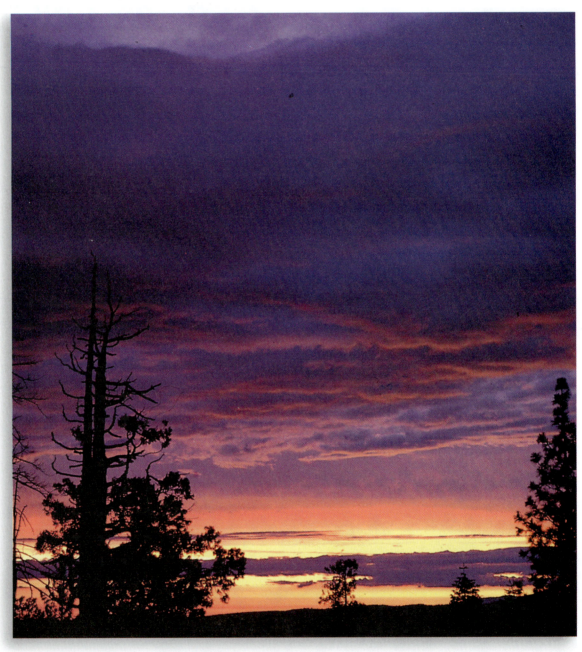

(Photo by Pat Badalmamente Zawadzki.)

SIXTH EDITION

METEOROLOGY TODAY

AN INTRODUCTION TO WEATHER, CLIMATE, AND THE ENVIRONMENT

C. DONALD AHRENS

Modesto Junior College

Brooks/Cole
Thomson Learning™

Pacific Grove • Albany • Belmont • Boston • Cincinnati • Johannesburg • London • Madrid •
Melbourne • Mexico City • New York • Scottsdale • Singapore • Tokyo • Toronto

Sponsoring Editor:	*Nina Horne*
Project Development Editor:	*Marie Carigma-Sambilay*
Marketing Team:	*Tami Cueny, Kelly Fielding, and Dena Donnelly*
Editorial Assistant:	*John-Paul Ramin*
Production Coordinator:	*Mary Anne Shahidi*
Production Service:	*Janet Bollow Associates*
Manuscript Editor:	*Stuart Kenter*
Permissions Editor:	*Mary Kay Hancharick*
Interior Design:	*Janet Bollow*
Cover Design:	*Irene Morris*
Cover Photo	*Fred Hirschmann*
Interior Illustration:	*Alexander Productions, Myrna Vladic, Richard Sheppard, George Kelvin, Carl Brown, Sue Sellars, Folium, House of Graphics, Joe Medeiros*
Photo Researcher:	*Stuart Kenter Associates*
Print Buyer:	*Vena Dyer*
Typesetting:	*Alphatype*
Cover Printing:	*R. R. Donnelley–Willard*
Printing and Binding:	*R. R. Donnelley–Willard*

LIBRARY OF CONGRESS CATALOGING-IN-PUBLICATION DATA
Ahrens, C. Donald.
    Meteorology today : an introduction to weather, climate, and the
environment / C. Donald Ahrens. — 6th ed.
        p.   cm.
    Includes bibliographical references and index.
    ISBN 0-534-37201-5 (alk. paper)
    1. Meteorology.   I. Title.
QC861.2.A3   1999   99-21123
551.5—dc21   CIP

C O N T E N T S

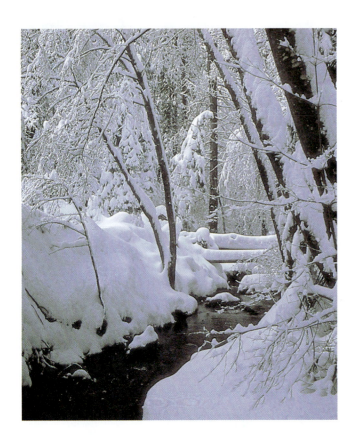

CHAPTER 4

LIGHT, COLOR, AND ATMOSPHERIC OPTICS   80

CHAPTER 3

SEASONAL AND DAILY TEMPERATURES   52

CHAPTER 5

ATMOSPHERIC MOISTURE   104

CHAPTER 6

CONDENSATION: DEW, FOG, AND CLOUDS   124

CHAPTER 7

STABILITY AND CLOUD DEVELOPMENT   158

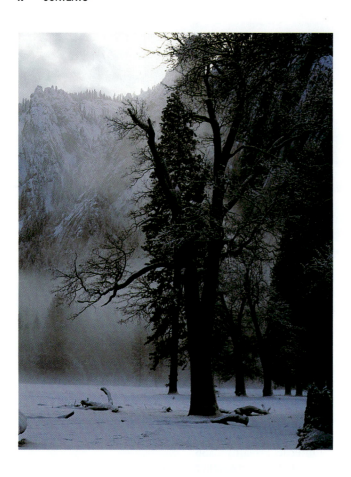

CHAPTER 13

MIDDLE LATITUDE CYCLONES   325

CHAPTER 14

WEATHER FORECASTING   350

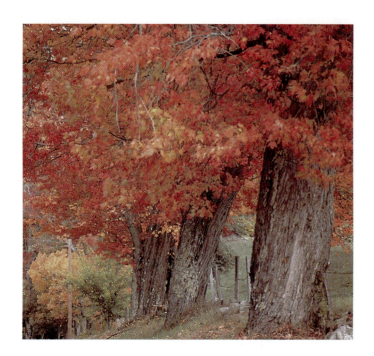

The world is an ever-changing picture of naturally occurring events. From drought and famine to devastating floods, some of the greatest challenges we face come in the form of natural disasters created by weather. Yet, dealing with weather and climate is an inevitable part of our lives. Sometimes it is as small as deciding what to wear for the day or how to plan a vacation. But it can also have life-shattering consequences, especially for those who are victims to a hurricane or tornado.

In recent years, weather and climate have become front page news from such environmental issues as greenhouse warming and ozone depletion in the stratosphere to the global weather influences of El Niño. The dynamic nature of the atmosphere seems to demand our attention and understanding more these days than ever before. Almost daily, there are newspaper articles describing some weather event or impending climate change. For this reason, and the fact that weather influences our daily lives in so many ways, interest in meteorology (the study of the atmosphere) has been growing. This rapidly developing and popular science is giving us more information about the workings of the atmosphere than ever before. Although the atmosphere will always provide challenges for us, as research and technology advance, our ability to understand our atmosphere improves, as well. The information available to you in this book, therefore, is intended to aid in your own personal understanding and appreciation of our earth's dynamic atmosphere.

ABOUT THIS BOOK

Meteorology Today is written for college-level students taking an introductory course on the atmospheric environment. The main purpose of the text is to convey meteorological concepts in a visual and practical manner, while simultaneously providing students with a comprehensive background in basic meteorology. This sixth edition includes up-to-date information on important topics, such as El Niño, ozone depletion, and global warming. Discussion of weather events, such as the destruction to Honduras wrought by Hurricane Mitch, the devastating floods of 1997 in California, and the horrific fires in Florida during 1998 (brought on by an extensive dry spell) also are included. As was the case with previous editions, no special prerequisites are necessary.

Written expressly for the student, this book emphasizes the understanding and application of meteorological principles. The text encourages watching the weather so that it becomes "alive," allowing readers to immediately apply textbook material to the world around them. To assist with this endeavor, a color Cloud Chart appears at the beginning of the text. The Cloud Chart can be separated from the book and used as a learning tool any place where one chooses to observe the sky. To strengthen points and clarify concepts, illustrations are rendered in full color throughout. Color photographs were carefully selected to illustrate features, stimulate interest, and show how exciting the study of weather can be.

This edition, organized into nineteen chapters, is designed to provide maximum flexibility to instructors of atmospheric science courses. Thus, chapters can be covered in any desired order. For example, Chapter 4, "Atmospheric Optics," is self-contained and can be covered later if so desired. Instructors, then, are able to tailor this text to their particular needs. This book basically follows a traditional approach. After an introductory chapter on the composition, origin, and structure of the atmosphere, it then covers energy, temperature, moisture, precipitation, and winds. Then come chapters that deal with air masses and middle-latitude cyclones. Weather prediction and severe storms are next. A chapter on air pollution is followed by a chapter on global climate. The final chapter deals with climate change.

Each chapter contains at least two Focus sections, which expand on material in the main text or explore a subject closely related to what is being discussed. Focus sections fall into one of three distinct categories: observations, special topics, and advanced topics. Some include material that is not always found in introductory meteorology textbooks, subjects such as the history of meteorology, aviation weather, and polar lows. Others help bridge theory and practice. Focus sections new to this edition include "The Changing Seasons and How We feel—It Could Be SAD," and "Florida's Fires and the Sea Breeze." Quantitative discussions of important equations, such as the geostrophic wind equation and the hydrostatic equation, are found in Focus sections on advanced topics.

Set apart as "Weather Watch" features in each chapter is weather information that may not be commonly known, yet pertains to the topic under discussion. Designed to bring the reader into the text, most of these weather highlights relate to some interesting weather fact or astonishing event.

Each chapter incorporates other effective learning aids:

- A major topic outline begins each chapter
- Interesting introductory pieces draw the reader naturally into the main text
- Important terms are boldfaced, with their definitions appearing in the glossary or in the text
- Key phrases are italicized
- English equivalents of metric units are immediately provided in parentheses
- A brief review of the main points is placed toward the middle of most chapters
- Summaries at the end of each chapter review the chapter's main ideas
- A list of key terms following each chapter allows students to review and reinforce their knowledge of the chief concepts they have encountered
- Questions for Review act to check how well students assimilate the material
- Questions for Thought require students to synthesize learned concepts for deeper understanding
- Problems and Exercises require mathematical calculations that provide a technical challenge to the student
- Questions for Exploration encourage students to integrate the text with the Blue Skies CD-ROM

Eleven appendices conclude the book, including one that is new to this edition—the adiabatic chart. The instant weather forecast chart, Appendix E, should especially interest students. In addition, at the end of the book, a compilation of supplementary material is presented, as is an extensive glossary.

SIXTH EDITION CHANGES

To help visualize how exciting meteorology can be, over 60 new color photographs and more than 100 new or revised color illustrations have been added to this edition. To complement the photographs and new art, the sixth edition of *Meteorology Today* has been extensively updated and revised to reflect the changing nature of the field. Chapter 1 still serves as a broad overview of the atmosphere, but the more advanced material on stratospheric ozone and the ozone hole has been placed in the chapter on air pollution (Chapter 17) next to the material on tropospheric ozone.

In Chapter 2, the section on temperature scales now appears in the main body of the text. A new Focus section, "Radiative Equilibrium and Forcing—The Ins and Outs," is in this chapter along with additional information on the atmospheric greenhouse effect. New material on atmospheric optics has been added to Chapter 4. Moreover, Chapter 5, "Atmospheric Moisture," has been slightly reorganized so that it flows better. Chapter 6 now contains the latest on METAR observations, and the Focus section on measuring cloud ceilings has been thoroughly updated and revised to include ASOS information.

The chapter on precipitation (Chapter 8) has been revised with added material on precipitation processes, precipitation types, and Doppler radar images. The section on isobaric charts has been rewritten in Chapter 9, "The Atmosphere in Motion: Air Pressure, Forces, and Winds." The chapter on small-scale and local winds (Chapter 10) now contains a Focus section on the Florida Fires of 1998 as well as a revised section on monsoon winds. In Chapter 11 (Wind: Global Systems), the section on El Niño and the Southern Oscillation has been completely updated and revised to include the latest major El Niño event of 1997–1998.

In Chapter 13, "Middle Latitude Cyclones," the section on developing lows has been restructured and rewritten for clarity. The chapter on weather forecasting (Chapter 14) has been reorganized and updated to include the modernization of the National Weather Service with new information on algorithms, meteograms, and soundings. Chapter 15, "Thunderstorms and Tornadoes," has been significantly revised with new art, photos, and Doppler radar images. Moreover, new to this chapter is a Focus section on Red Sprites and Blue Jets, as well as added information on microbursts, supercell storms, mesoscale convective complexes, and thunderstorm rotation.

Chapter 17, "Air Pollution," now contains information on stratospheric ozone and a Focus section on the ozone hole over Antarctica. The chapter on global climate (which is now Chapter 18) contains many new photos to illustrate the vegetation found in the various climatic regions. The chapter on climate change (Chapter 19) has been revised to include the latest material on global warming. Among the additions to this chapter is a Focus section on the ocean conveyor belt and climate change, and the latest assessment from the Intergovernmental Panel on Climate Change (IPCC).

ACKNOWLEDGMENTS

A special thank you to the many individuals who have contributed to the production of this sixth edition. Thanks go to my friends and colleagues who provided me with their beautiful photographs. A special thanks to Jan Null for his invaluable assistance. Thanks to Lita Teixeira, Mabel Labiack, and Ellen Deuel for their careful reading of the book and to

Aubrey Rhodes and Melissa Bouché for their assistance. I am most indebted to Janet Bollow who, once again, put a tremendous effort into ensuring a beautiful product. Thanks to Stuart Kenter for his careful editing and for his obtaining some of the photos.

My special thanks to the many professional people at Wadsworth and Brooks Cole Publishing who were instrumental in seeing that this project became a reality. Finally, for their comments, suggestions, and thoughtful input, I am indebted to those colleagues who were kind enough to review all or part of the manuscript, including

Jeffrey K. Lew
UCLA

Donald J. Wuebbles
University of Illinois

Karl E. Taylor
Lawrence Livermore National Laboratory

Alan Robock
Rutgers University

Wayne Wendland
University of Illinois (emeritus)

Robert Bornstein
San Jose State University

Douglas Yarger
Iowa State University

Paul Ruscher
Florida State University

Troy Kimmel, Jr.
University of Texas

John A. Ernst
Embry-Riddle Aeronautical University

Clifford Mass
University of Washington

Edward J. Zipser
Texas A & M University

Jan Null
Golden Gate Weather Service

Michael Richman
University of Oklahoma

Charles D. Weidman
University of Arizona

Steven Businger
University of Hawaii

Charles E. Konrad
University of North Carolina

Charles A. Doswell
National Severe Storms Laboratory

TO THE STUDENT

Learning about the atmosphere can be an enjoyable experience, especially if you become involved. This book is intended to give you some insight into the workings of the atmosphere, but for a real appreciation of your atmospheric environment, you must go outside and observe. Mountains take millions of years to form, while a cumulus cloud can develop into a raging thunderstorm in less than an hour. To help with your observations, a color Cloud Chart is bound toward the front of the book for easy reference. Remove it and keep it with you. And remember, all of the information in this book is out there—please, take the time to look.

Donald Ahrens

ANCILLARIES

A full suite of text and multimedia products accompanies *Meteorology Today*, Sixth Edition.

For Instructors

Instructor's Edition   A special edition for instructors containing the complete text, plus extra frontmatter material featuring a visual walk-through of the book's special pedagogical features and key ancillaries. (0-534-37207-4)

Instructor's Resource Manual   A resource guide for instructors that includes additional lecture resources, key terms, teaching suggestions, answers to critical thinking questions, and a test bank. (ISBN 0-534-37208-2)

Thomson Learning Testing Tools   This fully integrated suite of computerized testing tools allows instructors to deliver tests via print, LAN, or the Internet. With these tools, existing test questions can be used and edited, additional questions can be created, and feedback to objective test questions can be provided—enabling the system to work as a tutorial or an examination. Test types include essay, fill-in-the-blank, multiple-choice, true/false, and matching questions. Testing and tutorial results can be integrated into the classroom management tool, which offers scoring, gradebook, and reporting capabilities. (Dual-platform CD-ROM ISBN 0-534-37211-2)

Color Transparency Acetates   A set of 100 beautiful color transparencies featuring key images from the text. (0-534-37204-X)

Slide Set   A package of 75 color slides featuring key photos in the text. 0-534-37210-4.

The Brooks/Cole GeoLink 3.0   A lecture/presentation tool for instructors. This CD-ROM allows professors to customize lectures, assemble multimedia presentations, and export lecture notes for use on the World Wide Web. An extremely easy-to-use interface enables instructors to incorporate still graphics, digital video, animation, and audioclips into their lectures and presentations. Includes images from Ahrens and other Brooks-Cole geology titles. (0-534-52190-8)

CNN Today Video for Meteorology, Vol. 2   A 45-minute video showing dramatic, in-depth coverage of recent meteorological events from CNN's collection. (0-534-37206-6)

For Students

Blue Skies—Student Edition   This free CD-ROM, inserted in every instructor and student copy of the text, is an engaging, personal and interactive product encourages students to invest in active learning as opposed to passive reading. The Internet-activated CD-ROM provides real time data and images that are tied into simulations that allow students to take the information they have learned and make forecasts just as professional meteorologists do. (0-534-37197-3; bundle-only ISBN: 0-534-37198-1)

Workbook/Study Guide   A saleable student ancillary intended to facilitate learning of course content. Includes chapter overviews, self-tests with answers, discussion of key concepts, and other resources. (0-534-37209-0)

Meteorology Today for Scientists and Engineers, *by Roland Stull*   A supplemental text for more advanced coursework in meteorology where quantitative material is desirable. Contains equations omitted from *Meteorology Today*, Sixth Edition, and *Essentials of Meteorology*, Second Edition. No calculus is required. (0-534-37214-7)

Earth Online, *by Michael Ritter*   An inexpensive, hands-on Internet guide written for the novice. It provides a tool for students to get started on the Internet with homework exercises, lab exercises, web searches, and more. (0-534-51707-2)

InfoTrac College Edition   Free access to this extensive online library is provided to both instructors and students. The database contains full-length articles from *Weatherwise, Discover, Sky & Telescope, Science*, and 700 other scholarly and popular periodicals, updated daily and going back as far as 4 years.

Meteorology Today, *Sixth Edition, Website on the Brooks/Cole Earth Science Resource Center*   http://www.brookscole.com/geo An award-winning site that makes an encyclopedia's worth of online resources easy to find. It includes map resources and online field trips. On this website, students can access book-specific resources, such as hypercontents, critical thinking questions, and self-quizzes.

METEOROLOGY TODAY

AN INTRODUCTION TO WEATHER, CLIMATE, AND THE ENVIRONMENT

We live at the bottom of a swirling ocean of air. Here, air billowing up from the earth's surface forms into clouds and thunderstorms over the warm, tropical Pacific Ocean. (Photo: NASA)

THE EARTH AND ITS ATMOSPHERE

CONTENTS

I well remember a brilliant red balloon which kept me completely happy for a whole afternoon, until, while I was playing, a clumsy movement allowed it to escape. Spellbound, I gazed after it as it drifted silently away, gently swaying, growing smaller and smaller until it was only a red point in a blue sky. At that moment I realized, for the first time, the vastness above us: a huge space without visible limits. It was an apparent void, full of secrets, exerting an inexplicable power over all the earth's inhabitants. I believe that many people, consciously or unconsciously, have been filled with awe by the immensity of the atmosphere. All our knowledge about the air, gathered over hundreds of years, has not diminished this feeling.

Theo Loebsack, *Our Atmosphere*

Our *atmosphere* is a delicate life-giving blanket of air that surrounds the fragile earth. In one way or another, it influences everything we see and hear—it is intimately connected to our lives. Air is with us from birth, and we cannot detach ourselves from its presence. In the open air, we can travel for many thousands of kilometers in any horizontal direction, but should we move a mere eight kilometers above the surface, we would suffocate. We may be able to survive without food for a few weeks, or without water for a few days, but, without our atmosphere, we would not survive more than a few minutes. Just as fish are confined to an environment of water, so we are confined to an ocean of air. Anywhere we go, it must go with us.

The earth without an atmosphere would have no lakes or oceans. There would be no sounds, no clouds, no red sunsets. The beautiful pageantry of the sky would be absent. It would be unimaginably cold at night and unbearably hot during the day. All things on the earth would be at the mercy of an intense sun beating down upon a planet utterly parched.

Living on the surface of the earth, we have adapted so completely to our environment of air that we sometimes forget how truly remarkable this substance is. Even though air is tasteless, odorless, and (most of the time) invisible, it protects us from the scorching rays of the sun and provides us with a mixture of gases that allows life to flourish. Because we cannot see, smell, or taste air, it may seem surprising that between your eyes and the pages of this book are trillions of air molecules. Some of these may have been in a cloud only yesterday, or over another continent last week, or perhaps part of the life-giving breath of a person who lived hundreds of years ago.

In this chapter, we will examine a number of important concepts and ideas about the earth's atmosphere, many of which will be expanded in subsequent chapters.

OVERVIEW OF THE EARTH'S ATMOSPHERE

The universe contains billions of galaxies and each galaxy is made up of billions of stars. Stars are hot, glowing balls of gas that generate energy by converting hydrogen into helium near their centers. Our sun is an average size star situated near the edge of the Milky Way galaxy. Revolving around the sun is the earth and eight other planets (see Fig. 1.1). These planets, along with a host of other material (comets, asteroids, meteors, etc.), comprise our solar system.

Warmth for the planets is provided primarily by the sun's energy. At an average distance from the sun of nearly 150 million kilometers (km) or 93 million miles (mi), the earth intercepts only a very small fraction of the sun's total energy output. However, it is this *radiant energy (or radiation)** that drives the atmosphere into the patterns of everyday wind and weather and allows the earth to maintain an average surface temperature of about 15°C (59°F).† Although this temperature is mild, the earth experiences a wide range of temperatures, as readings can drop below –85°C (–121°F) during a frigid Antarctic night and climb, during the day, to above 50°C (122°F) on the oppressively hot subtropical desert.

The earth's **atmosphere** is a thin, gaseous envelope comprised mostly of nitrogen and oxygen, with small amounts of other gases, such as water vapor and carbon dioxide. Nested in the atmosphere are clouds of liquid water and ice crystals. The thin blue area near the horizon in Fig. 1.2 represents the most dense part of the atmosphere. Although our atmosphere extends upward for many hundreds of kilometers, almost 99 percent of the atmosphere lies within a mere 30 km (19 mi) of the earth's surface. In fact, if the earth were to shrink to the size of a beach ball, its inhabitable atmosphere would be thinner than a piece of paper. This thin blanket of air constantly shields the surface and its inhabitants from the sun's dangerous ultraviolet radiant energy, as well as from the onslaught of material from interplanetary space. There is no definite upper limit to the atmosphere; rather, it becomes thinner and thinner, eventually merging with empty space, which surrounds all the planets.

Composition of the Atmosphere   Table 1.1 shows the various gases present in a volume of air near the earth's surface. Notice that **nitrogen** (N_2) occupies about 78 percent and **oxygen** (O_2) about 21 percent of the total volume. If all the other gases are removed, these percentages for nitrogen and oxygen hold fairly constant up to an elevation of about 80 km (50 mi). (For a closer look at the composition of a breath of air at the earth's surface, read the Focus section on p. 5.)

---

*Radiation is energy transferred in the form of waves that have electrical and magnetic properties. The light that we see is radiation, as is ultraviolet light. More on this important topic is given in Chapter 2.

†The abbreviation °C is used when measuring temperature in degrees Celsius, and °F is the abbreviation for degrees Fahrenheit. More information about temperature scales is given in Appendix A and in Chapter 2.

FIGURE 1.1
The relative sizes and position of the planets in our solar system. (Positions are not to scale.)

FIGURE 1.2
The earth's atmosphere as viewed from space. The thin blue area near the horizon shows the shallowness of the earth's atmosphere.

At the surface, there is a balance between destruction (output) and production (input) of these gases. For example, nitrogen is removed from the atmosphere primarily by biological processes that involve soil bacteria. It is returned to the atmosphere mainly through the decaying of plant and animal matter. Oxygen, on the other hand, is removed from the atmosphere when organic matter decays and when oxygen combines with other substances, producing oxides. It is also taken from the atmosphere during breathing, as the lungs take in oxygen and release carbon dioxide (CO_2). The addition of oxygen to the atmosphere occurs during photosynthesis, as plants, in the presence of sunlight, combine carbon dioxide and water to produce sugar and oxygen.

The concentration of the invisible gas **water vapor** (H_2O), however, varies greatly from place to place, and from time to time. Close to the surface in warm, steamy, tropical locations, water vapor may account for up to 4 percent of the atmospheric gases, whereas in colder arctic areas, its concentration may dwindle to a mere fraction of a percent (see Table 1.1). Water vapor molecules are, of course, invisible. They become visible only when they transform into larger liquid or solid particles, such as cloud droplets and ice crystals. The changing of water vapor into liquid water is called *condensation*, whereas the process of liquid water becoming water vapor is called *evaporation*. In the lower atmosphere, water is everywhere. It is the only substance that exists as a gas, a liquid, and a solid at those temperatures and pressures normally found near the earth's surface (see Fig. 1.3).

Water vapor is an *extremely* important gas in our atmosphere. Not only does it form into both liquid and solid cloud particles that grow in size and fall to earth as precipitation, but it also releases large amounts of heat—called *latent heat*—when it changes from vapor into liquid water or ice. Latent heat is an important source of atmospheric energy, especially for storms, such as thunderstorms and hurricanes. Moreover, water vapor is a potent *greenhouse gas* because it strongly absorbs a portion of the earth's outgoing radiant energy (somewhat like the glass of a greenhouse prevents the heat

TABLE 1.1	Composition of the Atmosphere Near the Earth's Surface						
PERMANENT GASES				**VARIABLE GASES**			
Gas	Symbol	Percent (by Volume) Dry Air		Gas (and Particles)	Symbol	Percent (by Volume)	Parts per Million (ppm)*
Nitrogen	N_2	78.08		Water vapor	H_2O	0 to 4	
Oxygen	O_2	20.95		Carbon dioxide	CO_2	0.036	365*
Argon	Ar	0.93		Methane	CH_4	0.00017	1.7
Neon	Ne	0.0018		Nitrous oxide	N_2O	0.00003	0.3
Helium	He	0.0005		Ozone	O_3	0.000004	0.04**
Hydrogen	H_2	0.00006		Particles (dust, soot, etc.)		0.000001	0.01–0.15
Xenon	Xe	0.000009		Chlorofluorocarbons (CFCs)		0.00000002	0.0002

*For CO_2, 365 parts per million means that out of every million air molecules, 365 are CO_2 molecules.

**Stratospheric values are about 5 to 12 ppm.

FIGURE 1.3

The earth's atmosphere is a rich mixture of many gases, with clouds of condensed water vapor and ice crystals. Here, water evaporates from the ocean's surface. Rising air currents then transform the invisible water vapor into many billions of tiny liquid droplets that appear as puffy cumulus clouds. If the rising air in the cloud should extend to greater heights, where air temperatures are quite low, some of the liquid droplets would freeze into minute ice crystals.

inside from escaping and mixing with the outside air). Thus, water vapor plays a significant role in the earth's heat-energy balance.

Carbon dioxide (CO_2), a natural component of the atmosphere, occupies a small (but important) percent of a volume of air, about 0.036 percent. Carbon dioxide enters the atmosphere mainly from the decay of vegetation, but it also comes from volcanic eruptions, the exhalations of animal life, from the burning of fossil fuels (such as coal, oil, and natural gas), and from deforestation. The removal of CO_2 from the atmosphere takes place during *photosynthesis*, as plants con-

sume CO_2 to produce green matter. The CO_2 is then stored in roots, branches, and leaves. The oceans act as a huge reservoir for CO_2, as phytoplankton (tiny drifting plants) in surface water fix CO_2 into organic tissues. Carbon dioxide that dissolves directly into surface water mixes downward and circulates through greater depths. Estimates are that the oceans hold more than 50 times the total atmospheric CO_2 content.

Figure 1.4 reveals that the atmospheric concentration of CO_2 has risen more than 15 percent since 1958, when it was first measured at Mauna Loa Observatory in Hawaii. This increase means that CO_2 is entering the atmosphere at a

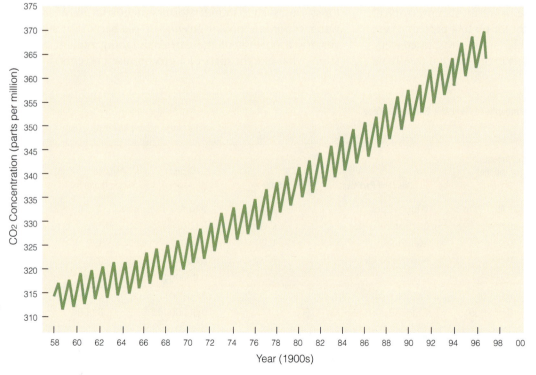

FIGURE 1.4

Measurements of CO_2 in parts per million (ppm) at Mauna Loa Observatory, Hawaii. Higher readings occur in winter when plants die and release CO_2 to the atmosphere. Lower readings occur in summer when more abundant vegetation absorbs CO_2 from the atmosphere.

FOCUS ON A SPECIAL TOPIC

A Breath of Fresh Air

If we could examine a breath of air, we would see that air (like everything else in the universe) is composed of incredibly tiny particles called *atoms*. We cannot see atoms individually. Yet, if we could see one, we would find electrons whirling at fantastic speeds about an extremely dense center, somewhat like hummingbirds darting and circling about a flower. At this center, or nucleus, are the protons and neutrons. Almost all of the atom's mass is concentrated here, in a trillionth of the atom's entire volume. In the nucleus, the proton carries a positive charge, whereas the neutron is electrically neutral. The circling electron carries a negative charge. As long as the total number of protons in the nucleus equals the number of orbiting electrons, the atom as a whole is electrically neutral (see Fig. 1).

Most of the air particles are *molecules*, combinations of two or more atoms (such as nitrogen, N_2, and oxygen, O_2), and most of the molecules are electrically neutral. A few, however, are electrically charged, having lost or gained electrons. These charged atoms and molecules are called *ions*.

An average breath of fresh air contains a tremendous number of molecules. With every deep breath, trillions of molecules from the atmosphere enter your body. Some of

these inhaled gases become a part of you, and others are exhaled.

The volume of an average size breath of air is about a liter.* Near sea level, there are roughly ten thousand million million million (10^{22})† air molecules in a liter. So,

1 breath of air ≈ 10^{22} molecules.

We can appreciate how large this number is when we compare it to the number of stars in the universe.

Astronomers have estimated that there are about 100 billion (10^{11}) stars in an average size galaxy and that there may be as many as 10^{11} galaxies in the universe. To determine the total number of stars in the universe, we multiply the number of stars in a galaxy by the total number of galaxies and obtain

$$10^{11} \times 10^{11} = 10^{22} \text{ stars in the universe.}$$

Therefore, each breath of air contains about as many molecules as there are stars in the known universe.

In the entire atmosphere, there are nearly 10^{44} molecules. The number 10^{44} is 10^{22} squared; consequently

$$10^{22} \times 10^{22} = 10^{44} \text{ molecules in the atmosphere.}$$

___

*One cubic centimeter is about the size of a sugar cube, and there are a thousand cubic centimeters in a liter.

†The notation 10^{22} means the number one followed by twenty-two zeros. For a further explanation of this system of notation see Appendix A.

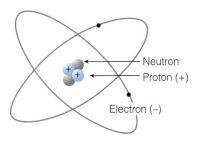

FIGURE 1
An atom has neutrons and protons at its center with electrons orbiting this center (or nucleus). Molecules are combinations of two or more atoms. The air we breathe is mainly molecular nitrogen (N_2) and molecular oxygen (O_2).

We thus conclude that there are about 10^{22} breaths of air in the entire atmosphere. In other words, there are as many molecules in a single breath as there are breaths in the atmosphere.

Each time we breathe, the molecules we exhale enter the turbulent atmosphere. If we wait a long time, those molecules will eventually become thoroughly mixed with all of the other air molecules. If none of the molecules were consumed in other processes, eventually there would be a molecule from that single breath in every breath that is out there. So, considering the many breaths people exhale in their lifetimes, it is possible that in our lungs are molecules that were once in the lungs of people who lived hundreds or even thousands of years ago. In a very real way then, we all share the same atmosphere.

___

greater rate than it is being removed. The increase appears to be due mainly to the burning of fossil fuels; however, deforestation also plays a role as cut timber, burned or left to rot, releases CO_2 directly into the air, perhaps accounting for about 20 percent of the observed increase. Measurements of CO_2 also come from ice cores. In Greenland and Antarctica, for example, tiny bubbles of air trapped within the ice sheets reveal that before the industrial revolution, CO_2 levels were stable at about 280 parts per million (ppm). Since the early 1800s, however, CO_2 levels have increased by as much as

25 percent. With CO_2 levels presently increasing by about 0.4 percent annually (1.5 ppm/year), scientists now estimate that the concentration of CO_2 will likely rise from its current value of about 365 ppm to a value near 500 ppm toward the end of the twenty-first century.

Carbon dioxide is another important greenhouse gas because, like water vapor, it traps a portion of the earth's outgoing energy. Consequently, with everything else being equal, as the atmospheric concentration of CO_2 increases, so should the average global surface air temperature. Most of the mathematical model experiments that predict future atmospheric conditions estimate that increasing levels of CO_2 (and other greenhouse gases) will result in a *global warming* of surface air between 1°C and 3.5°C (about 2°F to 6°F) by the year 2100. Such warming (as we will learn in more detail in Chapter 19) could result in a variety of consequences, such as increasing precipitation in certain areas and reducing it in others as the global air currents that guide the major storm systems across the earth begin to shift from their "normal" paths.

Carbon dioxide and water vapor are not the only greenhouse gases. Recently, others have been gaining notoriety, primarily because they, too, are becoming more concentrated. Such gases include *methane* (CH_4), *nitrous oxide* (N_2O), and *chlorofluorocarbons* (CFCs).*

Levels of methane, for example, have been rising over the past century, increasing recently by about one-half of one percent per year (see Fig. 1.5). Most methane appears to derive from the breakdown of plant material by certain bacteria in rice paddies, wet oxygen-poor soil, the biological activity of termites, and biochemical reactions in the stomachs of cows. Just why methane should be increasing so rapidly is currently under study. Levels of nitrous oxide—commonly known as laughing gas—have been rising annually at the rate of about one-quarter of a percent. Nitrous oxide forms in the soil through a chemical process involving bacteria and certain microbes. Ultraviolet light from the sun destroys it.

Chlorofluorocarbons represent a group of greenhouse gases that also have been increasing in concentration. These had been the most widely used propellants in spray cans. Today, however, they are mainly used as refrigerants, as propellants for the blowing of plastic-foam insulation, and as solvents for

---

*Because these gases (including CO_2) occupy only a small fraction of a percent in a volume of air near the surface, they are referred to collectively as *trace gases*.

cleaning electronic microcircuits. Although their average concentration in a volume of air is quite small (see Table 1.1), they have an important effect on our atmosphere as they not only have the potential for raising global temperatures, they also play a part in destroying the gas ozone in the stratosphere.

At the surface, **ozone** (O_3) is the primary ingredient of *photochemical smog*,* which irritates the eyes and throat and damages vegetation. But the majority of atmospheric ozone (about 97 percent) is found in the upper atmosphere—in the stratosphere—where it is formed naturally, as oxygen atoms combine with oxygen molecules. Here, the concentration of ozone averages less than 0.002 percent by volume. This small quantity is important, however, because it shields plants, animals, and humans from the sun's harmful ultraviolet rays. It is ironic that ozone, which damages plant life in a polluted environment, provides a natural protective shield in the upper atmosphere so that plants on the surface may survive. We will see in Chapter 17 that when CFCs enter the stratosphere ultraviolet rays break them apart, and the CFCs release ozone-destroying chlorine. Because of this effect, ozone concentration in the stratosphere has been decreasing over parts of the Northern and Southern Hemispheres. The reduction in stratospheric ozone levels over springtime Antarctica has plummeted at such an alarming rate that during September and October, there is an *ozone hole* over the region. (We will examine the ozone hole situation more closely in Chapter 17.)

Impurities from both natural and human sources are also present in the atmosphere: Wind picks up dust and soil from the earth's surface and carries it aloft; small saltwater drops from ocean waves are swept into the air (upon evaporating, these drops leave microscopic salt particles suspended in the atmosphere); smoke from forest fires is often carried high above the earth; and volcanoes spew many tons of fine ash particles and gases into the air (see Fig. 1.6). Collectively, these tiny solid or liquid suspended particles of various composition are called **aerosols**.

Some natural impurities found in the atmosphere are quite beneficial. Small, floating particles, for instance, act as surfaces on which water vapor condenses to form clouds. However, most human-made impurities (and some natural ones) are a nuisance, as well as a health hazard. These we call **pollutants**. For example, automobile engines emit copious amounts of *nitrogen dioxide* (NO_2), *carbon monoxide* (CO), and *hydrocarbons*. In sunlight, nitrogen dioxide reacts with hydrocarbons and other gases to produce ozone. Carbon monoxide is a major pollutant of city air. Colorless and odorless, this poisonous gas forms during the incomplete combustion of carbon-containing fuel. Hence, over 75 percent of carbon monoxide in urban areas comes from road vehicles.

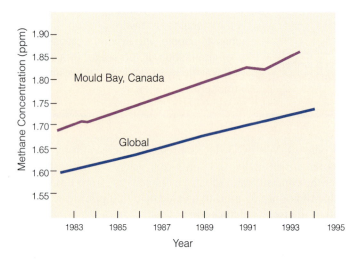

FIGURE 1.5
Global concentrations of atmospheric methane (CH_4) along with methane concentrations observed at Mould Bay, Canada.

---

*Originally the word *smog* meant the combining of smoke and fog. Today, however, the word usually refers to the type of smog that forms in large cities, such as Los Angeles, California. Because this type of smog forms when chemical reactions take place in the presence of sunlight, it is termed *photochemical smog*.

The burning of sulfur-containing fuels (such as coal and oil) releases the colorless gas *sulfur dioxide* (SO_2) into the air. When the atmosphere is sufficiently moist, the SO_2 may transform into tiny dilute drops of sulfuric acid. Rain containing sulfuric acid corrodes metals and painted surfaces and turns freshwater lakes acidic. *Acid rain* (thoroughly discussed in Chapter 17) is a major environmental problem, especially downwind from major industrial areas. In addition, high concentrations of SO_2 produce serious respiratory problems in humans, such as bronchitis and emphysema, and have an adverse effect on plant life. (More information on these and other pollutants is given in Chapter 17.)

The Early Atmosphere

The atmosphere that originally surrounded the earth was probably much different from the air we breathe today. The earth's first atmosphere (some 4.6 billion years ago) was most likely *hydrogen* and *helium*—the two most abundant gases found in the universe—as well as hydrogen compounds, such as methane and ammonia. Most scientists feel that this early atmosphere escaped into space from the earth's hot surface.

A second, more dense atmosphere, however, gradually enveloped the earth as gases from molten rock within its hot interior escaped through volcanoes and steam vents. We assume that volcanoes spewed out the same gases then as they do today: mostly water vapor (about 80 percent), carbon dioxide (about 10 percent), and up to a few percent nitrogen. These gases (mostly water vapor and carbon dioxide) probably created the earth's second atmosphere.

As millions of years passed, the constant outpouring of gases from the hot interior—known as *outgassing*—provided a rich supply of water vapor, which formed into clouds.* Rain fell upon the earth for many thousands of years, forming the rivers, lakes, and oceans of the world. During this time, large amounts of CO_2 were dissolved in the oceans. Through chemical and biological processes, much of the CO_2 became locked up in carbonate sedimentary rocks, such as limestone. With much of the water vapor already condensed and the concentration of CO_2 dwindling, the atmosphere gradually became rich in nitrogen (N_2), which is usually not chemically active.

It appears that oxygen (O_2), the second most abundant gas in today's atmosphere, probably began an extremely slow increase in concentration as energetic rays from the sun split water vapor (H_2O) into hydrogen and oxygen during a process called *photodissociation*. The hydrogen, being lighter, probably rose and escaped into space, while the oxygen remained in the atmosphere.

This slow increase in oxygen may have provided enough of this gas for primitive plants to evolve, perhaps 2 to 3 billion years ago. Or the plants may have evolved in an almost oxygen-free (anaerobic) environment. At any rate, plant

FIGURE 1.6
Erupting volcanoes can send tons of particles into the atmosphere, along with vast amounts of water vapor, carbon dioxide, and sulfur dioxide.

growth greatly enriched our atmosphere with oxygen. The reason for this enrichment is that, during photosynthesis, plants, in the presence of sunlight, combine carbon dioxide and water to produce oxygen. Hence, after plants evolved, the atmospheric oxygen content increased more rapidly, probably reaching its present composition about several hundred million years ago.

BRIEF REVIEW   Before going on to the next several sections, here is a review of some of the important concepts presented so far:

■ The earth's atmosphere is a mixture of many gases. In a volume of air near the surface, nitrogen (N_2) occupies about 78 percent and oxygen (O_2) about 21 percent.

■ Water vapor can condense into liquid cloud droplets or transform into delicate ice crystals. Water is the only substance in our atmosphere that is found naturally as a gas (water vapor), as a liquid (water), and as a solid (ice).

■ Both water vapor and carbon dioxide (CO_2) are important greenhouse gases.

■ The majority of water on our planet is believed to have come from its hot interior through outgassing.

---

*It is now believed that some of the earth's water may have originated from numerous collisions with small meteors and disintegrating comets when the earth was very young.

VERTICAL STRUCTURE OF THE ATMOSPHERE

A vertical profile of the atmosphere reveals that it can be divided into a series of layers. Each layer may be defined in a number of ways: by the manner in which the air temperature varies through it, by the gases that comprise it, or even by its electrical properties. At any rate, before we examine these various atmospheric layers, we need to look at the vertical profile of two important variables: air pressure and air density.

A Brief Look at Air Pressure and Air Density Earlier in this chapter we learned that most of our atmosphere is crowded close to the earth's surface. The reason for this fact is that air molecules (as well as everything else) are held near the earth by *gravity*. This strong invisible force pulling down on the air above squeezes (compresses) air molecules closer together, which causes their number in a given volume to increase. The more air above a level, the greater the squeezing effect or compression.

Gravity also has an effect on the weight of objects, including air. In fact, *weight* is the force upon an object due to gravity. Weight is defined as the mass of an object times the acceleration of gravity; thus

$$\text{Weight} = \text{mass} \times \text{gravity}.$$

An object's *mass* is the quantity of matter in the object. Consequently, the mass of air in a rigid container is the same everywhere in the universe. However, if you were to instantly travel to the moon, where the acceleration of gravity is one-sixth of what it is on earth, the mass of air in the container would be the same, but its weight would decrease by one-sixth.

When mass is given in grams (g) or kilograms (kg), volume is given in cubic centimeters (cm³) or cubic meters (m³). Near sea level, air density is about 1.2 kilograms per cubic meter (nearly 1.2 ounces per cubic foot).

The **density** of air (or any substance) is determined by the masses of atoms and molecules and the amount of space between them. In other words, density tells us how much matter is in a given space (that is, volume). We can express density in a variety of ways. The molecular density of air is the number of molecules in a given volume. Most commonly, however, density is given as the mass of air in a given volume; thus:

$$\text{Density} = \frac{\text{mass}}{\text{volume}}.$$

Because there are appreciably more molecules within the same size volume of air near the earth's surface than at higher levels, air density is greatest at the surface and decreases as we move up into the atmosphere. Notice in Fig. 1.7 that, because air near the surface is compressed, air density normally decreases rapidly at first, then more slowly as we move farther away from the surface.

Air molecules are in constant motion. On a mild spring day near the surface, an air molecule will collide about 10 billion times each second with other air molecules. It will also bump against objects around it—houses, trees, flowers, the ground, and even people. Each time an air molecule bounces against a person, it gives a tiny push. This small force (push) divided by the area on which it pushes is called **pressure**; thus

$$\text{Pressure} = \frac{\text{force}}{\text{area}}.$$

If we weigh a column of air 1 square inch in cross section, extending from the average height of the ocean surface (sea level) to the "top" of the atmosphere, it would weigh nearly 14.7 pounds. Thus, normal atmospheric pressure near sea level is close to 14.7 pounds per square inch. If more molecules are packed into the column, it becomes more dense, the air weighs more, and the surface pressure goes up. On the other hand, when fewer molecules are in the column, the air weighs less, and the surface pressure goes down. So, the surface air pressure can be changed by changing the mass of air above the surface.

Pounds per square inch is, of course, one way of expressing air pressure. However, the most common unit presently found on surface weather maps is the *millibar** (mb) although the *hectopascal* (hPa) is gradually replacing the millibar as the

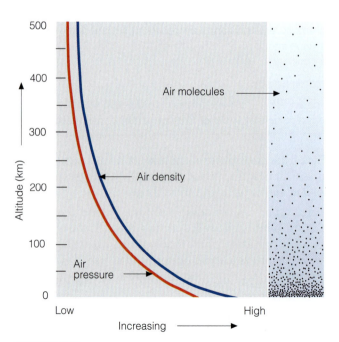

FIGURE 1.7
Both air pressure and air density decrease with increasing altitude.

---

*By definition, a *bar* is a force of 100,000 newtons (N) acting on a surface area of 1 square meter (m²). A *newton* is the amount of force required to move an object with a mass of 1 kilogram (kg) so that it increases its speed at a rate of 1 meter per second (m/sec) each second. Because the bar is a relatively large unit, and because surface pressure changes are usually small, the unit of pressure most commonly found on surface weather maps is the *millibar*, where 1 bar = 1000 mb. The unit of pressure designed by the International System (SI) of measurement is the *pascal* (Pa), where 1 pascal is the force of 1 newton acting on a surface of 1 square meter. A more common unit is the *hectopascal* (hPa), as 1 hectopascal equals 1 millibar.

preferred unit of pressure on surface charts. Another unit of pressure is inches of mercury (Hg), which is commonly used in the field of aviation and on television and radio weather broadcasts. At sea level, the *average*, or *standard*, value for atmospheric pressure is

1013.25 mb = 1013.25 hPa = 29.92 in. Hg.

Billions of air molecules push constantly on the human body. This force is exerted equally in all directions. We are not crushed by it because billions of molecules inside the body push outward just as hard. Even though we do not actually feel the constant bombardment of air, we can detect quick changes in it. For example, if we climb rapidly in elevation, our ears may "pop." This experience happens because air collisions outside the eardrum lessen. The popping comes about as air collisions between the inside and outside of the ear equalize. The drop in the number of collisions informs us that the pressure exerted by the air molecules decreases with height above the earth.

Air molecules not only take up space (freely darting, twisting, spinning, and colliding with everything around them), but—as we have seen—these same molecules have weight. In fact, air is surprisingly heavy. The weight of all the air around the earth is a staggering 5600 trillion tons. The weight of the air molecules acts as a force upon the earth. The amount of force exerted over an area of surface is called *atmospheric pressure* or, simply, **air pressure**. The pressure at any level in the atmosphere may be measured in terms of the total mass of air above any point. As we climb in elevation, fewer air molecules are above us; hence, *atmospheric pressure always decreases with increasing height*. Like air density, air pressure decreases rapidly at first, then more slowly at higher levels (see Fig. 1.7).

Figure 1.8 illustrates how rapidly air pressure decreases with height. Near sea level, atmospheric pressure is usually close to 1000 mb. Normally, just above sea level atmospheric pressure decreases by about 10 mb for every 100 meters (m) increase in elevation—about 1 inch of mercury for every 1000 feet (ft) of rise. At higher levels, air pressure decreases much more slowly with height. With a sea level pressure near 1000 mb, we can see in Fig. 1.8 that, at an altitude of only 5.5 km (3.5 mi), the air pressure is about 500 mb, or half of the sea level pressure, meaning that at a mere 18,000 ft above the surface, we are above one-half of all the molecules in the atmosphere.

At an elevation approaching the summit of Mt. Everest (about 9 km or 29,000 ft—the highest mountain peak on earth), the air pressure would be about 300 mb. The summit is above nearly 70 percent of all the molecules in the atmosphere. At an altitude of about 50 km, the air pressure is about 1 mb, which means that 99.9 percent of all the molecules are below this level. Yet the atmosphere extends upwards for many hundreds of kilometers, gradually becoming thinner and thinner until it ultimately merges with outer space. (Up to now, we have concentrated on the earth's atmosphere. For a brief look at the atmospheres of the other planets, read the Focus section on p. 10.)

WEATHER WATCH

Layers of the Atmosphere We have seen that both air pressure and density decrease with height above the earth—rapidly at first, then more slowly. *Air temperature*, however, has a more complicated vertical profile.*

Look closely at Fig. 1.9 and notice that air temperature normally decreases from the earth's surface up to an altitude of about 11 km, which is nearly 36,000 ft, or 7 mi. This decrease in air temperature with increasing height is due primarily to the fact (investigated further in Chapter 2) that sunlight warms the earth's surface, and the surface, in turn, warms the air above it. The rate at which the air temperature decreases with height is called the temperature **lapse rate**. The *average* (or *standard*) *lapse rate* in this region of the lower atmosphere is about 6.5°C

---

*Air temperature is the degree of hotness or coldness of the air and, as we will see in Chapter 2, it is also a measure of the average speed of the air molecules.

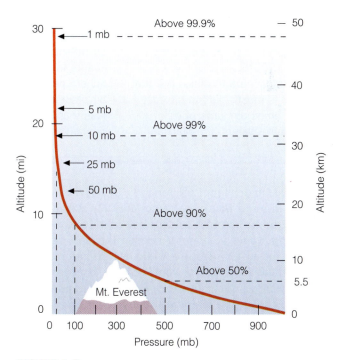

FIGURE 1.8

Atmospheric pressure decreases rapidly with height. Climbing to an altitude of only 5.5 km, where the pressure is 500 mb, would put you above one-half of the atmosphere's molecules.

The Atmospheres of Other Planets

Earth is unique. Not only does it lie at just the right distance from the sun so that life may flourish, it also provides its inhabitants with an atmosphere rich in nitrogen and oxygen—two gases that are not abundant in the atmospheres of either Venus or Mars, our closest planetary neighbors.

The Venusian atmosphere is mainly carbon dioxide (95 percent) with minor amounts of water vapor and nitrogen. An opaque acid-cloud deck encircles the planet, hiding its surface. The atmosphere is quite turbulent, as instruments reveal twisting eddies and fierce winds in excess of 200 km/hr. This thick dense atmosphere produces a surface air pressure of about 90,000 mb, which is 90 times greater than that on earth. To experience such a pressure on earth, one would have to descend in the ocean to a depth of about 900 m (2950 ft). Moreover, this thick atmosphere of CO_2 produces a strong greenhouse effect, with a scorching hot surface temperature of 480°C (900°F).

The atmosphere of Mars, like that of Venus, is mostly carbon dioxide, with only small amounts of other gases. Unlike Venus, the Martian

FIGURE 2
A portion of Jupiter extending from the equator to the southern polar latitudes. The Great Red Spot, as well as the smaller ones, are spinning eddies similar to storms that exist in the earth's atmosphere.

atmosphere is very thin, and heat escapes from the surface rapidly. Thus, surface temperatures on Mars are much lower, averaging around –60°C (–76°F). Because of its thin cold atmosphere, there is no liquid water on Mars and virtually no cloud cover—only a barren desertlike landscape. In addition, this thin atmosphere produces an average surface air pressure of about 7 mb, which is less than one-hundredth of that experienced at the surface of the earth.

FIGURE 3
The Great Dark Spot on Neptune. The white wispy clouds are similar to the high wispy cirrus clouds on earth. However, on Neptune, they are probably composed of methane ice crystals.

Such a pressure on earth would be observed above the surface at an altitude near 35 km (22 mi).

Occasionally, huge dust storms develop near the Martian surface. Such storms may be accompanied by winds of several hundreds of kilometers per hour. These winds carry fine dust around the entire planet. The dust gradually settles out, coating the landscape with a thin reddish veneer.

The atmosphere of the largest planet Jupiter is much different from

---

for every 1000 m or about 3.6°F for every 1000 ft rise in elevation. Keep in mind that these values are only averages. On some days, the air becomes colder more quickly as we move upward. This would increase or steepen the lapse rate. On other days, the air temperature would decrease more slowly with height, and the lapse rate would be less. Occasionally, the air temperature may actually *increase* with height, producing a condition known as a **temperature inversion**. So the lapse rate fluctuates, varying from day to day and season to season.

The region of the atmosphere (from the surface up to about 11 km) contains all of the weather we are familiar with on earth. Also, this region is kept well stirred by rising and descending air currents. Here, it is common for air molecules to circulate through a depth of more than 10 km in just a few

days. This region of circulating air extending upward from the earth's surface to where the air stops becoming colder with height is called the **troposphere**—from the Greek *tropein*, meaning to turn or change. The instrument used to measure the vertical profile of air temperature in the atmosphere up to an elevation sometimes exceeding 30 km (100,000 ft) is the **radiosonde**. More information on this instrument is given in the Focus section on p. 13.

Notice in Fig. 1.9 that just above 11 km the air temperature normally stops decreasing with height. Here, the lapse rate is zero. This region, where the air temperature remains constant with height, is referred to as an *isothermal* (equal temperature) zone. The bottom of this zone marks the top of the troposphere and the beginning of another layer, the

TABLE 1	Data on Planets and the Sun				
	DIAMETER	AVERAGE DISTANCE FROM SUN	AVERAGE SURFACE TEMPERATURE		MAIN ATMOSPHERIC COMPONENTS
	Kilometers	Millions of Kilometers	°C	°F	
Sun	$1,392 \times 10^3$		5,800	10,500	–
Mercury	4,880	58	260*	500	–
Venus	12,112	108	480	900	CO_2
Earth	12,742	150	15	59	N_2, O_2
Mars	6,800	228	–60	–76	CO_2
Jupiter	143,000	778	–110	–166	H_2, He
Saturn	121,000	1,427	–190	–310	H_2, He
Uranus	51,800	2,869	–215	–355	H_2, CH_4
Neptune	49,000	4,498	–225	–373	N_2, CH_4
Pluto	3,100	5,900	–235	–391	CH_4

*Sunlit side.

that of Venus and Mars. Jupiter's atmosphere is mainly hydrogen (H_2) and helium (He), with minor amounts of methane (CH_4) and ammonia (NH_3). A prominent feature on Jupiter is the Great Red Spot—a huge atmospheric storm about three times larger than earth—that spins counterclockwise in Jupiter's southern hemisphere (see Fig. 2). Large white ovals near the Great Red Spot are similar but smaller storm systems. Unlike the earth's weather machine, which is driven by the sun, Jupiter's massive swirling clouds appear to be driven by a collapsing core of hot hydrogen. Energy from this lower region rises toward the surface, then it (along with Jupiter's rapid rotation) stirs the cloud layer into more or less horizontal bands of various colors.

Swirling storms exist on other planets, too, such as on Saturn and Neptune. In fact, the large dark oval on Neptune (Fig. 3) appears to be a storm similar to Jupiter's Great Red Spot. The white wispy clouds in the photograph are probably composed of methane ice crystals. Studying the atmospheric behavior of other planets may give us added insight into the workings of our own atmosphere. (Additional information about size, surface temperature, and atmospheric composition of the other planets is given in Table 1.)

stratosphere. The boundary separating the troposphere from the stratosphere is called the **tropopause**. The height of the tropopause varies. It is normally found at higher elevations over equatorial regions, and it decreases in elevation as we travel poleward. Generally, the tropopause is higher in summer and lower in winter at all latitudes. In some regions, the tropopause "breaks" and is difficult to locate and, here, scientists have observed tropospheric air mixing with stratospheric air and vice versa. These breaks also mark the position of *jet streams*—high winds that meander in a narrow channel, like an old river, often at speeds exceeding 100 knots.*

From Fig. 1.9 we can see that, in the stratosphere at an altitude near 20 km (12 mi), the air temperature begins to increase with height, producing a *temperature inversion*. The inversion region, along with the lower isothermal layer, tends to keep the vertical currents of the troposphere from spreading into the stratosphere. The inversion also tends to reduce the amount of vertical motion in the stratosphere itself; hence, it is a stratified layer.

Even though the air temperature is increasing with height, the air at an altitude of 30 km is extremely cold, averaging less than –46°C. At this level above polar latitudes, air temperatures can change dramatically from one week to the next, as a *sudden warming* can raise the temperature in one week by more than 50°C. Such a rapid warming, although not well

---

*A knot is a nautical mile per hour. One knot is equal to 1.15 miles per hour (mi/hr), or 1.9 kilometers per hour (km/hr).

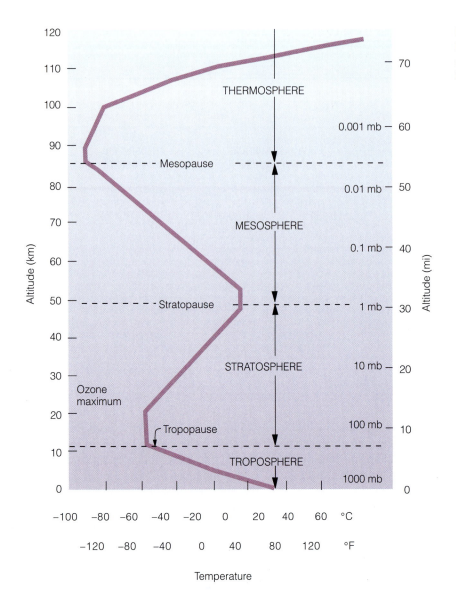

FIGURE 1.9
Layers of the atmosphere as related to the average profile of air temperature above the earth's surface. The heavy line illustrates how the average temperature varies in each layer.

understood, is probably due to sinking air associated with circulation changes that occur in late winter or early spring as well as with the poleward displacement of strong jet stream winds in the lower stratosphere.

The reason for the inversion in the stratosphere is that the gas ozone plays a major part in heating the air at this altitude. Recall that ozone is important because it absorbs energetic ultraviolet (UV) solar energy. Some of this absorbed energy warms the stratosphere, which explains why there is an inversion. If ozone were not present, the air probably would become colder with height, as it does in the troposphere.

Notice in Fig. 1.9 that the level of maximum ozone concentration is observed near 25 km (at middle latitudes), yet the stratospheric air temperature reaches a maximum near 50 km. The reason for this phenomenon is that the air at 50 km is less dense than at 25 km, and so the absorption of intense solar energy at 50 km raises the temperature of fewer molecules to a much greater degree. Moreover, much of the

solar energy responsible for the heating is absorbed in the upper part of the stratosphere and, therefore, does not reach down to the level of ozone maximum. And due to the low air density, the transfer of energy downward from the upper stratosphere is quite slow.

Above the stratosphere is the **mesosphere** (middle sphere). The boundary near 50 km, which separates these layers, is called the *stratopause*. The air at this level is extremely thin and the atmospheric pressure is quite low, averaging about 1 mb, which means that only one-thousandth of all the atmosphere's molecules are above this level and 99.9 percent of the atmosphere's mass is located below it.

The percentage of nitrogen and oxygen in the mesosphere is about the same as at sea level. Given the air's low density in this region, however, we would not survive very long breathing here, as each breath would contain far fewer oxygen molecules than it would at sea level. Consequently, without proper breathing equipment, the brain would soon become oxygen-

The Radiosonde

The vertical distribution of temperature, pressure, and humidity up to an altitude of about 30 km can be obtained with an instrument called a radiosonde.* The radiosonde is a small, lightweight box equipped with weather instruments and a radio transmitter. It is attached to a cord that has a parachute and a gas-filled balloon tied tightly at the end (see Fig. 4). As the balloon rises, the attached radiosonde measures air temperature with a small electrical thermometer—a thermistor—located just outside the box. The radiosonde measures humidity electrically by sending an electric current across a carbon-coated plate. Air pressure is obtained by a small barometer located inside the box. All of this information is transmitted to the surface by radio. Here, a computer rapidly reconverts the various frequencies into values of temperature,

pressure, and moisture. Special tracking equipment at the surface may also be used to provide a vertical profile of winds. (When winds are added, the observation is called a *rawinsonde*.) When plotted on a graph, the vertical distribution of temperature, humidity, and wind is called a *sounding*. Eventually, the balloon bursts and the radiosonde returns to earth, its descent being slowed by its parachute.

At most sites, radiosondes are released twice a day, usually at the time that corresponds to midnight and noon in Greenwich, England. Releasing radiosondes is an expensive operation because many of the instruments are never retrieved, and many of those that are retrieved are often in poor working condition. To complement the radiosonde, modern geostationary satellites (using instruments that measure radiant energy) are providing scientists with vertical temperature profiles in inaccessible regions.

*A radiosonde that is dropped by parachute from an aircraft is called a *dropsonde*.

FIGURE 4
The radiosonde with parachute and balloon.

starved—a condition known as *hypoxia*. Pilots who fly above 3 km (10,000 ft) for too long without oxygen-breathing apparatus experience this. With the first symptoms of hypoxia, there is usually no pain involved, just a feeling of exhaustion. Soon, visual impairment sets in and routine tasks become difficult to perform. Some people drift into an incoherent state, neither realizing nor caring what is happening to them. Of course, if this oxygen deficiency persists, a person will lapse into unconsciousness, and death may result. In fact, at this altitude in the mesosphere, we would suffocate in a matter of minutes.

There are other effects besides suffocating that could be experienced in the mesosphere. Exposure to ultraviolet solar energy, for example, could cause severe burns on exposed parts of the body. Also, given the low air pressure, the blood in one's veins would begin to boil at normal body temperatures.

The air temperature in the mesosphere decreases with height, a phenomenon due, in part, to the fact that there is little ozone in the air to absorb solar radiation. Consequently, the molecules (especially those near the top of the mesosphere) are able to lose more energy than they absorb, which

results in an energy deficit and cooling. So we find air in the mesosphere becoming colder with height up to an elevation near 85 km. At this altitude, the temperature of the atmosphere reaches its lowest average value, $-90°C$ ($-130°F$).

The "hot layer" above the mesosphere is the **thermosphere**. The boundary that separates the lower, colder mesosphere from the warmer thermosphere is the *mesopause*. In the thermosphere, oxygen molecules (O_2) absorb energetic solar rays, warming the air. Because there are relatively few atoms and molecules in the thermosphere, the absorption of a small amount of energetic solar energy can cause a large increase in air temperature. Furthermore, because the amount of solar energy affecting this region depends strongly on solar activity, temperatures in the thermosphere vary from day to day (see Fig. 1.10). The low density of the thermosphere also means that an air molecule will move an average distance (called *mean free path*) of over one kilometer before colliding with another molecule. A similar air molecule at the earth's surface will move an average distance of less than one millionth of a centimeter before it collides with another molecule.

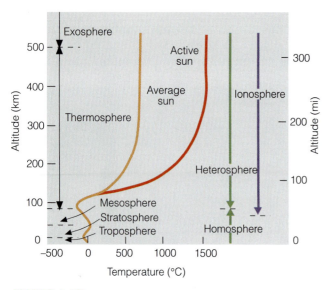

FIGURE 1.10

Layers of the atmosphere based on temperature (red line), composition (green line), and electrical properties (blue line). (An active sun is associated with large numbers of solar eruptions, described in Chapter 2.)

Because the air density in the upper thermosphere is so low, air temperatures there are not measured directly. They can, however, be determined by observing the orbital change of satellites caused by the drag of the atmosphere. Even though the air is extremely tenuous, enough air molecules strike a satellite to slow it down, making it drop into a slightly lower orbit. (For this reason, *Sky Lab* fell to earth in July, 1979, and the spacecraft *Solar Max* fell in December, 1989.) The amount of drag is related to the density of the air, and the density is related to the temperature. Therefore, by determining air density, scientists are able to construct a vertical profile of air temperature.

At the top of the thermosphere, about 500 km (300 mi) above the earth's surface, molecules can move distances of 10 km before they collide with other molecules. Here, many of the lighter, faster-moving molecules traveling in the right direction actually escape the earth's gravitational pull. The region where atoms and molecules shoot off into space is sometimes referred to as the **exosphere**, which represents the upper limit of our atmosphere.

Up to this point, we have examined the atmospheric layers based on the vertical profile of temperature. The atmosphere, however, may also be divided into layers based on its composition. For example, the composition of the atmosphere begins to slowly change in the lower part of the thermosphere. Below the thermosphere, the composition of air remains fairly uniform (78 percent nitrogen, 21 percent oxygen) by turbulent mixing. This lower, well-mixed region is known as the **homosphere** (Fig. 1.10). In the thermosphere, collisions between atoms and molecules are infrequent, and the air is unable to keep itself stirred. As a result, diffusion

takes over as heavier atoms and molecules (such as oxygen and nitrogen) tend to settle to the bottom of the layer, while lighter gases (such as hydrogen and helium) float to the top. The region from about the base of the thermosphere to the top of the atmosphere is often called the **heterosphere**.

The Ionosphere   The **ionosphere** is not really a layer, but rather an electrified region within the upper atmosphere where fairly large concentrations of ions and free electrons exist. *Ions* are atoms and molecules that have lost (or gained) one or more electrons. Atoms lose electrons and become positively charged when they cannot absorb all of the energy transferred to them by a colliding energetic particle or the sun's energy.

The lower region of the ionosphere is usually about 60 km above the earth's surface. From here (60 km), the ionosphere extends upward to the top of the atmosphere. Hence, the bulk of the ionosphere is in the thermosphere (see Fig. 1.10).

The ionosphere plays a major role in radio communications. The lower part (called the *D* region) reflects standard AM radio waves back to earth, but at the same time it seriously weakens them through absorption. At night, though, the *D* region gradually disappears and AM radio waves are able to penetrate higher into the ionosphere (into the *E* and *F* regions—see Fig. 1.11), where the waves are reflected back to earth. Because there is, at night, little absorption of radio waves in the higher reaches of the ionosphere, such waves bounce repeatedly from the ionosphere to the earth's surface and back to the ionosphere again. In this way, standard AM radio waves are able to travel for many hundreds of kilometers at night.

Around sunrise and sunset, AM radio stations usually make "necessary technical adjustments" to compensate for the changing electrical characteristics of the *D* region. Because they can broadcast over a greater distance at night, most AM stations reduce their output near sunset. This reduction prevents two stations—both transmitting at the same frequency but hundreds of kilometers apart—from interfering with each other's radio programs. At sunrise, as the *D* region intensifies, the power supplied to AM radio transmitters is normally increased. FM stations do not need to make these adjustments because FM radio waves are shorter than AM waves, and are able to penetrate through the ionosphere without being reflected.

BRIEF REVIEW   We have, in the last several sections, been examining our atmosphere from a vertical perspective. A few of the main points are listed below:

- Atmospheric pressure at any level represents the total mass of air above that level, and atmospheric pressure always decreases with increasing height above the surface.

- The atmosphere may be divided into layers (or regions) according to its vertical profile of temperature, its gaseous composition, or its electrical properties.

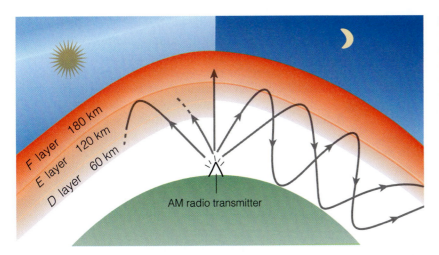

FIGURE 1.11
At night, the higher region of the ionosphere (F region) strongly reflects AM radio waves, allowing them to be sent over great distances. During the day, the lower D region strongly absorbs and weakens AM radio waves, preventing them from being picked up by distant receivers.

◾ Ozone at the earth's surface is the main ingredient of photochemical smog, while ozone in the stratosphere protects life on earth from the sun's harmful ultraviolet rays.

We will now turn our attention to weather events that take place in the lower atmosphere. As you read the remainder of this chapter, keep in mind that the content serves as a broad overview of material to come in later chapters, and that many of the concepts and ideas you encounter are designed to familiarize you with items you might read about in a newspaper or magazine, or see on television.

WEATHER AND CLIMATE

When we talk about the **weather**, we are talking about the condition of the atmosphere at any particular time and place. Weather—which is always changing—is comprised of the elements of:

1. *air temperature—the degree of hotness or coldness of the air*
2. *air pressure—the force of the air above an area*
3. *humidity—a measure of the amount of water vapor in the air*
4. *clouds—a visible mass of tiny water droplets and/or ice crystals that are above the earth's surface*
5. *precipitation—any form of water, either liquid or solid (rain or snow), that falls from clouds and reaches the ground*
6. *visibility—the greatest distance one can see*
7. *wind—the horizontal movement of air*

If we measure and observe these **weather elements** over a specified interval of time, say, for many years, we would obtain the "average weather" or the **climate** of a particular region. Climate, therefore, represents the accumulation of daily and seasonal weather events (the average range of weather) over a long period of time. The concept of climate is much more than this, for it also includes the extremes of weather—the heat waves of summer and the cold spells of winter—that occur in a particular region. The *frequency* of these extremes is what helps us distinguish among climates that have similar averages.

If we were able to watch the earth for many thousands of years, even the climate would change. We would see rivers of ice moving down stream-cut valleys and huge glaciers—sheets of moving snow and ice—spreading their icy fingers over large portions of North America. Advancing slowly from Canada, a single glacier might extend as far south as Kansas and Illinois, with ice several thousands of meters thick covering the region now occupied by Chicago. Over an interval of 2 million years or so, we would see the ice advance and retreat several times. Of course, for this phenomenon to happen, the average temperature of North America would have to decrease and then rise in a cyclic manner.

Suppose we could photograph the earth once every thousand years for many hundreds of millions of years. In time-lapse film sequence, these photos would show that not only is the climate altering, but the whole earth itself is changing as well: mountains would rise up only to be torn down by erosion; isolated puffs of smoke and steam would appear as volcanoes spew hot gases and fine dust into the atmosphere; and the entire surface of the earth would undergo a gradual transformation as some ocean basins widen and others shrink.*

In summary, the earth and its atmosphere are dynamic systems that are constantly changing. While major transformations of the earth's surface are completed only after long spans of time, the state of the atmosphere can change in a matter of minutes. Hence, a watchful eye turned skyward will be able to observe many of these changes.

Up to this point, we have looked at the concepts of weather and climate without discussing the word **meteorology**. What does this term actually mean and where did it originate?

---

*The movement of the ocean floor and continents is explained in the widely acclaimed theory of *plate tectonics*, formerly called the theory of continental drift.

Meteorology—A Brief History

Meteorology is the study of the atmosphere and its phenomena. The term itself goes back to the Greek philosopher Aristotle who, about 340 B.C., wrote a book on natural philosophy entitled *Meteorologica*. This work represented the sum of knowledge on weather and climate at that time, as well as material on astronomy, geography, and chemistry. Some of the topics covered included clouds, rain, snow, wind, hail, thunder, and hurricanes. In those days, all substances that fell from the sky, and anything seen in the air, were called meteors, hence the term *meteorology*, which actually comes from the Greek word *meteoros*, meaning "high in the air." Today, we differentiate between those meteors that come from extraterrestrial sources outside our atmosphere (meteoroids) and particles of water and ice observed in the atmosphere (hydrometeors).

In *Meteorologica*, Aristotle attempted to explain atmospheric phenomena in a philosophical and speculative manner. Even though many of his speculations were found to be erroneous, Aristotle's ideas were accepted without reservation for almost two thousand years. In fact, the birth of meteorology as a genuine natural science did not take place until the invention of weather instruments, such as the thermometer at the end of the sixteenth century, the barometer (for measuring air pressure) in 1643, and the hygrometer (for measuring humidity) in the late 1700s. With observations from instruments available, attempts were then made to explain certain weather phenomena employing scientific experimentation and the physical laws that were being developed at the time.

As more and better instruments were developed in the 1800s, the science of meteorology progressed. The invention of the telegraph in 1843 allowed for the transmission of routine weather observations. Ideas about winds and storms became, at least partially, understood, and crude weather maps were drawn. Around 1920, the concepts of air masses and weather fronts were formulated in Norway. By the 1940s, daily upper-air balloon observations of temperature, humidity, and pressure gave a three-dimensional view of the atmosphere.

Meteorology took another step forward in the 1950s, when high-speed computers were developed to solve the equations that describe the behavior of the atmosphere. Today, computers plot the observations, draw the lines on the map, and forecast the state of the atmosphere at some desired time in the future. After World War II, surplus military radars became available, and many were transformed into precipitation-measuring tools. In the mid-1990s, these conventional radars were replaced by the more sophisticated Doppler radars, which have the ability to peer into severe thunderstorms and unveil their winds.

In 1960, the first weather satellite, *Tiros I*, was launched, ushering in space-age meteorology. Subsequent satellites provided a wide range of useful information, ranging from day and night time-lapse images of clouds and storms to pictures that depict swirling ribbons of water vapor flowing around the globe. Throughout the 1990s, ever more sophisticated satellites were developed to supply computers with a far greater network of data so that more accurate forecasts—perhaps up to a week or more—will be available in the future. This trend will likely continue in the twenty-first century.

A Satellite's View of the Weather

A good view of the weather can be seen from a weather satellite. Figure 1.12 is a satellite photograph showing a portion of the Pacific Ocean and the North American continent. The photograph was obtained from a *geostationary satellite* situated about 36,000 km (22,300 mi) above the earth. At this elevation, the satellite travels at the same rate as the earth spins, which allows it to remain positioned above the same spot so it can continuously monitor what is taking place beneath it.

The dotted lines running from pole to pole on the satellite picture are called *meridians*. Since the zero meridian (or prime meridian) runs through Greenwich, England, the *longitude* of any place on earth is simply how far east or west, in degrees, it is from the prime meridian. North America is west of Great Britain and most of the United States lies between 75°W and 125°W longitude.

The dotted lines that parallel the equator are called *parallels of latitude*. The latitude of any place is how far north or south, in degrees, it is from the equator. The latitude of the equator is 0°, whereas the latitude of the North Pole is 90°N and that of the South Pole is 90°S. Most of the United States is located between latitude 30°N and 50°N, a region commonly referred to as the **middle latitudes**.

Storms of All Sizes Probably the most dramatic spectacle in Fig. 1.12 is the whirling cloud masses of all shapes and sizes. The clouds appear white because sunlight is reflected back to space from their tops. The dark areas show where skies are clear. The largest of the organized cloud masses are the sprawling storms. One such storm shows as an extensive band of clouds, over 2000 km long, west of the Great Lakes. This **middle latitude cyclonic storm** system (or *extratropical cyclone*) forms outside the tropics and, in the Northern Hemisphere, has winds spinning counterclockwise about its center, which is presently over Minnesota.

A slightly smaller but more vigorous storm is located over the Pacific Ocean near latitude 12°N and longitude 116°W. This tropical storm system, with its swirling band of rotating clouds and surface winds in excess of 64 knots* (74 mi/hr), is known as a **hurricane**. The diameter of the hurricane is about 800 km (500 mi). The tiny dot at its center is called the *eye*. In the eye, winds are light and skies are generally clear. Around the eye, however, is an extensive region where heavy rain and high surface winds are reaching peak gusts of 100 knots.

Smaller storms are seen as bright spots over the Gulf of Mexico. These spots represent clusters of towering *cumulus* clouds that have grown into **thunderstorms**; that is, tall churn-

---

*Recall from p. 11 that 1 knot equals 1.15 miles per hour.

ing clouds accompanied by lightning, thunder, strong gusty winds, and heavy rain. If you look closely at Fig. 1.12, you will see similar cloud forms in many regions. There were probably thousands of thunderstorms occurring throughout the world at that very moment. Although they cannot be seen individually, there are even some thunderstorms embedded in the cloud mass west of the Great Lakes. Later in the day on which this photograph was taken, a few of these storms spawned the most violent disturbance in the atmosphere—the **tornado**.

A tornado is an intense rotating column of air that extends downward from the base of a thunderstorm. Sometimes called *twisters*, or *cyclones*, they may appear as ropes or as a large circular cylinder. The majority are less than a kilometer wide and many are smaller than a football field. Tornado winds may exceed 200 knots but most probably peak at less than 125 knots. Some tornadoes never reach the ground, and often appear to hang from the base of a parent cloud as a rapidly rotating funnel. Often, they dip down then rise up before disappearing.

A Look at a Weather Map   We can obtain a better picture of the middle latitude storm system by examining a simplified surface weather map for the same day that the satellite picture was taken. The weight of the air above different regions varies and, hence, so does the atmospheric pressure. In Fig. 1.13, the letter L on the map indicates a region of low atmospheric pressure, often called a *low*, which marks the center of the middle latitude storm. The two letters H on the map represent regions of high atmospheric pressure, called *highs*, or *anticyclones*. The circles on the map represent individual weather stations. The **wind** is the horizontal movement of air. The **wind direction**—the direction *from which* the wind is blowing*—is given by lines that parallel the wind and extend outward from the center of the station. The *wind speed*—the rate at which the air is moving past a stationary observer—is indicated by barbs.

Notice how the wind blows around the highs and the lows. The horizontal pressure differences create a force that starts the air moving from higher pressure toward lower pressure. Because of the earth's rotation, the winds are deflected toward the right in the Northern Hemisphere.† This deflection causes the winds to blow *clockwise* and *outward* from the center of the highs, and *counterclockwise* and *inward* toward the center of the low.

As the surface air spins into the low, it flows together and rises, much like toothpaste does when its open tube is squeezed. The rising air cools, and the moisture in the air condenses into clouds. Notice in Fig. 1.13 that the area of precipitation (the shaded green area) in the vicinity of the low corresponds to an extensive cloudy region in the satellite photo (Fig. 1.12).

---

*If you are facing north and the wind is blowing in your face, the wind would be called a "north wind."

†This deflecting force, known as the *Coriolis force*, is discussed more completely in Chapter 9, as are the winds.

WEATHER WATCH

The blizzard of 1996 was an awesome event. Lasting three days in January, the storm dumped huge amounts of snow over the east coast of the United States. Winds swirled the snow into monstrous drifts that closed roads and left many people stranded in their homes. Nine months later, during October, a second blizzard of sorts took place—a blizzard baby boom. Hospitals began reporting a sudden surge in baby births. West Jersey Hospital in Pennsauken, New Jersey, for example, delivered 25 percent more babies in October, 1996, than during the same month in 1995. Other hospitals throughout the northeast reported similar increases.

Also notice by comparing Figs. 1.12 and 1.13 that, in the regions of high pressure, skies are generally clear. As the surface air flows outward away from the center of a high, air sinking from above must replace the laterally spreading air. Since sinking air does not usually produce clouds, we find generally clear skies and fair weather associated with the regions of high pressure.

The swirling air around the areas of high and low pressure are the major weather producers for the middle latitudes. Look at the middle latitude storm and the surface temperatures in Fig. 1.13 and notice that, to the southeast of the storm, southerly winds from the Gulf of Mexico are bringing warm, humid air northward over much of the southeastern portion of the nation. On the storm's western side, cool dry northerly breezes combine with sinking air to create generally clear weather over the Rocky Mountains. The boundary that separates the warm and cool air appears as a heavy, dark line on the map—a **front**, across which there is a sharp change in temperature, humidity, and wind direction.

Where the cool air from Canada replaces the warmer air from the Gulf of Mexico, a *cold front* is drawn in blue, with arrowheads showing its general direction of movement. Where the warm Gulf air is replacing cooler air to the north, a *warm front* is drawn in red, with half circles showing its general direction of movement. Where the cold front has caught up to the warm front and cold air is now replacing cool air, an *occluded front* is drawn in purple, with alternating arrowheads and half circles to show how it is moving. Along each of the fronts, warm air is rising, producing clouds and precipitation. In the satellite photo (Fig. 1.12), the occluded front and the cold front appear as an elongated, curling cloud band that stretches from the low pressure area over Minnesota into the northern part of Texas.

Notice in Fig. 1.13 that the weather front is to the west of Chicago. As the westerly winds aloft push the front eastward, a person on the outskirts of Chicago might observe the approaching front as a line of towering thunderstorms similar

FIGURE 1.12
This satellite image (taken in visible reflected light) shows a variety of cloud patterns and storms in the earth's atmosphere.

to those in Fig. 1.14. In a few hours, Chicago should experience heavy showers with thunder, lightning, and gusty winds as the front passes. All of this, however, should give way to clearing skies and surface winds from the west or northwest after the front has moved on by.

Observing storm systems, we see that not only do they move but they constantly change. Steered by the upper-level westerly winds, the middle latitude storm in Fig. 1.13 intensifies into a larger storm, which moves eastward, carrying its clouds and weather with it. In advance of this system, a sunny day in Ohio will gradually cloud over and yield heavy showers and thunderstorms by nightfall. Behind the storm, cool dry northerly winds rushing into eastern Colorado cause an overcast sky to give way to clearing conditions. Farther south, the thunderstorms presently over the Gulf of Mexico

(Fig. 1.12) expand a little, then dissipate as new storms appear over water and land areas. To the west, the hurricane over the Pacific Ocean drifts northwestward and encounters cooler water. Here, away from its warm energy source, it loses its punch; winds taper off, and the storm soon turns into an unorganized mass of clouds and tropical moisture.

Weather and Climate in Our Lives   Weather and climate play a major role in our lives. Weather, for example, often dictates the type of clothing we wear, while climate influences the type of clothing we buy. Climate determines when to plant crops as well as what type of crops can be planted. Weather determines if these same crops will grow to maturity. Although weather and climate affect our lives in many ways, perhaps their most immediate effect is on our comfort. In order to survive the

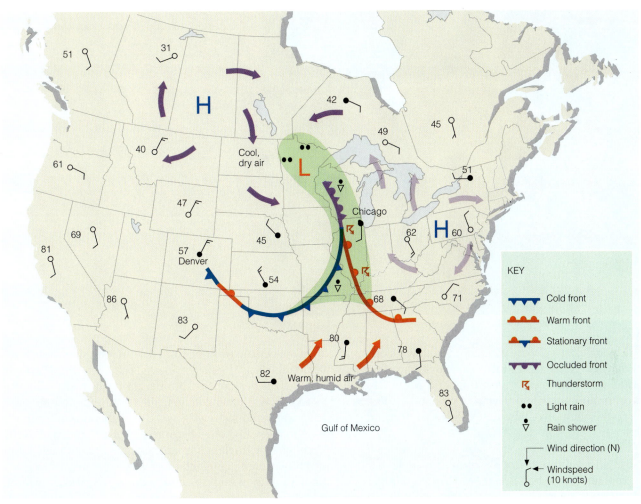

FIGURE 1.13
Simplified surface weather map that correlates with the satellite picture shown in Fig. 1.12. The shaded green area represents precipitation. The numbers on the map represent air temperatures in °F.

cold of winter and heat of summer, we build homes, heat them, air condition them, insulate them—only to find that when we leave our shelter, we are at the mercy of the weather elements.

Even when we are dressed for the weather properly, wind, humidity, and precipitation can change our perception of how cold or warm it feels. On a cold, windy day the effects of *wind chill* tell us that it feels much colder than it really is, and, if not properly dressed, we run the risk of *frostbite* or even *hypothermia* (the rapid, progressive mental and physical collapse that accompanies the lowering of human body temperature). On a hot, humid day we normally feel uncomfortably warm and blame it on the humidity. If we become too warm, our bodies overheat and *heat exhaustion* or *heat stroke* may result. Those most likely to suffer these maladies are the elderly with impaired circulatory systems and infants, whose heat regulatory mechanisms are not yet fully developed.

Weather affects how we feel in other ways, too. Arthritic pain is most likely to occur when rising humidity is accom-

WEATHER WATCH

When it rains, it rains pennies from heaven—sometimes. On July 17, 1940, a tornado reportedly picked up a treasure of over 1000 sixteenth-century silver coins, carried them into a thunderstorm, then dropped them on the village of Merchery in the Gorki region of Russia.

panied by falling pressures. In ways not well understood, weather does seem to affect our health. The incidence of heart attacks shows a statistical peak after the passage of warm fronts, when rain and wind are common, and after the passage of cold fronts, when an abrupt change takes place as showery precipitation is accompanied by cold gusty winds. Headaches are common on days when we are forced to squint, often due to hazy skies or a thin, bright overcast layer of high clouds.

FIGURE 1.14
Thunderstorms developing along an approaching cold front.

For some people, a warm dry wind blowing down-slope (a *chinook wind*) adversely affects their behavior (they often become irritable and depressed). Just how and why these winds impact humans physiologically is not well understood. We will take up the question of why these winds are warm and dry in Chapter 10.

When the weather turns colder or warmer than normal, it influences the lives and pocketbooks of many people. For example, the cool summer of 1992 over the eastern two-thirds of North America saved people billions of dollars in air-conditioning costs. On the other side of the coin, the bitter cold winter of 1986–1987 over Europe killed many hundreds of people and caused fuel rationing as demands for fuel exceeded supplies.

Major cold spells accompanied by heavy snow and ice can play havoc by snarling commuter traffic, curtailing airport services, closing schools, and downing power lines, thereby cutting off electricity to thousands of customers. For example, a huge ice storm during January, 1998, in northern New England and Canada left millions of people without power and caused over a billion dollars in damages, and a devastating snow storm during March, 1993, buried parts of the East Coast with 14-foot snow drifts and left Syracuse, New York, paralyzed with a snow depth of 36 inches. When the frigid air settles into the Deep South, many millions of dollars worth of temperature-sensitive fruits and vegetables may be ruined, the eventual consequence being higher produce prices in the supermarket.

Prolonged dry spells, especially when accompanied by high temperatures, can lead to a shortage of food and, in some places, widespread starvation. Parts of Africa, for example, have periodically suffered through major droughts and famine. In 1986, the southeastern section of the United States experienced a terrible drought as searing summer temperatures wilted crops, causing losses in excess of a billion dollars.

When the climate turns hot and dry, animals suffer too. Over 500,000 chickens perished in Georgia alone during a two-day period at the peak of the summer heat. Severe drought also has an effect on water reserves, often forcing communities to ration water and restrict its use. During periods of extended drought, vegetation often becomes tinder-dry and, sparked by lightning or a careless human, such a dried-up region can quickly become a raging inferno. During the summer of 1998, hundreds of thousands of acres in drought-stricken northern and central Florida were ravaged by wildfires.

Each year, the violent side of weather influences the lives of millions. It is amazing how many people whose family roots are in the Midwest know the story of someone who was severely injured or killed by a tornado. Tornadoes have not only taken many lives, but annually they cause damage to buildings and property totaling in the hundreds of millions of dollars, as a single large tornado can level an entire section of a town (see Fig. 1.15).

Although the gentle rains of a typical summer thunderstorm are welcome over much of North America, the heavy downpours, high winds, and hail of the *severe thunderstorms* are not. Cloudbursts from slowly moving, intense thunderstorms can provide too much rain too quickly, creating *flash floods* as small streams become raging rivers composed of mud and sand entangled with uprooted plants and trees (see Fig. 1.16). On the average, more people die in the United States from floods and flash floods than from any other natural disaster. Strong downdrafts originating inside a severe thunderstorm (a *downburst*) create turbulent winds that are capable of destroying crops and inflicting damage upon surface structures. Several airline crashes have been attributed to the turbulent *wind shear* zone within the downburst. Annually, hail damages crops worth millions of dollars, and light-

FIGURE 1.15
Tornadoes annually inflict widespread damage and cause the loss of many lives.

FIGURE 1.16
Flooding during April, 1997, inundates Grand Forks, North Dakota, as flood waters of the Red River extend over much of the city.

ning takes the lives of about eighty people in the United States and starts fires that destroy many thousands of acres of valuable timber (see Fig. 1.17).

Even the quiet side of weather has its influence. When winds die down and humid air becomes more tranquil, fog may form. Heavy fog can restrict visibility at airports, causing flight delays and cancellations. Every winter, deadly fog-related auto accidents occur along our busy highways and turnpikes. But fog has a positive side, too, especially during a dry spell, as fog moisture collects on tree branches and drips to the ground, where it provides water for the tree's root system.

Weather and climate have become so much a part of our lives that the first thing many of us do in the morning is to listen to the local weather forecast. For this reason, many radio and television newscasts have their own "weather person" to present weather information and give daily forecasts. More and more of these people are professionally trained in meteorology, and many stations require that the weathercaster obtain a seal of approval from the American Meteorological Society (AMS), or a certificate from the National Weather Association (NWA). To make their weather presentation as up-to-the-minute as possible, an increasing number of stations are taking advantage of the information provided by the National Weather Service (NWS), such as computerized weather forecasts, time-lapse satellite pictures, and color Doppler radar displays.

For many years now, a staff of trained professionals at "The Weather Channel" have provided weather information twenty-four hours a day on cable television. And finally, the National Oceanic and Atmospheric Administration (NOAA), in cooperation with the National Weather Service, sponsors weather radio broadcasts at selected locations across the

United States. Known as *NOAA weather radio* (and transmitted at VHF–FM frequencies), this service provides continuous weather information and regional forecasts (as well as special weather advisories, including watches and warnings) for over 90 percent of the nation.

FIGURE 1.17
Estimates are that lightning strikes the earth about 100 times every second. Consequently, it is a very common, and sometimes deadly, weather phenomenon.

SUMMARY

This chapter provides an overview of the earth's atmosphere. Our atmosphere is one rich in nitrogen and oxygen as well as smaller amounts of other gases, such as water vapor, carbon dioxide, and other greenhouse gases whose increasing levels may result in global warming. We examined the earth's early atmosphere and found it to be much different from the air we breathe today.

We investigated the various layers of the atmosphere and found that stratospheric ozone, which protects us from the sun's harmful ultraviolet rays, may be decreasing in concentration as gases such as chlorofluorocarbons in the stratosphere break apart and release ozone-destroying chlorine. We found that the coldest part of our atmosphere is in the mesosphere, that the warmest is in the thermosphere, and that all the weather we have come to know exists in the troposphere. We then looked at the ionosphere, the region of the upper atmosphere that contains large numbers of ions and free electrons.

We looked briefly at the weather map and a satellite photo and observed that dispersed throughout the atmosphere are storms and clouds of all sizes and shapes. The movement, intensification, and weakening of these systems, as well as the dynamic nature of air itself, produce a variety of weather events that we described in terms of weather elements. The sum total of weather and its extremes over a long period of time is what we call climate. Although sudden changes in weather may occur in a moment, climatic change takes place gradually over many years. The study of the atmosphere and all of its related phenomena is called *meteorology*, a term whose origin dates back to the days of Aristotle. Finally, we discussed some of many ways weather and climate influence our lives.

Key Terms

The following terms are listed in the order they appear in the text. Define each. Doing so will aid you in reviewing the material covered in this chapter.

atmosphere	lapse rate
nitrogen	temperature inversion
oxygen	troposphere
water vapor	radiosonde
carbon dioxide	stratosphere
ozone	tropopause
aerosol	mesosphere
pollutant	thermosphere
density	exosphere
pressure	homosphere
air pressure	heterosphere

ionosphere	hurricane
weather	thunderstorm
weather elements	tornado
climate	wind
meteorology	wind direction
middle latitudes	front
middle latitude cyclonic storm	

Questions for Review

1. List the four most abundant gases in today's atmosphere.
2. Of the four most abundant gases in our atmosphere, which one shows the greatest variation at the earth's surface?
3. What are some of the important roles that water plays in our atmosphere?
4. Briefly explain the production and natural destruction of carbon dioxide cycle near the Earth's surface. Why in the past 100 years, has CO_2 increased in the atmosphere?
5. List five greenhouse gases in the earth's atmosphere. What makes them greenhouse gases?
6. Explain how the atmosphere "protects" inhabitants at the earth's surface.
7. What are some of the aerosols in our atmosphere?
8. How has the composition of the earth's atmosphere changed over time? Briefly outline the evolution of the earth's atmosphere.
9. (a) Explain the concept of air pressure in terms of weight of air above some level. (b) Why does air pressure always decrease with increasing height above the surface?
10. What is standard atmospheric pressure at sea level in (a) inches of mercury, (b) millibars, and (c) hectopascals?
11. What is the average or standard lapse rate in the troposphere?
12. Briefly describe how the air temperature changes from the earth's surface to the lower thermosphere. What atmospheric layer contains all of our weather?
13. (a) In what atmospheric layer do we find the lowest average air temperature?
    (b) The highest average air temperature?
14. How does the ionosphere affect AM radio transmission during the day versus during the night?
15. Even though the actual concentration of oxygen is close to 21 percent (by volume) in the upper stratosphere, explain why you would not be able to survive there.
16. Define *meteorology* and discuss the origin of this word.
17. When someone says that "the wind direction today is south," what does that mean?
18. Describe some of the features observed on a surface weather map.

19. Explain how wind flow differs between low and high pressure systems.
20. Describe some of the ways weather and climate influence the lives of people.
21. How are fronts defined?
22. Rank the following storms in size from largest to smallest: hurricane, tornado, middle latitude storm, thunderstorm.

Questions for Thought

1. Which of the following statements relate more to weather and which relate more to climate?
    (a) The summers here are warm and humid.
    (b) Cumulus clouds presently cover the entire sky.
    (c) Our lowest temperature last winter was −29°C (−18°F).
    (d) The air temperature outside is 22°C (72°F).
    (e) December is our foggiest month.
    (f) The highest temperature ever recorded in Phoenix-ville, Pennsylvania, was 44°C (111°F) on July 10, 1936.
    (g) Snow is falling at the rate of 5 cm (2 in.) per hour.
    (h) The average temperature for the month of January in Chicago, Illinois, is −3°C (26°F).
2. How does weather differ from climate?
3. Why do you think that winds are named from the direction they are blowing?
4. A standard pressure of 1013.25 millibars is also known as one atmosphere (1 ATM). (a) Look at Fig. 1.8 and determine at approximately what levels you would record a pressure of 0.5 ATM and 0.1 ATM. (b) The surface air pressure on the planet Mars is about 0.007 ATM. If you were standing on Mars, the surface air pressure would be equivalent to a pressure observed at approximately what elevation in the Earth's atmosphere?

Problems and Exercises

1. Keep track of the weather. On an outline map of North America, mark the daily position of fronts and pressure systems for a period of several weeks or more. (This information can be obtained from newspapers, the TV news, or the Internet.) Plot the general upper-level flow pattern on the map. Observe how the surface systems move. Relate this information to the material on wind, fronts, and cyclones covered in later chapters.
2. Compose a one-week journal, including daily newspaper weather maps and weather forecasts from the newspaper or Internet. Provide a commentary for each day regarding the coincidence of actual and predicted weather.
3. Formulate a short-term climatology for your city for one month by recording maximum and minimum temperatures and precipitation amounts every day. You can get this information from television, newspapers, the Internet, or your own measurements. Compare it to the actual climatology for that month. How can you explain any large differences between the two?

 ## Questions for Exploration

On the Blue Skies: College Edition CD-ROM go to the **Atmospheric Basics** section of the CD and click on **"Layers of the Atmosphere."**

1. Explore the "Standard Atmosphere" by piloting up through this "average" atmosphere.
    (a) Does the temperature decrease or increase with height near the surface?
    (b) What heating process is responsible for this pattern?
    (c) The tropopause is marked by the point, usually above 8 km above sea level, where temperature systematically stops dropping with height and becomes either isothermal or increasing with height. At what height does this happen?
2. Explore the vertical profile of temperature for your location. Note the time of the sounding as either 00Z (GMT) or 12Z.
    (a) Does the temperature decrease or increase with height near the surface?
    (b) At what height does the tropopause happen over your location?
    (c) If the height over you is different from that in the standard atmosphere what might explain the difference?

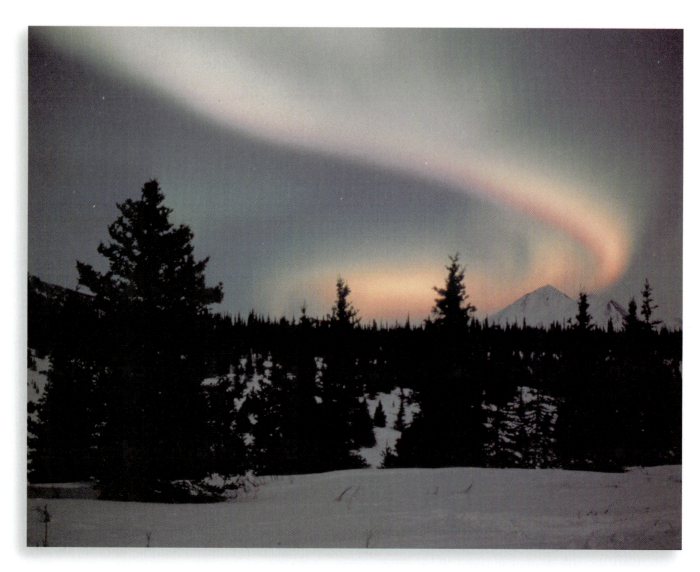

The aurora borealis, which forms
as energetic particles from the sun
interact with the earth's atmosphere.
(© Johnny Johnson, Allstock)

ENERGY: WARMING THE EARTH AND THE ATMOSPHERE

CONTENTS

At high latitudes after darkness has fallen, a faint, white glow may appear in the sky. Lasting from a few minutes to a few hours, the light may move across the sky as a yellow green arc much wider than a rainbow; or, it may faintly decorate the sky with flickering draperies of blue, green, and purple light that constantly change in form and location, as if blown by a gentle breeze.

For centuries curiosity and superstition have surrounded these eerie lights. Eskimo legend says they are the lights from demons' lanterns as they search the heavens for lost souls. Nordic sagas called them a reflection of fire that surrounds the seas of the north. Even today there are those who proclaim that the lights are reflected sunlight from polar ice fields. Actually, this light show in the Northern Hemisphere is the aurora borealis—the northern lights—which is caused by invisible energetic particles bombarding our upper atmosphere. Anyone who witnesses this, one of nature's spectacular color displays, will never forget it.

E nergy is everywhere. It is the basis for life. It comes in various forms: it can warm a house, melt ice, and drive the atmosphere, producing our everyday weather events. When the sun's energy interacts with our upper atmosphere we see energy at work in yet another form, a shimmering display of light from the sky—the aurora. What, precisely, is this common, yet mysterious, quantity we call "energy"? What is its primary source? How does it warm our earth and provide the driving force for our atmosphere? And in what form does it reach our atmosphere to produce a dazzling display like the aurora?

To answer these questions, we must first begin with the concept of energy itself. Then we will examine energy in its various forms and how energy is transferred from one form to another in our atmosphere. Finally, we will look more closely at the sun's energy and its influence on our atmosphere.

ENERGY, TEMPERATURE, AND HEAT

By definition, **energy** is the ability or capacity to do work on some form of matter. (Matter is anything that has mass and occupies space.) Work is done on matter when matter is either pushed, pulled, or lifted over some distance. When we lift a brick, for example, we exert a force against the pull of gravity—we "do work" on the brick. The higher we lift the brick, the more work we do. So, by doing work on something, we give it "energy," which it can, in turn, use to do work on other things. The brick that we lifted, for instance, can now do work on your toe—by falling on it.

The total amount of energy stored in any object (internal energy) determines how much work that object is capable of doing. A lake behind a dam contains energy by virtue of its position. This is called *gravitational potential energy* or simply **potential energy** because it represents the potential to do work—a great deal of destructive work if the dam were to break. The potential energy (PE) of any object is given as

$$PE = mgh,$$

where m is the object's mass, g is the acceleration of gravity, and h is the object's height above the ground.

A volume of air aloft has more potential energy than the same size volume of air just above the surface. This fact is so because the air aloft has the potential to sink and warm through a greater depth of atmosphere. A substance also possesses potential energy if it can do work when a chemical change takes place. Thus, coal, natural gas, and food all contain chemical potential energy.

Any moving substance possesses energy of motion, or **kinetic energy**. The kinetic energy (KE) of an object is equal to half its mass multiplied by its velocity squared; thus

$$KE = \tfrac{1}{2}\,mv^2.$$

Consequently, the faster something moves, the greater its kinetic energy; hence, a strong wind possesses more kinetic energy than a light breeze. Since kinetic energy also depends on the object's mass, a volume of water and an equal volume of air may be moving at the same speed, but, because the water has greater mass, it has more kinetic energy. The atoms and molecules that comprise all matter have kinetic energy due to their motion. This form of kinetic energy is often referred to as *heat energy*. Probably the most important form of energy in terms of weather and climate is the energy we receive from the sun—*radiant energy*.

Energy, therefore, takes on many forms, and it can change from one form into another. But the total amount of energy in the universe remains constant. *Energy cannot be created nor can it be destroyed.* It merely changes from one form to another in any ordinary physical or chemical process. In other words, the energy lost during one process must equal the energy gained during another. This is what we mean when we say that energy is conserved. This statement is known as the *law of conservation of energy*, and is also called the *first law of thermodynamics*.

We know that air is a mixture of countless billions of atoms and molecules. If they could be seen, they would appear to be moving about in all directions, freely darting, twisting, spinning, and colliding with one another like an angry swarm of bees. Close to the earth's surface, each individual molecule will travel only about a thousand times its diameter before colliding with another molecule. Moreover, we would see that all the atoms and molecules are not moving at the same speed, as some are moving faster than others. The temperature of the air (or any substance) is a measure of its average kinetic energy. Simply stated, **temperature** *is a measure of the average speed of the atoms and molecules*, where higher temperatures correspond to faster average speeds.

Suppose we examine a volume of surface air about the size of a large flexible balloon. If we warm the air inside, the molecules would move faster, but they also would move slightly farther apart—the air becomes less dense. Conversely, if we cool the air, the molecules would slow down, crowd closer together, and the air would become more dense. This molecular behavior is why, in many places throughout the book, we refer to surface air as either *warm, less-dense air* or as *cold, more-dense air*.

The atmosphere and oceans contain *internal energy*, which is the total energy (potential and kinetic) stored in their molecules. As we have just seen, the temperature of air and water is determined only by the *average* kinetic energy (average speed) of *all* their molecules. Since temperature only indicates how "hot" or "cold" something is relative to some set standard value, it does not always tell us how much internal energy that something possesses. For example, two identical mugs, each half-filled with water and each with the same temperature, contain the same internal energy. If the water from one mug is poured into the other, the total internal energy of the filled mug has doubled because its mass has doubled. Its temperature, however, has not changed, since the average speed of all of the molecules is still the same (see Fig. 2.1).

Now, imagine that you are sipping a hot cup of tea on a small raft in the middle of a lake. The tea has a much higher

FIGURE 2.1
Even though the temperature of the beverage in both mugs is the same, the mug on the left contains more internal energy because it contains more molecules.

temperature than the lake, yet the lake contains more internal energy because it is composed of many more molecules. If the cup of tea is allowed to float on top of the water, the tea would cool rapidly. The energy that would be transferred from the hot tea to the cool water (because of their temperature difference) is called *heat*.

In essence, **heat** *is energy in the process of being transferred from one object to another because of the temperature difference between them.* After heat is transferred, it is stored as internal energy. How is this energy transfer process accomplished? In the atmosphere, heat is transferred by *conduction*, *convection*, and *radiation*. We will examine these mechanisms of energy transfer after we look at temperature scales and at the important concepts of *specific heat* and *latent heat*.

Temperature Scales Suppose we take a small volume of air and allow it to cool. As the air slowly cools, its atoms and molecules would move slower and slower until the air reaches a temperature of −273°C (−459°F), which is the lowest temperature possible. At this temperature, called **absolute zero**, the atoms and molecules would possess a minimum amount of energy and theoretically no thermal motion. At absolute zero, we can begin a temperature scale called the *absolute* or **Kelvin scale** after Lord Kelvin (1824–1907), a famous British scientist who first introduced it. Since the Kelvin scale begins at absolute zero, it contains no negative numbers and is, therefore, quite convenient for scientific calculations.

Two other temperature scales commonly used today are the Fahrenheit and Celsius (formerly centigrade). The **Fahrenheit scale** was developed in the early 1700s by the physicist G. Daniel Fahrenheit, who assigned the number 32 to the temperature at which water freezes, and the number 212 to the temperature at which water boils. The zero point was simply the lowest temperature that he obtained with a mixture of ice, water, and salt. Between the freezing and boiling points are 180 equal divisions, each of which is called a degree. A thermometer calibrated with this scale is referred to as a Fahrenheit thermometer, for it measures an object's temperature in degrees Fahrenheit (°F).

The **Celsius scale** was introduced later in the eighteenth century. The number 0 (zero) on this scale is assigned to the

temperature at which pure water freezes, and the number 100 to the temperature at which pure water boils at sea level. The space between freezing and boiling is divided into 100 equal degrees. Therefore, each Celsius degree is 180/100 or 1.8 times bigger than a Fahrenheit degree. Put another way, an increase in temperature of 1°C equals an increase of 1.8°F. A formula for converting °C to °F is

$$°C = \tfrac{5}{9} \, (°F - 32).$$

On the Kelvin scale, degrees Kelvin are called *Kelvins* (abbreviated K). Each degree on the Kelvin scale is exactly the same size as a degree Celsius, and a temperature of 0 K is equal to −273°C. Converting from °C to K can be made by simply adding 273 to the Celsius temperature, as

$$K = °C + 273.$$

Figure 2.2 compares the Fahrenheit, Celsius, and Kelvin scales. Converting a temperature from one scale to another can be done by simply reading the corresponding temperature from the adjacent scale. Thus, 303 on the Kelvin scale is the equivalent of 30°C and 86°F.*

---

*A more complete table of conversions is given in Appendix A.

FIGURE 2.2
Comparison of Fahrenheit, Celsius, and Kelvin scales.

In most of the world, temperature readings are taken in °C. In the United States, however, temperatures above the surface are taken in °C, while temperatures at the surface are typically read in °F. Currently, then, temperatures on upper-level maps are plotted in °C, while, on surface weather maps, they are in °F. Since both scales are in use, temperature readings in this book will, in most cases, be given in °C followed by their equivalent in °F.

Specific Heat

A watched pot never boils, or so it seems. The reason for this is that water requires a relatively large amount of heat energy to bring about a small temperature change. The **heat capacity** of a substance is the ratio of the amount of heat energy absorbed by that substance to its corresponding temperature rise. The heat capacity of a substance per unit mass is called **specific heat**. In other words, specific heat is the amount of heat needed to raise the temperature of one gram (g) of a substance one degree Celsius.

If we heat 1 g of liquid water on a stove, it would take about 1 calorie* (cal) to raise its temperature by 1°C. So water has a specific heat of 1. If, however, we put the same amount (that is, same mass) of compact dry soil on the flame, we would see that it would take about one-fifth the heat (about 0.2 cal) to raise its temperature by 1°C. The specific heat of water is therefore 5 times greater than that of soil. In other words, water must absorb 5 times as much heat as the same quantity of soil in order to raise its temperature by the same amount. The specific heat of various substances is given in Table 2.1.

Not only does water heat slowly, it cools slowly as well. It has a much higher capacity for storing energy than other common substances, such as soil and air. A given volume of water can store a large amount of energy while undergoing only a small temperature change. Because of this attribute, water has a strong modifying effect on weather and climate. Near large bodies of water, for example, winters usually remain warmer

---

*By definition, a calorie is the amount of heat required to raise the temperature of 1 g of water from 14.5°C to 15.5°C. The kilocalorie is 1000 calories and is the heat required to raise 1 kg of water 1°C. In the SI system, the unit of energy is the joule (J), where 1 calorie = 4.186 J. (For pronunciation: joule rhymes with pool.)

TABLE 2.1 **Specific Heat of Various Substances**

SUBSTANCE	SPECIFIC HEAT (Cal/g × °C)	J/kg × °C
Water (pure)	1.00	4186
Wet mud	0.60	2512
Ice (0°C)	0.50	2093
Sandy clay	0.33	1381
Dry air (sea level)	0.24	1005
Quartz sand	0.19	795
Granite	0.19	794

and summers cooler than nearby inland regions—a fact well known to people who live adjacent to oceans or large lakes.

Latent Heat—The Hidden Warmth We know from Chapter 1 that water vapor is an invisible gas that becomes visible when it changes into larger liquid or solid (ice) particles. This process of transformation is known as a *change of state* or, simply, a *phase change*. The heat energy required to change a substance, such as water, from one state to another is called **latent heat**. But why is this heat referred to as "latent"? To answer this question, we will begin with something familiar to most of us—the cooling produced by evaporating water.

Suppose we microscopically examine a small drop of pure water. At the drop's surface, molecules are constantly escaping (evaporating). Because the more energetic, faster-moving molecules escape most easily, the average motion of all the molecules left behind decreases as each additional molecule evaporates. Since temperature is a measure of average molecular motion, the slower motion suggests a lower water temperature. *Evaporation is, therefore, a cooling process.* Stated another way, evaporation is a cooling process because the energy needed to evaporate the water—that is, to change its phase from a liquid to a gas—may come from the water or other sources, including the air.

In the everyday world, we experience evaporational cooling as we step out of a shower or swimming pool into a dry area. Because some of the energy used to evaporate the water comes from our skin, we may experience a rapid drop in skin temperature, even to the point where goose bumps form. In fact, on a hot, dry, windy day in Tucson, Arizona, cooling may be so rapid that we begin to shiver even though the air temperature is hovering around 38°C (100°F).

The energy lost by liquid water during evaporation can be thought of as carried away by, and "locked up" within, the water vapor molecule. The energy is thus in a "stored" or "hidden" condition and is, therefore, called *latent heat*. It is latent (hidden) in that the temperature of the substance changing from liquid to vapor is still the same. However, the heat energy will reappear as **sensible heat** (the heat we can feel and measure with a thermometer) when the vapor condenses back into liquid water. Therefore, *condensation* (the opposite of evaporation) *is a warming process*.

The heat energy released when water vapor condenses to form liquid droplets is called *latent heat of condensation*. Conversely, the heat energy used to change liquid into vapor at the same temperature is called *latent heat of evaporation* (vaporization). Nearly 600 cal (2500 J) are required to evaporate a single gram of water at room temperature. With many hundreds of grams of water evaporating from the body, it is no wonder that after a shower we feel cold before drying off.

In a way, latent heat is responsible for keeping a cold drink with ice colder than one without ice. As ice melts, its temperature does not change. The reason for this is that the heat added to the ice only breaks down the rigid crystal pattern, changing the ice to a liquid without changing its temperature.

The energy used in this process is called *latent heat of fusion* (melting). Roughly 80 cal (335 J) are required to melt a single gram of ice. Consequently, heat added to a cold drink with ice primarily melts the ice, while heat added to a cold drink without ice warms the beverage. If a gram of water at 0°C changes back into ice at 0°C, this same amount of heat (80 cal) would be released as sensible heat to the environment. Therefore, when ice melts, heat is taken in; when water freezes, heat is liberated.

The heat energy required to change ice into vapor (a process called *sublimation*) is referred to as *latent heat of sublimation*. For a single gram of ice to transform completely into vapor at 0°C requires nearly 680 cal—80 cal for the latent heat of fusion plus 600 cal for the latent heat of evaporation. If this same vapor transformed back into ice (a process called *deposition*), approximately 680 cal (2850 J) would be released.

Figure 2.3 summarizes the concepts examined so far. When the change of state is from left to right, heat is absorbed by the substance and taken away from the environment. The processes of melting, evaporation, and sublimation all cool the environment. When the change of state is from right to left, heat energy is given up by the substance and added to the environment. The process of freezing, condensation, and deposition all warm their surroundings.

Latent heat is an important source of atmospheric energy. Once vapor molecules become separated from the earth's surface, they are swept away by the wind, like dust before a broom. Rising to high altitudes where the air is cold, the vapor changes into liquid and ice cloud particles. During these processes, a tremendous amount of heat energy is released into the environment. This heat provides energy for storms, such as hurricanes, middle latitude cyclones, and thunderstorms (see Fig. 2.4).

FIGURE 2.3
Heat energy absorbed and released.

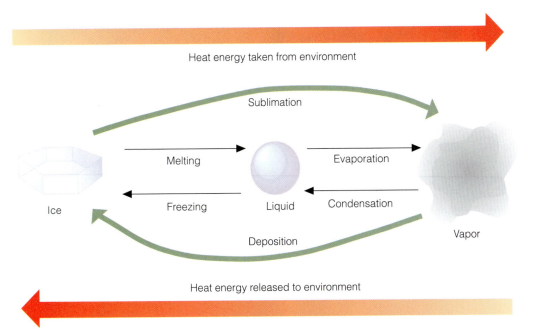

Heat energy taken from environment

Sublimation

Melting · Evaporation

Ice · Liquid · Vapor

Freezing · Condensation

Deposition

Heat energy released to environment

FIGURE 2.4
Every time a cloud forms, it warms the atmosphere. Inside this developing thunderstorm a vast amount of stored heat energy (latent heat) is given up to the air, as invisible water vapor becomes countless billions of water droplets and ice crystals. In fact, for the duration of this storm alone, more heat energy is released inside this cloud than is unleashed by a small nuclear bomb.

The Fate of a Sunbeam

Consider sunlight in the form of radiant energy striking a large lake. (See Fig. 1.) Part of the incoming energy heats the water, causing greater molecular motion and, hence, an increase in the water's kinetic energy. This greater kinetic energy allows more water molecules to evaporate from the surface. As each molecule escapes, work is done to break it away from the remaining water molecules. This energy becomes the latent heat energy that is carried with the water vapor.

Above the lake, a parcel of warm, moist air rises and expands. In order for this expansion to take place, the gas molecules inside the parcel must use some of their kinetic energy to do work against the parcel's sides. This results in a slower molecular speed and a lower temperature. Well above the surface, the water vapor in the rising, cooling parcel of moist air condenses into clouds. The condensation of water vapor releases latent heat energy into the atmosphere, warming the air. The tiny suspended cloud droplets possess potential energy, which becomes kinetic energy when these

FIGURE 1
Solar energy striking a large body of water goes through many transformations.

droplets grow into raindrops that fall earthward.

When the drops reach the surface, their kinetic energy erodes the land. As rain-swollen streams flow into a lake behind a dam, there is a buildup of potential energy, which can be transformed into kinetic energy as water is harnessed to flow down a chute. If the moving water drives a generator, kinetic energy is con-

verted into electrical energy, which is sent to cities. There, it heats, cools, and lights the buildings in which people work and live. Meanwhile, some of the water in the lake behind the dam evaporates and is free to repeat the cycle. Hence, the energy from the sunlight on a lake can undergo many transformations and help provide the moving force for many natural and human-made processes.

Water vapor evaporated from warm, tropical water can be carried into polar regions, where it condenses and gives up its heat energy. Thus, as we will see, evaporation–transportation– condensation is an extremely important mechanism for the relocation of heat energy (as well as water) in the atmosphere. (Before going on to the next section, you may wish to read the Focus section above, which summarizes some of the concepts considered thus far.)

HEAT TRANSFER IN THE ATMOSPHERE

Conduction   The transfer of heat from molecule to molecule within a substance is called **conduction**. Hold one end of a metal straight pin between your fingers and place a flaming candle under the other end (see Fig. 2.5). Because of the energy they absorb from the flame, the molecules in the pin vibrate faster. The faster-vibrating molecules cause adjoining mole-

cules to vibrate faster. These, in turn, pass vibrational energy on to their neighboring molecules, and so on, until the molecules at the finger-held end of the pin begin to vibrate rapidly. These fast-moving molecules eventually cause the molecules of your finger to vibrate more quickly. Heat is now being trans-

FIGURE 2.5
The transfer of heat from the hot end of the metal pin to the cool end by molecular contact is called *conduction*.

ferred from the pin to your finger, and both the pin and your finger feel hot. If enough heat is transferred, you will drop the pin. The transmission of heat from one end of the pin to the other, and from the pin to your finger, occurs by conduction. Heat transferred in this fashion always flows from *warmer to colder* regions. Generally, the greater the temperature difference, the more rapid the heat transfer.

When materials can easily pass energy from one molecule to another, they are considered to be good conductors of heat. How well they conduct heat depends upon how their molecules are structurally bonded together. Table 2.2 shows that solids, such as metals, are good heat conductors. It is often difficult, therefore, to judge the temperature of metal objects. For example, if you grab a metal pipe at room temperature, it will seem to be much colder than it actually is because the metal conducts heat away from the hand quite rapidly. Conversely, *air is an extremely poor conductor of heat*, which is why most insulating materials have a large number of air spaces trapped within them. Air is such a poor heat conductor that, in calm weather, the hot ground only warms a shallow layer of air a few centimeters thick by conduction. Yet, air can carry this energy rapidly from one region to another. How then does this phenomenon happen?

Convection
The transfer of heat by the mass movement of a fluid (such as water and air) is called **convection**. This type of heat transfer takes place in liquids and gases because they can move freely and it is possible to set up currents within them.

Convection happens naturally in the atmosphere. On a warm, sunny day, certain areas of the earth's surface absorb more heat from the sun than others; as a result, the air near the earth's surface is heated somewhat unevenly. Air molecules adjacent to these hot surfaces bounce against them, thereby gaining some extra energy by conduction. The heated air expands and becomes less dense than the surrounding cooler air. The expanded warm air is buoyed upward and rises. In this manner, large bubbles of warm air rise and transfer heat energy upward. Cooler, heavier air flows toward the surface to replace the rising air. This cooler air becomes heated in turn,

TABLE 2.2   Heat Conductivity* of Various Substances

SUBSTANCE	HEAT CONDUCTIVITY (Watts per meter per °C)
Still air	0.023 (at 20°C)
Wood	0.08
Dry soil	0.25
Water	0.60 (at 20°C)
Snow	0.63
Wet soil	2.1
Ice	2.1
Sandstone	2.6
Granite	2.7
Iron	80
Silver	427

*Heat (thermal) conductivity describes a substance's ability to conduct heat as a consequence of molecular motion.

rises, and the cycle is repeated. In meteorology, this vertical exchange of heat is called *convection*, and the rising air bubbles are known as **thermals** (see Fig. 2.6).

The rising air expands and gradually spreads outward. It then slowly begins to sink. Near the surface, it moves back into the heated region, replacing the rising air. In this way, a *convective circulation*, or thermal "cell," is produced in the atmosphere. In a convective circulation the warm, rising air cools. In our atmosphere, *any air that rises will expand and cool*, and *any air that sinks is compressed and warms*. This important concept is detailed in the Focus section on p. 32.

Although the entire process of heated air rising, spreading out, sinking, and finally flowing back toward its original location is known as a convective circulation, meteorologists usually restrict the term *convection* to the process of the rising and sinking part of the circulation.

The horizontally moving part of the circulation (called *wind*) carries properties of the air in that particular area with it. The transfer of these properties by horizontally moving air

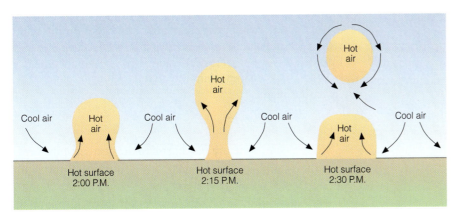

FIGURE 2.6
The development of a thermal. A thermal is a rising bubble of air that carries heat energy upward by *convection*.

Cool air   Hot air   Cool air   Cool air   Hot air   Cool air

Hot air   Hot air   Hot air

Hot surface 2:00 P.M.   Hot surface 2:15 P.M.   Hot surface 2:30 P.M.

Rising Air Cools and Sinking Air Warms

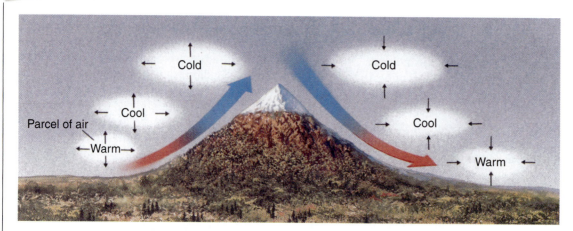

FIGURE 2
Rising air expands and cools; sinking air is compressed and warms.

To understand why rising air cools and sinking air warms we need to examine some air. Suppose we place air in an imaginary thin, elastic wrap about the size of a large balloon (see Fig. 2). This invisible balloonlike "blob" is called a *parcel*. The air parcel can expand and contract freely, but neither external air nor heat is able to mix with the air inside. By the same token, as the parcel moves, it does not break apart, but remains as a single unit.

At the earth's surface, the parcel has the same temperature and pressure as the air surrounding it. Suppose we lift the parcel. Recall from Chapter 1 that air pressure always decreases as we move up into the atmosphere. Consequently, as the parcel rises, it enters a region where the surrounding air pressure is lower. To equalize the pressure, the parcel molecules inside push the parcel walls outward, expanding it. Because there is no other energy source, the air molecules inside use some of their own energy to expand the parcel. This energy loss shows up as slower molecular speeds, which represent a lower parcel temperature. Hence, *any air that rises always expands and cools.*

If the parcel is lowered to the earth, it returns to a region where the air pressure is higher. The higher outside pressure squeezes (compresses) the parcel back to its original (smaller) shape. Because air molecules have a faster rebound velocity after striking the sides of a collapsing parcel, the average speed of the molecules inside goes up. (A Ping-pong ball moves faster after striking a paddle that is moving toward it.) This increase in molecular speed represents a warmer parcel temperature. Therefore, *any air that sinks (subsides), warms by compression.*

WEATHER WATCH

Although we can't see air, there are signs that tell us where the air is rising. One example: On a calm day, you can watch a hawk circle and climb high above level ground while its wings remain motionless. A rising thermal carries the hawk upward as it scans the terrain for prey. Another example: If the water vapor of a rising thermal condenses into liquid cloud droplets, the thermal becomes visible to us as a puffy cumulus cloud. Flying in a light aircraft beneath these clouds usually produces a bumpy ride, as passengers are jostled around by the rising and sinking air associated with convection.

is called **advection**. For example, wind blowing across a body of water will "pick up" water vapor from the evaporating surface and transport it elsewhere in the atmosphere. If the air cools, the water vapor may condense into cloud droplets and release latent heat. In a sense, then, heat is advected (carried) by the water vapor as it is swept along with the wind. Earlier we saw that this is an important way to redistribute heat energy in the atmosphere.

BRIEF REVIEW   Before moving on to the next section, here is a summary of some of the important concepts and facts we have covered:

■ The temperature of a substance is a measure of the average kinetic energy (average speed) of its atoms and molecules.

- Evaporation (the transformation of liquid into vapor) is a cooling process that can cool the air, whereas condensation (the transformation of vapor into liquid) is a warming process that can warm the air.

- Heat is energy in the process of being transferred from one object to another because of the temperature difference between them.

- In conduction, which is the transfer of heat by molecule-to-molecule contact, heat always flows from warmer to colder regions.

- Air is a poor conductor of heat.

- Convection is an important mechanism of heat transfer, as it represents the vertical movement of warmer air upward and cooler air downward.

There is yet another mechanism for the transfer of energy—radiation, or *radiant energy*, which is what we receive from the sun. In this method, energy may be transferred from one object to another without the space between them necessarily being heated.

RADIATION

On a summer day, you may have noticed how warm and flushed your face feels as you stand facing the sun. Sunlight travels through the surrounding air with little effect upon the air itself. Your face, however, absorbs this energy and converts it to thermal energy. Thus, sunlight warms your face without actually warming the air. The energy transferred from the sun to your face is called **radiant energy** or **radiation**. It travels in the form of waves that release energy when they are absorbed by an object. Because these waves have magnetic and electrical properties, we call them **electromagnetic waves**. Electromagnetic waves do not need molecules to propagate them. In a vacuum, they travel at a constant speed of nearly 300,000 km (186,000 mi) per second—the speed of light.

Figure 2.7 shows some of the different wavelengths of radiation. Notice that the **wavelength** (which is usually expressed by the Greek letter lambda, λ) is the distance measured along a wave from one crest to another. Also notice that some of the waves have exceedingly short lengths. For example, radiation that we can see (visible light) has an average wavelength of less than one-millionth of a meter—a distance nearly one-hundredth the diameter of a human hair. To measure these short lengths, we introduce a new unit of measurement called a **micrometer** (represented by the symbol μm), which is equal to one-millionth of a meter (m); thus

$$1 \text{ micrometer } (\mu m) = 0.000001 \text{ m} = 10^{-6} \text{ m}.$$

In Fig. 2.7, we can see that the average wavelength of visible light is about 0.0000005 m, which is the same as 0.5 μm. To give you a common object for comparison, the average height of a letter on this page is about 2000 μm, or 2 millimeters (2 mm), while the thickness of this page is about 100 μm.

We can also see in Fig. 2.7 that the longer waves carry less energy than do the shorter waves. When comparing the energy carried by various waves, it is useful to give electromagnetic radiation characteristics of particles in order to explain some of the waves' behavior. We can actually think of radiation as streams of particles or **photons** that are discrete packets of energy.*

---

*Packets of photons make up waves, and groups of waves make up a beam of radiation.

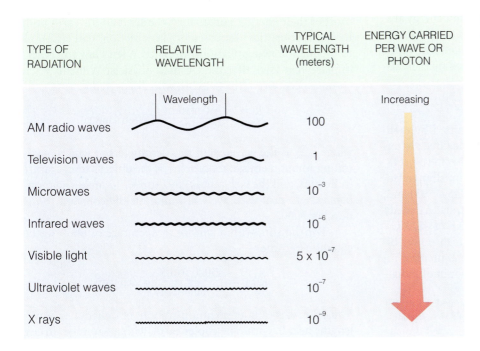

TYPE OF RADIATION	RELATIVE WAVELENGTH	TYPICAL WAVELENGTH (meters)	ENERGY CARRIED PER WAVE OR PHOTON
	Wavelength		Increasing
AM radio waves		100	
Television waves		1	
Microwaves		10^{-3}	
Infrared waves		10^{-6}	
Visible light		5×10^{-7}	
Ultraviolet waves		10^{-7}	
X rays		10^{-9}	

FIGURE 2.7
Radiation characterized according to wavelength. As the wavelength decreases, the energy carried per wave increases.

An ultraviolet photon carries more energy than a photon of visible light. In fact, certain ultraviolet photons have enough energy to produce sunburns and penetrate skin tissue, sometimes causing skin cancer. As we discussed in Chapter 1, it is ozone in the stratosphere that protects us from the vast majority of these harmful rays.

Radiation and Temperature

All things (whose temperature is above absolute zero), no matter how big or small, emit radiation. This book, your body, flowers, trees, air, the earth, the stars are all radiating a wide range of electromagnetic waves. The energy originates from rapidly vibrating electrons, billions of which exist in every object.

The wavelengths that each object emits depend primarily on the object's temperature. The higher the temperature, the faster the electrons vibrate, and the shorter are the wavelengths of the emitted radiation. This can be visualized by attaching one end of a rope to a post and holding the other end. If the rope is shaken rapidly (high temperature), numerous short waves travel along the rope; if the rope is shaken slowly (lower temperature), longer waves appear on the rope. Although objects at a temperature of about 500°C radiate waves with many lengths, some of them are short enough to stimulate the sensation of vision. We actually see these objects glow red. Objects cooler than this radiate at wavelengths that are too long for us to see. The page of this book, for example, is radiating electromagnetic waves. But because its temperature is only around 20°C (68°F), the waves emitted are much too long to stimulate vision. We are able to see the page, however, because light waves from other sources (such as light bulbs or the sun) are being *reflected* (bounced) off the paper. If this book were carried into a completely dark room, it would continue to radiate, but the pages would appear black because there are no visible light waves in the room to reflect off the pages.

Objects that have a very high temperature emit energy at a greater rate or intensity than objects at a lower temperature. Thus, *as the temperature of an object increases, more total radiation is emitted each second.* This can be expressed mathematically as

$$E = \sigma T^4 \qquad \text{(Stefan-Boltzmann law)},$$

where E is the maximum rate of radiation emitted by each square meter of surface area of the object, σ (the Greek letter sigma) is the Stefan-Boltzmann constant,* and T is the object's surface temperature in degrees Kelvin. This relationship, called the **Stefan-Boltzmann law** after Josef Stefan (1835–1893) and Ludwig Boltzmann (1844–1906), who derived it, states that all objects with temperatures above absolute zero (0 K or –273°C) emit radiation at a rate proportional to the fourth power of their absolute temperature. Consequently, a small increase in temperature results in a large increase in the amount of radia-

---

*The Stefan-Boltzmann constant σ in SI units is 5.67×10^{-8} W/m²k⁴. A watt (W) is a unit of power where one watt equals one joule (J) per second (J/s). One joule is equal to 0.24 cal. More conversions are given in Appendix A.

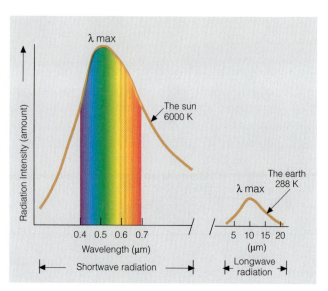

FIGURE 2.8

The hotter sun not only radiates more energy than that of the cooler earth (the area under the curve), but it also radiates the majority of its energy at much shorter wavelengths. (The scales for the two curves differ by a factor of 100,000.)

tion emitted because doubling the absolute temperature of an object increases the maximum energy output by a factor of 16, which is 2^4.

Radiation of the Sun and Earth

Most of the sun's energy is emitted from its surface, where the temperature is nearly 6000 K (10,500°F). The earth, on the other hand, has an average surface temperature of 288 K (15°C, 59°F). The sun, therefore, radiates a great deal more energy than does the earth (see Fig. 2.8). At what wavelengths do the sun and the earth radiate most of their energy? Fortunately, the sun and the earth both have characteristics (discussed in a later section) that enable us to use the following relationship called **Wien's law** (or *Wien's displacement law*) after the German physicist Wilhelm Wien (pronounced Wēēn, 1864–1928), who discovered it:

$$\lambda_{\text{max}} = \frac{\text{constant}}{T} \quad \text{(Wien's law)},$$

where λ_{max} is the wavelength in micrometers at which maximum radiation emission occurs, T is the object's temperature in Kelvins, and the constant is 2897 μm K. To make the numbers easy to deal with, we will round off the constant to the number 3000.

For the sun, with a surface temperature of 6000 K, the equation becomes

$$\lambda_{\text{max}} = \frac{3000 \ \mu\text{m K}}{6000 \ \text{K}} = 0.5 \ \mu\text{m}.$$

Thus, the sun emits a maximum amount of radiation at wavelengths near 0.5 μm. The cooler earth, with an average surface temperature of 288 K (rounded to 300 K), emits maximum radiation near wavelengths of 10 μm, since

$$\lambda_{max} = \frac{3000 \ \mu m \ K}{300 \ K} = 10 \ \mu m.$$

Thus, the earth emits most of its radiation at longer wavelengths between about 5 and 25 μm, while the sun emits the majority of its radiation at wavelengths less than 2 μm. For this reason, the earth's radiation is often called **longwave radiation**, whereas the sun's energy is referred to as **shortwave radiation**.

Wien's law demonstrates that, as the temperature of an object increases, the wavelength at which maximum emission occurs is shifted toward shorter values. For example, if the sun's surface temperature were to double to 12,000 K, its wavelength of maximum emission would be halved to about 0.25 μm. If, on the other hand, the sun's surface cooled to 3000 K, it would emit its maximum amount of radiation near 1.0 μm.

Even though the sun radiates at a maximum rate at a particular wavelength, it nonetheless emits some radiation at almost all other wavelengths. If we look at the amount of radiation given off by the sun at each wavelength, we obtain the sun's *electromagnetic spectrum*. A portion of this spectrum is shown in Fig. 2.9.

Since our eyes are sensitive to radiation between 0.4 and 0.7 μm, these waves reach the eye and stimulate the sensation of color. This portion of the spectrum is referred to as the **visible region**, and the light that reaches our eye is called *visible light*. The sun emits nearly 44 percent of its radiation in this zone, with the peak of energy output found at the wavelength corresponding to the color blue-green. The color violet is the shortest wavelength of visible light. Wavelengths shorter than violet (0.4 μm) are **ultraviolet (UV)**. X-rays and gamma rays with exceedingly short wavelengths also fall into this category. The sun emits only about 7 percent of its total energy at ultraviolet wavelengths.

The longest wavelengths of visible light correspond to the color red. Wavelengths longer than red (0.7 μm) are **infrared (IR)**. These waves cannot be seen by humans. Nearly 37 percent

WEATHER WATCH

The large ears of a jackrabbit are efficient emitters of infrared energy. Its ears help the rabbit survive the heat of a summer's day by radiating a great deal of infrared energy to the cooler sky above. Similarly, the large ears of the African elephant greatly increase its radiating surface area and promote cooling of its large mass.

of the sun's energy is radiated between 0.7 μm and 1.5 μm, with only 12 percent radiated at wavelengths longer than 1.5 μm.

Whereas the hot sun emits only a part of its energy in the infrared portion of the spectrum, the relatively cool earth emits almost all of its energy at infrared wavelengths. Although we cannot see infrared radiation, there are instruments called *infrared sensors* that can. Weather satellites that orbit the globe use these sensors to observe radiation emitted by the earth, the clouds, and the atmosphere. Since objects of different temperatures radiate their maximum energy at different wavelengths, infrared photographs can distinguish among objects of different temperatures. Clouds always radiate infrared energy; thus, cloud pictures using infrared sensors can be taken during both day and night.

In summary, both the sun and earth emit radiation. The *hot sun* (6000 K) radiates nearly 88 percent of its energy at wavelengths less than 1.5 μm, with maximum emission in the *visible region* near 0.5 μm. The *cooler earth* (288 K) radiates nearly all its energy between 5 and 25 μm with a peak intensity in the *infrared region* near 10 μm (see Fig. 2.8). The sun's surface is nearly 20 times hotter than the earth's surface. From the Stefan-Boltzmann relationship, this fact means that a unit area on the sun emits nearly 160,000 (20^4) times more energy during a given time period than the same size area on the earth. And since the sun has such a huge surface area from which to radiate, the total energy emitted by the sun each minute

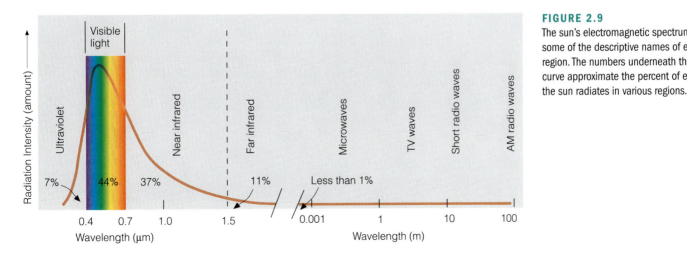

FIGURE 2.9

The sun's electromagnetic spectrum and some of the descriptive names of each region. The numbers underneath the curve approximate the percent of energy the sun radiates in various regions.

Wave Energy, Sun Burning, and UV Rays

Standing close to a fire makes us feel warmer than we do when we stand at a distance from it. Does this mean that, as we move away from a hot object, the waves carry less energy and are, therefore, weaker? Not really. The intensity of radiation decreases as we move away from a hot object because radiating energy spreads outward in all directions. Figure 3 illustrates that, as the distance from a radiating object increases, a given amount of energy is distributed over a larger area, so that the energy received per unit of area and per unit of time decreases. In fact, at twice the distance from the source, the radiation is spread over four times the area.

Another interesting fact about radiation that we learned earlier in this chapter is that shorter waves carry much more energy than do longer waves. Hence, a photon of ultraviolet light carries more energy than a photon of visible light. In fact, ultraviolet (UV) wavelengths in the range of 0.20 and 0.29 μm (known as *UV–C radiation*) are harmful to living things, as certain

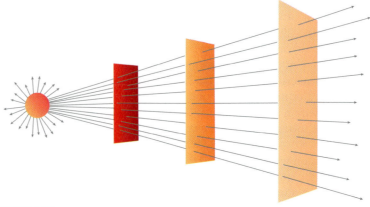

FIGURE 3

The intensity, or amount, of radiant energy transported by electromagnetic waves decreases as we move away from a radiating object because the same amount of energy is spread over a larger area.

waves can cause chromosome mutations, kill single-celled organisms, and damage the cornea of the eye. Fortunately, virtually all the ultraviolet radiation at wavelengths in the UV–C range is absorbed by ozone in the stratosphere.

Ultraviolet wavelengths between about 0.29 and 0.32 μm (known as *UV–B radiation*) reach the earth in

small amounts. Photons in this wavelength range have enough energy to produce sunburns and penetrate skin tissues, sometimes causing skin cancer. About 90 percent of all skin cancers are linked to sun exposure and UV–B radiation. Oddly enough, these same wavelengths activate provitamin D in the skin and convert it into vitamin D, which is essential to health.

amounts to a staggering 6 billion, billion, billion calories! (Additional information on radiation intensity and its effect on humans is given in the Focus section above.)

BALANCING ACT—ABSORPTION, EMISSION, AND EQUILIBRIUM

If the earth and all things on it are continually radiating energy, why doesn't everything get progressively colder? The answer is that all objects not only radiate energy, they absorb it as well. If an object radiates more energy than it absorbs, it gets colder; if it absorbs more energy than it emits, it gets warmer. On a sunny day, the earth's surface warms by absorbing more energy from the sun and the atmosphere than it radiates, while at night the earth cools by radiating more energy than it absorbs from its surroundings. When an object emits and absorbs energy at equal rates, its temperature remains constant.

The rate at which something radiates and absorbs energy depends strongly on its surface characteristics, such as color,

texture, and moisture, as well as temperature. For example, a black object in direct sunlight is a good absorber of visible radiation. It converts energy from the sun into internal energy, and its temperature ordinarily increases. You need only walk barefoot on a black asphalt road on a summer afternoon to experience this. At night, the blacktop road will cool quickly by emitting infrared radiation and, by early morning, it may be cooler than surrounding surfaces.

Any object that is a perfect absorber (that is, absorbs all the radiation that strikes it) and a perfect emitter (emits the maximum radiation possible at its given temperature) is called a **blackbody**. Blackbodies do not have to be colored black, they simply must absorb and emit all possible radiation. Since the earth's surface and the sun absorb and radiate with nearly 100 percent efficiency for their respective temperatures, they both behave as blackbodies. This is the reason we were able to use Wien's law and the Stefan-Boltzmann law to determine the characteristics of radiation emitted from the sun and the earth.

When we look at the earth from space, we see that half of it is in sunlight, the other half is in darkness. The outpouring of

Longer ultraviolet waves with lengths of about 0.32 to 0.40 μm (called *UV–A radiation*) are less energetic, but can still tan the skin. Although UV–B is mainly responsible for burning the skin, UV–A can cause skin redness. It can also interfere with the skin's immune system and cause long-term skin damage that shows up years later as accelerated aging and skin wrinkling. Moreover, recent studies indicate that longer UV–A exposures needed to create a tan pose about the same cancer risk as a UV–B tanning dose.

Upon striking the human body, ultraviolet radiation is absorbed beneath the outer layer of skin. To protect the skin from these harmful rays, the body's defense mechanism kicks in. Certain cells (when exposed to UV radiation) produce a dark pigment (*melanin*) that begins to absorb some of the UV radiation. (It is the production of melanin that produces a tan.) Consequently, a body that produces little melanin—one with pale skin—has little natural protection from UV–B.

Additional protection can come from a sunscreen. Unlike the old lotions that simply moisturized the skin before it baked in the sun, sunscreens today block UV rays from ever reaching the skin. Some contain chemicals (such as zinc oxide) that reflect UV radiation. (These are the white pastes seen on the noses of lifeguards.) Others consist of a mixture of chemicals (such as benzophenone and paraaminobenzoic acid, PABA) that actually absorb ultraviolet radiation, usually UV–B, although new products with UV–A-absorbing qualities are now on the market. The *Sun Protection Factor* (SPF) number on every container of sunscreen dictates how effective the product is in protecting from UV–B—the higher the number, the better the protection.

Protecting oneself from excessive exposure to the sun's energetic UV rays is certainly wise. Estimates are that, in a single year, over 30,000 Americans will be diagnosed with malignant melanoma, the most deadly form of skin cancer. And as the protective ozone shield diminishes, there is an ever-increasing risk of problems associated with UV–B. Using a good sunscreen and proper clothing can certainly help. The best way to protect yourself from too much sun, however, is to limit your time in direct sunlight, especially between the hours of 11 A.M. and 3 P.M. when the sun is highest in the sky and its rays are most direct.

Presently, the National Weather Service makes a daily prediction of UV radiation levels for selected cities throughout the United States. The forecast, known as the *Experimental Ultraviolet Index*, gives the UV level at its peak, around noon standard time or 1 P.M. daylight savings time. The 15-point index corresponds to five exposure categories set by the Environmental Protection Agency (EPA). An index value of between 0 and 2 is considered "minimal," whereas a value of 10 or greater is deemed "very high."

---

solar energy constantly bathes the earth with radiation, while the earth, in turn, constantly emits infrared radiation. If we assume that there is no other method of transferring heat, then, when the rate of absorption of solar radiation equals the rate of emission of infrared earth radiation, a state of *radiative equilibrium* is achieved. The average temperature at which this occurs is called the **radiative equilibrium temperature**. At this temperature, the earth (behaving as a blackbody) is absorbing solar radiation and emitting infrared radiation at equal rates, and its average temperature does not change. Because the earth is about 150 million km (93 million mi) from the sun, the earth's *radiative equilibrium temperature* is about 255 K (−18°C, 0°F). But this temperature is *much* lower than the earth's observed average surface temperature of 288 K (15°C, 59°F). Why is there such a large difference?

The answer lies in the fact that *the earth's atmosphere absorbs and emits infrared radiation.* Unlike the earth, the atmosphere does *not* behave like a blackbody, as it absorbs some wavelengths of radiation and is transparent to others. Objects that selectively absorb and emit radiation, such as

gases in our atmosphere, are known as **selective absorbers**. Let's examine this concept more closely.

Selective Absorbers and the Atmospheric Greenhouse Effect
Just as some people are selective eaters of certain foods, most substances in our environment are selective absorbers; that is, they absorb only certain wavelengths of radiation. Glass is a good example of a selective absorber in that it absorbs some of the infrared and ultraviolet radiation it receives, but not the visible radiation that is transmitted through the glass. As a result, it is difficult to get a sunburn through the windshield of your car, although you can see through it.

Objects that selectively absorb radiation also selectively emit radiation at the same wavelength. This phenomenon is called **Kirchhoff's law**. This law states that *good absorbers are good emitters at a particular wavelength, and poor absorbers are poor emitters at the same wavelength.*[*]

---

[*]Strictly speaking, this law only applies to gases.

FIGURE 2.10

The melting of snow outward from the trees causes small depressions to form. The melting is caused mainly by the snow's absorption of the infrared energy being emitted from the warmer tree and its branches. The trees are warmer because they are better absorbers of sunlight than is the snow.

Snow is a good absorber as well as a good emitter of infrared energy (white snow actually behaves as a blackbody in the infrared wavelengths). The bark of a tree absorbs sunlight and emits infrared energy, which the snow around it absorbs. During the absorption process, the infrared radiation is converted into internal energy, and the snow melts outward away from the tree trunk, producing a small depression that encircles the tree (see Fig. 2.10).

Figure 2.11 shows some of the most important selectively absorbing gases in our atmosphere. The shaded area represents the percent of radiation absorbed by each gas at various wavelengths. Notice that oxygen (O_2) and ozone (O_3) absorb almost 100 percent of the UV radiation whose lengths are shorter than about 0.3 μm. Notice, also that both water vapor (H_2O) and carbon dioxide (CO_2) are strong absorbers of infrared radiation and poor absorbers of visible solar radiation. Other, less important, selective absorbers include nitrous oxide (N_2O), methane (CH_4), and ozone (O_3), which is most abundant in the stratosphere. As these gases absorb infrared radiation emitted from the earth's surface, they gain kinetic energy (energy of motion). The gas molecules share this energy by colliding with neighboring air molecules, such as oxygen and nitrogen (both of which are poor absorbers of infrared energy). These collisions increase the average kinetic energy of the air, which results in an increase in air temperature. Thus, most of the infrared energy emitted from the earth's surface keeps the lower atmosphere warm.

Besides being selective absorbers, water vapor and CO_2 selectively emit radiation at infrared wavelengths.* This radiation travels away from these gases in all directions. A portion of this energy is radiated toward the earth's surface and absorbed,

---

*Nitrous oxide, methane, and ozone also emit infrared radiation, but their concentration in the atmosphere is much smaller than water vapor and carbon dioxide (see Table 1.1, p. 3).

FIGURE 2.11

Absorption of radiation by gases in the atmosphere. The shaded area represents the percent of radiation absorbed. The strongest absorbers of infrared radiation are water vapor and carbon dioxide.

thus heating the ground. The earth, in turn, radiates infrared energy upward, where it is absorbed and warms the lower atmosphere. In this way, water vapor and CO_2 absorb and radiate infrared energy and act as an insulating layer around the earth, keeping part of the earth's infrared radiation from escaping rapidly into space. Consequently, the earth's surface and the lower atmosphere are much warmer than they would be if these selectively absorbing gases were not present. In fact, as we saw earlier, the earth's mean radiative equilibrium temperature without CO_2 and water vapor would be around −18°C (0°F), or about 33°C (59°F) lower than at present.

The absorption characteristics of water vapor, CO_2, and other gases such as methane and nitrous oxide (see Fig. 2.11) were, at one time, thought to be similar to the glass of a florist's greenhouse. In a greenhouse, the glass allows visible radiation to come in, but inhibits to some degree the passage of outgoing infrared radiation. For this reason, the behavior of the water vapor and CO_2 in the atmosphere is popularly called the **greenhouse effect**. However, studies have shown that the warm air inside a greenhouse is probably caused more by the air's inability to circulate and mix with the cooler outside air, rather than by the entrapment of infrared energy. Because of these findings, some scientists insist that the greenhouse effect should be called the *atmosphere effect*. To accommodate everyone, we will usually use the term *atmospheric greenhouse effect* when describing the role that water vapor and CO_2 play in keeping the earth's mean surface temperature higher than it otherwise would be.

Look again at Fig. 2.11 and observe that, in the bottom diagram, there is a region between about 8 and 11 μm where neither water vapor nor CO_2 readily absorb infrared radiation. Because these wavelengths of emitted energy pass upward through the atmosphere and out into space, the wavelength range (between 8 and 11 μm) is known as the **atmospheric window**. At night, clouds can enhance the atmospheric greenhouse effect. Tiny liquid cloud droplets are selective absorbers in that they are good absorbers of infrared radiation but poor absorbers of visible solar radiation. Clouds even absorb the wavelengths between 8 and 11 μm, which are otherwise "passed up" by water vapor and CO_2. Thus, they have the effect of enhancing the atmospheric greenhouse effect by closing the atmospheric window.

Clouds, especially low thick ones, are also excellent emitters of infrared radiation. Their tops radiate infrared energy upward and their bases radiate energy back to the earth's surface where it is absorbed and, in a sense, re-radiated back to the clouds. This process keeps calm, cloudy nights warmer than calm, clear ones. If the clouds remain into the next day, they prevent much of the sunlight from reaching the ground by reflecting it back to space. Since the ground does not heat up as much as it would in full sunshine, cloudy, calm days are normally cooler than clear, calm days. Hence, the presence of clouds tends to keep nighttime temperatures higher and daytime temperatures lower.

In summary, the atmospheric greenhouse effect occurs because water vapor, CO_2, and other greenhouse gases are

WEATHER WATCH

The warming due to greenhouse gases was first recognized in 1827 by the French scientist Jean-Baptiste Fourier. He used the analogy that certain gases in the earth's atmosphere behaved in a similar manner to the glass of a florist's greenhouse. Hence, he is responsible for the name "greenhouse effect."

selective absorbers. They allow most of the sun's visible radiation to reach the surface, but they absorb a good portion of the earth's outgoing infrared radiation, preventing it from escaping into space (see Fig. 2.12). It is the atmospheric greenhouse effect, then, that keeps the temperature of our planet at a level where life can survive. The greenhouse effect is not just a "good thing;" it is essential to life on earth.

Enhancement of the Greenhouse Effect

In spite of the inaccuracies that plague temperature measurements, studies suggest that for the past 100 years or so, the earth's surface air temperature has been undergoing a slight warming of between 0.3°C and 0.6°C. There are scientific computer models, called *general circulation models* (GCMs) that mathematically simulate the physical processes of the atmosphere and oceans. These models (also referred to as *climate models*) predict that if such a warming should continue unabated, we would be irrevocably committed to some measure of climate change, notably a shift of the world's wind patterns that steer the rain-producing storms across the globe.

Many scientists believe that the main cause of this *global warming* is the greenhouse gas CO_2, whose concentration has been increasing primarily due to the burning of fossil fuels and deforestation. However, in recent years, increasing concentration of other greenhouse gases, such as methane (CH_4), nitrous oxide (N_2O), and chlorofluorocarbons (CFCs), has collectively been shown to have an effect almost equal to CO_2. Look at Fig. 2.11 and notice that both CH_4 and N_2O absorb strongly at infrared wavelengths. Moreover, a particular CFC (CFC-12) absorbs in the region of the atmospheric window between 8 and 11 μm. Thus, in terms of its absorption impact on infrared radiation, the addition of a single CFC-12 molecule to the atmosphere is equivalent to adding 10,000 molecules of CO_2. Overall, water vapor accounts for about 60 percent of the atmospheric greenhouse effect, CO_2 accounts for about 26 percent, and the remaining greenhouse gases contribute about 14 percent.

Presently, the concentration of CO_2 in a volume of air near the surface is about 0.036 percent. Computer models predict that doubling this amount could cause the earth's average surface temperature to rise on average 2.5 degrees Celsius by the end of the twenty-first century. How can doubling such a small quantity of CO_2 and adding miniscule amounts of other greenhouse gases bring about such a large temperature increase?

(a) Without greenhouse effect

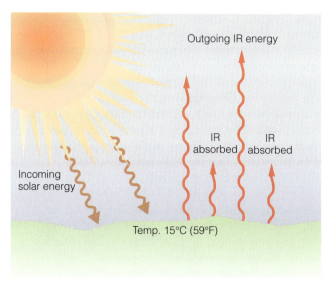

(b) With greenhouse effect

FIGURE 2.12

Sunlight warms the earth's surface only during the day, whereas the surface constantly emits infrared radiation upward during the day and at night. (a) Near the surface *without* water vapor, CO_2, and other greenhouse gases, the earth's surface would constantly emit infrared radiation (IR) energy; incoming energy from the sun would be equal to outgoing IR energy from the earth's surface. Since the earth would receive no IR energy from its lower atmosphere (no atmospheric greenhouse effect), the earth's average surface temperature would be a frigid $-18°C$ ($0°F$). (b) With greenhouse gases, the earth's surface receives energy from the sun and infrared energy from its atmosphere. Incoming energy still equals outgoing energy, but the added IR energy from the greenhouse gases raises the earth's average surface temperature about $33°C$, to a comfortable $15°C$ ($59°F$).

Mathematical climate models predict that rising ocean temperatures will cause an increase in evaporation rates. The added *water vapor*—the primary greenhouse gas—will enhance the atmospheric greenhouse effect and double the temperature rise, in what is known as a *positive feedback*. But there are other feedbacks to consider.*

The two potentially largest and least understood feedbacks in the climate system are the clouds and the oceans. Clouds can change area, depth, and radiation properties simultaneously with climatic changes. The net effect of all these changes is not well understood. Oceans, on the other hand, cover 70 percent of the planet. The response of ocean circulations, ocean temperatures, and sea ice to global warming will determine the global pattern and speed of climate change. Unfortunately, it is not now known how quickly or in what direction each of these will respond.

Satellite data from the *Earth Radiation Budget Experiment* (ERBE) suggest that clouds overall appear to cool the earth's climate, as they reflect and radiate away more energy than they retain. (The earth would be warmer if clouds were not present.) So an increase in global cloudiness (if it were to occur) might offset some of the global warming brought on by an enhanced atmospheric greenhouse effect. Therefore, if clouds were to act on the climate system in this manner, they would provide a *negative feedback* on climate change.†

Uncertainties unquestionably exist about the impact that increasing levels of CO_2 and other trace gases will have on enhancing the atmospheric greenhouse effect. Nonetheless, many (but not all) scientific studies suggest that increasing the concentration of these gases in our atmosphere will lead to global-scale climatic change by the end of the twenty-first century. Such change could adversely affect water resources and agricultural productivity. (We will examine this topic further in Chapter 19, when we cover climatic change in more detail.)

BRIEF REVIEW   In the last several sections, we have explored examples of some of the ways radiation is absorbed and emitted by various objects. Before reading the next several sections, let's review a few important facts and principles:

■ *All* objects with a temperature above absolute zero emit radiation.

---

*A feedback is a process whereby an initial change in a process will tend to either reinforce the process (positive feedback) or weaken the process (negative feedback). The *water vapor-temperature rise* feedback is a positive feedback because the initial increase in temperature is reinforced by the addition of more water vapor, which absorbs more of the earth's infrared energy, thus strengthening the greenhouse effect and enhancing the warming.

†Overall, current climate models tend to show that changes in clouds could provide either a net negative or a net positive feedback on climate change.

- The higher an object's temperature, the greater the amount of radiation emitted per unit surface area and the shorter the wavelength of maximum emission.

- The earth absorbs solar radiation only during the daylight hours; however, it emits infrared radiation continuously, both during the day and at night.

- The earth's surface behaves as a blackbody, making it a much better absorber and emitter of radiation than the atmosphere.

- Water vapor and carbon dioxide are important atmospheric greenhouse gases that selectively absorb and emit infrared radiation, thereby keeping the earth's average surface temperature warmer than it otherwise would be.

- Cloudy, calm nights are often warmer than clear, calm nights because clouds strongly absorb and emit infrared radiation.

- It is *not* the greenhouse effect itself that is of concern, but the *enhancement* of it due to increasing levels of greenhouse gases.

With these concepts in mind, we will first examine how the air near the ground warms, then we will consider how the earth and its atmosphere maintain a yearly energy balance. (But before reading on, you may wish to read the Focus section on p. 42 that describes, in more detail, some of the concepts presented so far.)

Warming the Air from Below   On a clear day, solar energy passes through the lower atmosphere with little effect upon the air. Ultimately it reaches the surface, warming it (see Fig. 2.13).

Air molecules in contact with the heated surface bounce against it, gain energy by *conduction*, then shoot upward like freshly popped kernels of corn, carrying their energy with them. Because the air near the ground is very dense, these molecules only travel a short distance (about 10^{-7} m) before they collide with other molecules. During the collision, these more rapidly moving molecules share their energy with less energetic molecules, raising the average temperature of the air. But air is such a poor heat conductor that this process is only important within a few centimeters of the ground.

As the surface air warms, it actually becomes less dense than the air directly above it. The warmer air rises and the cooler air sinks, setting up thermals, or *free convection cells* that transfer heat upward and distribute it through a deeper layer of air. The rising air expands and cools, and, if sufficiently moist, the water vapor condenses into cloud droplets, releasing latent heat that warms the air. Meanwhile, the earth constantly emits infrared energy. Some of this energy is absorbed by greenhouse gases (such as water vapor and carbon dioxide) that emit infrared energy upward and downward, back to the surface. Since the concentration of water vapor decreases rapidly above the earth, most of the absorption occurs in a layer near the surface. Hence, the lower atmosphere is mainly heated from below.

INCOMING SOLAR ENERGY

As the sun's radiant energy travels through space, essentially nothing interferes with it until it reaches the atmosphere. At the top of the atmosphere, solar energy received on a surface perpendicular to the sun's rays appears to remain fairly

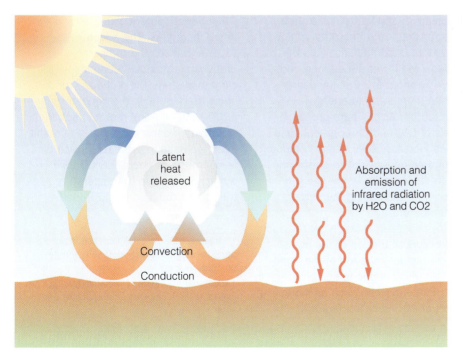

FIGURE 2.13

Air in the lower atmosphere is heated from below. Sunlight warms the ground, and the air above is warmed by conduction, convection, and radiation. Further warming occurs during condensation as latent heat is given up to the air inside the cloud.

Radiative Equilibrium and Forcing—The Ins and Outs

We know that all objects emit radiation, and that the hotter the object, the more radiant energy it emits. Earlier, we learned that this relationship between temperature and radiation is called the Stefan-Boltzmann law, which is written as

$$E = \sigma T^4$$

where E is the energy being emitted by each square meter of surface area of the object, T is the object's temperature in Kelvins, and σ is the Stephan-Boltzmann constant. The constant σ is 5.67×10^{-8} with units of Watts per square meter per Kelvin to the fourth power, or W/m^2K^4. We can obtain the units for E by plugging into the equation the units for σ and T; thus

$$E = \sigma T^4$$

$$E = \left(\frac{W}{m^2K^4}\right)(K^4)$$

$$E = \left(\frac{W}{m^2}\right).$$

So the units of energy E (called the *flux radiance* or *emittance*) are Watts per square meter.

In a previous section, we learned that the earth's radiative equilibrium temperature is about −18°C (0°F). How much energy would the earth be emitting at this temperature? Remember that we have to convert °C to Kelvins, otherwise at a temperature of −18°C the earth would be emitting negative amounts of energy, which is meaningless; consequently

$$K = °C + 273$$
$$K = -18 + 273$$
$$K = 255 \ K.$$

If we plug this temperature (255 K) into the Stefan-Boltzmann equation we obtain

$$E = \sigma T^4$$

$$E = (5.67 \times 10^{-8} \left(\frac{W}{m^2K^4}\right)(255 \ K)^4$$

$$E = 240 \left(\frac{W}{m^2}\right).$$

Thus the earth, in radiative equilibrium and behaving as a blackbody, would be emitting 240 watts of energy over each square meter of surface area.

Over the earth as a whole, outgoing infrared energy equals incoming solar energy (see Fig. 4). Consequently, without an atmospheric greenhouse effect, the earth's surface would (on average) emit 240 W/m^2 upward and, at the same time, receive 240 W/m^2 from the sun. But due to a greenhouse effect this equilibrium of 240 W/m^2 is only achieved at the *top* of the atmosphere. Moreover, as greenhouse gases slowly increase in concentration they alter this balance by gradually absorbing more and more of the earth's infrared radiation, thereby preventing this energy from escaping into space. So without any changes in the climate system, outgoing energy would gradually drop below a value of 240 W/m^2.

Climate models predict that, as long as everything else remains the same, a sudden doubling of the current levels of atmospheric CO_2 would result in a net radiation reduction (imbalance) of 4 watts per square meter at the top of the atmosphere. To restore this imbalance, the earth's surface and lower atmosphere must warm by about 1.2 Celsius degrees so that more infrared energy is directed upward.* Therefore, as levels of CO_2 and other greenhouse gases increase, they alter the amount of infrared energy lost to space and, in effect, *force* the atmosphere to respond by increasing the surface air temperature.

Any change in average net radiation that occurs at the top of the atmosphere (actually the top of the troposphere) which is due to some change in the climate system (such as increasing levels of CO_2) is called *radiative*

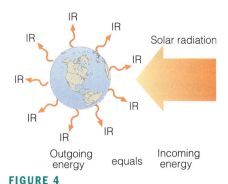

FIGURE 4

In radiative equilibrium, over the earth as a whole, incoming solar energy equals outgoing infrared (IR) energy.

forcing. Therefore, greenhouse gases (which are increasing in concentration) are referred to as *radiative forcing agents.*†

As levels of greenhouse gases increase, they alter the infrared radiation leaving the atmosphere. This process enhances the atmospheric greenhouse effect, which causes the surface air temperature to rise. As the surface air warms, more evaporation occurs from the oceans, and the water vapor content of the atmosphere increases. The added water vapor enhances the temperature rise, producing a positive feedback on the climate system. Most climate models show that doubling the concentration of CO_2 and allowing the atmospheric water vapor content to rise, will result in an increase in average surface air temperature of about 2.5°C (4.5°F)— a much larger rise in temperature than that produced by CO_2 alone.

---

*Remember that a small increase in temperature results in a great deal more energy emitted, as $E \sim T^4$.

†Additional examples of radiative forcing agents include changes in cloud cover and solar output.

constant at nearly two calories on each square centimeter each minute or 1367 W/m²—a value called the **solar constant**.*

Scattered and Reflected Light

Upon entering the atmosphere, a number of interactions take place between the atmosphere and solar radiation. For example, some of the energy is absorbed by gases, such as ozone, in the upper atmosphere. Moreover, when sunlight strikes very small objects, such as air molecules and dust particles, the light itself is deflected in all directions—forward, sideways, and backwards. The distribution of light in this manner is called **scattering**. (Scattered light is also called *diffuse light*.) Because air molecules are much smaller than the wavelengths of visible light, they are more effective scatterers of the shorter (blue) wavelengths than the longer (red) wavelengths. Hence, when we look away from the direct beam of sunlight, blue light strikes our eyes from all directions, turning the daytime sky blue. At midday, all the wavelengths of visible light from the sun strike our eyes, and the sun is perceived as white. At sunrise and sunset, when the white beam of sunlight must pass through a thick portion of the atmosphere, scattering by air molecules removes the blue light, leaving the longer wavelengths of red, orange, and yellow to pass on through, creating the image of a ruddy or yellowish sun (see Fig. 2.14).

Sunlight can be **reflected** from objects. Generally, reflection differs from scattering in that during the process of reflection more light is sent *backwards*. **Albedo** is the percent of radiation returning from a given surface compared to the amount of radiation initially striking that surface. Albedo, then, represents the *reflectivity* of the surface. In Table 2.3, notice that thick clouds have a higher albedo than thin clouds. On the average, the albedo of clouds is near 60 percent. When solar energy strikes a surface covered with snow, up to 95 percent of the sunlight may be reflected. Most of this energy is in the visible and ultraviolet wavelengths. Consequently, reflected radiation, coupled with direct sunlight, can produce severe sunburns on the exposed skin of unwary snow skiers, and unprotected eyes can suffer the agony of snow blindness. Water surfaces, on the other hand, reflect only a small amount of solar energy. For an entire day, a smooth water surface will have an average albedo of about 10 percent. Water has the highest albedo (and can therefore reflect sunlight best) when the sun is low on the horizon and the water is a little choppy. This may explain why people who wear brimmed hats while fishing from a boat in choppy water on a sunny day can still get sunburned during midmorning or midafternoon.

---

*By definition, the solar constant (which, in actuality, is *not* "constant") is the rate at which radiant energy from the sun is received on a surface at the outer edge of the atmosphere perpendicular to the sun's rays when the earth is at an average distance from the sun. Satellite measurements from the *Earth Radiation Budget Satellite* suggest the solar constant varies slightly as the sun's radiant output varies. The average is about 1.96 cal/cm²/min, or between 1365 W/m² and 1372 W/m² in the SI system of measurement.

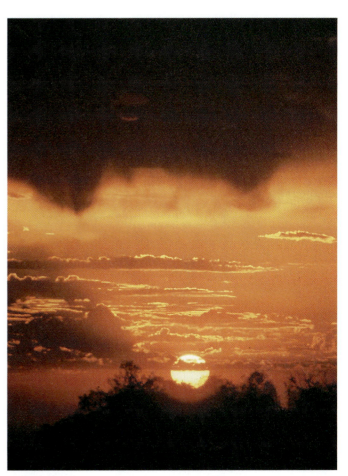

FIGURE 2.14
A brilliant red sunset produced by the process of scattering.

TABLE 2.3   Typical Albedo of Various Surfaces

SURFACE	ALBEDO (PERCENT)
Fresh snow	75 to 95
Clouds (thick)	60 to 90
Clouds (thin)	30 to 50
Venus	78
Ice	30 to 40
Sand	15 to 45
Earth and atmosphere	30
Mars	17
Grassy field	10 to 30
Dry, plowed field	5 to 20
Water	10*
Forest	3 to 10
Moon	7

*Daily average.

The Earth's Annual Energy Balance

Although the average temperature at any one place may vary considerably from year to year, the earth's overall average equilibrium temperature changes only slightly from one year to the next. This fact indicates that, each year, the earth and its atmosphere combined must send off into space just as much energy as they receive from the sun. The same type of energy balance must exist between the earth's surface and the atmosphere. That is, each year, the earth's surface must return to the atmosphere the same amount of energy that it absorbs. If this did not occur, the earth's average surface temperature would change. How do the earth and its atmosphere maintain this yearly energy balance?

Suppose 100 units of solar energy reach the top of the earth's atmosphere. We can see in Fig. 2.15 that, on the average, clouds, the earth, and the atmosphere reflect and scatter 30 units back to space and that the atmosphere and clouds together absorb 19 units, which leaves 51 units of direct and indirect solar radiation to be absorbed at the earth's surface.

Figure 2.16 shows approximately what happens to the solar radiation that is absorbed by the surface and the atmosphere. Out of 51 units reaching the surface, a large amount (23 units) is used to evaporate water and about 7 units are lost through conduction and convection, which leaves 21 units to be radiated away as infrared energy. Look closely at Fig. 2.16 and notice that the earth's surface actually radiates upward a whopping 117 units. It does so because, although it receives solar radiation only during the day, it constantly emits infrared energy both during the day and at night. Additionally, the atmosphere above only allows a small fraction of this energy (6 units) to pass through into space. The majority of it (111 units) is absorbed mainly by the greenhouse gases water vapor and CO_2, and by clouds. Much of this energy (96 units) is radi-

ated back to earth, producing the atmospheric greenhouse effect. Hence, the earth's surface receives nearly twice as much longwave infrared energy from its atmosphere as it does short-wave radiation from the sun. In all these exchanges, notice that the energy lost at the earth's surface (147 units) is exactly balanced by the energy gained there (147 units).

A similar balance exists between the earth's surface and its atmosphere. Again in Fig. 2.16 observe that the energy gained by the atmosphere (160 units) balances the energy lost. Moreover, averaged for an entire year, the solar energy received at the earth's surface (51 units) and that absorbed by the earth's atmosphere (19 units) balances the infrared energy lost to space by the earth's surface (6 units) and its atmosphere (64 units).

We can see the effect that conduction, convection, and latent heat play in the warming of the atmosphere if we look at the energy balance only in radiative terms. The earth's surface receives 147 units of radiant energy from the sun and its own atmosphere, while it radiates away 117 units, producing a *surplus* of 30 units. The atmosphere, on the other hand, receives 130 units (19 units from the sun and 111 from the earth), while it loses 160 units, producing a *deficit* of 30 units. The balance (30 units) is the warming of the atmosphere produced by the heat transfer processes of conduction and convection (7 units) and by the release of latent heat (23 units).

And so, the earth and the atmosphere absorb energy from the sun, as well as from each other. In all of the energy exchanges, a delicate balance is maintained. Essentially, there is no yearly gain or loss of total energy, and the average temperature of the earth and the atmosphere remains fairly constant from one year to the next. This equilibrium does not imply that the earth's average temperature does not change, but that the changes are small from year to year (usually less than one-

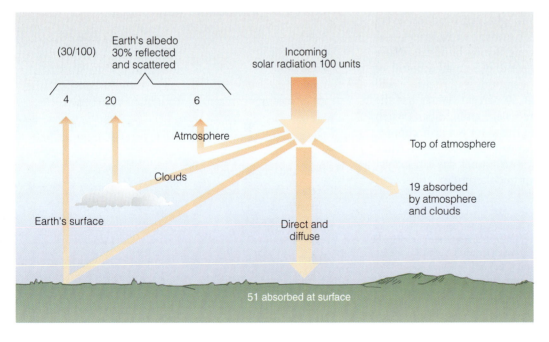

FIGURE 2.15
On the average, of all the solar energy that reaches the earth's atmosphere annually, about 30 percent ($^{30}/_{100}$) is reflected and scattered back to space, giving the earth and its atmosphere an albedo of 30 percent. Of the remaining solar energy, about 19 percent is absorbed by the atmosphere and clouds, and 51 percent is absorbed at the surface.

FIGURE 2.16

The earth-atmosphere energy balance. Numbers represent approximations based on surface observations and satellite data. While the actual value of each process may vary by several percent, it is the relative size of the numbers that is important.

tenth of a degree Celsius) and become significant only when measured over many years.

Up to this point we have considered radiant energy of the sun and earth. Before we turn our attention to how incoming solar energy, in the form of particles, produces a dazzling light show known as the aurora, you may wish to read the Focus section about the sun on p. 46.

Solar Particles and the Aurora   From the sun and its tenuous atmosphere comes a continuous discharge of particles. This discharge happens because, at extremely high temperatures, gases become stripped of electrons by violent collisions and acquire enough speed to escape the gravitational pull of the sun. As these charged particles (ions and electrons) travel through space, they are known as *plasma*, or **solar wind**. When the solar wind moves close enough to the earth, it interacts with the earth's magnetic field.

The magnetic field that surrounds the earth is much like the field around an ordinary bar magnet (see Fig. 2.17). Both have north and south magnetic poles, and both have invisible lines of force (field lines) that link the poles. On the earth, these field lines form closed loops as they enter near the magnetic north pole and leave near the magnetic south pole. Most scientists believe that an electric current coupled with fluid mo-

tions deep in the earth's hot molten core is responsible for its magnetic field. This field protects the earth, to some degree, from the onslaught of the solar wind.

Observe in Fig. 2.18 that, when the solar wind encounters the earth's magnetic field, it severely deforms it into a

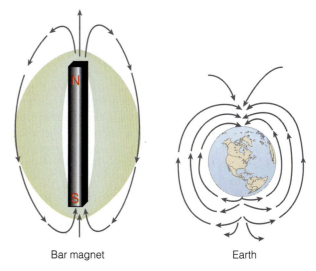

Bar magnet   Earth

FIGURE 2.17

Magnetic field about a bar magnet and about the earth.

Characteristics of the Sun

The sun is our nearest star. It is some 150 million km (93 million mi) from earth. The next star, Alpha Centauri, is more than 250,000 times further away. Even though the earth only receives about one two-billionths of the sun's total energy output, it is this energy that allows life to flourish. Sunlight determines the rate of photosynthesis in plants and strongly regulates the amount of evaporation from the oceans. It warms this planet and drives the atmosphere into the dynamic patterns we experience as everyday wind and weather. Without the sun's radiant energy, the earth would gradually cool, in time becoming encased in a layer of ice! Evidence of life on the cold, dark, and barren surface would be found only in fossils. Fortunately, the sun has been shining for billions of years, and it is likely to shine for at least several billion more.

The sun is a giant celestial furnace. Its core is extremely hot, with a temperature estimated to be near 15 million degrees Celsius. In the core, hydrogen nuclei (protons) collide at such fantastically high speeds that they fuse together to form helium nuclei. This thermonuclear process generates an enormous amount of energy, which gradually works its way

FIGURE 5
Various regions of the sun.

- Corona
- Prominence
- Core
- Chromosphere
- Photosphere
- Sun spot

FIGURE 6
A false-color image of the solar corona photographed aboard the *Skylab* spacecraft.

to the sun's outer luminous surface—the *photosphere* ("sphere of light"). Temperatures here are much cooler than in the interior, generally near 6000°C. We have noted already that a body with this surface temperature emits radiation at a maximum rate in the visible region of the spectrum.

teardrop-shaped cavity known as the *magnetosphere*. On the side facing the sun, the pressure of the solar wind compresses the field lines. On the opposite side, the magnetosphere stretches out into a long tail—the *magnetotail*—which reaches far beyond the moon's orbit. In a way, the magnetosphere acts as an obstacle to the solar wind by causing some of its particles to flow around the earth.

Inside the earth's magnetosphere are ionized gases. Some of these gases are solar wind particles, while others are ions from the earth's upper atmosphere that have moved upward along electric field lines into the magnetosphere.

Normally, the solar wind approaches the earth at an average speed of 400 km/sec. However, during periods of high solar activity (many sunspots and flares), the solar wind is more dense, travels much faster, and carries more energy. When these energized solar particles reach the earth, they cause a variety of effects, such as changing the shape of the magnetosphere and producing auroral displays.

The aurora is not reflected light from the polar ice fields, nor is it light from demons' lanterns as they search for lost souls. The *aurora* is produced by the solar wind disturbing the magnetosphere. The disturbance involves high-energy particles within the magnetosphere being ejected into the earth's upper atmosphere, where they excite atoms and molecules. The excited atmospheric gases emit visible radiation, which causes the sky to glow like a neon light. Let's examine this process more closely.

FIGURE 7
A spectacular solar flare spanning more than half a million kilometers across the solar surface.

The sun is, therefore, a shining example of such an object.

Dark blemishes on the photosphere called *sunspots* are huge, cooler regions that typically average more than five times the diameter of the earth. Although sunspots are not well understood, they are known to be regions of strong magnetic fields. They are cyclic, with the maximum number of spots occurring approximately every eleven years. Recent measurements from satellites reveal that during a sunspot maximum, more radiation may actually leave the pho-

tosphere. The reason for this lies in the fact that the bright areas around sunspots (known as *faculae*) radiate more energy as sunspot activity increases.

Above the photosphere are the *chromosphere* and the *corona* (see Fig. 5). The chromosphere ("color sphere") acts as a boundary between the relatively cool (6000°C) photosphere and the much hotter (2,000,000°C) corona, the outermost envelope of the solar atmosphere. During a solar eclipse, the corona is visible. It appears as a pale, milky cloud encircling the sun. Although much hotter than the photosphere, the corona radiates much less energy because its density is extremely low. This very thin solar atmosphere extends into space for many millions of kilometers* (see Fig. 6).

_____
*During a solar eclipse or at any other time, you should not look at the sun's corona either with sunglasses or through exposed negatives. Take this warning seriously. Much of the energy from this hot region is ultraviolet radiation. Viewing just a small area of the sun directly permits large amounts of UV radiation to enter the eye, causing serious and permanent damage to the retina. View the sun by projecting its image onto a sheet of paper, using a telescope or pinhole camera.

Violent solar activity occasionally occurs in the regions of sunspots. The most dramatic of these events are *prominences* and *flares*. Prominences are huge cloudlike jets of gas that often shoot up into the corona in the form of an arch. Solar flares are tremendous, but brief, eruptions (see Fig. 7). They emit large quantities of high-energy ultraviolet radiation, as well as energized charged particles, mainly protons and electrons, which stream outward away from the sun at extremely high speeds. A huge flare on May 21, 1980, lasted only 40 minutes but covered more than 5.2 billion square kilometers (2 billion square miles) of the sun's surface.

An intense solar flare can disturb the earth's magnetic field, producing a so-called *magnetic storm*. Because these storms can intensify the electrical properties of the upper atmosphere, they are often responsible for interruptions in radio and satellite communications. One such storm knocked out electricity throughout the province of Quebec, Canada, during March, 1989.

A high-energy particle from the magnetosphere will, upon colliding with an air molecule (or atom), transfer some of its energy to the molecule. The molecule then becomes excited (see Fig. 2.19). Just as excited football fans leap up when their favorite team scores the winning touchdown, electrons in an excited molecule jump into a higher energy level as they orbit its center. As the fans sit down after all the excitement is over, so electrons quickly return to their lower level. When molecules de-excite, they release the energy originally received from the energetic particle, either all at once (one big jump), or in steps (several smaller jumps). This emitted energy is given up as radiation. If its wavelength is in the visible range, we see it as visible light. In the Northern Hemisphere, we call this light show the **aurora borealis**, or *northern lights*; its counterpart in the Southern Hemisphere is the **aurora australis**, or *southern lights*.

Since each atmospheric gas has its own set of energy levels, each gas has its own characteristic color. For example, the de-excitation of atomic oxygen can emit green or red light. Molecular nitrogen gives off red and violet light. The shades of these colors can be spectacular as they brighten and fade, sometimes in the form of waving draperies, sometimes as unmoving, yet flickering, arcs and soft coronas. On a clear, quiet night the aurora is an eerie yet beautiful spectacle. (See the chapter-opening photograph.)

The aurora is most frequently seen in polar latitudes. Energetic particles trapped in the magnetosphere move along

FIGURE 2.18
The stream of charged particles from the sun—called
the *solar wind*—distorts the earth's magnetic field
into a teardrop shape known as the *magnetosphere*.

FIGURE 2.19
When an excited atom, ion, or molecule de-excites, it can
emit visible light. The electron in its normal orbit (a)
becomes excited by a charged particle and (b) jumps into
a higher energy level. When the electron returns to its normal orbit, it (c) emits a photon of light.

(a)   (b)   (c)

FIGURE 2.20
The aurora belt (solid line) represents the region where you would most likely
observe the aurora on a clear night. (The numbers represent the average number of nights per year on which you might see an aurora if the sky were clear.)

the earth's magnetic field lines. Because these lines emerge
from the earth near the magnetic poles, it is here that the particles interact with atmospheric gases to produce an aurora.
Notice in Fig. 2.20 that the zone of most frequent auroral
sightings (aurora belt) is not at the magnetic pole, but equatorward of it, where the field lines emerge from the earth's surface. At lower latitudes, where the field lines are oriented
almost horizontal to the earth's surface, the chances of seeing
an aurora diminish rapidly.

On rare occasions, however, the aurora is seen in the
southern United States. Such sightings happen only when the
sun is very active—as giant flares hurl electrons and protons
earthward at a fantastic rate. These particles move so fast that
some of them penetrate unusually deep into the earth's magnetic field before they are trapped by it. In a process not fully
understood, particles from the magnetosphere are accelerated
toward the earth along electrical field lines that parallel the
magnetic field lines. The acceleration of these particles gives

them sufficient energy so that when they enter the upper atmosphere they are capable of producing an auroral display much farther south than usual.

How high above the earth is the aurora? The exact height appears to vary. The base of an aurora is rarely lower than 80 km, and it averages about 105 km. Since the light of an aurora gradually fades, it is difficult to define an exact upper limit. Most auroras, however, are observed below 200 km.

In summary, energy for the aurora comes from the solar wind, which disturbs the earth's magnetosphere. This distur-bance causes energetic particles to enter the upper atmosphere, where they collide with atoms and molecules. The atmospheric gases become excited and emit energy in the form of visible light.

But there is other light coming from the atmosphere—a faint glow at night much weaker than the aurora. This feeble luminescence, called **airglow**, is detected at all latitudes and shows no correlation with solar wind activity. Apparently, this light comes from ionized oxygen and nitrogen and other gases that have been excited by solar radiation.

SUMMARY

In this chapter, we have seen how the concepts of heat and temperature differ and how heat is transferred in our environment. We learned that latent heat is an important source of atmospheric heat energy. We also learned that con-duction, the transfer of heat by molecular collisions, is most effective in solids. Because air is a poor heat conductor, con-duction in the atmosphere is only important in the shallow layer of air in contact with the earth's surface. A more impor-tant process of atmospheric heat transfer is convection, which involves the mass movement of air (or any fluid) with its energy from one region to another. Another significant heat transfer process is radiation—the transfer of energy by means of electromagnetic waves.

The hot sun emits most of its radiation as shortwave radi-ation. A portion of this energy heats the earth, and the earth, in turn, warms the air above. The cool earth emits most of its radiation as longwave infrared radiation. Selective absorbers in the atmosphere, such as water vapor and carbon dioxide, absorb some of the earth's infrared radiation and radiate a por-tion of it back to the surface, where it warms the surface, pro-ducing the atmospheric greenhouse effect. Because clouds are both good absorbers and good emitters of infrared radiation, they keep calm, cloudy nights warmer than calm, clear nights. The average equilibrium temperature of the earth and the atmosphere remains fairly constant from one year to the next because the amount of energy they absorb each year is equal to the amount of energy they lose.

Finally, we examined how the sun's energy in the form of solar wind particles interacts with our atmosphere to produce auroral displays.

Key Terms

energy	temperature
potential energy	heat
kinetic energy	absolute zero

Kelvin scale	visible region
Fahrenheit scale	ultraviolet (UV) radiation
Celsius scale	infrared (IR) radiation
heat capacity	blackbody
specific heat	radiative equilibrium
latent heat	temperature
sensible heat	selective absorbers
conduction	Kirchhoff's law
convection	greenhouse effect
thermals	atmospheric window
advection	solar constant
radiant energy (radiation)	scattering
electromagnetic waves	reflected (light)
wavelength	albedo
micrometer	solar wind
photons	aurora borealis
Stefan-Boltzmann law	aurora australis
Wien's law	airglow
longwave radiation	
shortwave radiation	

Questions for Review

1. Explain how the average speed of air molecules relates to the temperature of air.
2. Distinguish between temperature and heat.
3. (a) How does the Kelvin temperature scale differ from the Celsius scale? (b) Why is the Kelvin scale often used in scientific calculations? (c) Based on your experience, would a temperature of 250K be considered warm or cold? Explain.
4. Explain how heat is transferred by: (a) conduction; (b) convection; (c) radiation.
5. How is latent heat an important source of atmospheric energy?

6. In the atmosphere, how does advection differ from convection?

7. How does the temperature of an object influence the radiation that it emits?

8. How does the amount of radiation emitted by the earth differ from that emitted by the sun?

9. How do the wavelengths of most of the radiation emitted by the sun differ from those emitted by the surface of the earth?

10. Which photon carries the most energy—infrared, visible, or ultraviolet?

11. When a body reaches a radiative equilibrium temperature, what is taking place?

12. If the earth's surface continually radiates energy, why doesn't it become colder and colder?

13. Why are carbon dioxide and water vapor called selective absorbers?

14. Explain how the earth's atmospheric greenhouse effect works.

15. What gases appear to be responsible for the enhancement of the earth's greenhouse effect?

16. Why do some computer models predict that the earth's average surface temperature might increase by as much as 3.5°C toward the end of the twenty-first century?

17. What processes contribute to the earth's albedo being 30%?

18. Explain how the atmosphere near the earth's surface is warmed from below.

19. If a blackbody is a theoretical object, why can both the sun and earth be treated as blackbodies?

20. What is the solar wind?

21. Explain how the aurora is produced.

Questions for Thought

1. Explain why the bridge in the diagram is the first to become icy.

CAUTION! BRIDGE FREEZES BEFORE ROAD

2. Explain why the first snowfall of the winter usually "sticks" better to tree branches than to bare ground.

3. At night, why do materials that are poor heat conductors cool to temperatures less than the surrounding air?

4. Explain how ice can form on puddles (in shaded areas) when the temperature above and below the puddle is slightly above freezing.

5. In northern latitudes, the oceans are warmer in summer than they are in winter. In which season do the oceans lose heat most rapidly to the atmosphere by conduction? Explain.

6. How is heat transferred away from the surface of the moon? (Hint: The moon has no atmosphere.)

7. Why is ultraviolet radiation more successful in dislodging electrons from air atoms and molecules than is visible radiation?

8. Why must you stand closer to a small fire to experience the same warmth you get when standing further away from a large fire?

9. If water vapor were no longer present in the atmosphere, how would the earth's energy budget be affected?

10. Which will show the greatest increase in temperature when illuminated with direct sunlight: a plowed field or a blanket of snow? Explain.

11. Why does the surface temperature often increase on a clear, calm night as a low cloud moves overhead?

12. Which would have the greatest effect on the earth's greenhouse effect: Removing all of the CO_2 from the atmosphere or removing all of the water vapor? Explain why you chose your answer.

13. Explain why an increase in cloud cover surrounding the earth would increase the earth's albedo, yet not necessarily lead to a lower earth surface temperature.

14. Could a liquid thermometer register a temperature of −273°C when the air temperature is actually 1000°C? Where would this happen in the atmosphere, and why?

15. Why is it that auroral displays above Colorado can be forecasted several days in advance?

16. Why does the aurora usually occur more frequently above Maine than above Washington State?

Problems and Exercises

1. Suppose that 500 g of water vapor condense to make a cloud about the size of an average room. If we assume that the latent heat of condensation is 600 cal/g, how much heat would be released to the air? If the total mass of air before condensation is 100 kg, how much warmer would the air be after condensation? Assume that the air is not undergoing any pressure changes. (Hint: Use the specific heat of air in Table 2.1, p. 28.)

2. Suppose planet A is exactly twice the size (in surface area) of planet B. If both planets have the same exact surface temperature (1500 K), which planet would be emitting the most radiation? Determine the wavelength of maximum energy emission of both planets, using Wien's law.

3. Suppose, in question 2, the temperature of planet B doubles. (a) What would be its wavelength of maximum energy emission? (b) In what region of the electromagnetic spectrum would this wavelength be found? (c) If the temperature of planet A remained the same, determine which planet (A or B) would now be emitting the most radiation (use the Stefan-Boltzmann relationship). Explain your answer.

4. Suppose your surface body temperature averages 90°F. How much radiant energy in W/m² would be emitted from your body? (Hint: look at the Focus section on p. 42 and remember to convert °F to Kelvins.)

 ## Questions for Exploration

On the Blue Skies: College Edition CD-ROM go to the **Atmospheric Basics** section of the CD and click on **"Energy Balance."**

1. Explore the model of energy exchange. It mirrors the model presented in your text. Select one of the following parameters to change:
- Cloud cover
- Land characteristics
- CO_2 levels

(a) Before you begin, forecast how this change will affect: shortwave energy flows, longwave energy flows, and surface temperature.

(b) Explain your rationale for these forecasts.

(c) Change the parameter and watch the model adjust to the new environment.
- Did the changes occur as you expected?
- In hindsight can you explain what might explain the results you did not expect?

Shorter days and longer nights (created by a tilted earth revolving about the sun) promote cooler weather and the brilliant colors of autumn in New England. (Photo: Larry Ulrich, Tony Stone Images)

SEASONAL AND DAILY TEMPERATURES

CONTENTS

The sun doesn't rise or fall: it doesn't move, it just sits there, and we rotate in front of it. Dawn means that we are rotating around into sight of it, while dusk means we have turned another 180 degrees and are being carried into the shadow zone. The sun never "goes away from the sky." It's still there sharing the same sky with us; it's simply that there is a chunk of opaque earth between us and the sun which prevents our seeing it. Everyone knows that, but I really see it now. No longer do I drive down a highway and wish the blinding sun would set; instead I wish we could speed up our rotation a bit and swing around into the shadows more quickly.

Michael Collins, *Carrying the Fire*

s you sit quietly reading this book, you are part of a moving experience. The earth is speeding around the sun at thousands of kilometers per hour while, at the same time, it is spinning on its axis. When we look down upon the North Pole, we see that the direction of spin is counterclockwise, meaning that we are moving toward the east at hundreds of kilometers per hour. We normally don't think of it in that way, but, of course, this is what causes the sun, moon, and stars to rise in the east and set in the west. It is these motions coupled with the fact that the earth is tilted on its axis that causes our seasons. Therefore, we will begin this chapter by examining how the earth's motions and the sun's energy work together to produce temperature variations on a seasonal basis. Later, we will examine temperature variations on a daily basis.

WHY THE EARTH HAS SEASONS

The earth revolves completely around the sun in an elliptical path (not quite a circle) in slightly longer than 365 days (one year). As the earth revolves around the sun, it spins on its own axis, completing one spin in 24 hours (one day). The average distance from the earth to the sun is 150 million km (93 million mi). Because the earth's orbit is an ellipse instead of a circle, the actual distance from the earth to the sun varies during the year. The earth comes closer to the sun in January (147 million km) than it does in July (152 million km)* (see Fig. 3.1). From this we might conclude that our warmest weather should occur in January and our coldest weather in July. But, in the Northern Hemisphere, we normally experience cold weather in January when we are closer to the sun and warm weather in July when we are farther away. If nearness to the sun were the primary cause of the seasons then, indeed, January would be warmer than July. However, nearness to the sun is only a small part of the story.

Our seasons are regulated by the amount of solar energy received at the earth's surface. This amount is determined pri-

---

*The time around January 3rd when the earth is closest to the sun is called *perihelion* (from the Greek *peri*, meaning "near" and *helios*, meaning "sun"). The time when the earth is farthest from the sun (around July 4th) is called *aphelion* (from the Greek *ap*, "away from").

FIGURE 3.1
The elliptical path (highly exaggerated) of the earth about the sun brings the earth slightly closer to the sun in January than in July.

marily by the angle at which sunlight strikes the surface, and by how long the sun shines on any latitude (daylight hours). Let's look more closely at these factors.

Solar energy that strikes the earth's surface perpendicularly (directly) is much more intense than solar energy that strikes the same surface at an angle. Think of shining a flashlight straight at a wall—you get a small, circular spot of light (see Fig. 3.2). Now, tip the flashlight and notice how the spot of light spreads over a larger area. The same principle holds for sunlight. Sunlight striking the earth at an angle spreads out and must heat a larger region than sunlight impinging directly on the earth. Everything else being equal, an area experiencing more direct solar rays will receive more heat than the same size area being struck by sunlight at an angle. In addition, the more the sun's rays are slanted from the perpendicular, the more atmosphere they must penetrate. And the more atmosphere they penetrate, the more they can be scattered and absorbed (attenuated). As a consequence, when the sun is high in the sky, it can heat the ground to a much higher temperature than when it is low on the horizon.

The second important factor determining how warm the earth's surface becomes is the length of time the sun shines each day. Longer daylight hours, of course, mean that more energy is available from sunlight. In a given location, more solar energy reaches the earth's surface on a clear, long day than on a day that is clear but much shorter. Hence, more surface heating takes place.

From a casual observation, we know that summer days have more daylight hours than winter days. Also, the noontime summer sun is higher in the sky than is the noontime winter sun. Both of these events occur because our spinning planet is inclined on its axis (tilted) as it revolves around the sun. As Fig. 3.3 illustrates, the angle of tilt is $23\frac{1}{2}°$ from the perpendicular drawn to the plane of the earth's orbit. The earth's axis points to the same direction in space all year long; thus, the Northern Hemisphere is tilted toward the sun in summer (June), and away from the sun in winter (December). (If you have ever wondered what the "official" first day of a season really is, you may wish to read the Focus section on p. 56.)

Seasons in the Northern Hemisphere Let's first discuss the *warm summer* season. Note in Fig. 3.3 that, on June 21, the northern half of the world is directed toward the sun. At noon on this day, solar rays beat down upon the Northern Hemisphere more directly than during any other time of year. The sun is at its highest position in the noonday sky, directly above $23\frac{1}{2}°$ north (N) latitude (Tropic of Cancer). If you were standing at this latitude on June 21, the sun at noon would be directly overhead. This day, called the **summer solstice**, is the astronomical first day of summer in the Northern Hemisphere.†

---

†As we will see later in this chapter, the seasons are reversed in the Southern Hemisphere. Hence, in the Southern Hemisphere, this same day is the winter solstice, or the astronomical first day of winter.

FIGURE 3.2
Sunlight that strikes a surface at an angle is spread over a larger area than sunlight that strikes the surface directly. Oblique sun rays deliver less energy (are less intense) to a surface than direct sun rays.

High sun

Low sun

Earth

FIGURE 3.3
As the earth revolves about the sun, it is tilted on its axis by an angle of 23½°. The earth's axis always points to the same area in space (as viewed from a distant star). Thus, in June, when the Northern Hemisphere is tipped toward the sun, more direct sunlight and long hours of daylight cause warmer weather than in December, when the Northern Hemisphere is tipped away from the sun. (Diagram, of course, is not to scale.)

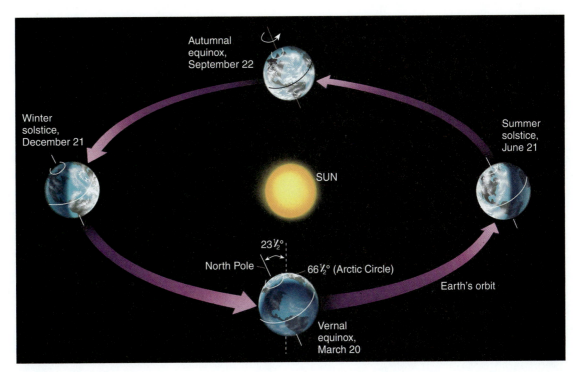

Autumnal equinox, September 22

Winter solstice, December 21

Summer solstice, June 21

SUN

23½°

North Pole

66½° (Arctic Circle)

Earth's orbit

Vernal equinox, March 20

Study Fig. 3.3 closely and notice that, as the earth spins on its axis, the side facing the sun is in sunshine and the other side is in darkness. Thus, half of the globe is always illuminated. If the earth's axis were not tilted, the noonday sun would always be directly overhead at the equator, and there would be 12 hours of daylight and 12 hours of darkness at each latitude every day of the year. However, the earth is tilted. Since the Northern Hemisphere faces toward the sun on June 21, each latitude in the Northern Hemisphere will have more than 12 hours of daylight. The farther north we go, the longer are the daylight hours. When we reach the Arctic Circle (66½°N), daylight lasts for 24 hours. Notice in Fig. 3.3 how the region above 66½°N never gets into the "shadow" zone as the earth spins. At the North Pole, the sun actually rises above the horizon on March 20 and has six months until it sets on September

22. No wonder this region is called the "Land of the Midnight Sun"! (See Fig. 3.4.)

Do longer days near polar latitudes mean that the highest daytime summer temperatures are experienced there? Not really. Nearly everyone knows that New York City (41°N) "enjoys" much hotter summer weather than Barrow, Alaska (71°N). The days in Barrow are much longer, so why isn't Barrow warmer? To figure this out, we must examine the *incoming solar radiation* (called *insolation*) on June 21. Figure 3.5 shows two curves: The upper curve represents the amount of insolation at the top of the earth's atmosphere on June 21, while the bottom curve represents the amount of radiation that eventually reaches the earth's surface on the same day.

The upper curve increases from the equator to the pole. This increase indicates that, during the entire day of June 21,

Is December 21 Really the First Day of Winter?

On December 21 (or 22, depending on the year) after nearly a month of cold weather, and perhaps a snowstorm or two, someone on the radio or TV has the audacity to proclaim that "today is the first official day of winter." If during the last several weeks it was not winter, then what season was it?

Actually December 21 marks the astronomical first day of winter in the Northern Hemisphere (NH), just as June 21 marks the *astronomical* first day of summer (NH). The earth is tilted on its axis by 23½° as it revolves around the sun. This fact causes the sun (as we view it from earth) to move in the sky from a point where it is directly above 23½° South latitude on December 21, to a point where it is directly above 23½° North latitude on June 21. The astronomical first day of spring (NH) occurs around March 20 as the sun crosses the equator moving northward and, likewise, the astronomical first day of autumn (NH) occurs around September 22 as the sun crosses the equator moving southward.

In the middle latitudes, summer is defined as the warmest season and winter the coldest season. If the year is divided into four seasons with each season consisting of three months, then the meteorological definition of summer over much of the Northern Hemisphere would be the three warmest months of June, July, and August. Winter would be the three coldest months of December, January, and February. Autumn would be September, October, and November—the transition between summer and winter. And spring would be March, April, and May—the transition between winter and summer.

So, the next time you hear someone remark on December 21 that "winter officially begins today," remember that this is the astronomical definition of the first day of winter. According to the meteorological definition, winter has been around for several weeks.

more solar radiation reaches the top of the earth's atmosphere above the poles than above the equator. True, the sun shines on these polar latitudes at a relatively large angle, but it does so for 24 hours, causing the maximum to occur there. The lower curve shows that the amount of solar radiation eventually reaching the earth's surface on June 21 is maximum near 30°N.

From there, the amount of insolation reaching the ground decreases as we move poleward.

The reason the two curves are different is that once sunlight enters the atmosphere, fine dust and air molecules scatter it, clouds reflect it, and some of it is absorbed by atmospheric gases. What remains reaches the surface. Generally, the greater

FIGURE 3.4
Land of the Midnight Sun. A series of exposures of the sun taken before, during, and after midnight in northern Alaska during July.

the thickness of atmosphere that sunlight must penetrate, the greater are the chances that it will be either scattered, reflected, or absorbed by the atmosphere. During the summer in far northern latitudes, the sun is never very high above the horizon, so its radiant energy must pass through a thick portion of atmosphere before it reaches the earth's surface (see Fig. 3.6). And because of the increased cloud cover during the arctic summer, much of the sunlight is reflected before it reaches the ground.

Solar energy that eventually reaches the surface in the far north does not heat the surface effectively. A portion of the sun's energy is reflected by ice and snow, while some of it melts frozen soil. The amount actually absorbed is spread over a large area. So, even though northern cities, such as Barrow, experience 24 hours of continuous sunlight on June 21, they are not warmer than cities farther south. Overall, they receive less radiation at the surface, and what radiation they do receive does not effectively heat the surface.

In our discussion of Fig. 3.5, we saw that, on June 21, solar energy incident on the earth's surface is maximum near latitude 30°N. On this day, the sun is shining directly above latitude 23½°N. Why, then, isn't the most sunlight received here? A quick look at a world map shows that the major deserts of the world are centered near 30°N. Cloudless skies and drier air predominate near this latitude. At latitude 23½°N, the climate is more moist and cloudy, causing more sunlight to be scattered and reflected before reaching the surface. In addition, day length is longer at 30°N than at 23½°N on June 21. For these reasons, more radiation falls on 30°N latitude than at the Tropic of Cancer (23½°N).

Each day past June 21, the noon sun is slightly lower in the sky. Summer days in the Northern Hemisphere begin to shorten. June eventually gives way to September, and fall begins.

Look at Fig. 3.3 again and notice that, by September 22, the earth will have moved so that the sun is directly above the equator. Except at the poles, the days and nights throughout the world are of equal length. This day is called the **autumnal** (fall) **equinox**, and it marks the astronomical beginning of fall in the Northern Hemisphere. At the North Pole, the sun appears on the horizon for 24 hours, due to the bending of light by the atmosphere. The following day (or at least within several days), the sun disappears from view, not to rise again for a long, cold six months. Throughout the northern half of the world on each successive day, there are fewer hours of daylight, and the noon sun is slightly lower in the sky. Less direct sunlight and shorter hours of daylight spell cooler weather for the Northern Hemisphere. Reduced radiation, lower air temperatures, and cooling breezes stimulate the beautiful pageantry of fall colors. (Does the first frost cause the beautiful colors? If you would like to know, read the Focus section on p. 58.)

In some years around the middle of autumn, there is an unseasonably warm spell, especially in the eastern two-thirds of the United States. This warm period, referred to as **Indian**

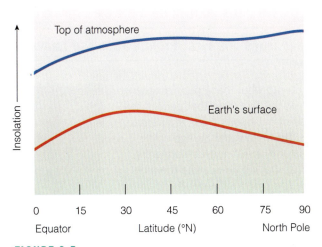

FIGURE 3.5

The relative amount of radiant energy received at the top of the earth's atmosphere and at the earth's surface on June 21—the summer solstice.

FIGURE 3.6

During the Northern Hemisphere summer, sunlight that reaches the earth's surface in far northern latitudes has passed through a thicker layer of absorbing, scattering, and reflecting atmosphere than sunlight that reaches the earth's surface farther south. Sunlight is lost through both the thickness of the pure atmosphere and by impurities in the atmosphere. As the sun's rays become more oblique, these effects become more pronounced.

summer,* may last from several days up to a week or more. It usually occurs when a large high-pressure area stalls near the southeast coast. The clockwise flow of air around this system moves warm air from the Gulf of Mexico into the central or eastern half of the nation. The warm, gentle breezes and smoke

---

*The origin of the term is uncertain, as it dates back to the eighteenth century. It may have originally referred to the good weather that allowed the Indians time to harvest their crops. Normally, a period of cool autumn weather must precede the warm weather period to be called Indian summer.

Does the First Frost Cause the Leaves to Change Color in Autumn?

Contrary to what many people believe, it is not the first frost that causes the leaves of deciduous trees to change color. The yellow and orange colors are actually in the leaves during the summer, but we do not see them because the green pigments—known as *chlorophylls*—are far more dominant.

Chlorophylls absorb and use the energy of red and violet wavelengths of sunlight to manufacture the simple sugars and starches that trees use as food. Most of the green wavelengths are not absorbed; instead, they are reflected. This, of course, causes the leaf to appear green.

The leaves remain green as long as the tree is able to replenish the chlorophylls it uses up. About several weeks before the first frost, however, a chemical change begins in the leaf, as shorter days and cooler nights cause leaf activity to diminish. As the chlorophylls decrease, the green slowly disappears. Other pigments, such as the yellow and orange carotenoid, which were hidden by the green of the leaf, now begin to show through. Addition-

FIGURE 1
The pageantry of fall colors along a country road in Vermont.

ally, the cooler weather stimulates the manufacture of other pigments, such as the anthocyanins, which turn the leaves red or purple. The combination of these pigments can produce a fiery display of autumn color (see Fig. 1).

The weather most suitable for an impressive display of fall colors is warm, sunny days followed by clear, cool nights, with temperatures dropping below 7°C (45°F), but remaining above freezing.

---

from a variety of sources respectively make for mild, hazy days. The warm weather ends abruptly when an outbreak of polar air reminds us that winter is not far away.

On December 21 (three months after the autumnal equinox), the Northern Hemisphere is tilted as far away from the sun as it will be all year (see Fig. 3.3). Nights are long and days are short. Notice in Table 3.1 that daylight decreases from 12 hours at the equator to 0 (zero) at latitudes above 66½°N. This is the shortest day of the year, called the **winter solstice,** the astronomical beginning of winter in the northern world. On this day, the sun shines directly above latitude 23½°S (Tropic of Capricorn). In the northern half of the world, the sun is at its lowest position in the noon sky. Its rays pass through a thick section of atmosphere and spread over a large area on the surface. Notice in Fig. 3.7 that on the average in the Northern Hemisphere, little (if any) solar energy reaches the earth's surface at high latitudes. With shorter days and less intense sunlight, we find ourselves bundled up for the cold winter.

With so little incident sunlight, the earth's surface cools quickly. A blanket of clean snow covering the ground aids in the cooling. The snow reflects much of the sunlight that reaches the surface and continually radiates away infrared energy during the long nights. In northern Canada and Alaska, the arctic air rapidly becomes extremely cold as it lies poised, ready to do battle with the milder air to the south. Periodically, this cold arctic air pushes down into the northern United States, producing a rapid drop in temperature called a *cold wave,* which occasionally reaches far into the south during the winter.

On each winter day after December 21, the sun climbs a bit higher in the midday sky. The periods of daylight grow longer until days and nights are of equal length, and we have another equinox.

The date of March 20, which marks the astronomical arrival of spring, is called the **vernal** (spring) **equinox.** At this equinox, the noonday sun is shining directly on the equator, while, at the North Pole, the sun (after hiding for six months) peeks above the horizon. Longer days and more direct solar radiation spell warmer weather for the northern world.

Three months after the vernal equinox, it is June again. The Northern Hemisphere is tilted toward the sun, which shines high in the noonday sky. The days have grown longer and warmer, and another summer season has begun.

At this point it is interesting to note that although sunlight is most intense in the Northern Hemisphere on June 21, the warmest weather in middle latitudes normally occurs weeks later, usually in July or August. This situation (called the *lag in seasonal temperature*) arises because although incoming energy from the sun is greatest in June, it still exceeds outgoing energy from the earth for a period of at least several weeks. When incoming solar energy and outgoing earth energy are in balance, the highest average temperature is attained. When outgoing energy exceeds incoming energy, the average temperature drops. Because outgoing earth energy exceeds incoming solar energy well past the winter solstice (December 21), we normally find our coldest weather occurring in January or February. As we will see later in this chapter, there is a similar lag in daily temperature between the time of most intense sunlight and the time of highest air temperature for the day.

In summary, the seasons are controlled by solar energy striking our tilted planet, as it makes its annual voyage about its star. This tilt of the earth causes a seasonal variation in both the length of daylight and the intensity of sunlight that reaches the surface. Because of these facts, high latitudes tend to lose more energy to space each year than they receive from the sun, while low latitudes tend to gain more energy during the course of a year than they lose. Does this mean that, on each successive year, high latitudes grow colder and low latitudes warmer? (For the answer, read the Focus section on p. 60.)

Seasons in the Southern Hemisphere

On June 21, the Southern Hemisphere is adjusting to an entirely different season. Because this part of the world is now tilted away from the sun, nights are long, days are short, and solar rays come in at an angle. All of these factors keep air temperatures fairly low. The June solstice marks the astronomical beginning of winter in the Southern Hemisphere. In this part of the world, summer will not "officially" begin until the sun is over the Tropic of Capricorn (23½°S)—remember that this occurs on December 21. So, when it is winter and June in the Southern Hemisphere, it is summer and June in the Northern Hemisphere. If you are tired of the hot June weather in your Northern Hemisphere city, travel to the winter half of the world and enjoy the cooler weather. The tilt of the earth as it revolves around the sun makes all this possible.

We know the earth comes nearer to the sun in January than in July. Even though this difference in distance amounts to only about 3 percent, the energy that strikes the top of the earth's atmosphere is almost 7 percent greater on January 3 than on July 4. These statistics might lead us to believe that

TABLE 3.1   Length of Time from Sunrise to Sunset for Various Latitudes on Different Dates

LATITUDE	NORTHERN HEMISPHERE (READ DOWN)			
	MARCH 20	JUNE 21	SEPT. 22	DEC. 21
0°	12 hr	12.0 hr	12 hr	12.0 hr
10°	12 hr	12.6 hr	12 hr	11.4 hr
20°	12 hr	13.2 hr	12 hr	10.8 hr
30°	12 hr	13.9 hr	12 hr	10.1 hr
40°	12 hr	14.9 hr	12 hr	9.1 hr
50°	12 hr	16.3 hr	12 hr	7.7 hr
60°	12 hr	18.4 hr	12 hr	5.6 hr
70°	12 hr	2 months	12 hr	0 hr
80°	12 hr	4 months	12 hr	0 hr
90°	12 hr	6 months	12 hr	0 hr
LATITUDE	SEPT. 22	DEC. 21	MARCH 20	JUNE 21
	SOUTHERN HEMISPHERE (READ UP)			

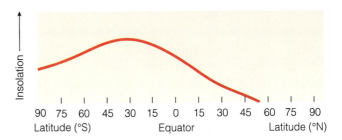

FIGURE 3.7

The relative amount of solar energy received at the earth's surface for various latitudes on December 21—the winter solstice.

summer should be warmer in the Southern Hemisphere than in the Northern Hemisphere, which, however, is not the case. A close examination of the Southern Hemisphere reveals that nearly 81 percent of the surface is water compared to 61 percent in the Northern Hemisphere. The added solar energy due to the closeness of the sun is absorbed by large bodies of water,

The Tilted Earth and Its Energy Balance

In Chapter 2, we saw that the amount of incoming energy averaged over the entire earth each year is roughly equal to the amount of outgoing energy. However, given the tilt of the earth, this balance is not maintained for each latitude.

As Fig. 2 shows, high latitudes lose more radiant energy to space each year than they receive. Low latitudes, on the other hand, gain more radiant energy from the sun than they return to space through infrared energy. Only at middle latitudes near 37° does the amount of radiation received each year balance the amount lost. Thus, in one year, high latitudes experience a net loss in radiant energy, while low latitudes experience a net gain. From this situation, we might conclude that polar regions are growing colder each year, while tropical regions are becoming warmer. But this obviously does not happen.

To compensate for these gains and losses of energy, winds in the atmosphere and currents in the oceans circulate warm air and water toward the poles, and cold air and water toward the equator. In part, the poleward transfer of heat occurs when water vapor

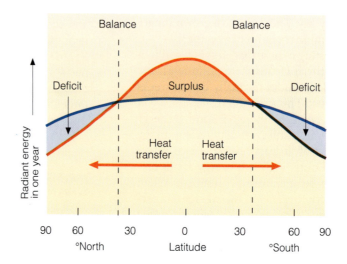

FIGURE 2
The average annual incoming solar radiation (red line) absorbed by the earth and the atmosphere along with the average annual infrared radiation (blue line) emitted by the earth and the atmosphere.

evaporates from tropical waters and is carried into high latitudes, where it condenses, releasing its latent heat. This process accounts for nearly 30 percent of the heat transported from equatorial to polar regions.

An additional 30 percent is carried northward in the form of sensible heat by winds associated with middle latitude cyclones and anticyclones, where on one side of each system warm air flows northward and on the other side

cold air moves southward. The remaining amount of heat (over 30 percent) travels poleward by ocean currents. Thus, the transfer of heat by atmospheric and oceanic circulations prevents low latitudes from steadily becoming warmer and high latitudes from steadily growing colder. These circulations are extremely important to weather and climate, and will be treated more completely in Chapter 11.

---

becoming well mixed and circulated within them. This keeps the average summer (January) temperatures in the Southern Hemisphere cooler than summer (July) temperatures in the Northern Hemisphere. Because of water's large heat capacity, it also tends to keep winters in the Southern Hemisphere warmer than we might expect.*

Another difference between the seasons of the two hemispheres concerns their length. Because the earth describes an ellipse as it journeys around the sun, the total number of days from the vernal (March 20) to the autumnal (September 22) equinox is about 7 days longer than from the autumnal to vernal equinox (see Fig. 3.8). This means that spring and summer in the Northern Hemisphere not only last about a week longer than northern fall and winter, but also about a week longer

---

*For a comparison of January and July temperatures, see Figs. 3.21 and 3.22, p. 69.

WEATHER WATCH

The position of the sun at noon above Barrow, Alaska, on June 21 (the summer solstice) is about 42° above the southern horizon. This angle represents the same position of the sun above the southern horizon at noon at Key West, Florida, on December 21 (the winter solstice).

than spring and summer in the Southern Hemisphere. Hence, the shorter spring and summer of the Southern Hemisphere somewhat offset the extra insolation received due to a closer proximity to the sun.

Up to this point, we have considered the seasons on a global scale. We will now shift to more local considerations. (Before going on to the next section you may wish to read the

The Changing Seasons and How We Feel—It Could Be SAD

Can a seasonal change affect how you feel? For example, when winter sets in, do you feel lethargic, down, or depressed? Do you have similar feelings when the heat of summer arrives? These psychological feelings may relate to the changing seasons—what scientists call *Seasonal Affective Disorder*, or SAD.

Studies suggest that SAD may relate to two main seasonal factors: a decrease in the amount of sunlight that accompanies the winter months and an increase in the temperature that occurs in summer. These changing seasonal events may affect one's eating patterns, sleeping patterns, and overall mental health. For example, as winter approaches and sunlight hours diminish, some people report feelings of listlessness, lack of motivation, antisocial behaviors, sadness, and even suicidal tendencies. People who report symptoms of winter SAD also tend to need extra sleep and crave large amounts of carbohydrates.

To alleviate these symptoms, doctors have experimented with light

therapy. In one study, patients with extreme winter SAD were subjected to one or more hours of extremely bright fluorescent light. In 80 percent of those treated, symptoms were relieved in 3 to 7 days. In another study, people with winter SAD were subjected to 30-minute exposures of bright light in the early morning in an attempt to trick the mind into believing the sun was rising earlier than it was. Nearly 60 percent of those receiving morning light therapy reported marked improvement in their SAD symptoms.

People who suffer from summer SAD report loss of energy, social withdrawal, oversleeping, sadness, and depression. Many of those reported that their environment (namely, increasing temperature and elevated relative humidity) seems to adversely influence their psychological well-being more than anything else.

The exact cause of winter and summer SAD is being investigated. The cause of winter SAD may be influenced by biological activity in the

brain. Some scientists feel that light either triggers, or at least brings into balance, certain neurotransmitters (chemical messengers involved in nerve cell communication), such as serotonin and dopamine. For winter SAD patients, reduction in sunlight may cause some form of imbalance in neurotransmitter activity and hence the onset of psychological discomfort. Recent studies tend to suggest that artificial morning bright light may change the brain's pattern of secreting melatonin, a hormone that regulates sleep and other daily biochemical rhythms. More and more studies are indicating that light therapy helps some people with winter SAD. So during the dark days of winter, when the blahs are abundant, a trip to a sunny tropical island may be just what the doctor orders.

Focus section above that relates to seasonal changes and their impact on how we feel.)

LOCAL SEASONAL VARIATIONS

Figure 3.9 shows how the sun's position changes in the middle latitudes of the Northern Hemisphere during the course of one year. Note that, during the winter, the sun rises in the southeast and sets in the southwest. During the summer, it rises in the northeast, reaches a much higher position in the sky at noon, and sets in the northwest. Clearly, objects facing south will receive more sunlight during a year than those facing north. This fact becomes strikingly apparent in hilly or mountainous country.

Hills that face south receive more sunshine and, hence, become warmer than the partially shielded north-facing hills. Higher temperatures usually mean greater rates of evaporation

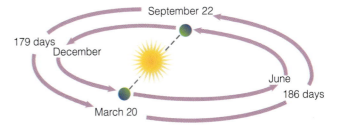

FIGURE 3.8
Because the earth travels more slowly when it is farther from the sun, it takes the earth a little more than 7 days longer to travel from March 20 to September 22 than from September 22 to March 20.

and slightly drier soil conditions. Thus, south-facing hillsides are usually warmer and drier as compared to north-facing slopes at the same elevation. In many areas of the far west, only sparse vegetation grows on south-facing slopes, while, on the

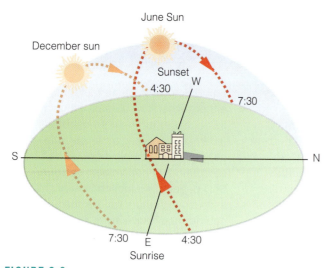

FIGURE 3.9

The changing position of the sun, as observed in middle latitudes in the Northern Hemisphere.

same hill, dense vegetation grows on the cool, moist hills that face north (see Fig. 3.10).

In northern latitudes, hillsides that face south usually have a longer growing season. Winemakers in western New York State do not plant grapes on the north side of hills. Grapes from vines grown on the warmer south side make better wine. Moreover, because air temperatures normally decrease with increasing height, trees found on the cooler north-facing side of mountains are often those that usually grow at higher elevations, while the warmer south-facing side of the mountain often supports trees usually found at lower elevations.

In the mountains, snow usually lingers on the ground for a longer time on north slopes than on the warmer south slopes.

For this reason, ski runs are built facing north wherever possible. Also, homes and cabins built on the north side of a hill usually have a steep pitched roof, as well as a reinforced deck to withstand the added weight of snow from successive winter storms.

The seasonal change in the sun's position during the year can have an effect on the vegetation around the home. In winter, a large two-story home can shade its own north side, keeping it much cooler than its south side. Trees that require warm, sunny weather should be planted on the south side, where sunlight reflected from the house can even add to the warmth.

The design of a home can be important in reducing heating and cooling costs. Large windows should face south, allowing sunshine to penetrate the home in winter. To block out excess sunlight during the summer, a small eave or overhang should be built (see Fig. 3.11). A kitchen with windows facing east will let in enough warm morning sunlight to help heat this area. Because the west side warms rapidly in the afternoon, rooms having small windows (such as garages) should be placed here to act as a thermal buffer. Deciduous trees planted on the west or south side of a home provide shade in the summer. In winter, they drop their leaves, allowing the winter sunshine to warm the house. If you like the bedroom slightly cooler than the rest of the home, face it toward the north. Let nature help with the heating and air conditioning. Proper house design, orientation, and landscaping can help cut the demand for electricity, as well as for natural gas and fossil fuels, which are rapidly being depleted.

From our reading of the last several sections, it should be apparent that, when solar heating a home, proper roof angle is important in capturing much of the winter sun's energy. (The information needed to determine the angle at which sunlight will strike a roof is given in the Focus section on p. 63.)

FIGURE 3.10

In areas where small temperature changes can cause major changes in soil moisture, sparse vegetation on the south-facing slopes will often contrast with lush vegetation on the north-facing slopes.

Solar Heating and the Noonday Sun

The amount of solar energy that falls on a typical American home each summer day is many times the energy needed to heat the inside for a year. Thus, many people are turning to the sun as a clean, safe, and virtually inexhaustible source of energy. If solar collectors are used to heat a home, they should be placed on south-facing roofs to take maximum advantage of the energy provided. The roof itself should be constructed as nearly perpendicular to winter sun rays as possible. To determine the proper roof angle at any latitude, we need to know how high the sun will be above the southern horizon at noon.

The noon angle of the sun can be calculated in the following manner:

1. Determine the number of degrees between your latitude and the latitude where the sun is currently directly overhead.

2. Subtract the number you calculated in step 1 from 90°. This will give you the sun's elevation above the southern horizon at noon at your latitude.

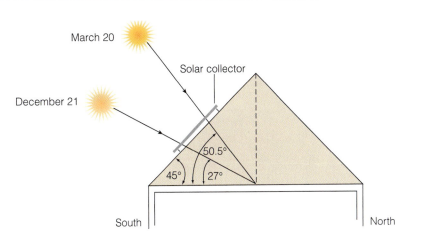

FIGURE 3

The roof of a solar-heated home constructed in Denver, Colorado, at an angle of 45° absorbs the sun's energy in midwinter at nearly right angles.

For example, suppose you live in Denver, Colorado (latitude 39½°N), and the date is December 21. The difference between your latitude and where the sun is currently overhead is 63° (39½°N to 23½°S), so the sun is 27° (90° − 63°) above the southern horizon at noon. On March 20 in Denver, the angle of the sun is 50½° (90° − 39½°). To determine a reasonable roof angle, we must consider the average altitude of the midwinter sun (about 39° for Denver), building costs, and snow loads. Figure 3 illustrates that a roof constructed in Denver, Colorado, at an angle of 45° will be nearly perpendicular to much of the winter sun's energy. Hence, the roofs of solar-heated homes in middle latitudes are generally built at an angle between 45° and 50°.

FIGURE 3.11

An overhang should be long enough to block out direct summer sunlight, yet short enough to allow winter sunlight to enter through the window.

DAILY TEMPERATURE VARIATIONS

In a way, each sunny day is like a tiny season as the air goes through a daily cycle of warming and cooling. The air warms during the morning hours, as the sun gradually rises higher in the sky, spreading a blanket of heat energy over the ground. The sun reaches its highest point around noon, after which it begins its slow journey toward the western horizon. It is around noon when the earth's surface receives the most intense solar rays. However, somewhat surprisingly, noontime is usually not the warmest part of the day. Rather, the air continues to be heated, often reaching a maximum temperature later in the afternoon. To find out why this *lag in temperature* occurs, we need to examine a shallow layer of air in contact with the ground.

Daytime Warming   As the sun rises in the morning, sunlight warms the ground, and the ground warms the air in contact

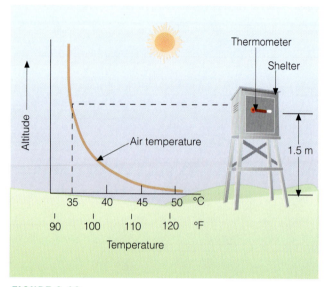

FIGURE 3.12
On a sunny, calm day, the air near the surface can be substantially warmer than the air a meter or so above the surface.

with it by conduction. However, air is such a poor heat conductor that this process only takes place within a few centimeters of the ground. As the sun rises higher in the sky, the air in contact with the ground becomes even warmer, and there exists a thermal boundary separating the hot surface air from the slightly cooler air above. Given their random motion, some air molecules will cross this boundary: The "hot" molecules below bring greater kinetic energy to the cooler air; the "cool" molecules above bring a deficit of energy to the hot, surface air. However, on a windless day, this form of heat exchange is slow,

and a substantial temperature difference usually exists just above the ground (see Fig. 3.12). This explains why joggers on a clear, windless, summer afternoon may experience air temperatures of over 50°C (122°F) at their feet and only 32°C (90°F) at their waist.

Near the surface, convection begins, and rising air bubbles (thermals) help to redistribute heat. In calm weather, these thermals are small and do not effectively mix the air near the surface. Thus, large vertical temperature gradients are able to exist. On windy days, however, turbulent eddies are able to mix hot surface air with the cooler air above. This form of mechanical stirring, sometimes called *forced convection*, helps the thermals to transfer heat away from the surface more efficiently. Therefore, on windy sunny days the molecules near the surface are more quickly carried away than on calm sunny days. Figure 3.13 shows a typical vertical profile of air temperature on windy days and on calm days in summer.

We can now see why the warmest part of the day is usually in the afternoon. Around noon, the sun's rays are most intense. However, even though incoming solar radiation decreases in intensity after noon, it still exceeds outgoing heat energy from the surface for a time. This yields an energy surplus for two to four hours after noon and substantially contributes to a lag between the time of maximum solar heating and the time of maximum air temperature several meters above the surface (see Fig. 3.14).

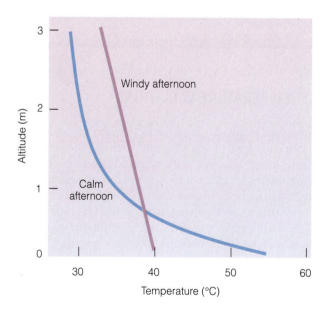

FIGURE 3.13
Vertical temperature profiles above an asphalt surface for a windy and a calm summer afternoon.

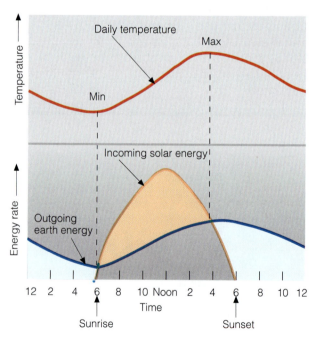

FIGURE 3.14
The daily variation in air temperature is controlled by incoming energy (primarily from the sun) and outgoing energy from the earth's surface. Where incoming energy exceeds outgoing energy (orange shade), the air temperature rises. Where outgoing energy exceeds incoming energy (blue shade), the air temperature falls.

The exact time of the highest temperature reading varies somewhat. Where the summer sky remains cloud-free all afternoon, the maximum temperature may occur sometime between 3:00 and 5:00 P.M. Where there is afternoon cloudiness or haze, the temperature maximum occurs an hour or two earlier. In Denver, afternoon clouds, which build over the mountains, drift eastward early in the afternoon. This sometimes causes the maximum temperature to occur as early as noon. If clouds persist throughout the day, the overall daytime temperatures are usually lower, as clouds reflect a great deal of incoming sunlight.

Adjacent to large bodies of water, cool air moving inland may modify the rhythm of temperature change such that the warmest part of the day occurs at noon or before. In winter, atmospheric storms circulating warm air northward can even cause the highest temperature to occur at night.

Just how warm the air becomes depends on such factors as the type of soil, its moisture content, and vegetation cover. When the soil is a poor heat conductor (as loosely packed sand is), heat energy does not readily transfer into the ground. This allows the surface layer to reach a higher temperature, availing more energy to warm the air above. On the other hand, if the soil is moist or covered with vegetation, much of the available energy evaporates water, leaving less to heat the air. As you might expect, the highest summer temperatures usually occur over desert regions, where clear skies coupled with low humidities and meager vegetation permit the surface and the air above to warm up rapidly.

Where the air is humid, haze and cloudiness lower the maximum temperature by preventing some of the sun's rays from reaching the ground. In humid Atlanta, Georgia, the average maximum temperature for July is 30.5°C (87°F). In contrast, Phoenix, Arizona—in the desert southwest at the same latitude as Atlanta—experiences an average July maximum of 40.5°C (105°F).

Nighttime Cooling

As the sun lowers, its energy is spread over a larger area, which reduces the heat available to warm the ground. Observe in Fig. 3.14 that sometime in late afternoon or early evening, the earth's surface and air above begin to lose more energy than they receive; hence, they start to cool.

Both the ground and air above cool by radiating infrared energy, a process called **radiational cooling**. The ground, being a much better radiator than air, is able to cool more quickly. Consequently, shortly after sunset, the earth's surface is slightly cooler than the air directly above it. The surface air transfers some energy to the ground by conduction, which the ground, in turn, quickly radiates away.

As the night progresses, the ground and the air in contact with it continue to cool more rapidly than the air a few meters higher. The warmer upper air does transfer *some* heat downward, a process that is slow due to the air's poor thermal conductivity. Therefore, by late night or early morning, the coldest air is found next to the ground, with slightly warmer air above (see Fig. 3.15).

FIGURE 3.15

On a clear, calm night, the air near the surface can be much colder than the air above. The increase in air temperature with increasing height above the surface is called a radiation temperature inversion.

This measured increase in air temperature just above the ground is known as a **radiation inversion** because it forms mainly through radiational cooling of the surface. Because radiation inversions occur on most clear calm nights, they are also called **nocturnal inversions**.

Radiation Inversions A strong radiation inversion occurs when the air near the ground is much colder than the air higher up. Ideal conditions for a strong inversion and, hence, very low nighttime temperatures exist when the air is calm, the night is long, and the air is fairly dry and cloud-free. Let's examine these ingredients one by one.

A windless night is essential for a strong radiation inversion because a stiff breeze tends to mix the colder air at the surface with the warmer air above. This mixing, along with the cooling of the warmer air as it comes in contact with the cold ground, causes a vertical temperature profile that is almost isothermal (constant temperature) in a layer several meters thick. In the absence of wind, the cooler, more dense surface air does not readily mix with the warmer, less dense air above, and the inversion is more strongly developed, as illustrated in Fig. 3.16.

A long night also contributes to a strong inversion. Generally, the longer the night, the longer the time of radiational cooling and the better are the chances that the air near the ground will be much colder than the air above. Consequently, winter nights provide the best conditions for a strong radiation inversion, other factors being equal.

Finally, radiation inversions are more likely with a clear sky and dry air. Under these conditions, the ground is able to

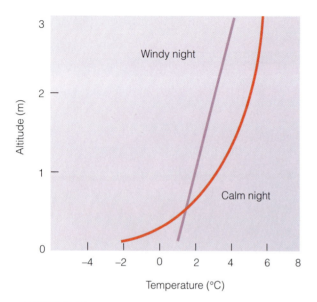

FIGURE 3.16
Vertical temperature profiles just above the ground on a windy night and on a calm night. Notice that the radiation inversion develops better on the calm night.

radiate its energy to outer space and thereby cool rapidly. However, with cloudy weather and moist air, much of the outgoing infrared energy is absorbed and radiated to the surface, retarding the rate of cooling. Also, on moist nights, condensation in the form of fog or dew will release latent heat, which warms the air. So, radiation inversions may occur on any night. But, during long winter nights, when the air is still, cloud-free, and relatively dry, these inversions can become strong and deep.

On winter nights in middle latitudes, it is common to experience below-freezing temperatures near the ground and air 5°C (9°F) warmer at your waist. In middle latitudes, the top of the inversion—the region where the air temperature stops increasing with height—is usually not more than 100 m (330 ft) above the ground. In dry, polar regions, where winter nights are measured in months, the top of the inversion is often 1000 m (about 3300 ft) above the surface. It may, however, extend to as high as 3000 m (about 10,000 ft).

It should now be apparent that how cold the night air becomes depends primarily on the length of the night, the moisture content of the air, cloudiness, and the wind. Even though wind may initially bring cold air into a region, the coldest nights usually occur when the air is clear and relatively calm.

There are, however, other factors that determine how cold the night air becomes. For example, a surface that is wet or covered with vegetation can add water vapor to the air, retarding nighttime cooling. Likewise, if the soil is a good heat conductor, heat ascending toward the surface during the night adds warmth to the air, which restricts cooling. On the other hand, snow covering the ground acts as an insulating blanket that prevents heat stored in the soil from reaching the air. Snow, a good emitter of infrared energy, radiates away energy rapidly at night, which helps keep the air temperature above a snow surface quite low.

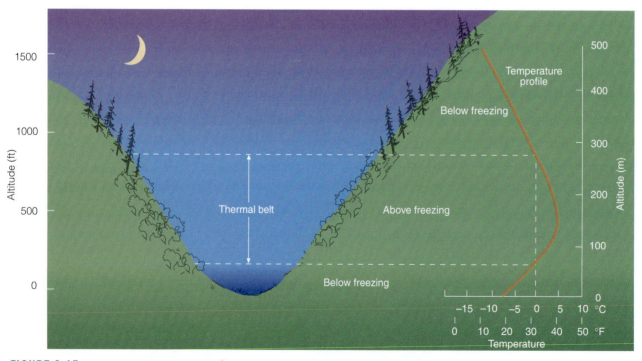

FIGURE 3.17
On cold, clear nights, the settling of cold air into valleys makes them colder than surrounding hillsides. The region along the side of the hill where the air temperature is above freezing is known as a *thermal belt*.

Look back at Fig. 3.14 and observe that the lowest temperature on any given day is usually observed around sunrise. However, the cooling of the ground and surface air may even continue beyond sunrise for a half hour or so, as outgoing energy can exceed incoming energy. This happens because light from the early morning sun passes through a thick section of atmosphere and strikes the ground at a low angle. Consequently, the sun's energy does not effectively heat the surface. Surface heating may be reduced further when the ground is moist and available energy is used for evaporation. (Any duck hunter lying flat in a marsh knows the sudden cooling that occurs as evaporation chills the air just after sunrise.) Hence, the lowest temperature may occur shortly after the sun has risen.

Cold, heavy surface air slowly drains downhill during the night and eventually settles in low-lying basins and valleys. Valley bottoms are thus colder than the surrounding hillsides (see Fig. 3.17). In middle latitudes, these warmer hillsides, called **thermal belts**, are less likely to experience freezing temperatures than the valley below. This encourages farmers to plant on hillsides those trees unable to survive the valley's low temperature.

On the valley floor, the cold, dense air is unable to rise. Smoke and other pollutants trapped in this heavy air restrict visibility. Therefore, valley bottoms are not only colder, but are also more frequently polluted than nearby hillsides. Even when the land is only gently sloped, cold air settles into lower-lying areas, such as river basins and floodplains. Because the flat floodplains are agriculturally rich areas, cold air drainage often forces farmers to seek protection for their crops.

Protecting Crops from the Cold On cold nights, many plants may be damaged by low temperatures. To protect small plants or shrubs, cover them with straw, cloth, or plastic sheeting. This prevents ground heat from being radiated away to the colder surroundings. If you are a household gardener concerned about outside flowers and plants during cold weather, simply wrap them in plastic or cover each with a paper cup.

Fruit trees are particularly vulnerable to cold weather in the spring when they are blossoming. The protection of such trees presents a serious problem to the farmer. Since the lowest temperatures on a clear, still night occur near the surface, the lower branches of a tree are the most susceptible to damage. Therefore, increasing the air temperature close to the ground may prevent damage. One way this can be done is to use **orchard heaters**, which warm the air around them by setting up convection currents close to the ground. Early forms of these heaters were called *smudge pots* because they produced large amounts of dense black smoke that caused severe pollution. People tolerated this condition only because they believed that the smoke acted like a blanket, trapping some of the earth's heat. Studies have shown this concept to be not as significant as previously thought. Orchard heaters are now designed to produce as little smoke as possible (see Fig. 3.18).

Another way to protect trees is to mix the cold air at the ground with the warmer air above, thus raising the tempera-

FIGURE 3.18
Orchard heaters circulate the air by setting up convection currents.

ture of the air next to the ground. Such mixing can be accomplished by using **wind machines** (see Fig. 3.19), which are power-driven fans that resemble airplane propellers. Farmers without their own wind machines can rent air mixers in the form of helicopters. Although helicopters are effective in mixing the air, they are expensive to operate.

If sufficient water is available, trees can be protected by irrigation. On potentially cold nights, the orchard may be flooded. Because water has a high heat capacity, it cools more slowly than dry soil. Consequently, the surface does not become as cold as it would if it were dry. Furthermore, wet soil has a higher thermal conductivity than dry soil. Hence, in wet

FIGURE 3.19
Wind machines mix cooler surface air with warmer air above.

FIGURE 3.20
A coating of ice protects these almond trees from damaging low temperatures, as an early spring freeze drops air temperatures well below freezing.

soil, heat is conducted upward from subsurface soil more rapidly, which helps to keep the surface warmer.

So far, we have discussed protecting trees against the cold air near the ground during a radiation inversion. Farmers often face another nighttime cooling problem. For instance, when subfreezing air blows into a region, the coldest air is not found at the surface; the air actually becomes colder with height. This condition is known as a **freeze.*** A single freeze in California or Florida can cause several million dollars damage to citrus crops.

Protecting an orchard from the damaging cold air blown by the wind can be a problem. Wind machines will not help because they would only mix cold air at the surface with the colder air above. Orchard heaters and irrigation are of little value as they would only protect the branches just above the ground. However, there is one form of protection that does work: An orchard's sprinkling system may be turned on so that it emits a fine spray of water. In the cold air, the water freezes around the branches and buds, coating them with a thin veneer of ice (see Fig. 3.20). As long as the spraying continues, the latent heat—given off as the water changes into ice—keeps the ice temperature at 0°C (32°F). The ice acts as a protective coating against the subfreezing air by keeping the buds (or fruit) at a temperature higher than their damaging point. Care must be taken since too much ice can cause the branches to break. The fruit may be saved from the cold air, while the tree itself may be damaged by too much protection.

Presently under research are tests using genetically altered microorganisms to inhibit frost on plants. For example, laboratory tests using a particular bacterium (*Pseudomonas syringae*),

---

*A freeze occurs over a widespread area when the surface air temperature remains below freezing for a long enough time to damage certain agricultural crops.

known as *frost ban*, have shown promise in protecting plants from frost damage at temperatures as low as −5°C (23°F).

BRIEF REVIEW   Up to this point we have examined temperature variations on a seasonal and daily basis. Before going on, here is a review of some of the important concepts and facts we have covered:

- The seasons are caused by the earth being tilted on its axis as it revolves around the sun. The tilt causes annual variations in the amount of sunlight that strikes the surface as well as variations in the length of time the sun shines at each latitude.

- During the day, the earth's surface and air above will continue to warm as long as incoming energy (mainly sunlight) exceeds outgoing energy from the surface.

- At night, the earth's surface cools, mainly by giving up more infrared radiation than it receives—a process called radiational cooling.

- The coldest nights of winter normally occur when the air is calm, fairly dry (low water-vapor content), and cloud free.

- The highest temperatures during the day and the lowest temperatures at night are normally observed at the earth's surface.

- Radiation inversions exist usually at night when the air near the ground is colder than the air above.

THE CONTROLS OF TEMPERATURE

The main factors that cause variations in temperature from one place to another are called the **controls of temperature**. Earlier we saw that the greatest factor in determining temperature is the amount of solar radiation that reaches the surface. This, of course, is determined by the length of daylight hours and the intensity of incoming solar radiation. Both of these factors are a function of latitude, hence, latitude is considered an important control of temperature. The main controls are listed as follows:

1. latitude
2. land and water
3. ocean currents
4. elevation

We can obtain a better picture of these controls by examining Figs. 3.21 and 3.22, which show the average monthly temperatures throughout the world for January and July. The lines on the map are **isotherms**—lines connecting places that have the same temperature. Because air temperature normally decreases with height, cities at very high elevations are much colder than their sea level counterparts. Consequently, the isotherms in Figs. 3.21 and 3.22 are corrected to read at the

FIGURE 3.21
Average air temperature near sea level in January (°F).

FIGURE 3.22
Average air temperature near sea level in July (°F).

WEATHER WATCH

Talk about extreme temperature ranges. One of the greatest temperature ranges ever recorded in the Northern Hemisphere (56°C or 100°F) occurred at Browning, Montana, on January 23, 1916, when the air temperature plummeted from 7°C (44°F) to –49°C (–56°F) in less than 24 hours. This huge temperature range, however, would represent a rather typical day on the planet Mars. On Mars, the average high temperature reaches about –12°C (10°F) and the average low drops to –79°C (–110°F), producing a daily temperature range of 67°C (120°F).

same horizontal level (sea level) by adding to each station above sea level an amount of temperature that would correspond to an average temperature change with height.*

Figures 3.21 and 3.22 show the importance of latitude on temperature. Note that, on the average, temperatures decrease poleward from the tropics and subtropics in both January and July. However, because there is a greater variation in solar radiation between low and high latitudes in winter than in summer, the isotherms in January are closer together (a tighter gradient)† than they are in July. This fact means that if you travel from New Orleans to Detroit in January, you are more likely to experience greater temperature variations than if you make the same trip in July. Notice also in Figs. 3.21 and 3.22 that the isotherms do not run horizontally; rather, in many places they bend, especially where they approach an ocean-continent boundary.

On the January map, the temperatures are much lower in the middle of continents than they are at the same latitude near the oceans; on the July map, the reverse is true. The reason for these temperature variations can be attributed to the unequal heating and cooling properties of land and water. For one thing, solar energy reaching land is absorbed in a thin layer of soil; reaching water, it penetrates deeply. Because water is able to circulate, it distributes its heat through a much deeper layer. Also, some of the solar energy striking the water is used to evaporate it rather than heat it.

Another important reason for the temperature contrasts is that water has a high *specific heat*. As we saw in Chapter 2, it takes a great deal more heat to raise the temperature of 1 gram of water 1°C than it does to raise the temperature of 1 gram of soil or rock by 1°C. Water not only heats more slowly than land, it cools more slowly as well, and so the oceans act like huge heat

---

*The amount of change is usually less than the standard temperature lapse rate of 6.5°C per 1000 m (3.6°F per 1000 ft). The reason is that the standard lapse rate is computed for altitudes above the earth's surface in the "free" atmosphere. In the less dense air at high elevations, the absorption of solar radiation by the ground causes an overall slightly higher temperature than that of the free atmosphere at the same level.

†Gradient represents the rate of change of some quantity (in this case, temperature) over a given distance.

reservoirs. Thus, mid-ocean surface temperatures change relatively little from summer to winter compared to the much larger annual temperature changes over the middle of continents.

Along the margin of continents, ocean currents often influence air temperatures. For example, along the eastern margins, warm ocean currents transport warm water poleward, while, along the western margins, they transport cold water equatorward. Some coastal areas also experience upwelling, which brings cold water from below to the surface (see Chapter 11).

Even large lakes can modify the temperature around them. In summer, the Great Lakes remain cooler than the land. As a result, refreshing breezes blow inland, bringing relief from the sometimes sweltering heat. As winter approaches, the water cools more slowly than the land. The first blast of cold air from Canada is modified as it crosses the lakes, and so the first freeze is delayed on the eastern shores of Lake Michigan.

AIR TEMPERATURE DATA

The careful recording and application of temperature data are tremendously important to us all. Without accurate information of this type, the work of farmers, power company engineers, weather analysts, and many others would be a great deal more difficult. In these next sections, we will study the ways temperature data are organized and used. We will also examine the significance of daily, monthly, and yearly temperature ranges and averages in terms of practical application to everyday living.

Daily, Monthly, and Yearly Temperatures   The greatest variation in daily temperature occurs right at the earth's surface. In fact, the difference between the daily maximum and minimum temperature—called the **daily** (or **diurnal**) **range of temperature**—is greatest next to the ground and becomes progressively smaller as we move away from the surface (see Fig. 3.23). This daily variation in temperature is also much larger on clear days than on cloudy ones.

The largest diurnal range of temperature occurs on high deserts, where the air is often cloud-free, and there is less CO_2 and water vapor above to radiate much infrared energy back to the surface. By day, clear summer skies allow the sun's energy to quickly warm the ground which, in turn, warms the air above to a temperature sometimes exceeding 35°C (95°F). At night, the ground cools rapidly by radiating infrared energy to space, and the minimum temperature in these regions occasionally dips below 5°C (41°F), thus giving a daily temperature range of 30°C (54°F).

A good example of a city with a large diurnal temperature range is Reno, Nevada, which is located on a plateau at an elevation of 1350 m (4400 ft) above sea level. Here, in the dry, thin summer air, the average daily maximum temperature for July is 33°C (92°F)—short-sleeve weather, indeed. But don't lose your shirt in Reno, for you will need it at night, as the average daily minimum temperature for July is 8°C (47°F). Reno has a daily range of 25°C (45°F)!

In humid regions, the diurnal temperature range is usually small. Here, haze and clouds lower the maximum temperature by preventing some of the sun's energy from reaching the surface. At night, the moist air keeps the minimum temperature high by absorbing the earth's infrared radiation and radiating a portion of it to the ground. An example of a humid city with a small summer diurnal temperature range is Charleston, South Carolina, where the average July maximum temperature is 32°C (90°F), the average minimum is 22°C (72°F), and the diurnal range is only 10°C (18°F).

Cities near large bodies of water typically have smaller diurnal temperature ranges than cities farther inland. This phenomenon is caused in part by the additional water vapor in the air and by the fact that water warms and cools much more slowly than land. Moreover, cities whose temperature readings are obtained at airports often have larger diurnal temperature ranges than those whose readings are obtained in downtown areas. The reason for this fact is that nighttime temperatures in cities tend to be warmer than those in outlying rural areas. This nighttime city warmth—called the *urban heat island*—is due to industrial and urban development, a topic that will be discussed more completely in Chapter 17.

The average of the highest and lowest temperature for a 24-hour period is known as the **mean daily temperature**. Most newspapers list the mean daily temperature along with the highest and lowest temperatures for the preceding day. The average of the mean daily temperatures for a particular date averaged for a 30-year period gives the average (or *"normal"*) temperatures for that date. The average temperature for each month is the average of the daily mean temperatures for that month.

At any location, the difference between the average temperature of the warmest and coldest months is called the **annual range of temperature**. Usually the largest annual ranges occur over land, the smallest over water. Hence, inland cities have larger annual ranges than coastal cities. Near the equator (because daylight length varies little and the sun is always high in the noon sky), annual temperature ranges are small, usually less than 3°C (5°F). Quito, Ecuador—on the equator at an elevation of 2850 m (9350 ft)—experiences an annual range of less than 1°C. In middle and high latitudes, large seasonal variations in the amount of sunlight reaching the surface produce large temperature contrasts between winter and summer. Here, annual ranges are large, especially in the middle of a continent. Yakutsk, in northeastern Siberia near the Arctic Circle, has an extremely large annual temperature range of 62°C (112°F)!

The average temperature of any station for the entire year is the **mean (average) annual temperature**, which represents the average of the twelve monthly average temperatures.* When two cities have the same mean annual temperature, it might first seem that their temperatures throughout the year are quite similar. However, often this is not the case. For example, San Francisco, California, and Richmond, Virginia, are at

*The mean annual temperature may be obtained by taking the sum of the 12 monthly means and dividing that total by 12, or by obtaining the sum of the daily means and dividing that total by 365.

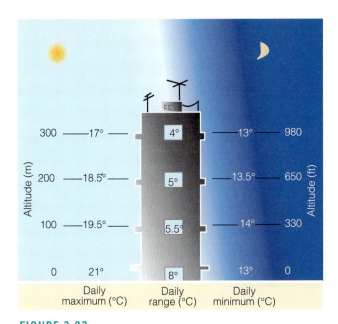

FIGURE 3.23
The daily range of temperature decreases as we climb away from the earth's surface. Hence, there is less day-to-night variation in air temperature near the top of a high-rise apartment complex than at the ground level.

the same latitude (37°N). Both have similar hours of daylight during the year; both have the same mean annual temperature—14°C (57°F). Here, the similarities end. The temperature differences between the two cities are apparent to anyone who has traveled to San Francisco during the summer with a suitcase full of clothes suitable for summer weather in Richmond.

Figure 3.24 summarizes the average temperatures for San Francisco and Richmond. Notice that the coldest month for both cities is January. Even though January in Richmond averages only 8°C (14°F) colder than January in San Francisco, people in Richmond awaken to an average January minimum temperature of –6°C (21°F), which is much colder than the lowest temperature ever recorded in San Francisco. Trees that thrive in San Francisco's weather would find it difficult surviving a winter in Richmond. So, even though San Francisco and Richmond have the same mean annual temperature, the behavior and range of their temperatures differ greatly.

The Use of Temperature Data An application of daily temperature developed by heating engineers in estimating energy needs is the **heating degree-day**. The heating degree-day is based on the assumption that people will begin to use their furnaces when the mean daily temperature drops below 65°F. Therefore, heating degree-days are determined by subtracting the mean temperature for the day from 65°F (18°C). Thus, if the mean temperature for a day is 64°F, there would be 1 heating degree-day on this day.*

*In the United States, the National Weather Service and the Department of Agriculture use degrees Fahrenheit in their computations.

FIGURE 3.24

Temperature data for San Francisco, California (37°N), and Richmond, Virginia (37°N)—two cities with the same mean annual temperature.

	SAN FRANCISO		RICHMOND	
	°C	°F	°C	°F
Mean annual temperature	14	57	14	57
Annual temperature range	6	11	22	40
Record high	39	103	41	105
Record low	−3	27	−24	−12

On days when the mean temperature is above 65°F, there are no heating degree-days. Hence, the lower the average daily temperature, the more heating degree-days and the greater the predicted consumption of fuel. When the number of heating degree-days for a whole year is calculated, the heating fuel requirements for any location can be estimated. Figure 3.25 shows the yearly average number of heating degree-days in various locations throughout the United States.

As the mean daily temperature climbs above 65°F, people begin to cool their indoor environment. Consequently, an index, called the **cooling degree-day**, is used during warm weather to estimate the energy needed to cool indoor air to a comfortable level. The forecast of mean daily temperature is converted to cooling degree-days by subtracting 65°F from the mean. The remaining value is the number of cooling degree-days for that day. For example, a day with a mean temperature of 70°F would correspond to 5 cooling degree-days (70 minus 65). High values indicate warm weather and high power production for cooling (see Fig. 3.26).

Knowledge of the number of cooling degree-days in an area allows a builder to plan the size and type of equipment that should be installed to provide adequate air conditioning. Also, the forecasting of cooling degree-days during the summer gives power companies a way of predicting the energy demand during peak energy periods. A composite of heating plus cooling degree-days would give a practical indication of the energy requirements over the year.

Farmers use an index, called **growing degree-days**, as a guide to planting and for determining the approximate dates when a crop will be ready for harvesting. A growing degree-day for a particular crop is defined as a day on which the mean daily temperature is one degree above the *base temperature* (also known as the *zero temperature*)—the minimum temperature required for growth of that crop. For sweet corn, the base temperature is 50°F and, for peas, it is 40°F.

On a summer day in Iowa, the mean temperature might be 80°F. From Table 3.2, we can see that, on this day, sweet corn would accumulate (80 − 50), or 30 growing degree-days. Theoretically, sweet corn can be harvested when it accumulates a total of 2200 growing degree-days. So, if sweet corn is planted in early April and each day thereafter averages about 20 growing degree-days, the corn would be ready for harvest about 110 days later, or around the middle of July.

At one time, corn varieties were rated in terms of "days to maturity." This rating system was unsuccessful because, in actual practice, corn took considerably longer in some areas than in others. This discrepancy was the reason for defining "growing degree-days." Hence, in humid Iowa, where summer nighttime temperatures are high, growing degree-days accumulate much faster. Consequently, the corn matures in considerably fewer days than in the drier west, where summer nighttime temperatures are lower, and each day accumulates fewer growing degree-days. Although moisture and other conditions are not taken into account, growing degree-days nevertheless serve as a useful guide in forecasting approximate dates of crop maturity.

AIR TEMPERATURE AND HUMAN COMFORT

Probably everyone realizes that the same air temperature can feel differently on different occasions. For example, a

TABLE 3.2   Estimated Growing Degree-Days for Certain Agricultural Crops to Reach Maturity

CROP (VARIETY, LOCATION)	BASE TEMPERATURE (°F)	GROWING DEGREE-DAYS TO MATURITY
Beans (Snap/South Carolina)	50	1200–1300
Corn (Sweet/Indiana)	50	2200–2800
Cotton (Delta Smooth Leaf/Arkansas)	60	1900–2500
Peas (Early/Indiana)	40	1100–1200
Rice (Vegold/Arkansas)	60	1700–2100
Wheat (Indiana)	40	2100–2400

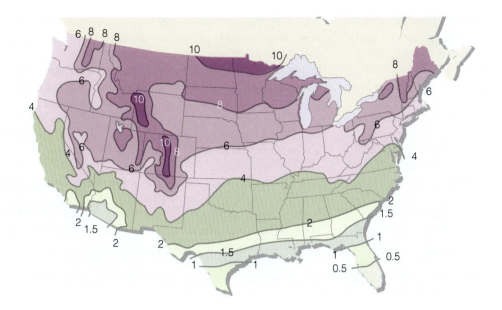

FIGURE 3.25
Mean annual total heating degree-days in thousands of °F, where the number 4 on the map represents 4000 (base 65°F).

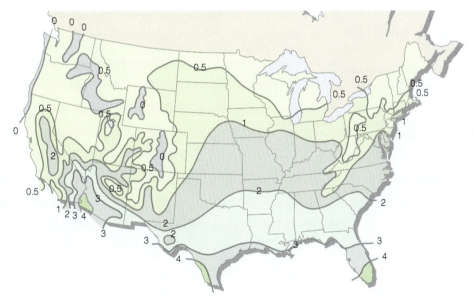

FIGURE 3.26
Mean annual total cooling degree-days in thousands of °F, where the number 1 on the map represents 1000 (base 65°F).

temperature of 20°C (68°F) on a clear windless March afternoon in New York City can almost feel balmy after a long hard winter. Yet, this same temperature may feel uncomfortably cool on a summer afternoon in a stiff breeze. The human body's perception of temperature obviously changes with varying atmospheric conditions. The reason for these changes is related to how we exchange heat energy with our environment.

The body stabilizes its temperature primarily by converting food into heat (*metabolism*). To maintain a constant temperature, the heat produced and absorbed by the body must be equal to the heat it loses to its surroundings. There is, therefore, a constant exchange of heat—especially at the surface of the skin—between the body and the environment.

One way the body loses heat is by emitting infrared energy. But we not only emit radiant energy, we absorb it as well. Another way the body loses and gains heat is by conduction and convection, which transfer heat to and from the body by air motions. On a cold day, a thin layer of warm air molecules forms close to the skin, protecting it from the surrounding cooler air and from the rapid transfer of heat. Thus, in cold weather, when the air is calm, the temperature we perceive—called the **sensible temperature**—is often higher than a thermometer might indicate. (Could the opposite effect occur where the air temperature is very high and a person might feel exceptionally cold? If you are unsure, read the Focus section on p. 75.)

Once the wind starts to blow, the insulating layer of warm air is swept away, and heat is rapidly removed from the skin by the constant bombardment of cold air. When all other factors are the same, the faster the wind blows, the greater the heat loss, and the colder we feel. How cold the wind makes us feel is usually expressed as a **wind-chill factor**.

The wind-chill charts (see Tables 3.3 and 3.4) translate the ability of the air to take heat away from the body with wind (its cooling power) into a wind-chill equivalent temperature.* For example, notice in Table 3.3 that an outside air temperature of 20°F with a wind speed of 30 mi/hr produces a wind-chill equivalent temperature of –18°F. This means that exposed skin would lose as much heat in one minute in air with a temperature of 20°F and a wind speed of 30 mi/hr as it would in calm air with a temperature of –18°F. Of course, how cold we feel actually depends on a number of factors, including the fit and type of clothing we wear, and the amount of exposed skin.

High winds, in below-freezing air, can remove heat from exposed skin so quickly that the skin may actually freeze and discolor. The freezing of skin, called *frostbite*, usually occurs on the body extremities first because they are the greatest distance from the source of body heat.

In cold weather, wet skin can be a factor in how cold we feel. A cold rainy day (drizzly, or even foggy) often feels colder

---

*Wind chill is also judged against a wind of 3 knots—the so-called "walking wind speed."

than a "dry" one because water on exposed skin conducts heat away from the body better than air does. In fact, in cold, wet, and windy weather a person may actually lose body heat faster than the body can produce it. This may even occur in relatively mild weather with air temperatures as high as 10°C (50°F). The rapid loss of body heat may lower the body temperature below its normal level and bring on a condition known as **hypothermia**—the rapid, progressive mental and physical collapse that accompanies the lowering of human body temperature.

The first symptom of hypothermia is exhaustion. If exposure continues, judgment and reasoning power begin to disappear. Prolonged exposure, especially at temperatures near or below freezing, produces stupor, collapse, and death when the internal body temperature drops to 26°C (79°F). Most cases of hypothermia occur when the air temperature is between 0°C and 10°C (between 32°F and 50°F). This may be because many people apparently do not realize that wet clothing in windy weather greatly enhances the loss of body heat, even when the temperature is well above freezing.

In cold weather, heat is more easily dissipated through the skin. To counteract this rapid heat loss, the peripheral blood vessels of the body constrict, cutting off the flow of blood to the outer layers of the skin. In hot weather, the blood vessels enlarge, allowing a greater loss of heat energy to the surround-

TABLE 3.3 Wind-Chill Equivalent Temperature (°F). A 20-mi/hr Wind Combined with an Air Temperature of 10°F Produces a Wind-Chill Equivalent Temperature of –24°F

					AIR TEMPERATURE (°F)													
	35	**30**	**25**	**20**	**15**	**10**	**5**	**0**	**-5**	**-10**	**-15**	**-20**	**-25**	**-30**	**-35**	**-40**	**-45**	
5	32	27	22	16	11	6	0	–5	–10	–15	–21	–26	–31	–36	–42	–47	–52	
10	22	16	10	3	–3	–9	–15	–22	–27	–34	–40	–46	–52	–58	–64	–71	–77	
15	16	9	2	–5	–11	–18	–25	–31	–38	–45	–51	–58	–65	–72	–78	–85	–92	
20	12	4	–3	–10	–17	–24	–31	–39	–46	–53	–60	–67	–74	–81	–88	–95	–103	
25	8	1	–7	–15	–22	–29	–36	–44	–51	–59	–66	–74	–81	–88	–96	–103	–110	
30	6	–2	–10	–18	–25	–33	–41	–49	–56	–64	–71	–79	–86	–93	–101	–109	–116	
35	4	–4	–12	–20	–27	–35	–43	–52	–58	–67	–74	–82	–89	–97	–105	–113	–120	
40	3	–5	–13	–21	–29	–37	–45	–53	–60	–69	–76	–84	–92	–100	–107	–115	–123	
45	2	–6	–14	–22	–30	–38	–46	–54	–62	–70	–78	–85	–93	–102	–109	–117	–125	

Wind Speed (mi/hr)

TABLE 3.4 Wind Chill Equivalent Temperature (°C)

					AIR TEMPERATURE (°C)											
	8	**4**	**0**	**-4**	**-8**	**-12**	**-16**	**-20**	**-24**	**-28**	**-32**	**-36**	**-40**	**-44**
Calm	8	4	0	–4	–8	–12	–16	–20	–24	–28	–32	–36	–40	–44
10	5	0	–4	–8	–13	–17	–22	–26	–31	–35	–40	–44	–49	–53
20	0	–5	–10	–15	–21	–26	–31	–36	–42	–47	–52	–57	–63	–68
30	–3	–8	–14	–20	–25	–31	–37	–43	–48	–54	–60	–65	–71	–77
40	–5	–11	–17	–23	–29	–35	–41	–47	–53	–59	–65	–71	–77	–83
50	–6	–12	–18	–25	–31	–37	–43	–49	–56	–62	–68	–74	–80	–87
60	–7	–13	–19	–26	–32	–39	–45	–51	–58	–64	–70	–77	–83	–89

Wind Speed (km/hr)

A Thousand Degrees and Freezing to Death

Is there somewhere in our atmosphere where the air temperature can be exceedingly high (say above 1000°C or 1800°F) yet a person might feel extremely cold? There is a region, but it's not at the earth's surface.

You may recall from Chapter 1, Fig. 1.10, that in the upper reaches of our atmosphere (in the middle and upper thermosphere), air temperatures may exceed 1000°C. However, a thermometer shielded from the sun in this region of the atmosphere would indicate an extremely low temperature. This apparent discrepancy lies in the meaning of air temperature and how we measure it.

In Chapter 2, we learned that the air temperature is directly related to the average speed at which the air molecules are moving—faster speeds correspond to higher temperatures. In the middle and upper thermosphere, air molecules are zipping about at speeds corresponding to extremely high temperatures. However, in order to transfer enough energy to heat something up by conduction (exposed skin or a thermometer bulb), an extremely large number of molecules must collide with the object. In the "thin" air of the upper atmosphere, air molecules are moving extraordinarily fast, but there

are simply not enough of them bouncing against the thermometer bulb for it to register a high temperature. In fact, when properly shielded from the sun, the thermometer bulb loses far more energy than it receives and indicates a temperature near absolute zero. This explains why an astronaut, when space walking, will not only survive temperatures exceeding 1000°C, but will also feel a profound coldness when shielded from the sun's radiant energy. At these high altitudes, the traditional meaning of air temperature (that is, regarding how "hot" or "cold" something feels) is no longer applicable.

---

ings. In addition to this, we perspire. As evaporation occurs, the skin cools because it supplies the large latent heat of vaporization (about 560 cal/g). When the air contains a great deal of water vapor and it is close to being saturated, perspiration does not readily evaporate from the skin. Less evaporational cooling causes most people to feel hotter than it really is, and a number of people start to complain about the "heat and humidity." (A closer look at how we feel in hot weather will be given in Chapter 5, after we have examined the concepts of relative humidity and wet-bulb temperature.)

MEASURING AIR TEMPERATURE

Thermometers were developed to measure air temperature. Each thermometer has a definite scale and is calibrated so that a thermometer reading of 0°C in Vermont will indicate the same temperature as a thermometer with the same reading in North Dakota. If a particular reading were to represent different degrees of hot or cold, depending on location, thermometers would be useless.

Liquid-in-glass thermometers are often used for measuring surface air temperature because they are easy to read and inexpensive to construct. These thermometers have a glass bulb attached to a sealed, graduated tube about 25 cm long. A very small opening, or bore, extends from the bulb to the end of the tube. A liquid in the bulb (usually mercury or red-colored alcohol) is free to move from the bulb up through the bore and into the tube. When the air temperature increases, the liquid in the bulb expands, and rises up the tube. When the air temperature decreases, the liquid contracts, and moves

down the tube. Hence, the length of the liquid in the tube represents the air temperature. Because the bore is very narrow, a small temperature change will show up as a relatively large change in the length of the liquid column.

Maximum and minimum thermometers are liquid-in-glass thermometers used for determining daily maximum and minimum temperatures. The **maximum thermometer** looks like any other liquid-in-glass thermometer with one exception: It has a small constriction within the bore just above the bulb (see Fig. 3.27). As the air temperature increases, the mercury expands and freely moves past the constriction up the tube, until the maximum temperature occurs. However, as the air temperature begins to drop, the small constriction prevents the mercury from flowing back into the bulb. Thus, the end of the stationary mercury column indicates the maximum temperature for the day. The mercury will stay at this position until either the air warms to a higher reading or the thermometer is reset by whirling it on a special holder and pivot. Usually, the whirling is sufficient to push the mercury back into the bulb past the constriction until the end of the column indicates the present air temperature.*

A **minimum thermometer** measures the lowest temperature reached during a given period. Most minimum thermometers use alcohol as a liquid, since it freezes at a temperature of −130°C compared to −39°C for mercury. The minimum thermometer is similar to other liquid-in-glass thermometers except that it contains a small barbell-shaped index marker in the bore (see Fig. 3.28). The small index marker is

---

*Thermometers that measure body temperature are maximum thermometers, which is why they are shaken both before and after you take your temperature.

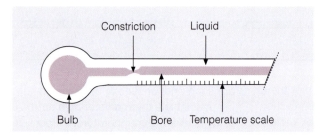

FIGURE 3.27
A section of a maximum thermometer.

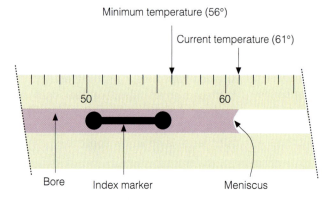

FIGURE 3.28
A section of a minimum thermometer showing both the current air temperature and the minimum temperature.

free to slide back and forth within the liquid. It cannot move out of the liquid because the surface tension at the end of the liquid column (the *meniscus*) holds it in.

A minimum thermometer is mounted horizontally. As the air temperature drops, the contracting liquid moves back into the bulb and brings the index marker down the bore with it. When the air temperature stops decreasing, the liquid and the index marker stop moving down the bore. As the air warms, the alcohol expands and moves freely up the tube past the stationary index marker. Because the index marker does not move as the air warms, the minimum temperature is read by observing the upper end of the marker.

To reset a minimum thermometer, simply tip it upside down. This allows the index marker to slide to the upper end of the alcohol column, which is indicating the current air temperature. The thermometer is then remounted horizontally, so that the marker will move toward the bulb as the air temperature decreases.

Highly accurate temperature measurements may be made with **electrical thermometers**. One type of electrical thermometer is the *electrical resistance thermometer*, which does not actually measure air temperature but rather the resistance of a wire, usually platinum or nickel, whose resistance increases as the temperature increases. An electrical meter measures the

resistance, and is calibrated to represent air temperature.* *Thermistors* are another type of electrical thermometer. They are made of ceramic material whose resistance increases as the temperature decreases. Although thermistors are not widely used for everyday surface observations, they are used extensively in remote sensing of air temperature, as radiosondes use them to measure the air temperature from the surface up to an altitude near 30 km.

Another electrical thermometer is the *thermocouple*. This device operates on the principle that the temperature difference between the junction of two dissimilar metals sets up a weak electrical current. When one end of the junction is maintained at a temperature different from that of the other end, an electrical current will flow in the circuit. This current is proportional to the temperature difference between the junctions.

Air temperature may also be obtained with instruments called *infrared sensors*, or **radiometers**. Radiometers do not measure temperature directly; rather, they measure emitted radiation (usually infrared). By measuring both the intensity of radiant energy and the wavelength of maximum emission of a particular gas, radiometers in orbiting satellites are now able to estimate the air temperature at selected levels in the atmosphere.

A **bimetallic thermometer** consists of two different pieces of metal (usually brass and iron) welded together to form a single strip. As the temperature changes, the brass expands more than the iron, causing the strip to bend. The small amount of bending is amplified through a system of levers to a pointer on a calibrated scale.

The bimetallic thermometer is usually the temperature-sensing part of the **thermograph**, an instrument that measures and records temperature. On a thermograph, the pointer is a pen that sits on a circular drum (see Fig. 3.29). The drum, which is covered with a piece of chart paper, is slowly rotated by a clock-drive. As the temperature of the air changes, the bimetallic thermometer bends accordingly, and the pen moves up or down, marking the chart paper with the current air temperature. Since the time is printed on the top of the chart paper, a continuous recording of temperature is obtained. Because bimetallic thermometers are less accurate than liquid-in-glass thermometers, thermographs must be checked periodically and corrected, using an accurate liquid thermometer. (Chances are, you may have heard someone exclaim something like, "Today the thermometer measured 90 degrees in the shade!" Does this mean that the air temperature was higher in the sun? If you are unsure of the answer, read the Focus section on p. 77 before reading the next section on instrument shelters.)

Thermometers and other instruments are usually housed in an **instrument shelter** (see Fig. 3.30). The shelter completely encloses the instruments, protecting them from rain, snow,

---

*Electrical resistance thermometers are the type of thermometer used in measuring air temperature in the over 900 fully automated surface weather stations (known as ASOS, for *Automated Surface Observing System*) that exist at airports and military facilities throughout the United States.

Thermometers Should Be Read in the Shade

When we measure air temperature with a common liquid thermometer, an incredible number of air molecules bombard the bulb, transferring energy either to or away from it. When the air is warmer than the thermometer, the liquid gains energy, expands, and rises up the tube; the opposite will happen when the air is colder than the thermometer. The liquid stops rising (or falling) when equilibrium between incoming and outgoing energy is established. At this point, we can read the temperature by observing the height of the liquid in the tube.

It is *impossible* to measure *air temperature* accurately in direct sunlight because the thermometer absorbs radiant energy from the sun in addition to energy from the air molecules. The thermometer gains energy at a much faster rate than it can radiate it away, and the liquid keeps expanding and rising until there is equilibrium between incoming and outgoing energy. Because of the direct absorption of solar energy, the level of the liquid in the thermometer indicates a temperature much higher than the actual air temperature, and so a statement that says "today the air temperature measured 100 degrees in the sun" has no meaning. Hence, a thermometer must be kept in a shady place to measure the temperature of the air accurately.

and the sun's direct rays. It is painted white to reflect sunlight, faces north to avoid direct exposure to sunlight, and has louvered sides, so that air is free to flow through it. This construction helps to keep the air inside the shelter at the same temperature as the air outside.

The thermometers inside a standard shelter are mounted about 1.5 to 2 m (5 to 6 ft) above the ground. As we saw in an earlier section, on a clear, calm night the air at ground level may be much colder than the air at the level of the shelter. As a result, on clear winter mornings it is possible to see ice or frost on the ground even though the minimum thermometer in the shelter did not reach the freezing point.

Because air temperatures vary considerably above different types of surfaces, shelters are usually placed over grass to ensure that the air temperature is measured at the same elevation over the same type of surface. Unfortunately, some shelters are placed on asphalt, others sit on concrete, while others are

located on the tops of tall buildings, making it difficult to compare air temperature measurements from different locations. In fact, if either the maximum or minimum air temperature in your area seems suspiciously different from those of nearby towns, find out where the instrument shelter is situated.

FIGURE 3.29
The thermograph with a bimetallic thermometer.

FIGURE 3.30
An instrument shelter protects the instruments inside from the weather elements.

SUMMARY

The earth has seasons because the earth is tilted on its axis as it revolves around the sun. The tilt of the earth causes a seasonal variation in both the length of daylight and the intensity of sunlight that reaches the surface. When the Northern Hemisphere is tilted toward the sun, the Southern Hemisphere is tilted away from the sun. Longer hours of daylight and more intense sunlight produce summer in the Northern Hemisphere, while, in the Southern Hemisphere, shorter daylight hours and less intense sunlight produce winter. On a more local setting, the earth's inclination influences the amount of solar energy received on the north and south side of a hill, as well as around a home.

The daily variation in air temperature near the earth's surface is controlled mainly by the input of energy from the sun and the output of energy from the surface. On a clear, calm day, the surface air warms, as long as heat input (mainly sunlight) exceeds heat output (mainly convection and radiated infrared energy). The surface air cools at night, as long as heat output exceeds input. Because the ground at night cools more quickly than the air above, the coldest air is normally found at the surface where a radiation inversion usually forms. When the air temperature in agricultural areas drops to dangerously low readings, fruit trees and grape vineyards can be protected from the cold by a variety of means, from mixing the air to spraying the trees and vines with water.

The greatest daily variation in air temperature occurs at the earth's surface. Both the diurnal and annual range of temperature are greater in dry climates than in humid ones. Even though two cities may have similar average annual temperatures, the range and extreme of their temperatures can differ greatly. Temperature information impacts our lives in many ways, from influencing decisions on what clothes to take on a trip to providing critical information for energy-use predictions and agricultural planning. We reviewed some of the many types of thermometers in use. Those designed to measure air temperatures near the surface are housed in instrument shelters to protect them from direct sunlight and precipitation.

Key Terms

summer solstice	freeze
autumnal equinox	controls of temperature
Indian summer	isotherms
winter solstice	daily (diurnal) range of temperature
vernal equinox	mean (average) daily temperature
radiational cooling	annual range of temperature
radiation inversion	mean annual temperature
nocturnal inversion	heating degree-day
thermal belts	cooling degree-day
orchard heaters	growing degree-days
wind machines	sensible temperature
wind-chill factor	electrical thermometers
hypothermia	radiometers
liquid-in-glass thermometers	bimetallic thermometer
maximum thermometer	thermograph
minimum thermometer	instrument shelter

Questions for Review

1. In the Northern Hemisphere, why are summers warmer than winters, even though the earth is actually closer to the sun in January?
2. What are the main factors that determine seasonal temperature variations?
3. During the Northern Hemisphere's summer, the daylight hours in northern latitudes are longer than in middle latitudes. Explain why northern latitudes are not warmer.
4. If it is winter and January in New York City, what is the season in Sydney, Australia?
5. Explain why Southern Hemisphere summers are not warmer than Northern Hemisphere summers.
6. Explain why the vegetation on the north-facing side of a hill is frequently different from the vegetation on the south-facing side of the same hill.
7. Look at Figures 3.13 and 3.16, which show vertical profiles of air temperature during different times of the day. Explain why the temperature curves are different.
8. What are some of the factors that determine the daily fluctuation of air temperature just above the ground?
9. Explain how incoming energy and outgoing energy regulate the daily variation in air temperature.
10. On a calm, sunny day, why is the air next to the ground normally much warmer than the air just above?
11. Explain why the warmest time of the day is usually in the afternoon, even though the sun's rays are most direct at noon.
12. Explain how radiational cooling at night produces a radiation temperature inversion.
13. What weather conditions are best suited for the formation of a cold night and a strong radiation inversion?
14. Explain why thermal belts are found along hillsides at night.
15. List some of the measures farmers use to protect their crops against the cold. Explain the physical principle behind each method.
16. Why are the lower tree branches most susceptible to damage from low temperatures?
17. Describe each of the controls of temperature.
18. Look at Fig. 3.21 (temperature map for January) and explain why the isotherms dip southward (equatorward) over the Northern Hemisphere continents.
19. Explain why the daily range of temperature is normally greater (a) in dry regions than in humid regions and (b) on clear days than on cloudy days.
20. Why is the largest annual range of temperatures normally observed over continents away from large bodies of water?

21. Two cities have the same mean annual temperature. Explain why this fact does not mean that their temperatures throughout the year are similar.

22. During a cold, calm, sunny day, why do we usually feel warmer than a thermometer indicates?

23. What atmospheric conditions can bring on hypothermia?

24. During the winter, frost can form on the ground when the minimum thermometer indicates a low temperature above freezing. Explain.

25. Why do daily temperature ranges decrease as you increase in altitude?

26. Why do the first freeze in autumn and the last freeze in spring occur in lowlying areas?

Questions for Thought

1. Explain (with the aid of a diagram) why the morning sun shines brightly through a south-facing bedroom window in December, but not in June.

2. Consider these two scenarios: (a) The tilt of the Earth decreased to 10°. (b) The tilt of the Earth increased to 40°. How would this change the summer and winter temperatures in your area? Explain, using a diagram.

3. At the top of the earth's atmosphere during the early summer (Northern Hemisphere), above what latitude would you expect to receive the most solar radiation in one day? During the same time of year, where would you expect to receive the most solar radiation at the surface? Explain why the two locations are different. (If you are having difficulty with this question, refer to Fig. 3.5, p. 57.)

4. If a construction company were to build a solar-heated home in middle latitudes in the Southern Hemisphere, in which direction should the solar panels on the roof be directed for maximum daytime heating?

5. Aside from the aesthetic appeal (or lack of such), explain why painting the outside north-facing wall of a middle latitude house one color and the south-facing wall another color is not a bad idea.

6. How would the lag in daily temperature experienced over land compare to the daily temperature lag over water?

7. Where would you expect to experience the smallest variation in temperature from year to year and from month to month? Why?

8. The average temperature in San Francisco, California, for December, January, and February is 11°C (52°F). During the same three-month period the average temperature in Richmond, Virginia, is 4°C (39°F). Yet, San Francisco and Richmond have nearly the same yearly total of heating degree-days. Explain why. (Hint: See Fig. 3.24, p. 72.)

9. On a warm summer day, one city experienced a daily range of 22°C (40°F), while another had a daily range of 10°C (18°F). One of these cities is located in New Jersey and the other in New Mexico. Which location most likely had the highest daily range, and which one had the smallest? Explain.

10. Minimum thermometers are usually read during the morning, yet they are reset in the afternoon. Explain why.

11. If clouds arrive at 2 A.M. in the middle of a calm, clear night it is quite common to see temperatures rise after 2 A.M. How does this happen?

12. Why does the wind chill temperature only decrease slightly as the wind speed increases above 30 mph?

13. In the Northern Hemisphere, south facing mountain slopes normally have a greater diurnal range in temperature than north facing slopes. Why?

14. If the poles have 24 hours of sunlight during the summer, why is the average summer temperature still below 0°F?

Problems and Exercises

1. Draw a graph similar to Figs. 3.5 (p. 57) and 3.7 (p. 59). Include in it the amount of solar radiation reaching the earth's surface at each latitude on the equinox.

2. Each day past the winter solstice the noon sun is a little higher above the southern horizon. (a) Determine how much change takes place each day at your latitude. (b) Does the same amount of change take place at each latitude in the Northern Hemisphere? Explain.

3. On approximately what dates will the sun be overhead at noon at latitudes: (a) 10°N? (b) 15°S?

4. Design a solar-heated home that sits on the north side of an east-west running street. If the home is located at 40°N, draw a proper roof angle for maximum solar heating. Design windows, doors, overhangs, and rooms with the intent of reducing heating and cooling costs. Place trees around the home that will block out excess summer sunlight and yet let winter sunlight inside. Choose a paint color for the house that will add to the home's energy efficiency.

5. Suppose peas are planted in Indiana on May 1. If the peas need 1200 growing degree-days before they can be picked, and if the average maximum temperature for May and June is 80°F and the average minimum is 60°F, on about what date will the peas be ready to pick? (Assume a base temperature of 55°F.)

6. (a) What is the wind-chill factor when the air temperature is 5°F and the wind speed is 35 mi/hr? (Use Table 3.3, p. 74.) (b) If the wind-chill equivalent temperature is –35°C and the wind speed is 40 km/hr, what is the actual air temperature? (Use Table 3.4, p. 74.)

Questions for Exploration

On the Blue Skies: College Edition CD-ROM go to the **Atmospheric Basics** section of the CD and click on "**Energy Balance.**"

1. Explore the model of energy exchange. This time modify only the time of day in three hour increments. (a) For each time step note the values of energy flow to and from the earth's surface and plot total gains and total losses versus time. (b) Based on this information when would you expect the temperature to reach its peak for the day?

Sunlight reflected and dispersed by falling
raindrops produces a dramatic rainbow
over southern Utah. (Photo by author.)

LIGHT, COLOR, AND ATMOSPHERIC OPTICS

CONTENTS

The sky is clear, the weather cold, and the year, 1818. Near Baffin Island in Canada, a ship with full sails enters unknown waters. On board are the English brothers James and John Ross, who are hoping to find the elusive "Northwest Passage," the waterway linking the Atlantic and Pacific oceans. On this morning, however, their hopes would be dashed, for directly in front of the vessel, blocking their path, is a huge towering mountain range. Disappointed, they turn back and report that the Northwest Passage does not exist. About seventy-five years later Admiral Perry met the same barrier and called it "Crocker land."

What type of treasures did this mountain conceal—gold, silver, precious gems? The curiosity of explorers from all over the world had been aroused. Speculation was the rule, until, in 1913, the American Museum of Natural History commissioned Donald MacMillan to lead an expedition to solve the mystery of Crocker land. At first, the journey was disappointing. Where Perry had seen mountains, MacMillan saw only vast stretches of open water. Finally, ahead of his ship was Crocker land, but it was more than two hundred miles farther west from where Perry had encountered it. MacMillan sailed on as far as possible. Then he dropped anchor and set out on foot with a small crew of men.

As the team moved toward the mountains, the mountains seemed to move away from them. If they stood still, the mountains stood still; if they started walking, the mountains receded again. Puzzled, they trekked onward over the glittering snow-fields until huge mountains surrounded them on three sides. At last the riches of Crocker land would be theirs. But in the next instant the sun disappeared below the horizon and, as if by magic, the mountains dissolved into the cold arctic twilight. Dumbfounded, the men looked around only to see ice in all directions—not a mountain was in sight. There they were, the victims of one of nature's greatest practical jokes, for Crocker land was a mirage.

The sky is full of visual events. Optical illusions (mirages) can appear as towering mountains or wet roadways. In clear weather, the sky can appear blue, while the horizon appears milky white. Sunrises and sunsets can fill the sky with brilliant shades of pink, red, orange, and purple. At night, the sky is black, except for the light from the stars, planets, and the moon. The moon's size and color seem to vary during the night, and the stars twinkle. To understand what we see in the sky, we will take a closer look at sunlight, examining how it interacts with the atmosphere to produce an array of atmospheric visuals.

WHITE AND COLORS

We know from Chapter 2 that nearly half of the solar radiation that reaches the atmosphere is in the form of visible light. As sunlight enters the atmosphere, it is either absorbed, reflected, scattered, or transmitted on through. How objects at the surface respond to this energy depends on their general nature (color, density, composition) and the wavelength of light that strikes them. How do we see? Why do we see various colors? What kind of visual effects do we observe because of the interaction between light and matter? In particular, what can we see when light interacts with our atmosphere?

We perceive light because electromagnetic waves stimulate antenna-like nerve endings in the retina of the human eye. These antennae are of two types—*rods* and *cones*. The rods respond to all wavelengths of visible light and give us the ability to distinguish light from dark. If people possessed rod-type receptors only, then only black and white vision would be possible. The cones respond to specific wavelengths of visible light. Radiation with a wavelength between 0.4 and 0.7 micrometers (μm) strikes the cones, which immediately fire an impulse through the nervous system to the brain, and we perceive this impulse as the sensation of color. (Color blindness is caused by missing or malfunctioning cones.) Wavelengths of radiation

shorter than 0.4 μm, or longer than 0.7 μm, do not stimulate color vision in humans.

White light is perceived when all visible wavelengths strike the cones of the eye with nearly equal intensity. Because the sun radiates almost half of its energy as visible light, all visible wavelengths from the midday sun reach the cones, and the sun usually appears white. A star that is cooler than our sun radiates most of its energy at slightly longer wavelengths; therefore, it appears redder. On the other hand, a star much hotter than our sun radiates more energy at shorter wavelengths and thus appears bluer. A star whose temperature is about the same as the sun's appears white.

Objects that are not hot enough to produce radiation at visible wavelengths can still have color. Everyday objects we see as red are those that absorb all visible radiation except red. The red light is reflected from the object to our eyes. Blue objects have blue light returning from them, since they absorb all visible wavelengths except blue. Some surfaces absorb all visible wavelengths and reflect no light at all. Since no radiation strikes the rods or cones, these surfaces appear black. Therefore, when we see colors, we know that light must be reaching our eyes.

WHITE CLOUDS AND SCATTERED LIGHT

One exciting feature of the atmosphere can be experienced when we watch the underside of a puffy, growing cumulus cloud change color from white to dark gray or black. When we see this change happen, our first thought is usually, "It's going to rain." Why is the cloud initially white? Why does it change color? To answer these questions, let's investigate the concept of *scattering*.

When sunlight bounces off a surface at the same angle at which it strikes the surface, we say that the light is **reflected**, and call this phenomenon *reflection* (see Fig. 4.1). There are various constituents of the atmosphere, however, that tend to deflect solar radiation from its path and send it out in all directions. We know from Chapter 2 that radiation reflected in this way is said to be **scattered**. (Scattered light is also called *diffuse light*.) During the scattering process, no energy is gained or lost and, therefore, no temperature changes occur. In the atmosphere, scattering is usually caused by small objects, such as air molecules, fine particles of dust, water molecules, and some pollutants. Just as the ball in a pinball machine bounces off the pins in many directions, so solar radiation is knocked about by small particles in the atmosphere.

Cloud droplets about 10 μm or so in diameter are large enough to effectively scatter all wavelengths of visible radiation more or less equally, a phenomenon we call *geometric scattering*. Clouds, even small ones, are optically thick, meaning that they are able to scatter vast amounts of sunlight and there is very little chance sunlight will pass through unscattered. These same clouds are poor absorbers of sunlight. Hence, when we look at a cloud, it appears white because countless cloud droplets scatter all wavelengths of visible sunlight in all directions (see Fig. 4.2).

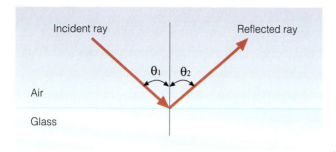

FIGURE 4.1

For a ray of light striking a flat, smooth surface, the angle at which the incident ray strikes the surface (the angle of incidence, or θ₁) is equal to the angle at which the reflected ray leaves the surface (the angle of reflection, or θ₂). This phenomenon is called *Snell's law*.

As a cloud grows larger and taller, more sunlight is reflected from it and less light can penetrate all the way through it (see Fig. 4.3). In fact, relatively little light penetrates a cloud whose thickness is 1000 m (3300 ft). Since little sunlight reaches the underside of the cloud, little light is scattered, and the cloud base appears dark. At the same time, if droplets near the cloud base grow larger, they become less effective scatterers and better absorbers. As a result, the meager amount of visible light that does reach this part of the cloud is absorbed rather than scattered, which makes the cloud appear even darker. These same cloud droplets may even grow large and heavy enough to fall to the earth as rain. From a casual observation of clouds, we know that dark, threatening ones frequently produce rain. Now, we know why they appear so dark.

BLUE SKIES AND HAZY DAYS

The sky appears blue because light that stimulates the sensation of blue color is reaching the retina of the eye. How does this happen?

Individual air molecules are much smaller than cloud droplets—their diameters are small even when compared with the wavelength of visible light. Each air molecule of oxygen and nitrogen is a *selective scatterer* in that each scatters shorter waves of visible light much more effectively than longer waves. This selective scattering is also known as *Rayleigh scattering* (see Table 4.1).

As sunlight enters the atmosphere, the shorter visible wavelengths of violet, blue, and green are scattered more by atmospheric gases than are the longer wavelengths of yellow, orange, and especially red.* In fact, violet light is scattered about 16 times more than red light. Consequently, as we view the sky, the scattered waves of violet, blue, and green strike the eye from all directions. Because our eyes are more sensitive to blue light, these waves, viewed together, produce the sensation of blue coming from all around us (see Fig. 4.4). Therefore, when we look at the sky it appears blue (see Fig. 4.5). (Earth, by the way,

---

*The reason for this fact is that the intensity of Rayleigh scattering varies as $\frac{1}{\lambda^4}$, where λ is the wavelength of radiation.

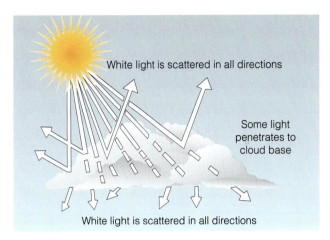

FIGURE 4.2
Since tiny cloud droplets scatter visible light in all directions, light from many billions of droplets turns a cloud white.

FIGURE 4.3
Average percent of radiation reflected, absorbed, and transmitted by clouds of various thickness.

is not the only planet with a colorful sky. On Mars, dust in the air turns the sky red at midday and purple at sunset.)

The selective scattering of blue light by air molecules and very small particles can make distant mountains appear blue, such as the Blue Ridge Mountains of Virginia and the Blue Mountains of Australia (see Fig. 4.6). In some places, a *blue haze* may cover the landscape, even in areas far removed from human contamination. Although its cause is still controversial, the blue haze appears to be the result of a particular process. Extremely tiny particles (hydrocarbons called *terpenes*) are released by vegetation to combine chemically with small amounts of ozone. This reaction produces tiny particles (about 0.2 μm in diameter) that selectively scatter blue light.

TABLE 4.1	The Various Types of Scattering of Visible Light		
TYPE OF PARTICLE	PARTICLE DIAMETER (MICROMETERS, μm)	TYPE OF SCATTERING	PHENOMENA
Air molecules	0.0001 to 0.001	Rayleigh	Blue sky, red sunsets
Aerosols (pollutants)	0.01 to 1.0	Mie	Brownish smog
Cloud droplets	10 to 100	Geometric	White clouds

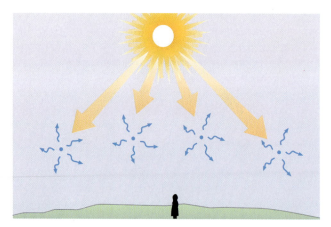

FIGURE 4.4
The sky appears blue because billions of air molecules selectively scatter the shorter wavelengths of visible light more effectively than the longer ones. This causes us to see blue light coming from all directions.

When small particles, such as fine dust and salt, become suspended in the atmosphere, the color of the sky begins to change from blue to milky white. Although these particles are small, they are large enough to scatter all wavelengths of visible light fairly evenly in all directions. (Recall that this is geometric scattering.) When our eyes are bombarded by all wavelengths of visible light, the sky appears milky white, the visibility lowers, and we call the day "hazy." If the humidity is high enough, soluble particles (nuclei) will "pick up" water vapor and grow into haze particles. Thus, the color of the sky gives us a hint about how much material is suspended in the air: the more particles,

the more scattering, and the whiter the sky becomes. Since most of the suspended particles are near the surface, the horizon often appears white. On top of a high mountain, when we are above many of these haze particles, the sky usually appears a deep blue.

Haze can scatter light from the rising or setting sun, so that we see bright lightbeams, or **crepuscular rays**, radiating across the sky. A similar effect occurs when the sun shines through a break in a layer of clouds (see Fig. 4.7). Dust, tiny water droplets, or haze in the air beneath the clouds scatters sunlight, making that region of the sky appear bright with rays. Because these rays seem to reach downward from clouds, some people will remark that the "sun is drawing up water." In England, this same phenomenon is referred to as "Jacob's ladder." No matter what these sunbeams are called, it is the scattering of sunlight by particles in the atmosphere that makes them visible.

RED SUNS AND BLUE MOONS

At midday, the sun seems a brilliant white, while at sunset it usually appears to be yellow, orange, or red. At noon, when the sun is high in the sky, light from the sun is most intense—all wavelengths of visible light are able to reach the eye with about equal intensity, and the sun appears white. (Looking directly at the sun, especially during this time of day, can cause irreparable damage to the eye. Normally, we get only glimpses or impressions of the sun out of the corner of our eye.)

Near sunrise or sunset, however, the rays coming directly from the sun strike the atmosphere at a low angle. They must pass through much more atmosphere than at any other time

FIGURE 4.5
Blue skies and white clouds. The selective scattering of blue light by air molecules produces the blue sky, while the scattering of all wavelengths of visible light by liquid cloud droplets produces the white clouds.

FIGURE 4.6
The Blue Ridge Mountains in Virginia. The blue haze is caused by the scattering of blue light by extremely small particles—smaller than the wavelengths of visible light. Notice that the scattered blue light causes the most distant mountains to become almost indistinguishable from the sky.

FIGURE 4.7
The scattering of sunlight by dust and haze produces these white bands of crepuscular rays.

WEATHER WATCH

The scattering of sunlight by our atmosphere prevents us from seeing the stars during the day. Since the moon has no atmosphere, its sky color is always black, the stars shine brightly all the time, and there is no twilight.

during the day. (When the sun is 4° above the horizon, sunlight must pass through an atmosphere more than 12 times thicker than when the sun is directly overhead.) By the time sunlight has penetrated this large amount of air, most of the shorter waves of visible light have been scattered away by the air molecules. Just about the only waves from a setting sun that make it on through the atmosphere on a fairly direct path are the yellow, orange, and red. Upon reaching the eye, these waves produce a bright yellow-orange sunset (see Fig. 4.8).

Bright, yellow-orange sunsets only occur when the atmosphere is fairly clean, such as after a recent rain. If the atmosphere contains many fine particles whose diameters are a little larger than air molecules, slightly longer (yellow) waves also would be scattered away. Only orange and red waves would penetrate through to the eye, and the sun would appear red-orange. When the atmosphere becomes loaded with particles, only the longest red wavelengths are able to penetrate the atmosphere, and we see a red sun.

Natural events may produce red sunrises and sunsets. Over the oceans, for example, the scattering characteristics of small, suspended salt particles and water vapor are responsible for the brilliant red suns observed from the beach (see Fig. 4.9).

Moreover, major volcanic eruptions send vast amounts of dust and ash high into the atmosphere. These fine particles, moved by the winds aloft, circle the globe, producing beautiful sunrises and sunsets for months and even years. There were beautiful ruddy sunsets in many parts of the Northern Hemisphere after the eruption of the Mexican volcano El Chichón in 1982 and the Philippine volcano Mt. Pinatubo in 1991.

Occasionally, the atmosphere becomes so laden with dust, smoke, and pollutants that even red waves are unable to pierce the filthy air. An eerie effect then occurs. Because no visible waves enter the eye, the sun literally disappears before it reaches the horizon.

The scattering of light by large quantities of atmospheric particles can cause some rather unusual sights. If the volcanic ash, dust, smoke particles, or pollutants are roughly uniform in size, they can selectively scatter the sun's rays. Even at noon, various colored suns have appeared: orange suns, green suns, and even blue suns. For blue suns to appear, the size of the suspended particles must be similar to the wavelength of visible light. (This situation produces a type of scattering called *Mie scattering*. Refer back to Table 4.1, p. 83.) When these particles are present they tend to scatter red light more than blue, which causes a bluing of the sun and a reddening of the sky. Although rare, the same phenomenon can happen to moonlight, making the moon appear blue; thus, the expression "once in a blue moon."

In summary, the scattering of light by small particles in the atmosphere causes many familiar effects: white clouds, blue skies, hazy skies, crepuscular rays, and colorful sunsets. In the absence of any scattering, we would simply see a white sun against a black sky—not an attractive alternative.

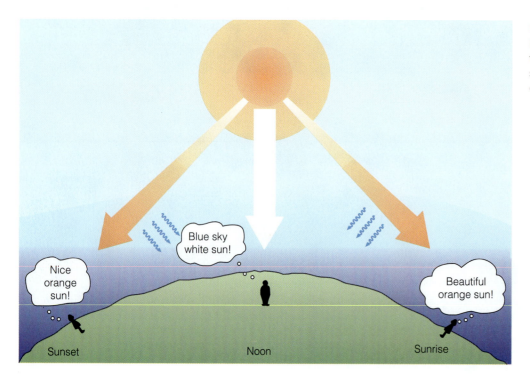

FIGURE 4.8

Because of the selective scattering by a thick section of atmosphere, the sun at sunrise and sunset appears either yellow, orange, or red. At noon, it is usually white.

FIGURE 4.10
The behavior of light as it enters and leaves a more-dense substance, such as water.

FIGURE 4.9
Red sunset near the coast of Iceland. The reflection of sunlight off the slightly rough water is producing a glitter path.

TWINKLING, TWILIGHT, AND THE GREEN FLASH

Light that passes through a substance is said to be *transmitted*. Upon entering a denser substance, transmitted light slows in speed. If it enters the substance at an angle, the light's path also bends. This bending is called **refraction**. The amount of refraction depends primarily on two factors: the density of the material and the angle at which the light enters the material.

Refraction can be demonstrated in a darkened room by shining a flashlight into a beaker of water (see Fig. 4.10). If the light is held directly above the water so that the beam strikes the surface of the water straight on, no bending occurs. But, if the light enters the water at some angle, it bends toward the *normal*, which is the dashed line in the diagram running perpendicular to the air-water boundary. (The normal is simply a line that intersects any surface at a right angle. We use it as a reference to see how much bending occurs as light enters and leaves various substances.) A small mirror on the bottom of the beaker reflects the light upward. This reflected light bends away from the normal as it re-enters the air. We can summarize these observations as follows: *Light that travels from a less-dense to a more-dense medium loses speed and bends toward the nor-*

mal, while light that enters a less-dense medium increases in speed and bends away from the normal.

The refraction of light within the atmosphere causes a variety of visual effects. At night, for example, the light from the stars that we see directly above us is not bent, but starlight that enters the earth's atmosphere at an angle is bent. In fact, a star whose light enters the atmosphere just above the horizon has more atmosphere to penetrate and is thus refracted the most. As we can see in Fig. 4.11, the bending is toward the

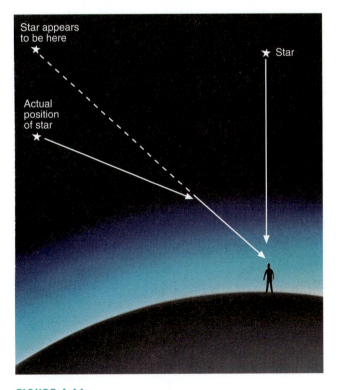

FIGURE 4.11
Due to the bending of starlight by the atmosphere, stars not directly overhead appear to be higher than they really are.

TABLE 4.2 The Amount of Atmospheric Refraction (Bending) in Minutes Viewed at Sea Level under Standard Atmospheric Conditions (60 minutes equals 1°)

ELEVATION ABOVE HORIZON (DEGREES)	REFRACTION (MINUTES)
0°	35.0
5°	10.0
20°	2.6
40°	1.2
60°	0.6
90°	0.0

normal as the light enters the more-dense atmosphere. By the time this "bent" starlight reaches our eyes, the star appears to be higher than it actually is because our eyes cannot detect that the light path is bent. We see light coming from a particular direction and interpret the star to be in that direction. So, the next time you take a midnight stroll, point to any star near the horizon and remember: this is where the star appears to be. To point to the star's true position, you would have to lower your arm just a bit (about one-half a degree, according to Table 4.2).

As starlight enters the atmosphere, it often passes through regions of differing air density. Each of these regions deflects and bends the tiny beam of starlight, constantly changing the apparent position of the star. This causes the star to appear to *twinkle* or flicker, a condition known as **scintillation**. Planets, being much closer to us, appear larger, and usually do not twinkle because their size is greater than the angle at which their light deviates as it penetrates the atmosphere. Planets sometimes twinkle, however, when they are near the horizon, where the bending of their light is greatest.

The refraction of light by the atmosphere has some other interesting consequences. For example, the atmosphere gradually bends the rays from a rising or setting sun or moon. Because light rays from the lower part of the sun (or moon) are bent more than those from the upper part, the sun appears to flatten out on the horizon, taking on an elliptical shape. Also, since light is bent most on the horizon, the sun and moon both appear to be higher than they really are. Consequently, they both rise about two minutes earlier and set about two minutes later than they would if there were no atmosphere (see Fig. 4.12).

You may have noticed that on clear days the sky is often bright for some time after the sun sets. The atmosphere refracts and scatters sunlight to our eyes, even though the sun itself has disappeared from our view. **Twilight** is the name given to the time after sunset (and immediately before sunrise) when the sky remains illuminated and allows outdoor activities to continue without artificial lighting. (*Civil twilight* lasts from sunset until the sun is 6° below the horizon, while *astronomical twilight* lasts until the sky is completely dark and the astronomical observation of the faintest stars is possible.)

The length of twilight depends on season and latitude. During the summer in middle latitudes, twilight adds about 30 minutes of light to each morning and evening for outdoor activities. The duration of twilight increases with increasing latitude, especially in summer. At high latitudes during the summer, morning and evening twilight may converge, producing a *white night*—a nightlong twilight.

The color of twilight can change when particles of volcanic ash and dust are present in the upper atmosphere. For example, after the volcanic eruptions of El Chichón in 1982 and Mt. Pinatubo in 1991, brilliant red twilights occurred over parts of North America as sulfur-rich volcanic particles scattered red light from the setting (and rising) sun (see Fig. 4.13).

In general, without the atmosphere, there would be no refraction or scattering, and the sun would rise later and set earlier than it now does. Instead of twilight, darkness would arrive immediately when the sun disappears below the horizon. Imagine the number of sandlot baseball games that would be called because of instant darkness.

Occasionally, a flash of green light—called the **green flash**—may be seen near the upper rim of a rising or setting sun (see Fig. 4.14). Remember from our earlier discussion that, when the sun is near the horizon, its light must penetrate a

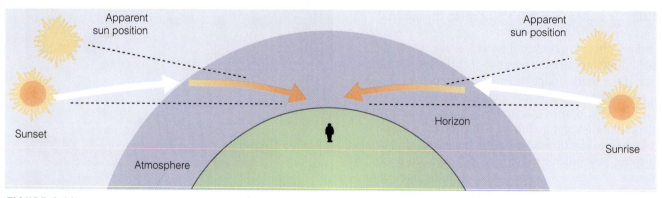

FIGURE 4.12
The bending of sunlight by the atmosphere causes the sun to rise about two minutes earlier, and set about two minutes later, than it would otherwise.

FIGURE 4.13
Bright twilight over California produced by the sulfur-rich particles from the volcano Mt. Pinatubo during September, 1992.

FIGURE 4.14
The very light green on the upper rim of the sun is the green flash. Also, observe how the atmosphere makes the sun appear to flatten on the horizon into an elliptical shape.

thick section of atmosphere. This thick atmosphere refracts sunlight, with purple and blue light bending the most, and red light the least. Because of this bending, more blue light should appear along the top of the sun. But because the atmosphere selectively scatters blue light, very little reaches us, and we see green light instead.

Usually, the green light is too faint to see with the human eye. However, under certain atmospheric conditions, such as when the surface air is very hot or when an upper-level inversion exists, the green light is magnified by the atmosphere. When this happens, a momentary flash of green light appears, often just before the sun disappears from view.

The flash usually lasts about a second, although in polar regions it can last longer. Here, the sun slowly changes in elevation and the flash may exist for many minutes. Members of Admiral Byrd's expedition in the south polar region reported seeing the green flash for 35 minutes in September as the sun slowly rose above the horizon, marking the end of the long winter.

BRIEF REVIEW Up to this point, we have examined how light can interact with our atmosphere. Before going on, here is a review of some of the important concepts and facts we have covered:

- When light is scattered, it is sent in all directions—forward, sideways, and backward.
- White clouds, blue skies, hazy skies, crepuscular rays, and colorful sunsets are the result of sunlight being scattered.
- The bending of light as it travels through regions of differing density is called *refraction*.

- As light travels from a less-dense substance (such as outer space) and enters a more-dense substance at an angle (such as our atmosphere), the light bends downward, toward the normal. This effect causes stars, the moon, and the sun to appear just a tiny bit higher than they actually are.

THE MIRAGE: SEEING IS NOT BELIEVING

In the atmosphere, when an object appears to be displaced from its true position, we call this phenomenon a **mirage**. A mirage is not a figment of the imagination—our minds are not playing tricks on us, but the atmosphere is.

Atmospheric mirages are created by light passing through and being bent by air layers of different densities. Such changes in air density are usually caused by sharp changes in air temperature. The greater the rate of temperature change, the greater the light rays are bent. For example, on a warm, sunny day, black road surfaces absorb a great deal of solar energy and become very hot. Air in contact with these hot surfaces warms by conduction and, because air is a poor thermal conductor, we find much cooler air only a few meters higher. On hot days, these road surfaces often appear wet (see Fig. 4.15). Such "puddles" disappear as we approach them, and advancing cars seem to swim in them. Yet, we know the road is dry. The apparent wet pavement above a road is the result of blue skylight refracting up into our eyes as it travels through air of different densities. A similar type of mirage occurs in deserts during the hot summer. Many thirsty travelers have been disappointed to find that what appeared to be a water hole was in actuality hot desert sand.

FIGURE 4.15
The road in the photo appears wet because blue skylight is bending up into the camera as the light passes through air of different densities.

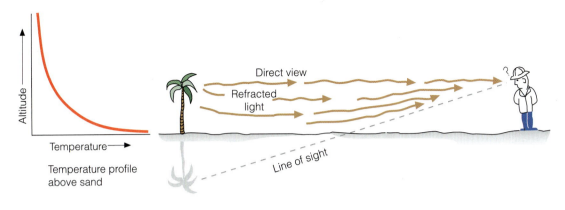

FIGURE 4.16
Inferior mirage over hot desert sand.

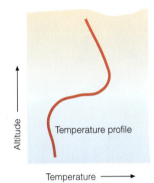

FIGURE 4.17
The formation of a superior mirage. When cold air lies close to the surface with warm air aloft, light from distant mountains is refracted toward the normal as it enters the cold air. This causes an observer on the ground to see mountains higher and closer than they really are.

Sometimes, these "watery" surfaces appear to *shimmer*. The shimmering results as rising and sinking air near the ground constantly change the air density. As light moves through these regions, its path also changes, causing the shimmering effect.

When the air near the ground is much warmer than the air above, objects may not only appear to be lower than they really are, but also (often) inverted. These mirages are called **inferior** (lower) **mirages**. The tree in Fig. 4.16 certainly doesn't grow upside down. So why does it look that way? It appears to be inverted because light reflected from the top of the tree moves outward in all directions. Rays that enter the hot, less-dense air above the sand are refracted upward, entering the eye from below. The brain is fooled into thinking that these rays came from below the ground, which makes the tree appear upside down. Some light from the top of the tree travels directly toward the eye through air of nearly constant density and, therefore, bends very little. These rays reach the eye "straight-on," and the tree appears upright. Hence, off in the distance, we see a tree and its upside-down image beneath it. (Some of the trees in Fig. 4.15 show this effect.)

The atmosphere can play optical jokes on us in extremely cold areas, too. In polar regions, air next to a snow surface can be much colder than the air many meters above. Because the air in this cold layer is very dense, light from distant objects entering it bends toward the normal in such a way that the objects can appear to be shifted upward. This phenomenon is

WEATHER WATCH

On December 14, 1890, a mirage—lasting several hours—gave the residents of Saint Vincent, Minnesota, a clear view of cattle nearly 13 km (8 mi) away.

called a **superior** (upward) **mirage**. Figure 4.17 shows the atmospheric conditions favorable for a superior mirage. (A special type of mirage, the **Fata Morgana**, is described in the Focus section on p. 92.)

HALOS, SUNDOGS, AND SUN PILLARS

A ring of light encircling and extending outward from the sun or moon is called a **halo**. Such a display is produced when sunlight or moonlight is refracted as it passes through ice crystals. Hence, the presence of a halo indicates that *cirriform clouds* are present (see Chapter 6).

The most common type of halo is the 22° halo—a ring of light 22° from the sun or moon.* Such a halo forms when tiny suspended column-type ice crystals (with diameters less than

---

*Extend your arm and spread your fingers apart. An angle of 22° is about the distance from the tip of the thumb to the tip of the little finger.

The *Fata Morgana*

A special type of superior mirage is the *Fata Morgana*, a mirage that transforms a fairly uniform horizon into one of vertical walls and columns with spires (see Fig. 1). According to legend, *Fata Morgana* (Italian for fairy Morgan) was the half-sister of King Arthur. Morgan, who was said to live in a crystal palace beneath the water, had magical powers that could build fantastic castles out of thin air. Looking across the Straits of Messina (between Italy and Sicily), residents of Reggio, Italy, on occasion would see buildings, castles, and sometimes whole cities appear, only to vanish again in minutes. The *Fata Morgana* is observed where the air tempera-ture increases with height above the surface, slowly at first, then more rapidly, then slowly again. Consequently, mirages like the *Fata Morgana* are frequently seen where warm air rests above a cold surface, such as above large bodies of water and in polar regions.

FIGURE 1

The *Fata Morgana* mirage over water. The mirage is the result of refraction—light from small islands and ships is bent in such a way as to make them appear to rise vertically above the water.

FIGURE 4.18

A 22° halo around the sun, produced by the refraction of sunlight through ice crystals.

20 μm) become randomly oriented as air molecules constantly bump against them. The refraction of light rays through these crystals forms a halo like the one shown in Fig. 4.18. Less common is the 46° halo, which forms in a similar fashion to the 22° halo (see Fig. 4.19). With the 46° halo, however, the light is refracted through column-type ice crystals that have diameters in a narrow range between about 15 and 25 μm.

Occasionally, a bright arc of light may be seen at the top of a 22° halo (see Fig. 4.20). Since the arc is tangent to the halo, it is called a **tangent arc**. Apparently, the arc forms as large six-sided (hexagonal), pencil-shaped ice crystals fall with their long axes horizontal to the ground. Refraction of sunlight through the ice crystals produces the bright arc of light. When the sun is on the horizon, the arc that forms at the top of the halo is called an *upper tangent arc*. When the sun is above the horizon, a *lower tangent arc* may form on the lower part of the halo beneath the sun. The shape of the arcs change greatly with the position of the sun.

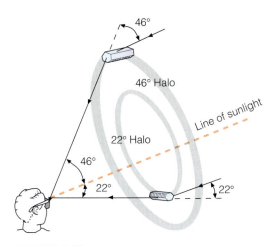

FIGURE 4.19
The formation of a 22° and a 46° halo.

A halo is usually seen as a bright, white ring, but there are refraction effects that can cause it to have color. To understand this, we must first examine refraction more closely.

When white light passes through a glass prism, it is refracted and split into a spectrum of visible colors (see Fig. 4.21). Each wavelength of light is slowed by the glass, but each is slowed a little differently. Because longer wavelengths (red) slow the least and shorter wavelengths (violet) slow the most, red light bends the least, and violet light bends the most. The breaking up of white light by "selective" refraction is called **dispersion**. As light passes through ice crystals, dispersion causes red light to be on the inside of the halo and blue light on the outside.

When hexagonal platelike ice crystals with diameters larger than about 30 μm are present in the air, they tend to fall slowly and orient themselves horizontally (see Fig. 4.22). (The horizontal orientation of these ice crystals prevents a ring halo.) In this position, the ice crystals act as small prisms, refracting and dispersing sunlight that passes through them. If the sun is near the horizon in such a configuration that it, ice crystals, and observer are all in the same horizontal plane, the observer will see a pair of brightly colored spots, one on either side of the sun. These colored spots are called **sundogs**, *mock suns*, or **parhelia** (meaning "with the sun") (see Fig. 4.23). The colors usually grade from red (bent least) on the inside closest to the sun to blue (bent more) on the outside.

While sundogs, tangent arcs, and halos are caused by *refraction* of sunlight *through* ice crystals, **sun pillars** are caused by *reflection* of sunlight *off* ice crystals. Sun pillars appear most often at sunrise or sunset as a vertical shaft of light extending upward or downward from the sun (see Fig. 4.24). Pillars may form as hexagonal platelike ice crystals fall with their flat bases oriented horizontally. As the tiny crystals fall in still air, they tilt from side to side like a falling leaf. This allows sunlight to reflect off the tipped surfaces of the crystals,

FIGURE 4.20
Halo with an upper tangent arc.

FIGURE 4.21
Refraction and dispersion of light through a glass prism.

FIGURE 4.22
Platelike ice crystals falling with their flat surfaces parallel to the earth produce sundogs.

FIGURE 4.23
The bright area on the left side of the photograph is a sundog photographed at Bartlett Cove, Glacier Bay, Alaska.

FIGURE 4.24
A sun pillar produced by the reflection of sunlight off ice crystals.

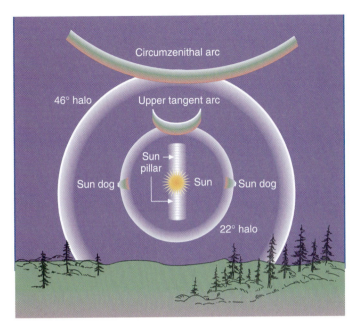

FIGURE 4.25
Ice crystal phenomena. (A picture of the circumzenithal arc is in Fig. 2 on p. 98.)

producing a relatively bright area in the sky above or below the sun. Pillars may also form as sunlight reflects off hexagonal pencil-shaped ice crystals that fall with their long axes oriented horizontally. As these crystals fall, they can rotate about their horizontal axes, producing many orientations that reflect sunlight. So, look for sun pillars when the sun is low on the horizon and cirriform (ice crystal) clouds are present. Figure 4.25 is a summary of some of the optical phenomena that form when cirriform clouds are present.

RAINBOWS

Now we come to one of the most spectacular light shows observed on the earth—the rainbow. **Rainbows** occur when rain is falling in one part of the sky, and the sun is shining in another. (Rainbows also may form by the sprays from waterfalls and water sprinklers.) To see the rainbow, we must face the falling rain with the sun at our backs. Look at Fig. 4.26 closely and note that, when we see a rainbow in the evening, we are facing east toward the rainshower. Behind us—in the west—it is clear. Because clouds tend to move from west to east in middle latitudes, the clear skies in the west suggest that the showers will give way to clearing. However, when we see a rainbow in the morning, we are facing west, toward the rainshower. It is a good bet that the clouds and showers will move toward us and it will rain soon. These observations explain why the following weather rhyme became popular:

Rainbow in morning, sailors take warning

Rainbow at night, a sailor's delight.*

When we look at a rainbow we are looking at sunlight that has entered the falling drops, and, in effect, has been redirected back toward our eyes. Exactly how this process happens requires some discussion.

As sunlight enters a raindrop, it slows and bends, with violet light refracting the most and red light the least (see Fig. 4.27). Although most of this light passes right on through the drop and is not seen by us, some of it strikes the backside of the drop at such an angle that it is reflected within the drop. The angle at which this occurs is called the *critical angle*. For water, this angle is 48°. Light that strikes the back of a raindrop at an angle exceeding the critical angle bounces off the back of the drop and is *internally reflected* toward our eyes. Because each light ray bends differently from the rest, each ray emerges from the drop at a slightly different angle. For red light, the angle is 42° from the beam of sunlight; for violet light, it is 40° (see Fig. 4.27b). The light leaving the drop is, therefore, dispersed into a spectrum of colors from red to violet. Since we see only a single color from each drop, it takes myriads of raindrops (each refracting and reflecting light back to our eyes at slightly

---

*This rhyme is often used with the words "red sky" in the place of rainbow. The red sky makes sense when we consider that it is the result of red light from a rising or setting sun being reflected from the underside of clouds above us. In the morning, a red sky indicates that it is clear to the east and cloudy to the west. A red sky in the evening suggests the opposite.

FIGURE 4.26
When you observe a rainbow, the sun is always to your back. In middle latitudes, a rainbow in the evening indicates that clearing weather is ahead.

WEATHER WATCH

During the summer, rainbows are occasionally seen after a thunderstorm. Because of this fact, the Shoshone Indians viewed the rainbow as a giant serpent that would sometimes rub its back on the icy sky and hurl pieces of ice (hail) to the ground.

different angles) to produce the brilliant colors of a *primary rainbow*.

Figure 4.27 might lead us erroneously to believe that red light should be at the bottom of the bow and violet at the top. A more careful observation of the behavior of light leaving two drops (see Fig. 4.28) shows us why the reverse is true. When violet light from the *lower drop* reaches an observer's eye, red light from the same drop is incident elsewhere, toward the waist. Notice that red light reaches the observer's eye from the *higher drop*. Because the color red comes from higher drops and the color violet from lower drops, the colors of a primary rainbow change from red on the outside (top) to violet on the inside (bottom).

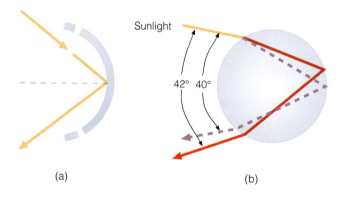

(a)   (b)

FIGURE 4.27

Sunlight internally reflected and dispersed by a raindrop. (a) The light ray is internally reflected only when it strikes the backside of the drop at an angle greater than the critical angle for water. (b) Refraction of the light as it enters the drop causes the point of reflection (on the back of the drop) to be different for each color. Hence, the colors are separated from each other when the light emerges from the raindrop.

Frequently, a larger second (secondary) rainbow with its colors reversed can be seen above the primary bow (see Fig. 4.29). Usually this secondary bow is much fainter than the primary one. The *secondary bow* is caused when sunlight enters the raindrops at an angle that allows the light to make two internal reflections in each drop. Each reflection weakens the light intensity and makes the bow dimmer. Figure 4.30 shows that the color reversals—with red now at the bottom and violet on top—are due to the way the light emerges from each drop after going through two internal reflections.

As you look at a rainbow, keep in mind that only one ray of light is able to enter your eye from each drop. Every time you move, whether it be up, down, or sideways, the rainbow moves with you. The reason why this happens is that, with every movement, light from different raindrops enters your eye. The bow you see is not exactly the same rainbow that the person standing next to you sees. In effect, each of us has a personal rainbow to ponder and enjoy! (Can a rainbow actually form on a day when it is not raining? The answer to this question is given in the Focus section on p. 98.)

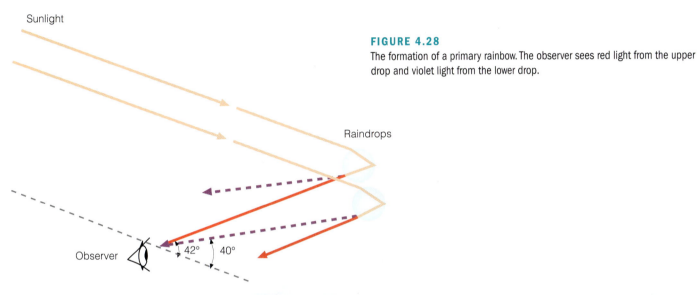

FIGURE 4.28
The formation of a primary rainbow. The observer sees red light from the upper drop and violet light from the lower drop.

FIGURE 4.29
A primary and a secondary rainbow.

Can It Be a Rainbow If It Is Not Raining?

Up to this point, we have seen that the sky is full of atmospheric visuals. One that we closely examined was the rainbow. Look back at Fig. 4.29, p. 97 and then look at Fig. 2. Is the color display in Fig. 2 a rainbow? The colors definitely show a rainbowlike brilliance. For this reason, some people will call this phenomenon a rainbow. But remember, for a rainbow to form it must be raining, and on this day it is not.

The color display in Fig. 2 is due to the refraction of light through ice crystals. Earlier in this chapter we saw that the refraction of light produces a variety of visuals, such as halos, tangent arcs, and sundogs. The color dis-play in Fig. 2 is a type of refraction phenomena called a *circumzenithal arc*. The photograph was taken in the afternoon, during late winter, when the sun was about 30° or so above the western horizon and the sky was full of ice crystal (cirrus) clouds. The arc was almost directly overhead, at the zenith; hence, its name—"circum-zenithal."

The circumzenithal arc forms about 45° above the sun as platelike ice crystals fall with their flat surfaces parallel to the ground. Remember that this is similar to the formation of a sundog (see Fig. 4.22, p. 94), except that in the formation of a cir-cumzenithal arc, sunlight enters the top of the crystal and exits one of its sides (see Fig. 3). This is why the short-lived circumzenithal arc can only form when the sun is lower than 32° above the horizon. When the sun is higher than this angle, the refracted light cannot be seen by the observer.

There are a wide variety of other refraction phenomena (too many to describe here) that may at first glance appear as a rainbow. Keep in mind, however, that rainbows only form when it is raining in one part of the sky and the sun is shining in the other.

FIGURE 2
Is this a rainbow? The photograph was taken looking almost straight up.

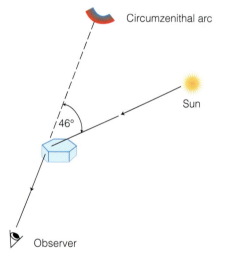

FIGURE 3
The formation of a circumzenithal arc.

CORONAS, GLORIES, AND *HEILIGENSCHEIN*

When the moon is seen through a thin veil of clouds composed of tiny spherical water droplets, a bright ring of light, called a **corona** (meaning crown), may appear to rest on the moon (see Fig. 4.31). The same effect can occur with the sun, but, due to the sun's brightness, it is usually difficult to see.

The corona is due to **diffraction**—the bending of light as it passes *around* objects. To understand the corona, imagine water waves moving around a small stone in a pond. As the waves spread around the stone, the trough of one wave may meet the crest of another wave. This situation results in the waves cancel-ing each other, thus producing calm water. (This is known as *destructive interference*.) Where two crests come together (*con-*

Falling raindrops

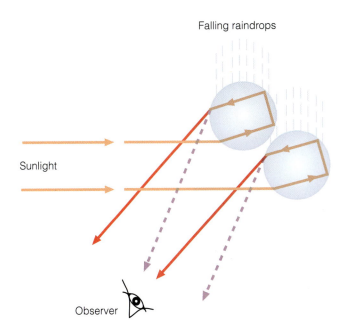

Sunlight

Observer

FIGURE 4.30
Two internal reflections are responsible for the weaker secondary rainbow.

FIGURE 4.31
The corona around the moon results from the diffraction of light by tiny liquid cloud droplets of uniform size.

structive interference), they produce a much larger wave. The same thing happens when light passes around tiny cloud droplets. Where light waves constructively interfere, we see bright light; where destructive interference occurs, we see darkness. Sometimes, the corona appears white, with alternating bands of light and dark. On other occasions, the rings have color.

The colors appear when the cloud droplets (or any kind of small particles) are of uniform size. Because the amount of bending due to diffraction depends upon the wavelength of light, the shorter wavelength blue light appears on the inside of a ring, while the longer wavelength red light appears on the outside. These colors may repeat over and over, becoming fainter as each ring is farther from the moon or sun (see Fig. 4.32). Also, the smaller the cloud droplets, the larger the ring diameter. Therefore, clouds that have recently formed (such as thin altostratus and altocumulus) are the best corona producers.

When different size droplets exist within a cloud, the corona becomes distorted and irregular. Sometimes the cloud exhibits patches of color, often pastel shades of pink, blue, or green. These bright areas produced by diffraction are called **iridescence** (see Fig. 4.33). Cloud iridescence is most often seen within 20° of the sun.

Like the corona, the **glory** is also a diffraction phenomenon. When an aircraft flies above a cloud layer composed of water droplets less than 50 μm in diameter, a set of colored rings, called the *glory*, may appear around the shadow of the aircraft (see Fig. 4.34). The same effect can happen when you stand with your back to the sun and look into a cloud or fog bank, as a bright ring of light may be seen around the shadow

of your head. In this case, the glory is called the *brocken bow*, after the Brocken Mountains in Germany, where it is particularly common.

For the glory and the brocken bow to occur, the sun must be to your back, so that sunlight can be returned to your eye

FIGURE 4.32
Corona around the sun. This type of corona, called *Bishop's ring*, is the result of diffraction of sunlight by tiny volcanic particles emitted from the volcano El Chichón in 1982.

FIGURE 4.33
Cloud iridescence.

FIGURE 4.34
The series of rings surrounding the shadow of the aircraft is called the glory.

from the water droplets. Sunlight that enters the small water droplet along its edge is refracted, then reflected off the backside of the droplet. The light then exits at the other side of the droplet, being refracted once again (see Fig. 4.35). However, in order for the light to be returned to your eyes, the light actually clings ever so slightly to the edge of the droplet—the light actually skims along the surface of the droplet as a *surface wave* for a short distance. Diffraction of light coming from the edges of the droplets produces the ring of light we see as the glory and the brocken bow. The colorful rings may be due to the various angles at which different colors leave the droplet.

Sunlight

Surface wave

FIGURE 4.35
Light that produces the glory follows this path in a water droplet.

FIGURE 4.36

The Heiligenschein is the ring of light around the shadow of the observer's head.

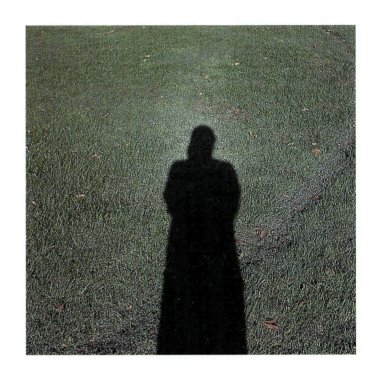

On a clear morning with dew on the grass, stand facing the dew with your back to the sun and observe that, around the shadow of your head, is a bright area—the **Heiligenschein** (German for halo). The *Heiligenschein* forms when sunlight, which falls on nearly spherical dew drops, is focused and reflected back toward the sun along nearly the same path that it took originally. (Light reflected in this manner is said to be *retroreflected*.) The light, however, does not travel along the exact path; it actually spreads out just enough to be seen as bright white light around the shadow of your head on a dew-covered lawn (see Fig. 4.36).

SUMMARY

The scattering of sunlight in the atmosphere can produce a variety of atmospheric visuals, from hazy days and blue skies to crepuscular rays and blue moons. Refraction of light by the atmosphere causes stars near the horizon to appear higher than they really are. It also causes the sun and moon to rise earlier and set later than they otherwise would. Under certain atmospheric conditions, the amplification of green light near the upper rim of a rising or setting sun produces the illusive green flash.

Mirages form when refraction of light displaces objects from their true positions. Inferior mirages cause objects to appear lower than they really are, while superior mirages displace objects upward.

Halos and sundogs form from the refraction of light through ice crystals. Sun pillars are the result of sunlight reflecting off gently falling ice crystals. The refraction, reflection, and dispersion of light in raindrops create a rainbow. To see a rainbow, the sun must be to your back, and rain must be falling in front of you. Diffraction of light produces coronas, glories, and cloud iridescence. We can see the *Heiligenschein* on a clear morning when sunlight falls on nearly spherical dew drops.

Key Terms

The following terms are listed in the order they appear in the text. Define each. Doing so will aid you in reviewing the material covered in this chapter.

reflected light	halo
scattered light	tangent arc
crepuscular rays	dispersion (of light)
refraction (of light)	sundog or parhelia
scintillation	sun pillars
twilight	rainbow
green flash	corona
mirage	diffraction
inferior mirage	iridescence
superior mirage	glory
Fata Morgana	*Heiligenschein*

Questions for Review

1. (a) Why are cumulus clouds normally white?
   (b) Why do the undersides of building cumulus clouds frequently change color from white to dark gray or even black?
2. Explain why the sky is blue during the day and black at night.
3. What can make a setting (or rising) sun appear red?
4. Why, on a hazy day, does the horizon appear white?
5. Why do stars "twinkle"?
6. How does light bend as it enters a more-dense substance at an angle? How does it bend upon leaving the more-dense substance? Make a sketch to illustrate your answer.

7. Since twilight occurs without the sun being visible, how does it tend to lengthen the day?

8. On a clear, dry, warm day, why do dark road surfaces frequently appear wet?

9. What atmospheric conditions are necessary for an inferior mirage? A superior mirage?

10. What process (refraction or scattering) produces crepuscular rays?

11. (a) Describe how a halo forms.
    (b) How is the formation of a halo different from that of a sundog?

12. At what time of day would you expect to observe the green flash?

13. Explain how sun pillars form.

14. Why is green the last possible color to appear in a sunset?

15. What process (refraction, scattering, diffraction) is responsible for lengthening the day?

16. Why are secondary rainbows higher and much dimmer than primary rainbows? Explain your answer with the aid of a diagram.

17. Explain how light is able to reach your eyes when you see
    (a) a corona
    (b) a glory
    (c) the Heiligenschein.

Questions for Thought

1. Explain why on a cloudless day the sky will usually appear milky white before it rains and a deep blue after it rains.

2. How long does twilight last on the moon? (Hint: The moon has no atmosphere.)

3. Why is it often difficult to see the road while driving on a foggy night with your high beam lights on?

4. What would be the color of the sky if air molecules scattered the longest wavelengths of visible light and passed the shorter wavelengths straight through? (Use a diagram to help explain your answer.)

5. Explain why the colors of the planets are not related to the temperatures of the planets, while the colors of the stars are related to the temperatures of the stars.

6. If there were no atmosphere surrounding the earth, what color would the sky be at sunrise? At sunset? What color would the sun be at noon? At sunrise? At sunset?

7. Why are rainbows seldom observed at noon?

8. On a cool, clear summer day, a blue haze often appears over the Great Smoky Mountains of Tennessee. Explain why the blue haze usually changes to a white haze as the humidity of the air increases.

9. During a lunar eclipse, the earth, sun, and moon are aligned as shown in Fig. 4.37. The earth blocks sunlight from directly reaching the moon's surface, yet the surface

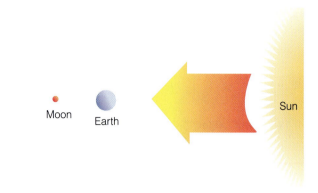

FIGURE 4.37

of the moon will often appear a pale red color during a lunar eclipse. How can you account for this phenomenon?

10. Explain why smoke rising from a cigarette often appears blue, yet appears white when blown from the mouth.

11. During Ernest Shackleton's last expedition to Antarctica, on May 8, 1915, seven days after the sun had set for the winter, he saw the sun reappear. Explain how this event—called the *Novaya Zemlya effect*—can occur.

12. Explain why it is easier to get sunburned on a high mountain than in the valley below. (The answer is not that you are closer to the sun on top of the mountain.)

13. Why are stars more visible on a clear night when there is no moon than on a clear night with a full moon?

Problems and Exercises

1. At least 5 times a day go outside and gaze at the sky. In a notebook record what you see. You will be amazed at what is there—crepuscular rays, halos, coronas, cloud iridescence, red sunsets, and more.

2. Make your own rainbow. On a sunny afternoon or morning, turn on the water sprinkler to create a spray of water drops. Stand as close to the spray as possible (without getting soaked) and observe that, as you move up, down, and sideways, the bow moves with you.
    (a) Explain why this happens.
    (b) Also, explain with the use of a diagram why the sun must be at your back in order to see the bow.

3. Take a large beaker or bottle and fill it with water. Add a small amount of nonfat powdered milk and stir until the water turns a faint milky white. Shine white light into the beaker, and, on the opposite side, hold a white piece of paper.
    (a) Explain why the milk has a blue cast to it and why the light shining on the paper appears ruddy.

(b) What do you know about the size of the milk particles?

(c) Is this a form of Rayleigh scattering or geometric scattering? Explain.

(d) How does this demonstration relate to the color of the sky and the color of the sun—at sunrise and sunset?

 ## Questions for Exploration

On the Blue Skies: College Edition CD-ROM go to the **Sky Identification** section of the CD and click on "**Atmospheric Optics**."

1. Explore the imagery of atmospheric optics. Define in your own words the physical process that leads to formation of
   (a) Sundogs
   (b) Rainbows
   (c) Halos.

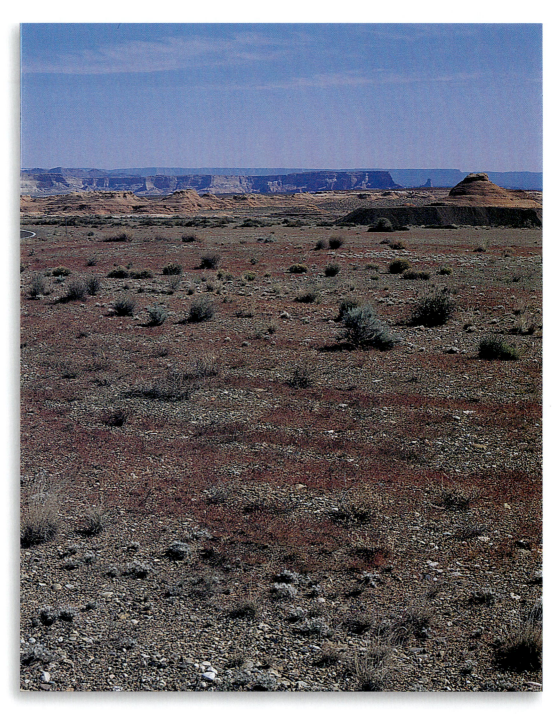

Drought-resistant vegetation struggles to survive another day in the arid climate of Arizona. Yet, on any summer day, there is actually more water vapor in the air of this desert than there is in the air of a wet New England snowstorm. (Photo by author.)

ATMOSPHERIC MOISTURE

CONTENTS

Sometimes it rains and still fails to moisten the desert—the falling water evaporates halfway down between cloud and earth. Then you see curtains of blue rain dangling out of reach in the sky while the living things wither below for want of water. Torture by tantalizing, hope without fulfillment. And the clouds disperse and dissipate into nothingness. . . . The sun climbed noon-high, the heat grew thick and heavy on our brains, the dust clouded our eyes and mixed with our sweat. My canteen is nearly empty and I'm afraid to drink what little water is left—there may never be any more. I'd like to cave in for a while, crawl under yonder cottonwood and die peacefully in the shade, drinking dust.

Edward Abbey, *Desert Solitaire—A Season in the Wilderness*

We know from Chapter 1 that, in our atmosphere, the concentration of the invisible gas water vapor is normally less than a few percent of all the atmospheric molecules. Yet water vapor is exceedingly important, for it transforms into cloud particles—particles that grow in size and fall to the earth as precipitation. The term *humidity* is used to describe the amount of water vapor in the air. To most of us, a moist day suggests high humidity. However, there is usually more water vapor in the hot, "dry" air of the Sahara Desert than in the cold, "damp" polar air in New England, which raises an interesting question: Does the desert air have a higher humidity? As we will see later in this chapter, the answer to this question is both yes and no, depending on the type of humidity we mean.

So that we may better understand the concept of humidity, we will begin this chapter by examining water vapor in the atmosphere. Then, we will look at different ways to express humidity. At the end of the chapter, we will investigate various ways to measure humidity.

WATER IN THE ATMOSPHERE

In the lower atmosphere, water is everywhere. As a gas, water vapor molecules move about quite freely, mixing well with neighboring atoms and molecules. As a liquid, the water molecules are closer together, and so they constantly jostle and bump each other. In the solid state (ice), the molecules arrange themselves into an orderly pattern, with each molecule more or less locked into a rigid position, able to vibrate, but not able to move about freely. Water vapor, an invisible gas, becomes visible only when many molecules join together to form tiny cloud droplets or ice particles. In this process—known as a *change of state* or, simply, a *phase change*—water only changes its disguise, it does not change its identity.

The Many Phases of Water To obtain a slightly different picture of water in the atmosphere, suppose we catch a falling snowflake in the palm of a glove. If we could magnify that tiny crystal of ice about a billion times, we would see H_2O molecules in the shape of tiny heads that resemble Mickey Mouse (see Fig. 5.1). The bulk of the "head" of the molecule is the oxy-

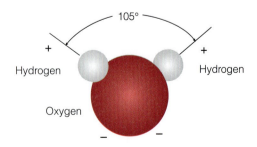

FIGURE 5.1
The water molecule.

FIGURE 5.2
The partially exposed hydrogen atom of one molecule is attracted to the negative oxygen atom of another molecule. Because the hydrogen atoms of each water molecule are separated by an angle of 105°, the joining of many billions of molecules produces a hexagonal-shaped ice crystal. In the atmosphere, many ice crystals may join together to form a snowflake.

gen atom. The "mouth" is a region of excess negative charge. The "ears" are partially exposed protons of the hydrogen atom, which are regions of excess positive charge.

When we look at many molecules (see Fig. 5.2), we can see that they are locked into specific positions and arranged as a six-sided (hexagonal) crystal form we call *ice*. Recall that the molecules are unable to move about freely, but they do vibrate. As we observe the ice crystal in freezing air, we see an occasional molecule gain enough energy to break away from its neighbors and enter into the air above. The molecule changes from an ice molecule directly into a vapor molecule without passing through the liquid state. This ice-to-vapor phase change is called **sublimation**. If a water vapor molecule should attach itself to the ice crystal, the vapor-to-ice phase change is called **deposition**.

If we apply warmth to the ice crystal, its molecules would vibrate faster. In fact, some of the molecules would actually vibrate out of their rigid crystal pattern into a disorderly condition—the ice melts. Suppose enough snow crystals melt to partially fill the beaker in Fig. 5.3.

If we were able to magnify the surface water in the beaker as we did with the ice crystal, we would see water molecules fairly close together, jiggling, bouncing, and moving about. We would also see that the molecules are not all moving at the same speed—some are moving much faster than others. At the surface, molecules with enough speed (and traveling in the right direction) would occasionally break away from the liquid surface and enter into the air above. These molecules, changing from the *liquid state into the vapor state* are **evaporating**. While some water molecules are leaving the liquid, others are returning. Those returning are **condensing** as they are changing from a *vapor state to a liquid state*.

When a cover is placed over the beaker (see Fig. 5.4), after a while the total number of molecules escaping from the liquid (evaporating) would be balanced by the number returning

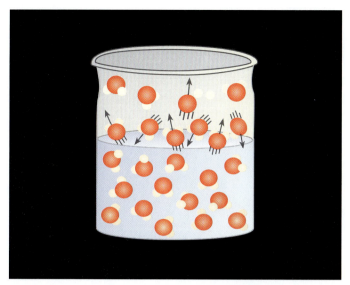

FIGURE 5.3
Water molecules at the surface of the water are evaporating (changing from liquid into vapor) and condensing (changing from vapor into liquid). Since more molecules are evaporating than condensing, net evaporation is occurring. (For clarity, only water molecules are illustrated.)

FIGURE 5.4
When the number of water molecules escaping from the liquid (evaporating) balances those returning (condensing), the air above the liquid is saturated with water vapor. (For clarity, only water molecules are illustrated.)

(condensing). When this condition exists, the air is said to be **saturated** with water vapor. For every molecule that evaporates, one must condense, and no net loss of liquid or vapor molecules results.

If we remove the cover and blow across the top of the water, some of the vapor molecules already in the air above would be blown away, creating a difference between the actual number of vapor molecules and the total number required for *saturation*. This would help prevent saturation from occurring and would allow for a greater amount of evaporation. Wind, therefore, enhances evaporation.

The temperature of the water also influences evaporation. All else being equal, warm water will evaporate more readily than cool water. The reason for this phenomenon is that, when heated, the water molecules will speed up. At higher temperatures, a greater fraction of the molecules have sufficient speed to break through the surface tension of the water and zip off into the air above. Consequently, the warmer the water, the greater the rate of evaporation.

If we could examine the air above the water in either Fig. 5.3 or Fig. 5.4, we would observe the water vapor molecules freely darting about and bumping into each other as well as neighboring molecules of oxygen and nitrogen. When these gas molecules collide, they tend to bounce off one another, constantly changing in speed and direction. However, the speed lost by one molecule is gained by another, and so the average speed of all the molecules does not change. Consequently, the temperature of the air does not change. Mixed in with all of the air molecules are microscopic bits of dust, smoke, salt, and other particles called *condensation nuclei* (so-

called because water vapor condenses on them). In the warm air above the water, fast-moving vapor molecules strike the nuclei with such impact that they simply bounce away. However, if the air is chilled, the molecules move more slowly and are more apt to stick and condense to the nuclei. When many billions of these vapor molecules condense onto the nuclei, tiny liquid cloud droplets form.

We can see then that condensation is more likely to happen as the air cools and the speed of the vapor molecules decreases. As the air temperature increases, condensation is less likely because most of the molecules have sufficient speed (sufficient energy) to remain as a vapor. As we will see in this and other chapters, *condensation occurs primarily when the air is cooled.*

Even though condensation is more likely to occur when the air cools, it is important to note that no matter how cold the air becomes, there will always be a few molecules with sufficient speed (sufficient energy) to remain as a vapor. It should be apparent, then, that with the same number of water vapor

WEATHER WATCH

Possibly the highest reported temperature a human has experienced (and lived to tell about it) occurred in a dry sauna with an air temperature of 127°C (260°F). Even though a roast would cook at this temperature, the person survived the ordeal for about 45 minutes because rapidly evaporating perspiration formed a protective layer of cool air around the individual's body.

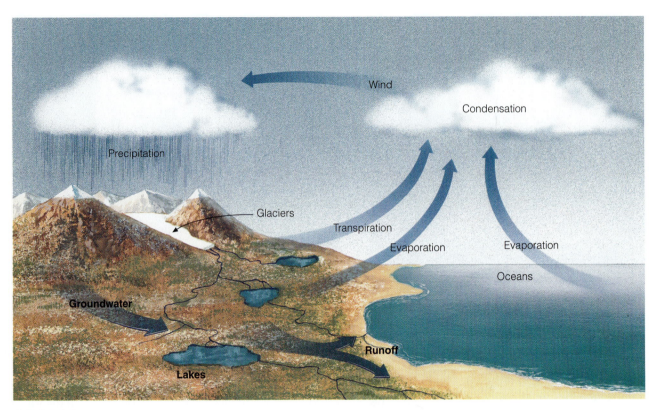

FIGURE 5.5
The hydrologic cycle.

molecules in the air, saturation is more likely to occur in cool air than in warm air. This idea often leads to the statement that "warm air can hold more water vapor molecules before becoming saturated than can cold air" or, simply, "warm air has a greater capacity for water vapor than does cold air." At this point, it is important to realize that although these statements are correct, the use of such words as "hold" and "capacity" are misleading when describing water vapor content, as air does not really "hold" water vapor in the sense of making "room" for it. (As we will see later, another way of explaining why cooling produces condensation is that the saturation vapor pressure decreases with lower temperatures.)

Circulation of Water in the Atmosphere   Within the atmosphere, an unending circulation of water begins as energy from the sun evaporates enormous quantities of water from the oceans. Winds transport the moist air to other regions, where it condenses into clouds, some of which produce rain and snow. If the precipitation falls into an ocean, the water is ready to begin its cycle again. If, on the other hand, the precipitation falls on a continent, a great deal of the water returns to the ocean in a complex journey. This cycle of moving and transforming water molecules from liquid to vapor and back to liquid again is called the **hydrologic** (water) **cycle**. In the form

with which we are most concerned, water molecules travel from ocean to atmosphere to land and then back to the ocean.

Figure 5.5 illustrates the complexities of the hydrologic cycle. For example, before falling rain ever reaches the ground, a portion of it evaporates back into the air. Some of the precipitation may be intercepted by vegetation, where it evaporates or drips to the ground long after a storm has ended. Once on the surface, a portion of the water soaks into the ground by percolating downward through small openings in the soil and rock, forming groundwater that can be tapped by wells. What does not soak in collects in puddles of standing water or runs off into streams and rivers, which find their way back to the ocean. Even the underground water moves slowly and eventually surfaces, only to evaporate or be carried seaward by rivers.

Over land, a considerable amount of vapor is added to the atmosphere through evaporation from the soil, lakes, and streams. Even plants give up moisture by a process called *transpiration*. The water absorbed by a plant's root system moves upward through the stem and emerges from the plant through numerous small openings on the underside of the leaf. In all, evaporation and transpiration from continental areas amount to only about 15 percent of the nearly 1.5 billion billion gallons of water that annually evaporate into the atmosphere; the remaining 85 percent evaporates from the oceans. If all of this vapor were to suddenly condense and fall as rain, it would be enough to cover the entire globe with 2.5 centimeters, or 1 inch of water.* The total mass of water vapor stored in the atmosphere at any moment adds up to only a little over a week's supply of the world's precipitation. Since this amount varies only slightly from day to day, the hydrologic cycle is exceedingly efficient in circulating water in the atmosphere.

ABSOLUTE HUMIDITY

Humidity refers to any one of a number of ways of specifying the amount of water vapor in the air. Since there are several ways to express atmospheric water vapor content, there are several meanings for the concept of humidity.

Suppose we enclose a volume of air in an imaginary thin elastic container—a *parcel*—about the size of a large balloon, as illustrated in Fig. 5.6. With a chemical drying agent, we can extract the water vapor from the air, weigh it, and obtain its mass. If we then compare the vapor's mass with the volume of air in the parcel, we would have determined the **absolute humidity** of the air—that is, the mass of water vapor in a given volume of air:

$$\text{Absolute humidity} = \frac{\text{mass of water vapor}}{\text{volume of air}}.$$

Absolute humidity represents the *water vapor density* (mass/volume) in the parcel and, normally, is expressed as grams of water vapor in a cubic meter of air. For example, if the water vapor in 1 cubic meter of air weighs 25 grams, the absolute humidity of the air is 25 grams per cubic meter (25 g/m^3).

We learned in Chapter 2 that a rising or descending parcel of air will experience a change in its volume because of the changes in surrounding air pressure. Consequently, when a volume of air fluctuates, the absolute humidity changes—even though the air's vapor content has remained constant (see Fig. 5.7). For this reason, the absolute humidity is not commonly used in atmospheric studies.

SPECIFIC HUMIDITY AND MIXING RATIO

Humidity, however, can be expressed in ways that are not influenced by changes in air volume. When the mass of the water vapor in the air parcel in Fig. 5.6 is compared with the mass of all the air in the parcel (including vapor), the result is called the **specific humidity**; thus

$$\text{Specific humidity} = \frac{\text{mass of water vapor}}{\text{total mass of air}}.$$

---

*If the water vapor in a column of air condenses and falls to the earth as rain, the depth of the rain on the surface is called *precipitable water*.

FIGURE 5.6
The water vapor content, or humidity, inside this air parcel can be expressed in a number of ways.

Another convenient way to express humidity is to compare the mass of the water vapor in the parcel to the mass of the remaining dry air. This is called the **mixing ratio**:

$$\text{Mixing ratio} = \frac{\text{mass of water vapor}}{\text{mass of dry air}}.$$

Both specific humidity and mixing ratio are expressed as grams of water vapor per kilogram of air (g/kg).

The specific humidity and mixing ratio of an air parcel remain constant *as long as water vapor is not added to or removed from the parcel*. This happens because the total number of molecules (and, hence, the mass of the parcel) remains constant, even as the parcel expands or contracts (see Fig. 5.8). Since changes in parcel size do not affect specific humidity and mixing ratio, these two concepts are used extensively in the study of the atmosphere.

Parcel Size	Mass of H2O Vapor	Absolute Humidity
2 m^3	10 g	5 g/m^3
1 m^3	10 g	10 g/m^3

FIGURE 5.7
With the same amount of water vapor in a parcel of air, an increase in volume decreases absolute humidity, while a decrease in volume increases absolute humidity.

	Mass of Parcel	Mass of H2O Vapor	Specific Humidity
	1 kg	1 g	1 g/kg
	1 kg	1 g	1 g/kg

FIGURE 5.8
The specific humidity does not change as air rises and descends.

Figure 5.9 shows how specific humidity varies with latitude. The average specific humidity is highest in the warm, muggy tropics. As we move away from the tropics, it decreases, reaching its lowest average value in the polar latitudes. Although the major deserts of the world are located near latitude 30°, Fig. 5.9 shows that, at this latitude, the average air contains nearly twice the water vapor that the air at latitude 50°N contains. Hence, the air of a desert is certainly not "dry," nor is the water vapor content very low. Since the hot, desert air of the Sahara often contains more moisture than the cold, polar air farther north, we can say that *summertime Sahara air has a higher specific humidity.* (We will see later in what sense we consider desert air to be "dry.")

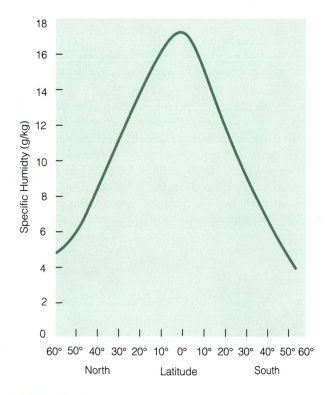

FIGURE 5.9
The average specific humidity for each latitude. The highest average values are observed in the tropics and the lowest values in polar regions.

VAPOR PRESSURE

The air's moisture content may also be described by measuring the pressure exerted by the water vapor in the air. Suppose the air parcel in Fig. 5.6, p. 109, is near sea level. The total pressure inside the parcel is due to the collision of all the molecules against the inside surface of the parcel. In other words, the total pressure inside the parcel is equal to the sum of the pressures of the individual gases. (This phenomena is known as *Dalton's law of partial pressure.*) If the total pressure inside the parcel is 1000 millibars (mb),* and the gases inside include nitrogen (78 percent), oxygen (21 percent), and water vapor (1 percent), then the partial pressure exerted by nitrogen would be 780 mb and that exerted by oxygen, 210 mb. The partial pressure of water vapor, called the **actual vapor pressure**, would be only 10 mb (1 percent of 1000).† It is evident, then, that because the number of water vapor molecules in any volume of air is small compared with the total number of air molecules in the volume, the actual vapor pressure is normally a small fraction of the total air pressure.

Everything else being equal, the more air molecules in a parcel, the greater the total air pressure. When you blow up a balloon, you increase its pressure by putting in more air. Similarly, an increase in the number of water vapor molecules will increase the total vapor pressure. Hence, the actual vapor pressure is a fairly good measure of the total amount of water vapor in the air: *High actual vapor pressure indicates large numbers of water vapor molecules, whereas low actual vapor pressure indicates comparatively small numbers of vapor molecules.***

Figure 5.10 shows the vapor pressures (expressed in millibars) over the United States for an average January. Notice that the amount of vapor in the air is greatest in the Gulf Coast states and lowest in the interior. Compare New Orleans with Omaha. Cold dry winds from northern Canada flow relentlessly into the Central Plains during the winter, keeping this area dry. But warm, moist air from the Gulf of Mexico helps maintain a higher vapor pressure in the South.

Figure 5.11 is a similar diagram showing the average vapor pressure for July. Again, the greatest vapor pressures are observed along the Gulf Coast. Note, too, that air over the eastern and central portion of the country in July contains between 3 and 6 times more moisture than it does in January. Compare St. Louis in July and January. Since cold air no longer moves in from Canada, the warm July air constantly receives water vapor

---

*You may recall from Chapter 1 that the millibar is the unit of pressure most commonly found on surface weather maps, and that it expresses atmospheric pressure as a force over a given area.

†When we use the percentages of various gases in a volume of air, Dalton's law only gives us an approximation of the actual vapor pressure. The point here is that, near the earth's surface, the actual vapor pressure is often close to 10 mb.

**Remember that actual vapor pressure is only an approximation of the total vapor content. A change in total air pressure will affect the actual vapor pressure even though the total amount of water vapor in the air remains the same.

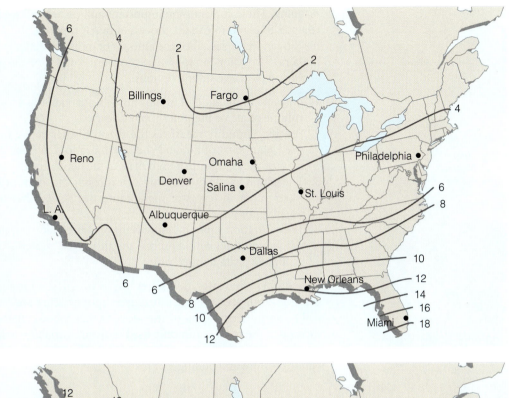

FIGURE 5.10
Average water vapor pressure in millibars near the earth's surface for the month of January.

FIGURE 5.11
Average water vapor pressure in millibars near the earth's surface for the month of July.

from the moist breezes that move northward from the Gulf of Mexico. The Great Basin region averages the lowest values with a minimum occurring over Nevada, an area surrounded by mountains that effectively shield it from significant amounts of moisture moving in from the southwest and northwest.

Actual vapor pressure indicates the air's total water vapor content, whereas **saturation vapor pressure** describes how much water vapor is necessary to make the air saturated at any given temperature. Put another way, *saturation vapor pressure is the pressure that the water vapor molecules would exert if the air were saturated with vapor at a given temperature.*

We can obtain a better picture of the concept of saturation vapor pressure by imagining molecules evaporating from a water surface. (Look back at Fig. 5.4, p. 107.) Recall that when the air is saturated, the number of molecules escaping from the water's surface equals the number returning. Since the number

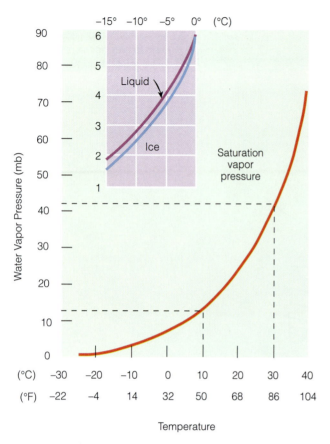

FIGURE 5.12

Saturation vapor pressure increases with increasing temperature. At a temperature of 10°C, the saturation vapor pressure is about 12 mb, whereas at 30°C it is about 42 mb. The insert illustrates that the saturation vapor pressure over water is greater than the saturation vapor pressure over ice.

of "fast-moving" molecules increases as the temperature increases, the number of water molecules escaping per second increases also. In order to maintain equilibrium, this situation causes an increase in the number of water vapor molecules in the air above the liquid. Consequently, at higher air temperatures, it takes more water vapor to saturate the air. And more vapor molecules exert a greater pressure. Saturation vapor pressure, then, depends primarily on the air temperature. From the graph in Fig. 5.12, we can see that at 10°C, the saturation vapor pressure is about 12 mb, whereas at 30°C it is about 42 mb.

The insert in Fig. 5.12 shows that, when both water and ice exist at the same temperature below freezing, *the saturation vapor pressure just above the water is greater than the saturation vapor pressure over the ice*. In other words, at any temperature below freezing, it takes more vapor molecules to saturate air directly above water than it does to saturate air directly above ice. This situation occurs because it is harder for molecules to escape an ice surface than a water surface. Consequently, fewer molecules escape the ice surface at a given temperature, requiring fewer in the vapor phase to maintain equilibrium. Likewise, salts in solution bind water molecules, reducing the number

escaping. These concepts are important and (as we will see in Chapter 8) play a role in the process of rain formation.

So far, we've described the amount of moisture actually in the air. If we want to report the moisture content of the air around us, we have several options:

1. Absolute humidity tells us the *mass* of water vapor in a fixed volume of air, or the *water vapor density*.
2. Specific humidity measures the *mass* of water vapor in a fixed *total mass* of air, and the mixing ratio describes the mass of water vapor in a fixed mass of the remaining dry air.
3. The actual vapor pressure of air expresses the amount of water vapor in terms of the amount of *pressure* that the water vapor molecules exert.
4. The saturation vapor pressure is the pressure that the water vapor molecules would exert if the air were saturated with vapor at a given temperature.

Each of these measures has its uses but, as we will see, the concepts of vapor pressure and saturation vapor pressure are critical to an understanding of the sections that follow. (Before looking at the most commonly used moisture variable—relative humidity—you may wish to read the Focus section on vapor pressure and boiling, p. 113.)

RELATIVE HUMIDITY

While relative humidity is the most commonly used way of describing atmospheric moisture, it is also, unfortunately, the most misunderstood. The concept of relative humidity may at first seem confusing because it does not indicate the actual amount of water vapor in the air. Instead, it tells us how close the air is to being saturated. The **relative humidity** *is the ratio of the amount of water vapor actually in the air compared to the maximum amount of water vapor required for saturation at that particular temperature (and pressure)*. It is the *ratio* of the air's water vapor *content* to its *capacity*; thus

$$\text{Relative humidity} = \frac{\text{water vapor content}}{\text{water vapor capacity}}.$$

We can think of the actual vapor pressure as a measure of the air's actual water vapor content, and the saturation vapor pressure as a measure of air's total capacity for water vapor. Hence, the relative humidity can be expressed as

$$\text{RH} = \frac{\text{actual vapor pressure}}{\text{saturation vapor pressure}} \times 100 \text{ percent.}^*$$

---

*Relative humidity may also be expressed as

$$\text{RH} = \frac{\text{actual mixing ratio}}{\text{saturation mixing ratio}} \times 100 \text{ percent,}$$

where the actual mixing ratio is the mixing ratio of the air and the saturation mixing ratio is the mixing ratio of saturated air at that particular temperature.

FOCUS ON A SPECIAL TOPIC

Vapor Pressure and Boiling—The Higher You Go, the Longer Cooking Takes

If you camp in the mountains, you may have noticed that, the higher you camp, the longer it takes vegetables to cook in boiling water. To understand this, we need to examine the relationship between vapor pressure and boiling. As water boils, bubbles of vapor rise to the top of the liquid and escape. For this to occur, the saturation vapor pressure exerted by the bubbles must equal the pressure of the atmosphere; otherwise, the bubbles would collapse. Boiling, therefore, occurs when the saturation vapor pressure of the escaping bubbles is equal to the total atmospheric pressure.

Because the saturation vapor pressure is directly related to the temperature of the liquid, higher water temperatures produce higher vapor pressures. Hence, any change in atmospheric pressure will change the temperature at which water boils: An increase in air pressure raises the boiling point, while a decrease in air pressure lowers it. Notice in Fig. 1 that, to make pure water boil at sea level, the water must be heated to a temperature of 100°C (212°F). At Denver, Colorado, which is situated about 1500 m (5000 ft) above sea level, the air pressure is near 850 millibars, and water boils at 95°C (203°F).

Once water starts to boil, its temperature remains constant, even if you continue to heat it. This happens because energy supplied to the water

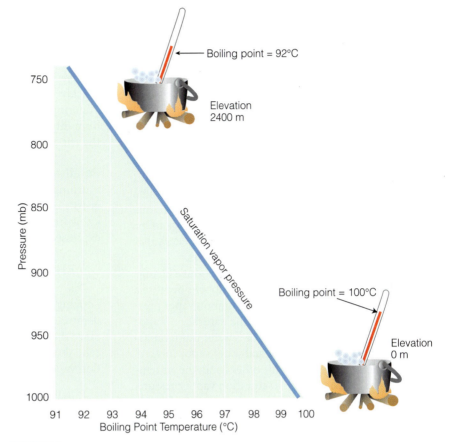

FIGURE 1
The lower the air pressure, the lower the saturation vapor pressure and, hence, the lower the boiling point temperature.

is used to convert the liquid to a gas (steam). Now we can see why vegetables take longer to cook in the mountains. To be thoroughly cooked, they must boil for a longer time because the boiling water is cooler than at lower levels. In New York City, which is near sea level, it takes about five minutes to hard boil an egg. An egg boiled for five minutes in the "mile high city" of Denver, Colorado, turns out to be runny.

Relative humidity is given as a percent. Air with a 50 percent relative humidity actually contains one-half the amount required for saturation. Air with a 100 percent relative humidity is said to be *saturated* because it is filled to capacity with water vapor. Air with a relative humidity greater than 100 percent is said to be **supersaturated**. Since relative humidity is used so much in the everyday world, let's examine it more closely.

A change in relative humidity can be brought about in two primary ways:

1. by changing the air's water vapor content
2. by changing the air temperature

A change in the air's water vapor content can change the air's actual vapor pressure. If the air temperature remains constant, an increase in the air's water vapor content increases the

WEATHER WATCH

Ninety days of dry weather, an afternoon dew point of –9°C (15°F), and a relative humidity of 8 percent left Bandon, Oregon, "bone dry" on September 26, 1936. To the east of Bandon, loggers burning brush started fires that got out of control and roared through the city, burning it to the ground. Out of nearly 500 buildings, only a handful were left standing.

air's actual vapor pressure and raises the relative humidity. The relative humidity increases as the actual vapor pressure approaches the saturation vapor pressure and the air approaches saturation. Conversely, if the air temperature remains constant, a decrease in the air's water vapor content decreases the air's actual vapor pressure and lowers the relative humidity. In summary, *as water vapor is added to the air (with no change in air temperature), the relative humidity increases, and, as water vapor is removed from the air, the relative humidity decreases.*

A change in the air temperature can bring about a change in the relative humidity. This phenomenon happens because a change in air temperature alters the air's saturation vapor pressure. If the air temperature increases, the saturation vapor pressure also increases, which raises the air's water vapor capacity. If there is no change in the air's actual water vapor content, the relative humidity lowers. If, on the other hand, the air temperature decreases, so does the air's saturation vapor pressure. As the saturation vapor pressure approaches the actual vapor pressure, the relative humidity increases as the air approaches saturation. In summary, *with no change in water vapor content, an increase in air temperature lowers the relative humidity, while a decrease in air temperature raises the relative humidity.*

In many places, the air's total vapor content varies only slightly during an entire day, and so it is the changing air temperature that primarily regulates the daily variation in relative humidity (see Fig. 5.13). As the air cools during the night, the

FIGURE 5.13

When the air is cool (morning), the relative humidity is high. When the air is warm (afternoon), the relative humidity is low. These conditions exist in clear weather when the air is calm or of constant wind speed.

relative humidity increases. Normally, the highest relative humidity occurs in the early morning, during the coolest part of the day. As the air warms during the day, the relative humidity decreases, with the lowest values usually occurring during the warmest part of the afternoon.

These changes in relative humidity are important in determining the amount of evaporation from vegetation and wet surfaces. If you water your lawn on a hot afternoon, when the relative humidity is low, much of the water will evaporate quickly from the lawn, instead of soaking into the ground. Watering the same lawn in the evening, when the relative humidity is higher, will cut down the evaporation and increase the effectiveness of the watering.

Relative Humidity and Dew Point Suppose it is early morning and the outside air is saturated. The air temperature is 10°C (50°F) and the relative humidity is 100 percent. We know from the previous section that relative humidity can be expressed as

$$\text{RH} = \frac{\text{actual vapor pressure}}{\text{saturation vapor pressure}} \times 100 \text{ percent.}$$

Looking back at Fig. 5.12, p. 112, we can see that air with a temperature of 10°C has a saturation vapor pressure of 12 mb. Since the air is saturated and the relative humidity is 100 percent, the actual vapor pressure *must* be the same as the saturation vapor pressure (12 mb), since

$$\text{RH} = \frac{12 \text{ mb}}{12 \text{ mb}} \times 100 \text{ \%} = 100 \text{ percent.}$$

Suppose during the day the air warms to 30°C (86°F), with no change in water vapor content (or air pressure). Because there is no change in water vapor content, the actual vapor pressure must be the same (12 mb) as it was in the early morning when the air was saturated. The saturation vapor pressure, however, has increased because the air temperature has increased. From Fig. 5.12, note that air with a temperature of 30°C has a saturation vapor pressure of 42 mb. The relative humidity of this unsaturated, warmer air is now much lower, as

$$\text{RH} = \frac{12 \text{ mb}}{42 \text{ mb}} \times 100 \text{ \%} = 29 \text{ percent.}$$

To what temperature must the outside air, with a temperature of 30°C, be cooled so that it is once again saturated? The answer, of course, is 10°C. For this amount of water vapor in the air, 10°C is called the **dew-point temperature** or, simply, the **dew point**. It represents *the temperature to which air would have to be cooled (with no change in air pressure or moisture content) for saturation to occur.* The dew point is determined with respect to a flat surface of water. When the dew point is determined with respect to a flat surface of ice, it is called the **frost point**.

The dew point is an important measurement used to predict the formation of dew, frost, fog, and even the minimum temperature. When used with an empirical formula (see Chapter 7), the dew point can help determine the height of the base of a cumulus cloud. Since atmospheric pressure varies only slightly at the earth's surface, *the dew point is a good indicator of the air's actual water vapor content. High dew points indicate high water vapor content; low dew points, low water vapor content.* Addition of water vapor to the air increases the dew point; removing water vapor lowers it.

The difference between air temperature and dew point can indicate whether the relative humidity is low or high. When the air temperature and dew point are far apart, the relative humidity is low; when they are close to the same value, the relative humidity is high. When the air temperature and dew point are equal, the air is saturated and the relative humidity is 100 percent. Even though the relative humidity may be 100 percent, the air, under certain conditions, may be considered "dry."

Observe, for example, in Fig. 5.14a that, because the air temperature and dew point are the same in the polar air, the relative humidity must be 100 percent. On the other hand, the desert air (Fig. 5.14b), with a large separation between air tem-

perature and dew point, has a much lower relative humidity—16 percent.* However, since dew point is a measure of the amount of water vapor in the air, the desert air (with a higher dew point) must contain *more* water vapor. So even though the polar air has a higher relative humidity, the desert air that contains more water vapor has a higher water vapor density, or *absolute humidity*, and a higher specific humidity and mixing ratio as well.

Now we can see why polar air is often described as being "dry" when the relative humidity is high (often close to 100 percent). In cold, polar air, the dew point and air temperature are normally close together. But the low dew-point temperature means that there is little water vapor in the air. Consequently, the air is said to be "dry" even though the relative humidity is high.

---

*The relative humidity can be computed from Fig. 5.12, p. 112. The desert air with an air temperature of 35°C has a saturation vapor pressure of about 56 mb. A dew-point temperature of 5°C gives the desert air an actual vapor pressure of about 9 mb. These values produce a relative humidity of $\frac{9}{56} \times 100$, or 16 percent.

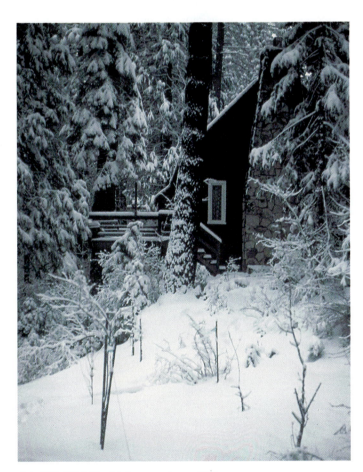

(a) POLAR AIR: Air temperature –2°C
Dew point –2°C
Relative humidity 100 percent

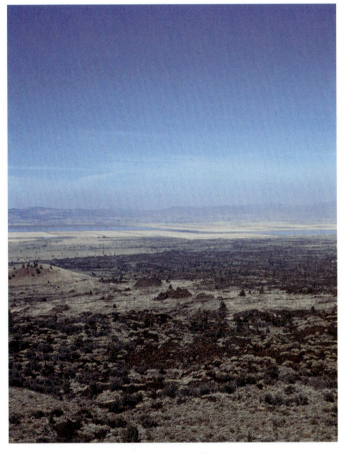

(b) DESERT AIR: Air temperature 35°C
Dew point 5°C
Relative humidity 16 percent

FIGURE 5.14
The polar air has the higher relative humidity, whereas the desert air, with the higher dew point, contains more water vapor.

BRIEF REVIEW   Up to this point we have looked at the different ways of describing humidity. Before going on, here is a review of some of the important concepts about relative humidity we have covered:

- Relative humidity does not tell us how much water vapor is actually in the air; rather, it tells us how close the air is to being saturated.

- Relative humidity can change when the air's water-vapor content changes, or when the air temperature changes.

- With a constant amount of water vapor, cooling the air raises the relative humidity and warming the air lowers it.

- The dew-point temperature is a good indicator of the air's water-vapor content.

- Dry air can have a high relative humidity. In polar air, the dew-point temperature is low and the air is considered dry. But the air temperature is close to the dew point, and so the relative humidity is high.

Comparing Humidities

Figure 5.15 shows how the average relative humidity varies from the equator to the poles. High relative humidities are normally found in the tropics and near the poles, where there is little separation between air temperature and dew point. The average relative humidity is low near latitude 30°—a latitude where we find the deserts of the world girdling the globe.

Of course, not all locations near 30°N are deserts. Take, for example, humid New Orleans, Louisiana. During July, the air

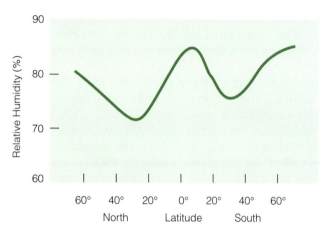

FIGURE 5.15
Relative humidity averaged for latitudes north and south of the equator.

in New Orleans contains a great deal of water vapor—nearly 50 percent more than does the air along the southern California coast. Since both locations are adjacent to large bodies of water, why is New Orleans more humid?

Figure 5.16 shows a summertime situation where air from the Pacific Ocean is moving into southern California and air from the Gulf of Mexico is moving into the southeastern states. Notice that the Pacific water is much cooler than the Gulf water. Westerly winds, blowing across the Pacific, cool to just about the same temperature as the water. Likewise, air over the warmer Gulf reaches a temperature near that of the water below it. Over the water, at both locations, the air is nearly saturated with water vapor. This means that the dew-point temperature of the air over the cooler Pacific Ocean is much lower than the dew-point temperature over the warmer Gulf. Conse-

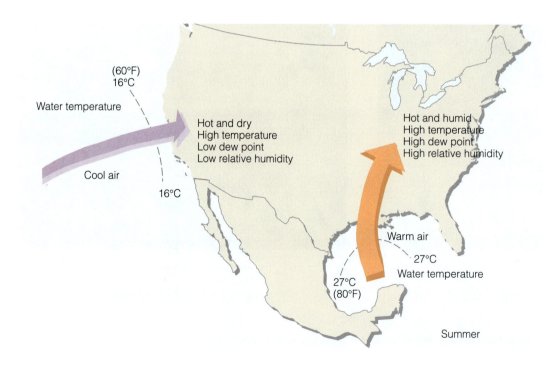

FIGURE 5.16
Air from the Pacific Ocean is hot and dry over land, whereas air from the Gulf of Mexico is hot and muggy over land.

Computing Relative Humidity and Dew Point

Suppose we want to compute the air's relative humidity and dew point from Table 1. Earlier, we learned that relative humidity may be expressed as the actual vapor pressure divided by the saturation vapor pressure times 100 percent. If the actual vapor pressure is designated by the letter e, and the saturation vapor pressure by e_s, then the expression for relative humidity becomes

$$RH = \frac{e}{e_s} \times 100\%.*$$

Let's look at a practical example of using vapor pressure to measure relative humidity and obtain dew point. Suppose the air temperature in a room is 27°C (80°F). Because the saturation vapor pressure (e_s) is dependent on the temperature of the air, to obtain e_s from Table 1 we simply read the value adjacent to the air temperature much like we did in Fig. 5.12. Hence, air with a temperature of 27°C has a saturation vapor pressure of 35 mb.

Now, suppose that the air in the room is cooled suddenly with no change in moisture content. At successively lower temperatures, the saturation vapor pressure decreases. As the lowering saturation vapor pressure (e_s) approaches the actual vapor pressure

*Relative humidity may also be expressed as $RH = w/w_s \times 100\%$, where w is the actual mixing ratio and w_s is the saturation mixing ratio. Relative humidity computations using mixing ratio and adiabatic charts are given in Chapter 7.

TABLE 1
Saturation Vapor Pressure over Water for Various Air Temperatures

AIR TEMPERATURE (°C)	(°F)	SATURATION VAPOR PRESSURE (MB)	AIR TEMPERATURE (°C)	(°F)	SATURATION VAPOR PRESSURE (MB)
−18	(0)	1.5	18	(65)	21.0
−15	(5)	1.9	21	(70)	25.0
−12	(10)	2.4	24	(75)	29.6
−9	(15)	3.0	27	(80)	35.0
−7	(20)	3.7	29	(85)	41.0
−4	(25)	4.6	32	(90)	48.1
−1	(30)	5.6	35	(95)	56.2
2	(35)	6.9	38	(100)	65.6
4	(40)	8.4	41	(105)	76.2
7	(45)	10.2	43	(110)	87.8
10	(50)	12.3	46	(115)	101.4
13	(55)	14.8	49	(120)	116.8
16	(60)	17.7	52	(125)	134.2

(e), the relative humidity increases. With an actual vapor pressure of 25 mb, 100 percent relative humidity will be reached at a temperature of 21°C (70°F). This temperature (21°C) must then be the *dew-point temperature* of the air. If, then, we know the actual vapor pressure in a room, we can determine the dew point by using Table 1 to locate the temperature at which air will be saturated with that amount of vapor. Similarly, if we are told that the dew point in the room has some value, we can look up that temperature in Table 1 and find the actual vapor pressure.

In essence, Table 1 can be used to obtain the saturation vapor pressure (e_s) and the actual vapor pressure (e) if the air temperature and dew point of the air are known. With this information we can calculate relative humidity. For example, what is the relative humidity of air with a temperature of 29°C and a dew point of 18°C? Answer: At 29°C, Table 1 shows $e_s = 41$ mb. For a dew point of 18°C, the actual vapor pressure (e) is 21 mb; therefore, the relative humidity is

$$RH = \frac{e}{e_s} = \frac{21}{41} \times 100\% = 51\%.$$

quently, the air from the Gulf of Mexico contains a great deal more water vapor because the Gulf of Mexico is much warmer than the Pacific Ocean.

As the air moves inland, away from the source of moisture, the air temperature in both cases increases. But the amount of water vapor in the air (and, hence, the dew-point temperature) hardly changes. Therefore, as the humid air moves into the southeastern states, high air temperatures along with high dew-point temperatures produce high relative humidities, often greater than 75 percent during the hottest part of the day.

On the other hand, over the southwestern part of the nation, high air temperatures and low dew-point temperatures produce low relative humidities, often less than 25 percent during the hottest part of the afternoon. Much of this inland area over the southwest is a desert. However, keep in mind that although considered "dry," this area, with a dew-point temperature above freezing, still contains more water vapor than does the cold, arctic air in polar regions. (For more information on the computation of relative humidity and dew point, read the Focus section above.)

WEATHER WATCH

It is possible for you to feel quite cool (and even begin to shiver) when the air temperature is 38°C (100°F) or above as long as your skin is wet, the wind is blowing, and the wet-bulb temperature is considerably below your exposed skin's temperature.

Relative Humidity in the Home   Question: How does the relative humidity of the winter air in your home compare with that in the Sahara Desert? Some homes actually have a lower relative humidity than the desert, and the inhabitants are usually unaware of it. Remember that cold arctic air contains only a little water vapor. Even when saturated, air with a temperature and dew point of −25°C (−13°F) has an actual vapor pressure of 0.8 mb. When this air is brought indoors and heated to 20°C (68°F), its saturation vapor pressure increases to 23.4 mb—about 30 times what it was outside. The relative humidity of the heated air inside the house drops to 3 percent.* This relative humidity is lower than what you would normally experience in a desert during the hottest time of the day!

Very low relative humidities in a house can have an adverse effect on things living inside. For example, house plants have a difficult time surviving because the moisture from their leaves and the soil evaporates rapidly. Hence, house plants usually need watering more frequently in winter than in summer. People suffer, too, when the relative humidity is quite low. The rapid evaporation of moisture from exposed flesh causes skin to crack, dry, flake, or itch. These low humidities also irritate the mucous membranes in the nose and throat, producing an "itchy" throat. Similarly, dry nasal passages permit inhaled bacteria to incubate, causing persistent infections. The remedy for most of these problems is simply to increase the relative humidity. But how?

The relative humidity in a home can be increased just by heating water and allowing it to evaporate into the air. This will raise the relative humidity to a more comfortable level. In modern homes, a humidifier, installed near the furnace, adds moisture to the air at a rate of about one gallon per room per day. The air, with its added moisture, is circulated throughout the home by a forced air heating system. In this way, all rooms get their fair share of moisture—not just the room where the vapor is added.

To lower the air's moisture content, as well as the air temperature, many homes are air conditioned. Outside air cools as it passes through a system of cold coils located in the air conditioning unit. The cooling increases the air's relative humidity, and the air reaches saturation. The water vapor condenses into liquid water, which is carried away. The cooler, dehumidified air is now forced into the home.

---

*$RH = \dfrac{0.8 \text{ mb}}{23.4 \text{ mb}} \times 100\% = 3\%$

In hot regions, where the relative humidity is low, *evaporative cooling systems* can be used to cool the air. These systems operate by having a fan blow hot, dry outside air across pads that are saturated with water. Evaporation cools the air, which is forced into the home, bringing some relief from the hot weather.

Evaporative coolers, also known as "swamp" coolers, work best when the relative humidity is low and the air is warm. They do not work well in hot, muggy weather because a high relative humidity greatly reduces the rate of evaporation. Besides, swamp coolers add water vapor to the air—something that is not needed when the air is already uncomfortably humid. That is why swamp coolers may be found on homes in Arizona, but not on homes in Alabama.

Relative Humidity and Human Discomfort   On a hot, muggy day when the relative humidity is high, it is common to hear someone exclaim (often in exasperation), "It's not so much the heat, it's the humidity." Actually, this statement is valid. In warm weather the main source of body cooling is through evaporation of perspiration. When the air temperature is high and the relative humidity low, perspiration on the skin evaporates quickly, often making us feel that the air temperature is lower than it really is. However, when both the air temperature and relative humidity are high and the air is nearly saturated with water vapor, body moisture does not readily evaporate; instead, it collects on the skin as beads of perspiration. Less evaporation means less cooling, and so we usually feel warmer than we did with a similar air temperature, but a lower relative humidity.

A good measure of how cool the skin can become is the **wet-bulb temperature**—the lowest temperature that can be reached by evaporating water into the air. On a hot day when the wet-bulb temperature is low, rapid evaporation (and, hence, cooling) takes place at the skin's surface. As the wet-bulb temperature approaches the air temperature, less cooling occurs, and the skin temperature may begin to rise. When the wet-bulb temperature exceeds the skin's temperature, no net evaporation occurs, and the body temperature can rise quite rapidly. Fortunately, most of the time, the wet-bulb temperature is considerably below the temperature of the skin.

When the weather is hot and muggy, a number of heat-related problems may occur. For example, in hot weather when the human body temperature rises, the *hypothalamus* gland (a gland in the brain that regulates body temperature) activates the body's heat-regulating mechanism, and over ten million sweat glands wet the body with as much as two liters of liquid per hour. As this perspiration evaporates, rapid loss of water and salt can result in a chemical imbalance that may lead to painful *heat cramps*. Excessive water loss through perspiring coupled with an increasing body temperature may result in *heat exhaustion*—fatigue, headache, nausea, and even fainting. If one's body temperature rises above about 41°C (106°F), *heatstroke* can occur, resulting in complete failure of the circulatory functions. If the body temperature continues to rise, death may result.

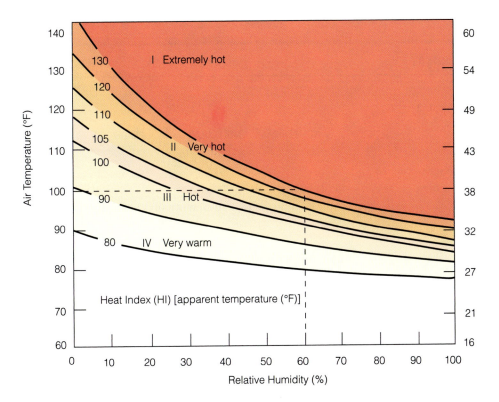

The heat index (HI). To calculate the apparent temperature, find the intersection of the air temperature and the relative humidity. (The thin dashed line represents the example given in the text.)

In an effort to draw attention to this serious weather-related health hazard, an index called the **heat index (HI)** is used by the National Weather Service. The index combines air temperature with relative humidity to determine an **apparent temperature**—what the air temperature "feels like" to the average person for various combinations of air temperature and relative humidity. For example, in Fig. 5.17, an air temperature of 38°C (100°F) and a relative humidity of 60 percent produce an apparent temperature of a little over 54°C (130°F). Heatstroke or sunstroke is imminent when the index reaches this level (see Table 5.1). (A Heat Index table is provided in Appendix G.)

During hot, humid weather some people remark about how "heavy" or how dense, the air feels. Is hot, humid air really more dense than hot, dry air? If you are interested in the answer, read the Focus section on p. 120.

MEASURING HUMIDITY

The common instrument used to obtain dew point and relative humidity is a **psychrometer**, which consists of two liquid-in-glass thermometers mounted side by side and attached to a piece of metal that has either a handle or chain at one end (see Fig. 5.18). The thermometers are exactly alike except that one

TABLE 5.1   The Heat Index (HI)

CATEGORY	APPARENT TEMPERATURE (°F)	HEAT SYNDROME
I	130° or higher	Heatstroke or sunstroke *imminent*
II	105°–130°	Sunstroke, heat cramps, or heat exhaustion *likely*, heatstroke *possible* with prolonged exposure and physical activity
III	90°–105°	Sunstroke, heat cramps, and heat exhaustion *possible* with prolonged exposure and physical activity
IV	80°–90°	Fatigue *possible* with prolonged exposure and physical activity

The sling psychrometer.

Is Humid Air or Dry Air More Dense?

Does a volume of hot, humid air really weigh more than a similar size volume of hot, dry air? The answer is no! At the same temperature and at the same level, humid air weighs *less* than dry air. (Keep in mind that we are referring strictly to water vapor—a gas—and not suspended liquid droplets.) To understand why, we must first see what determines the weight of atoms and molecules.

Almost all of the weight of an atom is concentrated in its nucleus, where the protons and neutrons are found. Neutrons weigh nearly the same as protons. To get some idea of how heavy an atom is, we simply add up the number of protons and neutrons in the nucleus. (Electrons are so light that we ignore them in comparing weights.) The larger this total, the heavier the atom. Now, we can compare one atom's weight with another's. For example, hydrogen, the lightest known atom, has only 1 proton in its center (no neutrons). Thus, it has an *atomic weight* of 1. Nitrogen, with 7 protons and 7 neutrons in its nucleus, has an atomic weight of 14. Oxygen, with 8 protons and 8 neutrons, weighs in at 16.

A molecule's weight is the sum of the atomic weights of its atoms. For example, molecular oxygen, with two oxygen atoms (O_2), has a molecular weight of 32. The most abundant atmospheric gas, molecular nitrogen (N_2), has a molecular weight of 28.

When we determine the weight of air, we are dealing with the weight of a mixture. As you might expect, a mixture's weight is a little more complex. We cannot just add the weights of all its atoms and molecules

because the mixture might contain more of one kind than another. Air, for example, has far more nitrogen (78 percent) than oxygen (21 percent). We allow for this by multiplying the molecule's weight by its share in the mixture. Since dry air is essentially composed of N_2 and O_2 (99 percent), we ignore the other parts of air for the rough average shown in Table 2.

The symbol $\approx$ means "is approximately equal to." Therefore, dry air has a molecular weight of about 29. How does this compare with humid air?

Water vapor is composed of two atoms of hydrogen and one atom of oxygen (H_2O). It is an invisible gas, just as oxygen and nitrogen are invisible. It has a molecular weight; its two atoms of hydrogen (each with atomic weight 1) and one atom of oxygen (atomic weight 16) give water vapor a molecular weight of 18. Obviously, air, at nearly 29, weighs appreciably more than water vapor.

Suppose we take a given volume of completely dry air and weigh it, then take exactly the same amount of water vapor at the same temperature and weigh it. We will find that the dry air weighs slightly more. If we replace dry air molecules one for one

with water vapor molecules, the total number of molecules remains the same, but the total weight of the drier air decreases. Since density is mass per unit volume, *hot, humid air at the surface is less dense (lighter) than hot dry air.*

This fact can have an important influence on our weather. The lighter the air becomes, the more likely it is to rise. All other factors being equal, hot, humid (less dense) air will rise more readily than hot, dry (more dense) air. It is of course the water vapor in the rising air that changes into liquid cloud droplets and ice crystals, which, in turn, grow large enough to fall to the earth as precipitation.

Of lesser importance to weather but of greater importance to sports is the fact that a baseball will "carry" farther in less dense air. Consequently, without the influence of wind, a ball should travel slightly farther on a hot humid day than on a hot dry day as long as the ball remains dry on the humid day.

TABLE 2

GAS	WEIGHT		NUMBER OF ATOMS		MOLECULAR WEIGHT		PERCENT BY VOLUME
Oxygen	16	×	2	=	32	×	21% ≈ 7
Nitrogen	14	×	2	=	28	×	78% ≈ 22

Molecular weight of dry air ≈ 29

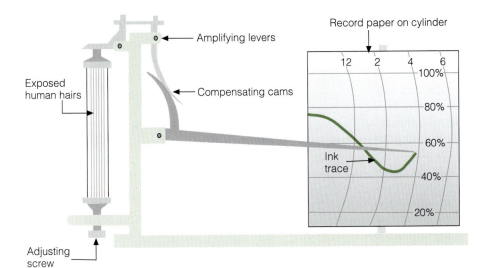

Exposed
human hairs

Amplifying levers

Compensating cams

Record paper on cylinder

Ink
trace

12 2 4 6
100%
80%
60%
40%
20%

Adjusting
screw

FIGURE 5.19
The hair hygrometer measures relative humidity
by amplifying and measuring changes in the
length of human (or horse) hair.

has a piece of cloth (wick) covering the bulb. The wick-covered thermometer—called the *wet bulb*—is dipped in clean (usually distilled) water, while the other thermometer is kept dry. Both thermometers are ventilated for a few minutes, either by whirling the instrument (*sling psychrometer*), or by drawing air past it with an electric fan (*aspirated psychrometer*). Water evaporates from the wick and that thermometer cools. The drier the air, the greater the amount of evaporation and cooling. After a few minutes, the wick-covered thermometer will cool to the lowest value possible. Recall from an earlier section that this is the *wet-bulb temperature*—the lowest temperature that can be attained by evaporating water into the air.

The dry thermometer (commonly called the *dry bulb*) gives the current air temperature, or *dry-bulb temperature*. The temperature difference between the dry bulb and the wet bulb is known as the *wet-bulb depression*. A large depression indicates that a great deal of water can evaporate into the air and that the relative humidity is low. A small depression indicates that little evaporation of water vapor is possible, so the air is close to saturation and the relative humidity is high. If there is no depression, the dry bulb, the wet bulb, and the dew point are the same; the air is saturated and the relative humidity is 100 percent. (Tables used to compute relative humidity and dew point are given in Appendix D.)

The length of human hair changes by about 2.5 percent between a relative humidity of 0 and 100 percent. Hair becomes shorter with low humidities and longer with high humidities. (In moist air, people with naturally curly hair experience the "frizzies" as their hair increases in length. Under the same conditions, people with long, straight hair find it going "limp.")

Instruments that measure humidity are commonly called **hygrometers**. One type—called the **hair hygrometer**—is constructed on the principle of hair length changing with the relative humidity. This instrument uses human (or horse) hair to measure relative humidity. A number of strands of hair (with

oils removed) are attached to a system of levers. A small change in hair length is magnified by a linkage system and transmitted to a dial (see Fig. 5.19) calibrated to show relative humidity, which can then be read directly or recorded on a chart. (Often, the chart is attached to a clock-driven rotating drum that gives a continuous record of relative humidity.) Because the hair hygrometer is not as accurate as the psychrometer (especially at very high and very low relative humidities and very low temperatures), it requires frequent calibration, principally in areas that experience large daily variations in relative humidity.

The **electrical hygrometer** is another instrument used to measure humidity. It consists of a flat plate coated with a film of carbon. An electric current is sent across the plate. As water vapor is absorbed, the electrical resistance of the carbon coating changes. These changes are translated into relative humidity. This instrument is commonly used in the radiosonde, which gathers atmospheric data at various levels above the earth. Still another instrument—the **infrared hygrometer**—measures atmospheric humidity by measuring the amount of infrared energy absorbed by water vapor in a sample of air.

The **dew-point hygrometer** measures the dew point temperature by cooling the surface of a mirror until condensation (dew) forms. This is the type of sensor that measures dew-point temperature in the over 900 fully automated weather stations—Automated Surface Observing System (ASOS)—that exist throughout the United States. Finally, the **dew cell** determines the amount of water vapor in the air by measuring the air's actual vapor pressure. This instrument consists of parallel electrical wires wound around an insulator covered with wicks that are coated with a water solution containing an excess of lithium chloride. A flow of alternating current through the solution raises the solution's temperature until all the water is evaporated. At this point, the "dry" vapor pressure of the lithium chloride equals the water vapor pressure of the air. The water vapor pressure then yields the dew-point temperature.

SUMMARY

This chapter examines water vapor in the atmosphere. It begins by looking at the different phases of water and how water circulates throughout our atmosphere. It then looks at the many ways of describing the amount of water vapor in the air. Here we learn that there are many ways of describing humidity. The absolute humidity represents the density of water vapor in a given volume of air. Specific humidity measures the mass of water vapor in a fixed mass of air, while the mixing ratio expresses humidity as the mass of water vapor in the fixed mass of remaining dry air. The actual vapor pressure indicates the air's total water vapor content by expressing the amount of water vapor in terms of the amount of pressure that the water vapor molecules exert. The saturation vapor pressure describes how much water vapor the air could hold at any given temperature in terms of how much pressure the water vapor molecules would exert if the air were saturated at that temperature. A good indicator of the air's actual water vapor content is the dew point—the temperature to which air would have to be cooled (at constant pressure) for saturation to occur.

Relative humidity is a measure of how close the air is to being saturated. Air with a high relative humidity does not necessarily contain a great deal of water vapor; it is simply close to being saturated. With a constant water vapor content, cooling the air causes the relative humidity to increase, while warming the air causes the relative humidity to decrease. When the air temperature and dew point are close together, the relative humidity is high, and, when they are far apart, the relative humidity is low. High relative humidity in hot weather makes us feel hotter than it really is by retarding the evaporation of perspiration. Although relative humidity can be confusing (because it can change with either air temperature or moisture content), it is nevertheless the most widely used way of describing the air's moisture content.

The chapter concludes by examining the various instruments that measure humidity, such as the psychrometer, hygrometer, and dew cell.

Key Terms

The following terms are listed in the order they appear in the text. Define each. Doing so will aid you in reviewing the material covered in this chapter.

sublimation	humidity
deposition	absolute humidity
evaporating (water)	specific humidity
condensing (water)	mixing ratio
saturation	actual vapor pressure
hydrologic cycle	saturation vapor pressure

relative humidity	hygrometer
supersaturation	hair hygrometer
dew-point temperature (dew point)	electrical hygrometer
frost point	infrared hygrometer
wet-bulb temperature	dew-point hygrometer
heat index (HI)	dew cell
apparent temperature	
psychrometer	

Questions for Review

1. Basically, how do the three states of water differ?
2. What are the primary factors that influence evaporation?
3. Explain why condensation occurs primarily when the air is cooled.
4. How are evaporation and condensation related to saturated air above a flat water surface?
5. Why are specific humidity and mixing ratio more commonly used in representing atmospheric moisture than absolute humidity? What is the only way to change the specific humidity or mixing ration of an air parcel?
6. In a volume of air, how does the actual vapor pressure differ from the saturation vapor pressure? When are they the same?
7. What does saturation vapor pressure primarily depend upon?
8. Explain why it takes longer to cook vegetables in the mountains than at sea level.
9. (a) What does the relative humidity represent?
   (b) When the relative humidity is given, why is it also important to know the air temperature?
   (c) Explain two ways the relative humidity may be changed.
10. Explain why, during a summer day, the relative humidity will change as shown in Fig. 5.13, p. 114.
11. Why do hot and humid summer days usually feel hotter than hot and dry summer days?
12. Why is the wet-bulb temperature a good measure of how cool human skin can become?
13. Explain why the air on a hot humid day is less dense than on a hot dry day.
14. (a) What is the dew-point temperature?
    (b) How is the difference between dew point and air temperature related to the relative humidity?
15. Why is cold polar air described as "dry" when the relative humidity of that air is very high?
16. How can a region have a high specific humidity and a low relative humidity? Give an example.

17. Why is the air from the Gulf of Mexico so much more humid than air from the Pacific Ocean at the same latitude?
18. How are the dew-point temperature and wet-bulb temperature different?

Questions for Thought

1. Would you expect water in a glass to evaporate more quickly on a windy, warm, dry summer day or on a calm, cold, dry winter day? Explain.
2. How can frozen clothes "dry" outside in subfreezing weather? What exactly is taking place?
3. Explain how and why each of the following will change as a parcel of air with an unchanging amount of water vapor rises, expands, and cools: (a) absolute humidity; (b) relative humidity; (c) actual vapor pressure; (d) saturation vapor pressure.
4. Where in the United States would you go to experience the *least* variation in dew point (actual moisture content) from January to July?
5. On a clear, calm morning, water condenses on the ground in a thick layer of dew. As the water slowly evaporates into the air, you measure a slow increase in dew point. Explain why.
6. Two cities have exactly the same amount of water vapor in the air. The 6:00 A.M. relative humidity in one city is 93 percent, while the 3:00 P.M. relative humidity in the other city is 28 percent. Explain how this can come about.
7. Suppose the dew point of cold outside air is the same as the dew point of warm air indoors. If the door is opened, and cold air replaces some of the warm inside air, would the new relative humidity indoors be (a) lower than before, (b) higher than before, or (c) the same as before? Explain your answer.
8. On a warm, muggy day, the air is described as "close." What are several plausible explanations for this expression?
9. Outside, on a very warm day, you swing a sling psychrometer for about a minute and read a dry-bulb temperature of 38°C and a wet-bulb temperature of 24°C. After swinging the instrument again, the dry bulb is still 38°C, but the wet bulb is now 26°C. Explain how this could happen.
10. Why are evaporative coolers used in Arizona, Nevada, and California but not in Florida, Georgia, or Indiana?
11. Devise a way of determining elevation above sea level if all you have is a thermometer and a pot of water.

Problems and Exercises

1. During July, how much more water vapor (in percent) is there normally in the air over New York City than in the air over Seattle, Washington? (Hint: See Fig. 5.11, p. 111.)
2. On a bitter cold, snowy morning, the air temperature and dew point of the outside air are –7°C. If this air is brought indoors and warmed to 21°C, with no change in vapor content, what is the relative humidity of the air inside the home? (Hint: See Table 1, p. 117.)
3. (a) With the aid of Fig. 5.11, p. 111 and Fig. 5.12, p. 112, determine the average summer dew points in St. Louis, MO; New Orleans, LA; and Los Angeles, CA. (b) If the high temperature on a particular summer day in all three cities is 32°C (90°F), then calculate the afternoon relative humidity at each of the three cities. (Hint: Either Fig. 5.12, p. 112 or Table 1, p. 117, will be helpful.)
4. Suppose with the aid of a sling psychrometer you obtain an air temperature of 30°C and a wet-bulb temperature of 25°C. What is (a) the wet-bulb depression, (b) the dew point, and (c) the relative humidity of the air? (Use the tables in Appendix D.)
5. If the air temperature is 35°C and the dew point is 21°C, determine the relative humidity using (a) Table 1, p. 117; (b) Fig. 5.12, p. 112; and (c) Tables D.1 and D.2 in Appendix D.
6. In Fig. 5.11, p. 111, the average vapor pressure in Nevada is about 8 mb. (a) Use Table 1, p. 117, to determine the average dew point of this air. (b) Much of the state is above an elevation of 1500 m (5000 ft). At 1500 m, the normal pressure is about 12.5 percent less than at sea level. If the air over Nevada were brought down to sea level, without any change in vapor content, what would be the new vapor pressure of the air?
7. In Yellowstone National Park, there are numerous ponds of boiling water. If Yellowstone is about 2200 m (7200 ft) above sea level (where the air pressure is normally about 775 mb), what is the normal boiling point of water in Yellowstone? (Hint: See Fig. 1, p. 113.)
8. What is the hottest cup of coffee you could get at a cafe located in the mountains at an elevation of 2000 m (6562 ft) above sea level? (Hint: The average air pressure at 2000 m is about 800 mb—see Fig. 1, p. 113.)

Questions for Exploration

On the Blue Skies: College Edition CD-ROM go to the **Atmospheric Moisture and Stability** section of the CD and click on "**Moisture Graph.**"

1. Using the current temperature and dew point in your location, investigate the vapor pressure, saturation vapor pressure, and relative humidity. (a) What is the amount of vapor pressure the atmosphere could hold at this temperature? (b) How much vapor pressure actually exists at the moment? (c) What is the relative humidity?
2. If your temperature rises 5°C without the addition of any more water vapor (i.e. the dew point temperature remains constant) what will be the resulting relative humidity?

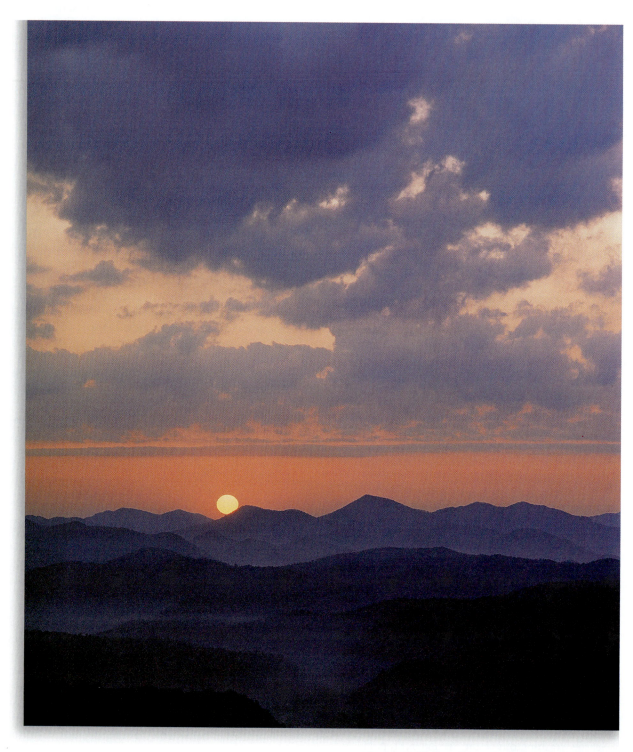

Haze slowly fills the valley, as clouds gradually dissipate over the Great Smoky Mountains on a warm, humid summer morning. (Photo © Carr Clifton, Minden Pictures.)

CONDENSATION: DEW, FOG, AND CLOUDS

CONTENTS

The weather is an ever-playing drama before which we are a captive audience. With the lower atmosphere as the stage, air and water as the principal characters, and clouds for costumes, the weather's acts are presented continuously somewhere about the globe. The script is written by the sun; the production is directed by the earth's rotation; and, just as no theater scene is staged exactly the same way twice, each weather episode is played a little differently, each is marked with a bit of individuality.

Clyde Orr, Jr., *Between Earth and Space*

ave you walked barefoot across a lawn on a summer morning and felt the wet grass under your feet? Did you ever wonder how those glistening droplets of dew could form on a clear summer night? Or why they formed on grass but not on bushes several meters above the ground? In this chapter, we will investigate first the formation of dew and frost. Then we will examine the different types of fog. The chapter concludes with the identification and observation of clouds.

THE FORMATION OF DEW AND FROST

On clear, calm nights, objects near the earth's surface cool rapidly by emitting infrared radiation. The ground and objects on it often become much colder than the surrounding air. Air that comes in contact with these cold surfaces cools by conduction. Eventually, the air cools to the *dew point*—the temperature at which saturation occurs. As surfaces such as twigs, leaves, and blades of grass cool below this temperature, water vapor begins to condense upon them, forming tiny visible specks of water called **dew** (see Fig. 6.1). If the air temperature should drop to freezing or below, the dew will freeze, becoming tiny beads of ice called **frozen dew**. Because the coolest air is usually at ground level, dew is more likely to form on blades of grass than on objects several meters above the surface. This thin coating of dew not only dampens bare feet, but is also a valuable source of moisture for many plants during periods of low rainfall. Averaged for an entire year in middle latitudes, dew yields a blanket of water between 12 and 50 mm (0.5 and 2 in.) thick.

Dew is more likely to form on nights that are clear and calm than on nights that are cloudy and windy. Clear nights allow objects near the ground to cool rapidly by emitting infrared radiation, and calm winds mean that the coldest air will be located at ground level. These atmospheric conditions are usually associated with large fair-weather, high-pressure systems. On the other hand, the cloudy, windy weather that inhibits rapid cooling near the ground and the forming of dew often signifies the approach of a rain-producing storm system. These observations inspired the following folk rhyme:

> When the dew is on the grass,
> rain will never come to pass.
> When grass is dry at morning light,
> look for rain before the night!

Visible white frost forms on cold, clear, calm mornings when the dew-point temperature is at or below freezing. When the air temperature cools to the dew point (now called the *frost point*) and further cooling occurs, water vapor can change directly to ice without becoming a liquid first—a process called *deposition*. The delicate, white crystals of ice that form in this manner are called *hoarfrost*, *white frost*, or simply **frost**. Frost has a treelike branching pattern that easily distinguishes it from the nearly spherical beads of frozen dew.

On cold winter mornings, frost may form on the inside of a windowpane in much the same way as it does outside, except that the cold glass chills the indoor air adjacent to it. When the temperature of the inside of the window drops below freezing, water vapor in the room forms a light, feathery deposit of frost (see Fig. 6.2).

In very dry weather, the air temperature may become quite cold and drop below freezing without ever reaching the frost point, and no visible frost forms. *Freeze* and *black frost* are

FIGURE 6.1
Dew forms on clear nights when objects on the surface cool to a temperature below the dew point. If these beads of water should freeze, they would become frozen dew.

FIGURE 6.2
These are the delicate ice-crystal patterns that frost exhibits on a window during a cold winter morning.

words denoting this situation. These conditions can severely damage crops.

So, dew, frozen dew, and frost form in the rather shallow layer of air near the ground on clear, calm nights. But what happens to air as a deeper layer adjacent to the ground is cooled? We have learned that if air cools without any change in water vapor content, the relative humidity increases. When air cools to the dew point, the relative humidity becomes 100 percent and the air is saturated. Continued cooling condenses some of the vapor into tiny cloud droplets.

CONDENSATION NUCLEI

Actually, the condensation process that produces clouds is not quite so simple. Just as dew and frost need a surface to form on, there must be airborne particles on which water vapor can condense to produce cloud droplets.

Although the air may look clean, it never really is. On an ordinary day, a volume of air about the size of your index finger contains between 1000 and 150,000 particles. Since many of these serve as surfaces on which water vapor can condense, they are called **condensation nuclei**. Without them, relative humidities of several hundred percent would be required before condensation could begin.

Some condensation nuclei are quite small and have a radius less than 0.2 μm; these are referred to as *Aitken nuclei*, after the British physicist who discovered that water vapor condenses on nuclei. Particles ranging in size from 0.2 to 1 μm are called *large nuclei*, while others, called *giant nuclei*, are much larger and have radii exceeding 1 μm (see Table 6.1). The condensation nuclei most favorable for producing clouds (called *cloud condensation nuclei*) have radii of 0.1 μm or more. Usually, between 100 and 1000 nuclei of this size exist in a cubic centimeter of air. These particles enter the atmosphere in a variety of ways: dust, volcanoes, factory smoke, forest fires, salt from ocean spray, and even sulfate particles emitted by phytoplankton in the oceans. In fact, studies show that sulfates provide the major source of cloud condensation nuclei in the marine atmosphere. Because most particles are released into the atmosphere near the ground, the largest concentrations of nuclei are observed in the lower atmosphere near the earth's surface.

Condensation nuclei are extremely light (many have a mass less than one-trillionth of a gram), so they can remain suspended in the air for many days. They are most abundant over industrial cities, where highly polluted air may contain nearly 1 million particles per cubic centimeter. They decrease in cleaner "country" air and over the oceans, where concentrations may dwindle to only a few nuclei per cubic centimeter.

Some particles are **hygroscopic** ("water-seeking"), and water vapor condenses upon these surfaces when the relative humidity is considerably lower than 100 percent. Ocean salt is hygroscopic, as is common table salt. In humid weather, it is

TABLE 6.1 Characteristic Sizes and Concentration of Condensation Nuclei and Cloud Droplets

TYPE OF PARTICLE	APPROXIMATE RADIUS (MICROMETERS)	NO. OF PARTICLES (per cm³)	
		RANGE	TYPICAL
Small (Aitken) condensation nuclei	<0.2	1000 to 10,000	1000
Large condensation nuclei	0.2 to 1.0	1 to 1000	100
Giant condensation nuclei	>1.0	<1 to 10	1
Fog and cloud droplets	>10	10 to 1000	300

difficult to pour salt from a shaker because water vapor condenses onto the salt crystals, sticking them together. Other hygroscopic nuclei include sulfuric and nitric acid particles. Not all particles serve as good condensation nuclei. Some are **hydrophobic** ("water-repelling")—such as oils, gasoline, and paraffin waxes*—and resist condensation even when the relative humidity is above 100 percent. As we can see, condensation may begin on some particles when the relative humidity is well below 100 percent and on others only when the relative humidity is much higher than 100 percent. However, at any given time there are usually many nuclei present, so that haze, fog, and clouds will form at relative humidities near or below 100 percent.

HAZE

Suppose you visit an area that has a layer of **haze** (that is, a layer of dust or salt particles) suspended above the region. There, you may notice that distant objects are usually more visible in the afternoon than in the morning, even when the concentration of particles in the air has not changed. Why? During the warm afternoon, the relative humidity of the air is often below the point where water vapor begins to condense, even on active hygroscopic nuclei. Therefore, the floating particles remain small—usually no larger than about one-tenth of a micrometer. These tiny, dry, haze particles selectively scatter some rays of sunlight, while allowing others to penetrate the

---

*A synthetic hydrophobic is PTFE, or Teflon—the material used in rain-repellent fabric.

FIGURE 6.3
The high relative humidity of the cold air above the lake is causing a layer of haze to form on a still winter morning.

air. The scattering effect of *dry haze* produces a bluish color when viewed against a dark background and a yellowish tint when viewed against a light-colored background.

As the air cools during the night, the relative humidity increases. When the relative humidity reaches about 75 percent, condensation may begin on the most active hygroscopic nuclei, producing a *wet haze*. As water collects on the nuclei, their size increases and the particles, although still small, become large enough to scatter light much more efficiently. (Recall from Chapter 4 that, when sunlight strikes wet haze, nearly all visible waves are scattered away evenly, causing the haze layer to appear white.) In fact, as the relative humidity increases from about 60 percent to 80 percent, the scattering effect increases by a factor of nearly 3. Since relative humidities are normally high during cool mornings, much of the light from distant objects is scattered away by the wet haze particles before reaching you; hence, it is difficult to see these distant objects.

Not only does wet haze restrict visibility more than dry haze, it also appears dull gray or white (see Fig. 6.3). Near sea-

shores and in clean air over the open ocean, large salt particles suspended in air with a high relative humidity often produce a thin white veil across the horizon.

FOG

By now, it should be apparent that condensation is a continuous process beginning when water vapor condenses onto hygroscopic nuclei at relative humidities as low as 75 percent. As the relative humidity of the air increases, the visibility decreases, and the landscape becomes masked with a grayish tint. As the relative humidity gradually approaches 100 percent, the haze particles grow larger, and condensation begins on the less-active nuclei. Now a large fraction of the available nuclei have water condensing onto them, causing the droplets to grow even bigger, until eventually they become visible to the naked eye. The increasing size and concentration of droplets further restrict visibility. When the visibility lowers to less than

1 km (0.62 mi), and the air is wet with countless millions of tiny floating water droplets, the wet haze becomes a cloud resting near the ground, which we call **fog**.*

With the same water content, fog that forms in dirty city air often is thicker than fog that forms over the ocean. Normally, the smaller number of condensation nuclei over the middle of the ocean produce fewer, but larger, fog droplets. City air with its abundant nuclei produces many tiny fog droplets, which greatly increase the thickness (or opaqueness) of the fog and reduce visibility. A dramatic example of a thick fog forming in air with abundant nuclei occurred in London, England, during the early 1950s. The fog became so thick, and the air so laden with smoke particles, that sunlight could not penetrate the smoggy air, requiring that street lights be left on at midday. Moreover, fog that forms in polluted air can turn acidic as the tiny liquid droplets combine with gaseous impurities, such as oxides of sulfur and nitrogen. **Acid fog** poses a threat to human health, especially to people with preexisting respiratory problems. We will examine this problem in more detail in Chapter 17.

As tiny fog droplets grow larger, they become heavier and tend to fall toward the earth. A fog droplet with a diameter of 25 μm settles toward the ground at about 5 cm (2 in.) each second. At this rate, most of the droplets in a fog layer 180 m (about 600 ft) thick would reach the ground in less than one hour. Therefore, two questions arise: How does fog form? How is fog maintained once it does form?

Fog, like any cloud, usually forms in one of two ways: (1) by cooling—air is cooled below its saturation point (dew point); and (2) by evaporation and mixing—water vapor is added to the air by evaporation, and the moist air mixes with relatively dry air. Once fog forms it is maintained by new fog droplets, which constantly form on available nuclei. In other words, the air must maintain its degree of saturation either by continual cooling or by evaporation and mixing of vapor into the air. Let's examine both processes.

Radiation Fog   How can the air cool so that a cloud will form near the surface? Radiation and conduction are the primary means for cooling nighttime air near the ground. Fog produced by the earth's radiational cooling is called **radiation fog**, or **ground fog**. It forms best on clear nights when a shallow layer of moist air near the ground is overlain by drier air. Since the moist layer is shallow, it does not absorb much of the earth's outgoing infrared radiation. The ground, therefore, cools rapidly and so does the air directly above it, and a surface inversion forms. The moist lower layer (chilled rapidly by the cold ground) quickly becomes saturated, and fog forms. The

longer the night, the longer the time of cooling and the greater the likelihood of fog. Hence, radiation fogs are most common over land in late fall and winter.

Another factor promoting the formation of radiation fog is a light breeze of less than 5 knots. Although radiation fog may form in calm air, slight air movement brings more of the moist air in direct contact with the cold ground, and the transfer of heat occurs more rapidly. A strong breeze would prevent a radiation fog from forming by mixing the air near the surface with the drier air above. The ingredients of clear skies and light winds are associated with large high-pressure areas (anticyclones). Consequently, during the winter, when a high becomes stagnant over an area, radiation fog may form on many consecutive days.

Because cold, heavy air drains downhill and collects in valley bottoms, we normally see radiation fog forming in low-lying areas. Hence, radiation fog is frequently called *valley fog*. The cold air and high moisture content in river valleys make them susceptible to radiation fog. Since radiation fog normally forms in lowlands, hills may be clear all day long, while adjacent valleys are fogged in (see Fig. 6.4).

Radiation fogs form upward from the ground as the night progresses and are usually deepest around sunrise. However, fog may occasionally form after sunrise, especially when evaporation and mixing take place near the surface. This usually occurs at the end of a clear, calm night as radiational cooling brings the air temperature close to the dew point in a rather shallow layer above the ground. At the surface, the air becomes saturated, forming a thick blanket of dew on the grass. At daybreak, the sun's rays evaporate the dew, adding water vapor to the air. A light breeze then stirs the moist air with the drier air above, causing saturation (and, hence, fog) to form in a shallow layer near the ground.

Often a shallow fog layer will dissipate or *burn off* by the afternoon. Of course, the fog does not "burn"; rather, sunlight penetrates the fog and warms the ground, causing the air temperature in contact with the ground to increase. The warm air rises and mixes with the foggy air above, which increases the temperature of the foggy air. In the slightly warmer air, some of the fog droplets evaporate, allowing more sunlight to reach the ground, which produces more heating, and soon the fog completely disappears (see Fig. 6.5).

Satellite photographs show that a blanket of radiation fog tends to "burn off" first around its periphery, where the fog is usually thinnest. Sunlight rapidly warms this region, causing the fog to dissipate as the warmer air mixes in toward the denser foggy area.

If the fog is thick, with little sunlight penetrating it, and there is little mixing along the outside edges, the fog may not dissipate. This is often the case in the Central Valley area of California during the late fall and winter. A fog layer over 500 m (1700 ft) thick settles between two mountain ranges, while a strong inversion normally keeps the warmest air above

---

*This is the official international definition of fog. The United States Weather Service reports fog as a restriction to visibility when fog restricts the visibility to 6 miles or less and the spread between the air temperature and dew point is 5°F or less. When the visibility is less than one-quarter of a mile, the fog is considered dense.

FIGURE 6.4
Radiation fog nestled in a valley.

FIGURE 6.5
Dissipation of radiation fog. As morning sunlight warms the ground, warm pockets of air mix with the foggy air above, causing the fog droplets to evaporate or "burn off."

Why Are Headlands Usually Foggier than Beaches?

If you drive along a highway that parallels an irregular coastline, you may have observed that advection fog is more likely to form in certain regions. For example, headlands that protrude seaward usually experience more fog than do beaches that are nestled in the mouths of bays. Why?

As air moves onshore, it crosses the coastline at nearly a right angle. This causes the air to flow together or converge in the vicinity of the headlands (see Fig. 1). This area of weak convergence causes the surface air to rise and cool just a little. If the rising air is close to being saturated, it will cool to its dew point, and fog will form.

Meanwhile, near the beach area, the surface air spreads apart or diverges as it crosses the coastline. This area of weak divergence creates sinking and slightly warmer air. Because the sinking of air increases the separation between air tempera-

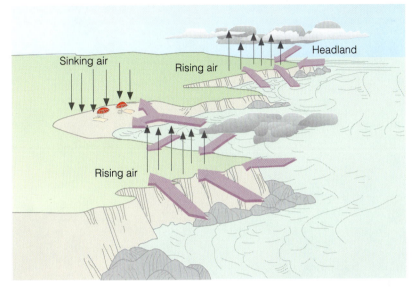

FIGURE 1

Along an irregular coastline, advection fog is more likely to form at the headland where moist surface air converges and rises than at the beach where air diverges and sinks.

ture and dew point, fog is less likely to form in this region. Hence, the headlands can be shrouded in fog while the beaches are basking in sunshine.

---

the top of the fog. During the day, much of the light from the low winter sun reflects off the top of the fog, allowing only a small amount of sunlight to penetrate the fog and warm the ground. As the air warms from below, the fog dissipates upward from the surface in a rather shallow layer less than 150 m (500 ft), creating the illusion that the fog is lifting. Since the fog no longer touches the ground, and a strong inversion exists above it, the fog is called a *high inversion fog*. (The low cloud above the ground is also called *stratus*, or, simply, *high fog*.) As soon as the sun sets, radiational cooling lowers the air temperature, and the fog once again forms on the ground. This daily lifting and lowering of the fog without the sun ever breaking through it may last for many days or even weeks during winter in California's Central Valley.

Advection Fog   Cooling surface air to its saturation point may be accomplished by warm moist air moving over a cold surface. The surface must be sufficiently cooler than the air above so that the transfer of heat from air to surface will cool the air to its dew point and produce fog. Fog that forms in this manner is called **advection fog**.

A good example of advection fog may be observed along the Pacific Coast during summer. The main reason fog forms in this region is that the surface water near the coast is much colder than the surface water farther offshore. Warm moist air from the Pacific Ocean is carried (advected) by westerly winds over the cold coastal waters. Chilled from below, the air temperature drops to the dew point, and fog is produced. Advection fog, unlike radiation fog, always involves the movement of air, so when there is a stiff summer breeze in San Francisco, it's common to watch advection fog roll in past the Golden Gate Bridge (see Fig. 6.6). It is also more common to see advection fog forming at headlands that protrude seaward than in the mouths of bays. If you are curious as to why, read the Focus section above.

As summer winds carry the fog inland over the warmer land, the fog near the ground dissipates, leaving a sheet of low-lying gray clouds that block out the sun. Further inland, the air is sufficiently warm so that even these low clouds evaporate and disappear. Since the fog is more likely to burn off during the warmer part of the day, a typical summertime weather forecast for coastal areas would read, "Fog and low cloudiness

FIGURE 6.6
Advection fog rolling in past the Golden Gate Bridge in San Francisco. As fog moves inland, the air warms and the fog lifts above the surface. Eventually, the air becomes warm enough to totally evaporate the fog.

along the coast extending locally inland both nights and mornings with sunny afternoons."

Because they provide moisture to the coastal redwood trees, advection fogs are important to the scenic beauty of the Pacific Coast. Much of the fog moisture collected by the needles and branches of the redwoods drips to the ground (*fog*

FIGURE 6.7
Tiny drops, each one made from many fog droplets, drip from the needles of this tree and provide a valuable source of moisture during the otherwise dry summer along the coast of California.

drip), where it is utilized by the tree's shallow root system (see Fig. 6.7). Without the summer fog, coastal redwood trees would have trouble surviving the dry California summers. Hence, we find them nestled in the fog belt along the coast.

Advection fogs also prevail where two ocean currents with different temperatures flow next to one another. Such is the case in the Atlantic Ocean off the coast of Newfoundland, where the cold southward-flowing Labrador Current lies almost parallel to the warm northward-flowing Gulf Stream. Warm southerly air moving over the cold water produces fog in that region—so frequently that fog occurs on about two out of three days during summer.

Advection fog also forms over land. In winter, warm moist air from the Gulf of Mexico moves northward over progressively colder and slightly elevated land. As the air cools to its saturation point, a fog forms in the southern or central United States. Because the cold ground is often the result of radiational cooling, fog that forms in this manner is sometimes called **advection-radiation fog**. During this same time of year, air moving across the warm Gulf Stream encounters the colder land of the British Isles and produces the thick fogs of England. Similarly, fog forms as marine air moves over an ice or snow surface. In extremely cold arctic air, ice crystals form instead of water droplets, producing an *ice fog*.

Upslope Fog   Fog that forms as moist air flows up along an elevated plain, hill, or mountain is called **upslope fog**. Typically, upslope fog forms during the winter and spring on the eastern side of the Rockies, where the eastward-sloping plains are nearly a kilometer higher than the land further east. Occasionally, cold air moves from the lower eastern plains westward. The air gradually rises, expands, becomes cooler, and—if sufficiently moist—a fog forms. Upslope fogs that form over an extensive area may last for many days.

Up to now, we have seen how the cooling of air produces fog. But remember that fog may also form by the mixing of two unsaturated masses of air. Fog that forms in this manner is usually called *evaporation fog* because evaporation initially enriches the air with water vapor. Probably, a more appropriate name for the fog is **evaporation (mixing) fog**. (For a better understanding of how mixing can produce fog, read the Focus section on p. 134.)

Evaporation (Mixing) Fog

On a cold day, you may have unknowingly produced evaporation fog. When moist air from your mouth or nose meets the cold air and mixes with it, the air becomes saturated, and a tiny cloud forms with each exhaled breath.

A common form of evaporation-mixing fog is **steam fog**, which forms when cold air moves over warm water. This type of fog forms above a heated outside swimming pool in winter. As long as the water is warmer than the unsaturated air above, water will evaporate from the pool into the air. The increase in water vapor raises the dew point, and, if mixing is sufficient, the air above becomes saturated. The colder air directly above the water is heated from below and becomes warmer than the air directly above it. This warmer air rises and, from a distance, the rising condensing vapor appears as "steam."

It is common to see steam fog forming over lakes on autumn mornings, as cold air settles over water still warm from the long

summer. On occasion, over the Great Lakes, columns of condensed vapor rise from the fog layer, forming whirling *steam devils*, which appear similar to the dust devils on land. If you travel to Yellowstone National Park, you will see steam fog forming above thermal ponds all year long (see Fig. 6.8). Over the ocean in polar regions, steam fog is referred to as *arctic sea smoke*.

Steam fog may form above a wet surface on a sunny day. This is commonly observed after a rainshower as sunlight shines on a wet road, heats the asphalt, and quickly evaporates the water. This added vapor mixes with the air above, producing steam fog. Fog that forms in this manner is short-lived and disappears as the road surface dries.

A warm rain falling through a layer of cold moist air can produce fog. Remember from Chapter 5 that the saturation vapor pressure depends on temperature: Higher temperatures correspond to higher saturation vapor pressures. When a warm raindrop falls into a cold layer of air, the saturation vapor pressure over the raindrop is greater than that of the air. This vapor pressure difference causes water to evaporate from the raindrop into the air. This process may saturate the air and, if mixing occurs, fog forms. Fog of this type is often associated with warm air riding up and over a mass of colder surface air. The fog

FIGURE 6.8
Even in summer, warm air rising above thermal pools in Yellowstone National Park condenses into a type of steam fog.

Fog that Forms by Mixing

How can unsaturated bodies of air mix together to produce fog (or a cloud)? To answer this question, let's first examine two unsaturated air parcels. (Later, we will look at the parcels mixed together.) The two parcels in Fig. 2 are essentially the same size and have a mass of 1 kg. Yet each has a different temperature and a different relative humidity. (We will assume that the parcels are near sea level where the atmospheric pressure is close to 1000 mb.)

Parcel A has an air temperature (t) of 20°C and a dew-point temperature (t_d) of 15°C. Remember from Chapter 5 that the air's relative humidity (RH) can be expressed as:

$$RH = \frac{\text{actual mixing ratio } (w)}{\text{saturation mixing ratio } (w_s)} \times 100 \text{ percent.*}$$

To obtain the *saturation mixing ratio*, we look at Table 1 and read the value that corresponds to the parcel's air temperature. For a temperature of

_____

*The actual mixing ratio (w) is the mass of water vapor per kilogram (kg) of dry air, usually expressed as grams per kilogram (g/kg). The saturation mixing ratio (w_s) is the mixing ratio of saturated air, also expressed as g/kg.

20°C, the saturation mixing ratio is 15.0 g/kg. Likewise, the *actual mixing ratio* is obtained by reading the value in Table 1 that corresponds to the parcel's dew-point temperature. For a dew-point temperature of 15°C, the actual mixing ratio is 10.8 g/kg. Hence, the relative humidity of the air in parcel A is

$$RH = \frac{w}{w_s} = \frac{10.8}{15.0} \times 100 \text{ percent}$$

RH = 72 percent.

Air parcel B is considerably colder than parcel A with a temperature of −10°C, and considerably drier with a dew-point temperature of −15°C. These temperatures yield a relative humidity of

$$RH = \frac{w}{w_s} = \frac{1.2}{1.8} \times 100 \text{ percent}$$

RH = 67 percent.

Suppose we now thoroughly mix the two parcels. After mixing, the new parcel's temperature will be close to the average of parcel A and parcel B, or about 5°C. The total water vapor content (the actual mix-

AIR TEMPERATURE (°C)	SATURATION MIXING RATIO (g/kg)
20	15.0
15	10.8
10	7.8
5	5.5
0	3.8
−5	2.6
−10	1.8
−15	1.2
−20	0.8

TABLE 1   Saturation Mixing Ratios of Water Vapor for Various Air Temperatures (Air Pressure Is 1000 mb)

ing ratio) of the mixed parcel will be the sum of the mixing ratios of parcel A and parcel B, or

$$10.8 \,{}^g\!/_{kg} + 1.2 \,{}^g\!/_{kg} = 12.0 \,{}^g\!/_{2\,kg} = 6.0 \,{}^g\!/_{kg}$$

Look at Table 1 and observe that the saturation mixing ratio for a saturated parcel at 5°C is only 5.5 g/kg. This means that the water vapor content of the mixed parcel, at 6.0 g/kg, is *above* that required for saturation and that the parcel is supersaturated with a relative humidity of 109 percent. Of course, such a high relative humidity is almost impossible to obtain, as water vapor would certainly condense on condensation nuclei, producing liquid water droplets as the two parcels mix together and the relative humidity approaches 100 percent. Hence, mixing two initially unsaturated masses of air can cause fog (or a cloud) to form. Fog that forms in this manner is listed under the heading of evaporation (mixing) fog. As you read that section, keep in mind that although evaporation has a part in fog formation, mixing plays the dominant role.

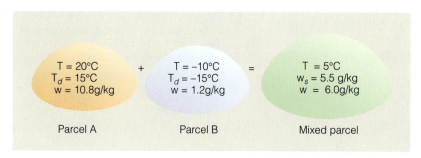

T = 20°C T_d = 15°C w = 10.8g/kg	T = −10°C T_d = −15°C w = 1.2g/kg	T = 5°C w_s = 5.5 g/kg w = 6.0g/kg
Parcel A	+ Parcel B =	Mixed parcel

FIGURE 2

The mixing of two unsaturated air parcels can produce fog. Notice in the saturated mixed parcel that the actual mixing ratio *(w)* is too high. As the mixed parcel cools below its saturation point, water vapor would condense onto nuclei, producing liquid droplets. This would keep the actual mixing ratio close to the saturation mixing ratio, and the relative humidity of the mixed parcel would remain close to 100 percent.

usually develops in the shallow layer of cold air just ahead of an approaching warm front or behind a cold front, which is why this type of evaporation fog is also known as *precipitation fog*, or **frontal fog**. Snow covering the ground is an especially favorable condition for the formation of frontal fog. The melting snow extracts heat from the environment, thereby cooling the already rain-saturated air.

FOGGY WEATHER

Notice in Fig. 6.9 that heavy fog is more prevalent in coastal margins—especially those regions lapped by cold currents—than in the center of the continent. In fact, three regions stand out as having the most days with heavy fog:

1. the Pacific Coast states
2. the Appalachian highland region
3. New England

Areas experiencing more than 20 days a year with heavy fog include the Gulf and Atlantic coastal states, portions of the Great Plains and Rocky Mountains, and the regions around the Great Lakes. The least number of foggy days occur in the Southwest.

The foggiest spot near sea level in the United States is Cape Disappointment, Washington. Located at the mouth of the Columbia River, it averages 2556 hours of heavy fog each year. Anyone who travels to this spot hoping to enjoy the sun during August and September would find its name appropriate indeed. Along the Atlantic Ocean just off the coast of Maine, heavy fog shrouds Mistake Island for an average of 1580 hours each year. And Mt. Washington, New Hampshire, nestled in the clouds at 1908 m (6262 ft) above sea level, experiences fog on more than 300 days each year!

Although fog is basically a nuisance, it has many positive aspects. For example, the California Central Valley fog that many people scorn is extremely important to the economy of that area. Fruit and nut trees that have finished growing during the summer and fall require **winter chilling**—a large number of hours with the air temperature below 7°C (45°F) before trees will begin to grow again. The winter fog blocks out the sun and helps keep daytime temperatures quite cool, while keeping nighttime temperatures above freezing: The more continuous the fog, the more effective the chilling. Consequently, the agricultural economy of the region depends heavily on the fog, for without it and the winter chill it stimulates, many of the fruit and nut trees would not grow well. During the spring, when trees are in bloom, fog prevents nighttime air temperatures from dipping to dangerously low readings by trapping infrared energy that is radiated by the earth and releasing latent heat to the air as fog droplets form.

Unfortunately, fog also has many negative aspects. Along a gently sloping highway, the elevated sections may have excellent visibility, while in lower regions—only a few kilometers away—fog may cause poor visibility. Driving from the clear area into the fog on a major freeway can be extremely dangerous. In fact, every winter many people are involved in fog-related auto accidents. These usually occur when a car enters the fog and, because of the reduced visibility, the driver puts on the brakes to slow down. The car behind then slams into the slowed vehicle, causing a chain-reaction accident with many cars involved. In fact, one such accident occurred near Chattanooga, Tennessee, during December, 1990, when 80 vehicles smashed into each other along a 3-km stretch of foggy Interstate 75. The accident left 12 dead, 51 injured, and a landscape that resembled a war zone.

Extremely limited visibility exists while driving at night in heavy fog with the high-beam lights on. The light scattered

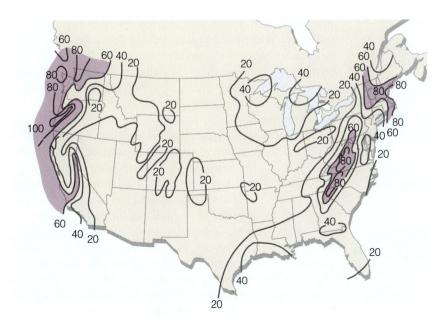

FIGURE 6.9

Average annual number of days with heavy fog throughout the United States.

back to the driver's eyes from the fog droplets makes it difficult to see very far down the road. However, even in thick fog, there is usually a drier and therefore clearer region extending about 35 cm above the road surface. People who drive a great deal in foggy weather take advantage of this by installing extra head lamps—called *fog lamps*—just above the front bumper. These lights are directed downward into the clear space where they provide improved visibility.

Fog-related problems are not confined to land. Even with sophisticated electronic equipment, dense fog in the open sea hampers navigation. A Swedish liner rammed the luxury liner *Andrea Doria* in thick fog off Nantucket Island on July 25, 1956, causing 52 casualties. On a fog-covered runway in the Canary Islands, two 747 jet airliners collided, taking the lives of over 570 people in March, 1977.

Airports suspend flight operations when fog causes visibility to drop below a prescribed minimum. The resulting delays and cancellations become costly to the airline industry and irritate passengers. With fog-caused problems such as these, it is no wonder that scientists have been seeking ways to disperse, or at least "thin," fog. (For more information on fog-thinning techniques, read the Focus section entitled "Fog Dispersal" on p. 137.)

Up to this point, we have looked at the different forms of condensation that occur on or near the earth's surface. In particular, we learned that fog is simply many millions of tiny liquid droplets (or ice crystals) that form near the ground. In the following sections we will see how these same particles, forming well above the ground, are classified and identified as clouds.

BRIEF REVIEW   However, before going on to the section on clouds, here is a brief review of some of the facts and concepts we covered so far:

- Dew, frost, and frozen dew generally form on clear nights when the temperature of objects on the surface cools below the air's dew-point temperature.

- Visible white frost forms in saturated air when the air temperature is at or below freezing. Under these conditions, water vapor can change directly to ice, in a process called *deposition*.

- Condensation nuclei act as surfaces on which water vapor condenses. Those nuclei that have an affinity for water vapor are called *hygroscopic*.

- Fog is a cloud resting on the ground.

- Radiation fog, advection fog, and upslope fog all form as the air cools. Evaporation (mixing) fog, such as steam fog and frontal fog, form as water evaporates and mixes with drier air.

CLOUDS

Clouds are aesthetically appealing and add excitement to the atmosphere. Without them, there would be no rain or snow, thunder or lightning, rainbows or halos. How monotonous if one had only a clear blue sky to look at. A cloud is a visible aggregate of tiny water droplets or ice crystals suspended in the air. Some are found only at high elevations, while others nearly touch the ground. Clouds can be thick or thin, big or little— they exist in a seemingly endless variety of forms. To impose order on this variety, we divide clouds into ten basic types. With a careful and practiced eye, you can become reasonably proficient in correctly identifying them.

Classification of Clouds   Although ancient astronomers named the major stellar constellations about 2000 years ago, clouds were not formally identified and classified until the early nineteenth century. The French naturalist Lamarck (1744–1829) proposed the first system for classifying clouds in 1802; however, his work did not receive wide acclaim. One year later, Luke Howard, an English naturalist, developed a cloud classification system that found general acceptance. In essence, Howard's innovative system employed Latin words to describe clouds as they appear to a ground observer. He named a sheet-like cloud *stratus* (Latin for "layer"); a puffy cloud *cumulus* ("heap"); a wispy cloud *cirrus* ("curl of hair"); and a rain cloud *nimbus* ("violent rain"). In Howard's system, these were the four basic cloud forms. Other clouds could be described by combining the basic types. For example, nimbostratus is a rain cloud that shows layering, whereas cumulonimbus is a rain cloud having pronounced vertical development.

In 1887, Abercromby and Hildebrandsson expanded Howard's original system and published a classification system that, with only slight modification, is still used today. Ten principal cloud forms are divided into four primary cloud groups. Each group is identified by the height of the cloud's base above the surface: high clouds, middle clouds, and low clouds. The fourth group contains clouds showing more vertical than horizontal development. Within each group, cloud types are identified by their appearance. Table 6.2 lists these four groups and their cloud types.

The approximate base height of each cloud group is given in Table 6.3. Note that the altitude separating the high and middle cloud groups overlaps and varies with latitude. Large temperature changes cause most of this latitudinal variation. For example, high cirriform clouds are composed almost entirely of ice crystals. In tropical regions, air temperatures low enough to freeze all liquid water usually occur only above 6000 m (about 20,000 ft). In polar regions, however, these same temperatures may be found at altitudes as low as 3000 m (about 10,000 ft). Hence, while you may observe cirrus clouds at 3600 m (about 12,000 ft) over northern Alaska, you will not see them at that elevation above southern Florida.

FOCUS ON A SPECIAL TOPIC

Fog Dispersal

In any airport fog-clearing operation the problem is to improve visibility so that aircraft can take off and land. Experts have tried various methods, which can be grouped into four categories: (1) increase the size of the fog droplets, so that they become heavy and settle to the ground as a light drizzle; (2) seed cold fog with dry ice (solid carbon dioxide), so that fog droplets are converted into ice crystals; (3) heat the air, so that the fog evaporates; and (4) mix the cooler saturated air near the surface with the warmer unsaturated air above.

To date, only one of these methods has been reasonably successful—the seeding of cold fog. *Cold fog* forms when the air temperature is below freezing, and most of the fog droplets remain as liquid water. (Liquid fog in below-freezing air is also called *supercooled fog.*) The fog can be cleared by injecting several hundred pounds of dry ice into it. As the tiny pieces of cold (−78°C) dry ice descend, they freeze some of the supercooled fog droplets in their path, producing ice crystals. As we will see in Chapter 8, these crystals then grow larger at the expense of the remaining liquid fog droplets. Hence, the fog droplets

evaporate and the larger ice crystals fall to the ground, which leaves a "hole" in the fog for aircraft takeoffs and landings.

Unfortunately, most of the fogs that close airports in the United States are *warm fogs* that form when the air temperature is above freezing. Since dry ice seeding does not work in warm fog, other techniques must be tried.

One method involves injecting hygroscopic particles into the fog. Large salt particles and other chemicals absorb the tiny fog droplets and form into larger drops. More large drops and fewer small drops improve the visibility; plus, the larger drops are more likely to fall as a light drizzle. Since the chemicals are expensive and the fog clears for only a short time, this method of fog dispersal is not economically feasible.

Another technique for fog dispersal is to warm the air enough so that the fog droplets evaporate and visibility improves. Tested at Los Angeles International Airport in the early 1950s, this technique was abandoned because it was smoky, expensive, and not very effective. In fact, the burning of hundreds of dollars worth

of fuel only cleared the runway for a short time. And the smoke particles, released during the burning of the fuel, provided abundant nuclei for the fog to recondense upon.

At large airports in France, a sophisticated fog-clearing technique called *turboclair* is being used. In this system, a series of jet engines installed underground along the edge of the runway pump hot exhaust gases into the air to warm and stir it, causing the fog droplets to evaporate. But the turbulence created by the exhaust makes takeoffs and landings unsafe for small aircraft.

A final method of warm fog dispersal uses helicopters to mix the air. The chopper flies across the fog layer, and the turbulent downwash created by the rotor blades brings drier air above the fog into contact with the moist fog layer (see Fig. 3). The aim, of course, is to evaporate the fog. Experiments show that this method works well, as long as the fog is a shallow radiation fog with a relatively low liquid water content. But many fogs are thick, have a high liquid water content, and form by other means. An inexpensive and practical method of dispersing warm fog has yet to be discovered.

(a)

(b)

FIGURE 3

Helicopters hovering above an area of shallow fog (diagram a) can produce a clear area (photograph b) by mixing the drier air into the foggy air below.

TABLE 6.2 The Four Major Cloud Groups and Their Types	

1. High clouds
    Cirrus (Ci)
    Cirrostratus (Cs)
    Cirrocumulus (Cc)
 	
2. Middle clouds
    Altostratus (As)
    Altocumulus (Ac)
	3. Low clouds
	    Stratus (St)
	    Stratocumulus (Sc)
	    Nimbostratus (Ns)
	
	4. Clouds with vertical
	    development
	       Cumulus (Cu)
	       Cumulonimbus (Cb)

Clouds cannot be accurately identified strictly on the basis of elevation. Other visual clues are necessary. Some of these are explained in the following section.

Cloud Identification

High Clouds   High clouds in middle and low latitudes generally form above 6000 m (20,000 ft). Because the air at these elevations is quite cold and "dry," high clouds are composed almost exclusively of ice crystals and are also rather thin.* High clouds usually appear white, except near sunrise and sunset, when the unscattered (red, orange, and yellow) components of sunlight are reflected from the underside of the clouds.

The most common high clouds are the **cirrus**, which are thin, wispy clouds blown by high winds into long streamers called *mares' tails*. Notice in Fig. 6.10 that they can look like a white, feathery patch with a faint wisp of a tail at one end. Cirrus clouds usually move across the sky from west to east, indicating the prevailing winds at their elevation, and they generally point to fair, pleasant weather.

Cirrocumulus clouds, seen less frequently than cirrus, appear as small, rounded, white puffs that may occur individually or in long rows (see Fig. 6.11). When in rows, the cirro-

FIGURE 6.10
Cirrus clouds.

cumulus cloud has a rippling appearance that distinguishes it from the silky look of the cirrus and the sheetlike cirrostratus. Cirrocumulus seldom cover more than a small portion of the sky. The dappled cloud elements that reflect the red or yellow light of a setting sun make this one of the most beautiful of all clouds. The small ripples in the cirrocumulus strongly resem-

---

*Studies conducted above Boulder, Colorado, during the early 1980s discovered small quantities of liquid water in cirrus clouds at temperatures as low as –36°C (–33°F).

TABLE 6.3 Approximate Height of Cloud Bases above the Surface for Various Locations			
CLOUD GROUP	**TROPICAL REGION**	**MIDDLE LATITUDE REGION**	**POLAR REGION**
High Ci, Cs, Cc	20,000 to 60,000 ft (6,000 to 18,000 m)	16,000 to 43,000 ft (5000 to 13,000 m)	10,000 to 26,000 ft (3000 to 8000 m)
Middle As, Ac	6500 to 26,000 ft (2000 to 8000 m)	6500 to 23,000 ft (2000 to 7000 m)	6500 to 13,000 ft (2000 to 4000 m)
Low St, Sc, Ns	surface to 6500 ft (0 to 2000 m)	surface to 6500 ft (0 to 2000 m)	surface to 6500 ft (0 to 2000 m)

ble the scales of a fish; hence, the expression *"mackerel sky"* commonly describes a sky full of cirrocumulus clouds.

The thin, sheetlike, high clouds that often cover the entire sky are **cirrostratus** (see Fig. 6.12), which are so thin that the sun and moon can be clearly seen through them. The ice crystals in these clouds refract the light passing through them and will often produce a halo. In fact, the veil of cirrostratus may be so thin that a halo is the only clue to its presence. Thick cirrostratus clouds give the sky a glary white appearance and frequently form ahead of an advancing storm; hence, they can be used to predict rain or snow within 12 to 24 hours, especially if they are followed by middle-type clouds.

Middle Clouds The middle clouds have bases between 2000 and 7000 m (6500 to 23,000 ft) in the middle latitudes. These clouds are composed of water droplets and—when the temperature becomes low enough—some ice crystals.

Altocumulus clouds are middle clouds that are composed mostly of water droplets and are rarely more than 1 km thick. They appear as gray, puffy masses, sometimes rolled out in parallel waves or bands (Fig. 6.13). Usually, one part of the cloud is darker than another, which helps to separate it from the higher cirrocumulus. Also, the individual puffs of the altocumulus appear larger than those of the cirrocumulus. A layer of altocumulus may sometimes be confused with

FIGURE 6.11
Cirrocumulus clouds.

FIGURE 6.12
Cirrostratus clouds with a halo.

FIGURE 6.13
Altocumulus clouds.

FIGURE 6.14
Altostratus cloud. The appearance of a dimly visible "watery sun" through a deck of gray clouds is usually a good indication that the clouds are altostratus.

altostratus; in case of doubt, clouds are called altocumulus if there are rounded masses or rolls present. Altocumulus clouds that look like "little castles" *(castellanus)* in the sky indicate the presence of rising air at cloud level. The appearance of these clouds on a warm, humid summer morning often portends thunderstorms by late afternoon.

The **altostratus** is a gray or blue-gray cloud composed of ice crystals and water droplets. Altostratus clouds often cover the entire sky across an area that extends over many hundreds of square kilometers. In the thinner section of the cloud, the sun (or moon) may be dimly visible as a round disk, as if the sun were shining through ground glass. This appearance is sometimes referred to as a "watery sun" (see Fig. 6.14). Thick cirrostratus clouds are occasionally confused with thin altostratus clouds. The gray color, height, and dimness of the sun are good clues to identifying an altostratus. The fact that halos only occur with cirriform clouds also helps one distinguish them. Another way to separate the two is to look at the ground for shadows. If there are none, it is a good bet that the cloud is altostratus because cirrostratus are usually transparent enough to produce them. Altostratus clouds often form ahead of storms having widespread and relatively continuous precipitation. If precipitation falls from an altostratus, its base usually lowers. If the precipitation reaches the ground, the cloud is then classified as *nimbostratus*.

Low Clouds Low clouds, with their bases lying below 2000 m (6500 ft), are almost always composed of water droplets; however, in cold weather, they may contain ice particles and snow.

FIGURE 6.15
The nimbostratus is the sheetlike cloud from which light rain is falling. The ragged-appearing cloud beneath the nimbostratus is stratus fractus, or scud.

FIGURE 6.16
Stratocumulus clouds. Notice that the rounded masses are larger than those of the altocumulus.

The **nimbostratus** is a dark gray, "wet"-looking cloudy layer associated with more or less continuously falling rain or snow (see Fig. 6.15). The intensity of this precipitation is usually light or moderate—it is never of the heavy, showery variety. The base of the nimbostratus cloud is normally impossible to identify clearly and its top may be over 3 km (10,000 ft) higher. Nimbostratus is easily confused with the altostratus. Thin nimbostratus is usually darker gray than thick altostratus, and you normally cannot see the sun or moon through a layer of nimbostratus. Visibility below a nimbostratus cloud deck is usually quite poor because rain will evaporate and mix with the air in this region. If this air becomes saturated, a lower layer of clouds or fog may form beneath the original cloud base. Since these lower clouds drift rapidly with the wind, they form irregular shreds with a ragged appearance that are called *stratus fractus*, or *scud*.

Stratocumulus are low lumpy clouds that appear in rows, in patches, or as rounded masses with blue sky visible between the individual cloud elements (see Fig. 6.16). Often they appear near sunset as the spreading remains of a much larger cumulus cloud. Occasionally, the sun will shine through the cloud breaks producing crepuscular rays (see Fig. 4.7, p. 85). The color of stratocumulus ranges from light to dark gray. It differs from altocumulus in that it has a lower base and larger individual cloud elements. (Compare Fig. 6.13 with Fig. 6.16.) To distinguish between the two, hold your hand at arm's length and point toward the cloud. Altocumulus cloud elements will generally be about the size of your thumbnail; stratocumulus cloud

WEATHER WATCH

On July 26, 1959, Colonel William A. Rankin took a wild ride inside a huge cumulonimbus cloud. Bailing out of his disabled military aircraft inside a thunderstorm at 14.5 km (about 47,500 ft), Rankin free-fell for about 3 km (10,000 ft). When his parachute opened, surging updrafts carried him higher into the cloud, where he was pelted by heavy rain and hail, and nearly struck by lightning.

elements will usually be about the size of your fist. Although precipitation rarely falls from stratocumulus, precipitation in the form of showers may occur in winter if the cloud elements develop vertically into much larger clouds and their tops grow colder than about –5°C (23°F).

Stratus is a uniform grayish cloud that often covers the entire sky. It resembles a fog that does not reach the ground (see Fig. 6.17). Actually, when a thick fog "lifts," the resulting cloud is a deck of low stratus. Normally, no precipitation falls from the stratus, but sometimes it is accompanied by a light mist or drizzle. This cloud commonly occurs over Pacific and Atlantic coastal waters in summer. A thick layer of stratus might be confused with nimbostratus, but the distinction between them can be made by observing the low base of the stratus cloud and remembering that light-to-moderate precipitation occurs with nimbostratus. Moreover, stratus often has a more uniform base than does nimbostratus. Also, a deck of stratus may be confused with a layer of altostratus. However, if you remember that stratus are lower and darker gray and that the sun normally appears "watery" through altostratus, the distinction can be made.

Clouds with Vertical Development Familiar to almost everyone, the puffy **cumulus** cloud takes on a variety of shapes, but most often it looks like a piece of floating cotton with sharp outlines and a flat base (see Fig. 6.18). The base appears white to light gray, and, on a humid day, may be only 1000 m (3300 ft) above the ground and a kilometer or so wide. The top of the cloud—often in the form of rounded towers—denotes the limit of rising air and is usually not very high. These clouds can be distinguished from stratocumulus by the fact that cumulus clouds are detached (usually a great deal of blue sky between each cloud) while stratocumulus usually occur in groups or patches. Also, the cumulus has a dome- or tower-shaped top as opposed to the generally flat tops of the stratocumulus. Cumulus clouds that show only slight vertical growth are called *cumulus humilis* and are associated with fair weather; therefore, we call these clouds "fair weather cumulus." Ragged-edge cumulus clouds that are smaller than cumulus humilis and scattered across the sky are called *cumulus fractus*.

Harmless-looking cumulus often develop on warm summer mornings and, by afternoon, become much larger and more vertically developed. When the growing cumulus resembles a head of cauliflower, it becomes a *cumulus congestus*, or *towering cumulus*. Most often, it is a single large cloud, but, occasionally, several grow into each other, forming a line of towering clouds, as shown in Fig. 6.19. Precipitation that falls from a cumulus congestus is always showery.

If a cumulus congestus continues to grow vertically, it develops into a giant **cumulonimbus**—a thunderstorm cloud (see Fig. 6.20). While its dark base may be no more than 600 m (2000 ft) above the earth's surface, its top may extend upward to the tropopause, over 12,000 m (39,000 ft) higher. A cumulonimbus can occur as an isolated cloud or as part of a line or "wall" of clouds.

Tremendous amounts of energy are released by the condensation of water vapor within a cumulonimbus and result in the development of violent up- and downdrafts, which may exceed 70 knots. The lower (warmer) part of the cloud is usu-

FIGURE 6.17
A layer of low-lying stratus clouds.

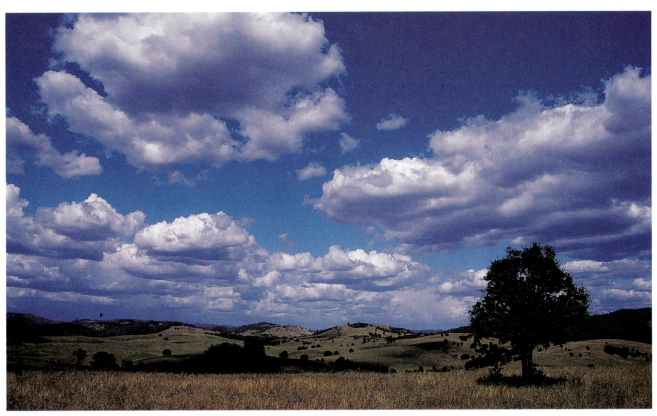

FIGURE 6.18
Cumulus clouds. Small cumulus clouds such as these are sometimes called *fair weather cumulus*, or *cumulus humilis*.

FIGURE 6.19
Cumulus congestus. This line of cumulus congestus clouds is building along Maryland's eastern shore.

FIGURE 6.20
A cumulonimbus cloud. Strong upper-level winds blowing from right to left produce a well-defined anvil. Sunlight scattered by falling ice crystals produces the white (bright) area beneath the anvil. Notice the heavy rain shower falling from the base of the cloud.

ally composed of only water droplets. Higher up in the cloud, water droplets and ice crystals both abound, while, toward the cold top, there are only ice crystals. Swift winds at these higher altitudes can reshape the top of the cloud into a huge flattened anvil *(cumulonimbus incus)*. These great thunderheads may contain all forms of precipitation—large raindrops, snowflakes, snow pellets, and sometimes hailstones—all of which can fall to earth in the form of heavy showers. Lightning, thunder, and even violent tornadoes are associated with the cumulonimbus. (More information on the violent nature of thunderstorms and tornadoes is given in Chapter 15.)

Cumulus congestus and cumulonimbus frequently look alike, making it difficult to distinguish between them. However, you can usually distinguish them by looking at the top of the cloud. If the sprouting upper part of the cloud is sharply defined and not fibrous, it is usually a cumulus congestus; conversely, if the top of the cloud loses its sharpness and becomes fibrous in texture, it is usually a cumulonimbus. Compare Fig. 6.19 with Fig. 6.20. The weather associated with these clouds also differs: lightning, thunder, and large hail only occur with cumulonimbus.

So far, we have discussed the ten primary cloud forms, summarized pictorially in Fig. 6.21. This figure, along with the cloud photographs and descriptions (and the cloud chart at the front of the book), should help you identify the more common cloud forms. Don't worry if you find it hard to estimate cloud heights. This is a difficult procedure, requiring much practice. You can use local objects (hills, mountains, tall buildings) of known height as references on which to base your height estimates.

To better describe a cloud's shape and form, a number of descriptive words may be used in conjunction with its name. We mentioned a few in the previous section; for example, a stratus cloud with a ragged appearance is a stratus fractus, and a cumulus cloud with marked vertical growth is a cumulus congestus. Table 6.4 lists some of the more common terms that are used in cloud identification.

Some Unusual Clouds   Although the ten basic cloud forms are the most frequently seen, there are some unusual clouds that deserve mentioning. For example, moist air crossing a mountain barrier often forms into waves. The clouds that form in the wave crest usually have a lens shape and are, therefore, called **lenticular clouds** (see Fig. 6.22). Frequently, they form one above the other like a stack of pancakes, and at a distance they may resemble a fleet of hovering spacecraft. Hence, it is no wonder a large number of UFO sightings take place when lenticular clouds are present. When a cloud forms over and extends downwind of an isolated mountain peak, as shown in Fig. 6.23, it is called a **banner cloud**.

Similar to the lenticular is the *cap cloud*, or **pileus**, that usually resembles a silken scarf capping the top of a sprouting cumulus cloud (Fig. 6.24). Pileus clouds form when moist winds are deflected up and over the top of a building cumulus

FIGURE 6.21
A generalized illustration of basic cloud types based on height above the surface and vertical development.

FIGURE 6.22
Lenticular clouds forming on the eastern side of the Sierra Nevada.

TABLE 6.4 Common Terms Used in Identifying Clouds

TERM	LATIN ROOT AND MEANING	DESCRIPTION
Lenticularis	(*lens, lenticula,* lentil)	Clouds having the shape of a lens or an almond, often elongated and usually with well-defined outlines. This term applies mainly to cirrocumulus, altocumulus, and stratocumulus
Fractus	(*frangere,* to break or fracture)	Clouds that have a ragged or torn appearance; applies only to stratus and cumulus
Humilis	(*humilis,* of small size)	Cumulus clouds with generally flattened bases and slight vertical growth
Congestus	(*congerere,* to bring together; to pile up)	Cumulus clouds of great vertical extent that from a distance may resemble a head of cauliflower
Calvus	(*calvus,* bald)	Cumulonimbus in which at least some of the upper part is beginning to lose its cumuliform outline
Capillatus	(*capillus,* hair; having hair)	Cumulonimbus characterized by the presence in the upper part of cirriform clouds with fibrous or striated structure
Undulatus	(*unda,* wave; having waves)	Clouds in patches, sheets, or layers showing undulations
Translucidus	(*translucere,* to shine through; transparent)	Clouds that cover a large part of the sky and are sufficiently translucent to reveal the position of the sun or moon
Incus	(*incus,* anvil)	The smooth cirriform mass of cloud in the upper part of a cumulonimbus that is anvil-shaped
Mammatus	(*mamma,* mammary)	Baglike clouds that hang like a cow's udder on the underside of a cloud; may occur with cirrus, altocumulus, altostratus, stratocumulus, and cumulonimbus
Pileus	(*pileus,* cap)	A cloud in the form of a cap or hood above or attached to the upper part of a cumuliform cloud, particularly during its developing stage
Castellanus	(*castellum,* a castle)	Clouds that show vertical development and produce towerlike extensions, often in the shape of small castles

FIGURE 6.23
The cloud forming over and downwind of Mt. Ranier is called a banner cloud.

FIGURE 6.24
A pileus cloud forming above a developing cumulus cloud.

congestus or cumulonimbus. If the air flowing over the top of the cloud condenses, a pileus often forms.

Most clouds form in rising air, but the mammatus forms in sinking air. **Mammatus clouds** derive their name from their appearance—baglike sacks that hang beneath the cloud and resembles a cow's udder (see Fig. 6.25). Although mammatus most frequently form on the underside of cumulonimbus, they may develop beneath cirrocumulus, altostratus, altocumulus, and stratocumulus. For mammatus to form, the sinking air must be cooler than the air around it and have a high liquid water or ice content. As saturated air sinks, it warms, but the warming is retarded because of the heat taken from the air to evaporate the liquid or melt ice particles. If the sinking air remains saturated and cooler than the air around it, the sinking air can extend below the cloud base appearing as rounded masses we call mammatus clouds.

Jet aircraft flying at high altitudes often produce a cirrus-like trail of condensed vapor called a *condensation trail* or **contrail** (see Fig. 6.26). The condensation may come directly from the water vapor added to the air from engine exhaust. In this case, there must be sufficient mixing of the hot exhaust gases with the cold air to produce saturation. The release of particles

FIGURE 6.25
Mammatus clouds.

FIGURE 6.26
A contrail forming behind a jet aircraft.

in the exhaust may even provide nuclei on which ice crystals form. Contrails evaporate rapidly when the relative humidity of the surrounding air is low. If the relative humidity is high, however, contrails may persist for many hours. Contrails may also form by a cooling process. The reduced pressure produced by air flowing over the wing causes the air to cool. This cooling may supersaturate the air, producing an *aerodynamic contrail*. This type of trail usually disappears quickly in the turbulent wake of the aircraft.

Aside from the cumulonimbus cloud that sometimes penetrates into the stratosphere, all of the clouds described so far are observed in the lower atmosphere—in the troposphere. Occasionally, however, clouds may be seen above the tropo-

sphere. For example, soft pearly looking clouds called **nacreous clouds**, or *mother-of-pearl clouds*, form in the stratosphere at altitudes above 30 km (see Fig. 6.27). They are best viewed in polar latitudes during the winter months when the sun, being just below the horizon, is able to illuminate them because of their high altitude. Their exact composition is not known, although they appear to be composed of water in either solid or liquid (supercooled) form.

Wavy bluish-white clouds, so thin that stars shine brightly through them, may sometimes be seen in the upper mesosphere, at altitudes above 75 km (46 mi). The best place to view these clouds is in polar regions at twilight. At this time, because of their altitude, the clouds are still in sunshine. To a ground

FIGURE 6.27
The clouds in this photograph are nacreous clouds. They form in the stratosphere and are most easily seen at high latitudes.

FIGURE 6.28
The wavy clouds in this photograph are noctilucent clouds. They are usually observed at high latitudes, at altitudes between 75 and 90 km above the earth's surface.

observer, they appear bright against a dark background and, for this reason, they are called **noctilucent clouds**, meaning "luminous night clouds" (see Fig. 6.28). Studies reveal that these clouds are composed of tiny ice crystals. The water to make the ice may originate in meteoroids that disintegrate when entering the upper atmosphere or from the chemical breakdown of methane gas at high levels in the atmosphere.

Cloud Observations

Determining Sky Conditions Often, a daily weather forecast will include a phrase such as, "overcast skies with clouds becoming scattered by evening." To the average person, this means that the cloudiness will diminish, but to the meteorologist, the terms *overcast* and *scattered* have a more specific meaning. In meteorology, descriptions of sky conditions are

defined by the fraction of sky covered by clouds. A *clear* sky, for example, is one where no clouds are present.* When there are between one-eighth and two-eighths clouds covering the sky, there are a *few* clouds present. When cloudiness increases to between three-eighths and four-eighths, the sky is described as being *scattered* with clouds. "Partly cloudy" also describes these sky conditions. Clouds covering between five-eighths and seven-eighths denote a sky with *broken* clouds ("mostly cloudy"), and *overcast* conditions exist when the sky is covered (eight-eighths) with clouds. Table 6.5 presents a summary of sky cover conditions.

Observing sky conditions far away can sometimes fool even the trained observer. A broken cloud deck near the

---

*In automated (ASOS) station usage, the phrase "clear sky" means that no clouds are reported whose bases are at or below 12,000 ft.

TABLE 6.5	Description of Sky Conditions			
		OBSERVATION		
DESCRIPTION	ASOS*	Human		MEANING
Clear (CLR or SKC)	0 to 5%	0		No clouds
Few	>5 to ≤25%	0 to $\frac{2}{8}$		Few clouds visible
Scattered (SCT)	>25 to ≤50%	$\frac{3}{8}$ to $\frac{4}{8}$		Partly cloudy
Broken (BKN)	>50 to ≤87%	$\frac{5}{8}$ to $\frac{7}{8}$		Mostly cloudy
Overcast (OVC)	>87 to 100%	$\frac{8}{8}$		Sky is covered by clouds
Sky obscured	—	—		Sky is hidden by surface-based phenomena, such as fog, blowing snow, smoke and so forth, rather than by cloud cover

*Automated Surface Observing System. Symbol > means greater than; < means less than; ≥ means equal to or greater than.

FOCUS ON AN OBSERVATION

Measuring Cloud Ceilings

In addition to knowing about sky conditions, it is usually important to have a good estimate of the height of cloud bases. Aircraft could not operate safely without accurate cloud height information, particularly at lower elevations.

The term *ceiling* is defined as the height of the lowest layer of clouds above the surface that are either broken or overcast, but not thin. Direct information on cloud height can be obtained from pilots who report the altitude at which they encounter the ceiling. Less directly, *ceiling balloons* are used to measure the height of clouds. A small balloon filled with a known amount of hydrogen or helium rises at a fairly constant and known rate. The ceiling is determined by measuring the time required for it to enter the lowest cloud layer. For example, if the balloon rises 125 m (about 400 ft) each minute, and it takes three minutes to enter a broken layer of stratocumulus, the ceiling would be 375 m (about 1200 ft). Ceiling balloon observations can be made at night simply by attaching a small battery-operated light to the balloon.

For many years, the *rotating-beam ceilometer* provided information on cloud ceiling, especially at airports. This instrument consists of a ground-

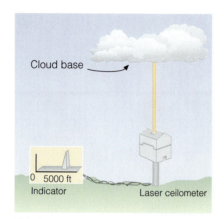

FIGURE 4

The laser-beam ceilometer sends pulses of infrared radiation up to the cloud. Part of this beam is reflected back to the ceilometer. The interval of time between pulse transmission and return is a measure of cloud height, as displayed on the indicator screen.

based projector that rotates vertically from horizon to horizon. As it rotates, it sends out a powerful light beam that moves along the base of the cloud. A light-sensitive detector, some known distance from the projector, points upward and picks up the light from the cloud base. By knowing the projector angle and its distance from the detector, the cloud height is determined mathematically.

Most of the rotating-beam ceilometers have been phased out and

replaced with *laser-beam ceilometers*. The laser ceilometer is a fixed-beam type whose transmitter and receiver point straight up at the cloud base (see Fig. 4). Short, intense pulses of infrared radiation from the transmitter strike the cloud base, and a portion of this radiation is reflected back to the receiver. The time interval between pulse transmission and its return from the cloud determines the cloud-base height.

The automated Surface Observing System (ASOS) uses a laser beam ceilometer to measure cloud height. The ceilometer measures the cloud height and then infers the amount of cloud cover by averaging the amount of clouds that have passed over the sensor during a duration of 30 minutes. The ASOS laser ceilometer is unable to measure clouds that are not above the sensor. To help remedy this situation, a second laser ceilometer may be located nearby. Another limitation of the ASOS ceilometer is that it does not report clouds above 12,000 ft.* A new laser ceilometer that will provide cloud height information up to 25,000 ft is being developed.

*The latest geostationary satellites above North America are equipped to measure cloud heights above 12,000 ft over ASOS stations.

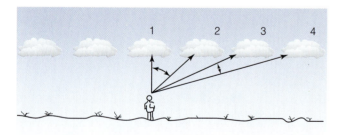

FIGURE 6.29

Clouds on the horizon appear closer together than clouds overhead. Note that the amount of clear space between each cloud is the same. To the observer, however, there appears to be more space between clouds 1 and 2 than between clouds 3 and 4.

horizon usually appears as overcast because the open spaces between the clouds are less visible at a distance (see Fig. 6.29). Therefore, cloudiness is usually over-estimated when clouds are near the horizon. Viewed from afar, clouds not normally associated with precipitation may appear darker and thicker than they actually are. The reason for this is that light from a distant cloud travels through more atmosphere and is more attenuated than the light from the same type of cloud closer to the observer. (Information on measuring the height of cloud bases is given in the Focus section above.)

Up to this point, we have seen how clouds look from the ground. We will now look at clouds from a different vantage point—the satellite view.

Satellite Observations The weather satellite is a cloud-observing platform in earth orbit. It provides extremely valuable cloud photographs of areas where there are no ground-based observations. Because water covers over 70 percent of the earth's surface, there are vast regions where few (if any) surface cloud observations are made. Before weather satellites were used, severe storms, such as hurricanes and typhoons, often went undetected until they moved dangerously near inhabited areas. Residents of the regions affected had little advance warning. Today, satellites spot these storms while they are still far out in the ocean and track them accurately.

There are two primary types of weather satellites in use for viewing clouds. The first are called **geostationary satellites** (or *geosynchronous satellites*) because they orbit the equator at the same rate the earth spins and, hence, remain at nearly 36,000 km (22,300 mi) above a fixed spot on the earth's surface (see Fig. 6.30). This positioning allows continuous monitoring of a specific region.

Geostationary satellites are also important because they use a "real time" data system, meaning that the satellites transmit photographs to the receiving system on the ground as soon as the camera takes the picture. Successive cloud photographs from these satellites can be put into a time-lapse movie sequence to show the cloud movement, dissipation, or development associated with weather fronts and storms. This information is a great help in forecasting the progress of large weather systems. Wind directions and speeds at various levels may also be approximated by monitoring cloud movement with the geostationary satellite.

To complement the geostationary satellites, there are **polar-orbiting satellites**, which closely parallel the earth's meridian lines. These satellites pass over the north and south polar regions on each revolution. As the earth rotates to the east beneath the satellite, each pass monitors an area to the west of the previous pass (see Fig. 6.31). Eventually, the satellite covers the entire earth.

Polar-orbiting satellites have the advantage of photographing clouds directly beneath them. Thus, they provide sharp pictures in polar regions, where photographs from a geostationary satellite are distorted because of the low angle at which the satellite "sees" this region. Polar orbiters also circle the earth at a much lower altitude (about 850 km) than geostationary satellites and provide detailed photographic information about objects, such as violent storms and cloud systems.

Continuously improved detection devices make weather observation by satellites more versatile than ever. Early satellites, such as *TIROS I*, launched on April 1, 1960, used television cameras to photograph clouds. Contemporary satellites use radiometers, which can observe clouds during both day and night by detecting radiation that emanates from the top of the clouds. Additionally, the new generation *Geostationary Operational Environmental Satellite (GOES)* series (see Fig. 6.32) has the capacity to obtain cloud images and, at the same time, provide vertical profiles of atmospheric temperature and moisture by detecting emitted radiation from atmospheric gases, such as water vapor. In modern satellites, a special type of advanced radiometer (called an *imager*) provides satellite pictures with much better resolution than did previous imagers. Moreover, another type of special radiometer (called a *sounder*) gives a more accurate profile of temperature and moisture at different levels in the atmosphere than did earlier

FIGURE 6.30

The geostationary satellite moves through space at the same rate that the earth rotates, so it remains above a fixed spot on the equator and monitors one area constantly.

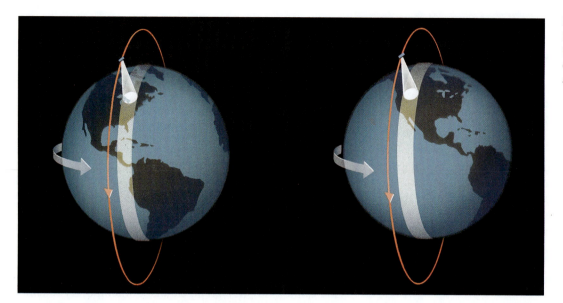

FIGURE 6.31
Polar-orbiting satellites scan from north to south, and on each successive orbit the satellite scans an area farther to the west.

instruments. In the latest *GOES* series, the imager and sounder are able to operate independent of each other.

Information on cloud thickness and height can be deduced from satellite photographs. Visible photographs show the sunlight reflected from a cloud's upper surface. Because thick clouds have a higher albedo (reflectivity) than thin clouds, they appear brighter on a visible satellite photograph. However, high, middle, and low clouds have just about the same albedo, so it is difficult to distinguish among them simply by using visible light photographs. To make this distinction, *infrared cloud pictures* are used. Such pictures produce a better image of the actual radiating surface because they do not show the strong visible reflected light. Since warm objects radiate more energy than cold objects, high temperature regions can

be artificially made to appear darker on an infrared photograph. Because the tops of low clouds are warmer than those of high clouds, cloud observations made in the infrared can distinguish between warm low clouds (dark) and cold high clouds (light) (see Fig. 6.33). Moreover, cloud temperatures can be converted by a computer into a three-dimensional

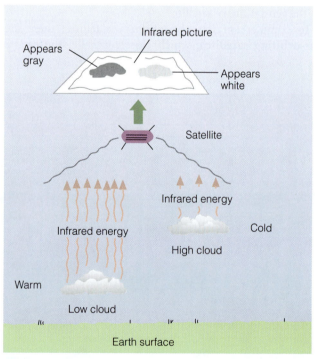

FIGURE 6.33

Generally, the lower the cloud, the warmer its top. Warm objects emit more infrared energy than do cold objects. Thus, an infrared satellite picture can distinguish warm, low (gray) clouds from cold, high (white) clouds.

FIGURE 6.32

Artist's view of the *GOES* meteorological satellite.

(a)   (b)

FIGURE 6.34
A visible image (a) and an infrared image (b) of the eastern Pacific taken on the same day at just about the same time.

image of the cloud. These are the 3-D cloud photos presented on television by many weathercasters.

Figure 6.34a shows a visible satellite image (from a geostationary satellite) of a storm system in the eastern Pacific. Notice that all of the clouds in the photo appear white. However, in the infrared photograph (see Fig. 6.34b), taken on the same day (and just about the same time), the clouds appear to have many shades of gray. In the visible photograph, the clouds covering part of Oregon and northern California appear relatively thin compared to the thicker, bright clouds to the west. Furthermore, these thin clouds must be high because they also appear bright in the infrared picture. The elongated band of clouds off the coast marks the position of an approaching weather front. Here, the clouds appear white and bright in both pictures, indicating a zone of thick, heavy clouds. Behind the front, the lumpy clouds are probably cumulus because they appear gray in the infrared photo, indicating that their tops are low and relatively warm.

When temperature differences are small, it is difficult to directly identify significant cloud and surface features on an infrared picture. Some way must be found to increase the contrast between features and their backgrounds. This can be done by a process called *computer enhancement*. Certain temperature ranges in the infrared photograph are assigned specific shades of gray—grading from black to white. Figure 6.35 is an infrared-enhanced picture for the same day and area as shown in Fig. 6.34. Note the dark and light contouring in the picture. Clouds with cold tops, and those with tops near freezing, are assigned the darkest gray color. Hence, the dark gray areas embedded along the front represent the region where the coldest and, therefore, highest and thickest clouds are found. It is here where the stormiest weather is probably occurring. Also notice that, near the southern tip of the picture, the dark gray blotches surrounded by areas of white are thunderstorms that have developed over warm tropical waters. They show up clearly as thick white clouds in both the visible and infrared

Satellites Do More than Observe Clouds

The use of satellites to monitor weather is not restricted to observing clouds. For example, there are satellites that relay data communications and television signals, and provide military surveillance. Moreover, satellites measure radiation from the earth's surface and atmosphere, giving us information about the earth-atmosphere energy balance, discussed in Chapter 2. The infrared radiation measurements, obtained by an atmospheric *sounder*, are transformed into vertical profiles of temperature and moisture, which are fed into National Weather Service computer forecast models.

Radiation intensities from the ocean surface are translated into sea-surface temperature readings. This information is valuable to the fishing industry, as well as to the meteorologist. In fact, the recently launched *Tropical Rainfall Measuring Mission (TRMM)* satellite obtains sea surface temperatures with a microwave scanner, even through clouds and atmospheric particles.

Satellites also monitor the amount of snow cover in winter, the extent of ice fields in the Arctic and Antarctic, the movement of large icebergs that drift into shipping lanes, and the height of the ocean's surface. One polar-orbiting satellite actually carries equipment that can detect faint distress signals anywhere on the globe, and relay them to rescue forces on the ground.

Infrared sensors on polar-orbiting satellites are able to assess conditions of crops, areas of deforestation, and regions of extensive drought. Satellites are also able to detect volcanic eruptions and follow the movement of ash clouds. During the winter, *GOES* satellites are able to monitor the southward progress of freezing air in

Florida and Texas, allowing forecasters to warn growers of impending low temperatures so that they can take necessary measures to protect sensitive crops.

The *Global Positioning System (GPS)* consists of 24 polar-orbiting satellites that transmit radio signals to ground receivers, which then use the signals for navigation and relative positioning on earth. The signal the satellites send to earth is slowed by the amount of water vapor in the air. Because of this effect, the GPS can estimate the atmosphere's *precipitable water vapor* (the total atmospheric water vapor contained in a vertical column of air).

Geostationary satellites, such as *GOES*, are equipped with systems that receive environmental information from remote data-collection platforms on the surface. These platforms include instrumented buoys, river gauges, automatic weather stations, seismic and tsunami ("tidal" wave) stations, and ships. This information is transmitted to the satellite, which relays it to a central receiving station at Wallop's Island, Virginia.

Normally, a network of five geostationary satellites positioned over the equator gives nearly complete global coverage from about latitude 60°N to 60°S. Japan's *Geostationary Meteorology Satellite (GMS)* monitors eastern Asia and the western Pacific. The United States' *GOES West*, positioned above the Pacific Ocean near 135°W longitude, observes the weather over western North America and much of the Pacific, whereas *GOES East* (another United States satellite), positioned over the central United States at about 75°W longitude, monitors North and South America and most of the Atlantic Ocean. The European

Space Agency's *METEOSAT* observes Europe, Africa, the Mediterranean, and the eastern Atlantic, whereas India's *INSAT* monitors the Indian Ocean and Asia.

Along with monitoring clouds and the atmosphere, the latest *GOES* series provides forecasters and researchers with data from Doppler radars and the network of automated surface-observing stations. The latest geostationary satellites detect pollution and haze, and provide accurate cloud-height measurements during the day. They even have the capacity to monitor the seasonal and daily trend in atmospheric ozone.

Satellites specifically designed to monitor the natural resources of the earth *(LandSat)* circle the earth every 103 minutes—14 times a day—in a near-polar circular orbit about 918 km (570 mi) above sea level. Each pass photographs a strip 185 km (115 mi) wide. These photographs, taken in several wavelength bands, provide valuable information about this planet's geology, hydrology, oceanography, and ecology. *LandSat* also collects data transmitted from remote ground stations in North America. These stations monitor a variety of environmental data, with water quality, rainfall amount, and snow depth of particular interest to the meteorologist and hydrologist.

Satellite information is not confined to the lower atmosphere. There are satellites that monitor the concentrations of ozone, air temperature, and winds in the upper atmosphere. And both geostationary and polar-orbiting satellites carry instruments that monitor solar activity. Even with all this information available, there are more sophisticated satellites on the drawing board that will provide more and improved data in the future.

FIGURE 6.35
An enhanced infrared image of the eastern Pacific taken on the same day as the images shown in Fig. 6.34(a) and (b).

WEATHER WATCH

Some of the tallest cumulonimbus clouds in the world, with tops sometimes higher than 18 km (59,000 ft), form during January and February over the tropical Pacific Ocean between northern Australia and Indonesia.

photographs. By examining the movement of these clouds on successive satellite photographs, forecasters can predict the arrival of clouds and storms, as well as the passage of weather fronts.

The shades of gray on enhanced infrared photos are often color-contoured to make specific features, such as deep cloud layers and the freezing level, more obvious. Usually, dark blue, red, or black is assigned to clouds with the coldest (highest) tops.

In regions where there are no clouds, it is difficult to observe the movement of the air. To help with this situation, the latest geostationary satellites are equipped with water-vapor sensors that can profile the distribution of atmospheric water vapor in the middle and upper troposphere (see Fig. 6.36). In time-lapse films, the swirling patterns of moisture clearly show wet regions and dry regions, as well as middle tropospheric swirling wind patterns and jet streams. (Additional information on satellites is provided in the Focus section on p. 154 entitled "Satellites Do More Than Observe Clouds.")

FIGURE 6.36
Infrared water vapor image. The darker areas represent dry air aloft, the brighter the gray, the more moist the air in the middle or upper troposphere. Bright white areas represent dense cirrus clouds or the tops of thunderstorms. The area in color represents the coldest cloud tops.

SUMMARY

In this chapter, we examined the different forms of condensation. We saw that dew forms when the air temperature cools to the dew point in a shallow layer of air near the surface. If the dew should freeze, it produces tiny beads of ice called frozen dew. Frost forms when the air cools to a dew point that is at 0°C or below.

As the air cools in a deeper layer near the surface, the relative humidity increases and water vapor begins to condense on hygroscopic condensation nuclei, forming wet haze. As the relative humidity approaches 100 percent, condensation occurs on most nuclei, and the air becomes filled with tiny liquid droplets (or ice crystals) called fog.

Fog forms in two primary ways: cooling of air and evaporating and mixing water into the air. Radiation fog, advection fog, and upslope fog form by the cooling of air, while steam fog and frontal fog are two forms of evaporation (mixing) fog. Although fog has some beneficial effects—providing winter chilling for fruit trees and water for thirsty redwoods—in many places it is a nuisance, for it disrupts air traffic and it is the primary cause of a number of auto accidents.

Condensation above the earth's surface produces clouds. When clouds are classified according to their height and physical appearance, they are divided into four main groups: high, middle, low, and clouds with vertical development. Since each cloud has physical characteristics that distinguish it from all the others, careful cloud observations normally lead to correct identification.

Satellites enable scientists to obtain a bird's-eye view of clouds on a global scale. Polar-orbiting satellites obtain data covering the earth from pole to pole, while geostationary satellites located above the equator continuously monitor a desired portion of the earth. Both types of satellites use radiometers (imagers) that detect emitted radiation. As a consequence, clouds can be observed both day and night.

Visible satellite pictures, which show sunlight reflected from a cloud's upper surface, can distinguish thick clouds from thin clouds. Infrared pictures show an image of the cloud's radiating top and can distinguish low clouds from high clouds. To increase the contrast between cloud features, infrared photographs are enhanced.

Satellites do a great deal more than simply photograph clouds. They provide us with a wealth of physical information about the earth and the atmosphere.

Key Terms

The following terms are listed in the order they appear in the text. Define each. Doing so will aid you in reviewing the material covered in this chapter.

dew	condensation nuclei
frozen dew	hygroscopic nuclei
frost	hydrophobic nuclei
haze	altostratus clouds
fog	nimbostratus clouds
acid fog	stratocumulus clouds
radiation (ground) fog	stratus clouds
advection fog	cumulus clouds
advection-radiation fog	cumulonimbus clouds
upslope fog	lenticular clouds
evaporation (mixing) fog	banner clouds
steam fog	pileus clouds
frontal fog	mammatus clouds
winter chilling	contrails
cirrus clouds	nacreous clouds
cirrocumulus clouds	noctilucent clouds
cirrostratus clouds	geostationary satellites
altocumulus clouds	polar-orbiting satellites

Questions for Review

1. Explain how dew, frozen dew, and visible frost each form.
2. Distinguish among dry haze, wet haze, and fog.
3. Why is fog that forms in industrial areas normally thick?
4. How can fog form when the air's relative humidity is less than 100 percent?
5. Name and describe the four types of fog. What conditions are necessary for the formation of radiation fog?
6. Why do ground fogs usually "burn off" by early afternoon?
7. List as many positive consequences of fog as you can.
8. List and describe three methods of fog dispersal.
9. What atmospheric conditions are necessary for the development of advection fog?
10. What is the name of the meteorological satellite that photographs the clouds shown on your local television news program?
11. Clouds are most generally classified by height. List the major height categories and the cloud types associated with each.
12. List at least two distinguishable characteristics of each of the ten basic clouds.
13. Why are high clouds normally thin? Why are they composed almost entirely of ice crystals?
14. How can you distinguish altostratus from cirrostratus?
15. Which clouds are associated with each of the following characteristics: (a) lightning; (b) heavy rain showers; (c) mackerel sky; (d) mares' tails; (e) halos; (f) light continuous rain or snow; (g) hailstones; (h) anvil top.
16. Why does a broken layer of clouds near the horizon often appear as overcast?
17. List and explain the various types of environmental information obtained from satellites.
18. How do geostationary satellites differ from polar orbiting satellites?
19. Explain why visible and infrared images can be used to dis-

tinguish: (a) high clouds from low clouds; (b) thick clouds from thin clouds.

20. Why are infrared images enhanced?
21. Name two clouds that form above the troposphere.

Questions for Thought

1. Explain the reasoning behind the wintertime expression, "Clear moon, frost soon."
2. Explain why icebergs are frequently surrounded by fog.
3. During a summer visit to New Orleans, you stay in an air-conditioned motel. One afternoon, you put on your sunglasses, step outside, and within no time your glasses are "fogged up." Explain what has apparently caused this.
4. While driving from cold air (well below freezing) into much warmer air (well above freezing), frost forms on the windshield of the car. Does the frost form on the inside or outside of the windshield? How can the frost form when the air is so warm?
5. Why are really clean atmospheres and really dirty atmospheres undesirable?
6. Why do relative humidities seldom reach 100 percent in polluted air?
7. Why are advection fogs rare over tropical water?
8. A January snowfall covers central Arkansas with 5 inches of snow. The following day, a south wind brings heavy fog to this region. Explain what has apparently happened.
9. If all fog droplets gradually settle earthward, explain how fog can last (without disappearing) for many days at a time.
10. Near the shore of an extremely large lake, explain why steam fog is more likely to form during the autumn and advection fog in early spring.
11. The air temperature during the night cools to the dew point in a deep layer, producing fog. Before the fog formed, the air temperature cooled each hour about 2°C. After the fog formed, the air temperature cooled by only 0.5°C each hour. Give *two* reasons why the air cooled more slowly after the fog formed.
12. On a winter night, the air temperature cooled to the dew point and fog formed. Before the formation of fog, the dew point remained almost constant. After the fog formed, the dew point began to decrease. Explain why.
13. Why can you see your breath on a cold morning? Does the air temperature have to be below freezing for this to occur?

14. Explain why altocumulus clouds might be observed at 6400 m (21,000 ft) above the surface in Mexico City, Mexico, but never at that altitude above Fairbanks, Alaska.
15. The sky is overcast and it is raining. Explain how you could tell if the cloud above you is a nimbostratus or a cumulonimbus.
16. Suppose it is raining lightly from a deck of nimbostratus clouds. Beneath the clouds are small, ragged, puffy clouds that are moving rapidly with the wind. What would you call these clouds? How did they probably form?

Problems and Exercises

1. The data in the chart below represent the dew-point temperature and expected minimum temperature near the ground for various clear winter mornings in a southeastern city. Assume that the dew point remains constant throughout the night. Answer the following questions about the data. (a) On which morning would there be the greatest likelihood of observing visible frost? Explain why. (b) On which morning would frozen dew most likely form? Explain why. (c) On which morning would there be black frost with no sign of visible frost, dew, or frozen dew? Explain. (d) On which morning would you probably only observe dew on the ground? Explain why.
2. If a ceiling balloon rises at 120 m (about 400 ft) each minute, what is the ceiling of an overcast deck of stratus clouds 1500 m (about 5000 ft) thick if the balloon disappears into the clouds in 5 minutes?
3. Compare the visible satellite picture (Fig. 6.34a) with the infrared picture (Fig. 6.34b). With the aid of the infrared photograph, label on the visible picture the regions of middle, high, and low clouds. On the enhanced infrared photograph (Fig. 6.35), label where the highest and thickest clouds appear to be located.

Questions for Exploration

1. On the Blue Skies: College Edition CD-ROM go to the **Sky Identification** section of the CD and click on "**Name That Cloud.**" Clouds will be presented in random order and you are challenged to identify the cloud type or cloud types for complex skies.

	MORNING 1	MORNING 2	MORNING 3	MORNING 4	MORNING 5
Dew-point temperature	2°C (35°F)	−7°C (20°F)	1°C (34°F)	−4°C (25°F)	3°C (38°F)
Expected minimum temperature	4°C (40°F)	−3°C (27°F)	0°C (32°F)	−4.5°C (24°F)	2°C (35°F)

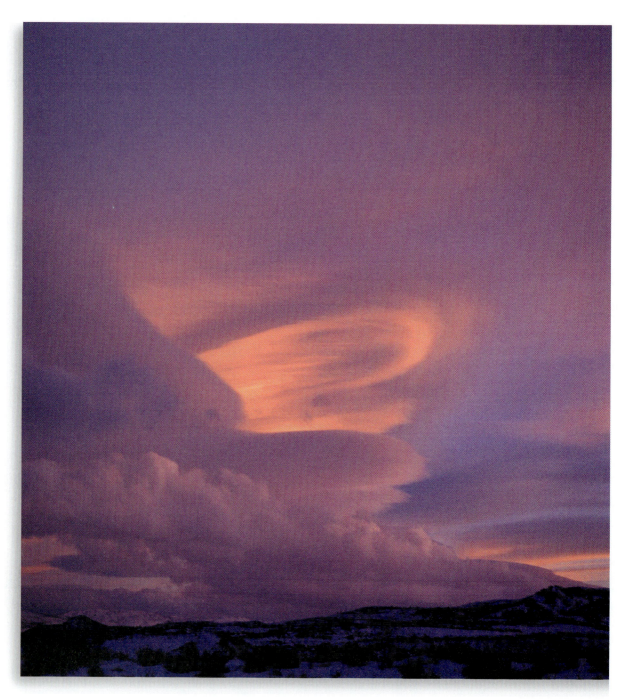

A mass of moist air gliding up and over the Sierra Nevada condenses into lenticular clouds on the leeward side. (Photo by author.)

STABILITY AND CLOUD DEVELOPMENT

CONTENTS

In July and August on the high desert the thunderstorms come. Mornings begin clear and dazzling bright, the sky as blue as the Virgin's cloak, unflawed by a trace of cloud in all that emptiness. . . . By noon, however, clouds begin to form over the mountains, coming it seems out of nowhere, out of nothing, a special creation.

The clouds multiply and merge, cumulonimbus piling up like whipped cream, like mashed potatoes, like sea foam, building upon one another into a second mountain range greater in magnitude than the terrestrial range below.

The massive forms jostle and grate, ions collide, and the sound of thunder is heard over the sun-drenched land. More clouds emerge from empty sky, anvil-headed giants with glints of lightning in their depths. An armada assembles and advances, floating on a plane of air that makes it appear, from below, as a fleet of ships must look to the fish in the sea.

Edward Abbey, *Desert Solitaire—A Season in the Wilderness*

Clouds, spectacular features in the sky, add beauty and color to the natural landscape. Yet, clouds are important for nonaesthetic reasons, too. As they form, vast quantities of heat are released into the atmosphere. Clouds help regulate the earth's energy balance by reflecting and scattering solar radiation and by absorbing the earth's infrared energy. And, of course, without clouds there would be no precipitation. But clouds are also significant because they visually indicate the physical processes taking place in the atmosphere; to a trained observer, they are signposts in the sky. This chapter examines the atmospheric processes these signposts point to, the first of which is atmospheric stability.

ATMOSPHERIC STABILITY

Most clouds form as air rises and cools. Why does air rise on some occasions and not on others? And why do the size and shape of clouds vary so much when the air does rise? Let's see how knowing about the air's stability will help us to answer these questions.

When we speak of atmospheric stability, we are referring to a condition of equilibrium. For example, rock A resting in the depression in Fig. 7.1 is in *stable* equilibrium. If the rock is pushed up along either side of the hill and then let go, it will quickly return to its original position. On the other hand, rock B, resting on the top of the hill, is in a state of *unstable* equilibrium, as a slight push will set it moving away from its original position. Applying these concepts to the atmosphere, we can see that air is in stable equilibrium when, after being lifted or lowered, it tends to return to its original position—it resists upward and downward air motions. Air that is in unstable equilibrium will, when given a little push, move farther away from its original position—it favors vertical air currents.

To explore the behavior of rising and sinking air, we must first put some air in an imaginary thin elastic wrap. This small volume of air is referred to as a **parcel of air**.* Although the air parcel can expand and contract freely, it does not break apart, but remains as a single unit. At the same time, neither external air nor heat can mix with the air inside the parcel. The space occupied by the air molecules within the parcel defines the air density. The average speed of the molecules is directly related to the air temperature, and the molecules colliding against the parcel walls determine the air pressure inside.

At the earth's surface, the parcel has the same temperature and pressure as the air surrounding it. Suppose we lift the air parcel up into the atmosphere. We know from Chapter 1 that air pressure decreases with height. Consequently, the air pressure surrounding the parcel lowers. The lower pressure outside allows the air molecules inside to push the parcel walls outward, expanding the parcel. Because there is no other energy source, the air molecules inside must use some of their own energy to expand the parcel. This shows up as slower average molecular speeds, which result in a lower parcel temperature. If the parcel is lowered to the surface, it returns to a region where the surrounding air pressure is higher. The higher pressure squeezes (compresses) the parcel back into its original (smaller) volume. This squeezing increases the average speed of the air molecules and the parcel temperature rises. Hence, *a rising parcel of air expands and cools, while a sinking parcel is compressed and warms.*

If a parcel of air expands and cools, or compresses and warms, with no interchange of heat with its surroundings, this situation is called an **adiabatic process**. As long as the air in the parcel is unsaturated (the relative humidity is less than 100 percent), the rate of adiabatic cooling or warming remains constant. This rate of heating or cooling is about 10°C for every 1000 m of change in elevation (5.5°F per 1000 ft) and applies only to unsaturated air. For this reason, it is called the **dry adiabatic rate**† (see Fig. 7.2).

As the rising air cools, its relative humidity increases as the air temperature approaches the dew-point temperature. If the air cools to its dew-point temperature, the relative humidity becomes 100 percent. Further lifting results in condensation, a cloud forms, and latent heat is released into the rising air. Because the heat added during condensation offsets some of the cooling due to expansion, the air no longer cools at the dry adiabatic rate but at a lesser rate called the **moist adiabatic rate**. If a saturated parcel containing water droplets were to sink, it would compress and warm at the moist adiabatic rate because evaporation of the liquid droplets would offset the rate of compressional warming. Hence, the rate at which rising or

FIGURE 7.1
When rock A is disturbed, it will return to its original position; rock B, however, will accelerate away from its original position.

---

*An air parcel is an imaginary body of air about the size of a large basketball. The concept of an air parcel is illustrated in Fig. 5.6, p. 109.

†For aviation purposes, the dry adiabatic rate is sometimes expressed as 3°C per 1000 ft.

sinking saturated air changes temperature—the moist adiabatic rate—is less than the dry adiabatic rate.*

Unlike the dry adiabatic rate, the moist adiabatic rate is not constant, but varies greatly with temperature and, hence, with moisture content—as warm saturated air produces more liquid water than cold saturated air. The added condensation in warm, saturated air liberates more latent heat. Consequently, the moist adiabatic rate is much less than the dry adiabatic rate when the rising air is warm; however, the two rates are nearly the same when the rising air is very cold (see Table 7.1). Although the moist adiabatic rate does vary, we will use an average of 6°C per 1000 m (3.3°F per 1000 ft) in most of our examples and calculations to make the numbers easy to deal with.

DETERMINING STABILITY

We determine the stability of the air by comparing the temperature of a rising parcel to that of its surroundings. If the rising air is colder than its environment, it will be more dense† (heavier) and tend to sink back to its original level. In this case, the air is *stable* because it resists upward displacement. If the rising air is warmer and, therefore, less dense (lighter) than the surrounding air, it will continue to rise until it reaches the same temperature as its environment. This is an example of *unstable* air. To figure out the air's stability, we need to measure the temperature both of the rising air and of its environment at various levels above the earth.

Stable Air   Suppose we release a balloon-borne instrument —a radiosonde (see Fig. 4, p. 13)—and it sends back temperature data as shown in Fig. 7.3. (Such a vertical profile of temperature is called a *sounding*.) We measure the air temperature in the vertical and find that it decreases by 4°C for every 1000 m (2°F per 1000 ft). Remember from Chapter 1 that the rate at which the air temperature changes with elevation is called the *lapse rate*. Because this is the rate at which the air temperature surrounding us will be changing if we were to climb upward into the atmosphere, we will refer to it as the **environmental lapse rate**. Now suppose in Fig. 7.3a that a parcel of unsaturated air with a temperature of 30°C is lifted

---

*Consider an air parcel initially at rest. Suppose the air parcel rises and cools, and a cloud forms. Further suppose that no precipitation (rain or snow) falls from the cloud (leaves the parcel). If the parcel should descend to its original level the latent heat released inside the parcel during condensation will be the same amount that is absorbed as the cloud evaporates. This process is called a *reversible adiabatic process*. If, on the other hand, rain or snow falls from the cloud during uplift and leaves the parcel, the sinking parcel will not recover during evaporation the same amount of latent heat released during condensation because the parcel's water content is lower. This process is known as an *irreversible pseudoadiabatic process*.

†When, at the same level in the atmosphere, we compare parcels of air that are equal in size but vary in temperature, we find that cold air parcels are more dense than warm air parcels; that is, in the cold parcel, there are more molecules that are crowded closer together.

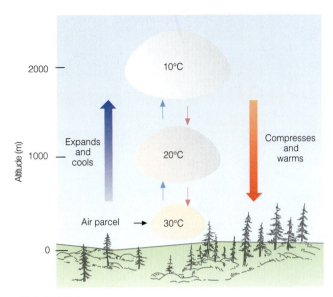

FIGURE 7.2
The dry adiabatic rate. As long as the air parcel remains unsaturated, it expands and cools by 10°C per 1000 m; the sinking parcel compresses and warms by 10°C per 1000 m.

TABLE 7.1   The Moist Adiabatic Rate for Different Temperatures and Pressures in °C/1000 m and °F/1000 ft

PRESSURE (MB)	TEMPERATURE (°C)					TEMPERATURE (°F)				
	-40	-20	0	20	40	-40	-5	30	65	100
1000	9.5	8.6	6.4	4.3	3.0	5.2	4.7	3.5	2.4	1.6
800	9.4	8.3	6.0	3.9		5.2	4.6	3.3	2.2	
600	9.3	7.9	5.4			5.1	4.4	3.0		
400	9.1	7.3				5.0	4.0			
200	8.6					4.7				

from the surface. As it rises, it cools at the dry adiabatic rate (10°C/ 1000 m), and the temperature inside the parcel at 1000 meters would be 20°C, or 6°C lower than the air surrounding it. Look at Fig. 7.3a closely and notice that, as the air parcel rises higher, the temperature difference between it and the surrounding air becomes even greater. Even if the parcel is initially saturated (see Fig. 7.3b), it will cool at the moist rate—6°C per 1000 m—and will be colder than its environment at all levels. In both cases, the rising air is colder and heavier than the air surrounding it. In this example, the atmosphere is **absolutely stable**. *The atmosphere is always absolutely stable when the environmental lapse rate is less than the moist adiabatic rate.*

Since absolutely stable air strongly resists upward vertical motion, it will, *if forced to rise*, tend to spread out horizontally.

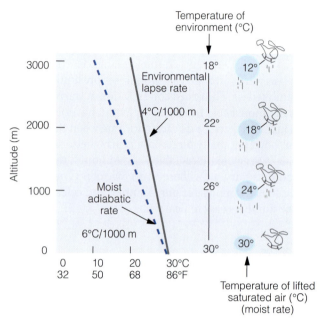

(a) Lifted, unsaturated air at each level is colder and heavier than the air around it. If given the chance, the parcel would return to its original position

(b) Lifted, saturated air at each position is colder and heavier than the air surrounding it. If released, the parcel would return to its original position.

FIGURE 7.3
An absolutely stable atmosphere occurs when the environmental lapse rate is less than the moist adiabatic rate.

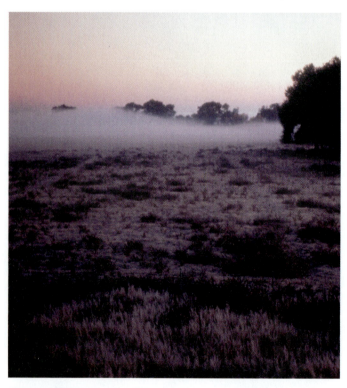

FIGURE 7.4
Cold surface air, on this morning, produces a stable atmosphere that inhibits vertical air motions and allows the fog and haze to linger close to the ground.

If clouds form in this rising air, they, too, will spread horizontally in relatively thin layers and usually have flat tops and bases. We might expect to see clouds—such as cirrostratus, altostratus, nimbostratus, or stratus —forming in stable air.

A Stable Atmosphere   What conditions are necessary to bring about a stable atmosphere? As we have just seen, the atmosphere is stable when the environmental lapse rate is small; that is, when the difference in temperature between the surface air and the air aloft is relatively small. This condition can occur when the air aloft warms, or the surface air cools. If the air aloft is being replaced by warmer air (warm advection), and the surface air is not changing appreciably, the environmental lapse rate decreases and the atmosphere becomes more stable. Similarly, the environmental lapse rate decreases and air becomes more stable when the lower layer cools. The cooling of the surface air may be due to:

1. nighttime radiational cooling of the surface
2. an influx of cold air brought in by the wind (cold advection)
3. air moving over a cold surface

Stable atmospheric conditions can cause a layer of fog or haze to persist close to the surface (see Fig. 7.4).

Another way the atmosphere becomes more stable is when an entire layer of air sinks. For example, if a layer of unsaturated air over 1000 m thick and covering a large area subsides, the entire layer will warm by adiabatic compression. As the layer subsides, it becomes compressed by the weight of the atmosphere and shrinks vertically. The upper part of the layer sinks farther, and, hence, warms more than the bottom part. This is illustrated in Fig. 7.5. After subsiding, the top of the layer is actually warmer than the bottom, and an inversion is formed. Inversions that form as air slowly sinks over a large area are called **subsidence inversions**. They sometimes occur at the surface, but more frequently, they are observed aloft and are often associated with large high-pressure areas because of the sinking air motions associated with these systems.

An inversion represents an atmosphere that is absolutely stable. Why? Within the inversion, warm air overlies cold air, and, if air rises into the inversion, it is becoming colder, while the air around it is getting warmer. Obviously, the colder air would tend to sink. Inversions, therefore, act as lids on vertical air motion. When an inversion exists near the ground, stratus, fog, haze, and pollutants are all kept close to the surface. In fact, as we will see in Chapter 17, most air pollution alerts occur with subsidence inversions.

Before we turn our attention to unstable air, let's first examine a condition known as **neutral stability**. If the lapse rate is exactly equal to the dry adiabatic rate, rising or sinking unsaturated air will cool or warm at the same rate as the air around it. At each level, it would have the same temperature and density as the surrounding air. Because this air tends neither to continue rising nor sinking, the atmosphere is said to be neutrally stable. For saturated air, *neutral stability* exists when the environmental lapse rate is equal to the moist adiabatic rate.

Unstable Air   Suppose a radiosonde sends back the temperatures above the earth as plotted in Fig. 7.6a. Once again, we determine the air's stability by comparing the environmental lapse rate to the moist and dry adiabatic rates. In this case, the environmental lapse rate is 11°C per 1000 m (6°F per 1000 ft). A rising parcel of unsaturated surface air will cool at the dry adiabatic rate. Because the dry adiabatic rate is less than the environmental lapse rate, the parcel will be warmer than the surrounding air and will continue to rise, constantly moving upward, away from its original position. The atmosphere is unstable. Of course, a parcel of saturated air cooling at the lower moist adiabatic rate will be even warmer than the air around it (see Fig. 7.6b). In both cases, the air parcels, once they start upward, will continue to rise on their own because they are warmer and less dense than the air around them. The atmosphere in this example is said to be **absolutely unstable**. *Absolute instability results when the environmental lapse rate is greater than the dry adiabatic rate.*

It should be noted, however, that deep layers in the atmosphere are seldom, if ever, absolutely unstable. Absolute instability is usually limited to a very shallow layer near the ground on hot, sunny days. Here the environmental lapse rate can exceed the dry adiabatic rate, and the lapse rate is called *superadiabatic*. On rare occasions when the environmental lapse rate exceeds about 3.4°C per 100 m (the *autoconvective lapse rate*), convection becomes spontaneous, resulting in the automatic overturning of the air.

So far, we have seen that the atmosphere is absolutely stable when the environmental lapse rate is less than the moist adiabatic rate and absolutely unstable when the environmental lapse rate is greater than the dry adiabatic rate. However, a typical type of instability exists when the lapse rate lies between the moist and dry adiabatic rates.

The environmental lapse rate in Fig. 7.7 is 7°C per 1000 m (4°F per 1000 ft). When a parcel of unsaturated air rises, it cools dry adiabatically and is colder at each level than the air around it (see Fig. 7.7a). It will, therefore, tend to sink back to its original level because it is in a stable atmosphere. Now, suppose the rising parcel is saturated. As we can see in Fig. 7.7b, the rising air is warmer than its environment at each level. Once the parcel is given a push upward, it will tend to move in that direction; the atmosphere is unstable for the saturated parcel. In this example, the atmosphere is said to be **conditionally unstable**. This type of stability depends upon whether or not the rising air is saturated. When the rising parcel of air is unsaturated, the atmosphere is stable; when the parcel of air is saturated, the atmosphere is unstable. Conditional instability means that, if unsaturated air could be lifted to a level where it becomes saturated, instability would result.

Conditional instability occurs whenever the environmental lapse rate is between the moist adiabatic rate and the dry

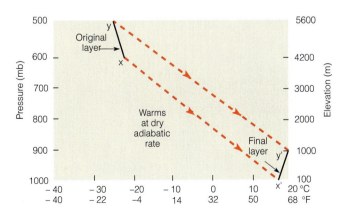

FIGURE 7.5

The layer x–y is initially 1400 m thick. If the entire layer slowly subsides, it shrinks in the more-dense air near the surface. As a result of the shrinking, the top of the layer warms more than the bottom, and the entire layer (x'–y') becomes more stable.

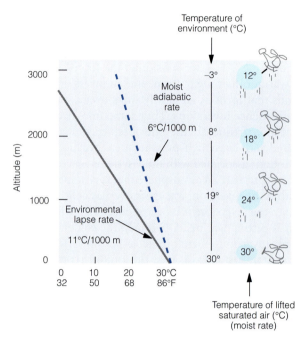

(a) The rising, unsaturated air parcel at each level is warmer and lighter than the air around it. If given the chance, the air parcel would accelerate away from its original position.

(b) The rising, saturated air parcel is warmer than its surroundings. If given the chance, it also would move away from its original position.

FIGURE 7.6

An absolutely unstable atmosphere occurs when the environmental lapse rate is greater than the dry adiabatic rate.

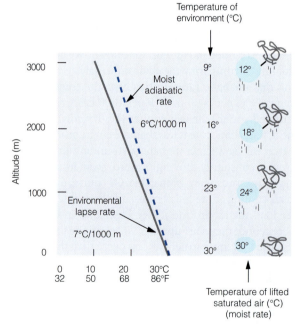

(a) The unsaturated parcel of air at each elevation is colder than its surroundings. The atmosphere is stable with respect to unsaturated, rising air.

(b) The lifted, saturated air parcel is warmer at each elevation than its surroundings. The atmosphere is unstable with respect to saturated, rising air.

FIGURE 7.7

Conditionally unstable air. The atmosphere is stable if the rising air is unsaturated (a), but unstable if the rising air is saturated (b).

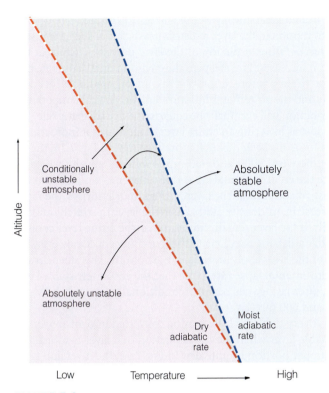

FIGURE 7.8
When the environmental lapse rate is greater than the dry adiabatic rate, the atmosphere is absolutely unstable. When the environmental lapse rate is less than the moist adiabatic rate, the atmosphere is absolutely stable. And when the environmental lapse rate lies between the dry adiabatic rate and the moist adiabatic rate (shaded gray area), the atmosphere is conditionally unstable.

adiabatic rate. Recall from Chapter 1 that the average lapse rate in the troposphere is about 6.5°C per 1000 m (3.6°F per 1000 ft). Since this value lies between the dry adiabatic rate and the average moist rate, *the atmosphere is ordinarily in a state of conditional instability.* (Figure 7.8 summarizes the concept of unstable, conditionally unstable, and stable atmospheres.)

As the atmosphere becomes more unstable, vertical motions increase, and thicker cumuliform clouds begin to form. Two questions arise: (1) What mechanisms cause the atmosphere to become more unstable? (2) What causes air to rise so that clouds are able to form? (Unstable air normally needs a "trigger" to start it moving upward.)

Causes of Instability   The atmosphere becomes more unstable as the environmental lapse rate steepens; that is, as the air temperature drops rapidly with increasing height. This circumstance may be brought on by either air aloft becoming colder or the surface air becoming warmer.

The cooling of the air aloft may be due to:

1. winds bringing in colder air (cold advection)
2. clouds (or the air) emitting infrared radiation to space (radiational cooling)

WEATHER WATCH

Nature's own fire extinguisher: Forest fires generate atmospheric instability by heating the air near the surface. The hot, rising air above the fire contains tons of tiny smoke particles that act as cloud condensation nuclei. As the air rises and cools, water vapor in the atmosphere as well as water vapor released during the burning of the timber, will often condense onto the nuclei, producing a cumuliform cloud. If the cloud builds high enough, and remains over the fire area, its heavy showers may actually help to extinguish the fire.

The warming of the surface air may be due to:

1. daytime solar heating of the surface
2. an influx of warm air brought in by the wind (warm advection)
3. air moving over a warm surface

The combination of cold air aloft and warm surface air can produce a steep lapse rate and an unstable atmosphere (see Fig. 7.9).

FIGURE 7.9
Unstable air. The warmth from the forest fire heats the air, causing instability near the surface. Warm, less-dense air (and smoke) bubbles upward, expanding and cooling as it rises. Eventually the rising air cools to its dew point, condensation begins, and a cumulus cloud forms.

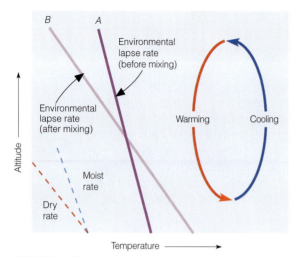

FIGURE 7.10
Mixing tends to steepen the lapse rate. Rising, cooling air lowers the temperature toward the top of the layer, while sinking, warming air increases the temperature near the bottom.

At this point, it should be apparent that the stability of the atmosphere changes during the course of a day. In clear, calm weather around sunrise, surface air is normally colder than the air above it, a radiation inversion exists, and the atmosphere is quite stable as indicated by smoke or haze lingering close to the ground. As the day progresses, sunlight warms the surface and the surface warms the air above. As the air temperature near the ground increases, the lower atmosphere gradually becomes unstable, with maximum instability usually occurring during the hottest part of the day.

Up to now, we have seen that a layer of air may become unstable by either cooling the air aloft or warming the air at the surface. A layer of air may also be made more unstable by either mixing or lifting. Let's look at mixing first. In Fig. 7.10, the environmental lapse rate before mixing is less than the moist rate, and the layer is stable (*A*). Now, suppose the air in the layer is mixed either by convection or by wind-induced turbulent eddies. Air is cooled adiabatically as it is brought up from below

and heated adiabatically as it is mixed downward. The up and down motion in the layer redistributes the air in such a way that the temperature at the top of the layer decreases, while, at the base, it increases. This steepens the environmental lapse rate and makes the layer more unstable. If this mixing continues for some time, and the air remains unsaturated, the vertical temperature distribution will eventually be equal to the dry adiabatic rate (*B*).

Just as lowering an entire layer of air makes it more stable, the lifting of a layer makes it more unstable. In Fig. 7.11, the air lying between 1000 and 900 mb is initially absolutely stable since the environmental lapse rate of layer *x–y* is less than the moist adiabatic rate. The layer is lifted, and, as it rises, the rapid decrease in air density aloft causes the layer to stretch out vertically. If the layer remains unsaturated, the entire layer cools at the dry adiabatic rate. Due to the stretching effect, however, the top of the layer cools more than the bottom. This steepens the environmental lapse rate. Note that the absolutely stable layer *x–y*, after rising, has become conditionally unstable between 500 and 600 mb (layer *x'–y'*).

A very stable air layer may be converted into an absolutely unstable layer when the lower portion of a layer is moist and the upper portion is quite dry. In Fig. 7.12, the inversion layer between 900 and 850 mb is absolutely stable. Suppose the bottom of the layer is saturated while the air at the top is unsaturated. If the layer is forced to rise, even a little, the upper portion of the layer cools at the dry adiabatic rate and grows cold quite rapidly, while the air near the bottom cools more slowly at the moist adiabatic rate. It does not take much lifting before the upper part of the layer is much colder than the bottom part; the environmental lapse rate steepens and the entire layer becomes absolutely unstable (layer *a'–b'*). The potential instability, brought about by the lifting of a stable layer whose surface is humid and whose top is "dry," is called **convective instability**. Convective instability is associated with the development of severe storms, such as thunderstorms and tornadoes, which are investigated more thoroughly in Chapter 15.

BRIEF REVIEW   Up to now, we have looked briefly at stability as it relates to cloud development. The next section describes how atmospheric stability influences the physical mechanisms responsible for the development of individual cloud types. However, before going on, here is a brief review of some of the facts and concepts concerning stability.

- The air temperature in a rising parcel of *unsaturated* air decreases at the dry adiabatic rate, whereas the air temperature in a rising parcel of *saturated* air decreases at the moist adiabatic rate.

- The dry adiabatic rate and moist adiabatic rate of cooling are different due to the fact that latent heat is released in a rising parcel of saturated air.

- In a stable atmosphere, a lifted parcel of air will be colder (heavier) than the air surrounding it. Because of this fact, the lifted parcel will tend to sink back to its original position.

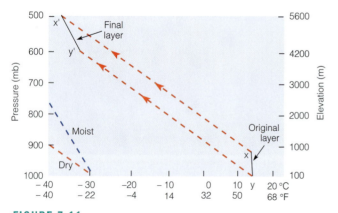

FIGURE 7.11
The lifting of an entire layer of air *(x–y)* tends to increase the instability of the layer.

- In an unstable atmosphere, a lifted parcel of air will be warmer (lighter) than the air surrounding it, and thus will continue to rise upward, away from its original position.

- The atmosphere becomes more stable as the surface air cools, the air aloft warms, or a layer of air sinks (subsides) over a vast area.

- The atmosphere becomes more unstable as the surface air warms, the air aloft cools, or a layer of air is either mixed or lifted.

- Layered clouds tend to form in a stable atmosphere, whereas cumuliform clouds tend to form in a conditionally unstable atmosphere.

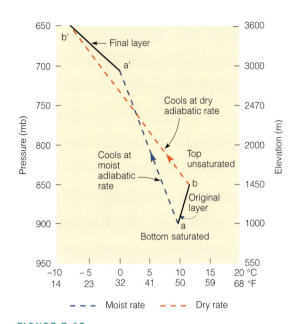

FIGURE 7.12
Convective instability. The layer a–b is initially absolutely stable. The lower part of the layer is saturated, and the upper part is "dry." After lifting, the entire layer (a'–b') becomes absolutely unstable.

CLOUD DEVELOPMENT

Most clouds form as air rises, expands, and cools. Basically, the following mechanisms are responsible for the development of the majority of clouds we observe:

1. surface heating and free convection
2. topography
3. widespread ascent due to convergence of surface air
4. uplift along weather fronts (see Fig. 7.13).

Convection and Clouds   Some areas of the earth's surface are better absorbers of sunlight than others and, therefore, heat up more quickly. The air in contact with these "hot spots" be-

comes warmer than its surroundings. A hot "bubble" of air—a *thermal*—breaks away from the warm surface and rises, expanding and cooling as it ascends. As the thermal rises, it mixes with the cooler, drier air around it and gradually loses its identity. Its upward movement now slows. Frequently, before it is

FIGURE 7.13
The primary ways clouds form: (a) surface heating and convection; (b) forced lifting along topographic barriers; (c) convergence of surface air; (d) forced lifting along weather fronts.

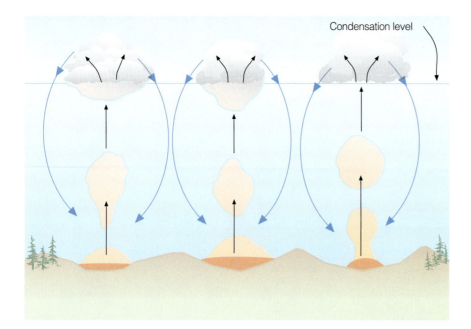

FIGURE 7.14
Cumulus clouds form as hot, invisible air bubbles detach themselves from the surface, then rise and cool to the condensation level. Below and within the cumulus clouds, the air is rising. Around the cloud, the air is sinking.

completely diluted, subsequent rising thermals penetrate it and help the air rise a little higher. If the rising air cools to its saturation point, the moisture will condense, and the thermal becomes visible to us as a cumulus cloud.

Observe in Fig. 7.14 that the air motions are downward on the outside of the cumulus cloud. The downward motions are caused in part by evaporation around the outer edge of the cloud, which cools the air, making it heavy. Another reason for the downward motion is the completion of the convection current started by the thermal. Cool air slowly descends to replace the rising warm air. Therefore, we have rising air in the cloud and sinking air around it. Since subsiding air greatly inhibits the growth of thermals beneath it, small cumulus clouds usually have a great deal of blue sky between them (see Fig. 7.15).

As the cumulus clouds grow, they shade the ground from the sun. This, of course, cuts off surface heating and upward convection. Without the continual supply of rising air, the cloud begins to erode as its droplets evaporate. Unlike the sharp outline of a growing cumulus, the cloud now has indistinct edges, with cloud fragments extending from its sides. As the cloud dissipates (or moves along with the wind), surface heating begins again and regenerates another thermal, which becomes a new cumulus. This is why you often see cumulus clouds form, gradually disappear, then reform in the same spot.

FIGURE 7.15
Cumulus clouds building on a warm summer afternoon. Each cloud represents a region where thermals are rising from the surface. The clear areas between the clouds are regions where the air is sinking.

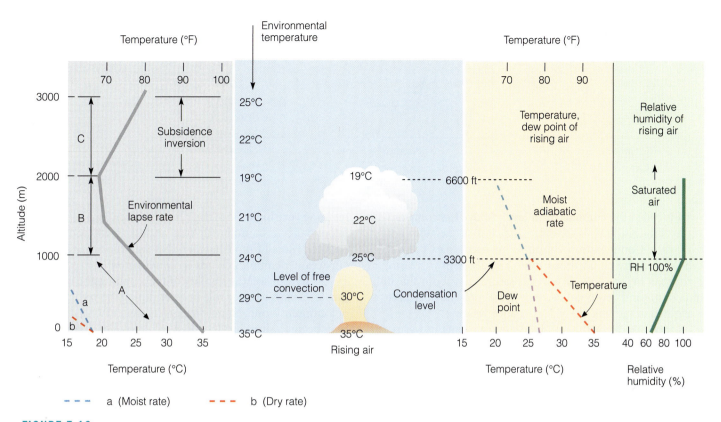

FIGURE 7.16
The development of a cumulus cloud.

Suppose that it is a warm, humid summer afternoon and the sky is full of cumulus clouds. The cloud bases are all at nearly the same level above the ground and the cloud tops extend only about a thousand meters higher. The development of these clouds depends primarily upon the air's stability and moisture content. To illustrate how these factors influence the formation of a convective cloud, we will examine the temperature and moisture characteristics within a rising bubble of air. Since the actual air motions that go into forming a cloud are rather complex, we will simplify matters by making these assumptions:

1. No mixing takes place between the rising air and its surroundings.
2. Only a single thermal produces the cumulus cloud.
3. The cloud forms when the relative humidity is 100 percent.
4. The rising air in the cloud remains saturated.

The environmental lapse rate on this particular day is plotted in Fig. 7.16 and is represented as a dark gray line on the far left of the illustration. The changing environmental lapse rate indicates changes in the air's stability. The environmental lapse rate in layer A is greater than the dry adiabatic rate, so the layer is absolutely unstable. The air layers above it—layer B and layer C—are both absolutely stable since the environmental lapse rate in each layer is less than the moist adiabatic rate. However, the overall environmental lapse rate from the surface up to the base of the inversion (2000 m) is 7.5°C per 1000 m (4.1°F per 1000 m), which indicates a conditionally unstable atmosphere.

Now, suppose that a warm bubble of air with an air temperature and dew-point temperature of 35°C and 27°C (95°F and 80.5°F), respectively, breaks away from the surface and begins to rise (which is illustrated in the middle of Fig. 7.16). Notice that, a short distance above the ground, the air inside the bubble is warmer than the air around it, so it is buoyant and rises freely. This level in the atmosphere where the rising air becomes warmer than the surrounding air is called the *level of free convection*. The rising bubble will continue to rise as long as it is warmer than the air surrounding it.

The rising air cools at the dry adiabatic rate and the dew point falls, but not as rapidly.* The rate at which the dew point drops varies with the moisture content of the rising air, but an approximation of 2°C per 1000 m (1°F per 1000 ft) is commonly used. So, as unsaturated rising air cools, the air temperature and dew point approach each other at the rate of 8°C per 1000 m (4.5°F per 1000 ft). This process causes an increase in the air's relative humidity (illustrated in the far right-hand side of Fig. 7.16, by the dark green line).

At an elevation of 1000 m (3300 ft), the air has cooled to the dew point, the relative humidity is 100 percent, condensation begins, and a cloud forms. The elevation where the cloud

---

*The decrease in dew-point temperature is caused by the rapid decrease in air pressure within the rising air. Since the dew point is directly related to the actual vapor pressure of the rising air, a decrease in total air pressure causes a corresponding decrease in vapor pressure and, hence, a lowering of the dew-point temperature.

(a) Cumulus humilis   (b) Cumulus congestus   (c) Cumulonimbus

FIGURE 7.17
The air's stability greatly influences the growth of cumulus clouds.

forms is called the **condensation level**. Above the condensation level the rising air is saturated and cools at the moist adiabatic rate. Condensation continues to occur, and the dew point within the cloud drops more rapidly with increasing height than before. The air remains saturated as both the air temperature and dew point decrease at the moist adiabatic rate (illustrated in the area of Fig. 7.16 shaded yellow).*

Notice that inside the cloud the rising air remains warmer than the environment and continues its spontaneous rise upward through layer B. The top of the bulging cloud at 2000 m (about 6600 ft) represents the top of the rising air, which has now cooled to a temperature equal to its surroundings. The air would have a difficult time rising much above this level because of the stable subsidence inversion directly above it. The subsidence inversion, associated with the downward air motions of a high-pressure system, prevents the clouds from

---

*A more sudden decrease in the dew-point temperature occurs above the condensation level because the water vapor content of the air lowers as liquid cloud droplets form.

building very high above their bases. Hence, an afternoon sky full of flat-base cumuli with little vertical growth indicates fair weather. (Recall from Chapter 6 that the proper name of these fair-weather cumulus clouds is *cumulus humilis*.)

As we can see, the stability of the air above the condensation level plays a major role in determining the vertical growth of a cumulus cloud. Notice in Fig. 7.17 that, when a deep stable layer begins a short distance above the cloud base, only cumulus humilis are able to form. If a deep conditionally unstable layer exists above the cloud base, cumulus congestus are likely to grow, with billowing cauliflowerlike tops. When the conditionally unstable layer is extremely deep—usually greater than 4 km (2.5 mi)—the cumulus congestus may even develop into a cumulonimbus.

Seldom do cumulonimbus clouds extend very far above the tropopause. The stratosphere is quite stable, so once a cloud penetrates the tropopause, it usually stops growing vertically and spreads horizontally. The low temperature at this altitude produces ice crystals in the upper section of the cloud. In the middle latitudes, high winds near the tropopause blow

FIGURE 7.18
Cumulus clouds developing into thunderstorms in unstable air over the Great Plains. Notice that, in the distance, the cumulonimbus with the anvil top has reached the stable part of the atmosphere.

the ice crystals horizontally, producing the flat anvil-shaped top so characteristic of cumulonimbus clouds (see Fig. 7.18).

The vertical development of a convective cloud also depends upon the mixing that takes place around its periphery. The rising, churning cloud mixes cooler air into it. Such mixing is called **entrainment**. If the environment around the cloud is very dry, the cloud droplets quickly evaporate. The effect of entrainment, then, is to increase the rate at which the rising air cools by the injection of cooler air into the cloud and the subsequent evaporation of the cloud droplets. If the rate of cooling approaches the dry adiabatic rate, the air stops rising and the cloud no longer builds, even though the lapse rate may indicate a conditionally unstable atmosphere.

Up to now, we have looked at convection over land. Convection and the development of cumulus clouds also occur over large bodies of water. As cool air flows over a body of relatively warm water, the lowest layer of the atmosphere becomes warm and moist. This induces instability—convection begins and cumulus clouds form. If the air moves over progressively warmer water, as is sometimes the case over the open ocean, more active convection occurs and a cumulus cloud can build into cumulus congestus and finally into cumulonimbus. This sequence of cloud development is observed from satellites as cold northerly winds move southward over the northern portions of the Atlantic and Pacific oceans.

Once a convective cloud forms, stability, humidity, and entrainment all play a part in its vertical development. The level at which the cloud initially forms, however, is determined primarily by the temperature and moisture content of the original thermals. (The Focus section on p. 172 uses this information and a simple formula to determine the bases of convective clouds.)

Topography and Clouds   Horizontally moving air obviously cannot go through a large obstacle, such as a mountain, so the air must go over it. Forced lifting along a topographic barrier is called **orographic uplift**. Often, large masses of air rise when they approach a long chain of mountains like the Sierra Nevada or Rockies. This lifting produces cooling, and, if the air is humid, clouds form. Clouds produced in this manner are called **orographic clouds**. The type of cloud that forms will depend on the air's stability and moisture content. On the leeward side of the mountain, as the air moves downhill, it warms. This sinking air is now drier, since much of its moisture was removed in the form of clouds and precipitation on the windward side. This region on the leeward side, where precipitation is noticeably less, is called a **rain shadow**.

An example of orographic uplift and cloud development is given in Fig. 7.19. Before rising up and over the barrier, the air at the base of the mountain (0 m) on the windward side has an air temperature of 20°C (68°F) and a dew-point temperature of 12°C (54°F). Notice that the atmosphere is conditionally unstable, as indicated by the environmental lapse rate of 8°C per 1000 m. (Remember from our earlier discussion that the atmosphere is conditionally unstable when the environmental lapse rate falls between the dry adiabatic rate and the moist adiabatic rate.)

As the unsaturated air rises, the air temperature decreases at the dry adiabatic rate (10°C per 1000 m) and the dew-point temperature decreases at 2°C per 1000 m. Notice that the rising, cooling air reaches its dew point and becomes saturated at 1000 m. This level (called the **lifting condensation level**, or **LCL**) marks the base of the cloud that has formed as air is lifted (in this case by the mountain). As the rising saturated air con-

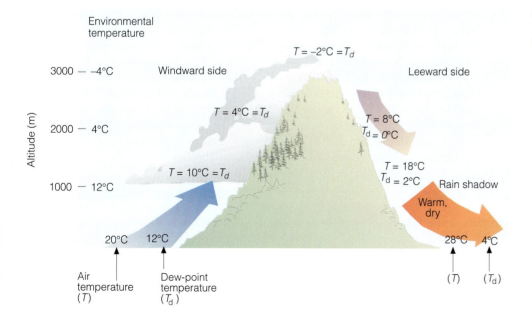

FIGURE 7.19
Orographic uplift, cloud development, and the formation of a rain shadow.

Determining Convective Cloud Bases

The bases of cumulus clouds that form by convection can be estimated quite easily when the surface air temperature and dew point are known. If the air is not too windy, we can assume that entrainment of air will not change the characteristics of a rising thermal. Since the rising air cools at the dry adiabatic rate of about 10°C per 1000 m, and the dew point drops at about 2°C per 1000 m, the air temperature and dew point approach each other at the rate of 8°C for every 1000 m of rise. Rising surface air with an air temperature and dew point spread of 8°C would produce saturation and a cloud at an elevation of 1000 m. Another way to look at it is that a 1°C difference between the surface air temperature and the dew point produces a cloud base at 125 m. Therefore, by finding the difference between surface air temperature (T) and dew point (T_d), and multiplying this value by 125, we can estimate the base of the convective cloud forming overhead, as

$$H_{meter} = 125 (T - T_d), \qquad (1)$$

where H is the height of the base of the cloud in meters above the surface, with both T and T_d measured in degrees Celsius. If T and T_d are in °F, H can be calculated with the formula

$$H_{feet} = 222 (T - T_d). \qquad (2)$$

To illustrate the use of formula (1), let's determine the base of the cumulus cloud in Fig. 7.16. Recall that the surface air temperature and dew point were 35°C and 27°C, respectively. The difference, $T - T_d$, is 8°C. This value multi-

	Elko, Nevada	Denver, Colorado	Hays, Kansas	Baltimore, Maryland
Surface temperature (T)	32°C (90°F)	32°C (90°F)	32°C (90°F)	32°C (90°F)
Dew point (T_d)	10°C (50°F)	15°C (59°F)	20°C (68°F)	24°C (75°F)

FIGURE 1

During the summer, cumulus cloud bases typically increase in elevation above the ground as one moves westward into the drier air of the Central Plains.

plied by 125 gives us a cumulus cloud with a base at 1000 m above the ground. This agrees with the condensation level we originally calculated.

Along the East Coast in summer, the air is warm and muggy, and the separation between air temperature and dew point may be smaller than 9°C (16°F). The bases of afternoon cumulus clouds over cities, such as Philadelphia and Baltimore, are typically about 1000 m (3300 ft) above the ground. Further west, in the Central Plains, where the air is drier and the spread between surface air temperature and dew point is greater, the cloud bases are higher. For example, west of Salina, Kansas, the

cumulus cloud bases are generally greater than 1500 m (about 5000 ft) above the surface. On a summer afternoon in central Nevada, it is not uncommon to observe cumulus forming at 2400 m (about 8000 ft). In the Central Valley of California, where the summer afternoon spread between air temperature and dew point usually exceeds 22°C (40°F), the air must rise to almost 2700 m (about 9000 ft) before a cloud forms. Due to sinking air aloft, thermals in this area are unable to rise to that elevation, and afternoon cumulus clouds are seldom observed forming overhead.

denses into many billions of liquid cloud droplets, and as latent heat is liberated by the condensing vapor, both the air temperature and dew-point temperature decrease at the moist adiabatic rate.

At the top of the mountain, the air temperature and dew point are both −2°C. Note in Fig. 7.19 that this temperature (−2°C) is higher than that of the surrounding air (−4°C). Conse-

quently, the rising air at this level is not only warmer, but unstable with respect to its surroundings. Therefore, the rising air should continue to rise and build into a much larger cumuliform cloud.

Suppose, however, that the air at the top of the mountain (temperature and dew point of −2°C) is forced to descend to the base of the mountain (0 m) on the leeward side. If we assume that the cloud remains on the windward side and does not extend beyond the mountain top, the temperature of the sinking air will increase at the dry adiabatic rate (10°C per 1000 m) all the way down to the base of the mountain. (The dew-point temperature increases at a much lower rate of 2°C per 1000 m.)

We can see in Fig. 7.19 that on the leeward side, after descending 3000 m, the air temperature is 28°C (82°F) and the dew-point temperature is 4°C (39°F). The air is now 8°C (14°F) warmer than it was before being lifted over the barrier. The higher air temperature on the leeward side is the result of latent heat being converted into sensible heat during condensation on the windward side. (In fact, the rising air at the *top* of the mountain is considerably warmer than it would have been had condensation not occurred.) The lower dew-point temperature and, hence, drier air on the leeward side are the result of water vapor condensing and then remaining as liquid cloud droplets and precipitation on the windward side. (A graphic representation of the preceding example is given in the Focus section on adiabatic charts, pp. 174–175.)

Although clouds are more prevalent on the windward side of mountains, they may, under certain atmospheric conditions, form on the leeward side as well. For example, stable air flowing over a mountain often moves in a series of waves that may extend for several hundred kilometers on the leeward side. These waves resemble the waves that form in a river downstream from a large boulder. Recall from Chapter 6 that wave clouds often have a characteristic lens shape and are called *lenticular clouds*. Lenticular clouds that form in the wave directly over the mountain are called *mountain wave clouds* (see Fig. 7.20), and those that form downwind of the mountain are called *lee wave clouds*.

The formation of lenticular clouds is shown in Fig. 7.21. As moist air rises on the upwind side of the wave, it cools and con-

WEATHER WATCH

The first modern "sighting" of a flying saucer was made over Mt. Rainer, Washington, a region where lenticular clouds commonly form.

denses, producing a cloud. On the downwind side of the wave, the air sinks and warms; the cloud evaporates. Viewed from the ground, the clouds appear motionless as the air rushes through them; hence, they are often referred to as *standing wave clouds*. Since they most frequently form at altitudes where middle clouds form, they are called *altocumulus standing lenticulars*.

When the air between the cloud-forming layers is too dry to produce clouds, lenticular clouds will form one above the other. Actually, when a strong wind blows almost perpendicular to a high mountain range, mountain waves may extend into the stratosphere, producing a spectacular display, sometimes resembling a fleet of hovering spacecraft (see Fig. 6.22, p. 145).

Notice in Fig. 7.21 that beneath the lee wave cloud, a large swirling eddy forms. The rising part of the eddy may cool enough to produce **rotor clouds**. The air in the rotor is extremely turbulent and presents a major hazard to aircraft in the vicinity. Dangerous flying conditions also exist near the leeside of the mountain, where strong downwind air motions are present. (These types of winds will be treated in more detail in Chapter 10.)

Widespread Ascent and Clouds   Just as mountains force air to rise, the convergence—flowing together—of air in the lower troposphere will initiate lifting and possibly the forma-

(Text continues on page 176.)

FIGURE 7.20
Lenticular clouds (mountain wave clouds) forming over Mt. Shasta in Northern California.

Adiabatic Charts

The adiabatic chart is a valuable tool for anyone who studies the atmosphere. The chart itself is a graph that shows how various atmospheric elements change with altitude (see Fig. 6). At first glance, the chart appears complicated because of its many lines. We will, therefore, construct these lines on the chart step by step.

Figure 2 shows horizontal lines of pressure decreasing with altitude, and vertical lines of temperature in °C increasing toward the right. The height values on the far right are approximate elevations that have been computed assuming that the air temperature decreases at a standard rate of 6.5°C per kilometer.

In Fig. 3, the slanted solid red lines are called **dry adiabats**. They show how the air temperature would change inside a rising or descending *unsaturated* air parcel. Suppose, for example, that an unsaturated air parcel at the surface (pressure 1013 mb) with a temperature of 10°C rises and cools at the dry adiabatic rate (10°C per km). What would be the parcel temperature at a pressure of 900 mb? To find out, simply follow the dry adiabat from the surface temperature of 10°C up to where it crosses the 900-mb line. Answer: about 0°C. If the same parcel returns to the surface, follow the dry adiabat back to the surface and read the temperature, 10°C.

On some charts, the dry adiabats are expressed as a potential temperature in Kelvins. The **potential temperature** is the temperature an air parcel would have if it were moved dry adiabatically to a pressure of 1000 mb. Moving parcels to the same level allows them to be observed under identical conditions. Thus, it can be determined which parcels are potentially warmer than others.

The sloping dashed blue lines in Fig. 4 are called **moist adiabats**. They show how the air temperature would change inside a rising or descending parcel of *saturated* air. In other words, they represent the moist adiabatic rate for a rising or sinking saturated air parcel, such as in a cloud.

The sloping gray lines in Fig. 5 are lines of constant *mixing ratio*. At any given temperature and pressure, they show how much water vapor the air could hold if it were saturated—the *saturation mixing ratio* (w_s) in grams of water vapor per kilogram of dry air (g/kg). At a given dew-point temperature, they show how much water vapor the air is actually holding—the *actual mixing ratio (w)* in g/kg. Hence, given the air temperature and dew-point temperature at some level, we can compute the relative humidity of the air.* For example, suppose at the surface (pressure 1013 mb) the air temperature and dew-point temperature are 29°C and 15°C, respectively. In Fig. 5, observe that at 29°C the saturation mixing ratio (w_s) is 26 g/kg, and with a dew-point temperature of 15°C, the actual mixing ratio *(w)* is 11 g/kg. This produces a relative humidity of $^{11}/_{26} \times 100$ percent, or 42 percent.

The mixing ratio lines also show how the dew-point temperature changes in a rising or sinking unsaturated air parcel. If an unsaturated air parcel with a dew point of 15°C rises from the surface (pressure 1013 mb) up to where the pressure is 700 mb (approximately 3 km), notice in Fig. 5 that the dew-point temperature inside the parcel would have dropped to a temperature near 10°C.

---

*The relative humidity (RH) of the air can be expressed as: $RH = w/w_s \times 100\%$.

FIGURE 2

FIGURE 3

FIGURE 4

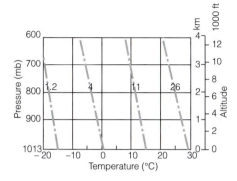

FIGURE 5

Figure 6 shows all of the lines described thus far on a single chart. We have already seen that the chart can be used to obtain graphically a number of atmospheric mathematical relationships. Therefore, let's use the chart to obtain information on air that rises up and over a mountain range.

Suppose we use the example given in Fig. 7.19 on p. 171. Air at an elevation of 0 m (pressure 1013 mb), with a temperature of 20°C (T_1) and a dew-point temperature of 12°C (D_1), first ascends then descends a 3000-meter-high mountain range. Look at Fig. 6 closely and observe that the surface air with a temperature of 20°C indicates a saturation mixing ratio of about 15 g/kg, and at 12°C the dew-point temperature indicates an actual mixing ratio of about 9 g/kg. Hence, the relative humidity of the air before rising over the mountain is $\frac{9}{15}$, or 60 percent.

Now, as the unsaturated air rises (as indicated by arrows in Fig. 6), the air temperature follows a dry adiabat (solid red line), and the dew-point temperature follows a line of constant mixing ratio (gray line). Carefully follow the mixing ratio line in Fig. 6 from 12°C up to where it intersects the dry adiabat that slopes upward from 20°C. Notice that the intersection occurs at an elevation near 1 km. This, of course, marks the base of the cloud—the *lifting condensation level (LCL)*—where the relative humidity is 100 percent and condensation begins. Above this level, the rising air is saturated. Consequently, the air temperature and dew-point temperature together follow a moist adiabat (dashed blue line) to the top of the mountain.

Notice in Fig. 6 that, at the top of the mountain (at 3 km or about 700 mb), both the air temperature and dew point are –2°C. If we assume that the cloud stays on the windward side, then from 3 km (700 mb) the descending air follows a dry adiabat all the way to the surface (1013 mb). Notice that, after descending, the air has a temperature of 28°C (T_2). From the mountaintop, the dew-point temperature follows a line of mixing ratio and reaches the surface (1013 mb) with a temperature of 4°C (D_2). Observe in Fig. 6 that, with an air temperature of 28°C, the saturation mixing ratio is about 25 g/kg and, with a dew point of 4°C, the actual mixing ratio is about 5 g/kg. Thus, the relative humidity of the air after descending is about $\frac{5}{25}$, or 20 percent. A more complete adiabatic chart is provided in Appendix J.

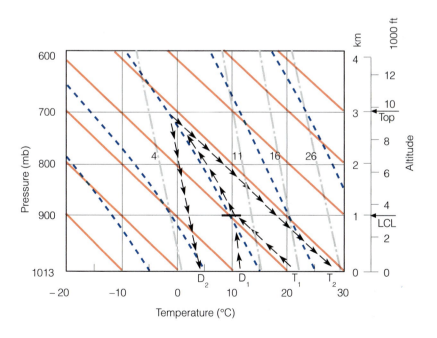

FIGURE 6
The adiabatic chart. The arrows in the chart illustrate the example given in the text.

FIGURE 7.21
The formation of lenticular clouds.

tion of clouds (see Fig. 7.13, p. 167). A cyclonic storm system is the most common cause of the widespread ascent of air. Associated with this kind of system are weather fronts that separate air with contrasting temperatures and densities. For example, along a warm front, warm air gently rides up and over colder surface air, producing clouds that may cover an area of many hundreds, and even thousands, of square kilometers. The clouds that form are usually layered, cirrus, cirrostratus, altostratus, or nimbostratus. Along a cold front, where usually warm moist air is forced upward, lines of cumuliform clouds often form. Even in the absence of fronts, clouds may form in regions of surface low pressure. Recall that surface air flows in toward the center of a low. This air, converging from

FIGURE 7.22
Satellite view of cloud streets, rows of stratocumulus clouds forming over the warm Georgia landscape.

all directions, must go somewhere. Since it cannot go into the ground, it rises, resulting in clouds.

Thus far, we have examined the primary ways air may be forced to rise so that clouds are able to form. We will now look at some of the other ways in which clouds develop.

Changing Cloud Forms   Under certain conditions, a layer of altostratus may change into altocumulus. This happens if the top of the original cloud deck cools while the bottom warms. Because clouds are such good absorbers and emitters of infrared radiation, the top of the cloud will often cool as it radiates infrared energy to space more rapidly than it absorbs solar energy. Meanwhile, the bottom of the cloud will warm as it absorbs infrared energy from below more quickly than it radiates this energy away. This process makes the cloud layer conditionally unstable to the point that small convection cells begin within the cloud itself. The up and down motions in a layered cloud produce globular elements that give the cloud a lumpy appearance. The cloud forms in the rising part of a cell, and clear spaces appear where descending currents occur (see Fig. 6.13, p. 140).

Cirrocumulus and stratocumulus may form in a similar way. When the wind is fairly uniform throughout a cloud layer, these new cloud elements appear evenly distributed across the sky. However, if the wind speed or direction changes with height, the horizontal axes of the convection cells align with the average direction of the wind. The new cloud elements then become arranged in rows and are given the name **cloud streets** (see Fig. 7.22). When the changes in wind speed and direction reach a critical value, and an inversion caps the cloud-forming layers, wavelike clouds called **billows** may form along the top of the cloud layer (see Fig. 7.23).

Occasionally, altocumulus show vertical development and produce towerlike extensions. The clouds often resemble floating castles and, for this reason, they are called *altocumulus castellanus* (see Fig. 7.24). They form when rising currents within the cloud extend into conditionally unstable air above the cloud. Apparently, the buoyancy for the rising air comes from the latent heat released during condensation within the cloud. This process can occur in cirrocumulus clouds, producing *cirrocumulus castellanus*. When altocumulus castellanus appear, they indicate that the mid-level of the troposphere is becoming more unstable (destabilizing). This destabilization is often the precursor to shower activity. So a morning sky full of altocumulus castellanus will likely become afternoon showers and even thunderstorms.

Mixing and Stratocumulus   Occasionally, the stirring of a moist layer of stable air will produce a deck of stratocumulus clouds. In Fig. 7.25, the air is stable and close to saturation. Suppose a strong wind mixes the layer from the surface up to an elevation of 600 m (2000 ft). As we saw earlier, the lapse rate will steepen as the upper part of the layer cools and the lower part warms. At the same time, mixing will make the moisture distribution in the layer more uniform. The warmer temperature and decreased moisture content cause the lower part of

FIGURE 7.23
Billow clouds forming in a region of rapidly changing wind speed, called wind shear.

FIGURE 7.24
An example of altocumulus castellanus.

the layer to dry out. On the other hand, the decrease in temperature and increase in moisture content saturate the top of the mixed layer, producing a layer of stratocumulus clouds. Figure 7.25 indicates that the air above the region of mixing is still stable and inhibits further mixing. In some cases, an inversion may actually form above the clouds. However, if the surface warms substantially, rising thermals may penetrate the stable region and the stratocumulus clouds may change into more widely separated clouds, such as cumulus or cumulus congestus. A stratocumulus layer changing to a sky dotted with growing cumulus clouds often occurs as surface heating increases on a warm, humid summer day.

FIGURE 7.25
The mixing of a moist layer of air near the surface can produce a deck of stratocumulus clouds.

Before mixing

Stable air

After mixing

Stable air

600 m

Mixed air

2000 ft

0

0

Altitude

Temperature

Temperature

Environmental lapse rate

Dew point lapse rate

Dry adiabatic rate

SUMMARY

In this chapter, we tied together the concepts of stability and the formation of clouds. We learned that rising unsaturated air cools at the dry adiabatic rate and, due to the release of latent heat, rising saturated air cools at the moist adiabatic rate. In a stable atmosphere, a lifted parcel of air will be colder (heavier) than the air surrounding it at each new level, and it will sink back to its original position. Because stable air tends to resist upward vertical motions, clouds forming in a stable atmosphere often spread horizontally and have a stratified appearance, such as cirrostratus and altostratus. A stable atmosphere may be caused by either cooling the surface air, warming the air aloft, or by the sinking (subsidence) of an entire layer of air, in which case a very stable subsidence inversion usually forms.

In an unstable atmosphere, a lifted parcel of air will be warmer (lighter) than the air surrounding it at each new level, and it will continue to rise upward away from its original position. In an unstable atmosphere, rising air tends to form clouds that develop vertically, such as cumulus congestus and cumulonimbus. Instability may be caused by warming the surface air, cooling the air aloft, or by the lifting or mixing of an entire layer of air.

On warm humid days the instability generated by surface heating can produce cumulus clouds at a height determined by the temperature and moisture content of the surface air. Instability may cause changes in existing clouds as convection changes an altostratus into an altocumulus. Also, mixing can change a clear day into a cloudy one.

In a conditionally unstable atmosphere, an unsaturated parcel of air can be lifted to a level where condensation begins, latent heat is released, and instability results, as the temperature inside the rising parcel becomes warmer than the air surrounding it.

Key Terms

The following terms are listed in the order they appear in the text. Define each. Doing so will aid you in reviewing the material covered in this chapter.

parcel of air
adiabatic process
dry adiabatic rate
moist adiabatic rate
environmental lapse rate
absolutely stable atmosphere
subsidence inversion
neutral stability
absolutely unstable
   atmosphere
conditionally unstable
   atmosphere
convective instability

condensation level
entrainment
orographic uplift
orographic clouds
rain shadow
lifting condensation level
   (LCL)
rotor clouds
dry adiabats
potential temperature
moist adiabats
cloud streets
billow clouds

Questions for Review

1. Why are moist and dry adiabatic rates of cooling different?
2. Under what conditions would the moist adiabatic rate of cooling be almost equal to the dry adiabatic rate?
3. Explain the difference between environmental lapse rate and dry adiabatic rate.
4. What is a stable atmosphere and how can it form?
5. Describe the general characteristics of clouds associated with stable and unstable air.
6. List and explain several processes by which a stable atmosphere can be made unstable.
7. If the atmosphere is conditionally unstable, what condition is necessary to bring on instability?
8. Explain why cumulus clouds are conspicuously absent over a cool water surface.
9. Why are cumulus clouds more frequently observed during the afternoon than at night?
10. Explain why an inversion represents an absolutely stable atmosphere.
11. How and why does lifting or lowering a layer of air change its stability?
12. Why do cumulonimbus clouds often have flat tops?
13. Describe the conditions necessary to produce stratocumulus clouds by mixing.
14. Why are there usually large spaces of blue sky between cumulus clouds?
15. List four primary ways clouds form, and describe the formation of one cloud type by each method.
16. (a) Why are lenticular clouds also called standing wave clouds? (b) On which side of a mountain (windward or leeward) would lenticular clouds most likely form?
17. List and explain several processes by which an unstable atmosphere can be made stable.
18. Explain why rain shadows form on the leeward side of mountains.
19. How can a layer of altostratus change into one of altocumulus?
20. Briefly describe how each of the following clouds form:
    (a) lenticular; (b) rotor; (c) billow; (d) castellanus

Questions for Thought

1. How is it possible for a layer of air to be convectively unstable and absolutely stable at the same time?
2. Are the bases of convective clouds generally higher during the day or the night? Explain.
3. Where would be the safest place to build an airport in a mountainous region? Why?
4. Use Fig. 5.11, p. 111 (Chapter 5), to help you explain why the bases of cumulus clouds, which form from rising thermals during the summer, increase in height above the surface as you move due west of a line that runs north-south through central Kansas.

5. For least polluted conditions, what would be the best time of day for a farmer to burn agricultural debris?

Problems and Exercises

1. Under which set of conditions would a cumulus cloud base be observed at the highest level above the surface? Surface air temperatures and dew points are as follows: (a) 35°C, 14°C; (b) 30°C, 19°C; (c) 34°C, 9°C; (d) 29°C, 7°C; (e) 32°C, 6°C.
2. If the height of a cumulus cloud is 1000 m above the surface, and the dew point at the earth's surface beneath the cloud is 20°C, determine the air temperature at the earth's surface beneath the cloud.
3. The condensation level over New Orleans, Louisiana, on a warm muggy afternoon is 2000 ft. If the dew point of the rising air at this level is 73°F, what is the approximate dew point and air temperature at the surface? Determine the surface relative humidity. (Hint: See Chapter 5, p. 117.)
4. Suppose the air pressure outside a conventional jet airliner flying at an altitude of 10 km (about 33,000 ft) is 250 mb. Further, suppose the air inside the aircraft is pressurized to 1000 mb. If the outside air temperature is –50°C (–58°F), what would be the temperature of this air if brought inside the aircraft and compressed at the dry adiabatic rate to a pressure of 1000 mb? (Assume that a pressure of 1000 mb is equivalent to an altitude of 0 m.)
5. In Fig. 7.26, above, a radiosonde is released and sends back temperature data as shown in the diagram. (This is the environment temperature.)
   (a) Calculate the environmental lapse rate from the surface up to 3000 m.
   (b) What type of atmospheric stability does the sounding indicate?
   Suppose the wind is blowing from the west and a parcel of surface air with a temperature of 10°C and a dew point of 2°C begins to rise upward along the western (windward) side of the mountain.
   (c) What is the relative humidity of the air parcel at 0 m (pressure 1013 mb) before rising? (Hint: See Chapter 5, p. 117.)
   (d) As the air parcel rises, at approximately what elevation would condensation begin and a cloud start to form?
   (e) What is the air temperature and dew point of the rising air at the base of the cloud?
   (f) What is the air temperature and dew point of the rising air inside the cloud at an elevation of 3000 m? (Use moist adiabatic rate of 6°C per 1000 m.)
   (g) At an altitude of 3000 m, how does the air temperature inside the cloud compare with the air temperature outside the cloud, as measured by the radiosonde? What type of atmospheric stability (stable or unstable) does this suggest? Explain.
   (h) At an elevation of 3000 m, would you expect the cloud to continue to develop vertically? Explain.

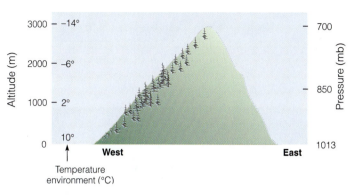

FIGURE 7.26
The information in this illustration is to be used in answering questions 5 and 6 in the Problems and Exercises section.

   (i) What would be the name of the cloud that is forming? Suppose that a parcel of air inside the cloud descends from the top of the mountain at 3000 m (pressure 700 mb) down the eastern (leeward) side of the mountain to an elevation of 0 m (pressure 1013 mb).
   (j) If the descending air warms at the dry adiabatic rate from the top of the mountain all the way down to 0 m, what is the sinking air's temperature and dew point when it reaches 0 m?
   (k) What would be the relative humidity of the sinking air at 0 m? (Hint: See Chapter 5, p. 117.)
   (l) What accounts for the sinking air being warmer at the base of the mountain on the eastern side?
   (m) Explain why the sinking air is drier (its dew point is lower) on the eastern side at 0 m.

Questions for Exploration

1. On the Blue Skies: College Edition CD-ROM go to the **Atmospheric Basics** section of the CD and click on "**Layers of the Atmosphere.**" (a) Explore the vertical profile of temperature and dew-point temperature for your location. Clouds will form in those layers where the dew-point temperature is equal to or near the temperature. Can you see such a layer in the profile? Are there clouds present over this location? (Hint: Look out the window) (b) What is the environmental lapse rate over the first 500 m of the atmosphere? Is this layer absolutely stable, conditionally unstable, or absolutely unstable?

2. On the Blue Skies: College Edition CD-ROM go to the **Atmospheric Moisture and Stability** section of the CD and click on "**Adiabatic.**" (a) Select a combination of temperature and dew point upwind. Record these in a log. Then traverse the mountain noting where the cloud forms. Determine the temperature and dew point on the downwind side. Record these in a log. (b) Repeat several times, keeping the initial temperature the same and changing the initial dew point. Form a relationship between the height at which the cloud forms and the difference between the initial temperature and dew point.

Heavy wet snowflakes that fall through the air with a temperature just below freezing cover everything with a blanket of white. (© Hans Strand, Tony Stone Images.)

PRECIPITATION

CONTENTS

By an unfortunate coincidence, as I write, the New Jersey countryside around me is in the grip of an ice storm—"the worst ice storm in a generation" so the papers tell me, and a look at my garden suffices to convince me. A 150-year-old tulip tree has already lost enough limbs to keep us in firewood for the rest of the winter; a number of black locusts stand beheaded; the silver birches are bent double to the ground; and almost every twig of every bush and tree is encased in a translucent cylinder of ice one to two inches in diameter. There is beauty in the sight, to be sure, for the sun has momentarily transmuted the virginal whites and grays into liquid gold. And there is hope, too, for some of the trees are still unbowed and look as though they had every intention of living to tell the tale.

George H. T. Kimble, *Our American Weather*

The young boy pushed his nose against the cold windowpane, hoping to see snowflakes glistening in the light of the street lamp across the way. Perhaps if it snowed, he thought, accumulations would be deep enough to cancel school—maybe for a day, possibly a week, or perhaps, forever. But a full moon with a halo gave little hope for snow on this evening. Nor did the voice from the back room that insisted, "Don't even think about snow. You know it won't snow tonight—it's too cold to snow."

Is it ever "too cold to snow"? Although many believe in this expression, the fact remains that it is *never* too cold to snow. True, colder air cannot "hold" as much water vapor as warmer air; but, no matter how cold the air becomes, it always contains some water vapor that could produce snow. At Fort Yellowstone, Wyoming, for example, 3 inches of snow fell on February 2, 1899, when the maximum temperature reached only –28°C (–18°F). In fact, tiny ice crystals have been observed falling at temperatures as low as –47°C (–53°F). We usually associate extremely cold air with "no snow" because the coldest winter weather occurs on clear, calm nights—conditions that normally prevail with strong high-pressure areas that have few if any clouds.

This chapter raises a number of interesting questions to consider regarding **precipitation**.* Why, for example, does the largest form of precipitation—hail—fall during the warmest time of the year? Why does it sometimes rain on one side of the street but not on the other? What is "sleet" and how does it differ from hail? First, we will examine the processes that produce rain and snow; then, we will look closely at the other forms of precipitation. Our discussion will conclude with a section on how precipitation is measured.

PRECIPITATION PROCESSES

As we all know, cloudy weather does not necessarily mean that it will rain or snow. In fact, clouds may form, linger for many days, and never produce precipitation. In Eureka, California, the August daytime sky is overcast more than 50 percent of the time, yet the average precipitation there for August is merely one-tenth of an inch. We know that clouds form by condensation, yet apparently condensation alone is not sufficient to produce rain. Why not? To answer this question we need to closely examine the tiny world of cloud droplets.

How Do Cloud Droplets Grow Larger?
An ordinary cloud droplet is extremely small, having an average diameter of 20 μm† or 0.002 cm. Notice in Fig. 8.1 that a typical cloud droplet is 100 times smaller than a typical raindrop. If a cloud droplet is in equilibrium with its surroundings, the size of the droplet does not change because the water molecules

---

*Precipitation is any form of water (liquid or solid) that falls from a cloud and reaches the ground.

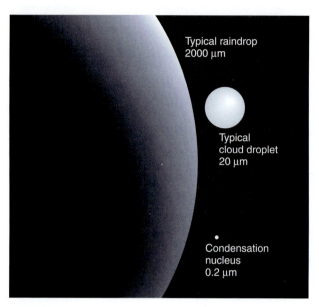

FIGURE 8.1
Relative sizes of raindrops, cloud droplets, and condensation nuclei.

condensing onto the droplet will be exactly balanced by those evaporating from it. If, however, it is not in equilibrium, the droplet size will either increase or decrease, depending on whether condensation or evaporation predominates.

Consider a cloud droplet in equilibrium with its environment. The total number of vapor molecules around the droplet remains fairly constant and defines the droplet's *saturation vapor pressure*. Since the droplet is in equilibrium, the saturation vapor pressure is also called the **equilibrium vapor pressure**. Figure 8.2 shows a cloud droplet and a flat water surface, both of which are in equilibrium. Because more vapor molecules surround the droplet, it has a greater equilibrium vapor pressure. The reason for this fact is that water molecules are less strongly attached to a curved (convex) water surface; hence, they evaporate more readily.

To keep the droplet in equilibrium, more vapor molecules are needed around it to replace those molecules that are constantly evaporating from its surface. Smaller cloud droplets exhibit a greater curvature, which causes a more rapid rate of evaporation. As a result of this process (called the **curvature effect**), smaller droplets require an even greater vapor pressure to keep them from evaporating away. Therefore, *when air is saturated with respect to a flat surface, it is unsaturated with respect to a curved droplet of pure water*, and the droplet evaporates. So, to keep cloud droplets in equilibrium with the surrounding air, the air must be *supersaturated*; that is, the relative humidity must be greater than 100 percent. The smaller the droplet, the greater its curvature, and the higher the supersaturation needed to keep it in equilibrium.

---

†Recall from Chapter 2 that one micrometer (μm) equals one-millionth of a meter.

Figure 8.3 shows the curvature effect for pure water. The dark blue line represents the relative humidity needed to keep a droplet with a given diameter in equilibrium with its environment. Note that when the droplet's size is less than 2 μm, the relative humidity (measured with respect to a flat surface) must be well above 100.1 percent for the droplet to survive. As droplets become larger, the effect of curvature lessens; for a droplet whose diameter is greater than 10 μm, the curvature effect is so small that the droplet behaves as if its surface were flat.

Just as relative humidities less than that required for equilibrium permit a water droplet to evaporate and shrink, those greater than the equilibrium value allow the droplet to grow by condensation. From Fig. 8.3, we can see that a droplet whose diameter is 1 μm will grow larger as the relative humidity approaches 101 percent. But relative humidities, even in clouds, rarely become greater than 101 percent. How, then, do tiny cloud droplets of less than 1 μm grow to the size of an average cloud droplet?

Recall from Chapter 6 (fog formation discussion) that condensation begins on tiny particles called *cloud condensation nuclei*. Because many of these nuclei are *hygroscopic* (that is, they have an affinity for water vapor), condensation may begin on such particles when the relative humidity is well below 100 percent. For example, when condensation begins on hygroscopic salt particles, they dissolve, forming a solution. Since the salt ions in solution bind closely with water molecules, it is more difficult for the water molecules to evaporate. This condition reduces the equilibrium vapor pressure, an effect known as the **solute effect**. Due to the solute effect, once an impurity (such as a salt particle) replaces a water molecule in the lattice structure of the droplet, the equilibrium vapor pressure surrounding the droplet is lowered. As a result of the solute effect, a droplet containing salt can be in equilibrium with its environment when the atmospheric relative humidity is much lower than 100 percent. Should the relative humidity of the air increase, water vapor molecules would attach themselves to the droplet at a faster rate than they would leave, and the droplet would grow larger in size.

Imagine that we place cloud condensation nuclei of varying sizes into moist but unsaturated air. As the air cools, the relative humidity increases. When the relative humidity reaches a value near 78 percent, condensation occurs on the majority of nuclei. As the air cools further, the relative humidity increases, with the nuclei containing the most salt reaching the largest sizes. As the relative humidity approaches 100 percent, the curvature effect becomes negligible for the larger nuclei but remains appreciable for the smaller nuclei.

The rising relative humidity of the air is countered by the growing particles that remove more water vapor as they become larger. When the particles are small, they cannot condense water vapor very rapidly; hence, the air's relative humidity continues to rise. As the particles grow bigger, they become more efficient in removing water vapor, until the rate at which

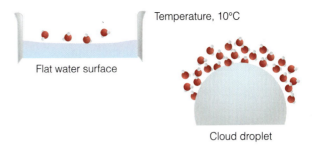

FIGURE 8.2

At equilibrium, the vapor pressure over a curved droplet of water is greater than that over a flat surface.

water vapor is removed equals the rate at which it is made available for condensation (from the updrafts in the cloud). After this point, the relative humidity decreases. Since the smaller nuclei are more affected by the curvature effect, only the larger nuclei are able to become cloud droplets.

Over land masses where large concentrations of nuclei exist, there may be many hundreds of droplets per cubic centimeter, all competing for the available supply of water vapor. Over the oceans where the concentration of nuclei is less, there are normally fewer (typically less than 100 per cubic centimeter) but larger cloud droplets.

We now have a cloud composed of many small droplets—too small to fall as rain. These minute droplets require only slight upward air currents to keep them suspended. Those droplets that do fall descend slowly and evaporate in the drier air beneath the cloud. It is evident, then, that most clouds cannot produce precipitation. The condensation process by itself is entirely too slow to produce rain. Even under ideal conditions, it would take several days for this process alone to create a raindrop. However, observations show that clouds can

FIGURE 8.3

The curved line represents the relative humidity needed to keep a droplet in equilibrium with its environment. For a given droplet size, the droplet will evaporate and shrink when the relative humidity is less than that given by the curve. The droplet will grow by condensation when the relative humidity is greater than the value on the curve.

WEATHER WATCH

Maybe it has never rained cats and dogs, but it has rained maggots. In Acapulco, Mexico, during October, 1968, swarms of maggots (about an inch in length) fell from the sky during a heavy rain shower, covering everything, even people who had gathered there to witness a yachting event. Apparently, the maggots were swept into a thunderstorm by strong vertical air currents.

develop and begin to produce rain in less than an hour. Since it takes about 1 million average size (20 μm) cloud droplets to make an average size (2000 μm) raindrop, there must be some other process by which cloud droplets grow large and heavy enough to fall as precipitation. Even though all of the intricacies of how rain is produced are not yet fully understood, two important processes stand out: (1) the collision-coalescence process and (2) the ice-crystal (Bergeron) process.

Collision and Coalescence Process   In relatively warm clouds with tops warmer than −15°C (5°F), collisions between droplets can play a significant role in producing precipitation. To produce the many collisions necessary to form a raindrop, some cloud droplets must be larger than others. Larger drops may form on large condensation nuclei, such as salt particles,

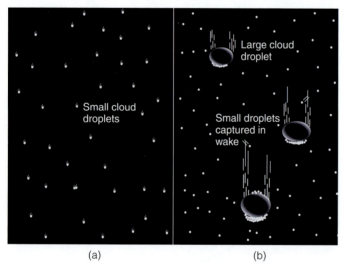

FIGURE 8.4

Collision and coalescence. (a) In a warm cloud composed only of small cloud droplets of uniform size, the droplets are less likely to collide as they all fall very slowly at about the same speed. Those droplets that do collide, frequently do not coalesce because of the strong surface tension that holds together each tiny droplet. (b) In a cloud composed of different size droplets, larger droplets fall faster than smaller droplets. Although some tiny droplets are swept aside, some collect on the larger droplet's forward edge, while others (captured in the wake of the larger droplet) coalesce on the droplet's backside.

or through random collisions of droplets. Recent studies also suggest that turbulent mixing between the cloud and its drier environment may play a role in producing larger droplets.

As cloud droplets fall, air retards the falling drops. The amount of air resistance depends on the size of the drop and on its rate of fall: The greater its speed, the more air molecules the drop encounters each second. The speed of the falling drop increases until the air resistance equals the pull of gravity. At this point, the drop continues to fall, but at a constant speed, which is called its **terminal velocity**. Because larger drops have a smaller surface-area-to-weight ratio, they must fall faster before reaching their terminal velocity. Thus *larger drops fall faster than smaller drops* (see Table 8.1). Note that, in calm air, a typical raindrop falls over 600 times faster than a typical cloud droplet!

Large droplets overtake and collide with smaller drops in their path. This merging of cloud droplets by collision is called **coalescence**. Laboratory studies show that collision does not always guarantee coalescence; sometimes the droplets actually bounce apart during collision. For example, the forces that hold a tiny droplet together (surface tension) are so strong that if the droplet were to collide with another tiny droplet, chances are they would not stick together (coalesce) (see Fig. 8.4). Coalescence appears to be enhanced if colliding droplets have opposite (and, hence, attractive) electrical charges.* An important factor influencing cloud droplet growth by the collision process is the amount of time the droplet spends in the cloud. A very large cloud droplet of 200 μm falling in still air takes about 12 minutes to travel through a cloud 500 m (1640 ft) thick and over an hour if the cloud thickness is 2500 m (8200 ft). Rising air currents in a forming cloud slow the rate at which droplets fall toward the ground. Consequently, a thick cloud with strong updrafts maximizes the time cloud droplets spend in the cloud and, hence, the size to which they can grow.

A warm stratus cloud is typically less than 500 m thick and has slow upward air movement (generally less than 0.1 m/sec). Under these conditions, a large droplet would be in the cloud for a relatively short time and grow by coalescence to only about 200 μm. If the air beneath the cloud is moist, the droplets may reach the ground as *drizzle*, the lightest form of rain. If, however, the stratus cloud base is fairly high above the ground, the drops will evaporate before reaching the surface, even when the relative humidity is 90 percent.

Clouds that have above-freezing temperatures at all levels are called *warm clouds*. In such clouds, precipitation forms by

---

*It was once thought that atmospheric electricity played a significant role in the production of rain. Today, many scientists feel that the difference in electrical charge that exists between cloud droplets results from the bouncing collisions between them. It is felt that the weak separation of charge and the weak electrical fields in developing, relatively warm clouds are not significant in initiating precipitation. However, studies show that coalescence is often enhanced in thunderstorms where strongly charged droplets exist in a strong electrical field.

the collision and coalescence process. For example, in tropical regions, where warm cumulus clouds build to great heights, convective updrafts of at least 1 m/sec (and some exceeding many tens of meters per second) occur. Look at the warm cumulus cloud in Fig. 8.5. Suppose a cloud droplet of 100 μm is caught in an updraft whose velocity is 6.5 m/sec. As the droplet rises, it collides with and captures smaller drops in its path and grows until it reaches a size of about 1000 μm. At this point, the updraft in the cloud is just able to balance the pull of gravity on the drop. Here, the drop remains suspended until it grows just a little bigger. Once the fall velocity of the drop is greater than the updraft velocity in the cloud, the drop slowly descends. As the drop falls, larger cloud droplets are captured by the falling drop, which then grows larger. By the time this drop reaches the bottom of the cloud, it will be a large raindrop with a diameter of over 5000 μm (5 mm). Because raindrops of this size fall faster and reach the ground first, they typically occur at the beginning of a rain shower originating in these warm, convective cumulus clouds.

Raindrops that reach the earth's surface are seldom larger than about 5 mm. The collisions between raindrops (whether glancing or head-on) tend to break them up into many smaller drops. Additionally, a large drop colliding with another large drop may result in oscillations within the combined drop. As the resultant drop grows, these oscillations may tear the drop apart into many fragments, all smaller than the original drop.

So far, we have examined the way cloud droplets in warm clouds (that is, those clouds with temperatures above freezing) grow large enough by the collision-coalescence process to fall as raindrops. Rain that falls from warm clouds is sometimes called *warm rain*. The most important factor in the production of raindrops is the cloud's *liquid water content*. In a cloud with sufficient water, other significant factors are:

1. the range of droplet sizes
2. the cloud thickness
3. the updrafts of the cloud
4. the electric charge of the droplets and the electric field in the cloud

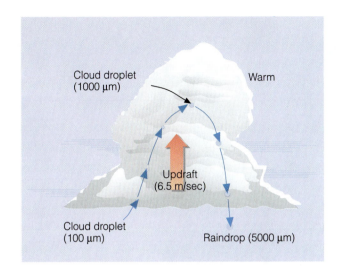

FIGURE 8.5
A cloud droplet rising then falling through a warm cumulus cloud can grow by collision and coalescence, and emerge from the cloud as a large raindrop.

Relatively thin stratus clouds with slow, upward air currents are, at best, only able to produce drizzle, whereas the towering cumulus clouds associated with rapidly rising air can cause heavy showers. Now, let's turn our attention to the ice-crystal process of rain formation.

Ice-Crystal Process   The **ice-crystal (or Bergeron)* process** of rain formation is extremely important in middle and high latitudes, where clouds extend upward into regions where the air temperature is well below freezing. Such clouds are called *cold clouds*. Figure 8.6 illustrates a typical cold cloud that has formed over the Great Plains, where the "cold" part is well above the 0°C isotherm.

---

*The ice-crystal process is also known as the *Bergeron process* after the Swedish meteorologist Tor Bergeron, who proposed that essentially all raindrops begin as ice crystals.

TABLE 8.1   Terminal Velocity of Different Size Particles Involved in Condensation and Precipitation Processes

| | TERMINAL VELOCITY | | |
DIAMETER (μm)	m/sec	ft/sec	TYPE OF PARTICLE
0.2	0.0000001	0.0000003	Condensation nuclei
20	0.01	0.03	Typical cloud droplet
100	0.27	0.9	Large cloud droplet
200	0.70	2.3	Large cloud droplet or drizzle
1000	4.0	13.1	Small raindrop
2000	6.5	21.4	Typical raindrop
5000	9.0	29.5	Large raindrop

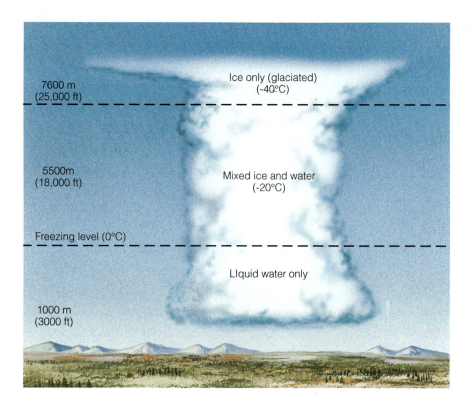

FIGURE 8.6
The distribution of ice and water in a cumulonimbus cloud.

7600 m
(25,000 ft)

Ice only (glaciated)
(-40°C)

5500m
(18,000 ft)

Mixed ice and water
(-20°C)

Freezing level (0°C)

Liquid water only

1000 m
(3000 ft)

Suppose we take an imaginary balloon flight up through the cumulonimbus cloud in Fig. 8.6. Entering the cloud, we observe cloud droplets growing larger by processes described in the previous section. As expected, only water droplets exist here, for the base of the cloud is warmer than 0°C. Surprisingly, in the cold air just above the 0°C isotherm, almost all of the cloud droplets are still composed of liquid water. Water droplets existing at temperatures below freezing are referred to as **supercooled**. Even at higher levels, where the air temperature is −10°C (14°F), there is only one ice crystal for every million liquid droplets. Near 5500 m (18,000 ft), where the temperature becomes −20°C (−4°F), ice crystals become more numerous, but are still outnumbered by water droplets. The distribution of ice crystals, however, is not uniform, as the downdrafts contain more ice than the updrafts. Clouds composed of both ice and water are called *mixed clouds*.

Not until we reach an elevation of 7600 m (25,000 ft), where temperatures drop below −40°C (also −40°F), do we find *only* ice crystals. The region of a cloud where only ice particles exist is called *glaciated*. Why are there so few ice crystals in the middle of the cloud, even though temperatures there are well below freezing?*

_____

*In continental clouds, such as the one in Fig. 8.6, where there are many small cloud droplets less than 20 μm in diameter, the onset of ice-crystal formation begins at temperatures between −9°C and −15°C. In clouds where larger but fewer cloud droplets are present, ice crystals begin to form at temperatures between −4°C and −8°C. In some of these clouds, glaciation can occur at −8°C, which may be only 2500 m (8200 ft) above the surface.

Pure ice melts when the temperature is raised to 0°C, but water does not necessarily freeze until temperatures fall well below that value. Over large bodies of fresh water, ice ordinarily forms when the temperature drops just slightly below 0°C. Yet, laboratory studies show that, the smaller the amount of pure water, the lower the temperature at which it freezes. A cloud droplet about 25 μm in diameter freezes spontaneously at −36°C (−33°F), while a droplet whose diameter is only a few micrometers does not freeze until the temperature is near −40°C (−40°F).

The freezing of pure water (without the benefit of some nucleus) is called **homogeneous** (*spontaneous*) **freezing**. For homogeneous freezing to occur, enough molecules within the water droplet must join together in a rigid pattern to form a tiny ice structure, or *ice embryo*. When the ice embryo grows to a critical size, it acts as a nucleus. Other molecules in the droplet then attach themselves to the nucleus of ice and the water droplet freezes.

Tiny ice embryos form in water at temperatures just below freezing, but at these temperatures thermal agitations are large enough to weaken their structure. The ice embryos simply form and then break apart. At lower temperatures, thermal motion is reduced, making it easier for bigger ice embryos to form. Hence, freezing is more likely. The chances of an ice embryo growing large enough to freeze water before the embryo is broken up by thermal agitation increase with bigger volumes of water. Consequently, only larger cloud droplets can freeze by homogeneous freezing at temperatures warmer than −40°C. In air colder than −40°C, however, it is almost certain that an ice embryo will grow to critical size in even the smallest cloud droplet. Thus, any cloud that forms in extremely cold air (below −40°C), such as cirrus clouds, will almost certainly be composed of ice, since any cloud droplets that form will freeze spontaneously.

Just as liquid cloud droplets form on condensation nuclei, ice crystals may form in subfreezing air on particles called **ice nuclei**. The number of ice-forming nuclei available in the atmosphere is small, especially at temperatures above −10°C (14°F). However, as the temperature decreases, more particles become active and promote freezing. Although some uncertainty exists regarding the principal source of ice nuclei, it is known that clay minerals, such as kaolinite, become effective nuclei at temperatures near −15°C (5°F). Some types of bacteria in decaying plant leaf material and ice crystals themselves are also excellent ice nuclei. Moreover, particles serve as excellent ice-forming nuclei if their geometry resembles that of an ice crystal. However, it is difficult to find substances in nature that have a lattice structure similar to ice, since there are so many possible lattice structures. In the atmosphere, it is easy to find hygroscopic ("water seeking") particles. Consequently, ice-forming nuclei are rare compared to cloud condensation nuclei.

In a cold cloud, there may be several types of ice-forming nuclei present. For example, certain ice nuclei allow water vapor to deposit as ice directly onto their surfaces in cold, saturated air. These are called *deposition nuclei* because, in this situation, water vapor changes directly into ice without going through the liquid phase. Ice nuclei that promote the freezing of supercooled liquid droplets are called *freezing nuclei*. Some freezing nuclei cause freezing after they are immersed in a liquid drop; some promote condensation, then freezing; yet others cause supercooled droplets to freeze if they collide with them. This last process is called **contact freezing**, and the particles involved are called *contact nuclei*. Recent studies suggest that contact nuclei can be just about any substance and that contact freezing may be the dominant force in the production of ice crystals in some clouds.

We can now understand why there are so few ice crystals in the cold mixed region of some clouds. Cloud droplets may freeze spontaneously, but only at the very low temperatures usually found at high altitudes. Ice nuclei may initiate the growth of ice crystals, but they do not abound in nature. Because there are many more cloud condensation nuclei than ice nuclei, we are left with a cold cloud that contains many more liquid droplets than ice particles, even at temperatures as low as −10°C (14°F). Neither the tiny liquid nor solid particles are large enough to fall as precipitation. How, then, does the ice-crystal process produce rain and snow?

In the subfreezing air of a cloud, many supercooled liquid droplets will surround each ice crystal. Suppose that the ice crystal and liquid droplet in Fig. 8.7 are part of a cold (−15°C) supercooled, saturated cloud. Since the air is saturated, both the liquid droplet and the ice crystal are in equilibrium, meaning that the number of molecules leaving the surface of both the droplet and the ice crystal must equal the number of mol-

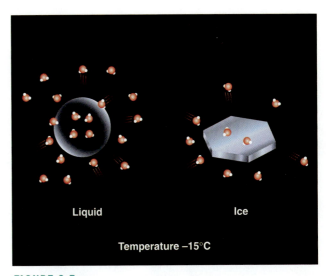

FIGURE 8.7

In a saturated environment, the water droplet and the ice crystal are in equilibrium, as the number of molecules leaving the surface of the droplet and ice crystal equals the number returning. The greater number of vapor molecules above the liquid indicates, however, that the saturation vapor pressure over water is greater than it is over ice.

ecules returning. Observe, however, that there are more vapor molecules above the liquid. The reason for this fact is that molecules escape the surface of water much easier than they escape the surface of ice. Consequently, more molecules escape the water surface at a given temperature, requiring more in the vapor phase to maintain saturation. This situation reflects the important fact discussed briefly in Chapter 5: At the same

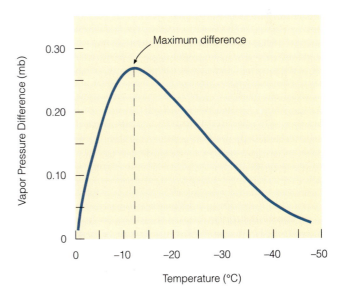

FIGURE 8.8

The difference in saturation vapor pressure between supercooled water and ice at different temperatures.

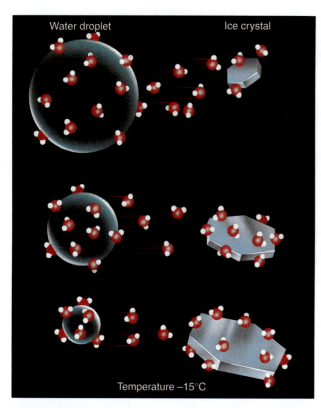

Water droplet   Ice crystal

Temperature −15°C

FIGURE 8.9

The ice-crystal process. The greater number of water vapor molecules around the liquid droplets causes water molecules to diffuse from the liquid drops toward the ice crystals. The ice crystals absorb the water vapor and grow larger, while the water droplets grow smaller.

subfreezing temperature, *the saturation vapor pressure just above a water surface is greater than the saturation vapor pressure above an ice surface.* This difference in saturation vapor pressure between water and ice is illustrated in Fig. 8.8.

This difference in vapor pressure causes water vapor molecules to move (diffuse) from the droplet toward the ice crystal. The removal of vapor molecules reduces the vapor pressure above the droplet. Since the droplet is now out of equilibrium with its surroundings, it evaporates to replenish the diminished supply of water vapor above it. This process provides a continuous source of moisture for the ice crystal, which absorbs the water vapor and grows rapidly (see Fig. 8.9). Hence, during the *ice-crystal process, ice crystals grow larger at the expense of the surrounding water droplets.*

The constant supply of moisture to the ice crystal allows it to enlarge rapidly. At some point, the ice crystal becomes heavy enough to overcome updrafts in the cloud and begins to fall. But a single falling ice crystal does not comprise a snowstorm; consequently, other ice crystals must quickly form.

In some clouds, especially those with relatively warm tops, ice crystals might collide with supercooled droplets, freezing

on contact and sticking together—a process called **accretion**, or *riming*. The icy matter (rime) that forms is called **graupel** (or *snow pellets*). As the graupel falls, it may fracture or splinter into tiny ice particles when it collides with cloud droplets. These splinters may grow to become new graupel, which, in turn, may produce more splinters. In colder clouds, the delicate ice crystals may collide with other crystals and fracture into smaller ice particles, or tiny seeds, which freeze hundreds of supercooled droplets on contact. In both cases a chain reaction may develop, producing many ice crystals. As the ice crystals fall, they may collide and stick to one another, in a process called **aggregation**,* forming an accumulation of ice crystals called a *snowflake* (see Fig. 8.10). If the snowflake melts before reaching the ground, it continues its fall as a raindrop. Therefore, much of the rain falling in middle and northern latitudes—even in summer—begins as snow.

For ice crystals to grow large enough to produce precipitation there must be many, many times more water droplets than ice crystals. Generally, the ratio of ice crystals to water droplets must be on the order of 1:100,000 to 1:1,000,000. When there are too few ice crystals in the cloud, each crystal grows large and falls out of the cloud, leaving the majority of cloud behind (unaffected). Since there are very few ice crystals, there is very little precipitation. If, on the other hand, there are too many ice crystals (such as an equal number of crystals and droplets), then each ice crystal receives the mass of one droplet. This would create a cloud of many tiny ice crystals, each too small to fall to the ground, and no precipitation. Now, if the ratio of crystals to droplets is on the order of 1:100,000, then each ice crystal would receive the mass of 100,000 droplets. Most of the cloud would convert to precipitation, as the majority of ice crystals would grow large enough to fall to the ground as precipitation.

The first person to formally propose the theory of ice-crystal growth due to differences in the vapor pressure between ice and supercooled water was Alfred Wegener (1880–1930), a German climatologist who also proposed the geological theory of continental drift. In the early 1930s, important additions to this theory were made by the Swedish meteorologist Tor Bergeron. Several years later, the German meteorologist Walter Findeisen made additional contributions to Bergeron's theory; hence, the ice-crystal theory of rain formation has come to be known as the *Wegener-Bergeron-Findeisen process*, or, simply, the *Bergeron process*.

Cloud Seeding   The primary goal in many **cloud seeding** experiments is to inject (or seed) a cloud with small particles that will act as nuclei, so that the cloud particles will grow large enough to fall to the surface as precipitation. The first ingredi-

---

*Significant aggregation seems possible only when the air is relatively warm, usually warmer than −10°C (14°F).

(a) Falling ice crystals may freeze supercooled droplets on contact (accretion), producing larger ice particles.

(b) Falling ice particles may collide and fracture into many tiny (secondary) ice particles.

(c) Falling ice crystals may collide and stick to other ice crystals (aggregation), producing snowflakes.

FIGURE 8.10
Ice particles in clouds.

ent in any seeding project is, of course, the presence of clouds, as seeding does not generate clouds. However, not just any cloud will do. For optimum results, the cloud must be cold; that is, at least a portion of it (preferably the upper part) must be supercooled because cloud seeding uses the ice-crystal process to cause the cloud particles to grow.

The idea in cloud seeding is to first find clouds that have too low a ratio of ice crystals to droplets and then to add enough artificial ice nuclei so that the ratio of crystals to droplets is about 1:100,000. However, it should be noted that the natural ratio of ice nuclei to cloud condensation nuclei in a typical cold cloud is about 1:100,000, just about optimal for producing precipitation.

Some of the first experiments in cloud seeding were conducted by Vincent Schaefer and Irving Langmuir during the late 1940s. To seed a cloud, they dropped crushed pellets of *dry ice* (solid carbon dioxide) from a plane. Because dry ice has a temperature of −78°C (−108°F), it acts as a cooling agent. Small pellets dropped into the cloud cool the air to the point where droplets are able to freeze in an instant (homogeneous freezing) at temperatures below −40°C (−40°F).* The newly formed ice crystals then grow larger by deposition at the expense of the

nearby liquid droplets and, upon reaching a sufficiently large size, fall as precipitation.

In 1947, Bernard Vonnegut demonstrated that silver iodide (AgI) could be used as a cloud-seeding agent. Because silver iodide has a crystalline structure similar to an ice crystal, it acts as an effective ice nucleus at temperatures of −4°C (25°F) and lower.

Silver iodide is much easier to handle than dry ice, since it can be supplied to the cloud from burners located either on the ground or on the wing of a small aircraft. Although other substances, such as lead iodide and cupric sulfide, are also effective ice nuclei, silver iodide still remains the most commonly used substance in cloud-seeding projects. (Additional information on the controversial topic, the effectiveness of cloud seeding, is given in the Focus section on p. 190.)

Under certain conditions, clouds may be seeded naturally. For example, when cirriform clouds lie directly above a lower cloud deck, ice crystals may descend from the higher cloud and seed the cloud below. As the ice crystals mix into the lower cloud, supercooled droplets are converted to ice crystals, and the precipitation process is enhanced. Sometimes the ice crystals in the lower cloud may settle out, leaving a clear area or "hole" in the cloud. When the cirrus clouds form waves downwind from a mountain chain (Fig. 8.11), bands of precipitation often form.

---

*Would the tiny dry-ice pellets freeze on contact with any supercooled liquid droplets that they encounter? (Answer: yes.)

FOCUS ON A SPECIAL TOPIC

Does Cloud Seeding Enhance Precipitation?

Just how effective is artificial seeding with silver iodide in increasing precipitation? This is a much-debated question among meteorologists. First of all, it is difficult to evaluate the results of a cloud-seeding experiment. When a seeded cloud produces precipitation, the question always remains as to how much precipitation would have fallen had the cloud not been seeded.

Other factors must be considered when evaluating cloud-seeding experiments: the type of cloud, its temperature, moisture content, droplet size distribution, and updraft velocities in the cloud.

Although some experiments suggest that cloud seeding does not increase precipitation, others seem to indicate that seeding *under the right conditions* may enhance precipitation between 5 percent and 20 percent. And so the controversy continues.

Some cumulus clouds show an "explosive" growth after being seeded. The latent heat given off when the droplets freeze functions to warm the cloud, causing it to become more buoyant. It grows rapidly and becomes a longer-lasting cloud, which may produce more precipitation.

The business of cloud seeding can be a bit tricky, since overseeding can produce too many ice crystals. When this phenomenon occurs, the cloud becomes glaciated (all liquid droplets become ice) and the ice particles, being very small, do not fall as precipitation. Since few liquid droplets exist, the ice crystals cannot grow by the ice-crystal (Bergeron) process; rather, they evaporate, leaving a clear area in a thin, stratified cloud. Because dry ice can produce the most ice crystals in a super-cooled cloud, it is the substance most suitable for deliberate overseeding. Hence, it is the substance most commonly used to dissipate cold fog at airports (see Chapter 6, p. 137).

Warm clouds with temperatures above freezing have also been seeded in an attempt to produce rain. Tiny water drops and particles of hygroscopic salt are injected into the base (or top) of the cloud. These particles (called *seed drops*), when carried into the cloud by updrafts, create large cloud droplets, which grow even larger by the collision-coalescence process. Apparently, the seed drop size plays a major role in determining the effectiveness of seeding with hygroscopic particles. To date, however, the results obtained using this method are inconclusive.

Cloud seeding may be inadvertent. Some industries emit large concentrations of condensation nuclei and ice nuclei into the air. Studies have shown that these particles are at least partly responsible for increasing precipitation in, and down-wind of, cities. On the other hand, studies have also indicated that the burning of certain types of agricultural waste may produce smoke containing many condensation nuclei. These produce clouds that yield less precipitation because they contain numerous, but very small, droplets.

In summary, cloud seeding in certain instances may lead to more precipitation; in others, to less precipitation, and, in still others, to no change in precipitation amounts. Many of the questions about cloud seeding have yet to be resolved.

Precipitation in Clouds   In cold, strongly convective clouds, precipitation may begin only minutes after the cloud forms and may be initiated by either the collision-coalescence or the ice-crystal process. Once either process begins, most precipitation growth is by accretion. Although precipitation is commonly absent in warm-layered clouds, such as stratus, it is often associated with such cold-layered clouds as nimbostratus and alto-stratus. This precipitation is thought to form principally by the ice-crystal process because the liquid water content of these clouds is generally lower than that in convective clouds, thus making the collision-coalescence process much less effective. Nimbostratus clouds are normally thick enough to extend to levels where air temperatures are quite low, and they usually last long enough for the ice-crystal process to initiate precipitation. Figure 8.12 illustrates how ice crystals produce precipitation in clouds of both low and high liquid-water content.

FIGURE 8.11
Natural seeding by cirrus clouds may form bands of precipitation downwind of a mountain chain.

FIGURE 8.12
How ice crystals grow and produce precipitation in clouds with a low liquid water content and a high liquid water content.

BRIEF REVIEW In the last few sections we encountered a number of important concepts and ideas about how cloud droplets can grow large enough to fall as precipitation. Before examining the various types of precipitation, here is a summary of some of the important ideas presented so far:

- Cloud droplets are very small, much too small to fall as rain.
- The smaller the cloud droplet, the greater its curvature, and the more likely it will evaporate.
- Cloud droplets form on cloud condensation nuclei. Hygroscopic nuclei, such as salt, allow condensation to begin when the relative humidity is less than 100 percent.
- Cloud droplets, in above-freezing air, can grow larger as faster-falling, bigger droplets collide and coalesce with smaller droplets in their path.
- In the ice-crystal (Bergeron) process of rain formation, both ice crystals and liquid cloud droplets must coexist at below-freezing temperatures. The difference in saturation vapor pressure between liquid and ice causes water vapor to diffuse from the liquid droplets (which shrink) toward the ice crystals (which grow).
- Most of the rain that falls over middle latitudes results from melted snow that formed from the ice crystal (Bergeron) process.
- Cloud seeding with silver iodide can only be effective in coaxing precipitation from clouds if the cloud is supercooled and the proper ratio of cloud droplets to ice crystals exists.

PRECIPITATION TYPES

Up to now, we have seen how cloud droplets are able to grow large enough to fall to the ground as rain or snow. While falling, raindrops and snowflakes may be altered by atmospheric conditions encountered beneath the cloud and transformed into other forms of precipitation that can profoundly influence our environment.

Rain Most people consider **rain** to be any falling drop of liquid water. To the meteorologist, however, that falling drop must have a diameter equal to, or greater than, 0.5 mm to be considered rain. Fine uniform drops of water whose diameters are smaller than 0.5 mm are called **drizzle**. Most drizzle falls from stratus clouds; however, small raindrops may fall through air that is unsaturated, partially evaporate, and reach the ground as drizzle. Occasionally, the rain falling from a cloud never reaches the surface because the low humidity causes rapid evaporation. As the drops become smaller, their rate of fall decreases, and they appear to hang in the air as a rain streamer. These evaporating streaks of precipitation are called **virga*** (see Fig. 8.13).

Raindrops may also fall from a cloud and not reach the ground, if they encounter rapidly rising air. Large raindrops have a terminal velocity of about 9 m/sec (20 mi/hr), and, if they encounter rising air whose speed is greater than 9 m/sec, they will not reach the surface. If the updraft weakens or changes direction and becomes a downdraft, the suspended drops will fall to the ground as a sudden rain **shower**. The showers falling from cumuliform clouds are usually brief and sporadic, as the cloud moves overhead and then drifts on by. If the shower is excessively heavy, it is termed a *cloudburst*. Beneath a cumulonimbus cloud, which normally contains large convection currents, it is entirely possible that one side of a street may be dry (updraft side), while a heavy shower is occurring across the street (downdraft side). Continuous rain, on the other hand, usually falls from a layered cloud that covers a large area and has smaller vertical air currents. These are the conditions normally associated with nimbostratus clouds.

Rain is usually transparent, but, occasionally, it takes on various colors. For example, brown rain is fairly common,

---

*Recent studies suggest that the "rain streamer" is actually caused by ice (which is more reflective) changing to water (which is less reflective). Apparently, most evaporation occurs below the virga line.

FIGURE 8.13
The streaks of falling precipitation that evaporate before reaching the ground are called *virga*.

especially during a drought when strong winds carry dirt and silt into a cloud. In extreme cases, where many tons of dust or volcanic ash are carried aloft, tiny mud or ash balls have actually fallen from clouds.

After a rainstorm, visibility usually improves primarily because precipitation removes (scavenges) many of the suspended particles. When rain combines with gaseous pollutants, such as oxides of sulfur and nitrogen, it becomes acidic. *Acid rain*, which has an adverse effect on plants and water resources, is becoming a major problem in many industrialized regions of the world (see Chapter 17).

It is important to know the interval of time over which rain falls. Did it fall over several days, gradually soaking into the soil? Or did it come all at once in a cloudburst, rapidly eroding the land, clogging city gutters, and causing floods along creeks and rivers unable to handle the sudden increased flow? The *intensity* of rain is the amount that falls in a given period; intensity of rain is always based on the accumulation during a certain interval of time. For example, *heavy rain* means that water is accumulating at a rate greater than 0.30 in. each hour (see Table 8.2). One can see that heavy rainfall reported for many consecutive hours can lead to substantial precipitation totals. (If you are curious as to the actual shape of a falling raindrop, read the Focus section on p. 193.)

FIGURE 8.14
Because snow scatters sunlight more effectively than rain, when you look toward the sun the sky appears darker above the melting level than below it.

Snow

We have learned that much of the precipitation reaching the ground actually begins as **snow**. In summer, the freezing level is usually above 3600 m (12,000 ft), and the snowflakes falling from a cloud melt before reaching the ground. However, in winter, the freezing level is much lower, and falling snowflakes have a better chance of survival. Snowflakes can generally fall about 300 m (1000 ft) below the freezing level before completely melting. Occasionally, you can spot the melting level when you look in the direction of the sun, if it is near the horizon. Because snow scatters incoming sunlight better than rain, the darker region beneath the cloud contains falling snow, while the lighter region is falling rain. The melting zone, then, is the transition between the light and dark areas (see Fig. 8.14).

The sky will look different, however, if you are looking directly up at the precipitation. Because snowflakes are such effective scatterers of light, they redirect the light beneath the cloud in all directions—some of it eventually reaching your

TABLE 8.2	Rainfall Intensity
RAINFALL DESCRIPTION	**RAINFALL RATE (in./hr)***
Light	0.01 to 0.10
Moderate	0.11 to 0.30
Heavy	>0.30

*In the United States, the National Weather Service measures rainfall in inches.

FOCUS ON A SPECIAL TOPIC

Are Raindrops Tear-Shaped?

As rain falls, the drops take on a characteristic shape. Choose the shape in Fig. 1 that you feel most accurately describes that of a falling raindrop. Did you pick number 1? The tear-shaped drop has been depicted by artists for many years. Unfortunately, *raindrops are not tear-shaped*. Actually, the shape depends on the drop size. Raindrops less than 2 mm in diameter are nearly spherical and look like raindrop number 2. The attraction among the molecules of the liquid (surface tension) tends to squeeze the drop into a shape that has

FIGURE 1
Which of the three drops drawn here represents the real shape of a falling raindrop?

the smallest surface area for its total volume—a sphere.

Large raindrops, with diameters exceeding 2 mm, take on a different shape as they fall. Believe it or not,

they look like number 3, slightly elongated, flattened on the bottom, and rounded on top. As the larger drop falls, the air pressure against the drop is greatest on the bottom and least on the sides. The pressure of the air on the bottom flattens the drop, while the lower pressure on its sides allows it to expand a little. This shape has been described as everything from a falling parachute to a loaf of bread, or even a hamburger bun. You may call it what you wish, but remember: it is not tear-shaped.

---

eyes, making the region beneath the cloud appear a lighter shade of gray. Falling raindrops, on the other hand, scatter very little light toward you, and the underside of the cloud appears dark. It is this change in shading that enables some observers to predict with uncanny accuracy whether precipitation will be in the form of rain or snow.

When ice crystals and snowflakes fall from high cirrus clouds they are called **fallstreaks**. Fallstreaks behave in much the same way as virga. As the ice particles fall into drier air, they usually sublimate (that is, change from ice into vapor). Because the winds at higher levels move the cloud and ice particles horizontally more quickly than do the slower winds at lower levels, fallstreaks appear as dangling white streamers (see Fig. 8.15). Moreover, fallstreaks descending into lower, supercooled clouds may actually seed them.

Snowflakes and Snowfall Snowflakes that fall through moist air that is slightly above freezing slowly melt as they descend.

FIGURE 8.15
The dangling white streamers of ice crystals beneath these cirrus clouds are known as *fallstreaks.* The bending of the streaks is due to the changing wind speed with height.

When Is It Too Warm to Snow?

In the beginning of this chapter, we learned that it is never too cold to snow. So when is it too warm to snow? A person who has never been in a snowstorm might answer, "When the air temperature rises above freezing." However, anyone who lives in a climate that experiences cold winters knows that snow may fall when the air temperature is considerably above freezing. In fact, in some areas, snowstorms often begin with a surface air temperature near 2°C (36°F). Why doesn't the falling snow melt in this air? Actually, it does melt, at least to some degree. Let's examine this in more detail.

In order for falling snowflakes to survive in air with temperatures much above freezing, the air must be unsaturated (relative humidity is less than 100 percent), and the wet-bulb temperature must be at freezing or below. You may recall from our discussion on humidity (Chapter 5) that the wet-bulb temperature is the lowest temperature that can be attained by

evaporating water into the air. Consequently, it is a measure of the amount of cooling that can occur in the atmosphere as water evaporates into the air. When rain falls into a layer of dry air with a low wet-bulb temperature, rapid evaporation and cooling occurs, which is why the air temperature often decreases when it begins to rain. During the winter, as raindrops evaporate in this dry air, rapid cooling may actually change a rainy day into a snowy one. This same type of cooling allows snowflakes to survive above freezing (melting) temperatures.

Suppose it is winter and the sky is overcast. At the surface, the air temperature is 2°C (36°F), the dew point is –6°C (21°F), and the wet-bulb temperature is 0°C (32°F).* The air temperature drops sharply with height,

*The wet-bulb temperature is always higher than the dew point, except when the air is saturated. At that point, the air temperature, wet-bulb temperature, and dew point are all the same.

from the surface up to the cloud deck. Soon, flakes of snow begin to fall from the clouds into the unsaturated layer below. In the air where temperatures are above freezing, the snowflakes begin to partially melt. The air, however, is dry, so the water quickly evaporates, cooling the air. In addition, evaporation cools the falling snowflake to the wet-bulb temperature, which retards its rate of melting. As snow continues to fall, evaporative cooling causes the air temperature to continue to drop. The addition of water vapor to the air increases the dew point, while the wet-bulb temperature remains essentially unchanged. Eventually, the entire layer of air cools to the wet-bulb temperature and becomes saturated at 0°C. As long as the wind doesn't bring in warmer air, the precipitation remains as snow.

We can see that when snow falls into warmer air (say at 8°C or 46°F), the air must be extremely dry in order to have a wet-bulb temperature at freezing or below. In fact, with an air temperature of 8°C (46°F) and a wet-bulb temperature of 0°C (32°F), the dew point would be –23°C (–9°F) and the relative humidity 11 percent. Conditions such as these are extremely unlikely at the surface before the onset of precipitation. Actually, the highest air temperature possible with a below-freezing wet-bulb temperature is about 10°C (50°F). Hence, snowflakes will melt rapidly in air with a temperature above this value. However, it is still possible to see flakes of snow at temperatures greater than 10°C (50°F), especially if the snowflakes are swept rapidly earthward by the cold, relatively dry downdraft of a thunderstorm.

FIGURE 2
It is snowing at 40°F during the middle of July near the summit of Beartooth Mountain, Montana.

TABLE 8.3
Ice Crystal Habits that Form at Various Temperatures

ENVIRONMENTAL TEMPERATURE (°C)	(°F)	CRYSTAL HABIT
0 to –4	32 to 25	Thin plates
–4 to –6	25 to 21	Needles
–6 to –10	21 to 14	Columns
–10 to –12	14 to 10	Plates
–12 to –16	10 to 3	Dendrites, plates
–16 to –22	3 to –8	Plates
–22 to –40	–8 to –40	Hollow columns

Note that at each temperature, the type of crystal that forms (e.g., hollow columns versus solid columns), will depend on the difference in saturation vapor pressure between ice and supercooled water.

TABLE 8.4   Snowfall Intensity

SNOWFALL DESCRIPTION	VISIBILITY
Light	Greater than $1/2$ mile*
Moderate	Greater than $1/4$ mile, less than or equal to $1/2$ mile
Heavy	Less than or equal to $1/4$ mile

*In the United States, the National Weather Service determines visibility (the greatest distance you can see) in miles.

FIGURE 8.16
The dendrite snow crystal.

Plate   Column   Dendrite   Needle

FIGURE 8.17
Common ice crystal forms (habits).

A thin film of water forms on the edge of the flakes, which acts like glue when other snowflakes come in contact with it. In this way, several flakes join to produce giant snowflakes often measuring several centimeters or more in diameter. These large, soggy snowflakes are associated with moist air and temperatures near freezing. However, when snowflakes fall through extremely cold air with a low moisture content, small, powdery flakes of "dry" snow accumulate on the ground. (To understand how snowflakes can survive in air that is above freezing, read the Focus section "When Is It Too Warm to Snow?", p. 194.)

If you catch falling snowflakes on a dark object and examine them closely, you will see that the most common snowflake form is a fernlike branching star shape called a *dendrite* (see Fig. 8.16). Since many types of ice crystals grow (see Fig. 8.17), why is the dendrite crystal the most common shape for snowflakes? The type of crystal that forms, as well as its growth rate, depends on the air temperature and relative humidity (the degree of supersaturation between water and ice). Table 8.3 summarizes the crystal forms (habits) that develop when supercooled water and ice coexist in a saturated environment. Note that dendrites are common at temperatures between –12°C and –16°C. The

maximum growth rate of ice crystals depends on the difference in saturation vapor pressure between water and ice, and this difference reaches a maximum in the temperature range where dendrite crystals are most likely to grow. (Look back at Fig. 8.8, p. 187.) Therefore, this type of crystal grows more rapidly than the other crystal forms. As ice crystals fall through a cloud, they are constantly exposed to changing temperature and moisture conditions. Since many ice crystals can join together (aggregate) to form a much larger snowflake, snow crystals may assume many complex patterns (see Fig. 8.18).

Snow falling from developing cumulus clouds is often in the form of **flurries**. These are usually light showers that fall intermittently for short durations and produce only light accumulations. A more intense snow shower is called a **snow squall**. These brief but heavy falls of snow are comparable to summer rain showers and, like snow flurries, usually fall from cumuliform clouds. A more continuous snowfall (sometimes steadily, for several hours) accompanies nimbostratus and altostratus clouds.

The intensity of snow is based on its reduction of horizontal visibility at the time of observation (see Table 8.4).

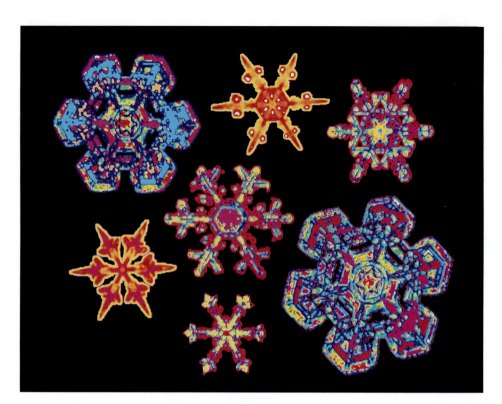

FIGURE 8.18
Computer color-enhanced image of dendrite snowflakes.

Heavy snow means the visibility due to falling snow is less than 0.5 km (¼ mi). As a general rule, heavy snow means a total accumulation of 10 or more centimeters (4 in.) during a 12-hour period, or a fall of 15 cm (6 in.) or more during a 24-hour period. There are, however, some variations to this general rule. For example, where average 10-cm snowfalls are common, a heavy snowfall is typically 15 cm or more. Where snow is infrequent, a heavy snow may be an accumulation of 5 or 6 cm (about 2 or 3 inches).

When a strong wind is blowing at the surface, snow can be picked up and deposited into huge drifts. Drifting snow is usually accompanied by *blowing snow*; that is, snow lifted from the surface by the wind and blown about in such quantities that horizontal visibility is greatly restricted. The combination of drifting and blowing snow, after falling snow has ended, is called a *ground blizzard*. A true **blizzard** is a weather condition characterized by low temperatures and strong winds (greater than 30 knots) bearing large amounts of fine, dry, powdery particles of snow, which can reduce visibility to only a few meters.

A Blanket of Snow   A mantle of snow covering the landscape is much more than a beautiful setting—it is a valuable resource provided by nature. A blanket of snow is a good insulator (poor heat conductor). In fact, the more air spaces there are between the individual snowflake crystals, the better insulator they become. A light, fluffy covering of snow protects sensitive plants and their root systems from damaging low temperatures by retarding the loss of ground heat.

On winter nights, ground that is covered with dry snow maintains a warmer temperature than ground that is exposed to the cold air. In this way, snow can prevent the ground from freezing downward to great depths. In cold climates that receive little snow, it is often difficult to grow certain crops because the frozen soil makes spring cultivation almost impossible. Frozen ground also prevents early spring rains from percolating downward into the soil, leading to rapid water runoff and flooding. If subsequent rains do not fall, the soil could even become moisture-deficient. If you become lost in a cold and windy snowstorm, build a snow cave and climb inside. It not only will protect you from the wind, but it also will protect you from the extreme cold by slowing the escape of heat your body generates.

The accumulation of snow in mountains provides for winter recreation, and the melting snow in spring and summer is of great economic value in that it supplies streams and reservoirs with much-needed water.

Winter snows may be beautiful, but they are not without hardships and potential hazards. As spring approaches, rapid melting of the snowpack may flood low-lying areas. Too much snow on the side of a steep hill or mountain may become an avalanche as the spring thaw approaches. The added weight of snow on the roof of a building may cause it to collapse, leading to costly repairs and even loss of life. Each winter, heavy snows clog streets and disrupt transportation. To keep traffic moving, streets must be plowed and sanded, or salted to lower the temperature at which the snow freezes (melts). This effort can

Sounds of Silence

Once snow settles on the ground, it can affect the way sound propagates. You may have noticed that, after a snowfall, it seems quieter than usual: Freshly fallen snow can absorb sound—just like acoustic tiles. As the snow gets deeper, this absorption increases. Anyone who has walked through the woods on a snowy evening knows the quiet created by a thick blanket of snow. As snow becomes older and more densely packed, its ability to absorb sound is reduced.

New snow covering a pavement will sometimes squeak as you walk in it. The sound produced is related to the snow's temperature. When the air and snow are only slightly below freezing, the pressure from the heel of a boot partially melts the snow. The snow can then flow under the weight of the boot and no sound is made. However, on cold days when the snow temperature drops below –10°C (14°F), the heel of the boot will not melt the snow, and the ice crystals are crushed. The crunching of the crystals produces the creaking sound.

---

be expensive, especially if the snow is heavy and wet. Cities unaccustomed to snow are usually harder hit by a moderate snowstorm than cities that frequently experience snow. A January snowfall of several centimeters in New Orleans, Louisiana, can bring traffic to a standstill, while a snowfall of several centimeters in Buffalo, New York, would go practically unnoticed. (A blanket of snow also has an effect on the way sounds are transmitted. More on this subject is given in the Focus section above.)

Sleet and Freezing Rain   Consider the falling snowflake in Fig. 8.19. As it falls into warmer air, it begins to melt. When it falls through the deep subfreezing surface layer of air, the partially melted snowflake or cold raindrop turns back into ice, not as a snowflake, but as a *tiny ice pellet* called **sleet**.* Generally, these ice pellets are transparent (or translucent), with diameters of 5 mm (0.2 in.) or less. They bounce when striking the ground and produce a tapping sound when they hit a window or piece of metal.

The cold surface layer beneath a cloud may be too shallow to freeze raindrops as they fall. In this case, they reach the surface as supercooled liquid drops. Upon striking a cold object, the drops spread out and almost immediately freeze, forming a thin veneer of ice. This form of precipitation is called **freezing rain**, or *glaze*. If the drops are small (less than 0.5 mm in diameter), the precipitation is called **freezing drizzle**. When small supercooled cloud or fog droplets strike an object whose temperature is below freezing, the tiny droplets freeze, forming an accumulation of white or milky granular ice called **rime** (see Fig. 8.20).

Freezing rain can create a beautiful winter wonderland by coating everything with silvery, glistening ice. At the same time,

FIGURE 8.19
Sleet forms when a partially melted snowflake or a cold raindrop freezes into a pellet of ice before reaching the ground.

FIGURE 8.20
An accumulation of rime forms on tree branches as supercooled fog droplets freeze on contact in the below-freezing air.

---

*Occasionally, the news media incorrectly use the term *sleet* to represent a mixture of rain and snow. The term used this way is, however, the British meaning.

FIGURE 8.21
A heavy coating of freezing rain during this ice storm caused tree limbs to break and power lines to sag.

highways turn into skating rinks for automobiles, and the destructive weight of the ice—which can be many tons on a single tree—breaks tree branches, power lines, and telephone cables (see Fig. 8.21). A case in point is the huge ice storm of January, 1998, which left millions of people without power in Northern New England and Canada and caused over $1 billion in damages. In another storm, as much as 4 cm (1.6 in.) of ice covered sections of Oregon and Washington in January, 1970, causing $6 million in damage to fruit trees in the Hood River Valley. The area most frequently hit by these storms extends over a broad region from Texas into Minnesota and eastward into the middle Atlantic states and New England. Such storms are extremely rare in southern California and Florida. (For additional information on freezing rain and its effect on air-craft, read the Focus section on p. 199.)

In summary, Fig. 8.22 shows various winter temperature profiles and the type of precipitation associated with each. In profile (a), the air temperature is below freezing at all levels, and snowflakes reach the surface. In (b), a zone of above-freezing air causes snowflakes to partially melt; then, in the deep, subfreez-

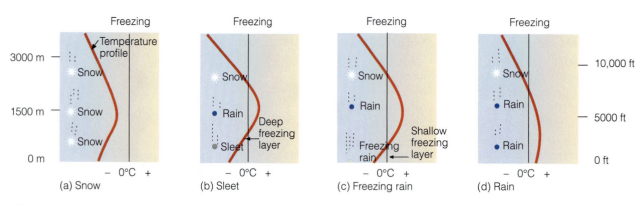

FIGURE 8.22
Vertical temperature profiles (solid red line) associated with different forms of precipitation.

Aircraft Icing

Consider an aircraft flying through an area of freezing rain or through a region of large supercooled droplets in a cumuliform cloud. As the large, supercooled drops strike the leading edge of the wing, they break apart and form a film of water, which quickly freezes into a solid sheet of ice. This smooth, transparent ice—called *clear ice*—is similar to the freezing rain or glaze that coats trees during ice storms. Clear ice can build up quickly; it is heavy and difficult to remove, even with modern de-icers.

When an aircraft flies through a cloud composed of tiny, supercooled liquid droplets, *rime ice* may form. Rime ice forms when some of the cloud droplets strike the wing and freeze before they have time to spread, thus leaving a rough and brittle coating of ice on the wing. Because the small, frozen droplets trap air between them, rime ice usually appears white (see Fig. 8.20). Even though rime ice redistributes the flow of air over the wing more than clear ice does, it is lighter in weight and is more easily removed with de-icers.

Because the raindrops and cloud droplets in most clouds vary in size, a mixture of clear and rime ice usually forms on aircraft. Also, because concentrations of liquid water tend to be greatest in warm air, icing is usually heaviest and most severe when the air temperature is between 0°C and −10°C (32°F and 14°F).

A major hazard to aviation, icing reduces aircraft efficiency by increasing weight. Icing has other adverse effects, depending on where it forms. On a wing or fuselage, ice can disrupt the air flow and decrease the plane's flying capability. When ice forms in the air intake of the engine, it robs the engine of air, causing a reduction in power. Icing may also affect the operation of brakes, landing gear, and instruments. Because of the hazards of ice on an aircraft, its wings are usually sprayed with a type of antifreeze before taking off during cold, inclement weather.

---

ing air at the surface, the liquid freezes into sleet. In the shallow subfreezing surface air in (c), the melted snowflakes freeze on contact, producing freezing rain. In (d), the air temperature is above freezing in a sufficiently deep layer so that precipitation reaches the surface as rain. (Weather symbols for these and other forms of precipitation are presented in Appendix B.)

Snow Grains and Snow Pellets   **Snow grains** are small, opaque grains of ice, the solid equivalent of drizzle. They are fairly flat or elongated, with diameters generally less than 1 mm (0.04 in.). They fall in small quantities from stratus clouds, and never in the form of a shower. Upon striking a hard surface, they neither bounce nor shatter. **Snow pellets**, on the other hand, are white, opaque grains of ice, with diameters less than 5 mm (0.2 in.). They are sometimes confused with snow grains. The distinction is easily made, however, by remembering that, unlike snow grains, snow pellets are brittle, crunchy, and bounce (or break apart) upon hitting a hard surface. They usually fall as showers, especially from cumulus congestus clouds.

To understand how snow pellets form consider the cumulus congestus cloud with a high liquid water content in Fig. 8.23. The freezing level is near the surface and, since the atmosphere is unstable, the air temperature drops quickly with height. An ice crystal falling into the cold (−23°C) middle region of the cloud would be surrounded by many supercooled cloud droplets and ice crystals. In the very cold air, the crystals tend to rebound after colliding rather than sticking to one another. However, when the ice crystals collide with the supercooled water droplets, they immedi-

ately freeze the droplets, producing a spherical accumulation of icy matter (rime) containing many tiny air spaces. These small air bubbles have two effects on the growing ice particle: (1) they keep its density low; and (2) they scatter light, making the particle opaque. By the time the ice particle reaches the lower half of the cloud, it has grown in size and its original shape is gone. When the ice particle accumulates so much rime that it can no longer be recognized as an ice crystal (or snowflake), it is called *graupel*. Since

FIGURE 8.23

The formation of snow pellets. In the cold air of a convective cloud, with a high liquid water content, ice particles collide with supercooled cloud droplets, freezing them into clumps of icy matter called *graupel*. Upon reaching the relatively cold surface, the graupel is classified as snow pellets.

WEATHER WATCH

Each year between 500 and 700 hailstorms produce hail large enough to cause damage or injury in the United States. It is rare, however, for a person to be killed by hail from one of these storms. But hail can be lethal—a hailstone the size of a baseball will whiz earthward at a speed close to 87 knots (100 mi/hr). During April, 1996, for instance, 92 people tragically perished in a hailstorm near Gopolganj, Bangladesh. Some of the hailstones in that storm reportedly weighed about 1 kg (2.2 lbs.). In another example, 200 sheep caught in an open field during a violent hailstorm were killed in Montana during April, 1978, as baseball-size hailstones became lethal missiles.

the freezing level is at a low elevation, the graupel reaches the surface as a light, round clump of snowlike ice called a *snow pellet*. In a thunderstorm, when the freezing level is well above the surface, graupel that reaches the ground is sometimes called *soft hail*. During summer, the graupel may melt and reach the surface as a large raindrop. In vigorously convective clouds, however, the graupel may develop into full-fledged hailstones (see Fig. 8.24).

Hail   **Hailstones** are pieces of ice, either transparent or partially opaque, ranging in size from that of small peas to that of golf balls or larger. Some are round, others take on irregular

FIGURE 8.24
The accumulation of small hail after a thunderstorm. The hail formed as supercooled cloud droplets collected on ice particles called *graupel* inside a cumulonimbus cloud.

shapes. The largest authenticated hailstone in the United States fell on Coffeyville, Kansas, in September, 1970. This giant weighed 757 grams (1.67 lb) and had a measured diameter of over 14 cm (5.5 in.). Canada's record hailstone fell on Cedoux, Saskatchewan, during August, 1973. It weighed 290 grams (0.6 lb) and measured about 10 cm (4 in.) in diameter. Needless to say, large hailstones are quite destructive. They can break windows, dent cars, batter roofs of homes, and cause extensive damage to livestock and crops. In fact, a single hailstorm can destroy a farmer's crop in a matter of minutes, which is why farmers sometimes call it "the white plague."

Estimates are that, in the United States alone, hail damage amounts to hundreds of millions of dollars annually. The costliest hailstorm on record in the United States battered the Front Range of the Rocky Mountains in Colorado with golf ball- and baseball-size hail on July 11, 1990. The storm damaged thousands of roofs and tens of thousands of cars, trucks, and streetlights, causing an estimated $625 million in damage. Although hailstones are potentially lethal, only two fatalities due to falling hail have been documented in the United States during the twentieth century.

Hail is produced in a cumulonimbus cloud when graupel, large frozen raindrops, or just about any particles (even insects) act as embryos that grow by accumulating supercooled liquid droplets—*accretion*. It takes a million cloud droplets to form a single raindrop, but it takes about 10 billion cloud droplets to form a golf ball–size hailstone. For a hailstone to grow to this size, it must remain in the cloud for between 5 and 10 minutes. Violent, upsurging air currents within the cloud carry small embryos high above the freezing level. When the updrafts are tilted, the embryos are swept laterally through the cloud. Studies reveal that the width and tilt of the main updraft are very important to hailstone growth, with the best trajectory being one that is nearly horizontal through the cloud (see Fig. 8.25).

As the embryos pass through regions of varying liquid water content, a coating of ice forms around them and they grow larger and larger. When the ice particles are appreciable size, they become too large and heavy to be supported by the rising air, and they then begin to fall as hail. As they slowly descend, the hailstones may get caught in a violent updraft only to be carried upward once again to repeat the cycle. Or, they may fall through the cloud and begin to melt in the warmer air below. Small hailstones often melt before reaching the ground, but, in the violent thunderstorms of summer, hailstones may grow large enough to reach the ground before completely melting. Strangely, then, we find the largest form of frozen precipitation occurring during the warmest time of the year.

As the cumulonimbus cloud moves along, it may deposit its hail in a long narrow band (often several kilometers wide and about 10 kilometers long) known as a **hailstreak**. If the cloud should remain almost stationary for a period of time, substantial accumulation of hail is possible. For example, in

June, 1984, a devastating hailstorm lasting over an hour dumped knee-deep hail on the suburbs of Denver, Colorado. In addition to its destructive effect, accumulation of hail on a roadway is a hazard to traffic, as when, for example, four people lost their lives near Soda Springs, California, in a 15-vehicle pileup on a hail-covered freeway during September, 1989.

Figure 8.26 shows a cut section of the giant Coffeyville hailstone. Notice that it has concentric layers of milky white and clear ice. We have already seen that as a hailstone travels through regions of a cloud with varying water content, it grows larger by accumulating supercooled water droplets. If the air is very cold and the liquid water content relatively low, supercooled droplets freeze immediately on the stone, producing a coating of white or opaque rime ice containing many air bubbles. As the water freezes, heat is released, which keeps the surface of the hailstone warmer than its environment.

If the stone enters the wetter part of the cloud, where the liquid water content is higher and many more supercooled water droplets exist, liquid droplets may collect so rapidly that the hailstone's surface temperature remains above 0°C. Now, the supercooled droplets no longer freeze on impact; instead, they spread a coating of water around the hailstone. As the hailstone re-enters a colder section of the cloud, the surface water cools and slowly freezes. Trapped air escapes, and the ice that forms is clear. Therefore, as a hailstone passes through a cloud of changing temperatures and liquid water content, alternating layers of opaque and clear ice form. Observing these layers can help us understand much about the hailstone's environment during its formation.

Because hailstones are so damaging, various methods have been tried to prevent them from forming in thunderstorms. One method employs the seeding of clouds with large quantities of silver iodide. These nuclei freeze supercooled water droplets and convert them into ice crystals. The ice crystals

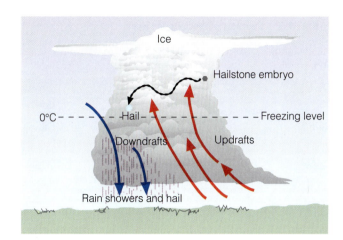

FIGURE 8.25

Hailstones begin as embryos (usually ice particles) that remain suspended in the cloud by violent updrafts. When the updrafts are tilted, the ice particles are swept horizontally through the cloud, producing the optimal trajectory for hailstone growth. Along their path, the ice particles collide with supercooled liquid droplets, which freeze on contact. The ice particles eventually grow large enough and heavy enough to fall toward the ground as hailstones.

grow larger as they come in contact with additional supercooled cloud droplets. In time, the ice crystals grow large enough to be called graupel, which then becomes a hailstone embryo. Large numbers of embryos are produced by seeding in hopes that competition for the remaining supercooled droplets may be so great that none of the embryos would be able to grow into large and destructive hailstones. Russian scientists claim great success in suppressing hail using ice nuclei, such as silver iodide and lead iodide. In the United States, the results of most hail-suppression experiments have been inconclusive.

FIGURE 8.26

The giant Coffeyville hailstone first cut then photographed under regular light *(left)* and polarized light *(right)*. This procedure reveals its layered structure.

TABLE 8.5 Summary of Precipitation Types

PRECIPITATION TYPE	WEATHER SYMBOL	DESCRIPTION
Drizzle	" (light)	Tiny water drops with diameters less than 0.5 mm that fall slowly, usually from a stratus cloud
Rain	•• (light)	Falling liquid drops that have diameters greater than 0.5 mm
Snow	✳✳ (light)	White (or translucent) ice crystals in complex hexagonal (six-sided) shapes that often join together to form snowflakes
Sleet (ice pellets)	△	Frozen raindrops that form as cold raindrops (or partially melted snowflakes) refreeze while falling through a relatively deep subfreezing layer
Freezing rain	∼ (light)	Supercooled raindrops that fall through a relatively shallow subfreezing layer and freeze upon contact with cold objects at the surface
Snow grains (granular snow)	△	White or opaque particles of ice less than 1 mm in diameter that usually fall from stratus clouds, and are the solid equivalent of drizzle
Snow pellets (graupel)	⚇ (light showers)	Brittle, soft white (or opaque), usually round particles of ice with diameters less than 5 mm that generally fall as showers from cumuliform clouds; they are softer and larger than snow grains
Hail	⬦ (moderate or heavy showers)	Transparent or partially opaque ice particles in the shape of balls or irregular lumps that range in size from that of a pea to that of a softball; the largest form of precipitation. *Large hail* has a diameter of $3/4$ in. or greater; hail almost always is produced in a thunderstorm

Up to this point, we have examined the various types of precipitation. The different types (from drizzle to hail) are summarized in Table 8.5.

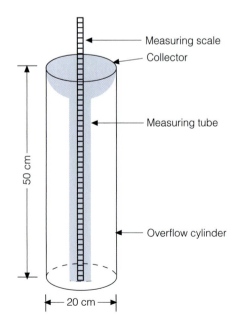

Measuring scale
Collector
Measuring tube

50 cm

Overflow cylinder

FIGURE 8.27
Components of the standard rain gauge.

20 cm

MEASURING PRECIPITATION

Instruments Any instrument that can be used to collect and measure rainfall is called a *rain gauge*. A **standard rain gauge** consists of a funnel-shaped collector attached to a long measuring tube (see Fig. 8.27). The cross-sectional area of the collector is 10 times that of the tube. Hence, rain falling into the collector is amplified tenfold in the tube, permitting measurements of great precision. A wooden scale, calibrated to allow for the vertical exaggeration, is inserted into the tube and withdrawn. The wet portion of the scale indicates the depth of water. So, 10 inches of water in the tube would be measured as 1 inch of rainfall. Because of this amplification, rainfall measurements can be made when the amount is as small as one-hundredth (0.01) of an inch. An amount less than this is called a **trace**.

The measuring tube can only collect 5 cm or 2 in. of rain. Rainfall of more than this amount causes an overflow into an outer cylinder. Here, the excess rainfall is stored and protected from appreciable evaporation. When the gauge is emptied, the overflow is carefully poured into the tube and measured.

Another instrument that measures rainfall is the **tipping bucket rain gauge**. In Fig. 8.28, notice that this gauge has a receiving funnel leading to two small metal collectors (buckets). The bucket beneath the funnel collects the rain water. When it accumulates the equivalent of one-hundredth of an inch of rain, the weight of the water causes it to tip and empty itself. The second bucket immediately moves under the funnel to catch the water. When it fills, it also tips and empties itself, while the original bucket moves back beneath the funnel. Each time a bucket tips, an electric contact is made, causing a pen to register a mark on a remote recording chart. Adding up the total number of marks gives the rainfall for a certain time period.

A problem with the tipping bucket rain gauge is that during each "tip" it loses some rainfall and, therefore, undermeasures rainfall amounts, especially during heavy downpours. The tipping bucket is the rain gauge used in the automated (ASOS) weather stations.

Remote recording of precipitation can also be made with a **weighing-type rain gauge**. With this gauge, precipitation is caught in a cylinder and accumulates in a bucket. The bucket sits on a sensitive weighing platform. Special gears translate the accumulated weight of rain or snow into millimeters or inches of precipitation. The precipitation totals are recorded by a pen on chart paper, which covers a clock-driven drum. By using special electronic equipment, this information can be transmitted from rain gauges in remote areas to satellites or land-based stations, thus providing precipitation totals from previously inaccessible regions.

The depth of snow in a region is determined by measuring its depth at three or more representative areas. The amount of snowfall is defined as the average of these measurements. Since snow often blows around and accumulates into drifts, finding a representative area can be a problem. Determining the actual depth of snow can include considerable educated guesswork. Snow depth may also be measured by removing the collector and inner cylinder of a standard rain gauge and allowing snow to accumulate in the outer tube. Turbulent air around the edge of the tube often blows flakes away from the gauge. This makes the amount of snow collected less than the actual snowfall. To remedy this, slatted windshields are placed around the cylinder to block the wind and ensure a more correct catch.

The depth of water that would result from the melting of a snow sample is called the **water equivalent**. In a typical fresh snowpack, about 10 cm of snow will melt down to about 1 cm of water, giving a water equivalent ratio of 10:1. This ratio, however, will vary greatly, depending on the texture and packing of the snow. Very wet snow falling in air near freezing may have a water equivalent of 6:1. On the other hand, in dry powdery snow, the ratio may be as high as 30:1. Toward the end of the winter, large compacted drifts representing the accumulation of many storms may have a water equivalent of less than 2:1.

Determining the water equivalent of snow is a fairly straightforward process: The snow accumulated in a rain gauge is melted and its depth is measured. Another method

FIGURE 8.28
The tipping bucket rain gauge. Each time the bucket fills with one-hundredth of an inch of rain, it tips, sending an electric signal to the remote recorder.

uses a long, hollow tube pushed into the snow to a desired depth. This snow sample is then melted and poured into a rain gauge for measuring its depth. Knowing the water equivalent of snow can provide valuable information about spring runoff and the potential for flooding, especially in mountain areas.

Precipitation is a highly variable weather element. A huge thunderstorm may drench one section of a town while leaving another section completely dry. Given this variability, it should be apparent that a single rain gauge on top of a building cannot represent the total precipitation for any particular region.

Weather Radar and Precipitation After it showers, ask someone what time it began to rain. More than likely the answer will be, "When the first raindrops hit the ground." Did the rain really begin at this time? Actually, raindrops begin to fall within clouds as soon as they are large enough to overcome any updrafts. Raindrops falling from the middle of a cloud 6100 m (about 20,000 ft) thick take over 9 minutes to reach the surface. And what about updrafts? Falling drops can become caught in rising currents and remain suspended for a long time. So, when *does* rain begin to fall? Until atmospheric scientists began using weather radar to study precipitation, the answer to this question was indeed difficult to obtain.

Radar (*radio detection and ranging*) has become an essential tool of the atmospheric scientist, for it gathers information about storms and precipitation in previously inaccessible regions. Atmospheric scientists use weather radar to examine the inside of a cloud much like physicians use X-rays to examine the inside of a human body. Essentially, the radar unit consists of a transmitter that sends out short, powerful microwave pulses having wavelengths between 1 and 20 cm. When this energy encounters a foreign object—called a *target*—a fraction of the energy is scattered back toward the transmitter and is detected by a receiver. The returning signal is amplified and displayed on a screen, producing an image, or *echo*, from the target. The elapsed time between transmission and reception indicates the target's distance.

(a)

(b)

FIGURE 8.29
(a) Doppler radar display showing precipitation intensity over Oklahoma for April 24, 1999.
(b) Doppler radar display showing 1-hour rainfall over Oklahoma for April 24, 1999 amounts.

Smaller targets require detection by shorter wavelengths. Cloud droplets are detected by radar using wavelengths of 1 cm, whereas longer wavelengths (between 3 and 10 cm) are only weakly scattered by tiny cloud droplets, but are strongly scattered by larger precipitation particles. The brightness of the echo is directly related to the amount (intensity) of rain falling in the cloud. So, the radar screen shows not only where precipitation is occurring, but also how intense it is. The radar image typically is displayed using various colors to denote the intensity of precipitation, with light green representing the lightest precipitation and orange and red representing the heaviest precipitation.

In recent years, **Doppler radar** has replaced the conventional radar units that were put into service shortly after World War II. Doppler radar is like conventional radar in that it can detect areas of precipitation and measure rainfall intensity (see Fig. 8.29a). Using special computer programs called *algorithms,* the rainfall intensity, over a given area for a given time, can be computed and displayed as an estimate of total rainfall over that particular area (see Fig. 8.29b). But the Doppler radar can do more than conventional radar.

Because the Doppler radar uses the principle called *Doppler shift,** it has the capacity to measure the speed at which

*The Doppler shift (or effect) is the change in the frequency of waves that occurs when the emitter or the observer is moving toward or away from the other. As an example, suppose a high-speed train is approaching you. The higher-pitched (higher frequency) whistle you hear as the train approaches will shift to a lower pitch (lower frequency) after the train passes.

falling rain is moving horizontally toward or away from the radar antenna. Falling rain moves with the wind. Consequently, Doppler radar allows scientists to peer into a tornado-generating thunderstorm and observe its wind. We will investigate these ideas further in Chapter 15, when we consider the formation of severe thunderstorms and tornadoes.

In some instances, radar displays indicate precipitation where there is none reaching the surface. This situation happens because the radar beam travels in a straight line and the earth curves away from it. Hence, the return echo is not necessarily that of precipitation reaching the ground, but is that of raindrops in the cloud.

SUMMARY

In this chapter, we have seen that cloud droplets are too small and light to reach the ground as rain. Cloud droplets do grow larger by condensation, but this process by itself is much too slow to produce substantial precipitation. Because larger cloud droplets fall faster and farther than smaller ones, they grow larger as they fall by coalescing with drops in their path. If the temperature in a cloud drops below freezing, then ice crystals play an important role in producing precipitation. Some ice crystals may form directly on ice nuclei, or they may result when an ice nucleus makes contact with and freezes a water droplet. Because of differences in vapor pressures between water and ice, an ice crystal surrounded by water droplets grows larger at the expense of the droplets. As it begins to fall, it grows even larger by colliding with liquid droplets, which freeze on contact. In an attempt to coax more precipitation from them, some clouds are seeded.

Precipitation can reach the surface in a variety of forms. In winter, raindrops may freeze on impact, producing freezing rain that can disrupt electrical service by downing power lines. Raindrops may freeze into tiny pellets of ice above the ground and reach the surface as sleet. Depending on conditions, snow may fall as pellets, grains, or flakes, all of which can influence how far we see and hear. Strong updrafts in a cumulonimbus cloud may carry ice particles high above the freezing level, where they acquire a further coating of ice and form destructive hailstones. Although the rain gauge is still the most commonly used method of measuring precipitation, Doppler radar has become an important instrument for determining precipitation intensity and estimating rainfall amount.

Key Terms

The following terms are listed in the order they appear in the text. Define each. Doing so will aid you in reviewing the material covered in this chapter.

precipitation	coalescence
equilibrium vapor pressure	ice-crystal (Bergeron) process
curvature effect	supercooled (water droplets)
solute effect	homogeneous (spontaneous) freezing
terminal velocity	

ice nuclei	sleet (ice pellets)
contact freezing	freezing rain
accretion	freezing drizzle
graupel	rime
aggregation	snow grains
cloud seeding	snow pellets
rain	hailstones
drizzle	hailstreak
virga	standard rain gauge
shower (rain)	trace (of precipitation)
snow	tipping bucket rain gauge
fallstreaks	weighing-type rain gauge
flurries (of snow)	water equivalent
snow squall	radar
blizzard	Doppler radar

Questions for Review

1. What is the primary difference between a cloud droplet and a raindrop?
2. Why do typical cloud droplets seldom reach the ground as rain?
3. Describe how the process of collision and coalescence produces precipitation.
4. Would the collision-and-coalescence process work better at producing rain in (a) a warm, thick nimbostratus cloud or (b) a warm, towering cumulus congestus cloud? Explain.
5. List and describe three ways in which ice crystals can form in a cloud.
6. When the temperature in a cloud is –30°C, why are larger cloud droplets more likely to freeze than smaller cloud droplets?
7. In a cloud where the air temperature is –10°C, why are there many more cloud droplets than ice crystals?
8. How does the ice-crystal (Bergeron) process produce precipitation? What is the *main* premise describing this process?
9. Why is it foolish to seed a clear sky with silver iodide?
10. List the advantages and disadvantages of heavy snowfall.

11. Why do heavy showers usually fall from cumuliform clouds? Why does steady precipitation normally fall from stratiform clouds?

12. Why are large snowflakes usually observed when the air temperature near the ground is just below freezing?

13. In a cloud composed of water droplets and ice crystals, is the saturation vapor pressure greater over the droplets or over the ice?

14. When seeding a cloud to promote rainfall, is it possible to overseed the cloud so that it prevents rainfall? Explain.

15. What atmospheric conditions are necessary for snow to fall when the air temperature is considerably above freezing?

16. How do the atmospheric conditions that produce sleet differ from those that produce hail?

17. What is the difference between freezing rain and sleet?

18. Why is hail more common in summer than in winter?

19. How does radar measure the intensity of precipitation?

20. List the common precipitation gauges that measure rain and snow.

Questions for Thought

1. Ice crystals that form by accretion are fairly large. Explain why they fall slowly.

2. Why is a warm, tropical cumulus cloud more likely to produce precipitation than a cold, stratus cloud?

3. Explain why very small cloud droplets of pure water evaporate even when the relative humidity is 100 percent.

4. Suppose a thick nimbostratus cloud contains ice crystals and cloud droplets all about the same size. Which precipitation process will be most important in producing rain from this cloud? Why?

5. Clouds that form over water are usually more efficient in producing precipitation than clouds that form over land. Why?

6. Everyday in summer a blizzard occurs over the Great Plains. Explain where and why.

7. During a recent snowstorm, Denver, Colorado, received 7 cm (3 in.) of snow. Sixty kilometers east of Denver, a city received no measurable snowfall, while 150 km east of Denver another city received 10 cm (4 in.) of snow. Since Denver is located to the east of the Rockies, and the upper-level winds were westerly during the snowstorm, give an explanation as to what *could* account for this snowfall pattern.

8. Raindrops rarely grow larger than 5 mm. Two reasons were given on p. 185. Can you think of a third? (Hint: See the Focus section on p. 193, and look at the shape of a large drop.)

9. Lead iodide is an effective ice-forming nucleus. Why do you think it has not been used for that purpose?

10. When cirrus clouds are above a deck of altocumulus clouds, occasionally a clear area, or "hole," will appear in the altocumulus cloud layer. What do you suppose could cause this to happen?

11. It is −12°C (10°F) in Albany, New York, and freezing rain is falling. Can you explain why?

12. When falling snowflakes become mixed with sleet, why is this condition often followed by the snowflakes changing into rain?

Problems and Exercises

1. In the daily newspaper, a city is reported as receiving 1.32 cm (0.52 in.) of precipitation over a 24-hour period. If all the precipitation fell as snow, and if we assume a normal water equivalent ratio of 10:1, how much snow did this city receive?

2. How many times faster does a large raindrop (diameter 5000 μm) fall than a cloud droplet (diameter 20 μm), if both are falling at their terminal velocity in still air?

3. (a) How many minutes would it take drizzle with a diameter of 200 μm to reach the surface if it falls at its terminal velocity from the base of a cloud 1000 m (about 3300 ft) above the ground? (Assume the air is saturated beneath the cloud, the drizzle does not evaporate, and the air is still.)
   (b) Suppose the drizzle in problem 3a evaporates on its way to the ground. If the drop size is 200 μm for the first 450 m of descent, 100 μm for the next 450 m, and 20 μm for the final 100 m, how long will it take the drizzle to reach the ground if it falls in still air?

4. Suppose a large raindrop (diameter 5000 μm) falls at its terminal velocity from the base of a cloud 1500 m (about 5000 ft) above the ground.
   (a) If we assume the raindrop does not evaporate, how long would it take the drop to reach the surface?
   (b) What would be the shape of the falling raindrop just before it reaches the ground?
   (c) What type of cloud would you expect this raindrop to fall from? Explain.

5. Describe a plausible life history for the large hailstone in the figure below. Explain how the areas of clear ice and rime ice formed.

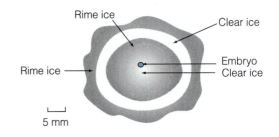

Questions for Exploration

1. On the Blue Skies: College Edition CD-ROM go to the **Weather Analysis** section of the CD and click on "**Weather Maps**." Use the interactive weather maps to locate regions of the country where precipitation is being reported.

   (a) What is the relationship between relative humidity and precipitation?

   (b) What is the relationship between pressure patterns and precipitation?

2. On the Blue Skies: College Edition CD-ROM go to the **Atmospheric Basics** section of the CD and click on "**Layers of the Atmosphere**." Choose a sounding from an area where precipitation is occurring. Can you see the layer that is most likely producing the precipitation in the profile?

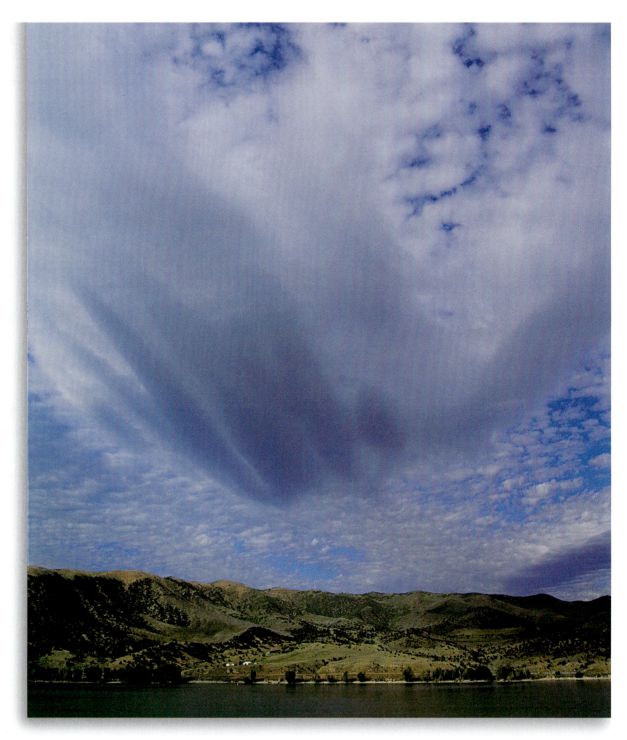

High winds blowing over irregular terrain in
Wyoming whip these altocumulus clouds
into waves. (Photo by author.)

THE ATMOSPHERE IN MOTION: AIR PRESSURE, FORCES, AND WINDS

CONTENTS

December 19, 1980, was a cool day in Lynn, Massachusetts, but not cool enough to dampen the spirits of more than 2000 people who gathered in Central Square—all hoping to catch at least one of the 1500 dollar bills that would be dropped from a small airplane at noon. Right on schedule, the aircraft circled the city and dumped the money onto the people below. However, to the dismay of the onlookers, a westerly wind caught the currency before it reached the ground and carried it out over the cold Atlantic Ocean. Had the pilot or the sponsoring leather manufacturer examined the weather charts beforehand, it might have been able to predict that the wind would ruin the advertising scheme.

This opening scenario raises two questions: (1) Why does the wind blow? and (2) How can one tell its direction by looking at weather charts? Chapter 1 has already answered the first question: Air moves in response to horizontal differences in pressure. This phenomenon happens when we open a vacuum-packed can—air rushes from the higher pressure region outside the can toward the region of lower pressure inside. In the atmosphere, the wind blows in an attempt to equalize imbalances in air pressure. Does this mean that the wind always blows directly from high to low pressure? Not really, because the movement of air is controlled not only by pressure differences but by other forces as well.

In this chapter, we will first consider how and why atmospheric pressure varies, then we will look at the forces that influence atmospheric motions aloft and at the surface. Through studying these forces, we will be able to tell how the wind should blow in a particular region by examining surface and upper-air charts.

ATMOSPHERIC PRESSURE

In Chapter 1, we learned several important concepts about atmospheric pressure. One stated that **air pressure** is simply the weight of air above a given level. As we climb in elevation above the earth's surface, there are fewer air molecules above us; hence, atmospheric pressure always decreases with increasing height. Another concept we learned was that most of our atmosphere is crowded close to the earth's surface, which causes air pressure to decrease with height, rapidly at first, then more slowly at higher altitudes. But there are other important concepts to consider. For example, air pressure, air density (the mass of air in a given volume), and air temperature are all interrelated. If one of these variables changes, the other two usually change as well. The relationship among these three variables is expressed by the gas law, which is described in the Focus section on p. 212.

To help eliminate some of the complexities of the atmosphere, scientists construct *models*. Figure 9.1 shows a simple atmospheric model—a column of air, extending well up into the atmosphere. In the column, the dots represent air molecules. Our model assumes: (1) that the air molecules are not crowded close to the surface and, unlike the real atmosphere, the air density remains constant from the surface up to the top of the column, and (2) that the width of the column does not change with height.

Suppose we somehow force more air into the column in Fig. 9.1. What would happen? If the air temperature in the column does not change, the added air would make the column more dense, and the added mass of the air in the column would increase the surface air pressure. Likewise, if a great deal of air were removed from the column, the surface air pressure would decrease. With the same assumptions for our model in Fig. 9.1, let's look at Fig. 9.2.

Air column

FIGURE 9.1

A model of the atmosphere where air density remains constant with height. The air pressure at the surface is related to the number of molecules above. When air of the same temperature is stuffed into the column, the surface air pressure rises. When air is removed from the column, the surface pressure falls.

Suppose the two air columns in Fig. 9.2a are located at the same elevation and have identical surface air pressures. This condition, of course, means that there must be the same number of molecules (same mass of air) in each column above both cities. Further suppose that the surface air pressure for both cities remains the same, while the air above city 1 cools and the air above city 2 warms (see Fig. 9.2b).

As the air in column 1 cools, the molecules move more slowly and crowd closer together—the air becomes more dense. In the warm air above city 2, the molecules move faster and spread farther apart—the air becomes less dense. Since the width of the columns does not change, (and if we assume an invisible barrier exists between the columns), the surface pressure does not vary and the total number of molecules above each city must remain the same. Therefore, in the more dense cold air above city 1, the column shrinks, while the column rises in the less-dense, warm air above city 2.

We now have a cold shorter column of air above city 1 and a warm taller air column above city 2. From this situation, we can conclude that *it takes a shorter column of cold, more dense air to exert the same surface pressure as a taller column of warm, less dense air.* This concept has a great deal of meteorological significance.

Atmospheric pressure decreases more rapidly with elevation in the cold column of air. In the cold air above city 1 (Fig. 9.2b), move up the column and observe how quickly you pass through the densely packed molecules. This activity indicates a rapid change in pressure. In the warmer, less dense air, the pressure does not decrease as rapidly with height, simply because you climb above fewer molecules in the same vertical distance.

In Fig. 9.2c, move up the warm column until you come to the letter H. Now move up the cold column the same distance until you reach the letter L. Notice that there are more molecules above the letter H in the warm column than above the letter L in the cold column. The fact that the number of molecules above any level is a measure of the atmospheric pressure leads to an important concept: *Warm air aloft is normally associated with high atmospheric pressure, and cold air aloft is associated with low atmospheric pressure.*

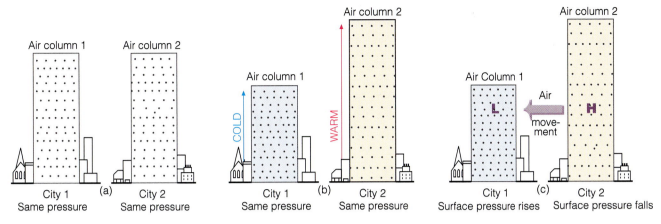

FIGURE 9.2

It takes a shorter column of cold air to exert the same pressure as a taller column of warm air. Because of this fact, cold air aloft is associated with low pressure and warm air aloft with high pressure. The pressure differences aloft create a force that causes the air to move from a region of higher pressure toward a region of lower pressure. The removal of air from column 2 causes its surface pressure to drop, whereas the addition of air into column 1 causes its surface pressure to rise. (The difference in height between the two columns is greatly exaggerated.)

In Fig. 9.2c, the horizontal difference in temperature creates a horizontal difference in pressure. The pressure difference establishes a force (called the *pressure gradient force*) that causes the air to move from higher pressure toward lower pressure. Consequently, if we remove the invisible barrier between the two columns and allow the air aloft to move horizontally, the air will move from column 2 toward column 1. As the air aloft leaves column 2, the mass of the air in the column decreases, and so does the surface air pressure. Meanwhile, the accumulation of air in column 1 causes the surface air pressure to increase.

In summary, heating or cooling a column of air can establish horizontal variations in pressure that cause the air to move. The net accumulation of air above the surface causes the surface air pressure to rise, whereas a decrease in the amount of air above the surface causes the surface air pressure to fall.

From what we have learned so far, we might expect to see the surface pressure dropping as the air temperature rises, and vice versa. Over large continental areas, especially the southwestern United States in summer, hot surface air is accompanied by surface low pressure. Likewise, bitter cold arctic air in winter is often accompanied by surface high pressure. Yet, on a daily basis, any cyclic change in surface pressure brought on by daily temperature changes is concealed by the pressure changes created by the warming of the upper atmosphere.

In the tropics, for example, pressure rises and falls in a regular pattern twice a day (see Fig. 9.3). Maximum pressures occur around 10:00 A.M. and 10:00 P.M., minimum near

Sea level pressure changes in middle latitudes on a given day

Average sea level pressure changes in the tropics

FIGURE 9.3

Daily sea level pressure variations in middle and low latitudes. Middle-latitude pressure changes are primarily the result of large high- and low-pressure areas that move toward or away from a region. Diurnal surface pressure changes in the tropics reflect the influence of the atmospheric tides.

The Atmosphere Obeys the Gas Law

The relationship among the pressure, temperature, and density of air can be expressed by

Pressure = temperature × density × constant.

This simple relationship, often referred to as the *gas law* (or *equation of state*), tells us that the pressure of a gas is equal to its temperature times its density times a constant. When we ignore the constant and look at the gas law in symbolic form, it becomes

$$p \sim T \times \rho$$

where, of course, p is pressure, T is temperature, and ρ (the Greek letter rho, pronounced "row") represents air density.* The line $\sim$ is a symbol meaning "is proportional to." A change in one variable causes a corresponding change in the other two variables. Thus, it will be easier to understand

---

*The gas law may also be written as $p \times v = T \times$ constant. Consequently, pressure and temperature changes are also related to changes in volume.

FIGURE 1

Air above a region of surface high pressure is more dense than air above a region of surface low pressure (at the same temperature). (The dots in each column represent air molecules.)

the behavior of a gas if we keep one variable from changing and observe the behavior of the other two.

Suppose, for example, we hold the temperature constant. The relationship then becomes

$$p \sim \rho \text{ (temperature constant)}.$$

This expression says that the pressure of the gas is proportional to its density, as long as its temperature does not change. Consequently, if the

temperature of a gas (such as air) is held constant, as the pressure increases the density increases, and as the pressure decreases the density decreases. In other words, at *the same temperature, air at a higher pressure is more dense than air at a lower pressure*. If we apply this concept to the atmosphere, then with nearly the same temperature and elevation, air above a region of surface high pressure is more dense than air above a region of surface low pressure (see Fig. 1).

We can see, then, that for surface high-pressure areas (anticyclones) and surface low-pressure areas (mid-latitude storms) to form, the air density (mass of air) above these systems must change. As we will see later in this chapter, as well as in other chapters, surface air pressure increases when the wind causes more air to move into a column of air than is able to leave (called *net convergence*), and surface air pressure decreases when the wind causes more air to move out

---

4:00 A.M. and 4:00 P.M. The largest pressure difference, about 2.5 mb, occurs near the equator. It also shows up in higher latitudes, but with a much smaller amplitude. This daily (*diurnal*) fluctuation of pressure appears to be due primarily to the absorption of solar energy by ozone in the upper atmosphere and by water vapor in the lower atmosphere. The warming and cooling of the air creates density oscillations known as *thermal* (or *atmospheric*) *tides* that show up as small pressure changes near the earth's surface.

Pressure Measurements Instruments that detect and measure pressure changes are called **barometers**, which literally means an instrument that measures bars. You may recall from Chapter 1 that a *bar* is a unit of pressure that describes a force over a given area.* Because the bar is a relatively large

---

*A bar is a force of 100,000 newtons acting on a surface area of 1 square meter. A *newton* (N) is the amount of force required to move an object with a mass of 1 kilogram so that it increases its speed at a rate of 1 meter per second each second. Additional pressure units and conversions are given in Appendix A.

unit, and because surface pressure changes are normally small, the unit of pressure commonly found on surface weather maps is, as we saw in Chapter 1, the **millibar** (mb), where 1 mb = 1/1000 bar or

1 bar = 1000 mb.

A common pressure unit used in aviation is *inches of mercury* (Hg). At sea level, *standard atmospheric pressure* is

1013.25 mb = 29.92 in. Hg = 76 cm.

As a reference, Fig. 9.4 compares pressure readings in millibars and in inches of mercury.

The unit of pressure designated by the International System (SI) of measurement is the *pascal*, named in honor of Blaise Pascal (1632–1662), whose experiments on atmospheric pressure greatly increased our knowledge of the atmosphere. A pascal (Pa) is the force of 1 newton acting on a surface area of 1 square meter. Thus, 100 pascals equals 1 millibar. The scientific community often uses the *kilopascal* (kPa) as the preferred of a column of air than is able to enter (called *net divergence*).

Earlier, we considered how pressure and density are related when the temperature is not changing. What happens to the gas law when the pressure of a gas remains constant? In shorthand notation, the law becomes

(Constant pressure) $\times$ constant $= T \times \rho$.

This relationship tells us that when the pressure of a gas is held constant, the gas becomes less dense as the temperature goes up, and more dense as the temperature goes down. Therefore, *at a given atmospheric pressure, air that is cold is more dense than air that is warm.* Keep in mind that the idea that cold air is more dense than warm air applies only when we compare volumes of air at the same level, where pressure changes are small in any horizontal direction.

We can use the gas law to obtain information about the atmosphere. For example, at an altitude of about 5600 m (18,400 ft) above sea level, the atmospheric pressure is normally close to 500 mb. If we obtain the average density at this level, with the aid of the gas law we can calculate the average air temperature.

Recall that the gas law is written as

$$p = T \times \rho \times C.$$

With the pressure (p) in millibars (mb), the temperature (T) in Kelvins, and the density (ρ) in kilograms per cubic meter (kg/m^3), the numerical value of the constant (C) is about 2.87.*

At an altitude of 5600 m above sea level, where the average (or standard) air pressure is about 500 mb and the average air density is 0.690 kg/m^3, the average air temperature becomes

$$p = T \times \rho \times C$$
$$500 = T \times 0.690 \times 2.87$$
$$\frac{500}{0.690 \times 2.87} = T$$
$$252.5 \text{ K} = T$$

To convert Kelvins into degrees Celsius, we subtract 273 from the Kelvin

---

*The constant is usually expressed as 2.87 $\times$ 10^6 erg/g K, or, in the SI system, as 287 J/kg K. (See Appendix A for information regarding the units used here.)

temperature and obtain a temperature of –20.5°C, which is the same as 5°F.

If we know the numerical values of temperature and density, with the aid of the gas law we can obtain the air pressure. For example, in Chapter 1 we saw that the average global temperature near sea level is 15°C (59°F), which is the same as 288 K. If the average air density at sea level is 1.226 kg/m^3, what would be the average (standard) sea level pressure?

Using the gas law, we obtain:

$$p = T \times \rho \times C$$
$$p = 288 \times 1.226 \times 2.87$$
$$p = 1013 \text{ mb}$$

Since the air pressure is related to both temperature and density, a small change in either or both of these variables can bring about a change in pressure.

---

pressure unit, where 1 kPa = 10 mb. However, a more convenient unit is the **hectopascal** (hPa), as

1 hPa = 1 mb.

Currently, pressure readings on all surface weather maps in the United States are expressed in millibars. However, when the National Weather Service converts entirely to the metric system, the hectopascal, because of its ease of conversion with the millibar, will probably be the unit of pressure displayed on surface charts.

Because we measure atmospheric pressure with an instrument called a *barometer*, atmospheric pressure is also referred to as *barometric pressure*. Evangelista Torricelli, a student of Galileo, invented the **mercury barometer** in 1643. His barometer, similar to those in use today, consisted of a long glass tube open at one end and closed at the other (see Fig. 9.5). Removing air from the tube and covering the open end, Torricelli immersed the lower portion into a dish of mercury. He removed the cover, and the mercury rose up the tube to nearly

WEATHER WATCH

Mercury is 13.6 times more dense than water. Consequently, a water barometer resting on the ground near sea level would have to be read from a ladder over 10.3 m (34 ft) tall.

76 cm (or about 30 in.) above the level in the dish. Torricelli correctly concluded that the column of mercury in the tube was balancing the weight of the air above the dish, and hence its height was a measure of atmospheric pressure.

Why is mercury rather than water used in the barometer? The primary reason is convenience. (Also, water can evaporate in the tube.) Mercury seldom rises to a height above 80 cm (31.5 in.). A water barometer, however, presents a problem. Because water is 13.6 times less dense than mercury, an atmospheric pressure of 76 cm (30 in.) of mercury would be equivalent to 1034 cm (408 in.) of water.

in. Hg   **mb**

32.78 — — 1110
32.48 — — 1100
32.19 — — 1090
31.89 — — 1080   ← 1084 mb (32.01 in.) Highest recorded sea level pressure:
31.60 — — 1070   Agata, Siberia (December, 1968)
31.30 — — 1060   ← 1064 mb (31.42 in.) Highest recorded sea level
31.00 — — 1050   pressure in United States: Miles City, Montana (December,1983)
30.71 — — 1040   ← Strong high pressure system
30.42 — — 1030
30.12 — — 1020
29.82 — — 1010   ← 1013.25 mb (29.92 in.) Average sea level pressure
29.53 — — 1000
29.24 — — 990
28.94 — — 980   ← Deep low pressure system
28.64 — — 970
28.35 — — 960
28.05 — — 950
27.76 — — 940
27.46 — — 930
27.17 — — 920
26.87 — — 910
26.58 — — 900
26.28 — — 890   ← 888 mb (26.22 in.) Hurricane Gilbert (September, 1988)
25.99 — — 880
25.69 — — 870   ← 870 mb (25.70 in.) Lowest recorded sea level pressure:
25.40 — — 860   Typhoon Tip (October, 1979)
25.10 — — 850

FIGURE 9.4
Atmospheric pressure in inches of mercury and in millibars.

FIGURE 9.6
The aneroid barometer.

STORMY RAIN CHANGE FAIR DRY
980 990 1000 1010 1020 1030 1040

Lever system
Aneroid cell

FIGURE 9.5
The mercury barometer. The height of the mercury column is a measure of atmospheric pressure.

Vacuum
Mercury column
Glass tube
Height 76 cm (29.92 in.)
Air pressure
Air pressure
Mercury in dish

The most common type of home barometer—the **aneroid barometer**—contains no fluid. Inside this instrument is a small, flexible metal box called an aneroid cell. Before the cell is tightly sealed, air is partially removed, so that small changes in external air pressure cause the cell to expand or contract. The size of the cell is calibrated to represent different pressures, and any change in its size is amplified by levers and transmitted to an indicating arm, which points to the current atmospheric pressure (see Fig. 9.6).

Notice that the aneroid barometer often has descriptive weather-related words printed above specific pressure values. These adjectives indicate the most likely weather conditions when the needle is pointing to that particular pressure read-ing. Generally, the higher the reading, the more likely clear weather will occur, and the lower the reading, the better are the chances for inclement weather. This phenomenon occurs because surface high-pressure areas are associated with sink-ing air and normally fair weather, whereas surface low-pres-sure areas are associated with rising air and usually cloudy, wet weather.

The *altimeter* and *barograph* are two types of aneroid barometers. Altimeters are aneroid barometers that measure pressure, but are calibrated to indicate altitude. Barographs are recording aneroid barometers. Basically, the barograph con-sists of a pen attached to an indicating arm that marks a con-tinuous record of pressure on chart paper. The chart paper is attached to a drum rotated slowly by an internal mechanical clock (see Fig. 9.7).

Pressure Readings   The seemingly simple task of reading the height of the mercury column to obtain the air pressure is actually not all that simple. Being a fluid, mercury is sensitive to changes in temperature; it will expand when heated and contract when cooled. Consequently, to obtain accurate pressure readings without the influence of temperature, all mercury barometers are corrected as if they were read at the same temperature. Because the earth is not a perfect sphere, the force of gravity is not a constant. Since small gravity differences influence the height of the mercury column, they must be considered when reading the barometer. Finally, each barometer has its own "built-in" error, called *instrument error*, which is caused, in part, by the surface tension of the mercury against the glass tube. After being corrected for temperature, gravity, and instrument error, the barometer reading at a particular location and elevation is termed **station pressure**.

Figure 9.8a gives the station pressure measured at four locations only a few hundred kilometers apart. The different station pressures of the four cities are due primarily to the cities being at different altitudes. This fact becomes even clearer when we realize that atmospheric pressure changes much more

FIGURE 9.7
A recording barograph.

quickly when we move upward than it does when we move sideways. As an example, the vertical change in air pressure from the base to the top of the Empire State Building—a distance of a little more than ½ km—is typically much greater than the horizontal difference in air pressure from New York City to Miami, Florida—a distance of over 1600 km. Therefore, we can see that a small vertical difference between two observation sites can yield a large difference in station pressure. Thus,

FIGURE 9.8
The top diagram (a) shows four cities (A, B, C, and D) at varying elevations above sea level, all with different station pressures. The middle diagram (b) represents sea level pressures of the four cities plotted on a sea-level chart. The bottom diagram (c) shows isobars drawn on the chart (dark lines) at intervals of 4 mb.

to properly monitor horizontal changes in pressure, barometer readings must be corrected for altitude.

Altitude corrections are made so that a barometer reading taken at one elevation can be compared with a barometer reading taken at another. Station pressure observations are normally adjusted to an altitude of 0 meters, which is referred to as *mean sea level (MSL)*—the level representing the average surface of the ocean. The adjusted pressure reading is called **sea level pressure**. The size of the correction depends primarily on how high the station is above sea level.

Near the earth's surface, atmospheric pressure decreases by about 10 mb for every 100 m increase in elevation (about 1 in. of mercury for each 1000-ft rise) in an atmosphere where the air temperature decreases at the standard lapse rate of 6.5°C/1000 m. Therefore, we generally add 10 mb for every 100 m of altitude to obtain sea-level pressure.

Because a standard atmosphere seldom exists, this value (10 mb per 100 m) only approximates actual conditions. For example, as we saw earlier in this chapter, atmospheric pressure decreases more rapidly with height in cold (more-dense) air than it does in warm (less-dense) air. Hence, in cold air, the vertical rate of pressure change is typically greater than 10 mb per 100 m, whereas in warm air, it is less. Moreover, since the vertical lapse rate changes slightly from day to day, temperature and moisture corrections made for "average" conditions may introduce additional errors. However, if we simplify matters and assume a standard lapse rate, we can correct each of the station pressures in Fig. 9.8a to sea level simply by adding 10 mb per 100 m to each station pressure. When we plot these corrected values on the sea-level pressure chart (Fig. 9.8b), we are able to see the horizontal variation in sea-level pressure. Notice that the pressure varies from higher pressure on the left side of the map to lower pressure on the right—something impossible to see from the uncorrected station pressures we initially examined in Fig. 9.8a.

When more pressure data are added (see Fig. 9.8c), the chart can be analyzed and the pressure pattern visualized. **Isobars** (lines connecting points of equal pressure) are drawn at intervals of 4 mb,* with 1000 mb being the base value. Note that the isobars do not pass through each point, but, rather, between many of them, with the exact values being interpolated from the data given on the chart. For example, follow the 1008-mb line from the top of the chart southward and observe that there is no plotted pressure of 1008 mb. The 1008-mb isobar, however, comes closer to the station with a sea level pressure of 1007 mb than it does to the station with a pressure of 1010 mb. With its isobars, the bottom chart (Fig. 9.8c) is now called a *sea-level pressure chart* or simply a *surface map*. When weather data are plotted on the map it becomes a *surface weather map*.

---

*An interval of 2 mb would put the lines too close together, and an 8-mb interval would spread them too far apart.

SURFACE AND UPPER-LEVEL CHARTS

The isobars on the surface map in Fig. 9.9a are drawn precisely, with each individual observation taken into account. Notice that many of the lines are irregular, especially in mountainous regions over the Rockies. The reason for the wiggle is due, in part, to small-scale local variations in pressure and to errors introduced by correcting observations that were taken at high-altitude stations. An extreme case of this type of error occurs at Leadville, Colorado (elevation 3096 m), the highest city in the United States. Here, the station pressure is typically near 700 mb. This means that nearly 300 mb must be added to obtain a sea level pressure reading! A mere 1 percent error in estimating the exact correction would result in a 3-mb error in sea-level pressure. For this reason, isobars are smoothed through readings from high-altitude stations and from stations that might have small observational errors. Figure 9.9b shows how the isobars appear on the surface map after they are smoothed.

The sea-level pressure chart described so far is called a **constant height chart** because it represents the atmospheric pressure at a constant level—in this case, sea level. The same type of chart could be drawn to show the horizontal variations in pressure at any level in the atmosphere; for example, at 3000 m (see Fig. 9.10).

Another type of chart is commonly used in studying the weather, namely, the **constant pressure (isobaric) chart**. Instead of showing pressure variations at a constant altitude, these charts are constructed to show height variations along an equal pressure *(isobaric)* surface. Constant pressure charts are convenient to use because the height variables they show are easier to deal with in meteorological equations than the variables of pressure. Since isobaric charts are in common use, let's examine them in detail.

Imagine the light blue shaded area inside the air column in Fig. 9.11 represents tightly packed air molecules from the surface up to the tropopause. Assume that the air density is constant throughout the entire air layer and that all of the air molecules are squeezed into this layer. If we climb halfway up the air column and stop, then draw a sheetlike surface representing this level, we will have made a constant height surface. This altitude (5600 m) is where we would, under standard conditions, measure a pressure of 500 mb. Observe that everywhere along this surface (shaded gray in the diagram) there are an equal number of molecules above it. This condition means that the level of constant height also represents a level of constant pressure. At every point on this **isobaric surface**, the height is 5600 m above sea level and the pressure is 500 mb. Within the air column, we could cut any number of horizontal slices, each one at a different altitude, and each slice would represent both an isobaric and constant height surface. A map of any one of these surfaces would be blank, since there are no horizontal variations in either pressure or altitude.

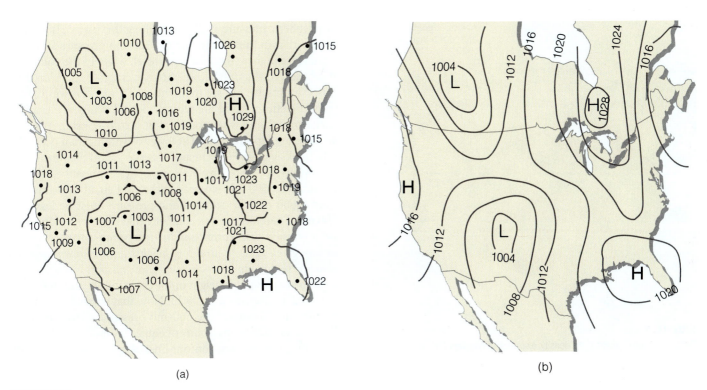

(a)

(b)

FIGURE 9.9

(a) Sea-level isobars drawn so that each observation is taken into account. Not all observations are plotted. *(b)* Sea-level isobars after smoothing.

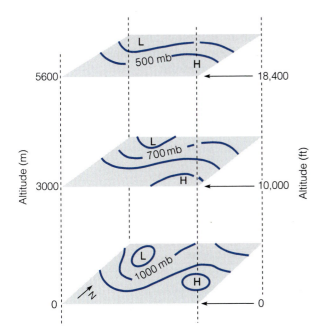

FIGURE 9.10

Each map shows isobars on a constant height chart. The isobars represent variations in horizontal pressure at that altitude. An average isobar at sea level would be about 1000 mb; at 3000 m, about 700 mb; and at 5600 m, about 500 mb.

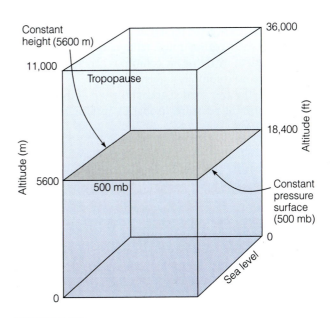

FIGURE 9.11

When there are no horizontal variations in pressure, constant pressure surfaces are parallel to constant height surfaces. In the diagram, a measured pressure of 500 mb is 5600 m above sea level everywhere.

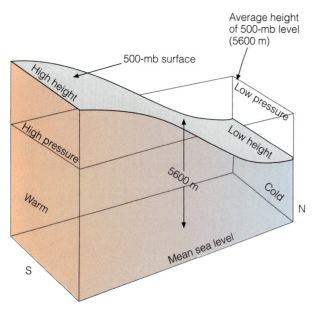

FIGURE 9.12

Because of the changes in air density, a surface of constant pressure (shaded gray area) rises in warm, less-dense air and lowers in cold, more-dense air. Where the horizontal temperature changes most quickly, the constant pressure surface changes elevation most rapidly.

If the air temperature should change in any portion of the column, the air density and pressure would change along with it (see Fig. 9.12). Note that we have colder air to the north and warmer air to the south. To simplify this situation, we will assume that the atmospheric pressure at the earth's surface remains constant. Hence, the total number of molecules in the column above each region must remain constant. Consequently, the cold, more dense air column is shorter, while the warm, less dense air column is higher.

In Fig. 9.12, the area shaded gray at the top of the column represents a constant pressure (isobaric) surface, where the atmospheric pressure at all points along this surface is 500 mb. Notice that in the warmer air the 500-mb pressure surface is found at a higher (than average) level, while in the colder air, it is observed at a much lower (than average) level. From these observations, we can see that when the air aloft is warm, constant pressure surfaces are typically found at higher elevations than normal, and when the air aloft is cold, constant pressure surfaces are typically found at lower elevations than normal.

Look again at Fig. 9.12 and observe that in the warm air at an altitude of 5600 m, the atmospheric pressure must be greater than 500 mb, whereas in the cold air, at the same altitude (5600 m), the atmospheric pressure must be less than 500 mb. Therefore, we can conclude that *high heights on a constant pressure chart correspond to higher-than-normal pressures at any given altitude, and low heights on a constant pressure chart correspond to lower-than-normal pressures.*

The variations in height of the constant pressure surface in Fig. 9.12 are shown in Fig. 9.13. Note that where the constant altitude lines intersect the 500-mb pressure surface, **contour lines** (lines connecting points of equal elevation) are drawn on the 500-mb map. Each contour line, of course, tells us the altitude above sea level at which we can obtain a pressure reading of 500 mb. In the warmer air to the south, the elevations are high, while in the cold air to the north, the elevations are low. The contour lines are crowded together in the middle of the chart, where the pressure surface dips rapidly due to the changing air temperatures. Where there is little horizontal temperature change, there are also few contour lines. Although contour lines are height lines, keep in mind that they illustrate pressure in the manner of isobars, as contour lines of low height represent a region of lower pressure and contour lines of high height represent a region of higher pressure.

Since cold air aloft is normally associated with low heights or low pressures, and warm air aloft with high heights or high pressures, on upper-air charts representing the Northern Hemisphere, contour lines and isobars usually decrease in value from south to north because the air is typically warmer to the south

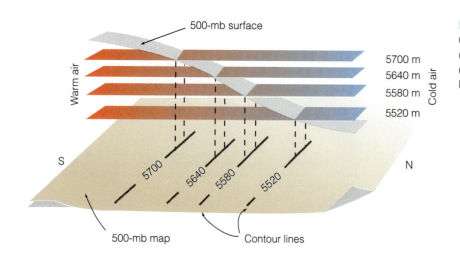

FIGURE 9.13

Changes in elevation of a constant pressure surface (500 mb) show up as contour lines on a constant pressure (500 mb) map. Where the surface dips most rapidly, the lines are closer together.

TABLE 9.1   Common Isobaric Charts and Their Approximate Elevation above Sea Level

ISOBARIC SURFACE (MB) CHARTS	APPROXIMATE ELEVATION	
	(m)	(ft)
1000	120	400
850	1,460	4,800
700	3,000	9,800
500	5,600	18,400
300	9,180	30,100
200	11,800	38,700
100	16,200	53,200

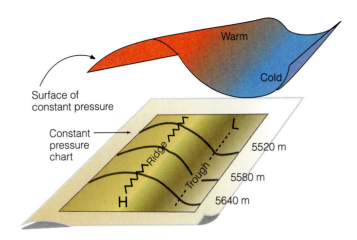

FIGURE 9.14
The wavelike patterns of a constant pressure surface reflect the changes in air temperature. An elongated region of warm air aloft shows up on a constant pressure chart (isobaric map) as higher heights and a ridge; the colder air shows as lower heights and a trough.

and colder to the north. The lines, however, are not straight; they bend and turn, indicating **ridges** *(elongated highs)* where the air is warm and indicating depressions, or **troughs** *(elongated lows)*, where the air is cold. In Fig. 9.14, we can see how the wavy contours on the map relate to the changes in altitude of the pressure surface.

Although we have examined only the 500-mb chart, other isobaric charts are commonly used. Table 9.1 lists these charts and their approximate heights above sea level.

Upper-level charts are a valuable tool. As we will see, they show wind-flow patterns that are extremely important in forecasting the weather. They can also be used to determine the movement of weather systems and to predict the behavior of surface pressure areas. To the pilot of a small aircraft, a constant pressure chart can help determine whether the plane is

flying at an altitude either higher or lower than its altimeter indicates. (For more information on this topic, read the Focus section "Flying on a Constant Pressure Surface—High to Low, Look Out Below," p. 220.)

Figure 9.15a is a simplified surface map that shows areas of high and low pressure and arrows that indicate *wind direction*—the direction from which the wind is blowing. The large blue H's on the map indicate the centers of high pressure, which are also called **anticyclones**. The large L's represent centers of low pressure, also known as *depressions* or **mid-latitude**

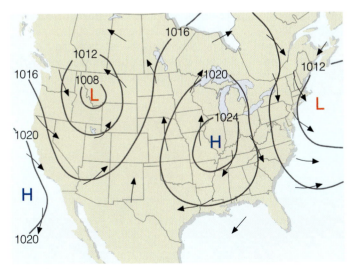

(a) Surface map

(b) 500-millibar map

FIGURE 9.15
(a) Surface map showing areas of high and low pressure. The solid lines are isobars drawn at 4-mb intervals. The arrows represent wind direction. (b) The 500-mb map for the same day as the surface map. Solid dark lines on the map are contour lines in meters above sea level. Dashed red lines are isotherms in °C. Arrows show wind direction.

Flying on a Constant Pressure Surface—High to Low, Look Out Below

Aircraft that use pressure altimeters typically fly along a constant pressure surface rather than a constant altitude surface. They do this because the *altimeter*, as we saw earlier, is simply an aneroid barometer calibrated to convert atmospheric pressure to an approximate elevation. The altimeter elevation indicated by an altimeter assumes a standard atmosphere where the air temperature decreases at the rate of 6.5°C/1000 m (3.6°F/1000 ft). Since the air temperature seldom, if ever, decreases at this rate, altimeters generally indicate an altitude different from their true elevation.

Figure 2 shows a standard column of air bounded on each side by air with a different temperature and density. On the left side, the air is warm; on the right, it is cold. The heavy dashed line represents a constant pressure surface of 700 mb as seen from the side. In the standard air, the 700-mb surface is located at 10,000 ft above sea level.

In the warm air, the 700-mb surface rises; in the cold air, it descends. An aircraft flying along the 700-mb surface would be at an altitude less than 10,000 ft in the cold air, equal to 10,000 ft in the standard air, and greater than 10,000 ft in the warmer air. With no corrections for temperature, the altimeter would indicate the same altitude at all three positions because the air pressure does not change. We can see that, if no temperature corrections are made, an aircraft flying into warm air will increase in altitude and fly higher than its altimeter indicates. Put another way: The altimeter inside the plane will read an altitude lower than the plane's true elevation.

Flying from standard air into cold air represents a potentially dangerous situation. As an aircraft flies into cold air, it flies along a lowering pressure surface. If no correction for temperature is made, the altimeter shows no change in elevation even though the aircraft is losing altitude; hence, the plane will be flying lower

than the altimeter indicates. This problem can be serious, especially for planes flying above mountainous terrain with poor visibility and where high winds and turbulence can reduce the air pressure drastically. To ensure adequate clearance under these conditions, pilots fly their aircraft higher than they normally would, consider air temperature, and compute a more realistic altitude by resetting their altimeters to reflect these conditions.

Even without sharp temperature changes, pressure surfaces may dip suddenly. This event is especially likely close to the ground (see Fig. 3). An aircraft flying into an area of decreasing pressure will lose altitude unless corrections are made. For example, suppose a pilot has set the altimeter for sea-level pressure above station A. At this location, the plane is flying along an isobaric surface at a true altitude of 500 ft. As the plane flies toward station B, the pressure surface (and the plane) dips but the altime-

ter continues to read 500 ft, which is too high. To correct for such changes in pressure, a pilot can obtain a current altimeter setting from ground control. With this additional information, the altimeter reading will more closely match the aircraft's actual altitude.

Because of the inaccuracies inherent in the pressure altimeter, many high performance and commercial aircraft are now equipped with a radio altimeter. This device is like a small radar unit that measures the altitude of the aircraft by sending out radio waves, which bounce off the terrain. The time it takes these waves to reach the surface and return is a measure of the aircraft's altitude. If used in conjunction with a pressure altimeter, a pilot can determine the variations in a constant pressure surface simply by flying along that surface and observing how the true elevation measured by the radio altimeter changes.

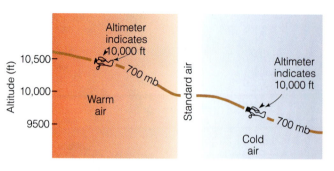

FIGURE 2

An aircraft flying along a surface of constant pressure may change altitude as the air temperature changes. Without being corrected for the temperature change, a pressure altimeter will continue to read the same elevation.

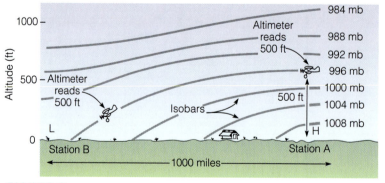

FIGURE 3

In the absence of horizontal temperature changes, pressure surfaces can dip toward the earth's surface. An aircraft flying along the pressure surface will either lose or gain altitude, depending on the direction of flight.

cyclones because they form in the middle latitudes, outside of the tropics. The solid dark lines are isobars with units in millibars. Notice that the surface winds tend to blow across the isobars toward regions of lower pressure. In fact, as we briefly observed in Chapter 1, in the Northern Hemisphere the winds blow counterclockwise and inward toward the center of the lows and clockwise and outward from the center of the highs.

Figure 9.15b shows an upper-air chart (a 500-mb map) for the same day as the surface map in Fig. 9.15a. The solid dark lines on the map are contour lines given in meters above sea level. The difference in elevation between each contour line (called the *contour interval*) is 60 m. Superimposed on this map are dashed red lines, which represent lines of equal temperature (isotherms). Observe how the contour lines tend to parallel the isotherms. As we would expect, the contour lines tend to decrease in value from south to north.

The arrows on the 500-mb map show the wind direction. Notice that, unlike the surface winds that cross the isobars in Fig. 9.15a, the winds on the 500-mb chart tend to flow parallel to the contour lines in a more or less west-to-east direction. Why does the wind tend to cross the isobars on a surface map, yet blow parallel to the contour lines (or isobars) on an upper-air chart? To answer this question we will now examine the forces that affect winds.

NEWTON'S LAWS OF MOTION

Our understanding of why the wind blows stretches back through several centuries, with many scientists contributing to our knowledge. When we think of the movement of air, however, one great scholar stands out—Isaac Newton (1642–1727), who formulated several fundamental laws of motion.

Newton's first law of motion states that *an object at rest will remain at rest and an object in motion will remain in motion (and travel at a constant velocity along a straight line) as long as no force is exerted on the object*. For example, a baseball in a pitcher's hand will remain there until a force (a push) acts upon the ball. Once the ball is pushed (thrown), it would continue to move in that direction forever if it were not for the force of air friction (which slows it down), the force of gravity (which pulls it toward the ground), and the catcher's mitt (which exerts an equal but opposite force to bring it to a halt). Similarly, to start air moving, to speed it up, to slow it down, or even to change its direction requires the action of an external force. This brings us to Newton's second law.

Newton's second law states that *the force exerted on an object equals its mass times the acceleration produced*. In symbolic form, this law is written as

$$F = ma.$$

From this relationship we can see that, when the mass of an object is constant, the force acting on the object is directly related to the acceleration that is produced. A force in its simplest form is a push or a pull. *Acceleration is the speeding up, the slowing down,*

or the changing of direction of an object. (More precisely, acceleration is the change in velocity* over a period of time.)

Because more than one force may act upon an object, Newton's second law always refers to the *net*, or total, force that results. An object will always accelerate in the direction of the total force acting on it. Therefore, to determine in which direction the wind will blow, we must identify and examine all of the forces that affect the horizontal movement of air. These forces include:

1. pressure gradient force
2. Coriolis force
3. centripetal force
4. friction

We will first study the forces that influence the flow of air aloft. Then we will see which forces modify winds near the ground.

FORCES THAT INFLUENCE THE WINDS

We already know that horizontal differences in atmospheric pressure cause air to move and, hence, the wind to blow. Since air is an invisible gas, it may be easier to see how pressure differences cause motion if we examine a visible fluid, such as water.

In Fig. 9.16, the two large tanks are connected by a pipe. Tank A is two-thirds full and tank B is only one-half full. Since the water pressure at the bottom of each tank is proportional to the weight of water above, the pressure at the bottom of tank A is greater than the pressure at the bottom of tank B. Moreover, since fluid pressure is exerted equally in all directions, there is a greater pressure in the pipe directed from tank A toward tank B than from B toward A.

Since pressure is force per unit area, there must also be a net force directed from tank A toward tank B. This force causes the water to flow from left to right, from higher pressure toward

---

*Velocity specifies both the speed of an object and its direction of motion.

FIGURE 9.16

The higher water level creates higher fluid pressure at the bottom of tank A and a net force directed toward the lower fluid pressure at the bottom of tank B. This net force causes water to move from higher pressure toward lower pressure.

FIGURE 9.17
The pressure gradient between point 1 and point 2 is 4 mb per 100 km. The net force directed from higher toward lower pressure is the pressure gradient force.

FIGURE 9.18
The closer the spacing of the isobars, the greater the pressure gradient. The greater the pressure gradient, the stronger the pressure gradient force (*PGF*). The stronger the *PGF*, the greater the wind speed. The red arrows represent the relative magnitude of the force, which is always directed from higher toward lower pressure.

lower pressure. The greater the pressure difference, the stronger the force, and the faster the water moves. In a similar way, horizontal differences in atmospheric pressure cause air to move.

Pressure Gradient Force Figure 9.17 shows a region of higher pressure on the map's left side, lower pressure on the right. The isobars show how the horizontal pressure is changing. If we compute the amount of pressure change that occurs over a given distance, we have the **pressure gradient**; thus

$$\text{Pressure gradient} = \frac{\text{difference in pressure}}{\text{distance}}.$$

If we let the symbol delta (Δ) mean "a change in," we can simplify the expression and write the pressure gradient as

$$PG = \frac{\Delta p}{d},$$

where Δp is the pressure difference between two places some horizontal distance (d) apart. In Fig. 9.17 the pressure gradient between points 1 and 2 is 4 mb per 100 km.

Suppose the pressure in Fig. 9.17 were to change and the isobars become closer together. This condition would produce a rapid change in pressure over a relatively short distance, or what is called a *steep* (or *strong*) *pressure gradient*. However, if the pressure were to change such that the isobars spread farther apart, then the difference in pressure would be small over a relatively large distance. This condition is called a *gentle* (or *weak*) *pressure gradient*.

Notice in Fig. 9.17 that when differences in horizontal air pressure exist there is a net force acting on the air. This force, called the **pressure gradient force** (*PGF*), is *directed from higher toward lower pressure at right angles to the isobars*. The magnitude of the force is directly related to the pressure gradient. Steep pressure gradients correspond to strong pressure gradient forces and vice versa. Figure 9.18 shows the relationship between pressure gradient and pressure gradient force.

The *pressure gradient force is the force that causes the wind to blow*. Because of this effect, closely spaced isobars on a weather chart indicate steep pressure gradients, strong forces, and high winds. On the other hand, widely spaced isobars indicate gentle pressure gradients, weak forces, and light winds. These rules also apply to surface weather maps. An example of a steep pressure gradient producing strong winds is given in Fig. 9.19. Notice that the tightly packed isobars along the green line are producing a steep pressure gradient of 32 mb per 500 km and surface winds of 40 knots.

If the pressure gradient force were the only force acting upon air, we would always find winds blowing directly from higher toward lower pressure. However, the moment air starts to move, it is deflected in its path by the *Coriolis force*.

Coriolis Force The Coriolis force describes an apparent force that is due to the rotation of the earth. To understand how it works, consider two people playing catch as they sit opposite one another on the rim of a merry-go-round (see Fig. 9.20, platform A). If the merry-go-round is not moving, each time the ball is thrown, it moves in a straight line to the other person.

Suppose the merry-go-round starts turning counterclockwise—the same direction the earth spins as viewed from above the North Pole. If we watch the game of catch from above, we see that the ball moves in a straight-line path just as before. However, to the people playing catch on the merry-go-round, the ball seems to veer to its right each time it is thrown, always landing to the right of the point intended by the thrower (see Fig. 9.20, platform B). This perception is due to the fact that, while the ball moves in a straight-line path, the merry-go-round rotates beneath it; by the time the ball reaches the opposite side, the catcher has moved. To anyone on the merry-go-round, it seems as if there is some force causing the ball to deflect to the right. This apparent force is called the **Coriolis force** after Gaspard Coriolis, a nineteenth-century French scientist who worked it out mathematically. (Because it is an

	Miles (statute) per hour	Knots
◎	Calm	Calm
	1–2	1–2
	3–8	3–7
	9–14	8–12
	15–20	13–17
	21–25	18–22
	26–31	23–27
	32–37	28–32
	38–43	33–37
	44–49	38–42
	50–54	43–47
	55–60	48–52
	61–66	53–57
	67–71	58–62
	72–77	63–67
	78–83	68–72
	84–89	73–77
	119–123	103–107

FIGURE 9.19

Surface weather map for 6 A.M. (CST), Tuesday, November 10, 1998. Dark gray lines are isobars with units in millibars. The interval between isobars is 4 mb. A deep low with a central pressure of 972 mb (28.70 in.) is moving over southeastern Minnesota. The distance along the green line X–X' is 500 km. The difference in pressure between X and X' is 32 mb, producing a pressure gradient of 32 mb/500 km. The tightly packed isobars along the green line are producing strong northwesterly winds of 40 knots, with gusts even higher. Wind directions are given by lines that parallel the wind. Wind speeds are indicated by barbs and flags. (The solid blue line is a cold front, the solid red line a warm front, and the solid purple line an occluded front. The heavy dashed line is a trough.)

apparent force due to the rotation of the earth, it is also called the *Coriolis effect.*) This effect occurs on the rotating earth, too. All free-moving objects, such as ocean currents, aircraft, artillery projectiles, and air molecules seem to deflect from a straight-line path because the earth rotates under them.

The Coriolis force *causes the wind to deflect to the right of its path in the Northern Hemisphere and to the left of its path in the Southern Hemisphere.* To illustrate this, consider a satellite in polar circular orbit. If the earth were not rotating, the path of the satellite would be observed to move directly from north to south, parallel to the earth's meridian lines. However, the earth *does* rotate, carrying us and meridians eastward with it. Because of this, in the Northern Hemisphere, we see the satellite moving southwest instead of due south; it seems to veer off its path and move toward *its right*. In the Southern Hemisphere, the earth's direction of rotation is clockwise as viewed

WEATHER WATCH

The deep, low-pressure area illustrated in Fig. 9.19 was quite a storm. The intense low with its tightly packed isobars and strong pressure gradient produced extremely high winds that gusted over 90 knots in Wisconsin and in Michigan's Mackinac Island. The extreme winds caused blizzard conditions over the Dakotas, closed many Interstate highways, shut down airports, and overturned trucks. The winds pushed a school bus off the road near Albert Lea, Minnesota, injuring two children, and blew the roofs off homes in Wisconsin. This notorious deep storm set an all-time record low pressure of 963 mb (28.43 in.) for Minnesota on November 10, 1998.

from above the South Pole. Consequently, a satellite moving northward from the South Pole would appear to move northwest and, hence, would veer to the *left* of its path.

The magnitude of the Coriolis force varies with the speed of the moving object and the latitude. Figure 9.21 shows this variation for various wind speeds at different latitudes. In each case, as the wind speed increases, the Coriolis force increases; hence, *the stronger the wind speed, the greater the deflection.* Also, note that the Coriolis force increases for all wind speeds from a value of *zero at the equator* to *a maximum at the poles.* We can see this latitude effect better by examining Fig. 9.22.

Imagine in Fig. 9.22 that there are three aircraft, each at a different latitude and each flying along a straight-line path, with no external forces acting on them. The destination of each aircraft is due east and is marked on the diagram (see Fig.

FIGURE 9.20

On nonrotating platform A, the thrown ball moves in a straight line. On platform B, which rotates counterclockwise, the ball continues to move in a straight line. However, platform B is rotating while the ball is in flight; thus, to anyone on platform B, the ball appears to deflect to the right of its intended path.

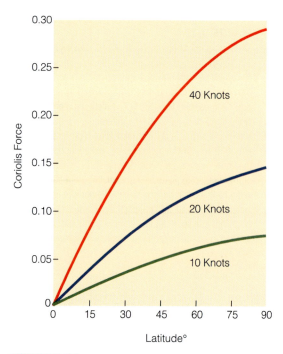

FIGURE 9.21
The relative variation of the Coriolis force at different latitudes with different wind speeds.

effect on the plane at high latitudes (large deviation) than on the plane at low latitudes (small deviation). On the equator, it has no effect at all. The same is true of its effect on winds.

In summary, to an observer on the earth, objects moving in *any direction* (north, south, east, or west) are deflected to the *right* of their intended path in the Northern Hemisphere and to the *left* of their intended path in the Southern Hemisphere. The amount of deflection depends upon:

1. the rotation of the earth
2. the latitude
3. the object's speed*

In addition, the Coriolis force acts at right angles to the wind, only influencing wind direction and never wind speed.

The Coriolis "force" behaves as a real force, constantly tending to "pull" the wind to its right in the Northern Hemisphere and to its left in the Southern Hemisphere. Moreover, this effect is present in all motions relative to the earth's surface. However, in most of our everyday experiences, the Coriolis force is so small (compared to other forces involved in those experiences) that it is negligible and, contrary to popular belief, does not cause water to turn clockwise or counterclockwise when draining from a sink. The Coriolis force is also minimal on small-scale winds, such as those that blow inland along coasts in summer. Here, the Coriolis force might be strong because of high winds, but the force cannot produce much deflection over the relatively short distances. Only where winds blow over vast regions is the effect significant.

---

*These three factors are grouped together and shown in the expression

$$\text{Coriolis force} = 2m\,\Omega V \sin \phi,$$

where m is the object's mass, Ω is the earth's angular rate of spin (a constant), V is the speed of the object, and ϕ is the latitude.

9.22a). Each plane travels in a straight path relative to an observer positioned at a fixed spot in space. The earth rotates beneath the moving planes, causing the destination points at latitudes 30° and 60° to change direction slightly (to the observer in space) (see Fig. 9.22b). To an observer standing on the earth, however, it is the plane that appears to deviate. The amount of deviation is greatest toward the pole and nonexistent at the equator. Therefore, the Coriolis force has a far greater

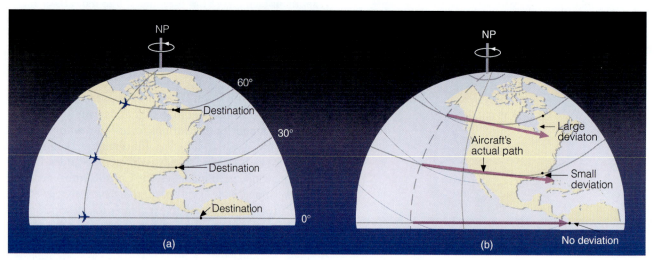

FIGURE 9.22
Except at the equator, a free-moving object heading either east or west (or any other direction) will appear from the earth to deviate from its path as the earth rotates beneath it. The deviation (Coriolis force) is greatest at the poles and decreases to zero at the equator.

BRIEF REVIEW   In summary, we know that:

- The pressure gradient force is always directed from higher pressure toward lower pressure and it is the pressure gradient force that causes the air to move.
- Steep pressure gradients (tightly packed isobars) produce strong pressure gradient forces and high winds; gentle pressure gradients (widely spaced isobars) produce weak pressure gradient forces and light winds.
- Once the wind starts to blow, the Coriolis force causes it to bend to the right of its intended path in the Northern Hemisphere and to the left of its intended path in the Southern Hemisphere.

With this information in mind, we will first examine how aloft (that is, above the friction layer) the pressure gradient force and the Coriolis force produce straight-line winds. We will then see what influence the centripetal force has on winds that blow along a curved path.

Straight-line Flow Aloft—Geostrophic Winds   Earlier in this chapter, we saw that the winds aloft on an upper-level chart blow more or less parallel to the isobars or contour lines. We can see why this phenomenon happens by carefully looking at Fig. 9.23, which shows a map of the Northern Hemisphere, above the earth's frictional influence,* with horizontal pressure variations at an altitude of about 1 km. The evenly spaced isobars indicate a constant pressure gradient force (PGF) directed from south toward north as indicated by the arrow at the left. Why, then, does the map show a west wind? We can answer this question by placing a parcel of air at position 1 in the diagram and watching its behavior.

At position 1, the *PGF* acts immediately upon the air parcel, accelerating it northward toward lower pressure. However, the instant the air begins to move, the Coriolis force deflects the air toward its right, curving its path. As the parcel of air increases in speed (positions 2, 3, and 4), the magnitude of the Coriolis force increases (as shown by the longer arrows), bending the wind more and more to its right. Eventually, the wind speed increases to a point where the Coriolis force just balances the *PGF*. At this point (position 5), the wind no longer accelerates because the net force is zero. Here the wind flows in a straight path, parallel to the isobars at a constant speed.† This flow of air is called a **geostrophic** (*geo*: earth; *strophic*: turning) **wind**. Notice that the

---

*We will see later that the friction layer (the layer where the wind is influenced by frictional interaction with objects on the earth's surface) usually extends from the surface up to about 1000 m (3300 ft) above the ground.

†At first, it may seem odd that the wind blows at a constant speed with no net force acting on it. But when we remember that the net force is necessary only to accelerate (*F = ma*) the wind, it makes more sense. For example, it takes a considerable net force to push a car and get it rolling from rest. But once the car is moving, it only takes a force large enough to counterbalance friction to keep it going. There is no net force acting on the car, yet it rolls along at a constant speed.

geostrophic wind blows in the Northern Hemisphere with lower pressure to its left and higher pressure to its right.

When the flow of air is purely geostrophic, the isobars are straight and evenly spaced, and the wind speed is constant. In the atmosphere, isobars are rarely straight or evenly spaced, and the wind normally changes speed as it flows along. So, the geostrophic wind is usually only an approximation of the real wind. However, the approximation is generally close enough to help us more clearly understand the behavior of the winds aloft.

As we would expect from our previous discussion of winds, the speed of the geostrophic wind is directly related to the pressure gradient. In Fig. 9.24, we can see that a geostrophic wind flowing parallel to the isobars is similar to water in a stream flowing parallel to its banks. At position 1, the wind is blowing at a low speed; at position 2, the pressure gradient increases and the wind speed picks up. Notice also that at position 2, where the wind speed is greater, the Coriolis force is greater and balances the stronger pressure gradient force. (A more mathematical approach to the concept of geostrophic wind is given in the Focus section on p. 226.)

In Fig. 9.25, we can see that the geostrophic wind direction can be determined by studying the orientation of the isobars; its speed can be estimated from the spacing of the isobars. On an isobaric chart, the geostrophic wind direction and speed are

FIGURE 9.23

Above the level of friction, air initially at rest will accelerate until it flows parallel to the isobars at a steady speed with the pressure gradient force (PGF) balanced by the Coriolis force (CF). Wind blowing under these conditions is called geostrophic.

A Mathematical Look at the Geostrophic Wind

We know from an earlier discussion that the geostrophic wind gives us a good approximation of the real wind above the level of friction, about 500 to 1000 m above the earth's surface. Above the friction layer, the winds tend to blow parallel to the isobars, or contours. We know that, for any given latitude, the speed of the geostrophic wind is proportional to the pressure gradient. This may be represented as:

$$V_g \sim \frac{\Delta p}{d},$$

where V_g is the geostrophic wind and Δp is the pressure difference between two places some horizontal distance (d) apart. From this, we can see that the greater the pressure gradient, the stronger the geostrophic wind.

When we consider a unit mass of moving air, we must take into account the air density (mass per unit volume) expressed by the symbol ρ. The geostrophic wind is now directly proportional to the pressure gradient force; thus

$$V_g \sim \frac{1}{\rho} \frac{\Delta p}{d},$$

We can see from this expression that, with the same pressure gradient (at the same latitude), the geostrophic wind will increase with increasing elevation because air density decreases with height.

In a previous section, we saw that the geostrophic wind represents a balance of forces between the Coriolis force and the pressure gradient force. Here, it should be noted that the Coriolis force can be expressed as

Coriolis force = $2\Omega V \sin \phi$,

FIGURE 4

A portion of an upper-air chart for part of the Northern Hemisphere at an altitude of 5600 meters above sea level. The lines on the chart are isobars where 500 equals 500 millibars. The air temperature is −25°C and the air density is 0.70 kg/m³.

where Ω is the earth's angular spin (a constant), V is the speed of the wind, and ϕ is the latitude. The sin ϕ is a trigonometric function that takes into account the variation of the Coriolis force with latitude. At the equator (0°), sin ϕ is 0; at 30° latitude, sin ϕ is 0.5, and, at the poles (90°), sin ϕ is 1.

This balance between the Coriolis force and the pressure gradient force can be written as

$$CF = PGF$$

$$2\Omega V_g \sin \phi = \frac{1}{\rho} \frac{\Delta p}{d}.$$

Solving for V_g, the geostrophic wind, the equation becomes

$$V_g = \frac{1}{2\Omega \sin \phi \rho} \frac{\Delta p}{d}. \quad (1)$$

Customarily, the rotational (2Ω) and latitudinal (sin ϕ) factors are combined into a single value f, called the *Coriolis parameter*. Thus, we have the geostrophic wind equation written as

$$V_g \sim \frac{1}{f\rho} \frac{\Delta p}{d}. \quad (2)$$

Suppose we compute the geostrophic wind for the example given in Fig. 4. Here the wind is blowing parallel to the isobars in the Northern Hemisphere at latitude 40°. The spacing between the isobars is 200 km and the pressure difference is 4 mb. The altitude is 5600 m above sea level, where the air temperature is −25°C (−13°F) and the air density is 0.70 kg/m³. First, we list our data and put them in the proper units, as

$$\Delta p = 4 \text{ mb} = 400 \text{ Newtons/m}^2$$
$$d = 200 \text{ km} = 2 \times 10^5 \text{ m}$$
$$\sin \phi = \sin(40°) = 0.64$$
$$\rho = 0.70 \text{ kg/m}^3$$
$$2\Omega = 14.6 \times 10^{-5} \text{ radian/sec.}*$$

When we use equation (1) to compute the geostrophic wind, we obtain

$$V_g = \frac{1}{2\Omega \sin \phi \rho} \frac{\Delta p}{d},$$

$$V_g = \frac{400}{14.6 \times 10^{-5} \times 0.64 \times 0.70 \times 2 \times 10^5},$$

$$V_g = 30.6 \text{ m/sec, or } 59.4 \text{ knots.}$$

---

*The rate of the earth's rotation (Ω) is 360° in one day, actually a sidereal day consisting of 23 hr, 56 min, 4 sec, or 86,164 seconds. This gives a rate of rotation of 4.18×10^{-3} degree per second. Most often, Ω is given in radians, where 2π radians equals 360° ($\pi = 3.14$). Therefore, the rate of the earth's rotation can be expressed as 2π radians/86,164 sec, or 7.29×10^{-5} radian/sec, and the constant 2Ω becomes 14.6×10^{-5} radian/sec.

FIGURE 9.24

The isobars and contours on an upper-level chart are like the banks along a flowing stream. When they are widely spaced, the flow is weak; when they are narrowly spaced, the flow is stronger. The increase in winds on the chart results in a stronger Coriolis force *(CF)*, which balances a larger pressure gradient force *(PGF)*.

related in a similar way to the contour lines. Therefore, if we know the isobar or contour patterns on an upper-level chart, we also know the direction and relative speed of the geostrophic wind, even for regions where no direct wind measurements have been made. Similarly, if we know the geostrophic wind direction and speed, we can estimate the orientation and spacing of the isobars, even if we don't have a current weather map. (It is also possible to estimate the wind flow and pressure patterns aloft by watching the movement of clouds. The Focus section on p. 228 illustrates this further.)

We know that the winds aloft do not always blow in a straight line; frequently, they curve and bend into meandering loops as they tend to follow the patterns of the isobars. In the Northern Hemisphere, winds blow counterclockwise around lows and clockwise around highs. The next section explains why.

Curved Winds Around Lows and Highs Aloft—Gradient

Winds   Because lows are also known as cyclones, the counterclockwise flow of air around them is often called *cyclonic flow*. Likewise, the clockwise flow of air around a high, or anticyclone, is called *anticyclonic flow*. Look at the wind flow around the upper-level low (Northern Hemisphere) in Fig. 9.26a. At first, it appears as though the wind is defying the Coriolis force by bending to the left as it moves counterclockwise around the system. Let's see why the wind blows in this manner.

Suppose we consider a parcel of air initially at rest at position 1 in Fig. 9.26a. The pressure gradient force accelerates the air inward toward the center of the low and the Coriolis force deflects the moving air to its right, until the air is moving parallel to the isobars at position 2. If the wind were geostrophic, at position 3 the air would move northward parallel to straight-line isobars at a constant speed. The wind is blowing at a constant speed, but parallel to *curved isobars*. A wind that

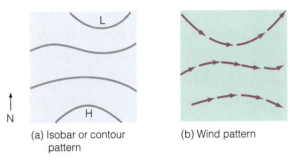

(a) Isobar or contour pattern   (b) Wind pattern

FIGURE 9.25

By observing the orientation and spacing of the isobars (or contours) in diagram (a), the geostrophic wind direction and speed can be determined for diagram (b).

blows at a constant speed parallel to curved isobars above the level of frictional influence is termed a **gradient wind**.

Earlier in this chapter we learned that an object accelerates when there is a change in its speed or direction (or both). Therefore, the gradient wind blowing *around* the low-pressure center is constantly accelerating because it is constantly changing direction. This acceleration, called the **centripetal acceleration**, is directed at right angles to the wind, inward toward the low center.

Remember from Newton's second law that, if an object is accelerating, there must be a *net force* acting on it. In this case, the net force acting on the wind must be directed toward the center of the low, so that the air will keep moving in a circular path. This inward-directed force is called the **centripetal force** (*centri*: center; *petal*: to push toward). The magnitude of the centripetal force is related to the wind velocity (*V*) and the radius of the wind's path (*r*) by the formula

$$\text{Centripetal force} = \frac{V^2}{r}.$$

FOCUS ON AN OBSERVATION

Estimating Wind Direction and Pressure Patterns Aloft

Both the wind direction and the orientation of the isobars aloft can be estimated by observing middle- and high-level clouds from the earth's surface. Suppose, for example, we are in the Northern Hemisphere watching clouds directly above us move from southwest to northeast at an elevation of 3000 m (about 10,000 ft) (see Fig. 5a). This indicates that the geostrophic wind at this level is southwesterly. Looking downwind, the geostrophic wind blows parallel to the isobars with lower pressure on the left and higher pressure on the right. Thus, if we *stand with our backs to the direction from which the clouds are moving, lower pressure aloft will always be to our left and higher pressure to our right.* From this observation, we can draw a rough upper-level chart (see Fig. 5b), which shows isobars and wind direction for an elevation of approximately 3000 m.

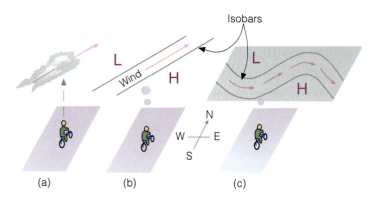

FIGURE 5
This drawing of a simplified upper-level chart is based on cloud observations. Upper-level clouds moving from the southwest (a) indicate isobars and winds aloft (b). When extended horizontally, the upper-level chart appears as in (c), where lower pressure is to the northwest and higher pressure is to the southeast.

The isobars aloft will not continue in a southwest-northeast direction indefinitely; rather, they will often bend into wavy patterns. We may carry our observation one step farther, then, by assuming a bending of the lines (see Fig. 5c). Thus, with a southwesterly geostrophic wind aloft, a trough of low pressure will be found to our west and a ridge of high pressure to our east. What would be the pressure pattern if the winds aloft were blowing *from* the northwest? Answer: A trough would be to the east and a ridge to the west.

Where wind speeds are light and there is little curvature (large radius), the centripetal force is weak and, compared to other forces, may be considered insignificant. However, where the wind is strong and blows in a tight curve (small radius), as in the case of tornadoes and tropical hurricanes, the centripetal force is large and becomes quite important.

The centripetal force results from an imbalance between the Coriolis force and the pressure gradient force.* Again, look

*In some cases, it is more convenient to express the centripetal force (and the centripetal acceleration) as the *centrifugal force*, an apparent force that is equal in magnitude to the centripetal force, but directed outward from the center of rotation. The gradient wind is then described as a balance of forces between the centrifugal force

$$\frac{V^2}{r},$$

the pressure gradient force

$$\frac{1}{\rho}\frac{\Delta p}{d},$$

and the Coriolis force $2\Omega V \sin \phi$. Under these conditions, the *gradient wind equation* is expressed as

$$\frac{V^2}{r} + \frac{1}{\rho}\frac{\Delta p}{d} + 2\Omega V \sin \phi = 0.$$

closely at position 3 (Fig. 9.26a) and observe that the inward-direction pressure gradient force *(PGF)* is greater than the outward-directed Coriolis force *(CF)*. The difference between these two forces—the net force—is the inward-directed centripetal force. In Fig. 9.26b, the wind blows clockwise around the center of the high. The spacing of the isobars tells us that the magnitude of the *PGF* is the same as in Fig. 9.26a. However, to keep the wind blowing in a circle, the inward-directed Coriolis force must now be greater in magnitude than the outward-directed pressure gradient force, so that the centripetal force (again, the net force) is directed inward.

The greater Coriolis force around the high results in an interesting observation. Because the Coriolis force (at any given latitude) can increase only when the wind speed increases, we can see that for the same pressure gradient (the same spacing of the isobars), the winds around a high-pressure area (or a ridge) must be greater than the winds around a low-pressure area (or a trough). Actually, however, the winds normally blow much faster around a low (cyclonic storm) than they do around a high, because the isobars around a low are usually spaced much closer together, resulting in a much stronger pressure gradient.

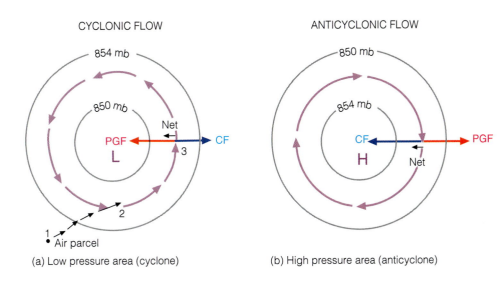

CYCLONIC FLOW

ANTICYCLONIC FLOW

(a) Low pressure area (cyclone)

(b) High pressure area (anticyclone)

FIGURE 9.26
Winds and related forces around areas of low and high pressure above the friction level in the Northern Hemisphere. Notice that the pressure gradient force *(PGF)* is in red, while the Coriolis force *(CF)* is in blue.

In the Southern Hemisphere, the pressure gradient force starts the air moving and the Coriolis force deflects it to the *left*, thereby causing the wind to blow *clockwise around lows* and *counterclockwise around highs*. Figure 9.27 shows a satellite image of clouds and wind-flow around a low-pressure area in the Northern Hemisphere (9.27a) and in the Southern Hemisphere (9.27b).

Near the equator, where the Coriolis force is minimum, winds may blow around intense tropical storms with the centripetal force being almost as large as the pressure gradient force. In this type of flow, the Coriolis force is considered negligible, and the wind is called *cyclostrophic*.

So far we have seen how winds blow in theory, but how do they appear on an actual map?

Winds on Upper-Level Charts   Figure 9.28 shows a 500-mb chart with contour lines (heavy gray lines), isotherms (dashed red lines), and winds. As we would expect at this level, the regions of lowest elevations (lowest pressures) are associated with the coolest air, and the highest elevations (highest pressures) with the warmest air. As an example, compare the cold −25°C isotherm within the low near the Washington-Oregon coast with the relatively warm −10°C isotherm within the ridge farther to the west.

(a)

(b)

FIGURE 9.27
Clouds and related wind flow patterns around low-pressure areas in (a) the Northern Hemisphere and in (b) the Southern Hemisphere.

	Miles (statute) per hour	Knots
◎	Calm	Calm
—	1–2	1–2
	3–8	3–7
	9–14	8–12
	15–20	13–17
	21–25	18–22
	26–31	23–27
	32–37	28–32
	38–43	33–37
	44–49	38–42
	50–54	43–47
	55–60	48–52
	61–66	53–57
	67–71	58–62
	72–77	63–67
	78–83	68–72
	84–89	73–77
	119–123	103–107

FIGURE 9.28

An upper-level 500-mb map showing wind direction, as indicated by lines that parallel the wind. Wind speeds are indicated by barbs and flags. (See the green insert.) Solid gray lines are contours in meters above sea level. Dashed red lines are isotherms in °C.

The wind directions on the map are given by lines that parallel the wind, whereas the wind speeds are indicated by barbs and flags. Note that the wind generally blows parallel to the contour lines, counterclockwise around the low and its trough and clockwise around the high and the ridge. The strength of the wind is directly related to the magnitude of the contour gradient. (This has the same meaning as the pressure gradient.) Observe the close spacing of the contour lines off the west coast of North America and the corresponding high winds. Near the high-pressure area, where gradients are weak, the winds are light.

In a region where the winds blow in a west-to-east direction, paralleling lines of latitude, the wind flow is termed **zonal**. The winds in Fig. 9.28 are approximately zonal over the eastern third of the United States and Canada. Because the winds aloft generally blow from west to east, planes flying in this direction have a beneficial tail wind, explaining why a flight from San Francisco to New York City takes about 30 minutes less than the return flight. If the flow aloft is zonal, clouds, storms, and surface anticyclones tend to move rapidly from west to east.

When the wind flows in large, looping meanders, following a more north-south trajectory paralleling the meridian lines, the flow is termed **meridional**. Meridional flow can be seen in Fig. 9.28 along the west coast of North America. Here, cold polar air moves southward on the western side of the low, while relatively warm subtropical air moves northward on its eastern side. As we will see in Chapter 13, during periods of meridional flow, surface storms tend to move slowly, often intensifying into major storm systems.

Moving from south to north, we can see that the contour lines decrease in elevation. We expect this, since the average air temperature is warmer to the south than it is farther north. Where horizontal temperature contrasts are large, there are also large height gradients and strong winds. We can also see that when the flow is zonal, cold air and lower heights are observed to the north of the wind. Where there is meridional flow (such as off the west coast of North America), there is a gradient of height as well as temperature that runs in a west-east direction. In general, the north-south temperature gradient is strongest in winter. This is why the winds aloft are usually much stronger in winter than in summer.

It is clear from Fig. 9.28 that the winds aloft in the middle latitudes of the Northern Hemisphere tend to move in a wavy west-to-east pattern. Because the winds in the Southern Hemisphere blow clockwise around lows and counterclockwise around highs, does this mean that the winds aloft in the middle latitudes of the Southern Hemisphere blow from east to west? The answer is found in the Focus section on p. 231.

Take a minute and look back at Fig. 9.19 on p. 223. Observe that the winds on this surface map tend to cross the isobars, blowing from higher pressure toward lower pressure. Observe also that along the green line, the tightly packed isobars (with a pressure gradient of 16 mb per 500 km) are producing a steady surface wind of 40 knots. However, this same pressure gradient (with the same air temperature) would, on an upper-level chart, produce a geostrophic wind of 55 knots. Why do surface winds normally cross the isobars and why do

FOCUS ON A SPECIAL TOPIC

Winds Aloft in the Southern Hemisphere

In the Southern Hemisphere, just as in the Northern Hemisphere, the winds aloft blow because of horizontal differences in pressure. The pressure differences, in turn, are due to variations in temperature. Recall from an earlier discussion of pressure that warm air aloft is associated with high pressure and cold air aloft with low pressure. Look at Fig. 6. It shows an upper-level chart that extends from the Northern Hemisphere into the Southern Hemisphere. Over the equator, where the air is warmer, the pressure aloft is high. North and south of the equator, where the air is colder, the pressure aloft is lower.

Let's assume, to begin with, that there is no wind on the chart. In the Northern Hemisphere, the pressure gradient force directed northward starts the air moving toward lower pressure. Once the air is set in motion, the Coriolis force bends it to

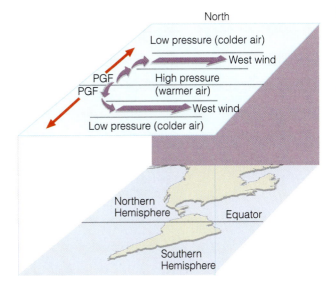

FIGURE 6
Upper-level chart that extends over the Northern and Southern Hemispheres. Solid gray lines on the chart are isobars.

the right until it is a *west wind*, blowing parallel to the isobars. In the Southern Hemisphere, the pressure gradient force directed southward starts the air moving south. But notice that the Coriolis force in the Southern Hemisphere bends the moving air to its *left*, until the wind is blowing parallel to the isobars *from the west*. Hence, in the middle and high latitudes of both hemispheres, we generally find westerly winds aloft.

---

they blow more slowly than the winds aloft? The answer to both of these questions is *friction*.*

Surface Winds   The frictional drag of the ground slows the wind down. Because the effect of friction decreases as we move away from the earth's surface, wind speeds tend to increase with height above the ground. The atmospheric layer that is influenced by friction, called the **friction layer** (or *planetary boundary layer*), usually extends upward to an altitude near 1000 m

---
* The geostrophic wind equation given in the Focus section on p. 226 is

$$V_g = \frac{1}{2\Omega \sin \phi \rho} \frac{\Delta p}{d},$$

when

$$
\begin{aligned}
p &= 16 \text{ mb} = 1600 \text{ Newtons/m}^2 \\
d &= 500 \text{ km} = 5 \times 10^5 \text{ m} \\
\sin \phi &= \sin(40°) = 0.64 \\
\rho &= 1.2 \text{ kg/m}^3 \\
2\Omega &= 14.6 \times 10^{-5} \text{ radian/sec.}
\end{aligned}
$$

The geostrophic wind is

$$V_g = \frac{1600}{14.6 \times 10^{-5} \times 0.64 \times 1.2 \times 5 \times 10^{-5}},$$

$$V_g = 28.6 \text{ m/sec, or 55 knots.}$$

(3300 ft) above the surface, but this altitude may vary due to strong winds or irregular terrain. (We will examine the planetary boundary layer winds more thoroughly in Chapter 10.)

In Fig. 9.29a, the wind aloft is blowing at a level above the frictional influence of the ground. At this level, the wind is approximately geostrophic and blows parallel to the isobars, with the pressure gradient force (*PGF*) on its left balanced by the Coriolis force on its right. At the earth's surface, the same pressure gradient will not produce the same wind speed, and the wind will not blow in the same direction.

Near the surface, *friction reduces the wind speed, which in turn reduces the Coriolis force.* Consequently, the weaker Coriolis force no longer balances the PGF, and the wind blows across the isobars toward lower pressure. The PGF is now balanced by the sum of the frictional force and the Coriolis force. Therefore, in the Northern Hemisphere, we find surface winds blowing counterclockwise and *into* a low; they flow clockwise and *out* of a high (see Fig. 9.29b). In the Southern Hemisphere, winds blow clockwise and inward around surface lows; counterclockwise and outward around surface highs. See the surface weather map and the general wind flow pattern for South America (Fig. 9.30).

(a)

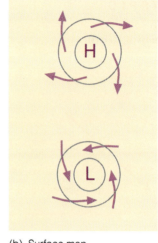

FIGURE 9.29
The effect of surface friction is to slow down the wind so that, near the ground, the wind crosses the isobars and blows toward lower pressure. This phenomenon produces an outflow of air around a high and an inflow around a low.

(b) Surface map
Northern Hemisphere

FIGURE 9.30
Surface weather map showing isobars and winds on a day in December in South America. The boxed area shows the idealized flow around surface-pressure systems in the Southern Hemisphere.

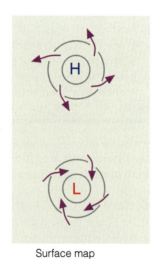

Surface map
Southern Hemisphere

In Fig. 9.29a, the angle (α) at which the wind crosses the isobars to a large degree depends upon the roughness of the terrain. Everything else being equal, the rougher the surface, the larger the angle. Over hilly land, for example, the angle might average between 35° and 40°, while over an open body of relatively smooth water it may average between 10° and 15°. Taking into account all types of surfaces, the average is near 30°. This angle also depends on the wind speed. Typically, the angle is smallest for high winds and largest for gentle breezes. As we move upward through the friction layer, the wind becomes more and more parallel to the isobars.

So far, we have seen that, because of friction, surface winds move more slowly than geostrophic winds with the same pressure gradient. Surface winds also blow across the isobars toward lower pressure. The angle at which the winds cross the isobars depends upon surface friction, wind speed, and the height above the surface. As we have seen, *if we stand with the wind aloft to our backs, lower pressure will be to our left and higher pressure to our right in the Northern Hemisphere* (see Fig. 9.31a). The same rule applies to the surface wind but with a slight modification due to the fact that here the wind crosses the isobars. Look at Fig. 9.31b and notice that, at the surface, *if*

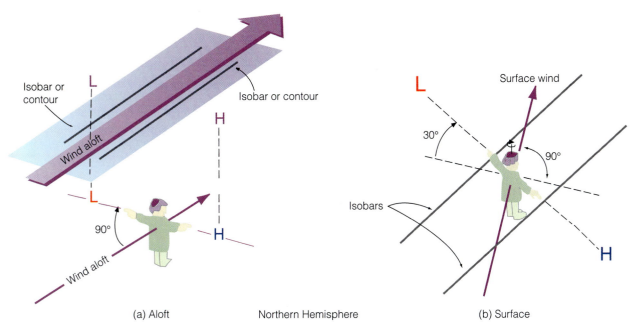

FIGURE 9.31
In the Northern Hemisphere, if you stand with the wind aloft at your back, lower pressure aloft will be to your left and higher pressure to your right (a). At the surface, the same relationship holds if, with your back to the surface wind, you turn clockwise about 30° (b).

we stand with our backs to the wind, then turn clockwise about 30°, lower pressure will be to our left. (In the Southern Hemisphere, if we stand with our backs to the wind, then turn counterclockwise about 30°, lower pressure will be to our right.) This relationship between wind and pressure is often called **Buys-Ballot's law**, after the Dutch meteorologist Christoph Buys-Ballot (1817–1890), who formulated it.

WINDS AND VERTICAL AIR MOTIONS

Up to this point, we have seen that surface winds blow in toward the center of low pressure and outward away from the center of high pressure. As air moves inward toward the center of a low-pressure area (see Fig. 9.32), it must go somewhere. Since this converging air cannot go into the ground, it slowly rises. Above the surface low (at about 6 km or so), the air begins to diverge (that is, move horizontally outward more quickly than it is taken in) to compensate for the converging surface air. As long as the upper-level diverging air balances the converging surface air, the central pressure in the low does not change. However, the surface pressure *will change* if upper-level divergence and surface convergence are not in balance. For example, if upper-level divergence exceeds surface convergence (that is, more air is removed at the top than is taken in at the surface), the central pressure of the low will decrease, and isobars around the low will become more tightly packed. This increases the pressure gradient (and,

hence, the pressure gradient force), which, in turn, increases the surface winds.

Surface winds move outward (diverge), away from the center of a high-pressure area. To replace this laterally spreading air, the air aloft converges and slowly descends (see Fig. 9.32). Again, as long as upper-level converging air balances surface diverging air, the central pressure in the high will not change. (Convergence and divergence of air are so important to the development or weakening of surface pressure systems that we will examine this topic again when we look more closely at the vertical structure of pressure systems in Chapter 13.)

FIGURE 9.32
Winds and air motions associated with surface highs and lows in the Northern Hemisphere.

FOCUS ON AN ADVANCED TOPIC

The Hydrostatic Equation

Air is in hydrostatic equilibrium when the upward-directed pressure gradient force is exactly balanced by the downward force of gravity. Figure 7 shows air in hydrostatic equilibrium. Since there is no net vertical force acting on the air, there is no net vertical acceleration, and the sum of the forces is equal to zero, all of which is represented by

$$PGF_{vertical} + g = 0$$

$$\frac{1}{\rho}\frac{\Delta p}{\Delta z} + g = 0,$$

where ρ is the air density, Δp is the decrease in pressure along a small change in height (Δz), and g is the force of gravity. This expression is usually given as

$$\frac{\Delta p}{\Delta z} = -\rho g.$$

This equation is called the *hydrostatic equation*. The hydrostatic equation tells us that the rate at which the pressure decreases with height is equal to the air density times the acceleration of gravity (where ρg is actually the force of gravity per unit

FIGURE 7

When the vertical pressure gradient force (PGF) is in balance with the force of gravity (g), the air is in hydrostatic equilibrium.

volume). The minus sign indicates that, as the air pressure decreases, the height increases. When the hydrostatic equation is given as

$$\Delta p = -\rho g\,\Delta z,$$

it tells us something important about the atmosphere that we learned earlier: The air pressure decreases more rapidly with height in cold (more-dense) air than it does in warm (less-dense) air. In addition, we can use the hydrostatic equation to determine how rapidly the air pressure decreases with increasing height above the surface. For example, suppose at the surface a 1000 m-thick

layer of air (under standard conditions) has an average density of 1.1 kg/m^3 and an acceleration of gravity of 9.8 m/sec^2. Therefore, we have:

$\rho = 1.1$ kg/m^3
$g = 9.8$ m/sec^2
$\Delta z = 1000$ m. (This is the height difference from the surface [0 m] to an altitude of 1000 m.)

Using the hydrostatic equation to compute Δp, the difference in pressure in a 1000 m-thick layer of air, we obtain

$\Delta p = \rho g\,\Delta z$
$\Delta p = (1.1)\,(9.8)\,(1000)$
$\Delta p = 10,780$ Newtons/m^2.

Since 1 mb = 100 Newtons/m^2,

$\Delta p = 108$ mb.

Hence, air pressure decreases by about 108 mb in a standard 1000-m layer of air near the surface. This closely approximates the pressure change of 10 mb per 100 m we used in converting station pressure to sea-level pressure earlier in this chapter.

The rate at which air rises above a low or descends above a high is small compared to the horizontal winds that spiral about these systems. Generally, the vertical motions are usually only about several centimeters per second, or about 1.5 km per day.

Earlier in this chapter we learned that air moves in response to pressure differences. Because air pressure decreases rapidly with increasing height above the surface, there is always a strong pressure gradient force (*PGF*) directed upward, much stronger than in the horizontal. Why, then, doesn't the air rush off into space?

Air does not rush off into space because the upward-directed *PGF* is nearly always exactly balanced by the downward

force of gravity. When these two forces are in exact balance, the air is said to be in **hydrostatic equilibrium**. When air is in hydrostatic equilibrium, there is no net vertical force acting on it, and so there is no net vertical acceleration. Most of the time, the atmosphere approximates hydrostatic balance, even when air slowly rises or descends at a constant speed. However, this balance does not exist in violent thunderstorms and tornadoes, where the air shows appreciable vertical acceleration. But these occur over relatively small vertical distances, considering the total vertical extent of the atmosphere. (A more mathematical look at hydrostatic equilibrium, expressed by the **hydrostatic equation**, is given in the Focus section above.)

SUMMARY

This chapter gives us a broad view of how and why the wind blows. We examined constant pressure charts and found that low heights correspond to low pressure and high heights to high pressure. In regions where the air aloft is cold, the air pressure is normally lower than average; where the air aloft is warm, the air pressure is normally higher than average. Where horizontal variations in temperature exist, there is a corresponding horizontal change in pressure. The difference in pressure establishes a force, the pressure gradient force (*PGF*), which starts the air moving from higher toward lower pressure.

Once the air is set in motion, the Coriolis force bends the moving air to the right of its intended path in the Northern Hemisphere and to the left in the Southern Hemisphere. Above the level of surface friction, the wind is bent enough so that it blows nearly parallel to the isobars, or contours. Where the wind blows in a straight-line path, and a balance exists between the *PGF* and the Coriolis force, the wind is termed geostrophic. Where the wind blows parallel to curved isobars (or contours), the centripetal acceleration becomes important, and the wind is called a gradient wind.

The interaction of the forces causes the winds aloft in the Northern Hemisphere to blow clockwise around regions of high pressure and counterclockwise around areas of low pressure. In the Southern Hemisphere, the winds aloft blow counterclockwise around highs and clockwise around lows. The effect of surface friction is to slow down the wind. This causes the surface air to blow across the isobars from higher pressure toward lower pressure. Consequently, in both hemispheres, surface winds blow outward, away from the center of a high, and inward, toward the center of a low.

Key Terms

The following terms are listed in the order they appear in the text. Define each. Doing so will aid you in reviewing the material covered in this chapter.

air pressure
barometer
millibar
hectopascal
mercury barometer
aneroid barometer
station pressure
sea level pressure
isobars
constant height chart
constant pressure
  (isobaric) chart

isobaric surface
contour lines (on isobaric
  charts)
ridges
troughs
anticyclones
mid-latitude cyclones
pressure gradient
pressure gradient force (*PGF*)
Coriolis force
geostrophic wind
gradient wind

centripetal acceleration
centripetal force
zonal flow
meridional flow

friction layer
Buys-Ballot's Law
hydrostatic equilibrium
hydrostatic equation

Questions for Review

1. Why does air pressure decrease with height more rapidly in cold air than in warm air?
2. What is considered standard atmospheric pressure?
3. How does an aneroid barometer differ from a mercury barometer?
4. Explain how sea-level pressure differs from station pressure.
5. On an upper-level chart, is cold air aloft generally associated with low or high pressure? What about warm air aloft?
6. What do Newton's first and second laws of motion tell us?
7. Explain why, in the Northern Hemisphere, the average height of contour lines on an upper-level isobaric chart tend to decrease northward.
8. What is the force that initially sets the air in motion?
9. What does the Coriolis force do to moving air (a) in the Northern Hemisphere? (b) in the Southern Hemisphere?
10. Explain how each of the following influences the Coriolis force: (a) rotation of the earth; (b) wind speed; (c) latitude.
11. Explain why on a map, closely spaced isobars (or contours) indicate strong winds, and widely spaced isobars (or contours) indicate weak winds.
12. Why would you not expect to observe a geostrophic wind at the equator?
13. Why do upper-level winds in the middle latitudes of both hemispheres generally blow from the west?
14. Describe how the wind blows around highs and lows aloft and near the surface (a) in the Northern Hemisphere and (b) in the Southern Hemisphere.
15. What are the forces that affect the horizontal movement of air?
16. What factors influence the angle at which surface winds cross the isobars?
17. Describe the type of vertical air motions associated with high- and low-pressure areas.
18. Since there is always an upward-directed pressure gradient force, why doesn't the air rush off into space?
19. How does Buys-Ballot's law help to locate regions of high and low pressure aloft and at the surface?

20. Explain the effect friction has on wind speed and direction.

21. Why are highs associated with ridges and lows associated with troughs on the 500 mb surface?

Questions for Thought

1. Explain why, on a sunny day, an aneroid barometer would indicate "stormy" weather when carried to the top of a hill or mountain.

2. Suppose the air column above city Q is completely saturated with water vapor, and the air column above city T is completely dry. If the temperature of the air in both columns is the same, which column will have the highest atmospheric pressure at the surface? Explain. (Hint: Refer back to the Focus section on p. 120 in Chapter 5, "Is Humid Air or Dry Air More Dense?")

3. Suppose you are in the Northern Hemisphere watching altocumulus clouds 4000 m (13,000 ft) above you drift from the northeast. Draw the orientation of the isobars above you. Locate and mark regions of lowest and highest pressure on this map. Finish the map by drawing isobars and the upper-level wind flow pattern hundreds of kilometers in all directions from your position. Would this type of flow be zonal or meridional? Explain.

4. Pilots often use the expression "high to low, look out below." In terms of upper-level temperature and pressure, explain what this can mean.

5. Suppose an aircraft using a pressure altimeter flies along a constant pressure surface from standard temperature into warmer-than-standard air without any corrections. Explain why the altimeter would indicate an altitude lower than the aircraft's true altitude.

6. If the earth were not rotating, how would the wind blow with respect to centers of high and low pressure?

7. Why are surface winds that blow over the ocean closer to being geostrophic than those that blow over the land?

8. If the wind aloft is blowing parallel to curved isobars, with the horizontal pressure gradient force being of greater magnitude than the Coriolis force, would the wind flow be cyclonic or anticyclonic? In this example, what would be the relative magnitude of the centripetal force, and how would it be directed?

9. With your present outside surface wind, use Buys-Ballot's law to determine where regions of surface high- and low-pressure areas are located. If clouds are moving overhead, use the relationship to locate regions of higher and lower pressure aloft.

10. If you live in the Northern Hemisphere and a region of surface low pressure is directly west of you, what would probably be the surface wind direction at your home? If an upper-level low is also directly west of your location, describe the probable wind direction aloft and the direction in which middle-type clouds would move. How would the wind direction and speed change from the surface to where the middle clouds are located?

11. In the Northern Hemisphere, you observe surface winds shift from N to NE to E, then to SE. From this observation, you determine that a west-to-east moving high-pressure area (anticyclone) has passed north of your location. Describe with the aid of a diagram how you were able to come to this conclusion.

12. The Coriolis force causes winds to deflect to the right of their intended path in the Northern Hemisphere, yet around a surface low-pressure area, winds blow counterclockwise, appearing to bend to their left. Explain why.

13. Why is it that, on the equator, winds may blow either counterclockwise or clockwise with respect to an area of low pressure?

14. Use the gas law in the Focus section on p. 213 to explain why a car with tightly closed windows will occasionally have a window "blow out" or crack when exposed to the sun on a hot day.

15. Consider wind blowing over a land surface that crosses a coastline and then blows over a lake. How will the wind speed and direction change as it moves from the land surface to the lake surface?

Problems and Exercises

1. Convert a sea-level pressure of 1013.25 mb to:
  (a) inches of mercury
  (b) kilopascals
  (c) hectopascals

2. A station 300 m above sea level reports a station pressure of 994 mb. What would be the sea-level pressure for this station, assuming standard atmospheric conditions? If the observation were taken on a hot summer afternoon, would the sea-level pressure be greater or less than that obtained during standard conditions? Explain.

3. The map you see on p. 237 is a sea-level pressure chart (Northern Hemisphere), with isobars drawn for every 4 mb. Answer the following questions, which refer to this map.

(a) What is the lowest possible pressure in whole millibars that there can be in the center of the closed low? What is the highest pressure possible?

(b) Place a dashed line through the ridge and a dotted line through the trough.

(c) What would be the wind direction at point A and at point B?

(d) Where would the stronger wind be blowing, at point A or B? Explain.

(e) Compute the pressure gradient between points 1 and 2, and between points 3 and 4. How do the computed pressure gradients relate to the pressure gradient force?

(f) If point A and point B are located at 30°N, and if the air density is 1.2 kg/m³, use the geostrophic wind equation in the Focus section on p. 226 to compute the geostrophic wind at point A and point B. (Hint: Be sure to convert km to m and mb to Newtons/m², where 1 mb = 100 Newtons/m².)

(g) Would the actual winds at point A and point B be greater than, less than, or equal to the wind speeds computed in problem f? Explain.

4. (a) Suppose the atmospheric pressure at the bottom of a deep air column 5.6 km thick is 1000 mb. If the average air density of the column is 0.91 kg/m³, and the acceleration of gravity is 9.8 m/sec², use the hydrostatic equation on p. 234 to determine the atmospheric pressure at the top of the column. (Hint: Be sure to convert km to m and mb to Newtons/m², where 1 mb = 100 N/m².)

(b) If the air in the column of problem (a) becomes much colder than average, would the atmospheric pressure at the top of the new column be greater than, less than, or equal to the pressure computed in problem (a)? Explain.

(c) Determine the atmospheric pressure at the top of the air column in problem a if the air in the column is quite cold and has an average density of 0.97 kg/m³.

5. Use the gas law in the Focus section on p. 213 to calculate the air pressure in millibars when the air temperature is −23°C and the air density is 0.700 kg/m³. (Hint: Be sure to use the Kelvin temperature.) At approximately what elevation would you expect to observe this pressure?

6. Suppose air in a closed container has a pressure of 1000 mb and a temperature of 20°C.

(a) Use the gas law to determine the air density in the container.

(b) If the density in the container remains constant, but the pressure doubles, what would be the new temperature?

7. A large balloon is filled with air so that the air pressure inside just equals the air pressure outside. The volume of

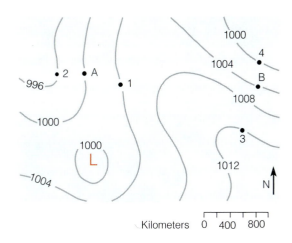

FIGURE 9.33
Map for problem 3.

the filled balloon is 3 m³, the mass of air inside is 3.6 kg, and the temperature inside is 20°C. What is the air pressure? (Hint: Density = mass/volume.)

8. If the clouds overhead are moving from north to south, would the upper-level center of low pressure be to the east or west of you?

Questions for Exploration

1. On the Blue Skies: College Edition CD-ROM go to the **Atmospheric Forces** section of the CD and click on "**Coriolis Force**."

(a) Fly a plane from the North Pole to a city of your choice. Note the path the plane maps out, though it never changes direction. In which direction is the plane pulled relative to its motion? Explain, in your own words, a reason for its trajectory.

(b) Try the same task for a city in the southern hemisphere. How does this differ from the Northern Hemisphere?

2. On the Blue Skies: College Edition CD-ROM go to the **Atmospheric Forces** section of the CD and click on "**Winds in Two Hemispheres**."

(a) Observe the wind pattern around a low pressure system in the United States, Canada, Europe or Asia (northern hemisphere). Describe the circulation pattern.

(b) Observe the wind pattern around a low pressure system in southern South America, Southern Africa or Australia (southern hemisphere). Describe the circulation pattern. What differences did you find?

As air sweeps over the open ocean it ultimately encounters obstacles such as the Canary Islands, which, on their downwind side, disrupt the flow into swirling eddies. (NASA photo.)

WIND: SMALL-SCALE AND LOCAL SYSTEMS

CONTENTS

On January 22, 1983, a DC-8 jetliner carrying 138 passengers and a crew of 6 was heading for San Francisco from Chicago. Some mild bumpiness over Colorado prompted the "fasten your seat belt" sign to go on. But other than that it was a routine, uneventful flight until suddenly over Utah a tremendous vibration ran through the plane. In an instant the aircraft nosed up, then dropped several thousand feet before stabilizing. Inside, screaming passengers were flung up against the walls and ceilings, then dropped. Those not fastened into their seats were sent hurling against overhead luggage compartments. Small bags and luggage, having slipped out from under the seats, were flying about inside the plane. Within seconds the entire ordeal was over. Twelve people suffered injuries and one passenger was knocked unconscious for about two minutes.

The aircraft in our opening vignette encountered a turbulent eddy—an "air pocket"—in perfectly clear air. Such violent eddies are not uncommon, especially in the vicinity of jet streams. In this chapter, we will examine a variety of eddy circulations. First, we will see how eddies form and how eddies and other small-scale winds interact with our environment. Then, we will examine slightly larger circulations—local winds—such as the sea breeze and the chinook, describing how they form and the type of weather they generally bring.

SMALL-SCALE WINDS INTERACTING WITH THE ENVIRONMENT

The air in motion—what we commonly call *wind*—is invisible, yet we see evidence of it nearly everywhere we look. It sculptures rocks, moves leaves, blows smoke, and lifts water vapor upward to where it can condense into clouds. The wind is with us wherever we go. On a hot day, it can cool us off; on a cold day, it can make us shiver. A breeze can sharpen our appetite when it blows the aroma from the local bakery in our direction. The wind is a powerful element. The workhorse of weather, it moves storms and large fair-weather systems around the globe. It transports heat, moisture, dust, insects, bacteria, and pollens from one area to another.

Circulations of all sizes exist within the atmosphere. Little whirls form inside bigger whirls, which encompass even larger whirls—one huge mass of turbulent, twisting *eddies*.* For clarity, meteorologists arrange circulations according to their size. This hierarchy of motion from tiny gusts to giant storms is called the **scales of motion**.

Scales of Motion   Consider smoke rising into the otherwise clean air from a chimney in the industrial section of a large city (see Fig. 10.1a). Within the smoke, small chaotic motions—

*Eddies are spinning globs of air that have a life history of their own.

tiny eddies—cause it to tumble and turn. These eddies constitute the smallest scale of motion—the **microscale**. At the microscale, eddies with diameters of a few meters or less not only disperse smoke, they also sway branches and swirl dust and papers into the air. They form by convection or by the wind blowing past obstructions and are usually short-lived, lasting only a few minutes at best.

In Fig. 10.1b observe that, as the smoke rises, it drifts toward the center of town. Here the smoke rises even higher and is carried back toward the industrial section. This circulation of city air constitutes the next larger scale—the **mesoscale** (meaning middle scale). Typical mesoscale winds range from a few kilometers to about a hundred kilometers in diameter. Generally, they last longer than microscale motions, often many minutes, hours, or in some cases as long as a day. Mesoscale circulations include local winds (which form along shorelines and mountains), as well as thunderstorms, tornadoes, and small tropical storms.

When we look at the smoke stack on a surface weather map (see Fig. 10.1c), neither the smoke stack nor the circulation of city air shows up. All that we see are the circulations around high- and low-pressure areas. We are now looking at the **synoptic scale**, or weather map scale. Circulations of this magnitude dominate regions of hundreds to even thousands of square kilometers and, although the life spans of these features vary, they typically last for days and sometimes weeks.

The largest wind patterns are seen at the **planetary** (*global*) **scale**. Here, we have wind patterns ranging over the entire earth. Sometimes, the synoptic and global scales are combined and referred to as the **macroscale**. (Table 10.1 summarizes the various scales of motion and their average life span.)

In the next several sections, we will concentrate primarily on microscale winds and the effect they have on our environment.

Friction and Turbulence in the Boundary Layer   We are all familiar with friction. If we rub our hand over the top of a table, friction tends to slow its movement because of irregularities in the table's surface. On a microscopic level, friction arises

(a) Microscale

(b) Mesoscale

(c) Synoptic scale

FIGURE 10.1

Scales of atmospheric motion. The tiny microscale motions constitute a part of the larger mesoscale motions, which, in turn, are part of the much larger synoptic scale. Notice that as the scale becomes larger, motions observed at the smaller scale are no longer visible.

TABLE 10.1   The Scales of Atmospheric Motion with the Phenomena's Average Size and Life Span*

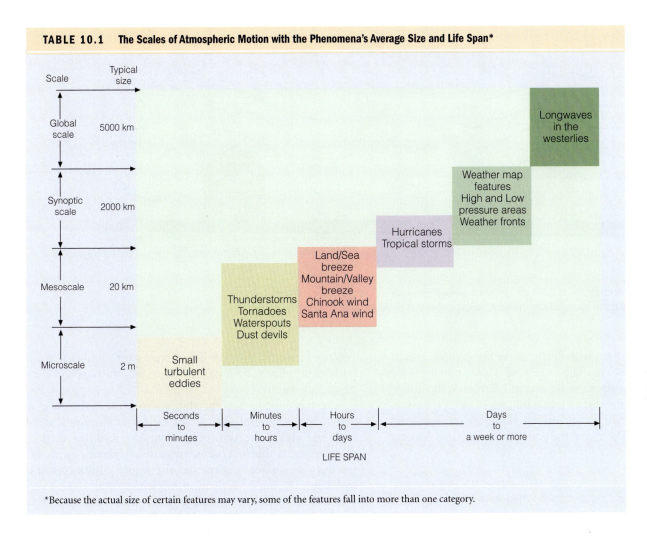

*Because the actual size of certain features may vary, some of the features fall into more than one category.

as atoms and molecules of the two surfaces seem to adhere, then snap apart as the hand slides over the table. Friction is not restricted to solid objects; it occurs in moving fluids as well. Consider, for example, a steady flow of water in a stream. When a paddle is placed in the stream, turbulent whirls (*eddies*) form behind it. These eddies create fluid friction by draining energy from the main stream flow, slowing it down. Let's examine the idea of fluid friction in more detail.

The friction of fluid flow is called **viscosity**. When the slowing of a fluid—such as air—is due to the random motion of the gas molecules, the viscosity is referred to as *molecular viscosity*. Consider a mass of air gliding horizontally and smoothly (*laminar flow*) over a stationary mass of air. Even though the molecules in the stationary air are not moving horizontally, they are darting about and colliding with each other. At the boundary separating the air layers, there is a constant exchange of molecules between the stationary air and flowing air. The overall effect of this molecular exchange is to slow down the moving air. If molecular viscosity were the only type of friction acting on moving air, the effect of friction would disappear in a thin layer just above the surface. There is, how-

ever, another frictional effect that is far more important in reducing wind speeds.

When laminar flow gives way to irregular turbulent motion, there is an effect similar to molecular viscosity, but which occurs throughout a much larger portion of the moving air. The internal friction produced by turbulent whirling eddies is called *eddy viscosity*. Near the surface, it is related to the roughness of the ground. As wind blows over a landscape dotted with trees and buildings, it breaks into a series of irregular, twisting eddies that can influence the air flow for hundreds of meters above the surface. Within each eddy, the wind speed and direction fluctuate rapidly, producing the irregular air motion known as *wind gusts*. Eddy motions created by obstructions are commonly referred to as **mechanical turbulence**.* Mechanical turbulence creates a drag on the flow of air, one far greater than that caused by molecular viscosity.

The frictional drag of the ground normally decreases as we move away from the earth's surface. Because of this, wind

---

*Keep in mind that *turbulence* represents any irregular or disturbed flow in the atmosphere that produces gusts and eddies.

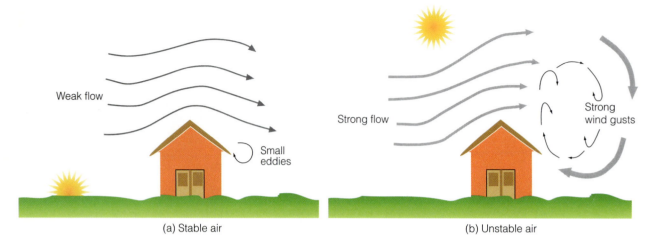

FIGURE 10.2

Winds flowing past an obstacle. (a) In stable air, light winds produce small eddies and little vertical mixing. (b) Greater winds in unstable air create deep, vertically mixing eddies that produce strong, gusty surface winds.

speeds tend to increase with height above the ground. In fact, at a height of only 10 m (33 ft), the wind is often moving twice as fast as at the surface. As we saw in Chapter 9, the atmospheric layer near the surface that is influenced by friction (turbulence) is called the *friction layer* or **planetary** *(atmospheric)* **boundary layer**. The top of the boundary layer is usually near

FIGURE 10.3

When the air is stable and the terrain fairly smooth (a), vertical mixing is at a minimum, and the effect of surface friction only extends upward a relatively short distance above the surface. When the air is unstable and the terrain rough (b), vertical mixing is at a maximum, and the effect of surface friction extends upward through a much greater depth of atmosphere. Within the region of frictional influence, vertical mixing increases the wind speed near the ground and decreases it aloft. (Wind at the surface is measured at 10 m above the surface.)

1000 m (3300 ft), but this height may vary somewhat since both strong winds and rough terrain extend the region of frictional influence.

Surface heating and instability also cause turbulence to extend to greater altitudes. As the earth's surface heats, thermals rise and convection cells form. The resulting vertical motion creates **thermal turbulence**, which increases with the intensity of surface heating and the degree of atmospheric instability. During the early morning, when the air is most stable, thermal turbulence is normally at a minimum. As surface heating increases, instability is induced and thermal turbulence becomes more intense. If this heating produces convective clouds that rise to great heights, there may be turbulence from the earth's surface to the base of the stratosphere.

Although we have treated thermal and mechanical turbulence separately, they occur together in the atmosphere—each magnifying the influence of the other. Let's consider a simple example: the eddy forming behind the barn in Fig. 10.2. In stable air with weak winds, the eddy is nonexistent or small and the air flow is often around the barn. As wind speed and surface heating increase, instability develops, and the eddy becomes larger and extends through a greater depth. The rising side of the eddy carries slow-moving surface air upward, causing a frictional drag on the faster flow of air aloft. Some of the faster moving air is brought down with the descending part of the eddy, producing a momentary gust of wind. Because of the increased depth of circulating eddies in unstable air, strong, gusty surface winds are more likely to occur when the atmosphere is unstable. Greater instability also leads to a greater exchange of faster moving air from upper levels with slower moving air at lower levels. In general, this exchange increases the average wind speed near the surface and decreases it aloft, producing the distribution of wind speed with height shown in Fig. 10.3.

We can now see why surface winds are usually stronger in the afternoon. Vertical mixing during the middle of the day links surface air with the faster moving air aloft. The result is that the surface air is pulled along more quickly. At night, when convection is reduced, the interchange between the air at the surface and the air aloft is at a minimum. Hence, the wind near the ground is less affected by the faster wind flow above, and so it blows more slowly.

In summary, the friction of air flow (viscosity) is a result of the exchange of air molecules moving at different speeds. The exchange brought about by random molecular motions (molecular viscosity) is quite small in comparison with the exchange brought about by turbulent motions (eddy viscosity). Therefore, the frictional effect of the surface on moving air depends largely upon mechanical and thermal turbulent mixing. The depth of mixing and, hence, frictional influence (in the boundary layer) depend primarily upon three factors:

1. surface heating—producing a steep lapse rate and strong thermal turbulence
2. strong wind speeds—producing strong mechanical turbulent motions
3. rough or hilly landscape—producing strong mechanical turbulence

When these three factors occur simultaneously, the frictional effect of the ground is transferred upward to considerable heights, and the wind at the surface is typically strong and gusty.

Eddies—Big and Small   When the wind encounters a solid object, a whirl of air—an eddy (see Fig. 10.4)—forms on the object's leeward side. The size and shape of the eddy depend upon the size and shape of the obstacle and on the speed of the wind. Light winds produce small stationary eddies. Wind moving past trees, shrubs, and even your body produces small eddies. (You may have had the experience of dropping a piece of paper on a windy day only to have it carried away by a swirling eddy as you bend down to pick it up.) Air flowing over a building produces larger eddies that will, at best, be about the size of the building. Strong winds blowing past an open sports stadium can produce eddies that may rotate in such a way as to create surface winds on the playing field. These surface winds may move in a direction opposite to the wind flow above the stadium. Wind blowing over a fairly smooth surface produces few eddies, but when the surface is rough, many eddies form.

The eddies that form downwind from obstacles can produce a variety of interesting effects. For instance, wind moving over a mountain range in a stable atmosphere with a speed greater than 40 knots usually produces waves and eddies, such as those shown in Fig. 10.5. We can see that eddies form both close to the mountain and beneath each wave crest. These are called *roll eddies*, or **rotors**, and have violent vertical motions that produce extreme turbulence and hazardous flying conditions. Strong winds blowing over a mountain in stable air sometimes provide a *mountain wave eddy* on the downwind side, with a reverse flow near the ground. On a much smaller scale, the howling of wind on a blustery night is believed to be caused by eddies that are constantly being shed around obstructions, such as chimneys and roof corners. Whirling eddies create small areas of high air density. These tightly packed molecules act like a tiny pulse of compressed air, which

FIGURE 10.4

As the wind blows over the mountain peak, the direction of air flow is disturbed and circular eddies form on the mountain's leeward (downward) side. The shallow cloud pattern indicates that some eddies are circulating clockwise while others are circulating counterclockwise. Eddies such as these are commonly called *von Karman vortices*.

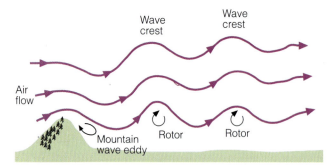

FIGURE 10.5
Under stable conditions, air flowing past a mountain range can create eddies many kilometers downwind from the mountain itself.

pushes on the surrounding air molecules. These, in turn, push on other molecules and so on until the pulse of compressed air reaches your eardrums, producing a sound. On a windy night, one eddy after another forms around sharp corners with such regularity that the wind sounds as if it were howling.

The largest atmospheric eddies form as the flow of air becomes organized into huge spiraling whirls—the cyclones and anticyclones of middle latitudes—which can have diameters greater than 1000 km (600 mi). Since it is these migrating systems that make our middle latitude weather so changeable, we will examine the formation and movement of these systems in Chapters 12 and 13.

Turbulent eddies form aloft as well as near the surface. Turbulence aloft can occur suddenly and unexpectedly, especially where the wind changes its speed or direction (or both) abruptly. Such a change is called a **wind shear**. The shearing creates forces that produce eddies along a mixing zone. If the eddies form in clear air, this form of turbulence is called **clear air turbulence** (or **CAT**). (Additional information on this topic is given in the Focus section on p. 245.)

The Force of the Wind

Because a small increase in wind speed can greatly increase the wind force on an object, strong winds may blow down trees, overturn mobile homes, and even move railroad cars. Here are just a few examples to illustrate how powerful the wind really is. On November 17, 1869, in Columbia County, New York, high winds lifted a moving train from the tracks and hurled it down a 6-m embankment. In Nebraska, in 1894, strong winds propelled six fully loaded coal cars over 160 km (100 mi) in a little more than three hours. Similarly, in February, 1965, the wind presented people in North Dakota with a "ghost train" as it pushed five railroad cars from Portal to Minot (about 125 km) without a locomotive. And, while the people in Mount Pleasant, Iowa, awaited the 1979 Fourth of July celebration, a phenomenally strong wind—estimated at 90 knots—blew the Goodyear blimp from its mooring and rolled it 300 m into a corn field, where it came to rest in ruins.

Wind blowing with sufficient force to demolish the Goodyear blimp is uncommon. However, wind blowing with enough force to move an automobile is very common, especially when the automobile is exposed to a strong crosswind. On a normal road, the force of a crosswind is usually insufficient to move a car sideways because of the reduced wind flow near the ground. However, when the car crosses a high bridge, where the frictional influence of the ground is reduced, the increased wind speed can be felt by the driver. Near the top of a high bridge, where the wind flow is typically strongest, complicated eddies pound against the car's side as the air moves past obstructions, such as guard railings and posts. In a strong wind, these eddies may even break into extremely turbulent whirls that buffet the car, causing difficult handling as it moves from side to side. If there is a wall on the bridge, the wind may swirl around and strike the car from the side opposite the wind direction.

A similar effect occurs where the wind moves over low hills paralleling a highway (see Fig. 10.6). When the vehicle moves by the obstruction, a wind gust from the opposite direction can suddenly and without warning push it to the opposite side. This

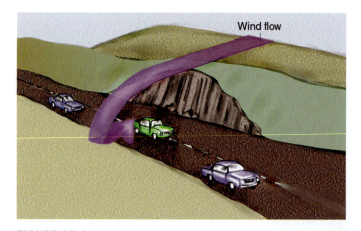

FIGURE 10.6
Strong winds flowing past an obstruction, such as these hills, can produce a reverse flow of air that strikes an object from the side opposite the general wind direction.

Eddies and "Air Pockets"

To better understand how eddies form along a zone of wind shear, imagine that, high in the atmosphere, there is a stable layer of air having vertical wind speed shear (changing wind speed with height) as depicted in Fig. 1. The top half of the layer slowly slides over the bottom half, and the relative speed of both halves is low. As long as the wind shear between the top and bottom of the layer is small, few if any eddies form. However, if the shear and the corresponding relative speed of these layers increase, wavelike undulations may form. When the shearing exceeds a certain value, the waves break into large swirls, with significant vertical movement. Eddies such as these often form in the upper troposphere near jet streams, where large wind speed shears

exist. They also occur in conjunction with mountain waves, which may extend upward into the stratosphere (see Fig. 2). As we learned earlier, when these huge eddies develop in clear air, this form of turbulence is referred to as *clear air turbulence*, or *CAT*.

The eddies that form in clear air may have diameters ranging from a couple of meters to several hundred meters. An unsuspecting aircraft entering such a region may be in for more than just a bumpy ride. If the aircraft flies into a zone of descending air, it may drop suddenly, producing the sensation that there is no air to support the wings. Consequently, these regions have come to be known as *air pockets*.

Commercial aircraft entering an air pocket have dropped hundreds of

meters, injuring passengers and flight attendants not strapped into their seats. For example, on April 4, 1981, a DC-10 jetliner flying at 11,300 m (37,000 ft) over central Illinois encountered a region of severe clear air turbulence and reportedly plunged about 600 m (2000 ft) toward the earth before stabilizing. Twenty-one of the 154 people aboard were injured; one person sustained a fractured hip and another person, after hitting the ceiling, jabbed himself in the nose with a fork, then landed in the seat in front of him. Clear air turbulence has occasionally caused structural damage to aircraft by breaking off vertical stabilizers and tail structures. Fortunately, the effects are usually not this dramatic.

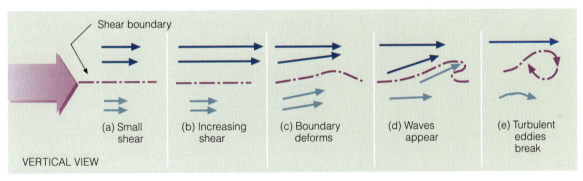

Shear boundary

(a) Small shear   (b) Increasing shear   (c) Boundary deforms   (d) Waves appear   (e) Turbulent eddies break

VERTICAL VIEW

FIGURE 1

The formation of clear air turbulence (CAT) along a boundary of increasing wind speed shear. The view of wind is from the side, with a top layer of air moving over a layer below.

FIGURE 2

Turbulent eddies forming downwind of a mountain chain in a wind shear zone produce these billow clouds, which are also known as Kelvin-Helmholtz billows.

wind hazard is a special problem for trucks, campers, and trailers, and highway signs warning of gusty wind areas are often posted.

Up to now we have seen that, when the wind meets a barrier, it exerts a force upon it. If the barrier doesn't move, the wind moves around, up, and over it. When the barrier is long and low like a water wave, the slight updrafts created on the windward side support the wings of birds, allowing them to skim the water in search of food without having to flap their wings. Elongated hills and cliffs that face into the wind create upward air motions that can support a hang glider in the air for a long time. (The cliffs in the Hawaiian Islands and along the California coast with their steep escarpments are especially fine hang-gliding areas.) Wind speeds greater than about 15 knots blowing over a smooth yet moderately sloping ridge may provide excellent *ridge-soaring* for the sailplane enthusiast.

As stable air flows over a ridge, it increases in speed. Thus, winds blowing over mountains tend to be stronger than winds blowing at the same level on either side. In fact, the greatest wind speed ever recorded near the ground occurred at the summit of Mt. Washington, New Hampshire, elevation 1909 m (6262 ft), where the wind gusted to 200 knots on April 12, 1934. A similar increase in wind speed occurs where air accelerates as it funnels through a narrow constriction, such as a low pass or saddle in a mountain crest.

Microscale Winds Blowing over the Earth's Surface
We have already seen that winds can move objects on the ground. However, strange as it may seem, winds capable of moving large boulders are not capable of lifting fine clay and dust particles.* The reason is that, near the ground in a thin layer less than 0.1 mm thick, the wind speed is practically zero even when high wind speeds exist only a few meters higher. How, then, do these very fine particles become airborne?

Sand grains large enough to protrude into the region of wind flow are pushed along when the wind speeds are sufficiently strong. Particles with diameters larger than 0.2 mm begin to move in a 13-knot wind, and larger sand particles, whose diameters are about 2.0 mm, are put in motion by a 26-knot wind. Once the grains begin to move, they roll and slide over the ground, with the smaller ones rolling faster than the larger ones. After moving a short distance, a grain may collide with a larger, stationary particle. This collision causes the smaller particle to jump into the air to a height of between 1 mm and several meters, depending upon the wind speed and the weight of the particle. This bouncing and skipping motion of sand grains is called *saltation*. Once airborne, the particles are swept along even more quickly by the wind. Eventually, they drop back to the ground, where they dislodge other particles, some being fine clay and dust too small to be picked up by the wind alone. If these tiny particles are caught in the windstream, they remain suspended for a long time and, in some cases, are carried over great distances.

*Diameters of clay- and dust-size particles are typically less than 0.002 mm.

Wind and Exposed Soil Where the wind blows over exposed soil, it takes an active role in shaping the landscape. This is especially noticeable in deserts. As tiny, loose particles of sand, silt, and dust are lifted from the surface and carried away by the wind, the ground level gradually lowers. The removal of this fine material leaves the surface covered with gravel and pebbles, which are too large to be transported by the wind. Such a landscape is termed *desert pavement*.

Because sand grains usually travel close to the ground, strong winds armed with sand can erode and shape rocks lying near the surface. If the rocks are soft and the winds are exceptionally strong (and prevail from one direction), the constant impact of hard quartz sand grains can weather solid rocks, notching the base of a cliff or a boulder. The constant sandblasting of rock fragments may cause a flattened and pitted surface on the windward side. Rocks shaped by the abrasive action of the wind are called *ventifacts*. Even telephone poles have no immunity to desert winds and are often tapered near the base, where windblown sand has cut away the wood. And, finally, a motorist traveling through a wind-driven sandstorm knows all too well the disastrous effects of sandblasting on a car's paint job and windshield.

Blowing sand eventually comes to rest *behind* obstacles, which can be anything from a rock to a clump of vegetation. As the sand grains accumulate, they pile into a larger heap that, when high enough, acts as an obstacle itself. If the wind speed is strong and continues to blow in the same direction for a sufficient time, the sand piles up higher and eventually becomes a *sand dune*. On the dune's surface, the sand rolls, slides, and gradually creeps along, producing wavelike patterns called *sand ripples*. Each ripple forms perpendicular to the wind direction, with a gentle slope on the windward side and a steeper slope on the leeward side. (If the wind direction frequently changes, the ripple becomes more symmetric.) On a larger scale, the dune itself may take on this shape. Sand is carried forward and up the dune until it reaches the top. Here, the air flow is strongest, and the sand continues its forward movement and cascades down the backside of the dune into quieter air. The effect of this migration is to create a dune whose windward slope is more gentle than its leeward slope. Therefore, the shape of a sand dune reveals the prevailing wind direction during its formation (see Fig. 10.7).

Wind and Snow Surfaces Wind blowing over a snow-covered landscape may also create wavelike patterns several centimeters high and oriented at right angles to the wind. These *snow ripples* are similar to sand ripples. On a larger scale, winds may create *snow dunes*, which are quite similar to sand dunes. Irregularities at the surface can cause a strong wind (40 knots) to break into turbulent eddies. If the snow on the ground is moist and sticky, some of it may be picked up by the wind and sent rolling. As it rolls along, it collects more snow and grows bigger. If the wind is sufficiently strong, the moving clump of snow becomes cylindrical, often with a hole extend-

FIGURE 10.7
The shape of this sand dune reveals that the wind was blowing from left to right when it formed. Note also the shape of the sand ripples on the dune.

FIGURE 10.8
Snow rollers—natural cylindrical rolls of snow—grow larger as the wind blows them down a hillside.

ing through it lengthwise. These **snow rollers** range from the size of eggs to that of small barrels. The tracks they make in the snow are typically less than 1 centimeter deep and several meters long. Snow rollers are rare, but, when they occur, they create a striking winter scene (see Fig. 10.8). In populated areas, they may escape notice as they are often mistaken as having been made by children rather than by nature.

Strong winds blowing over a vast region of open plains can alter the landscape in a different way. Consider, for example, a light snowfall several centimeters deep covering a large portion of central South Dakota. After the snow stops falling, strong winds may whip it into the air, leaving fields barren of snow. The cold, dry wind also robs the soil of any remaining moisture and freezes it solid. Meanwhile, the snow settles out of the air when the wind encounters obstacles. Since the greatest density of such obstructions is normally in towns, municipal snowfall measurements may show an accumulation of many centimeters, while the surrounding countryside, which may desperately need the snow, has practically none.

To help remedy this situation, *snow fences* are constructed in open spaces (see Fig. 10.9). Behind the snow fence, the wind speed is reduced because the air is broken into small eddies, which allow the snow to settle to the ground. Added snow cover is important for open areas because it acts like an insulating blanket that protects the ground from the bitter cold air, which often follows in the wake of a major snowstorm. In regions of low rainfall, moisture from the spring snowmelt can be a critical factor during long, dry summers. Snow fences are also built to protect major highways in these areas. Hopefully, the snow will accumulate behind the fence rather than in huge drifts on the road.

Wind and Vegetation We have already examined how wind affects the landscape in regions where the amount of exposed vegetation is limited. What can happen to vegetation in regions

FIGURE 10.9
Snow accumulating behind snow fences in Wyoming.

of high winds? Strong, prevailing winds can bend and twist the branches of conifers toward the leeward side, producing **wind-sculptured trees** (see Fig. 10.10). Strong winds can unexpectedly prune trees by breaking off their branches. Armed with sand, strong winds can damage or destroy tender new vegetation, decreasing crop productivity. Most plants increase their rate of transpiration as wind speed increases.* This leads to rapid water loss, especially in warmer areas having low humidities, and may actually dry out plants. If sustained, this drying-out effect may stunt plant growth, and, in some windy, dry regions, mature trees that should be many meters tall grow only to the height of a small shrub.

---

*This effect actually drops above a certain wind speed and varies greatly among plant species.

FIGURE 10.10
In the high country, trees standing unprotected from the wind are often sculpted into "flag" trees.

Wind-dried vegetation can result in an area of high fire danger. If a fire should begin here, any additional wind helps it along, directing its movement, adding oxygen for combustion, and carrying burning embers elsewhere to start new fires. On the open plains, where the wind blows practically unimpeded, wind-whipped prairie fires can imperil homes and livestock as the fires burn out of control over large areas.

Wind erosion is greatly reduced by a continuous cover of vegetation. The vegetation screens the surface from the direct force of the wind and anchors the soil. Soil moisture also helps to resist wind erosion by cementing particles together, which increases their cohesiveness. From this fact, we can see that land where natural vegetation has been removed for farming purposes—followed by several years of drought—is ripe for wind erosion. This situation happened in parts of the Great

FIGURE 10.11
A properly designed shelterbelt can reduce the air flow downwind for a distance of 25 times the height of the belt. The minimum wind flow behind the belt is typically measured downwind at a distance of about 4 times the belt's height.

Plains in the middle 1930s, when winds carried millions of tons of dust into the air, creating vast duststorms that buried whole farmhouses, reduced millions of acres to unproductive wasteland, and financially ruined thousands of families. Because of these disastrous effects of the wind, portions of the western plains became known as the "dust bowl."

To protect crops and soil, *windbreaks*—commonly called **shelterbelts**—are planted. Shelterbelts usually consist of a series of mixed conifer and deciduous trees or shrubs planted in rows perpendicular to the prevailing wind flow. They greatly reduce the wind speed behind them (see Fig. 10.11). As air filters through the belt, the flow is broken into small eddies, which have little mixing effect on the air near the surface. However, if trees are planted too close together, several unwanted effects may result. For one thing, the air moving past the belt may be broken into larger, more turbulent eddies, which swirl soil about. Furthermore, in high winds, strong downdrafts may damage the crops.

The use of properly designed shelterbelts has benefited agriculture. In some parts of the Central Plains, these belts have stabilized the soil and increased wheat yield. Despite their advantages, many of the shelterbelts planted during the mid-1930s drought years have been removed. Some are economically unfeasible because they occupy valuable crop land. Others interfere with the large center pivot sprinkler systems now in use. At any rate, one wonders how the absence of these shelterbelts would affect this region if it were struck by a drought similar to that experienced in the 1930s.

Wind and Waves   The impact of the wind on the earth's surface is not limited to land; wind also influences water—it makes waves. Waves forming by wind blowing over the surface of the water are known as **wind waves**. Just as air blowing over the top of a water-filled pan creates tiny ripples, so waves are created as the frictional drag of the wind transfers energy to the water. In general, the greater the wind speed, the greater the amount of energy added, and the higher will be the waves. Actually, the amount of energy transferred to the water (and thus the height to which a wave can build) depends upon three factors:

1. the wind speed
2. the length of time that the wind blows over the water
3. the *fetch*, or distance, of deep water over which the wind blows

A sustained 50-knot wind blowing steadily for nearly three days over a minimum distance of 2600 km (1600 mi) can generate waves with an average height of 15 m (49 ft). Thus, a stationary storm system centered somewhere over the open sea is capable of creating large waves with waveheights occasionally measuring over 31 m (100 ft).

Microscale winds actually help waves grow taller. Consider, for example, the wind blowing over the small wave depicted in Fig. 10.12. Observe that both the wind and the

Pedaling into the Wind

Anyone who rides a bicycle knows that it is much easier to pedal with the wind than against it. The reason is obvious: As the wind blows against an object, it exerts a force upon it. The amount of force exerted by the wind over an area increases as the square of the wind velocity. This is shown by the relationship

$$F \sim V^2,$$

where F is the wind force and V is the wind velocity. From this we can see that, if the wind velocity doubles, the force goes up by a factor of 2^2, or 4, which means that pedaling into a 40-knot wind requires 4 times as much effort as pedaling into a 20-knot wind.

Wind striking an object exerts a pressure on it. The amount of pressure depends upon the object's shape and size, as well as on the amount of reduced pressure that exists on the object's downwind side. Without concern for all the complications, we can approximate the wind pressure on an object with a simple formula. For example, if the wind velocity (V) is in miles per hour, and the wind force (F) is in pounds, and the object's surface area (A) is measured in square feet, the wind pressure (P), in pounds per square feet, is

$$\frac{F}{A} = P = 0.004 \ V^2.$$

We can look at a practical example of this expression if we consider a bicycle rider going 10 mi/hr into a head wind of 40 mi/hr. With the total velocity of the wind against the rider (wind speed plus bicycle speed) being 50 mi/hr, the pressure of the wind is

$$P = 0.004 \ V^2$$
$$= 0.004 \ (50^2)$$
$$= 10 \ \text{lb/ft}^2.$$

If the rider has a surface body area of 5 ft², the total force exerted by the wind becomes

$$F = P \times A$$
$$= 10 \ \text{lb/ft}^2 \times 5 \ \text{ft}^2$$
$$= 50 \ \text{lb}.$$

This force is enough to make pedaling into the wind extremely difficult. To remedy this adverse effect, cyclists—especially racers—bend forward as low as possible in order to expose a minimum surface area to the wind. It should also be obvious why track records are asterisked with "wind-aided" when the runners race with a tail wind of more than 3 mi/hr.

---

wave are moving in the same direction, and that the wave crest deflects the wind upward, producing an undulation in the air flow just above the water. This looping air motion establishes a small eddy of air between the two crests. The upward and downward motion of the eddy reinforces the upward and downward motion of the water. Consequently, the eddy helps the wave to build in height.

Traveling in the open ocean, waves represent a form of energy. As they move into a region of weaker winds, they gradually change: Their crests become lower and more rounded, forming what are commonly called *swells*. When waves reach a shoreline they transfer their energy—sometimes catastrophically—to the coast and structures along it. High, storm-induced waves can hurl thousands of tons of water against the shore. If this happens during an unusually high tide, resort homes overlooking the ocean can be pounded into a twisted mass of board and nails by the surf. Bear in mind that the storms creating these waves may be thousands of kilometers away and, in fact, may never reach the shore. Some of the largest, most damaging waves ever to strike the beach communities of Southern California arrived on what was described as "one of the clearest days imaginable." On the more positive

side, in the Hawaiian Islands these high waves are excellent for surfboarding.

Up to this point, we have seen how the wind blowing over the surface can produce a variety of features, from snow rollers to large ocean waves. How the force of the wind can influence someone riding a bicycle is found in the Focus section above.

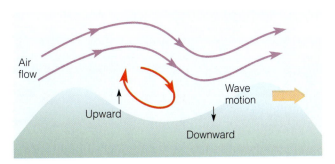

FIGURE 10.12

Wind blowing over a wave creates a small eddy of air that helps to reinforce the up-and-down motion of the water.

DETERMINING WIND DIRECTION AND SPEED

Wind is characterized by its direction, speed, and gustiness. Because air is invisible, we cannot really see it. Rather, we see things being moved by it. Therefore, we can determine wind direction by watching the movement of objects as air passes them. For example, the rustling of small leaves, smoke drifting near the ground, and flags waving on a pole all indicate wind direction. In a light breeze, a tried and true method of determining wind direction is to raise a wet finger into the air. The dampness quickly evaporates on the windward side, cooling the skin. Traffic sounds carried from nearby railroads or airports can be used to help figure out the direction of the wind. Even your nose can alert you to the wind direction as the smell of fried chicken or broiled hamburgers drifts with the wind from a local restaurant.

We already know that *wind direction* is given as the direction from which it is blowing—a north wind blows from the north toward the south. However, near large bodies of water and in hilly regions, wind direction may be expressed differently. For example, wind blowing from the water onto the land is referred to as an **onshore wind**, while wind blowing from land to water is called an **offshore wind** (see Fig. 10.13). Air moving uphill is an *upslope wind*; air moving downhill is a *downslope wind*. The wind direction may also be given as degrees about a 360° circle. These directions are expressed by the numbers shown in Fig. 10.14. For example: A wind direction of 360° is a north wind; an east wind is 90°; a south wind is 180°; and calm is expressed as zero. It is also common practice to express the wind direction in terms of compass points, such as N, NW, NE, and so on.

The Influence of Prevailing Winds
At many locations, the wind blows more frequently from one direction than from any other. The **prevailing wind** is the name given to the wind direction most often observed during a given time period. Prevailing winds can greatly affect the climate of a region. For example,

where the prevailing winds are upslope, clouds, fog, and precipitation are more likely than where the winds are downslope. Prevailing onshore winds in summer carry moisture, cool air, and fog into coastal regions, while prevailing offshore breezes carry warmer and drier air into the same locations. In city planning, the prevailing wind can help decide where industrial centers, factories, and city dumps should be built. All of these, of course, must be located so that the wind will not carry pollutants into populated areas. Sewage disposal plants must be situated downwind from large housing developments, and major runways at airports must be aligned with the prevailing wind to assist aircraft in taking off or landing.

The prevailing wind can even be a significant factor in building an individual home. In the northeastern half of the United States, the prevailing wind in winter is northwest and in summer it is southwest. Thus, houses built in the northeastern United States should have windows facing southwest to provide summertime ventilation and few, if any, windows facing the cold winter winds from the northwest. The northwest side of the house should be thoroughly insulated and even protected by a windbreak.

From the prevailing wind, biologists can predict the direction disease-carrying insects and plant spores will move and, hence, how a disease may spread. Geologists use the prevailing wind to predict where ejected debris from potentially active volcanoes will land.

Many local ground and landscape features show the effect of a prevailing wind. For example, smoke particles from an industrial stack settle to the ground on its downwind side. From the air, the prevailing wind direction can be seen as a discolored landscape on the leeward side of the stack. Wind blowing over surfaces of snow and sand produces ripples with a more gentle slope facing into the wind. As previously mentioned, sand dunes have similar shapes and, thus, show the prevailing wind direction. Trees with branches twisted and broken on their windward side appear gnarled as the remaining branches point away from the prevailing wind. Look at Fig.

FIGURE 10.13

An onshore wind blows from water to land, whereas an offshore wind blows from land to water.

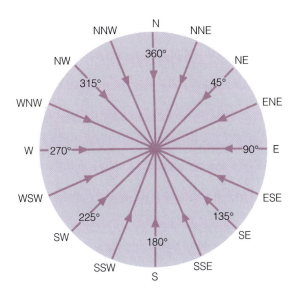

FIGURE 10.14
Wind direction can be expressed in degrees about a circle or as compass points.

FIGURE 10.15
Which way was the wind blowing when this cinder cone erupted? (Answer given at the bottom of this page.)

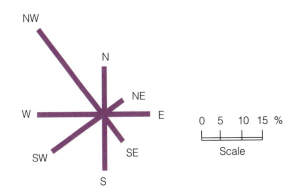

FIGURE 10.16
This wind rose represents the percent of time the wind blew from different directions at a given site during the month of January for the past ten years. The prevailing wind is NW and the wind direction of least frequency is NE.

10.15 and see if you can determine the prevailing wind when this cinder cone erupted.*

The prevailing wind can be represented by a **wind rose**, which indicates the percentage of time the wind blows from different directions. Extensions from the center of a circle point to the wind direction, and the length of each extension indicates the percentage of time the wind blew from that direction. Figure 10.16 shows a wind rose for a city averaged over a period of ten years during the month of January. Observe that the longest extension points toward the northwest and that the wind blew from this direction 25 percent of the time. This is the prevailing wind for this time period. Of course, a wind rose can be made for any time of the day, and it can represent the wind direction for any month or season of the year.

The prevailing wind in a town does not always represent the prevailing wind of an entire region. In mountainous regions, the wind is usually guided by topography and is often deflected by obstructions that cause its direction to change abruptly. Within this region, the wind may be blowing from one direction on one side of a valley and from an entirely different direction on the other side.

Wind Instruments   A very old, yet reliable, weather instrument for determining wind direction is the **wind vane**. Most wind vanes consist of a long arrow with a tail, which is allowed to move freely about a vertical post (see Fig. 10.17). The arrow always points into the wind and, hence, always gives the wind direction. Wind vanes can be made of almost any material. At

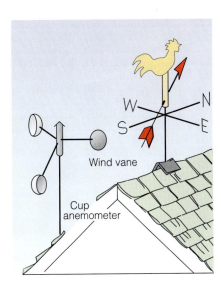

FIGURE 10.17
A wind vane and a cup anemometer.

Wind vane
Cup anemometer

*During eruption, the prevailing wind was from left to right. We can tell this by the volcano's shape. Material ejected from the volcano was blown by the wind to the right, where it accumulated, producing a more gentle slope.

airports, a cone-shaped bag opened at both ends so that it extends horizontally as the wind blows through it sits near the runway. This form of wind vane, called a *wind sock*, enables pilots to tell the surface wind direction when landing.

The instrument that measures wind speed is the **anemometer**. The oldest type of anemometer is the *pressure plate anemometer* developed by Robert Hooke in 1667. It consists of a rectangular metal plate, which is free to swing in the vertical. As the speed of the wind increases, the force of the wind on the plate pushes it outward at a greater angle. The wind speed is read from a scale mounted adjacent to the arm of the swinging plate. Most anemometers today consist of three (or more) hemispherical cups (*cup anemometer*) mounted on a vertical shaft (see Fig. 10.17). The difference in wind pressure from one side of a cup to the other causes the cups to spin about the shaft. The rate at which they rotate is directly proportional to the speed of the wind. The spinning of the cups is usually translated into wind speed through a system of gears, and may be read from a dial or transmitted to a recorder.

The **aerovane** (*skyvane*) is an instrument commonly used to indicate both wind speed and direction. It consists of a bladed propeller that rotates at a rate proportional to the wind speed. Its streamlined shape and a vertical fin keep the blades facing into the wind (see Fig. 10.18). When attached to a recorder, a continuous record of both wind speed and direction is obtained.

The wind-measuring instruments described thus far are "ground-based" and only give wind speed or direction at a particular fixed location. But the wind is influenced by local conditions, such as buildings, trees, and so on. Also, wind speed normally increases rapidly with height above the ground. Thus, wind instruments should be exposed to freely flowing air at a height of at least 10 m above the surface and well above the roofs of buildings. In practice, unfortunately, anemometers are

FIGURE 10.18
The aerovane (skyvane).

placed at various levels; the result, then, is often erratic wind observations.

A simple way to obtain wind data above the surface is with a **pilot balloon**. A small balloon filled with helium is released from the surface. The balloon rises at a known rate, but drifts freely with the wind. It is manually tracked with a small telescope called a *theodolite*. Every minute (or half minute), the balloon's vertical angle (height) and horizontal angle (direction) are measured. The data from the observations are fed into a computer or plotted on a special board, and the wind speed and direction are computed at specific intervals—usually every 300 m (1000 ft)—above the surface.

The pilot balloon principle can be used to obtain wind information during a radiosonde observation. During this type of observation, a balloon rises from the surface carrying a *radiosonde* (an instrument package designed to measure the vertical profile of temperature, pressure, and humidity—see Chapter 1, p. 13). Equipment located on the ground constantly tracks the balloon, measuring its vertical and horizontal angles, as well as its height above the ground. From this information, a computer determines and prints the vertical profile of wind from the surface up to where the balloon normally pops, typically in the stratosphere near 30 km (19 mi). The observation of winds using a radiosonde balloon is called a *rawinsonde observation*.

Above an altitude of 30 km, rockets and radar can inform us about the wind flow. One type of rocket ejects an instrument attached to a parachute that drifts with the wind as it slowly falls to earth. While descending, the instrument is tracked by a ground-based radar unit that determines wind information for that region of the atmosphere. Other rockets eject metal strips at some desired level. Again, radar tracks these drifting pieces of chaff, which provide valuable wind speed and direction data for elevations outside the normal radiosonde range.

A device similar to radar called **lidar** (*light detection and ranging*) uses infrared or visible light in the form of a laser beam to determine wind information. Basically, it sends out a narrow beam of light that is reflected from particles, such as smoke or dust—it measures wind velocity by measuring the movement of these particles.

Another instrument presently under study as a possible wind information gatherer is the *infrared radiometer*. Radiometers do not measure wind directly; instead, they measure the intensity of infrared radiation emitted from an object. Studies show that strong variations in wind speed and wind direction are accompanied by sharp variations in air temperature, especially near the underside of a severe thunderstorm, where a cold downdraft pours out of the cloud. Infrared radiometers can detect these temperature variations and relate them to the wind field. Studies also show that the amount of water vapor changes drastically in regions of strong wind shear related to clear air turbulence. Because water vapor is an excellent emitter of infrared radiation, radiometers can measure these sudden changes in radiation intensity. Installed in an airplane, this instrument could warn pilots of possible wind turbulence as far as 100 km (60 mi) in front of the aircraft.

In remote regions of the world where upper-air observations are lacking, wind speed and direction can be obtained from satellites. So far, the most reliable wind data have come from geostationary satellites positioned above a particular location. From this vantage point, the satellite shows the movement of clouds. The direction of cloud movement indicates wind direction, and the horizontal distance the cloud moves during a given time period indicates the wind speed.

Recently, *Doppler radar* has been employed to obtain a vertical profile of wind speed and direction up to an altitude of 16 km or so above the ground. Such a profile is called a *wind sounding*, and the radar, a **wind profiler** (or simply a *profiler*). Doppler radar, like conventional radar, emits pulses of microwave radiation that are returned (backscattered) from a target, in this case the irregularities in moisture and temperature created by turbulent, twisting eddies that move with the wind. Doppler radar works on the principle that, as these eddies move toward or away from the receiving antenna, the returning radar pulse will change in frequency. The Doppler radar wind profilers are so sensitive that they can translate the backscattered energy from these eddies into a vertical picture of wind speed and direction in a column of air 16 km (10 mi) thick. Presently, there is a network of wind profilers scattered across the central United States.

Up to now, we have seen that the wind is a powerful weather element that affects our environment in many ways. It can shape the landscape, influence crop production, transport material from one area to another, and generate waves. In an attempt to harness the wind's energy and turn it into electricity, many countries are building wind generators. (More information on this topic is given in the Focus section on p. 254.)

BRIEF REVIEW   Up to now we have been examining winds at the microscale. Before we turn our attention to winds on a larger scale, here is a brief review of some of the main points presented so far:

- Viscosity is the friction of fluid flow. The small-scale fluid friction that is due to the random motion of the molecules is called *molecular viscosity*. The larger scale internal friction produced by turbulent flow is called *eddy viscosity*.
- Mechanical turbulence is created by twisting eddies that form as the wind blows past obstructions. Thermal turbulence results as rising and sinking air forms when the earth's surface is heated unevenly by the sun.
- The planetary boundary layer (or friction layer) is usually given as the first 1000 m (3300 ft) above the surface.
- Wind shear is a sudden change in wind speed or wind direction (or both).
- The prevailing wind is the wind direction most frequently observed during a given time.

LOCAL WIND SYSTEMS

Every summer, millions of people flock to the New Jersey shore, hoping to escape the oppressive heat and humidity of the inland region. On hot, humid afternoons, these travelers often encounter thunderstorms about 30 km or so from the ocean, thunderstorms that invariably last for only a few minutes. In fact, by the time the vacationers arrive at the beach, skies are generally clear and air temperatures are much lower, as cool ocean breezes greet them. If the travelers return home in the afternoon, these "mysterious" showers often occur at just about the same location as before.

The showers are not really mysterious. Actually, they are caused by a local wind system—the sea breeze. As cooler ocean air pours inland, it forces the warmer, unstable humid air to rise and condense, producing majestic clouds and rainshowers along a line where the wind systems meet.

The sea breeze forms as part of a thermally driven circulation. Consequently, we will begin our study of local winds by examining the formation of thermal circulations.

Thermal Circulations   Consider the vertical distribution of pressure shown in Fig. 10.19a. The isobaric surfaces all lie parallel to the earth's surface; thus, there is no horizontal variation in pressure (or temperature), and there is no pressure gradient and no wind. Suppose the atmosphere is cooled to the north and warmed to the south (see Fig. 10.19b). In the cold, dense air above the surface, the isobars bunch closer together, while in the warm, less-dense air, they spread farther apart. This dipping of the isobars produces a horizontal pressure gradient force aloft that causes the air to move from higher pressure (warm air) toward lower pressure (cold air).

At the surface, the air pressure changes as the air aloft begins to move. As the air aloft moves from south to north, air leaves the southern area and "piles up" above the northern area. This redistribution of air reduces the surface air pressure to the south and raises it to the north. Consequently, a pressure gradient force is established at the earth's surface from north to south and, hence, surface winds begin to blow from north to south.

We now have a distribution of pressure and temperature and a circulation of air, as shown in Fig. 10.19c. As the cool surface air flows southward, it warms and becomes less dense. In the region of surface low pressure, the warm air slowly rises, expands, cools, and flows out the top at an elevation of about 1 km above the surface. At this level, the air flows horizontally northward toward lower pressure, where it completes the circulation by slowly sinking and flowing out the bottom of the surface high. Circulations brought on by changes in air temperature, in which warmer air rises and colder air sinks, are termed **thermal circulations**.

The regions of surface high and low atmospheric pressure created as the atmosphere either cools or warms are called **thermal** (*cold-core*) **highs** and **thermal** (*warm-core*) **lows**. In

Wind Power

Thousands of small windmills, their arms spinning in a stiff breeze, have pumped water, sawed wood, and even supplemented the electrical needs of small farms for many decades. It was not until the energy crisis of the early 1970s, however, that we seriously considered wind-driven turbines to run generators. In 1980, wind energy got a boost from Congress in the form of a bill—the Wind Energy Research, Demonstration, and Utilization Act—that authorized $900 million to define the nation's wind-energy potential and provide support for the research and development of wind generators called *wind turbines.*

Wind power seems an attractive way of producing energy—it is nonpolluting and, unlike solar power, is not restricted to daytime use. It does, however, pose some problems. Large wind machines may interfere with television reception in their immediate vicinity. Also, the financial aspects are another problem, as the cost of a single wind turbine can exceed $1 million. In addition, a region dotted with large wind machines is unaesthetic. (Probably, though, it is no more of an eyesore than the parades of huge electrical towers marching across many open areas.) And, unfortunately, each year the blades of spinning turbines kill countless birds.

If the wind turbine is to produce electricity, there must be wind, not just any wind, but a steady flow of air neither too weak nor too strong. A slight breeze will not turn the blades, and a powerful wind gust could severely damage the machine. Thus, regions with the greatest potential for wind-generated power would have

FIGURE 3

A portion of a wind farm near the summit of Altamont Pass, California. With over 7000 wind turbines, this is the world's largest wind energy development project.

moderate, steady winds. Examples of such areas include the central Great Plains, high mountain passes, and the coastal areas of New England and the Pacific Northwest.

The efficiency of a wind machine increases with both the wind speed and the size of the rotor blade. The available energy is proportional to the cube of the wind speed. A wind speed of 20 knots, for example, can provide 8 times more energy than a wind of 10 knots, meaning that a small increase in wind speed can result in a large increase in electrical energy. Large-blade wind turbines are more efficient than small-blade models because the power output of the machine increases with the square of the blade diameter.

With fossil fuels rapidly diminishing, the wind can provide a pollution-free alternative form of energy. For example, in 1994, over 15,000 wind machines were generating almost 3.4 billion kilowatt-hours of electricity in the United States. This quantity is enough energy to supply the annual needs of more than one million people. In California alone, there are thousands of wind turbines, many of which are on *wind farms*—clusters of 50 or more wind turbines. (See Fig. 3.) In 1997, California's wind farms were producing a little less than 2 percent of the state's total electrical needs. Present estimates are that wind power may be able to furnish up to a few percent of the nation's total energy needs during the first half of the twenty-first century.

general, they are shallow systems, usually extending no more than a few kilometers above the ground. These systems weaken with height. For example, at the surface, atmospheric pressure is lowest in the center of the warm thermal low in Fig. 10.20. In the warm air above the low, the isobars spread apart, and, at some intermediate level, the thermal low disappears and actually changes into a high. A similar phenomenon happens above the cold thermal high. The surface pressure is greatest in its center, but because the isobars aloft are crowded together due to the cold dense air, the surface thermal high becomes a low a kilometer or so above the ground. We can summarize the typical characteristics of thermal pressure systems as being shallow, weakening with height, and being maintained, for the most part, by local heating and cooling.

Sea and Land Breezes

The sea breeze is a type of thermal circulation. The uneven heating rates of land and water (described in Chapter 3) cause these mesoscale coastal winds. During the day, the land heats more quickly than the adjacent water, and the intensive heating of the air above produces a shallow thermal low. The air over the water remains cooler than the air over the land; hence, a shallow thermal high exists above the water. The overall effect of this pressure distribution is a **sea breeze** that blows from the sea toward the land (see Fig. 10.21a). Since the strongest gradients of temperature and pressure occur near the land-water boundary, the strongest winds typically occur right near the beach and diminish inland. Further, since the greatest contrast in temperature between land and water usually occurs in the afternoon, sea breezes are strongest at this time. (The same type of breeze that develops along the shore of a large lake is called a **lake breeze**.)

At night, the land cools more quickly than the water. The air above the land becomes cooler than the air over the water, producing a distribution of pressure such as the one shown in Fig. 10.21b. With higher surface pressure now over the land, the wind reverses itself and becomes a **land breeze**—a breeze that flows from the land toward the water. Temperature contrasts between land and water are generally much smaller at night, hence, land breezes are usually weaker than their daytime counterpart, the sea breeze. In regions where greater nighttime temperature contrasts exist, stronger land breezes occur over the water, off the coast. They are not usually noticed much on shore, but are frequently observed by ships in coastal waters.

Look at Fig. 10.21 again and observe that the rising air is over the land during the day and over the water during the night. Therefore, along the humid East Coast, daytime clouds tend to form over land and nighttime clouds over water. This explains why, at night, distant lightning flashes are sometimes seen over the ocean.

Sea breezes are best developed where large temperature differences exist between land and water. Such conditions prevail year-round in many tropical regions. In middle latitudes, however, sea breezes are invariably spring and summer phenomena.

FIGURE 10.19

A thermal circulation produced by the heating and cooling of the atmosphere near the ground. The H's and L's refer to atmospheric pressure. The lines represent surfaces of constant pressure (isobaric surfaces). (For more information on isobaric surfaces, see Chapter 9, p. 216.)

FIGURE 10.20

The vertical distribution of pressure with thermal highs and thermal lows.

During the summer, a sea breeze usually sets in about mid-morning after the land has been warmed. By early afternoon, the breeze has increased in strength and depth. By late afternoon, the cool ocean air may reach a depth of more than 300 m (1000 ft) and extend inland for more than 20 km (12 mi).

(a) Sea breeze

(b) Land breeze

FIGURE 10.21
Development of a sea breeze and a land breeze. (a) At the surface, a sea breeze blows from the water onto the land, whereas (b) the land breeze blows from the land out over the water. Notice that the pressure at the surface changes more rapidly with the sea breeze. This situation indicates a stronger pressure gradient force and higher winds with a sea breeze.

The leading edge of the sea breeze is called the **sea breeze front**. As the front moves inland, a rapid drop in temperature occurs just behind it. In some locations, this temperature change may be 5°C (9°F) or more during the first hours—a refreshing experience on a hot, sultry day. In regions where the water temperature is warm, the cooling effect of the sea breeze is hardly evident. Since cities near the ocean usually experience the sea breeze by noon, their highest temperature usually occurs much earlier than in inland cities. Along the East Coast, the passage of the sea breeze front is marked by a wind shift, usually from west to east. In the cool ocean air, the relative humidity rises as the temperature drops. If the relative humidity increases to above 70 percent, water vapor begins to condense upon particles of sea salt or industrial smoke, producing haze. When the ocean air is highly concentrated with pollutants, the sea breeze front may meet relatively clear air and thus appear as a *smoke front*, or a *smog front*. If the ocean air becomes saturated, a mass of low clouds and fog will mark the leading edge of the marine air.

When there is a sharp contrast in air temperature across the frontal boundary, the warmer, lighter air will converge and rise. In many regions, this makes for good sea breeze glider soaring. If this rising air is sufficiently moist, a line of cumulus clouds will form along the sea breeze front, and, if the air is also conditionally unstable, thunderstorms may form. As previously mentioned, on a hot, humid day one can drive toward the shore, encounter heavy showers several kilometers from the ocean, and arrive at the beach to find it sunny with a steady onshore breeze.

A sea breeze moving over a forest fire can be dangerous. First of all, gusty surface winds often make the fire difficult to control. Another problem is the return flow aloft. Along the sea breeze frontal boundary, air can rise to elevations where it becomes part of the return flow. Should burning embers drift seaward with this flow and drop to the ground behind the fire, they could start new fires. Flames from these fires pushed on by surface winds can trap firefighters between the two blazes.

When cool, dense, stable marine air encounters an obstacle, such as a row of hills, the heavy air tends to flow around

Florida's Fires and the Sea Breeze

The months of June and July, 1998, were certainly memorable over much of Florida. During this period, thunderstorms became less numerous, record heat and drought set in, and wild fires ravaged the northern half of the state. The weather pattern responsible for these conditions began when in June a ridge of high pressure became entrenched over the Gulf of Mexico. Sinking air from the high formed a stable atmospheric layer over Florida. This situation cut off one of the necessary conditions for thunderstorm development: a conditionally unstable atmosphere. At the surface, the dominance of westerly winds prevented the convergence of the two sea breezes and the necessary uplift for thunderstorm development (see Fig. 4). The thunderstorms that did form were mostly isolated storms that provided only spotty precipitation across the state.

During June, record high temperatures were set or tied in many cities. Melbourne, Florida, for example, set high temperature records on 22 days in June. Tallahassee had 12 days with temperatures at or above 38°C (100°F). Daytona Beach had record highs on 14 days, with June being the warmest month on record. Extremely dry conditions prevailed during the heat wave as total rainfall for June in

FIGURE 4

Typically, during the summer over Florida, converging sea breezes in the afternoon produce uplift that enhance thunderstorm development and rainfall. However, during June and July, 1998, westerly surface winds dominated the flow, a ridge of high pressure formed over the area, and thunderstorm activity diminished.

Melbourne amounted to a record low 0.41 cm (0.16 in.).

Extreme drought and low relative humidities left much of the north state ripe for fire. Tragically, wild fires, fed by strong westerly winds, swept across northern Florida. During June and early July, the fires damaged or destroyed over 100 homes, forced many thousands of people to evacuate, burned hundreds of thousands of acres of brush and timber, and left a thick blanket of smoke over much of the north state. At times, the smoke became so thick that stretches of Interstate 95 (the major north-south thoroughfare along Florida's east

coast) had to be closed. Moreover, thick smoke and poor visibility in some areas forced motorists to turn their headlights on during the day.

At one point in June, an average of 80 new fires a day were being set, mostly by lightning and burning embers that drifted with the wind and settled on the parched landscape. Fortunately, by mid-July the ridge weakened and a more normal rainfall pattern developed as the converging sea breezes once again provided uplift and abundant precipitation that helped extinguish the fires.

---

them rather than over them. When the opposing breezes meet on the opposite side of the obstruction, they form what is called a *sea breeze convergence zone.* Such conditions are common along the Pacific Coast.

Sea breezes in Florida help produce that state's abundant summertime rainfall. On the Atlantic side of the state, the sea breeze blows in from the east; on the Gulf shore, it moves in from the west. The convergence of these two moist wind systems, coupled with daytime convection, produces cloudy conditions and showery weather over the land. Over the water (where cooler, more stable air lies close to the surface), skies often remain cloud-free.

On many days during June and July of 1998, however, Florida's converging wind system did not materialize. The lack of converging surface air and its accompanying showers left much of the state parched. Huge fires broke out over northern and central Florida, which left hundreds of people homeless and burned many thousands of acres of grass and woodlands. More on the devastating fires is given in the Focus section above.

Convergence of coastal breezes is not restricted to ocean areas, as large lakes are capable of producing well-defined lake breezes. For example, both Lake Superior and Lake Michigan can produce strong lake breezes. In upper Michigan, these large bodies of water are separated by a narrow strip of land

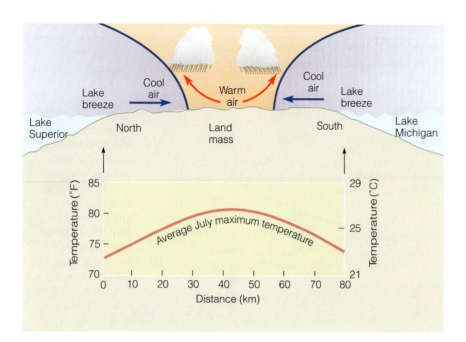

about 80 km (50 mi) wide. As can be seen from Fig. 10.22, the two breezes push inland and converge near the center of the peninsula, creating afternoon clouds and showers, while the lakeshore area remains sunny, pleasantly cool, and dry.

Local Winds and Water

Frequently, local winds will change speed and direction as they cross a large body of water. Figure 10.23 shows the wind speed and direction as air flows over a large lake. At position A, on the upwind side, the wind is blowing at 10 knots from the WNW; at position B, the wind speed is 15 knots from the NW; at position C, the wind is again blowing at 10 knots from the WNW. Why does the wind blow faster and from a slightly different direction in the center of the lake? As the air moves from the rough land over the relatively smooth lake, friction with the surface lessens, and the wind

speed increases. The increase in wind speed, however, increases the Coriolis force, which turns the wind flow to the right, as shown by the report at position B. When the air reaches the opposite side of the lake, it again encounters rough land, and its speed slows. This process reduces the Coriolis force, and the wind responds by shifting to a more westerly direction, as shown by the report at position C.

Changes in wind speed along the shore of a large lake can inhibit cloud formation on one side and enhance it on the other. Suppose warm, moist air flows over a lake, as illustrated in Fig. 10.24. Observe that clouds are forming on the downwind side, but not on the upwind side. The lake is slightly cooler than the air. Consequently, by the time the air reaches the lee side of the lake, it will be cooler, denser, and less likely to rise. Why, then, are clouds forming on this side of the lake? As air moves from the land over the water, it travels from a region of greater friction into a region of less friction, so it increases in speed, which causes the surface air to diverge—to spread apart. Such spreading of air forces air from above to slowly sink, which, of course, inhibits the formation of clouds. Hence, there are no clouds on the upwind side of the lake. Out over the lake, the separation between air temperature and dew point lessens. As this nearly saturated air moves onshore, friction with the rougher ground slows it down, causing it to "bunch up" or converge (which forces the air upward). This slight upward motion coupled with surface heating is often sufficient to initiate the formation of clouds along the downwind side of the lake.

Strong winds blowing over an open body of water, such as a lake, can cause the water to slosh back and forth rhythmically. This sloshing causes the water level to periodically rise and fall, much like water does at both ends of a bathtub when the water is disturbed. Such water waves that oscillate back and forth are

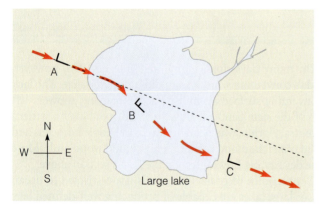

FIGURE 10.23
Wind can change in both speed and direction when crossing a large lake.

FIGURE 10.24
Sinking air develops where surface winds move offshore, speed up, and diverge. Rising air develops as surface winds move onshore, slow down, and converge.

standing waves called **seiches** (pronounced "sayshes"). In addition to strong winds, seiches may also be generated by sudden changes in atmospheric pressure or by earthquakes. Around the Great Lakes, seiche applies to any sudden rise in water level whether or not it oscillates. During December, 1986, seiches generated by strong easterly winds caused extensive coastal flooding along the southwestern shores of Lake Michigan.

Seasonally Changing Winds—The Monsoon The word *monsoon* derives from the Arabic *mausim*, which means seasons. A **monsoon wind system** is one that *changes direction seasonally*, blowing from one direction in summer and from the opposite direction in winter. This seasonal reversal of winds is especially well developed in eastern and southern Asia.

In some ways, the monsoon is similar to a huge sea breeze. During the winter, the air over the continent becomes much colder than the air over the ocean (see Fig. 3.21, p. 69). A large, shallow high-pressure area develops over continental Siberia, producing a *clockwise* circulation of air that flows out over the Indian Ocean and South China Sea (see Fig. 10.25a). Subsiding

air of the anticyclone and the downslope movement of northeasterly winds from the inland plateau provide eastern and southern Asia with generally fair weather. Hence, the *winter monsoon*, which lasts from about December through February, means clear skies (dry season), with winds that blow from land to sea.

In summer, the wind flow pattern reverses itself as air over the continents becomes much warmer than air above the water (see Fig. 3.22, p. 69). A shallow thermal low develops over the continental interior. The heated air within the low rises, and the surrounding air responds by flowing *counterclockwise* into the low center. This condition results in moisture-bearing winds sweeping into the continent from the ocean. The humid air converges with a drier westerly flow, causing it to rise; further lifting is provided by hills and mountains. Lifting cools the air to its saturation point, resulting in heavy showers and thunderstorms. Thus, the *summer monsoon* of southeastern Asia, which lasts from about June through September, means wet, rainy weather (wet season) with winds that blow from sea to land (see Fig. 10.25b). Although the majority of rain falls during the wet season, it does not rain all the time. In fact, rainy

N (a) Winter Monsoon

N (b) Summer Monsoon

FIGURE 10.25
Changing annual wind flow patterns associated with the winter and summer Asian monsoon.

WEATHER WATCH

Cherrapunji, India, received 2647 cm (87 ft) of rain in 1861, most of which fell between April and October—the summer monsoon. In fact, during July, 1861, Cherrapunji recorded a whopping 930 cm (30.5 ft) of rain.

periods of between 15 to 40 days are often followed by several weeks of hot, sunny weather.

Many factors help create the monsoon wind system. The latent heat given off during condensation aids in the warming of the air over the continent and strengthens the summer monsoon circulation. Rainfall is enhanced by weak, westward moving low-pressure areas called *monsoon depressions*. The formation of these depressions is aided by an upper-level jet stream. Where winds in the jet diverge, surface pressures drop, the monsoon depressions intensify, and surface winds increase. The greater inflow of moist air supplies larger quantities of latent heat, which, in turn, intensifies the summer monsoon circulation. Findings of the Monsoon Experiment (MONEX), a joint research project of the United States and other nations, suggests that surges of rainfall over western India might be related to a strong, moist, low-level jet stream that flows along the eastern hills of Kenya, then swings out over the Indian Ocean, where it feeds the monsoon winds.

The strength of the Indian monsoon appears to be related to the reversal of surface air pressure that occurs at irregular intervals about every 2 to 7 years at opposite ends of the tropical South Pacific Ocean. As we will see in Chapter 11, this reversal of pressure (which is known as the *Southern Oscillation*) is linked to an ocean warming phenomenon known as *El Niño*. During a major El Niño event, surface water near the equator becomes much warmer over the central and eastern Pacific. Over the region of warm water we find rising air, convection, and heavy rain. Meanwhile, to the west of the warm water (over the region influenced by the summer monsoon), sinking air inhibits cloud formation and convection. Hence, during El Niño years, monsoon rainfall is extremely deficient.

Summer monsoon rains over southern Asia can reach record amounts. Located about 300 km inland on the southern slopes of the Khasi Hills in northeastern India, Cherrapunji receives an average of 1080 cm (425 in.) of rainfall each year, most of it during the summer monsoon between April and October. The summer monsoon rains are essential to the agriculture of that part of the world. With a population of over 600 million people, India depends heavily on the summer rains so that food crops will grow. The people also depend on the rains for drinking water. Unfortunately, the monsoon can be unreliable in both duration and intensity. Since the monsoon is vital to the survival of so many people, it is no wonder that meteorologists have investigated it extensively. They have tried to develop methods of accurately forecasting the intensity and duration of the monsoon. With the aid of current research projects and the latest climate models (which tie in the interaction of ocean and atmosphere), there is hope that monsoon forecasts will begin to improve in accuracy.

Monsoon wind systems exist in other regions of the world, where large contrasts in temperature develop between oceans and continents. (Usually, however, these systems are not as pronounced as in southeast Asia.) Monsoon winds can, for example, be found in Australia, Africa, and South America. In the United States, a weak monsoonlike flow occurs in the Mississippi Valley, as winter winds tend to be cold, dry, and more northerly and summer winds warm, humid, and southerly. Along the Pacific Coast, the strong onshore surface flow that begins in spring and lasts until fall is so persistent it is called the *Pacific monsoon*.

A monsoonlike circulation also exists in the southwestern United States, especially in Arizona and New Mexico, where spring and early summer are normally dry, as warm westerly winds sweep over the region. By mid-July, however, moist southerly winds are more common, and so are afternoon showers and thunderstorms.

Mountain and Valley Breezes   Mountain and valley breezes develop along mountain slopes. Observe in Fig. 10.26 that, during the day, sunlight warms the valley walls, which in turn warm the air in contact with them. The heated air, being less dense than the air of the same altitude above the valley, rises as a gentle upslope wind known as a **valley breeze**. At night, the flow reverses. The mountain slopes cool quickly, chilling the air in contact with them. The cooler, more dense air glides downslope into the valley, providing a **mountain breeze**. (Because gravity is the force that directs these winds downhill, they are also referred to as *gravity winds*, or *drainage winds*.) This daily cycle of wind flow is best developed in clear summer weather when prevailing winds are light.

In many areas, the upslope winds begin early in the morning, reach a peak speed of about 6 knots by midday, and reverse direction by late evening. The downslope mountain breeze increases in intensity, reaching its peak in the early morning hours, usually just before sunrise. In the Northern Hemisphere, valley breezes are particularly well developed on south-facing slopes, where sunlight is most intense. On partially shaded north-facing slopes, the upslope breeze may be weak or absent. Since upslope winds begin soon after the sun's rays strike a hill, valley breezes typically begin first on the hill's east-facing side. In the late afternoon, this side of the mountain goes into shade first, producing the onset of downslope winds at an earlier time than experienced on west-facing slopes. Hence, it is possible for campfire smoke to drift downslope on one side of a mountain and upslope on the other side.

When the upslope winds are well developed and have sufficient moisture, they can reveal themselves as building cumulus clouds above mountain summits (see Fig. 10.27). Since

Valley Breeze

Mountain Breeze

FIGURE 10.26
Valley breezes blow uphill during the day; mountain breezes blow downhill at night. (The L's and H's represent pressure, whereas the purple lines represent surfaces of constant pressure.)

valley breezes usually reach their maximum strength in the early afternoon, cloudiness, showers, and even thunderstorms are common over mountains during the warmest part of the day—a fact well known to climbers, hikers, and seasoned mountain picnickers.

Katabatic Winds Although any downslope wind is technically a **katabatic wind**, the name is usually reserved for downslope winds that are much stronger than mountain breezes.

Katabatic (or *fall*) winds can rush down elevated slopes at hurricane speeds, but most are not that intense and many are on the order of 10 knots or less.

The ideal setting for a katabatic wind is an elevated plateau surrounded by mountains, with an opening that slopes rapidly downhill. When winter snows accumulate on the plateau, the overlying air grows extremely cold and a shallow dome of high pressure forms near the surface (see Fig. 10.28). Along the edge of the plateau, the horizontal pressure gradient force is usually

FIGURE 10.27
As mountain slopes warm during the day, air rises and often condenses into cumuliform clouds, such as these.

FIGURE 10.28
Strong katabatic winds can form where cold winds rush downhill from an elevated plateau covered with snow.

strong enough to cause the cold air to flow across the isobars through gaps and saddles in the hills. Along the slopes of the plateau, the wind continues downhill as a gentle or moderate cold breeze. If the horizontal pressure gradient increases substantially, such as when a storm approaches, or if the wind is confined to a narrow canyon or channel, the flow of air can increase, often destructively, as cold air rushes downslope like water flowing over a fall.

Katabatic winds are observed in various regions of the world. For example, along the northern Adriatic coast in the former Yugoslavia, a polar invasion of cold air from Russia descends the slopes from a high plateau and reaches the lowlands as the *bora*—a cold, gusty, northeasterly wind with speeds sometimes in excess of 100 knots. A similar, but often less violent, cold wind known as the *mistral* descends the western mountains into the Rhone Valley of France, and then out over the Mediterranean Sea. It frequently causes frost damage to exposed vineyards and makes people bundle up in the otherwise mild climate along the Riviera. Strong, cold katabatic winds also blow downslope off the icecaps in Greenland and Antarctica, occasionally with speeds greater than 100 knots.

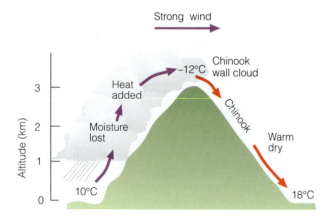

FIGURE 10.29
Conditions that may enhance a chinook.

In North America, when cold air accumulates over the Columbia plateau, it may flow westward through the Columbia River Gorge as a strong, gusty, and sometimes violent wind. Even though the sinking air warms by compression, it is so cold to begin with that it reaches the ocean side of the Cascade Mountains much colder than the marine air it replaces. The *Columbia Gorge wind* is often the harbinger of a prolonged cold spell.

Strong downslope katabatic-type winds funneled through a mountain canyon can do extensive damage. For example, during January, 1984, a ferocious downslope wind blew through Yosemite National Park at speeds estimated at 100 knots. The wind toppled trees and, unfortunately, caused a fatality when a tree fell on a park employee sleeping in a tent.

Chinook (Foehn) Winds

The **chinook wind** is a warm dry wind that descends the eastern slope of the Rocky Mountains. The region of the chinook is rather narrow (only several hundred kilometers wide) and extends from northeastern New Mexico northward into Canada. Similar winds occur along the leeward slopes of mountains in other regions of the world. In the European Alps, for example, such a wind is called a **foehn** and, in Argentina, a *zonda*. When these winds move through an area, the temperature rises sharply, sometimes over 20°C (36°F) in one hour, and a corresponding sharp drop in the relative humidity occurs, occasionally to less than 5 percent. (More information on temperature changes associated with chinooks is given in the Focus section on p. 263.)

Chinooks occur when strong westerly winds aloft flow over a north-south-trending mountain range, such as the Rockies and Cascades. Such conditions (described in Chapter 13) can produce a trough of low pressure on the mountain's eastern side, a trough that tends to force the air downslope. As the air descends, it is compressed and warms at the dry adiabatic rate (10°C/km). So the main source of warmth for a chinook is *compressional heating*, as potentially warmer (and drier) air is brought down from aloft.

When clouds and precipitation occur on the mountain's windward side, they can enhance the chinook. For example, as the cloud forms on the windward side of the mountain in Fig. 10.29, the conversion of latent heat to sensible heat supplements the compressional heating on the leeward side. This phenomenon makes the descending air at the base of the mountain on the leeward side warmer than it was before it started its upward journey on the windward side. The air is also drier, since much of its moisture was removed as precipitation on the windward side.

Along the front range of the Rockies, a bank of clouds forming over the mountains is a telltale sign of an impending chinook. This *chinook wall cloud* (which looks like a wall of clouds) usually remains stationary as air rises, condenses, and then rapidly descends the leeward slopes, often causing strong winds in foothill communities. Figure 10.30 shows how a chinook wall cloud appears as one looks west toward the Rockies

FOCUS ON A SPECIAL TOPIC

Snow Eaters and Rapid Temperature Changes

Chinooks are thirsty winds. As they move over a heavy snow cover, they can melt and evaporate a foot of snow in less than a day. This situation has led to some tall tales about these so-called "snow eaters." Canadian folklore has it that a sled-driving traveler once tried to outrun a chinook. During the entire ordeal his front runners were in snow while his back runners were on bare soil.

Actually, the chinook is important economically. It not only brings relief from the winter cold, but it uncovers prairie grass, so that livestock can graze on the open range. Also, these warm winds have kept railroad tracks clear of snow, so that trains can keep running. On the other hand, the drying effect of a chinook can create an extreme fire hazard. And when a chinook follows spring planting, the seeds may die in the parched soil. Along with the dry air comes a buildup of static electricity, making a simple handshake a shocking experience. These warm dry winds have sometimes adversely affected human behavior. During periods of chinook winds some people feel irritable and depressed and others become ill. The

FIGURE 5
Cities near the warm air–cold air boundary can experience sharp temperature changes, if cold air should rock up and down like water in a bowl.

exact reason for this phenomenon is not clearly understood.

Chinook winds have been associated with rapid temperature changes. Figure 5 shows a shallow layer of extremely cold air that has moved southward out of Canada and is now resting against the Rocky Mountains. In the cold air, temperatures are near –15°C (5°F), while just a short distance up the mountain a warm chinook wind raises the air temperature to 7°C (45°F). The cold air behaves just as any fluid, and, in some cases, atmospheric conditions may cause the air to move up and down much like water does when a bowl is rocked back and forth. This rocking motion can cause extreme

temperature variations for cities located at the base of the hills along the periphery of the cold air–warm air boundary, as they are alternately in and then out of the cold air. Such a situation is held to be responsible for the unbelievable two-minute temperature change of 27°C (49°F) recorded at Spearfish, South Dakota, during the morning of January 22, 1943. On the same morning, in nearby Rapid City, the temperature fluctuated from –20°C (–4°F) at 5:30 a.m. to 12°C (54°F) at 9:40 a.m., then down to –12°C (11°F) at 10:30 a.m. and up to 13°C (55°F) just 15 minutes later. At nearby cities, the undulating cold air produced similar temperature variations that lasted for several hours.

---

from the Colorado plains. The photograph was taken on a winter afternoon with the air temperature about –7°C (20°F). That evening, the chinook moved downslope at high speeds through foothill valleys, picking up sand and pebbles (which dented cars and cracked windshields). The chinook spread out over the plains like a warm blanket, raising the air temperature the following day to a mild 15°C (59°F). The chinook and its wall of clouds remained for several days, bringing with it a welcomed break from the cold grasp of winter.

Santa Ana Winds   A warm dry wind that blows from the east or northeast into Southern California is the **Santa Ana wind**. As the air descends from the elevated desert plateau, it funnels through mountain canyons in the San Gabriel and San

Bernardino mountains, finally spreading over the Los Angeles basin and San Fernando Valley. The wind often blows with exceptional speed in the Santa Ana Canyon (the canyon from which it derives its name).

These warm, dry winds develop as a region of high pressure builds over the Great Basin. The clockwise circulation around the anticyclone forces air downslope from the high plateau. Thus, *compressional heating* provides the primary source of warming. The air is dry, since it originated in the desert, and it dries out even more as it is heated. Figure 10.31 shows a typical wintertime Santa Ana situation.

As the wind rushes through canyon passes, it lifts dust and sand and dries out vegetation. Such conditions set the stage for serious brush fires, especially in autumn, when chaparral-

FIGURE 10.30
A chinook wall cloud forming over the Colorado Rockies (viewed from the plains).

FIGURE 10.31
Surface weather map showing Santa Ana conditions in January. Maximum temperatures for this particular day are given in °F. Observe that the downslope winds blowing into Southern California raised temperatures into the upper 80s, while elsewhere temperature readings were much lower.

covered hills are already parched from the dry summer.* One such fire in November of 1961—the infamous *Bel Air fire*—burned for three days, destroying 484 homes and causing over $25 million in damage. During October, 1993, 15 wildfires driven by Santa Ana winds swept through wealthy suburbs and rural communities near Los Angeles. The fires charred over 200,000 acres, destroyed over 1000 homes, injured hundreds of people, and caused an estimated $1 billion in damages. Four hundred miles to the north in Oakland, California, a ferocious Santa Ana–type wind was responsible for the disastrous *Oakland hills fire* during October, 1991. The fire started in the parched Oakland hills, just east of San Francisco, where a firestorm driven by strong northeast winds blackened almost 2000 acres, damaged or destroyed over 3000 dwellings, caused almost $5 billion in damage, and took 25 lives. With the protective vegetation cover removed, the land is ripe for erosion, as winter rains may wash away topsoil and, in some areas, create serious mudslides. The adverse effects of a wind-driven Santa Ana fire may be felt long after the fire itself has been put out.

A similar downslope-type wind called a *California norther* can produce unbearably high temperatures in the northern half of California's Central Valley. On August 8, 1978, for example, a ridge of high pressure formed to the north of this region, while a thermal low was well entrenched to the south. This pressure pattern produced a north wind in the area. A summertime north wind in most parts of the country means cooler weather and a welcome relief from a hot spell, but not in Red Bluff, California, where the winds moved downslope off the

---

*Chaparral denotes a shrubby environment, in which many of the plant species contain highly flammable oils.

mountains. Heated by compression, these winds increased the air temperature in Red Bluff to an unbelievable 48°C (119°F) for two consecutive days—amazing when you realize that Red Bluff is located at about the same latitude as Philadelphia!

Desert Winds   Local winds form in deserts, too. *Dust storms* form in dry regions, where strong winds are able to lift and fill the air with particles of fine dust. In desert areas where loose sand is more prevalent, *sandstorms* develop, as high winds enhanced by surface heating rapidly carry sand particles close to the ground. A spectacular example of a storm composed of dust or sand is the **haboob** (from Arabic *hebbe*: blown). The haboob forms as cold downdrafts along the leading edge of a thunderstorm lift dust or sand into a huge, tumbling dark cloud that may extend horizontally for over 150 km and rise vertically to the base of the thunderstorm (see Fig. 10.32). Spinning whirlwinds of dust frequently form along the turbulent cold air boundary, giving rise to sightings of huge *dust devils* and even tornadoes. Haboobs are most common in the African Sudan (where about 24 occur each year) and in the desert southwest of the United States, especially in southern Arizona.

The spinning vortices so commonly seen on hot days in dry areas are called **whirlwinds**, or **dust devils**. (In Australia, the Aboriginal word *willy-willy* is used to refer to a dust devil.) Generally, dust devils form on clear, hot days over a dry surface where most of the sunlight goes into heating the surface, rather than evaporating water from vegetation. The air directly above

the hot surface becomes absolutely unstable (see Chapter 7), convection sets in, and the heated air rises. Wind, often deflected by small topographic barriers, flows into this region, rotating the rising air (see Fig. 10.33). Depending on the nature of the topographic feature, the spin of a dust devil around its central eye may be cyclonic or anticyclonic, and both directions occur with about equal frequency.

Having diameters of only a few meters and heights of less than a hundred meters, most dust devils are small and last only a short time (see Fig. 10.34). There are, however, some dust devils of sizable dimension, extending upward from the surface for many hundreds of meters. Such whirlwinds are capable of considerable damage; winds exceeding 75 knots may overturn mobile homes and tear the roofs off buildings. Fortunately, the majority of dust devils are small. Also keep in mind that dust

FIGURE 10.32
A haboob approaching Phoenix, Arizona. The dust cloud is rising to a height of about 450 m (1475 ft) above the valley floor.

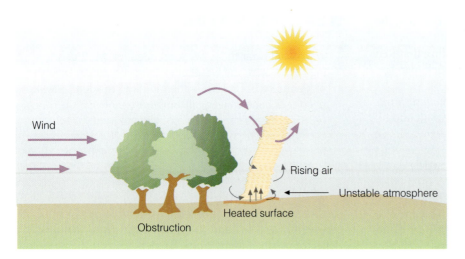

FIGURE 10.33
The formation of a dust devil. On a hot, dry day, the atmosphere next to the ground becomes unstable. As the heated air rises, wind blowing past an obstruction twists the rising air, forming a rotating column, or *dust devil*. Air from the sides rushes into the rising column, lifting sand, dust, leaves, or any other loose material from the surface.

FIGURE 10.34
A dust devil forming on a clear, hot summer day just south of Phoenix, Arizona.

devils *are not* tornadoes. The circulation of a tornado descends downward from the base of a thunderstorm, whereas the circulation of a dust devil begins at the surface, normally in sunny weather, although some form beneath convective-type clouds.

There are other desert winds that should be mentioned. Winds originating over the Sahara Desert are given local names as they move into different regions. For example, the normal flow of surface air over North Africa is from the north; however, when a storm system is located west of Africa or southern Spain (position 1, Fig. 10.35), a hot, dry, and dusty easterly or southeasterly wind—the *leste*—blows over Morocco and out into the Atlantic. If the wind crosses the Mediterranean, it becomes the *leveche* when it enters southern Spain. Because of the short time it is over water, the leveche remains hot and dry.

When a low-pressure center is located at position 2, a warm, dry, dust-laden south or southeast wind originates over the Sahara Desert and blows across North Africa. This wind is known as the *sirocco*. As it moves over the Mediterranean, it picks up moisture and arrives in Sicily and southern Italy as a warm but more humid wind.

A storm located still farther to the east (position 3) can cause a dry, hot southerly wind—the *khamsin*—to blow over Egypt, the Red Sea, and Saudi Arabia. In Israel, this wind is called the *sharav*. These winds are exceedingly hot and can raise the air temperature to 50°C (122°F), while lowering the relative humidity to less than 10 percent. Because storm systems are not common over the Mediterranean in summer, scorching breezes such as these occur in spring or fall.

Other Local Winds of Interest   Up to now, we have examined the causes of some mesoscale wind systems recognized more than just locally. Other local winds are known by various names in different locales. Let's look at some examples.

In winter, when an intense storm tracks east across the Great Plains of North America, cold northerly winds often

FIGURE 10.35
Local winds that originate over North Africa.

plunge southward behind it. As the cold air moves through Texas, it may drop temperatures tens of degrees in a few hours. Such a cold wind is called a **Texas norther**, or *blue norther*, especially if accompanied by snow. If the cold air penetrates into Central America, it is known as a *norte*. Meanwhile, if the strong, cold winds over the plains states are accompanied by drifting, blowing, or falling snow, the term *blizzard* is applied to this weather situation.

When a storm develops or intensifies off the eastern seaboard of North America and moves northeastward along the coast, it causes strong northeasterly winds along coastal areas. These **northeasters** (or *nor'easters*) usually bring heavy snow or sleet and gale force winds, which frequently attain maximum intensity off the coast of New England. The ferocious northeaster of December, 1992, produced strong northeasterly winds from Maryland to Massachusetts (see Fig. 10.36). Huge waves accompanied by hurricane-force winds, that reached 78 knots (90 mi/hr) in Wildwood, New Jersey, pounded the shoreline, causing extensive damage to beaches, beach front homes, sea walls, and boardwalks. Heavy snow and rain, which lasted for several days, coupled with high winds and high tides, put many coastal areas and highways under water, including parts of the New York City subway.

Studies suggest that some of the northeasters, which batter the coastline in winter, may actually possess some of the characteristics of a tropical hurricane, such as very high winds and an "eye-like" center. The northeaster of December, 1992, actually developed something like a hurricane's "eye" as the winds at its center went calm when it moved over Atlantic City, New Jersey. (We will examine hurricanes and their characteristics in more detail in Chapter 16.)

FIGURE 10.36
The surface weather map for 7:00 A.M. (EST) December 11, 1992, shows an intense low-pressure area (central pressure 988 mb, or 29.18 in.), which is generating strong northeasterly winds and heavy precipitation (area shaded green) from the mid-Atlantic states into New England. This northeaster devastated a wide area of the eastern seaboard, causing damage in the hundreds of millions of dollars.

TABLE 10.2   Local Winds of the World

NAME	DESCRIPTION
COLD WINDS	
Buran	A strong, cold wind that blows over Russia and central Asia
Purga	A buran accompanied by strong winds and blowing snow
Pampero	A cold wind blowing from the south over Argentina, Uruguay, and into the Amazon Basin
Burga	A cold northeasterly wind in Alaska usually accompanied by snow; similar to the buran and purga of Russia
Bise	Generally a cold north or northeast wind that blows over southern France; often brings damaging spring frosts
Papagayo	A cold northeasterly wind along the Pacific coast of Nicaragua and Guatemala; occurs when a cold air mass overrides the mountains of Central America
Tehuantepecer	A strong wind from the north or northwest funneled through the gap between the Mexican and Guatemalan mountains and out into the Gulf of Tehuantepec
MILD WINDS	
Levanter	A mild, humid, and often rainy east or northeast wind that blows across southern Spain
Harmattan	A dry, dusty but mild wind from the northeast or east that originates over the cool Sahara in winter and blows over the west coast of Africa; brings relief from the hot, humid weather along the coastal region
HOT WINDS	
Simoom	A strong, dry, and dusty desert wind that blows over the African and Arabian deserts; name means "poison wind" because it is often accompanied by temperatures in excess of 52°C (125°F), which may cause heat stroke

Local winds that also deserve attention are the strong down-mountain winds that blow along the eastern slopes of the Rockies during chinook conditions. Such winds are especially notorious in winter in Boulder, Colorado, where the average yearly windstorm damage is about $1 million. These *Boulder winds* have been recorded at over 100 knots, damaging roofs, uprooting trees, overturning mobile homes and trucks, and sandblasting car windows. Although the causes of these high winds are not completely understood, some meteorologists believe that they may be associated with large vertically oriented spinning whirls of air that some scientists call *mountainadoes*. How these rapidly rotating vortices form is presently being investigated.

Table 10.2 lists winds of local significance observed in other regions of the world.

SUMMARY

In this chapter, we concentrated on microscale and mesoscale winds. In the beginning of the chapter, we considered both how our environment influences the wind and how the wind influences our environment. We saw that the friction of airflow—viscosity—can be brought about by the random motion of air molecules (molecular viscosity) or by turbulent whirling eddies of air (eddy viscosity). The depth of the atmospheric layer near the surface that is influenced by surface friction (the boundary layer) depends upon atmospheric stability, the wind speed, and the roughness of the terrain. Although it may vary, the top of the boundary layer is usually near 1000 meters.

Winds blowing past obstructions can produce a number of effects, from gusty winds at a sports stadium to howling winds on a blustery night. Aloft, winds blowing over a mountain range may generate hazardous rotors downwind of the range. And the eddies that form in a region of strong wind shear, especially in the vicinity of a jet stream, can produce extreme turbulence, even in clear air.

Wind blowing over the earth's surface can create a variety of features. In deserts, we see sand dunes, ventifacts, and desert pavement. Over a snow surface, the wind produces snow ripples and snow rollers. Where high winds blow over a ridge, trees may be sculpted into "flag" trees. In unprotected areas, shelterbelts are planted to protect crops and soil from damaging winds.

We also examined winds on a slightly larger scale. Land and sea breezes are true mesoscale winds that blow in response to local pressure differences created by the uneven heating and cooling rates of land and water. When winds move across a large body of water, they often change in speed and direction. Where the winds change direction seasonally, they are termed monsoon winds. Monsoon winds exist in many parts of the world, including North America, Asia, Australia, and Africa.

Local winds that blow uphill during the day are called valley breezes and those that blow downhill at night, mountain breezes. A strong, cold downslope wind is the katabatic (or fall) wind. Examples of katabatic winds are the bora in Yugoslavia and the mistral in France.

A warm, dry wind that descends the eastern side of the Rocky Mountains is the chinook. The same type of wind in the Alps is called a foehn. A warm, dry downslope wind that blows into Southern California from the east or northeast is the Santa Ana wind.

Local intense heating of the surface can produce small rotating winds, such as the dust devil, while downdrafts in a thunderstorm are responsible for the desert haboob. Some winds, such as the blizzard, are snow-bearing, while others, such as the sirocco, are dust-bearing. Finally, the wind's name may express the direction from which it blows (the Texas norther), or it may represent the region that it blows from (the Santa Ana).

Key Terms

The following terms are listed in the order they appear in the text. Define each. Doing so will aid you in reviewing the material covered in this chapter.

scales of motion	aerovane (skyvane)
microscale	pilot balloon
mesoscale	lidar
synoptic scale	wind profiler
planetary (global) scale	thermal circulations
macroscale	thermal highs
viscosity	thermal lows
mechanical turbulence	sea breeze
planetary boundary layer	lake breeze
thermal turbulence	land breeze
rotors	sea breeze front
wind shear	seiches
clear air turbulence (CAT)	monsoon wind system
snow rollers	valley breeze
wind-sculptured trees	mountain breeze
shelterbelts (windbreaks)	katabatic (or fall) wind
wind waves	chinook wind
onshore wind	foehn wind
offshore wind	Santa Ana wind
prevailing wind	haboob
wind rose	dust devils (whirlwind)
wind vane	Texas norther
anemometer	northeaster

Questions for Review

1. Describe the various scales of motion, and give an example of each.
2. How does the earth's surface influence the flow of air above it?
3. What causes wind gusts?
4. Why are winds near the surface typically stronger and more gusty in the afternoon?
5. A friend has just returned from a trans-Atlantic jet aircraft flight and reported that the plane dropped about 1000 m when it entered an "air pocket." Explain to your friend what apparently happened to cause this drop.
6. Explain why the car in the diagram below may experience a west wind as it travels past the wall.

7. What is an air pocket?
8. How do sand dunes form in a desert?
9. Explain how shelterbelts protect sensitive crops from wind damage.
10. With the same wind speed, explain why a camper is more easily moved by the wind than a car.
11. What are the necessary conditions for the development of large wind waves?
12. How can a coastal area have heavy waves on a clear, non-stormy day?
13. If you are standing directly south of a smoke stack and the wind from the stack is blowing over your head, what would be the wind direction?
14. An upper wind direction is reported as 315°. From what compass direction is the wind blowing?
15. List as many ways as you can of determining the wind direction and the wind speed.
16 Name and describe three instruments used to measure wind speed and direction.
17 Using a diagram, explain how a thermal circulation develops.
18. Why do winds usually change direction and speed when moving over a large body of water?

19. Discuss the factors that contribute to the formation of the summer monsoon and the winter monsoon in India. Where are the locations of other monsoons around the globe?

20. You are fly fishing in a mountain stream during the early morning; would you expect the wind to be blowing upstream or downstream? Explain.

21. Which wind will most likely produce clouds: a valley breeze or a mountain breeze? Why?

22. Explain why chinook winds are warm and dry.

23. Name some of the benefits of a chinook wind.

24. What atmospheric conditions contribute to the development of a strong Santa Ana condition?

25. How do strong katabatic winds form?

26. Explain how a Texas norther (or blue norther) differs from a northeaster.

27. Why are haboobs more prevalent in Arizona than in Oklahoma?

28. In what part of the world would you expect to encounter each of the following winds, and what type of weather would each wind bring?
  (a) bora   (e) buran
  (b) mistral   (f) bise
  (c) blizzard   (g) sirocco
  (d) zonda   (h) leveche

Questions for Thought

1. A pilot enters the weather service office and wants to know what time of the day she can expect to encounter the least turbulent winds at 760 m (2500 ft) above central Kansas. If you were the weather forecaster, what would you tell her?

2. Why is it dangerous during hang gliding to enter the leeward side of the hill when the wind speed is strong?

3. After a winter snowstorm, Cheyenne, Wyoming, reports a total snow accumulation of 48 cm (19 in.), while the maximum depth in the surrounding countryside is only 28 cm (11 in.). If the storm's intensity and duration were practically the same for a radius of 50 km around Cheyenne, explain why Cheyenne received so much more snow.

4. Why is the difference in surface wind speed between morning and afternoon typically greater on a clear, sunny day than on a cloudy, overcast day?

5. Might it be possible to have a city/suburb breeze? If so, would you expect it to be more prominent during the day or night? Describe how it would form. Use a diagram to help you.

6. Average annual wind speed information in knots is given here for two cities located on the Great Plains. Which city would probably be the best site for a wind turbine? Why?

										Average Annual Wind Speed (Knots)
	Midnight	**3**	**6**	**9**	**Noon**	**3**	**6**	**9**		
City A:	12	7	8	13	15	18	14	13		12.5
City B:	8	6	6	13	20	22	15	10		12.5

TIME header spans the time columns.

7. Which of the sites in the drawing found in the next column (A, B, or C) would probably be the best place to construct a wind turbine? Which would be the worst? Explain.

8. Explain why cities near large bodies of cold water in summer experience well-developed sea breezes, but only poorly-developed land breezes.

9. Why do clouds tend to form over land with a sea breeze and over water with a land breeze?

10. The convergence of two sea breezes in Florida frequently produces rain showers; the convergence of two sea breezes in California does not. Explain.

11. If campfire smoke is blowing uphill along the east-facing side of the hill and downhill along the west-facing side of the same hill, are the fires cooking breakfast or dinner? From the drift of the smoke, how were you able to tell?

12. Why don't chinook winds form on the east side of the Appalachians?

13. Show, with the aid of a diagram, what atmospheric and topographic conditions are necessary for an area in the Northern Hemisphere to experience hot summer breezes from the north.

14. The prevailing winds in southern Florida are northeasterly. Knowing this, would you expect the strongest sea breezes to be along the east or west coast of southern Florida? What about the strongest land breezes?

Problems and Exercises

1. A model city is to be constructed in the middle of an uninhabited region. The wind rose seen here shows the annual frequency of wind directions for this region. With the aid of the wind rose, on a square piece of paper determine where the following should be located:
  (a) industry
  (b) parks

(c) schools
(d) shopping centers
(e) sewage disposal plants
(f) housing development
(g) an airport with two runways

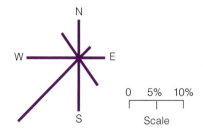

2. What would be the total force exerted on a camper 15 ft long and 8 ft high, if a wind of 40 mi/hr blows perpendicular to one of its sides?

3. On the map of the United States, label where each of the following winds might be observed, then show with arrows the general direction of air flow that occurs with each of the winds.
   (a) Santa Ana wind
   (b) chinook wind

(c) California norther
(d) northeaster
(e) Columbia Gorge wind (downslope)
(f) Texas norther (blue norther)
(g) sea breeze along the New Jersey shore
(h) sea breeze in Los Angeles, California

4. On the same map of the United States, show where the centers of atmospheric pressure should be located in order to produce the following winds. Place a large *L* on the map for the center of low pressure and a large *H* for the center of high pressure. (Be sure to place the letter representing the wind next to the *L* or the *H*.)
   (a) high-pressure area for a Santa Ana wind
   (b) low-pressure area for a chinook wind
   (c) high-pressure area for a California norther
   (d) low-pressure area for a northeaster
   (e) high-pressure area for a Columbia Gorge wind (downslope wind)
   (f) high- and low-pressure areas for a Texas norther (blue norther)
   (g) high-pressure area for a sea breeze along the New Jersey shore
   (h) low-pressure area for a sea breeze in Los Angeles, California

Advanced image processing techniques convert satellite data into a global picture of winds. Arrows show wind direction and colors represent wind speed, with yellow being the highest wind speed and blue the lowest. (Photo: NASA. The image was created by Peter Woiceshyn and Andrew Pursh of the Jet Propulsion Lab, Professor Morton Wurtele of UCLA, and Dr. Stephen Peteherych of the Canadian Atmospheric Environment Service.)

WIND: GLOBAL SYSTEMS

CONTENTS

One sunny August morning, three vessels set sail from Palos, Spain, on an established course. After halting at the Canary Islands (about 30°N) for a few days, the vessels sailed due west. For most of their Atlantic journey, steady winds from the northeast filled their sails and blew them along at a hundred miles a day. They were quite fortunate. The steady northeast winds had edged unusually far north this year, and only for about ten days or so did the vessels encounter the light, variable winds more typical of this region—a notorious region where, years later, ships were frequently becalmed under a blistering hot sun. And on October 12, 1492, the flotilla landed on a small island in the Bahamas. Christopher Columbus had not only found a route to the New World, he had discovered the trade winds—the steady northeast winds that are an integral part of the world's global wind system.

*I*n Chapter 10, we learned that local winds vary considerably from day to day and from season to season. As you may suspect, these winds are part of a much larger circulation—the little whirls within larger whirls that we spoke of before. Indeed, if the rotating high- and low-pressure areas are like spinning eddies in a huge river, then the flow of air around the globe is like the meandering river itself. When winds throughout the world are averaged over a long period, the local wind patterns vanish, and what we see is a picture of the winds on a global scale—what is commonly called the **general circulation of the atmosphere**. Just as the eddies in a river are carried along by the overall flow of water, so the highs and lows in the atmosphere are swept along by the general circulation. We will examine this large-scale circulation of air, its effects and its features, in this chapter.

GENERAL CIRCULATION OF THE ATMOSPHERE

Before we study the general circulation, we must remember that it only represents the *average* air flow around the world. Actual winds at any one place and at any given time may vary considerably from this average. Nevertheless, the average can answer why and how the winds blow around the world the way they do—why, for example, prevailing surface winds are northeasterly in Honolulu and westerly in New York City. The average can also give a picture of the driving mechanism behind these winds, as well as a model of how heat and momentum are transported from equatorial regions poleward, keeping the climate in middle latitudes tolerable.

The underlying cause of the general circulation is the unequal heating of the earth's surface. We learned in Chapter 3 that, averaged over the entire earth, incoming solar radiation is roughly equal to outgoing earth radiation. However, we also know that this energy balance is not maintained for each latitude, since the tropics experience a net gain in energy, while polar regions suffer a net loss. To balance these inequities, the atmosphere transports warm air poleward and cool air equatorward. Although seemingly simple, the actual flow of air is complex; certainly not everything is known about it. In order to better understand it, we will first look at some models (that is, artificially constructed simulations) that eliminate some of the complexities of the general circulation.

Single-Cell Model   The first model is the single-cell model, in which we assume that:

1. The earth's surface is uniformly covered with water (so that differential heating between land and water does not come into play).
2. The sun is always directly over the equator (so that the winds will not shift seasonally).
3. The earth does not rotate (so that the only force we need deal with is the pressure gradient force).

With these assumptions, the general circulation of the atmosphere would look much like the representation in Fig. 11.1a, a huge thermally driven convection cell in each hemisphere.* (For reference, the names of the different regions of the world and their approximate latitude is given in Figure 11.1b.)

The circulation of air described in Fig. 11.1a is the **Hadley cell** (named after the eighteenth-century English meteorologist George Hadley, who first proposed the idea). It is referred to as a *thermally direct cell* because it is driven by energy from the sun. Excessive heating of the equatorial area produces a broad region of surface low pressure, while at the poles excessive cooling creates a region of surface high pressure. In response to the horizontal pressure gradient, cold surface polar air flows equatorward, while at higher levels air flows toward the poles. The entire circulation consists of a closed loop with rising air near the equator, sinking air over the poles, an equatorward flow of air near the surface, and a return flow aloft. In this manner, some of the excess energy of the tropics is transported as sensible and latent heat to the regions of energy deficit at the poles.

Such a simple cellular circulation as this does not actually exist on the earth. For one thing, the earth rotates, so the Coriolis force would deflect the southward-moving surface air in the Northern Hemisphere to the right, producing easterly surface winds at practically all latitudes. These winds would be moving in a direction opposite to that of the earth's rotation and, due to friction with the surface, would slow down the earth's spin. We know that this does not happen and that prevailing winds in middle latitudes actually blow from the west. Therefore, observations alone tell us that a closed circulation of air between the equator and the poles is not the proper model for a rotating earth. (Models that simulate air flow around the globe have also verified this.) How, then, does the wind blow on a rotating planet? To answer, we will keep our model simple by retaining our first two assumptions—that is, that the earth is covered with water and that the sun is always directly above the equator.

Three-Cell Model   If we allow the earth to spin, the simple convection system breaks into a series of cells as shown in Fig. 11.2a. Although this model is considerably more complex than the single-cell model, there are some similarities. The tropical regions still receive an excess of heat and the poles a deficit. In each hemisphere, three cells instead of one have the task of energy redistribution. A surface high-pressure area is located at the poles, and a broad trough of surface low pressure still exists at the equator. From the equator to latitude 30°, the circulation closely resembles that of a Hadley cell, as does the circulation from the poles to about latitude 60°. Let's look at this model more closely by examining what happens to the air above the equator. (Refer to Fig. 11.2 as you read the following section.)

Over equatorial waters, the air is warm, horizontal pressure gradients are weak, and winds are light. This region is

---

*Additional information on thermal circulations is found on p. 253 in Chapter 10.

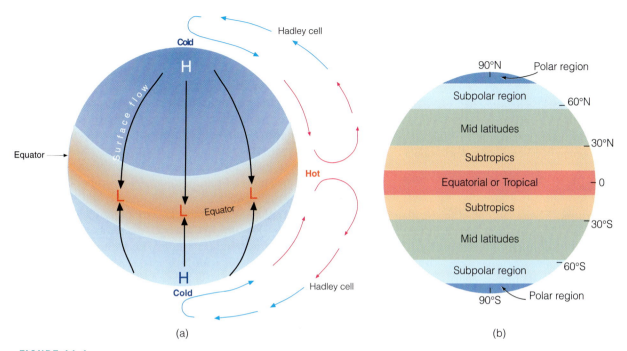

FIGURE 11.1
Diagram (a) shows the general circulation of air on a nonrotating earth uniformly covered with water and with the sun directly above the equator. (Vertical air motions are highly exaggerated in the vertical.) Diagram (b) shows the names that apply to the different regions of the world and their approximate latitudes.

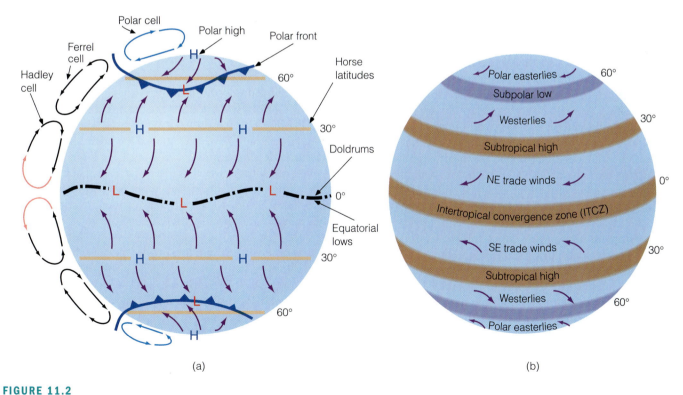

FIGURE 11.2
Diagram (a) shows the idealized wind and surface-pressure distribution over a uniformly water-covered rotating earth. Diagram (b) gives the names of surface winds and pressure systems over a uniformly water-covered rotating earth.

WEATHER WATCH

Does the African Sahara desert have an influence on South America's Amazon basin? Annually, millions of tons of soil and trace elements from the Sahara drift westward with the northeast trades thousands of kilometers across the Atlantic where they settle upon the Amazon basin. Some scientists suggest that this is one of the major sources of soil nutrients for the poor soil of the Amazon.

referred to as the **doldrums**. (The monotony of the weather in this area has given rise to the expression "down in the doldrums.") Here, warm air rises, often condensing into huge cumulus clouds and thunderstorms called *convective "hot" towers* because of the enormous amount of latent heat they liberate. This heat makes the air more buoyant and provides energy to drive the Hadley cell. The rising air reaches the tropopause, which acts like a barrier, causing the air to move laterally toward the poles. The Coriolis force deflects this poleward flow toward the right in the Northern Hemisphere and to the left in the Southern Hemisphere, providing westerly winds aloft in both hemispheres. (We will see later that these westerly winds reach maximum velocity and produce jet streams near 30° and 60° latitudes.)

As air moves poleward from the tropics, it constantly cools by radiation, and at the same time it also begins to converge, especially as it approaches the middle latitudes.* This convergence (piling up) of air aloft increases the mass of air above the surface, which in turn causes the air pressure at the surface to increase. Hence, at latitudes near 30°, the convergence of air aloft produces belts of high pressure called **subtropical highs** (or anticyclones). As the converging, relatively dry air above the highs slowly descends, it warms by compression. This subsiding air produces generally clear skies and warm surface temperatures; hence, it is here that we find the major deserts of the world, such as the Sahara. Over the ocean, the weak pressure gradients in the center of the high produce only weak winds. According to legend, sailing ships traveling to the New World were frequently becalmed in this region, and, as food and supplies dwindled, horses were either thrown overboard or eaten. As a consequence, this region is sometimes called the **horse latitudes**.

From the horse latitudes, some of the surface air moves back toward the equator. It does not flow straight back, however, because the Coriolis force deflects the air, causing it to blow from the northeast in the Northern Hemisphere and from the southeast in the Southern Hemisphere. These steady winds provided sailing ships with an ocean route to the New

World; hence, these winds are called the **trade winds**. Near the equator, the *northeast trades* converge with the *southeast trades* along a boundary called the **intertropical convergence zone** (ITCZ). In this region of surface convergence, air rises and continues its cellular journey.

Meanwhile, at latitude 30°, not all of the surface air moves equatorward. Some air moves toward the poles and deflects toward the east, resulting in a more or less westerly air flow—called the *prevailing westerlies*, or, simply, **westerlies**—in both hemispheres. Consequently, from Texas northward into Canada, it is much more common to experience winds blowing out of the west than from the east. The westerly flow in the real world is not constant; migrating areas of high and low pressure break up the surface flow pattern from time to time. In the middle latitudes of the Southern Hemisphere, where the surface is mostly water, winds blow more steadily from the west.

As this mild air travels poleward, it encounters cold air moving down from the poles. These two air masses of contrasting temperature do not readily mix. They are separated by a boundary called the **polar front**, a zone of low pressure—the **subpolar low**—where surface air converges and rises and storms develop. Some of the rising air returns at high levels to the horse latitudes, where it sinks back to the surface in the vicinity of the subtropical high. In this model, the middle cell (a *thermally indirect cell* called the *Ferrel cell*, after the American meteorologist William Ferrel) is completed when surface air from the horse latitudes flows poleward toward the polar front.

Behind the polar front, the cold air from the poles is deflected by the Coriolis force, so that the general flow of air is northeasterly. Hence, this is the region of the **polar easterlies**. In winter, the polar front with its cold air can move into middle and subtropical latitudes, producing a cold polar outbreak. Along the front, a portion of the rising air moves poleward, and the Coriolis force deflects the air into a westerly wind at high levels. Air aloft eventually reaches the poles, slowly sinks to the surface, and flows back toward the polar front, completing the weak *polar cell*.

We can summarize all of this by referring back to Fig. 11.2 and noting that, at the surface, there are two major areas of high pressure and two major areas of low pressure. Areas of high pressure exist near latitude 30° and the poles; areas of low pressure exist over the equator and near 60° latitude in the vicinity of the polar front. By knowing the way the winds blow around these systems, we have a generalized picture of surface winds throughout the world. The trade winds extend from the subtropical high to the equator, the westerlies from the subtropical high to the polar front, and the polar easterlies from the poles to the polar front.

How does this three-cell model compare with actual observations of winds and pressure? We know, for example, that upper-level winds at middle latitudes generally blow from the west. The middle Ferrel cell, however, suggests an east wind aloft as air flows equatorward. Hence, discrepancies exist between this model and atmospheric observations. This model

---

*You can see why the air converges if you have a globe of the world. Put your fingers on meridian lines at the equator and then follow the meridians poleward. Notice how the lines and your fingers bunch together in the middle latitudes.

Dynamic General Circulation Models

Dynamic models that simulate the physical processes of the atmosphere (and the oceans) are called *General Circulation Models*, or *GCMs* for short. When an atmospheric component of a GCM is linked to an ocean component, the model is said to be "coupled" and the model is called a *Coupled General Circulation Model*, or *CGCM*. General circulation models use mathematics and the laws of physics to describe the atmosphere. To reduce some of the atmosphere's complexities, the models make simplified assumptions about the atmosphere and describe the atmosphere in more simplified physical terms. They also reduce many of the small-scale atmospheric processes (such as those due to clouds) into a single approximation, or parameter, which is known as *parameterization*. The GCMs represent the atmosphere by dividing it up into grid squares, usually several hundred kilometers on a side. General circulation models simulate the behavior of the real atmosphere and describe the major

FIGURE 1

A computer-generated picture of sea level pressure and cloudiness from a general circulation model developed at the National Center for Atmospheric Research.

circulation features as well as the seasonal and latitudinal temperature patterns. The weather patterns generated by a GCM appear very similar to the weather patterns of the real atmosphere (see Fig. 1).

We will look at atmosphere models again in Chapter 14 (weather forecasting) and in Chapter 19 (climate

change). Weather prediction models begin with a detailed mathematical description of the atmosphere. This information is then fed into a computer that forecasts the movement and development of large-scale weather systems, such as high and low pressure areas, for up to ten days into the future. Climate models, on the other hand, are GCMs that are first run for a few decades to make sure that the model simulates the real atmosphere. Then, to see how some variables (such as increasing levels of CO_2) might influence the atmosphere, the model is repeatedly run with increasing concentrations of CO_2. In this manner, the GCMs reveal how the atmosphere and its circulation might change with time, due to increasing levels of greenhouse gases.

Although, as we will see in Chapter 19, there are drawbacks to the GCMs (as they do not take into account all of the human and natural factors that affect climate), they still remain the most sophisticated tools available for estimating climate change.

does, however, agree closely with the winds and pressure distribution at the *surface*, and so we will examine this next. (Before going on, you may wish to read the Focus section above that describes dynamic general circulation models.)

Average Surface Winds and Pressure: The Real World

When we examine the real world with its continents and oceans, mountains and ice fields, we obtain an average distribution of sea-level pressure and winds for January and July, as shown in Fig. 11.3. Even though these data are based on sparse observations, especially in unpopulated areas, we can see that there are regions where pressure systems appear to persist throughout the year. These systems are referred to as **semipermanent highs and lows** because they move only slightly during the course of a year.

In Fig. 11.3a, we can see that there are four semipermanent pressure systems in the Northern Hemisphere during January.

In the eastern Atlantic, between latitudes 25° and 35°N is the *Bermuda–Azores high*, often called the **Bermuda high**, and, in the Pacific Ocean, its counterpart, the **Pacific high**. These are the subtropical anticyclones that develop in response to the convergence of air aloft near an upper-level jet stream. Since surface winds blow clockwise around these systems, we find the trade winds to the south and the prevailing westerlies to the north. In the Southern Hemisphere, where there is relatively less land area, there is less contrast between land and water, and the subtropical highs show up as well-developed systems with a clearly defined circulation.

Where we would expect to observe the polar front (between latitudes 40° and 65°), there are two semipermanent subpolar lows. In the North Atlantic, there is the *Greenland-Icelandic low*, or simply **Icelandic low**, which covers Iceland and southern Greenland, while the **Aleutian low** sits over the Aleutian Islands in the North Pacific. These zones of cyclonic activity actually represent regions where numerous storms, having traveled eastward, tend to converge, especially in winter. In the Southern Hemisphere, the subpolar low forms a continuous trough that completely encircles the globe.

On the January map (Fig. 11.3a), there are other pressure systems, which are not semipermanent in nature. Over Asia, for example, there is a huge (but shallow) thermal anticyclone called the **Siberian high**, which forms because of the intense cooling of the land. South of this system, the winter monsoon shows up clearly, as air flows away from the high across Asia and out over the ocean. A similar (but less intense) anticyclone (called the *Canadian high*) is evident over North America.

As summer approaches, the land warms and the cold shallow highs disappear. In some regions, areas of surface low pressure replace areas of high pressure. The lows that form over the warm land are *thermal lows*. On the July map (Fig. 11.3b),

warm thermal lows are found over the desert southwest of the United States and over the plateau of Iran. Notice that these systems are located at the same latitudes as the subtropical highs. We can understand why they form when we realize that, during the summer, the subtropical high-pressure belt girdles the world aloft near 30° latitude (see Fig. 11.8b). Within this system, the air subsides and warms, producing clear skies (which allow intense surface heating by the sun). This air near the ground warms rapidly, rises only slightly, then flows laterally several hundred meters above the surface. The outflow lowers the surface pressure and, as we saw in Chapter 10, a shallow thermal low forms. The thermal low over India, also called the *monsoon low*, develops when the continent of Asia warms. As the low intensifies, warm, moist air from the ocean is drawn into it, producing the wet summer monsoon so characteristic of India and Southeast Asia. Where these surface

(a) January

Longitude

FIGURE 11.3

Average sea-level pressure distribution and surface wind-flow patterns for January (a) and for July (b). The heavy dashed line represents the position of the ITCZ.

winds converge with the general westerly flow, rather weak monsoon depressions form. These enhance the position of the monsoon low on the July map.

When we compare the January and July maps, we can see several changes in the semipermanent pressure systems. The strong subpolar lows so well developed in January over the Northern Hemisphere are hardly discernible on the July map. The subtropical highs, however, remain dominant in both seasons. Because the sun is overhead in the Northern Hemisphere in July and overhead in the Southern Hemisphere in January, the zone of maximum surface heating shifts seasonally. In response to this shift, the major pressure systems, wind belts, and ITCZ (heavy dashed line—see Fig. 11.3) shift toward the north in July and toward the south in January. Figure 11.4 illustrates a winter weather map where the main features of the general circulation have been displaced southward.

The General Circulation and Precipitation Patterns   The position of the major features of the general circulation and their latitudinal displacement (which annually averages about 10° to 15°) strongly influence the precipitation of many areas. For example, on the global scale, we would expect abundant rainfall where the air rises and very little where the air sinks. Consequently, areas of high rainfall exist in the tropics, where humid air rises in conjunction with the ITCZ, and between 40° and 55° latitude, where middle latitude storms and the polar front force air upward. Areas of low rainfall are found near 30° latitude in the vicinity of the subtropical highs and in polar regions where the air is cold and dry (see Fig. 11.5).

Poleward of the equator, between the doldrums and the horse latitudes, the area is influenced by both the ITCZ and the subtropical high. In summer (high sun period), the subtropical high moves poleward and the ITCZ invades this area, bringing

(b) July

FIGURE 11.4
A winter weather map depicting the main features of the general circulation over North America. Notice that the Canadian high, polar front, and subpolar lows have all moved southward into the United States, and that the prevailing westerlies exist south of the polar front. The arrows on the map illustrate wind direction.

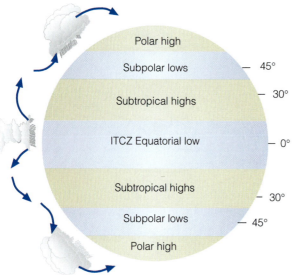

FIGURE 11.5
Major pressure systems and idealized air motions (heavy blue arrows) and precipitation patterns of the general circulation. (Areas shaded light blue represent abundant rainfall.)

with it ample rainfall. In winter (low sun period), the subtropical high moves equatorward, bringing with it clear, dry weather.

During the summer, the Pacific high drifts northward to a position off the California coast (see Fig. 11.6). Sinking air on its eastern side produces a strong upper-level subsidence inversion, which tends to keep summer weather along the West Coast relatively dry. The rainy season typically occurs in winter when the high moves south and the polar front and storms are able to penetrate the region. Along the East Coast, the clockwise circulation of winds around the Bermuda high (see Fig. 11.6) brings warm tropical air northward into the United States and southern Canada from the Gulf of Mexico. Because subsiding air is not as well developed on this side of the high,

FIGURE 11.6
During the summer, the Pacific high moves northward. Sinking air along its eastern margin produces a strong subsidence inversion, which causes relatively dry weather to prevail. Along the western margin of the Bermuda high, southerly winds bring in humid air, which rises, condenses, and produces abundant rainfall.

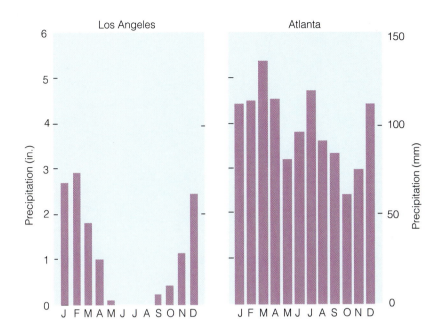

FIGURE 11.7
Average annual precipitation for Los Angeles, California, and Atlanta, Georgia.

the humid air can rise and condense into towering cumulus clouds and thunderstorms. So, in part, it is the air motions associated with the subtropical highs that keep summer weather dry in California and moist in Georgia. (Compare the rainfall patterns for Los Angeles, California, and Atlanta, Georgia—see Fig. 11.7.)

Average Wind Flow and Pressure Patterns Aloft Figures 11.8a and 11.8b are average global 500-mb charts for the months of January and July, respectively. Look at both charts carefully and observe that some of the surface features of the general circulation are reflected on these upper-air charts. On the January map, for example, both the Icelandic low and Aleutian low are located to the west of their surface counterparts. On the July map, the subtropical high-pressure areas of the Northern Hemisphere appear as belts of high height (pressure) that tend to circle the globe south of 30°N. In both hemispheres, the air is warmer over low latitudes and colder over high latitudes. This horizontal temperature gradient establishes a horizontal pressure (contour) gradient that causes the winds to blow from the west, especially in middle and high latitudes.* Notice that the temperature gradients and the contour gradients are steeper in winter than in summer. Consequently, the winds aloft are stronger in winter than in summer. The westerly winds, however, do not extend all the way to the equator, as easterly winds appear on the equatorward side of the upper-level subtropical highs.

---

*Remember that, at this level (about 5600 m or 18,000 ft above sea level), the winds are approximately geostrophic, and tend to blow more or less parallel to the contour lines.

In middle and high latitudes, the westerly winds continue to increase in speed above the 500-mb level. We already know that the wind speed increases up through the friction layer, but why should it continue to increase at higher levels? You may remember from Chapter 9 that the geostrophic wind at any latitude is directly related to the pressure gradient and inversely related to the air density. Therefore, a greater pressure gradient will result in stronger winds, and so will a decrease in air density. Owing to the fact that air density decreases with height, the same pressure gradient will produce stronger winds at higher levels. In addition, the north-to-south temperature gradient causes the horizontal pressure (contour) gradient to increase with height up to the tropopause. As a result, the winds increase in speed up to the tropopause. Above the tropopause, the temperature gradients reverse. This changes the pressure gradients and reduces the strength of the westerly winds. Where strong winds tend to concentrate into narrow bands at the tropopause, we find rivers of fast-flowing air—*jet streams*. (In the following section, you will read about a wavy jet stream. The Focus section on p. 284 describes an experiment that illustrates how these waves form.)

JET STREAMS

Atmospheric **jet streams** are swiftly flowing air currents thousands of kilometers long, a few hundred kilometers wide, and only a few kilometers thick (see Fig. 11.9). Wind speeds in the central core of a jet stream often exceed 100 knots and occasionally exceed 200 knots. Jet streams are usually found at the tropopause at elevations between 10 and 15 km (6 and 9 mi), although they may occur at both higher and lower altitudes.

(a) January

FIGURE 11.8

Average 500-mb chart for the month of January (a) and for July (b). Solid lines are contour lines in meters above sea level. Dashed lines are isotherms in °C. Arrowheads illustrate wind direction.

FIGURE 11.9

A jet stream is a swiftly flowing current of air that moves in a wavy west-to-east direction. In the Northern Hemisphere, it forms along a boundary where colder air lies to the north and warmer air to the south. In the Southern Hemisphere, it forms where colder air lies to the south and warmer air to the north.

Jet streams were first encountered by high-flying military aircraft during World War II, but their existence was suspected before the war. Ground-based observations of fast-moving cirrus clouds had revealed that westerly winds aloft must be moving rapidly indeed.

Figure 11.10 illustrates the average position of the jet streams, tropopause, and general air flow for the Northern Hemisphere in winter. From this diagram, we can see that there are two jet streams, both located in tropopause gaps, where mixing between tropospheric and stratospheric air takes place. The jet stream situated at nearly 13 km (43,000 ft) above the subtropical high is the **subtropical jet stream**. The jet stream situated at about 10 km (33,000 ft) near the polar front is known as the **polar front jet stream** or, simply, the *polar jet*. Since both are found at the tropopause, they are referred to as *tropopause jets*.

In Fig. 11.10, the wind in the jet core would be flowing as a westerly wind away from the viewer. This direction, of course,

(b) July

Longitude

is only an average, as jet streams often meander into broad loops that sweep north and south. When the polar jet develops this pattern, it may even merge with the subtropical jet. Occasionally, the polar jet splits into two jet streams. The jet stream to the north is often called the *northern branch* of the polar jet, whereas the one to the south is called the *southern branch*.

We can see the looping pattern of the jet by studying Fig. 11.11. This diagram shows the position of the polar jet stream at the 300-mb level (near 9 km or 30,000 ft) on March 10, 1998. The air flow and jet core are represented by the colored arrows; the solid gray lines represent lines of equal wind speed (isotachs) in nautical miles per hour (knots). The map shows a strong subtropical jet stream over the Gulf states with a much weaker polar jet to the north. Notice how the subtropical jet swings northward, parallel to the East Coast. Observe also that the weaker polar jet displays a number of

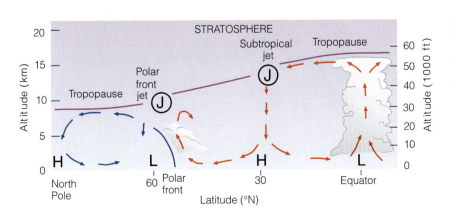

FIGURE 11.10

Average position of the polar jet stream and the subtropical jet stream, with respect to a model of the general circulation in winter. Both jet streams are flowing into the page, away from the viewer, which would be from west to east.

FOCUS ON AN OBSERVATION

The "Dishpan" Experiment

We know that the primary cause of the atmosphere's general circulation is the unequal heating that occurs between tropical and polar regions. A laboratory demonstration that tries to replicate this situation is the *"dishpan"* experiment.

The dishpan experiment consists of a flat, circular pan, filled with water several centimeters deep. The pan is positioned on a rotating table (see Fig. 2a). Around the edge of the pan, a heating coil supplies heat to the pan's "equator." In the center of the pan, a cooling cylinder represents the "pole," and ice water is continually supplied here. When the pan is rotated counterclockwise, the temperature difference between "equator" and "pole" produces a thermally driven circulation that transports heat poleward.

Aluminum powder (or dye) is added to the water so that the motions of the fluid can be seen. If the pan rotates at a speed that corresponds to the rotation of the earth, the flow develops into a series of waves and rotating eddies similar to those shown in Fig. 2b. The atmospheric counterpart of these eddies are the cyclones and anticyclones of the middle latitudes. At the earth's surface,

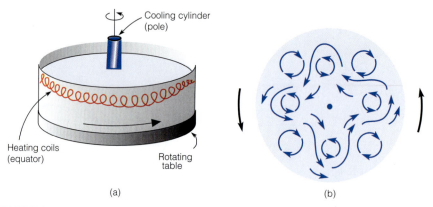

FIGURE 2

(a) A "dishpan" with a hot "equator" and a cold "pole" rotating at a speed corresponding to that of the earth (b) produces troughs, ridges, and eddies, which appear (when viewed from above) very similar to the patterns we see on an upper-level chart.

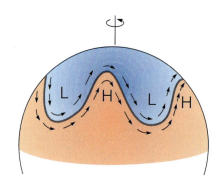

FIGURE 3

The circulation of the air aloft is in the form of waves—troughs and ridges—that encircle the globe.

they occur as winds circulating around centers of low and high pressure. Aloft, the waves appear as a series of troughs and ridges that encircle the globe and slowly migrate from west to east (see Fig. 3). The waves represent a fundamental feature of the atmosphere's circulation. Later in this chapter we will see how they transfer momentum, allowing the atmosphere to maintain its circulation. In Chapter 13, we will see that these waves are instrumental in the development of surface mid-latitude storms.

WEATHER WATCH

Time flies when traveling from west to east. Because of the jet stream, a plane flight from New York City to Europe takes about an hour less time than the return flight.

loops and actually merges with the subtropical jet over the central United States. Since the wind flow at the 300-mb level pretty much parallels the contour lines, a trough of low pressure exists off the northwest coast and over central Canada, while a ridge of high pressure extends northward over the

Atlantic, just east of Canada. The mesoscale region of maximum winds is called a *jet maximum* or **jet streak**. In Chapter 13, we will see how this region of strong winds is an important factor in the development and intensification of surface storm systems.

The looping (meridional) pattern of the jet stream has an important function. Observe in Fig. 11.11 that on the eastern side of the troughs, the red arrows indicate that swiftly moving air is directing warmer air poleward, while, on the trough's western side, the more northerly flow (indicated by blue arrows) directs cold air equatorward. Jet streams, therefore, play a major role in the global transfer of heat. Since jet streams tend to meander around the world, we can easily understand

FIGURE 11.11
Position of the polar jet stream and the subtropical jet stream at the 300-mb level during the morning of March 10, 1998. Solid gray lines are lines of equal wind speed (isotachs) in knots. Heavy lines show the position of the jet streams. Heavy blue lines show where the jet stream directs cold air southward, while heavy red arrows show where the jet stream directs warm air northward.

how pollutants or volcanic ash injected into the atmosphere in one part of the globe could eventually settle to the ground many thousands of kilometers to the west.

The ultimate cause of jet streams is the energy imbalance that exists between high and low latitudes. How, then, do jet streams actually form?

The Formation of the Polar Front Jet and the Subtropical Jet

Horizontal variations in temperature and pressure offer clues to the existence of the polar front jet. Figure 11.12(a) is a model that shows a side view of the atmosphere in the region of the polar front. Since the polar front is a boundary separating the cold polar air to the north from the warm subtropical air to the south, the greatest contrast in air temperature occurs along the frontal zone. We can see this contrast as the −20°C isotherm dips sharply crossing the front. This rapid change in temperature produces a rapid change in pressure (as shown by the sharp bending of the constant pressure surfaces as they pass through the front). In Figure 11.12(b) we can see that *the sudden change in pressure sets up a steep pressure (contour) gradient that intensifies the wind speed and causes the jet stream.* Observe in Figure 11.12(b) that the wind is blowing along the front (from the west), parallel to the contour lines, with cold air on its left side.* Diagram (c) describes how the polar front,

*Recall from Chapter 6 that any horizontal change in temperature causes the isobaric surfaces to tip or slant. The greater the temperature difference, the greater the slanting, and the stronger the winds. The changing wind speed with height due to horizontal temperature variations is referred to as the *thermal wind*. The thermal wind always blows with cold air on its left side in the Northern Hemisphere and on its right side in the Southern Hemisphere.

FIGURE 11.12
Diagram (a) is a model that shows a vertical (north-to-south) view of the polar front in association with sharply dipping pressure surfaces (solid lines) and an isotherm (dashed line). It is highly exaggerated in the vertical. Diagram (b) represents an upper-level chart along the green line x-x' in diagram (a). Solid lines are isobars. Sharp temperature contrasts along the front produce lightly packed isobars and stronger winds. Diagram (c) describes a 3-D perspective of the polar front, dipping 500-mb surface, cold and warm air, and the position of the polar-front jet stream. Notice that in both diagram (b) and (c) the winds are westerly, parallel to the polar front.

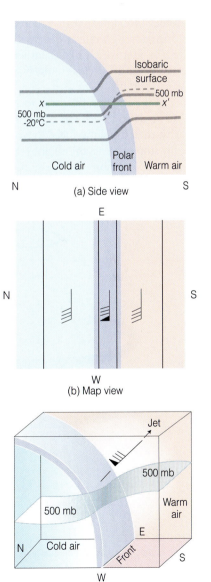

dipping pressure surface, and polar jet stream all fit together in a 3-D perspective. Because the north-south temperature contrast along the front is strongest in winter and weakest in summer, the polar-front jet shows seasonal variations. In winter, the winds blow stronger and the jet moves farther south as the leading edge of the cold air extends into subtropical regions. In summer, the jet is weaker, and is usually found over more northerly latitudes.

The subtropical jet stream, which is usually strongest slightly above the 200-mb level, tends to form along the poleward side of the Hadley cell (see Fig. 11.10). Here, warm air carried poleward by the Hadley cell produces sharp temperature contrasts along a boundary sometimes called the *subtropical front*. In the vicinity of the subtropical front (which does not have a frontal structure extending to the surface), sharp contrasts in temperature produce sharp contrasts in pressure and strong winds.

When we examine jet streams carefully, we see that another mechanism (other than a steep temperature gradient) causes a strong westerly flow aloft. The cause appears to be the same as that which makes an ice skater spin faster when the arms are pulled in close to the body—the *conservation of angular momentum*.

At the equator, the earth rotates toward the east at a speed close to 1000 knots. On a windless day, the air above moves eastward at the same speed. If somehow the earth should suddenly stop rotating, the air above would continue to move eastward until friction with the surface brought it to a halt; the air keeps moving because it has momentum.

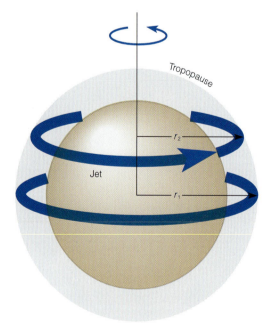

FIGURE 11.13
Air flowing poleward at the tropopause moves closer to the rotational axis of the earth (r_2 is less than r_1). This decrease in radius is compensated for by an increase in velocity and the formation of a jet stream.

Straight-line momentum—called *linear momentum* —is the product of the mass of the object times its velocity. An increase in either the mass or the velocity (or both) produces an increase in momentum. Air on a spinning planet moves about an axis in a circular path and has angular momentum. Along with the mass and the speed, angular momentum depends upon the distance (r) between the mass of air and the axis about which it rotates. *Angular momentum* is defined as the product of the mass (m) times the velocity (v) times the radial distance (r). Thus

$$\text{Angular momentum} = mvr.$$

As long as there are no external twisting forces (torques) acting on the rotating system, the angular momentum of the system does not change. We say that angular momentum is *conserved*; that is, the product of the quantity mvr at one time will equal the numerical quantity mvr at some later time. Hence, a decrease in radius must produce an increase in speed and vice versa. An ice skater, for instance, with arms fully extended rotates quite slowly. As the arms are drawn in close to the body, the radius of the circular path (r) decreases, which causes an increase in rotational velocity (v), and the skater spins faster. As arms become fully extended again, the skater's speed decreases. The conservation of angular momentum, when applied to moving air, will help us to understand the formation of a jet stream.

Consider heated air parcels rising from the equatorial surface on a calm day. As the parcels approach the tropopause, they spread laterally and begin to move poleward. If we follow the air moving northward (Fig. 11.13) we see that, because of the curvature of the earth, air constantly moves closer to its axis of rotation (r decreases). Because angular momentum is conserved (and since the mass of air is unchanged), the decrease in radius must be compensated for by an increase in speed. The air must, therefore, move faster to the east than a point on the earth's surface does. To an observer, this is a west wind. Hence, the conservation of angular momentum of northward-flowing air leads to the generation of strong westerly winds and the formation of a jet stream. (A more detailed look at the general circulation and the exchange of momentum between the earth and the atmosphere is provided in the Focus section on p. 287.)

Other Jet Streams There is another jet stream that forms in summer near the tropopause above Southeast Asia, India, and Africa. Here, the altitude of the summer tropopause and the jet stream is near 15 km. Because the jet forms on the equatorward side of the upper-level subtropical high, its winds are easterly and, hence, it is known as the **tropical easterly jet stream**. Although the exact causes of this jet have yet to be completely resolved, its formation appears to be, at least in part, related to the warming of the air over large elevated land masses, such as Tibet. During the summer, the air above this region (even at high elevations) is warmer than the air above the ocean to the south. This contrast in temperature produces

Momentum—A Case of Give and Take

We know that energy from the sun drives the atmospheric circulation. Although the circulation pattern is complex, the general flow aloft is westerly. However, the westerly flow is not constant, for it breaks into eddies, cyclones, and anticyclones that transfer heat and momentum poleward. This transfer process feeds energy to the jet stream and maintains the westerly winds aloft. Therefore, this transfer mechanism is responsible for maintaining the general circulation of the atmosphere. How, then, does the momentum exchange take place?

There is a constant exchange of momentum between the earth and the atmosphere. Since momentum is simply mass times velocity, the momentum of an object whose mass is 1 is represented by velocity only. For such an object (a mass of air, for example), a change in momentum represents a change in velocity.

Consider the earth to be a rotating globe and the atmosphere to be your hand. When your hand rests on the globe and rotates with it at the same speed, there is, for all practical purposes, no transfer of momentum. But if your hand is on the equator and moves more slowly to the east than the rotating globe, then the friction between your hand and the globe will reduce the globe's rate of spin. Hence, there is a transfer of momentum from your hand to the globe. Because the actual winds in the tropics are easterly, there is a transfer of momentum from the moving air to the earth, which should slow down the earth's rotation in a similar manner. But the earth's spin does not slow because of the westerly winds in middle latitudes.

To see what effect westerly winds have on the earth, place your hand on a rotating globe, then move it faster to the east than the globe rotates. Momentum transferred from your hand to the globe will cause the globe to spin faster. Similarly, the westerly winds of middle latitudes should increase the earth's rate of

FIGURE 4

A well-developed surface storm usually shows up as a wave with a tilted trough (dashed line) on a 500-mb chart. The wave transports westerly momentum poleward because the winds east of the trough have a greater westerly component than do the winds west of the trough.

spin, but they do not because they are compensated for by the easterly winds in the tropics.

The rotating earth affects the momentum of the atmosphere as well. We know that the northeast trades blow from about 30°N latitude toward the equator. As the air moves closer to the equator, it also moves farther away from the earth's axis of rotation (*r* increases). Therefore, to conserve *angular momentum*, the increase in *r* must be offset by a decrease in velocity (*v*). However, the slower the air moves eastward on the rotating earth, the faster the wind appears to be blowing from the east to an observer on the earth's surface. (For a calm wind, the air is moving eastward at the same rate that the earth spins.) As a consequence, the trades should be strong easterly winds near the equator. In fact, however, the trades are fairly steady, but weak. The reason is that friction with the earth drags the air along more rapidly toward the east; thus, the apparent westward motion of the air decreases. The net result of this frictional interaction is that the earth imparts some of its momentum to the tropical air above.

So, in low latitudes, the atmosphere gains momentum from the earth.

Meanwhile, in middle latitudes, the prevailing westerlies curve slightly northward, away from the subtropical high. As they move northward, they also move closer to the earth's axis of rotation (*r* decreases). The conservation of angular momentum requires that, as *r* decreases, surface wind velocity should increase and eventually reach the speed of a fast-flowing westerly jet. Surface winds do not blow that fast; again, the reason is surface friction, which slows the westerlies and reduces the air's angular momentum. Therefore, in middle latitudes, the air near the surface loses momentum to the earth.

If the atmosphere were to continually lose momentum in middle latitudes and gain momentum in low latitudes, both the prevailing westerlies and northeast trades would slow until the air is calm. Since we know this does not happen, there must be a net transfer of momentum from low latitudes toward high latitudes in order to maintain our wind systems. It is the large low- and high-pressure areas, the cyclones and anticyclones of the middle latitudes, that are primarily responsible for this transfer of momentum.

For instance, a storm in the upper troposphere might look like a trough similar to the one seen in Fig. 4. Notice that it has an asymmetric shape and tilts in a southwest-northeast direction. The winds on the east side of the trough are generally stronger than the winds on the west side. Also, the northward-moving winds on the east side have a stronger west-to-east component than do the southward-moving winds on the west side. This means that there is a net west-to-east transfer of momentum from lower latitudes to the middle latitude westerlies. Therefore, the next time a storm causes you some discomfort, remember that, without such storms, the atmosphere could not maintain its circulation.

a north-to-south pressure gradient and strong easterly winds that usually reach a maximum speed near 15°N latitude.

Not all jet streams form at the tropopause. For example, there is a jet stream that forms near the top of the stratosphere over polar latitudes. Because little, if any, sunlight reaches the polar region during the winter, air in the upper stratosphere is able to cool to low temperatures. By comparison, in equatorial regions, sunlight prevails all year long, allowing stratospheric ozone to absorb solar energy and warm the air. The horizontal temperature gradients between the cold poles and the warm tropics create steep horizontal pressure gradients, and a strong westerly jet forms in polar regions at an elevation near 50 km (30 mi). Because this wind maximum occurs in the stratosphere during the dark polar winter, it is known as the **stratospheric polar night jet stream**.

In summer, the polar regions experience more hours of sunlight than do tropical areas. Stratospheric temperatures over the poles increase more than at the same altitude above the equator, which causes the horizontal temperature gradient to reverse itself. The jet stream disappears, and in its place there are weaker easterly winds.

Jet streams also form in the upper mesosphere and in the thermosphere. Not much is known about the winds at these high levels, but they are probably related to the onslaught of charged particles that constantly bombard this region of the atmosphere.

Jet streams form near the earth's surface as well. One such jet develops over the central plains of the United States, where it occasionally attains speeds of 60 knots several hundred meters above the surface. This wind speed maximum, which usually flows from the south or southwest, is known as a **low-level jet**. It typically forms at night above a temperature inversion. Apparently, the stable air reduces the interaction between the air within the inversion and the air directly above. Consequently, the air in the vicinity of the jet is able to flow faster because it is not being slowed by the lighter winds below. Also, the north-south trending Rocky Mountains tend to funnel the air northward. However, these factors alone are unable to account for the strong winds in the jet core.

Another important element contributing to the formation of the low-level jet is the downward sloping of the land from the Rockies to the Mississippi Valley, which causes nighttime air above regions to the west to be cooler than air at the same elevation to the east. This horizontal contrast in temperature causes pressure surfaces to dip toward the west. The dipping of pressure surfaces produces strong pressure gradient forces directed from east to west which, in turn, cause strong southerly winds.

During the summer, these strong southerly winds carry moist air from the Gulf of Mexico into the Central Plains. This moisture, coupled with converging, rising air of the low-level jet, enhances thunderstorm formation. Therefore, on warm, moist, summer nights, when the low-level jet is present, it is common to have nighttime thunderstorms over the plains.

BRIEF REVIEW   Before going on to the next section, which describes the many interactions between the atmosphere and the ocean, here is a review of some of the important concepts presented so far:

- The two major semipermanent subtropical highs that influence the weather of North America are the Pacific high situated off the west coast and the Bermuda high situated off the southeast coast.
- The polar front is a zone of low pressure where storms often form. It separates the mild westerlies of the middle latitudes from the cold, polar easterlies of the high latitudes.
- In equatorial regions, the intertropical convergence zone (ITCZ) is a boundary where air rises in response to the convergence of the northeast trades and the southwest trades.
- In the Northern Hemisphere, the major global pressure systems and wind belts shift northward in summer and southward in winter.
- The northward movement of the Pacific high in summer tends to keep summer weather along the west coast of North America relatively dry.
- Jet streams exist where strong winds become concentrated in narrow bands. The polar-front jet stream meanders in a wavy, west-to-east pattern, becoming strongest in winter when the contrast in temperature between high and low latitudes is greatest.

ATMOSPHERE–OCEAN INTERACTIONS

Although scientific understanding of all the interactions between the oceans and the atmosphere is far from complete, there are some relationships that deserve mentioning here.

Global Wind Patterns and Surface Ocean Currents   As the wind blows over the oceans, it causes the surface water to drift along with it. The moving water gradually piles up, creating pressure differences within the water itself. This leads to further motion several hundreds of meters down into the water. In this manner, the general wind flow around the globe starts the major surface ocean currents moving. The relationship between the general circulation and ocean currents can be seen by comparing Fig. 11.3 (p. 278) and Fig. 11.14, p. 289.

Because of the larger frictional drag in water, ocean currents move more slowly than the prevailing wind. Typically, these currents range in speed from several kilometers per day to several kilometers per hour. As we can see in Fig. 11.14, major ocean currents do not follow the wind pattern exactly; rather, they spiral in semiclosed circular whirls called *gyres*. In the North Atlantic, the prevailing winds blow clockwise and outward from the subtropical highs, while the ocean currents

FIGURE 11.14
Average position and extent of the major surface ocean currents. Cold currents are shown in blue; warm currents are shown in red. Names of the ocean currents are given in Table 11.1.

TABLE 11.1   Major Ocean Currents

1. Gulf Stream	9. South Equatorial Current	17. Peru or Humbolt Current
2. North Atlantic Drift	10. South Equatorial Countercurrent	18. Brazil Current
3. Labrador Current	11. Equatorial Countercurrent	19. Falkland Current
4. West Greenland Drift	12. Kuroshio Current	20. Benguela Current
5. East Greenland Drift	13. North Pacific Drift	21. Agulhas Current
6. Canary Current	14. Alaska Current	22. West Wind Drift
7. North Equatorial Current	15. Oyashio Current	
8. North Equatorial Countercurrent	16. California Current	

move in a more or less circular, but still clockwise, pattern. As the water moves beneath the wind, the Coriolis force deflects the water to the right in the Northern Hemisphere and to the left in the Southern Hemisphere. This deflection causes the surface water to move at an angle between 20° and 45° to the direction of the wind. Hence, surface water tends to move in a circular pattern as winds blow outward, away from the center of the subtropical high.

Important interactions between the atmosphere and the ocean can be seen by examining the huge gyre in the North Atlantic. Flowing northward along the east coast of the United States is a tremendous warm water current called the *Gulf Stream*. The Gulf Stream carries vast quantities of warm tropical water into higher latitudes. To the north, on the western side of

the smaller subpolar gyre, cold water moves southward along the Atlantic coast of North America. This *Labrador Current* brings cold water as far south as Massachusetts in summer and North Carolina in winter. In the vicinity of the Grand Banks of Newfoundland, where the two opposing currents flow side by side, there is a sharp temperature gradient. When warm Gulf Stream air blows over the cold Labrador Current water, the stage is set for the formation of the fog so common to this region.

Meanwhile, steered by the prevailing westerlies, the Gulf Stream swings away from the coast of North America and moves eastward toward Europe. Gradually, it widens and slows as it merges into the broader *North Atlantic Drift*. As this current approaches Europe, it divides into two currents. A portion flows northward along the coasts of Great Britain and Norway, bringing with it warm water (which helps keep winter temperatures much warmer than one would expect this far north). The other part of the North Atlantic Drift flows southward as the *Canary Current*, which transports cool northern water equatorward. Eventually, the Atlantic gyre is completed as the Canary Current merges with the westward-moving *North Equatorial Current*, which derives its energy from the northeast trades.

The ocean circulation in the North Pacific is similar to that in the North Atlantic. On the western side of the ocean is the Gulf Stream's counterpart, the warm, northward-flowing *Kuroshio Current*, which gradually merges into the slower-moving *North Pacific Drift*. A portion of this current flows southward along the coastline of the western United States as the cool *California Current*. In the Southern Hemisphere, surface ocean circulations are much the same except that the gyres move counterclockwise in response to the winds around the subtropical highs. Notice that the ocean currents at higher latitudes tend to move in a more west-to-east pattern than do the currents of the Northern Hemisphere. This zonal pattern limits, to some extent, the poleward transfer of warm tropical water. Hence, there is a much smaller temperature difference between the ocean's surface and the atmosphere than exists over Northern Hemisphere oceans. This situation tends to limit the development of vigorous convective activity over the oceans of the Southern Hemisphere. In the Indian Ocean, monsoon circulations tend to complicate the general pattern of ocean currents.

Up to now, we have seen that atmospheric circulations and ocean circulations are closely linked; wind blowing over the oceans produces surface ocean currents. The currents, along with the wind, transfer heat from tropical areas, where there is a surplus of energy, to polar regions, where there is a deficit. This helps to equalize the latitudinal energy imbalance with about 40 percent of the total heat transport in the Northern Hemisphere coming from surface ocean currents. The environmental implications of this heat transfer are tremendous. If the energy imbalance were to go unchecked, yearly temperature differences between low and high latitudes would increase greatly, and the climate would gradually change.

Satellite pictures reveal that distinct temperature gradients exist along the boundaries of surface ocean currents. For example, off the east coast of the United States, where the warm Gulf Stream meets cold waters to the north, sharp temperature contrasts often exist. The boundary separating the two masses of water with contrasting temperatures and densities is called an **oceanic front**. Along this frontal boundary, a portion of the meandering Gulf Stream occasionally breaks away and develops into a closed circulation of either cold or warm water—a whirling eddy (see Fig. 11.15). Because these eddies transport heat and momentum from one region to another, they may have a far-reaching effect upon the climate and a more immediate impact upon coastal waters. Scientists are investigating the effects of these eddies.

FIGURE 11.15

The Gulf Stream (dark red band) and its eddies are revealed in this satellite mosaic of sea surface temperatures of the western North Atlantic during June, 1984. Bright red shows the warmest water (about 27°C or 80°F), followed by orange and yellow. Green, blue, and purple represent the coldest water.

Upwelling   Earlier, we saw that the cool California Current flows roughly parallel to the west coast of North America. From this, we might conclude that summer surface water temperatures would be cool along the coast of Washington and gradually warm as we move south. A quick glance at the water temperatures along the west coast of the United States during August (see Fig. 11.16) quickly alters that notion. The coldest water is observed along the northern California coast near Cape Mendocino. Why there? To answer this, we need to examine how the wind influences the movement of surface water.

As the wind blows over an open stretch of ocean, the surface water beneath it is set in motion. The Coriolis force bends the moving water to the right. Thus, if we look at a shallow surface layer of water, we see that it moves at an average angle of about 45° to the direction of the wind (see Fig. 11.17). If we imagine the top layer of ocean water to be broken into a series of layers, then each layer will exert a frictional drag on the layer below. Each successive layer will not only move a little slower than the one above, but (because of the Coriolis effect) each layer will also rotate slightly to the right of the layer above. (The rotation of each layer is to the right in the Northern Hemisphere and to the left in the Southern Hemisphere.) Consequently, descending from the surface, we would find water slowing and turning to the right until, at some depth (usually about 100 m), the water actually moves in a direction opposite to the flow of water at the surface. This turning of water with depth is known as the **Ekman Spiral.*** The Ekman Spiral in Fig. 11.17 shows us that the average movement of surface water down to a depth of about 100 m is at right angles (90°) to the surface wind direction. The Ekman Spiral helps to explain why, in summer, surface water is cold along the west coast of North America.

The summertime position of the Pacific high and the low coastal mountains cause winds to blow parallel to the California coastline (see Fig. 11.18). The net transport of surface water is at right angles to the wind, in this case, out to sea. As

---

*The Ekman Spiral is also present in the atmospheric boundary layer, from the surface up to the top of the friction layer, which is usually about 1000 m above the surface.

WEATHER WATCH

What summer weather condition could make the water temperature a lot warmer along the coast of northern California? Actually, if the summertime winds along the northern California coast stopped blowing from the northwest, upwelling would diminish and the average coastal water temperature would rise between 5.5°C and 8°C (10°F and 14°F).

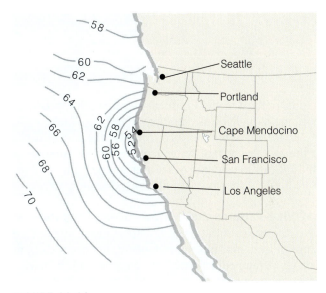

FIGURE 11.16
Average sea surface temperatures (°F) along the west coast of the United States during August.

surface water drifts away from the coast, cold, nutrient-rich water from below rises to replace it. The rising of cold water is known as **upwelling**. Upwelling is strongest and surface water is coolest in this area because here the wind parallels the coast.

Summertime weather along the West Coast often consists of low clouds and fog, as the air over the water is chilled to its

FIGURE 11.17
The Ekman Spiral. Winds move the water, and the Coriolis force deflects the water to the right (Northern Hemisphere). Below the surface each successive layer of water moves more slowly and is deflected to the right of the layer above. The average transport of surface water in the Ekman layer is at right angles to the prevailing winds.

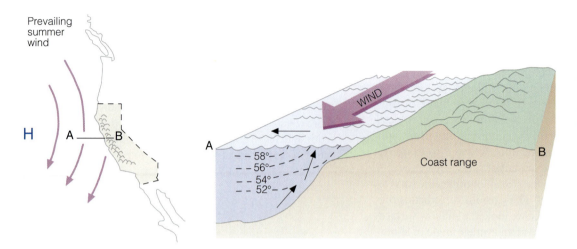

FIGURE 11.18
As winds blow parallel to the west coast of North America, surface water is transported to the right (out to sea). Cold water moves up from below (upwells) to replace the surface water.

saturation point. On the brighter side, upwelling produces good fishing, as higher concentrations of nutrients are brought to the surface. But swimming is only for the hardiest of souls, since the average surface water temperature in summer is nearly 10°C (18°F) colder than the average coastal water temperature found at the same latitude along the Atlantic Coast.

Global Wind Patterns and Surface Ocean Temperatures

Between the ocean surface and the atmosphere, there is an exchange of heat, moisture, and momentum that depends, in part, on temperature differences between water and air. In winter, when air-water temperature contrasts are greatest, there is a substantial transfer of sensible and latent heat from the ocean surface into the atmosphere. This energy helps to maintain the global air flow. Because of the difference in heat capacity between water and air, even a relatively small change in surface ocean temperatures could modify atmospheric circulations. This could have far-reaching effects on global weather patterns.

Meteorologists have tried to correlate surface water temperatures in the Pacific Ocean with weather in the United States. Between 1971 and 1975, the surface temperatures in the central Pacific were warmer than normal, while surface temperatures were cooler than normal along the west coast of North America. During this period, the east coast experienced warm, dry winters, while the weather in the West was relatively wet. In contrast, the winter of 1976–1977 produced record-breaking cold temperatures in the East and drought in the West. At about this time, surface water temperatures in the central Pacific became cooler, while those along the west coast became warmer.

During the winter of 1982–1983, the water temperatures off the west coast became much warmer than normal. Fish accustomed to colder water migrated northward. As the westerly jet stream intensified, massive storms, one after another, pounded the California coast, causing extensive damage and record-breaking rains. Heavy snow fell over the Rocky Mountains and widespread flooding occurred in the Gulf states.

Meanwhile, the northeastern part of the nation experienced an exceptionally mild winter. The cause of these weather events has been linked to an atmospheric-oceanic phenomenon referred to as El Niño.

EL NIÑO AND THE SOUTHERN OSCILLATION

Along the west coast of South America, where the cool Peru Current sweeps northward, southerly winds promote upwelling of cold, nutrient-rich water that gives rise to large fish populations, especially anchovies. The abundance of fish supports a large population of sea birds whose droppings (called *guano*) produce huge phosphate-rich deposits, which support the fertilizer industry. Near the end of the calendar year, a warm current of nutrient-poor tropical water often moves southward, replacing the cold, nutrient-rich surface water. Because this condition frequently occurs around Christmas, local residents call it *El Niño* (Spanish for boy child), referring to the Christ child.

In most years, the warming lasts for only a few weeks to a month or more, after which weather patterns usually return to normal and fishing improves. However, when El Niño conditions last for many months, and a more extensive ocean warming occurs, the economic results can be catastrophic. This extremely warm episode, which occurs at irregular intervals of 2 to 7 years and covers a large area of the tropical Pacific Ocean, is now referred to as a *major El Niño event*, or simply **El Niño**.*

During a major El Niño event, large numbers of fish and marine plants may die. Dead fish and birds may litter the water and beaches of Peru; their decomposing carcasses deplete the

---

*It was thought that El Niño was a local event that occurs along the west coast of Peru and Ecuador. It is now known that the ocean-warming associated with a major El Niño can cover an area of the tropical Pacific much larger than the continental United States.

water's oxygen supply, which leads to the bacterial production of huge amounts of smelly hydrogen sulfide. The El Niño of 1972–1973 reduced the annual Peruvian anchovy catch from 10.3 million metric tons in 1971 to 4.6 million metric tons in 1972. Since much of the harvest of this fish is converted into fishmeal and exported for use in feeding livestock and poultry, the world's fishmeal production in 1972 was greatly reduced. Countries such as the United States that rely on fishmeal for animal feed had to use soybeans as an alternative. This raised poultry prices in the United States by more than 40 percent.

Why does the ocean become so warm over the eastern tropical Pacific? Normally, in the tropical Pacific Ocean, the trades are persistent winds that blow westward from a region of higher pressure over the eastern Pacific toward a region of lower pressure centered near Indonesia (see Fig. 11.19a). The trades create upwelling that brings cold water to the surface. As this water moves westward, it is heated by sunlight and the atmosphere. Consequently, in the Pacific Ocean, surface water along the equator usually is cool in the east and warm in the west. In addi-

tion, the dragging of surface water by the trades raises sea level in the western Pacific and lowers it in the eastern Pacific, which produces a thick layer of warm water over the tropical western Pacific Ocean and a weak ocean current (called the *countercurrent*) that flows slowly eastward toward South America.

Every few years, the surface atmospheric pressure patterns break down, as air pressure rises over the region of the western Pacific and falls over the eastern Pacific (see Fig. 11.19b). This change in pressure weakens the trades, and, during strong pressure reversals, east winds are replaced by west winds that strengthen the countercurrent. Surface water warms over a broad area of the tropical Pacific and heads eastward toward South America in a surge known as a *Kelvin wave*, which is an enormous wave perhaps 15 centimeters high but extending for hundreds of kilometers north and south of the equator (see Fig. 11.20). Toward the end of the warming period, which may last between one and two years, atmospheric pressure over the eastern Pacific reverses and begins to rise, whereas, over the western Pacific, it falls. This seesaw pattern of reversing surface

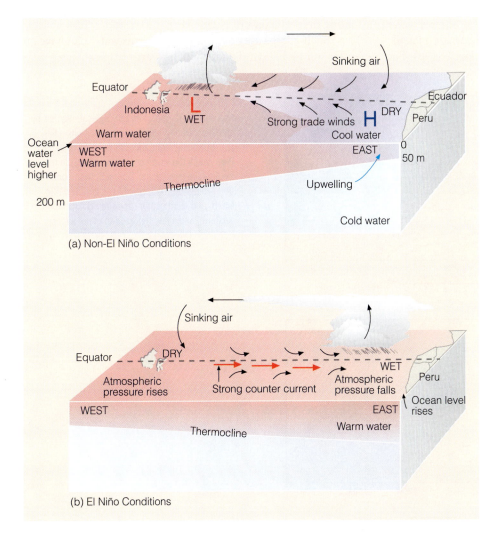

(a) Non-El Niño Conditions

(b) El Niño Conditions

FIGURE 11.19

In diagram (a), under ordinary conditions higher pressure over the southeastern Pacific and lower pressure near Indonesia produce easterly trade winds along the equator. These winds promote upwelling and cooler ocean water in the eastern Pacific, while warmer water prevails in the western Pacific. The trades are part of a circulation (called the *Walker circulation*) that typically finds rising air and heavy rain over the western Pacific and sinking air and generally dry weather over the eastern Pacific. When the trades are exceptionally strong, water along the equator in the eastern Pacific becomes quite cool. This cool event is called *La Niña*. During El Niño conditions—diagram (b)— atmospheric pressure decreases over the eastern Pacific and rises over the western Pacific. This change in pressure causes the trades to weaken or reverse direction. This situation enhances the countercurrent that carries warm water from the west over a vast region of the eastern tropical Pacific. The thermocline, which separates the warm water of the upper ocean from the cold water below, changes as the ocean conditions change from non-El Niño to El Niño.

FIGURE 11.20

These three images depict the evolution of a warm water Kelvin wave moving eastward in the equatorial Pacific ocean during March and April, 1997. The white areas represent ocean levels about 20 cm (8 in.) higher than average, while the red areas represent ocean levels about 10 cm (4 in.) higher than average. These data were collected by the altimeter on board the joint United States/French *TOPEX/Poseidon* satellite.

air pressure at opposite ends of the Pacific Ocean is called the **Southern Oscillation**. Because the pressure reversals and ocean warming are more or less simultaneous, scientists call this phenomenon the *El Niño/Southern Oscillation* or **ENSO** for short. Although most ENSO episodes follow a similar evolution, each event has its own personality, differing in both strength and behavior.

During especially strong ENSO events (such as in 1982–83 and 1997–98) the easterly trades may actually become westerly winds. As these winds push eastward, they drag surface water with them. This dragging raises sea level in the eastern Pacific and lowers sea level in the western Pacific (see Fig. 11.19b). The

eastward-moving water gradually warms under the tropical sun, becoming as much as 6°C (11°F) warmer than normal in the eastern equatorial Pacific. Gradually, a thick layer of warm water pushes into coastal areas of Ecuador and Peru, choking off the upwelling that supplies cold, nutrient-rich water to South America's coastal region. The unusually warm water may extend from South America's coastal region for many thousands of kilometers westward along the equator (see Fig. 11.21). The warm tropical water may even spread northward along the west coast of North America.

Such a large area of abnormally warm water can have an effect on global wind patterns. The warm tropical water fuels

(a)

(b)

FIGURE 11.21

Sea surface temperature (SSTs) as measured by satellites. During non-El Niño conditions—diagram (a)—upwelling along the equator and coast of Peru keeps the water cool (blue colors) in the tropical eastern Pacific. During El Niño conditions—diagram (b)—upwelling is greatly diminished, and warm water (deep red color) from the western Pacific has replaced the cool water.

FIGURE 11.22
The numbers on the left side of the diagram represent a multivariable ENSO index which includes six parameters over the tropical Pacific ocean. These variables include: air temperature, surface water temperature, sea-level pressure, cloudiness, and wind speed and direction. Warm episodes are in red, cold episodes in blue. An index value greater than +1 represents an El Niño. A value less than –1 represents a La Niña.

the atmosphere with additional warmth and moisture, which the atmosphere turns into additional storminess and rainfall. The added warmth from the oceans and the release of latent heat during condensation apparently influence the westerly winds aloft in such a way that certain regions of the world experience too much rainfall, whereas others have too little. Meanwhile, over the warm tropical central Pacific, the frequency of typhoons usually increases. However, over the tropical Atlantic, between Africa and Central America, the winds aloft tend to disrupt the organization of thunderstorms that is necessary for hurricane development; hence, there are fewer hurricanes in this region during strong El Niño events. And, as we saw in Chapter 10, during El Niño events there is a tendency for monsoon conditions over India to weaken.

Although the actual mechanism by which changes in surface ocean temperatures influence global wind patterns is not fully understood, the by-products are plain to see. For example, during exceptionally warm El Niños, drought is normally felt in Indonesia, southern Africa, and Australia, while heavy rains and flooding often occur in Ecuador and Peru. In the Northern Hemisphere, a strong subtropical westerly jet stream normally directs storms into California and heavy rain into the Gulf Coast states. The total damage worldwide due to flooding, winds, and drought may exceed $8 billion.

Following an ENSO event, the trade winds usually return to normal. However, if the trades are exceptionally strong, unusually cold surface water moves over the central and eastern Pacific, and the warm water and rainy weather is confined mainly to the western tropical Pacific. This cold-water episode, which is the opposite of El Niño conditions, has been termed **La Niña** (the girl child).

Figure 11.22 shows warm events, or El Niño years, in red and cold events, or La Niña years, in blue. Notice that the two strongest El Niños were 1982–83 and 1997–98. The weaker El Niño events also have an effect on the Northern Hemisphere's weather patterns. For example, during the El Niño of 1986–87, the subtropical jet stream (being fueled by the warm tropical waters and huge thunderstorms) curved its way over the southeastern United States, where it brought abundant rainfall to a region that, during the previous summer, had suffered through a devastating drought. Moreover, the cold La Niña of 1988 may have contributed to the summer drought over North America that year. During the El Niño of 1991–92, the subtropical jet stream once again swung over North America. Water evaporating from the warm tropical oceans fueled huge thunderstorms. The subtropical jet stream initially swept this moisture into Texas, where it caused extensive flooding. This El Niño continued into early 1995, becoming the longest lasting one in the past 50 years. Notice in Figure 11.22 that since 1976, El Niño events have been stronger and more numerous. Is this a natural cycle or is it somehow associated with greenhouse gases and global warming? The answer is presently unknown.

Figure 11.23a illustrates typical winter weather patterns over North America during El Niño conditions. Notice that a

(a) El Niño

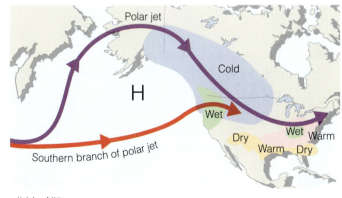

(b) La Niña

FIGURE 11.23
Typical winter weather patterns across North America during an El Niño warm event (a) and during a La Niña cold event (b).

persistent trough of low pressure forms over the north Pacific and, to the south of the low, the jet stream (from off the Pacific) steers wet weather and storms into California and the southern part of the United States. A weak polar jet stream forms over eastern Canada allowing warmer than normal weather to prevail over a large part of North America. (Compare the jet stream pattern in Fig. 11.23a with the jet stream pattern during an El Niño event in Fig. 11.11.)

Figure 11.23b shows typical winter weather patterns with a La Niña. Notice that a persistent (called blocking) high-pressure area forms south of Alaska forcing the polar jet stream into Alaska, then southward into Canada and the western United States. The southern branch of the polar jet, which forms south of the high, directs moist air from the ocean into the Pacific northwest, producing a wet winter for that region. Meanwhile, winter months in the southern part of the United States tend to be warmer and drier than normal.

As we have seen, El Niño and the Southern Oscillation are part of a large-scale ocean-atmosphere interaction that can take several years to run its course. During this time, there are certain regions in the world where significant climatic responses to an ENSO event are likely. Using data from previous ENSO episodes, scientists at the National Oceanic and Atmospheric Administration's Climatic Prediction Center have obtained a global picture of where climatic abnormalities are most likely (see Fig. 11.24).

Some scientists feel that the trigger necessary to start an ENSO event lies within the changing of the seasons, especially the transition periods of spring and fall. Others feel that the winter monsoon plays a major role in triggering a major El Niño event. As noted earlier, it appears that an ENSO episode and the monsoon system are intricately linked, so that a change in one brings about a change in the other.

Presently, scientists (with the aid of coupled general circulation models) are trying to simulate atmospheric and oceanic

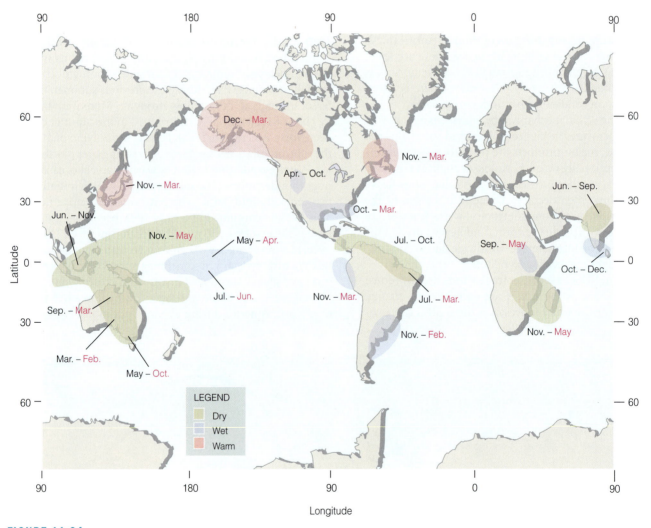

FIGURE 11.24

Regions of climatic abnormalities associated with El Niño–Southern Oscillation conditions. A strong ENSO event may trigger a response in nearly all indicated areas, whereas a weak event will likely play a role in only some areas. Note that the months in black type indicate months during the same years the major warming began; months in red type indicate the following year. (After NOAA Climatic Prediction Center.)

conditions, so that El Niño and the Southern Oscillation can be anticipated. At this point, several models have been formulated that show promise in predicting the onset and life history of an ENSO event. In addition, an in-depth study known as TOGA (*Tropical Ocean and Global Atmosphere*), which began in 1985 and ended in 1994, is providing scientists with valuable information about the interactions that occur between the ocean and the atmosphere. The primary aim of TOGA, a major component of the *World Climate Research Program* (WCRP), is to provide enough scientific information so that researchers can better predict climatic fluctuations (such as ENSO) that occur over periods of months and years. The hope is that a better understanding of El Niño and the Southern Oscillation will provide improved long-range forecasts of weather and climate.

Is there a similar pattern in the Atlantic that compares to the southern oscillation in the Pacific? Over the Atlantic there is a reversal of pressure (called the *North Atlantic Oscillation*) that has an effect on the weather in Europe and along the east coast of North America. For example, in winter when atmospheric pressure in the vicinity of the Icelandic low rises, and the pressure in the region of the Bermuda-Azores high lowers (see Fig. 11.3a for reference), northern Europe's winter turns extremely cold, and storms that normally take a more northerly route slam into Spain. In the northeastern United States, extremely cold arctic air invades the region. There is also an event similar to the tropical Pacific's El Niño that occurs over the tropical Atlantic. Typically in the eastern Atlantic (off the coast of Africa), the water is cooler and the weather drier than in the tropical western Atlantic. Periodically, the cool water along the African coast is replaced by warm water and heavy rainfall. This Atlantic warming, however, occurs more sporadically, and is not as strong as the warming in the tropical Pacific.

SUMMARY

In this chapter, we described the large-scale patterns of wind and pressure that persist around the world. We found that, at the surface in both hemispheres, the trade winds blow equatorward from the semipermanent high-pressure areas centered near 30° latitude. Near the equator, the trade winds converge along a boundary known as the intertropical convergence zone (ITCZ). On the poleward side of the subtropical highs are the prevailing westerly winds. The westerlies meet cold polar easterly winds along a boundary called the polar front, a zone of low pressure where middle latitude storms often form. The annual shifting of the major pressure areas and wind belts—northward in July and southward in January—strongly influences the annual precipitation of many regions.

Warm air aloft (high pressure) over low latitudes and cold air aloft (low pressure) over high latitudes produce westerly winds in both hemispheres, especially at middle and high latitudes. Near the equator, easterly winds exist. The jet streams are located where strong winds concentrate into narrow bands. The polar jet stream forms in response to temperature contrasts along the polar front, while the subtropical jet stream forms at higher elevations above the subtropics, along an upper-level boundary called the subtropical front.

Near the surface, we examined the interaction between the atmosphere and oceans. We found the interaction to be an ongoing process where everything, in one way or another, seems to influence everything else. On a large scale, winds blowing over the surface of the water drive the major ocean currents; the oceans, in turn, release energy to the atmosphere, which helps to maintain the general circulation. Where winds and the Ekman Spiral move surface water away from a coastline, cold, nutrient-rich water upwells to replace it, creating good fishing. When atmospheric circulation patterns change and the trade winds weaken or reverse direction, warm tropical water is able to flow eastward toward South America, where it chokes off upwelling and produces disastrous economic conditions. When the warm water extends over a vast area of the tropical Pacific, the warming is called a major El Niño event and the associated reversal of pressure over the Pacific Ocean is called the Southern Oscillation. The large-scale interaction between the atmosphere and the ocean during El Niño and the Southern Oscillation (ENSO) affects global atmospheric circulation patterns. The sweeping winds aloft provide too much rain in some areas and not enough in others. Studies now in progress are designed to determine how the interchange between atmosphere and ocean can produce such events.

Key Terms

The following terms are listed in the order they appear in the text. Define each. Doing so will aid you in reviewing the material covered in this chapter.

general circulation (of the atmosphere)	intertropical convergence zone (ITCZ)
Hadley cell	westerlies
doldrums	polar front
subtropical highs	subpolar low
horse latitudes	polar easterlies
trade winds	semipermanent highs and lows

Bermuda high
Pacific high
Icelandic low
Aleutian low
Siberian high
jet streams
subtropical jet stream
polar front jet stream
jet streak
    (or jet maximum)
tropical easterly jet stream

stratospheric polar night jet
    stream
low-level jet
oceanic front
Ekman Spiral
upwelling
El Niño
Southern Oscillation
ENSO
La Niña

Questions for Review

1. Draw a large circle. Now, place the major surface pressure and wind belts of the world at their appropriate latitudes.
2. Explain how and why the average surface pressure features shift from summer to winter.
3. Why is it impossible on the earth for a Hadley cell to extend from the pole to the equator?
4. Along a meridian line running from the equator to the poles, how does the general circulation help to explain zones of abundant and sparse precipitation?
5. Explain why the winds in the middle and upper troposphere tend to blow from west to east in both the Northern and the Southern Hemispheres.
6. How does the polar front influence the development of the polar-front jet stream?
7. Describe how the conservation of angular momentum plays a role in the formation of a jet stream.
8. Why is the polar-front jet stream stronger in winter than in summer?
9. What conditions lead to the development of a low-level jet? During the summer, how does the low-level jet affect weather on the Great Plains?
10. Explain the relationship between the circulation of air and the circulation of ocean currents.
11. List at least four important interactions that exist between the ocean and the atmosphere.
12. Describe how the Ekman Spiral forms.
13. What conditions are necessary for upwelling to occur along the west coast of North America? The east coast of North America?
14. (a) What is a major El Niño event?
    (b) What happens to the surface pressure at opposite ends of the Pacific Ocean during the Southern Oscillation?
    (c) Describe how the Southern Oscillation influences a major El Niño event.
15. What are the conditions over the tropical eastern and central Pacific Ocean during the phenomenon known as La Niña?

16. What type of weather (cold/warm, wet/dry) would you expect over North America during a strong El Niño? During a strong La Niña?

Questions for Thought

1. What effect would continents have on the circulation of air in the single-cell model?
2. How would the general circulation of air appear in summer and winter if the earth were tilted on its axis at an angle of 45° instead of 23½°?
3. Summer weather in the southwestern section of the United States is influenced by a subtropical high-pressure cell, yet Fig. 11.3b shows an area of low pressure at the surface. Explain.
4. Explain why icebergs tend to move at right angles to the direction of the wind.
5. Over the open ocean in the vicinity of the Pacific high, observations have indicated that ozone concentrations hundreds of meters above the surface are greater than expected. Give a possible explanation for this.
6. Give *two* reasons why pilots would prefer to fly in the core of a jet stream rather than just above or below it.
7. Why do the major ocean currents in the North Indian Ocean reverse direction between summer and winter?
8. Explain why the surface water temperature along the northern California coast is warmer in winter than it is in summer.
9. You are given an upper-level map that shows the position of two jet streams. If one is the polar front jet and the other the subtropical jet, how would you be able to tell which is which?
10. The Coriolis force deflects moving water to the right in the Northern Hemisphere and to the left in the Southern Hemisphere. Why, then, does upwelling tend to occur along the western margin of continents in both hemispheres?

Problems and Exercises

1. Locate the following cities on a world map. Then, based on the general circulation of surface winds, predict the prevailing wind for each one during July and January.
   (a) Nashville, Tennessee
   (b) Oklahoma City, Oklahoma
   (c) Melbourne, Australia
   (d) London, England
   (e) Paris, France
   (f) Reykjavik, Iceland
   (g) Fairbanks, Alaska
   (h) Seattle, Washington

2. Suppose a balloon is released over southern Baja California on the morning of March 10, 1998. (Refer to Fig. 11.11, p. 285.) If the balloon rises to the 300-mb level and is "caught" by the jet stream, calculate about how long it would take the balloon to exit the United States off the coast of Georgia.

3. Below is a list of average weather conditions that prevail during the month of July at San Francisco, California, and Atlantic City, New Jersey. Both cities lie adjacent to an ocean at nearly the same latitude; however, the average weather conditions vary greatly. In terms of the average surface winds and pressure systems (see Fig. 11.3b) and the interaction between the atmosphere and ocean, explain what accounts for the variation between the two cities of each weather element.

Atlantic City, New Jersey
(latitude 39°N)

Average weather, July

Temperature maximum	84°F
minimum	66°F
Dew point	64°F
Precipitation	3.72 in.
Prevailing wind	S
Weather: Clear to partly cloudy with the possibility of afternoon showers	
Water temperature	70°F

San Francisco, California
(latitude 37°N)

Average weather, July

Temperature maximum	64°F
minimum	53°F
Dew point	53°F
Precipitation	0.01 in.
Prevailing wind	NW
Wind speed (average)	11 mi/hr
Weather: Fog and low clouds during the night and morning with afternoon clearing	
Water temperature	53°F

Questions for Exploration

1. On the Blue Skies: College Edition CD-ROM go to the **Atmospheric Circulation** section of the CD and click on "**Global Circulation.**" Loop the animation and identify the location of the InterTropical Convergence Zone (ITCZ), the belt of trade winds, and the prevailing westerlies. Where do the deserts of the world lie relative to these wind zones?

2. On the Blue Skies: College Edition CD-ROM go to the **Atmospheric Circulation** section of the CD and click on "**Global Oceans.**" Is the mixing of warm water with cold evenly distributed around the ocean or focused on certain regions? What features can you observe that may be important to the exchange of heat from the tropics to the polar regions?

3. On the Blue Skies: College Edition CD-ROM go to the **Atmospheric Circulation** section of the CD and click on "**Southern Oscillation.**" What relationships can you see between the movement of warm water in the Pacific Ocean and wet and dry patterns on the continents?

Clouds developing along an approaching cold front. (Photo by author.)

AIR MASSES AND FRONTS

CONTENTS

About two o'clock in the afternoon it began to grow dark from a heavy, black cloud which was seen in the northwest. Almost instantly the strong wind, traveling at the rate of 70 miles an hour, accompanied by a deep bellowing sound, with its icy blast, swept over the land, and everything was frozen hard. The water in the little ponds in the roads froze in waves, sharp edged and pointed, as the gale had blown it. The chickens, pigs and other small animals were frozen in their tracks. Wagon wheels ceased to roll, froze to the ground. Men, going from their barns or fields a short distance from their homes, in slush and water, returned a few minutes later walking on the ice. Those caught out on horseback were frozen to their saddles, and had to be lifted off and carried to the fire to be thawed apart. Two young men were frozen to death near Rushville. One of them was found with his back against a tree, with his horse's bridle over his arm and his horse frozen in front of him. The other was partly in a kneeling position, with a tinder box in one hand and a flint in the other, with both eyes wide open as if intent on trying to strike a light. Many other casualties were reported. As to the exact temperature, however, no instrument has left any record; but the ice was frozen in the stream, as variously reported, from six inches to a foot in thickness in a few hours.

John Moses, *Illinois: Historical and Statistical*

The opening details the passage of a spectacular cold front as it moved through Illinois on December 21, 1836. Although no reliable temperature records are available, estimates are that, as the front swept through, air temperatures dropped almost instantly from the balmy 40s (°F) to 0 degrees. Fortunately, temperature changes of this magnitude are quite rare with cold fronts.

In this chapter, we will examine the more typical weather associated with cold fronts and warm fronts. We will address questions such as: Why are cold fronts usually associated with showery weather? How can warm fronts cause freezing rain and sleet to form over a vast area during the winter? And how can one read the story of an approaching warm front by observing its clouds? But, first, so that we may better understand fronts, we will examine air masses. We will look at where and how they form and the type of weather usually associated with them.

AIR MASSES

An **air mass** is an extremely large body of air whose properties of temperature and humidity are fairly similar in any horizontal direction at any given altitude. Air masses may cover many thousands of square kilometers. In Fig. 12.1, a large winter air mass, associated with a high-pressure area, covers over half of the United States. Note that, although the surface air temperature and dew point vary somewhat, everywhere the air is cold and dry, with the exception of the zone of snow showers on the eastern shores of the Great Lakes. This cold, shallow anticyclone

will drift eastward, carrying with it the temperature and moisture characteristic of the region where the air mass formed; hence, in a day or two, cold air will be located over the central Atlantic Ocean. Part of weather forecasting is, then, a matter of determining air mass characteristics, predicting how and why they change, and in what direction the systems will move.

Source Regions   Regions where air masses originate are known as **source regions**. In order for a huge mass of air to develop uniform characteristics, its source region should be generally flat and of uniform composition, with light surface winds. The longer the air remains stagnant over its source region, the more likely it will acquire properties of the surface below. Consequently, ideal source regions are usually those areas dominated by high pressure. They include the ice- and snow-covered arctic plains in winter and subtropical oceans and desert regions in summer. The middle latitudes, where surface temperatures and moisture characteristics vary considerably, are not good source regions. Instead, this region is a transition zone where air masses with different physical properties move in, clash, and produce an exciting array of weather activity.

Classification   Air masses are grouped into four general categories according to their source region. Air masses that originate in polar latitudes are designated by the capital letter "P" (for *polar*); those that form in warm tropical regions are designated by the capital letter "T" (for *tropical*). If the source region is land, the air mass will be dry and the lowercase letter "c" (for *continental*) precedes the P or T. If the air mass originates over

FIGURE 12.1
Here, a large, extremely cold winter air mass is dominating the weather over much of the United States. At almost all cities, the air is cold and dry. Upper number is air temperature (°F); bottom number is dew point (°F).

water, it will be moist—at least in the lower layers—and the lowercase letter "m" (for *maritime*) precedes the P or T. We can now see that polar air originating over land will be classified cP on a surface weather chart, while tropical air originating over water will be marked as mT. In winter, an extremely cold cP air mass is designated as cA, *continental arctic*. Often, however, it is difficult to distinguish between arctic and polar air masses, especially when the arctic air mass has traveled over warmer terrain. By the same token, an extremely hot, humid air mass originating over equatorial waters is sometimes designated as mE, for *maritime equatorial*. Distinguishing between equatorial and tropical air masses is usually difficult. Table 12.1 lists the four basic air masses.

After the air mass spends some time over its source region, it usually begins to move in response to the winds aloft. As it moves away from its source region, it encounters surfaces that may be warmer or colder than itself. The relationship between the air mass and its present environment leads to an additional classification. If the air mass is colder than the surface over which it is moving, the lowercase letter "k" (for cold) is added. When the air mass is warmer than the underlying surface, the lowercase letter "w" is added. Hence, a cPk air mass is polar air that originated over land and is presently colder than the surface directly beneath it. What is an mTw air mass? (Answer: Moist, tropical air that is warmer than the surface over which it is moving.)

When the air mass is colder than the underlying surface, it is warmed from below, which results in a steeper lapse rate and instability at low levels. In this case, increased convection and turbulent mixing near the surface usually produce good visibility, cumuliform clouds, and showers of rain or snow. On the other hand, when the air mass is warmer than the surface below, the lower layers are chilled by contact with the cold earth. Warm air above cooler air produces a stable lapse rate with little vertical mixing. This causes the accumulation of dust, smoke, and pollutants, which restricts surface visibilities. In moist air, stratiform clouds accompanied by drizzle or fog may form.

Air Masses of North America

The principal air masses (with their source regions) that invade the United States are shown in Fig. 12.2. We are now in a position to study the formation and modification of each of these air masses and the variety of weather that accompanies them.

cP (Continental Polar) and cA (Continental Arctic) Air Masses

The bitterly cold weather that enters the United States in winter is associated with **continental polar** and **continental arctic** air masses. These originate over the ice- and snow-covered regions of northern Canada and Alaska where long, clear nights allow for strong radiational cooling of the surface. Air in contact with the surface becomes quite cold and stable. Since little moisture is added to the air, it is also quite dry, and specific humidities are often less than 0.3 gram per kilogram. Eventually a portion of this cold air breaks away and, under the influence of the air flow aloft, moves southward as an enormous shallow high-pressure area.

TABLE 12.1
Air Mass Classification and Characteristics

SOURCE REGION	POLAR (P)	TROPICAL (T)
Land continental (c)	cP Cold, dry, stable	cT Hot, dry, stable air aloft; unstable surface air
Water maritime (m)	mP Cool, moist, unstable	mT Warm, moist; usually unstable

As the cold air moves into the interior plains, there are no topographic barriers to restrain it, so it continues southward, bringing with it cold wave warnings and frigid temperatures. The infamous Texas norther is associated with cP (and cA) air. As the air mass moves over warmer land to the south, the air temperature moderates slightly and becomes cPk as it is heated from below. However, even during the afternoon, when the

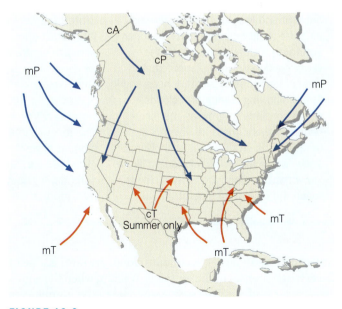

FIGURE 12.2
Air mass source regions and their paths.

Lake-Effect Snows

During the winter, when the weather in the Midwest is dominated by clear, brisk cP (or cA) air, people living on the eastern shores of the Great Lakes brace themselves for heavy snow showers. Snowstorms that form on the downwind side of one of these lakes are known as *lake-effect snows*. These storms are highly localized, extending from just a few kilometers to more than 50 km inland. The snow usually falls as a heavy shower or squall in a concentrated zone. So centralized is the region of snowfall, that one part of a city may accumulate many centimeters of snow, while, in another part, the ground is bare.

Lake-effect snows are most numerous from November to January. During these months, cP air moves over the lakes when they are relatively warm and not quite frozen. The contrast in temperature between water and air can be as much as 25°C (45°F). Studies show that the greater the contrast in temperature, the greater the potential for snow showers. In Fig. 1, we can see that, as the cold air moves over the warmer water, the air mass is quickly warmed from below, making it more buoyant and less stable. Rapidly, the air sweeps up moisture, soon becoming saturated. Out over the water, the vapor condenses into steam fog. As the air continues to warm, it rises and forms billowing cumuliform clouds, which continue to grow as the air becomes more unstable. Eventually, these clouds produce heavy showers of snow, which make the lake seem like a snow factory. Once the air and clouds reach the downwind side of the lake,

FIGURE 1

The formation of lake-effect snows. Cold, dry air crossing the lake gains moisture and warmth from the water. The more buoyant air now rises, forming clouds that deposit large quantities of snow on the lake's leeward shores.

FIGURE 2

Areas shaded purple show regions that experience heavy lake-effect snows.

additional lifting is provided by low hills and the convergence of air as it slows down over the rougher terrain. In late winter, the frequency and intensity of lake-effect snows taper off as the temperature contrast between water and air diminishes and larger portions of the lakes freeze.

Generally, the longer the stretch of water over which the air mass travels (the longer the fetch), the greater the amount of warmth and moisture derived from the lake, and the greater the potential for heavy snow showers. In fact, studies show that, for significant

snowfall to occur, the air must move across 80 km (50 mi) of open water. Consequently, forecasting lake-effect snowfalls depends to a large degree on determining the trajectory of the air as it flows over the lake. Regions that experience heavy lake-effect snowfalls are shown in Fig. 2.

As the cP air moves farther east, the heavy snow showers usually taper off; however, the western slope of the Appalachian Mountains produces further lifting, enhancing the possibility of more and heavier showers. The heat given off during condensation warms the air and, as the air descends the eastern slope, compressional heating warms it even more. Snowfall ceases, and by the time the air arrives at Philadelphia, New York, or Boston, the only remaining trace of the snow showers occurring on the other side of the mountains are the puffy cumulus clouds drifting overhead.

Lake-effect snows are not confined to the Great Lakes. In fact, any large unfrozen lake (such as the Great Salt Lake) can enhance snowfall when cold, relatively dry air sweeps over it.

surface air is most unstable, cumulus clouds are rare because of the extreme dryness of the air mass. At night, when the winds die down, rapid surface cooling and clear skies combine to produce low minimum temperatures. If the cold air moves as far south as central or southern Florida, the winter vegetable crop may be severely damaged. When the cold, dry air mass moves over a relatively warm body of water, such as the Great Lakes, heavy snow showers—called **lake-effect snows**—often form on the eastern shores. (More information on lake-effect snows is provided in the Focus section above.)

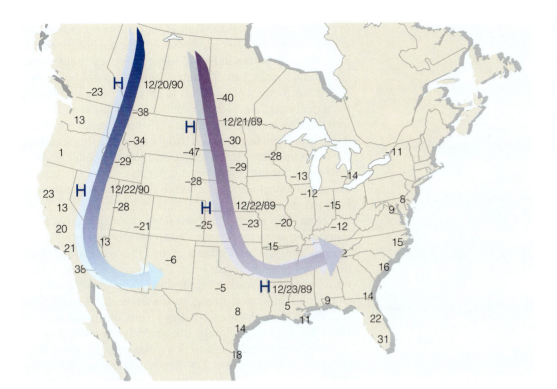

FIGURE 12.3
Average upper-level wind flow (heavy arrows) and surface position of anticyclones (H) associated with two extremely cold outbreaks of arctic air during December. Numbers on the map represent minimum temperatures (°F) measured during each cold snap.

In winter, the generally fair weather accompanying cP air is due to the stable nature of the atmosphere aloft. Sinking air develops above the large dome of high pressure. The subsiding air warms by compression and creates warmer air, which lies above colder surface air. Therefore, a strong upper-level subsidence inversion often forms. Should the anticyclone stagnate over a region for several days, the visibility gradually drops as pollutants become trapped in the cold air near the ground. Usually, however, winds aloft move the cold air mass either eastward or southeastward.

The Rockies, Sierra Nevada, and Cascades normally protect the Pacific Northwest from the onslaught of cP air, but, occasionally, cP air masses do invade these regions. When the upper-level winds over Washington and Oregon blow from the north or northeast on a trajectory beginning over northern Canada or Alaska, cold cP (and cA) air can slip over the mountains and extend its icy fingers all the way to the Pacific Ocean. As the air moves off the high plateau, over the mountains, and on into the lower valleys, compressional heating of the sinking air causes its temperature to rise, so that by the time it reaches the lowlands, it is considerably warmer than it was originally. However, in no way would this air be considered warm. In some cases, the subfreezing temperatures slip over the Cascades and extend southward into the coastal areas of southern California.

A similar but less dramatic warming of cP and cA air occurs along the eastern coast of the United States. Air rides up and over the lower Appalachian Mountains. Turbulent mixing and compressional heating increase the air temperatures on the downwind side. Consequently, cities located to the east of the Appalachian Mountains usually do not experience temperatures as low as those on the west side. In Fig. 12.1, notice that for the same time of day—in this case 7 a.m. EST—Philadelphia, with an air temperature of 14°F, is 16°F (9°C) warmer than Pittsburgh, with an air temperature of –2°F.

Figure 12.3 shows two upper-air wind patterns that led to extremely cold outbreaks of arctic air during December 1989 and 1990. Upper-level winds typically blow from west to east, but, in both of these cases, the flow, as given by the heavy, dark arrows, had a strong north-south (meridional) trajectory. The H represents the positions of the cold surface anticyclones. Numbers on the map represent minimum temperatures (°F) recorded during the cold spells. East of the Rocky Mountains, over 350 record low temperatures were set between December 21 and 24, 1989, with the arctic outbreak causing an estimated $480 million in damage to the fruit and vegetable crops in Texas and Florida. Along the West Coast, the frigid air during December, 1990, caused over $300 million in damage to the vegetable and citrus crops, as temperatures over parts of California plummeted to their lowest readings in more than fifty years. Notice in both cases how the upper-level wind directs the paths of the air masses.

The cP air that moves into the United States in summer has properties much different from its winter counterpart. The source region remains the same but is now characterized by long summer days that melt snow and warm the land. The air is only moderately cool, and surface evaporation adds water vapor to the air. A summertime cP air mass usually brings relief from the oppressive heat in the central and eastern states, as cooler air lowers the air temperature to more comfortable levels. Daytime heating warms the lower layers, producing surface instability. The water vapor in the rising air may condense and create a sky

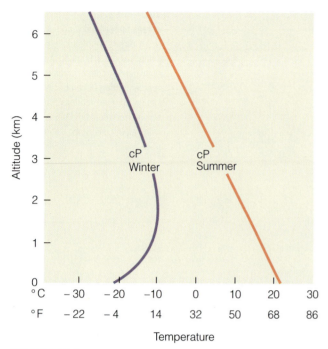

FIGURE 12.4

Typical vertical temperature profile over land for a summer and a winter cP air mass.

dotted with fair weather cumulus clouds (cumulus humilis). A typical profile of temperatures for a summer and a winter cP air mass is given in Fig. 12.4. Notice that the strong inversion so prevalent in winter is absent in summer.

When an air mass moves over a large body of water, its original properties may change considerably. For instance, cold dry cP air moving over the Gulf of Mexico warms rapidly and gains moisture. The air quickly assumes the qualities of a maritime air mass. Notice in Fig. 12.5 that rows of cumulus clouds (called *cloud streets*) are forming over the Gulf of Mexico parallel to northerly surface winds as cP air is being warmed by the water beneath it. As the air continues its journey southward into Mexico and Central America, strong, moist northerly winds build into heavy clouds (bright area) and showers along the northern coast. Hence, a once cold, dry, and stable air mass can be modified to such an extent that its original characteristics are no longer discernible. When this happens, the air mass is given a new designation.

Notice also in Fig. 12.5 that a similar modification of cP air is occurring along the Atlantic Coast, as northwesterly winds are blowing over the mild Atlantic. When this air encounters the much warmer Gulf Stream water, it warms rapidly and becomes unstable. Vertical mixing brings down faster-flowing cold air from aloft. This creates strong, gusty surface winds and choppy seas, which can be hazardous to ships. Thus, the modification of cP air by a large body of water can adversely affect shipping.

In summary, polar and arctic air masses are responsible for the bitter cold winter weather that can cover wide sections of North America. When the air mass originates over the Canadian Northwest Territories, frigid air can bring record-breaking low temperatures. Such was the case on Christmas Eve, 1983, when arctic air covered most of North America. (A detailed look at this air mass and its accompanying record-setting low temperatures is given in the Focus section beginning on p. 307.)

mP (Maritime Polar) Air Masses During the winter, cP air originating over Asia and frozen polar regions is carried east-

FIGURE 12.5

Visible satellite image showing the modification of cP air as it moves over the warmer Gulf of Mexico and the Atlantic Ocean.

The Return of the Siberian Express

The winter of 1983–1984 was one of the coldest on record across North America. Unseasonably cold weather arrived in December, which, for much of the country, was one of the coldest Decembers since records have been kept. During the first part of the month, continental polar air covered most of the northern and central plains. As the cold air moderated slightly, far to the north a huge mass of bitter cold arctic air was forming over the frozen reaches of the Canadian Northwest Territories.

By midmonth, the frigid air, associated with a massive high-pressure area, covered all of northwest Canada. Meanwhile, an upper-level ridge was forming over Alaska. On the eastern side of the ridge, strong northerly winds associated with the polar jet stream directed the frigid air southward over the prairie provinces of Canada. A portion of the extraordinarily cold air broke away, and, like a large swirling bubble, moved as a cold, shallow anticyclone southward into the United States. Because the frigid air was accompanied in some regions by winds gusting to 45 knots, at least one news reporter dubbed the onslaught of this arctic blast, "the Siberian Express."

The express dropped temperatures to some of the lowest readings ever recorded during the month of December. On December 22, Elk Park, Montana, recorded an unofficial low of –53°C (–64°F), only 4°C higher than the all-time low of –57°C (–70°F) for the nation (excluding Alaska) recorded at Rogers Pass, Montana, on January 20, 1954.

The center of the massive anticyclone gradually pushed southward out of Canada. By December 24, its center was over eastern Montana (Fig. 3), where the sea level pressure

FIGURE 3

Surface weather map for 7 A.M., EST, December 24, 1983. Solid lines are isobars. Areas shaded green represent precipitation. An extremely cold arctic air mass covers nearly 90 percent of the United States. (Weather symbols for the surface map are given in Appendix B.)

at Miles City reached an incredible 1064 mb (31.40 in.)—a new United States record that topped the old mark of 1063 mb set in Helena, Montana, on January 10, 1962. An enormous ridge of high pressure stretched from the Canadian arctic coast to the Gulf of Mexico. On the east side of the ridge, cold westerly winds brought lake-effect snows to the eastern shores of the Great Lakes. To the south of the high-pressure center, cold easterly winds, rising along the elevated plains, brought light amounts of *upslope snow** to sections of the Rocky Mountain states. Notice in Fig. 3 that, on Christmas Eve, arctic air covered almost 90 percent of the United States. As the cold air swept east-

*Upslope snow forms as cold air moving from east to west gradually rises (and cools even more) as it approaches the Rocky Mountains.

ward and southward, a hard freeze caused hundreds of millions of dollars in damage to the fruit and vegetable crops in Texas, Louisiana, and Florida. On Christmas Day, 125 record low temperature readings were set in twenty-four states. That afternoon, at 1:00 p.m., it was actually colder in Atlanta, Georgia, at –13°C than it was in Fairbanks, Alaska (–12°C). One of the worst cold waves to occur in December during the twentieth century continued through the week, as many new record lows were established in the deep south from Texas to Louisiana.

By January 1, the extreme cold had moderated, as the upper-level winds became more westerly. These winds brought milder mP Pacific air eastward into the Great Plains. The warmer pattern continued until about January 10, when the Siberian

(*continued on page 308*)

The Return of the Siberian Express (continued)

Express decided to make a return visit. Driven by strong upper-level northerly winds, impulse after impulse of arctic air from Canada swept across the United States. On January 18, an all-time record low of –54°C (–65°F) was recorded for the state of Utah at Middle Sinks. On January 19, temperatures plummeted to a new low of –22°C (–7°F) for the airports in Philadelphia and Baltimore. Toward the end of the month, the upper-level winds once again became more westerly. Over much of the nation, the cold

air moderated. But the express was to return at least one more time.

The beginning of February saw relatively warm air covering much of the nation from California to the Atlantic coast. On February 4, an arctic outbreak of cA air spread southward and eastward across the nation. Although freezing air extended southward into central Florida, the express ran out of steam, and a February heat wave soon engulfed most of the United States east of the Rocky Mountains. Maritime tropical air from the Gulf of

Mexico brought record warmth to much of the eastern two-thirds of the nation. Near the middle of the month, Louisville, Kentucky, reported 23°C (73°F) and Columbus, Ohio, 21°C (69°F). Even though February was one of the warmest months on record over parts of the United States, the winter of 1983–1984 (December, January, and February) will go down in the record books as one of the coldest winters for the United States as a whole since reliable record keeping began in 1931.

ward and southward over the Pacific Ocean by the circulation around the Aleutian low. The ocean water modifies the cP air by adding warmth and moisture to it. Since this air has a fetch of many hundreds or even thousands of kilometers, it gradually changes into a **maritime polar** air mass.

FIGURE 12.6

A winter upper-air pattern that brings mP air into the west coast of North America. The large arrow represents the upper-level flow. Note the trough of low pressure along the coast. The small arrows show the trajectory of the mP air at the surface. Regions that normally experience precipitation under these conditions are also shown on the map. Showers are most prevalent along the coastal mountains and in the Sierra Nevada.

By the time this air mass reaches the Pacific Coast it is cool, moist, and unstable. The ocean's effect is to keep air near the surface warmer than the air aloft. Temperature readings in the 40s and 50s (°F) are common near the surface, while air at an altitude of about a kilometer or so above the surface may be at the freezing point. Within this colder air, characteristics of the original cP air mass may still prevail. As the air moves inland, coastal mountains force it to rise, and much of its water vapor condenses into rain-producing clouds. In the colder air aloft, the rain changes to snow, with heavy amounts accumulating in mountain regions. A typical upper-level wind-flow pattern that brings mP air onto the west coast of North America is shown in Fig. 12.6.

When the mP air moves inland, it loses much of its moisture as it crosses a series of mountain ranges. Beyond these mountains, it travels over a cold, elevated plateau that chills the surface air and slowly transforms the lower level into dry, stable cP air. East of the Rockies this air mass is referred to as *Pacific air* (see Fig. 12.7). Here, it often brings fair weather and temperatures that are cool but not nearly as cold as the cP air that invades this region from northern Canada. In fact, when mP air from the west replaces retreating cP air from the north, chinook winds often develop. Furthermore, when the modified mP air replaces moist tropical air, storms can form along the boundary separating the two air masses.

Along the East Coast, mP air originates in the North Atlantic as cP air moves southward some distance off the Atlantic Coast. Steered by northeasterly winds, mP air then swings southwestward toward the northeastern states. Because the water of the North Atlantic is very cold and the air mass travels only a short distance, wintertime Atlantic mP air masses are usu-

FIGURE 12.7
After crossing several mountain ranges, cool moist mP air from off the Pacific ocean descends the eastern side of the Rockies as modified, relatively dry Pacific air.

ally much colder than their Pacific counterparts. Because the prevailing winds aloft are westerly, Atlantic mP air masses are also much less common.

Figure 12.8 illustrates a typical late winter or early spring surface weather pattern that carries mP air from the Atlantic into the New England and middle Atlantic states. A slow-moving, cold anticyclone drifting to the east (north of New England) causes a northeasterly flow of mP air to the south. The boundary separating this invading colder air from warmer air even farther south is marked by a stationary front. North of this front, northeasterly winds provide generally undesirable weather, consisting of damp air and low, thick clouds from which light precipitation falls in the form of rain, drizzle, or snow. When upper atmospheric conditions are right, storms may develop along the stationary front, move eastward, and intensify near the shores of Cape Hatteras. Such storms, called *Hatteras lows*, sometimes swing northeastward along the coast, where they become *northeasters* (commonly called *nor'easters*) bringing with them strong northeasterly winds, heavy rain or snow, and coastal flooding. (Such developing storms will be treated in detail in Chapter 13.)

mT (Maritime Tropical) Air Masses The wintertime source region for Pacific **maritime tropical** air masses is the subtropical east Pacific Ocean. Air from this region must travel over 1600 km of water before it reaches the southern California coast. Consequently, these air masses are very warm and moist by the time they arrive along the West Coast. The warm air produces heavy precipitation usually in the form of rain, even at high elevations. Melting snow and rain quickly fill rivers, which overflow into the low-lying valleys. The rapid snowmelt leaves local ski slopes barren, and the heavy rain can cause disastrous mud slides in the steep canyons.

Figure 12.9 shows maritime tropical air (usually referred to as *subtropical air*) streaming into northern California on January 1, 1997. The humid, subtropical air, which originated near the Hawaiian Islands, was termed by at least one weather-caster as *"the pineapple connection."* After battering the Pacific Northwest with heavy rain, the pineapple connection roared into northern and central California, causing catastrophic

floods that sent over 100,000 people fleeing from their homes, mud slides that closed roads, property damage (including crop losses) that amounted to more than $1.5 billion, and eight fatalities. Yosemite National Park, which sustained over $170 million in damages due mainly to flooding, was forced to close for more than two months.

The mT air that influences much of the weather east of the Rockies originates over the Gulf of Mexico and Caribbean Sea. In winter, cold polar air tends to dominate the continental weather scene, so mT air is usually confined to the Gulf and extreme southern states. Occasionally, a slow-moving

FIGURE 12.8
Winter and early spring surface weather patterns that usually prevail during the invasion of mP air into the mid-Atlantic and New England states. (Green-shaded area represents precipitation.)

FIGURE 12.9
An infrared satellite image that shows maritime tropical air (heavy red arrow) moving into northern California on January 1, 1997. The warm, humid airflow (sometimes called "the pineapple connection") produced heavy rain and extensive flooding in northern and central California.

storm system over the Central Plains draws mT air northward. Gentle, moist south or southwesterly winds blow into the central and eastern parts of the nation in advance of the system. Since the land is still extremely cold, air near the surface is chilled to its dew point. Fog and low clouds form in the early morning, dissipate by midday, and reform in the evening. This mild winter weather in the Mississippi and Ohio valleys lasts, at best, only a few days. Soon cold polar air will move down from the north behind the eastward-moving storm system. Along the boundary between the two air masses, the mT air is lifted above the more dense cP air. This often leads to heavy and widespread precipitation and storminess.

When a storm system stalls over the Central Plains, a constant supply of mT air from the Gulf of Mexico can bring record-breaking maximum temperatures to the eastern half of the country. Sometimes the air temperatures are higher in the mid-Atlantic states than they are in the Deep South, as compressional heating warms the air even more as it moves downslope after crossing the Appalachian Mountains. Figure 12.10 shows a surface weather map and the associated upper airflow (heavy arrow) that brought unseasonably warm mT air into the central and eastern states during April, 1976. A large high centered off the southeast coast coupled with a strong southwesterly flow aloft carried warm moist air into the Midwest and East, causing a

record-breaking April heat wave. The flow aloft prevented the surface low and the cP air behind it from making much eastward progress, so that the warm spell lasted for five days. Providence, Rhode Island, experienced a record-breaking high temperature for April of 36°C (96°F). Note that, on the west side of the surface low, the winds aloft funneled cold cP air from the north into the western states, creating unseasonably cold weather from California to the Rockies. Hence, while people in the Southwest were huddled around heaters, others several thousand kilometers away in the Northeast were turning on air conditioners. We can see that it is the upper-level meridional flow, directing cP air southward and mT air northward, which makes these contrasts in temperature possible.

As maritime air moves inland over the hot continent, it warms, rises, and frequently causes cumuliform clouds, which produce afternoon showers and thunderstorms. You can almost count on thunderstorms developing along the Gulf Coast each afternoon in summer. As evening approaches, thunderstorm activity typically dies off. Nighttime cooling lowers the air temperature and, if the air becomes saturated, fog or low clouds form. These, of course, dissipate by late morning as surface heating warms the air again.

A weak, but often persistent, flow around an upper-level anticyclone will spread mT air from the Gulf of Mexico into

Minimum temperatures (°F)

Maximum temperatures (°F)

FIGURE 12.10
Weather conditions during an unseasonably hot spell in the eastern portion of the United States that occurred between the 15th and 20th of April, 1976. The surface low-pressure area and fronts are shown for April 17. Numbers to the east of the surface low (in red) are maximum temperatures recorded during the hot spell, while those to the west of the low (in blue) are minimums reached during the same time period. The heavy arrow is the average upper-level flow during the period. The faint L and H show average positions of the upper-level trough and ridge.

the southern and central Rockies, where it causes afternoon thunderstorms. Occasionally, this easterly flow may work its way even farther west, producing shower activity in the otherwise dry southwestern desert.

During the summer, humid mT air originating over the tropical eastern Pacific normally remains south of California. Occasionally, a weak upper-level southerly flow will spread this humid air northward into the southwestern United States, most often Arizona, Nevada, and the southern part of Califor-

nia. In many places, the moist, unstable air aloft only shows up as middle and high cloudiness, especially altocumulus and cirrocumulus castellanus. However, where the moist flow meets a mountain barrier, it usually rises and condenses into towering, shower-producing clouds as seen in Fig. 12.11.

cT (Continental Tropical) Air Masses   The only real source region for hot, dry **continental tropical** air masses in North America is found during the summer in northern Mexico and

FIGURE 12.11
Summertime humid maritime tropical air produces towering cumulus clouds and thunderstorms as it is lifted up along the Sierra Nevada.

WEATHER WATCH

A continental tropical air mass, stretching from southern California to the heart of Texas, brought record warmth to the desert southwest during the last week of June, 1990. The temperature, which soared to a sweltering peak of 50°C (122°F) in Phoenix, Arizona, on June 26, caused officials to suspend aircraft takeoffs at Sky Harbor Airport. The extreme heat had lowered air density to the point where it reduced aircraft lift.

the adjacent arid southwestern United States. Here, the air mass is hot, dry, and unstable at low levels, with frequent dust devils forming during the day. Because of the low relative humidity (typically less than 10 percent during the afternoon), air must rise over 3000 m (10,000 ft) before condensation begins. Furthermore, an upper-level ridge usually produces weak subsidence over the region, tending to make the air aloft rather stable and the surface air even warmer. Consequently, skies are generally clear, the weather is hot, and rainfall is practically nonexistent where cT air masses prevail. If this air mass moves outside its source region and into the Great Plains and stagnates over that region for any length of time, a severe drought may result. Figure 12.12 shows a weather map situation where cT air covers a large portion of the western United States and produces hot, dry weather northward to Canada.

So far, we have examined the various air masses that enter North America annually. The characteristics of each depend upon the air mass source region and the type of surface over which the air mass moves. The winds aloft determine the trajectories of these air masses. Occasionally, an air mass will control the weather in a region for some time. These persistent weather conditions are known as **air mass weather**.

Air mass weather is especially common in the southeastern United States during summer as, day after day, mT air from the Gulf brings sultry conditions and afternoon thunderstorms. It is also common in the Pacific Northwest in winter when unstable, cool mP air accompanied by widely scattered showers dominates the weather for several days or more. The real weather action, however, usually occurs not within air masses but at their margins, where air masses with sharply contrasting properties meet—in the zone marked by weather fronts.*

BRIEF REVIEW　Before we examine fronts, here is a review of some of the important facts about air masses:

- An air mass is a large body of air whose properties of temperature and humidity are fairly similar in any horizontal direction.
- Source regions for air masses tend to be generally flat, of uniform composition, and in an area of light winds.
- Continental air masses form over land. Maritime air masses form over water. Polar air masses originate in cold, polar regions, and extremely cold air masses form over arctic regions. Tropical air masses originate in warm, tropical regions.
- Continental polar (cP) air masses are cold and dry; continental arctic (cA) air masses are extremely cold and dry; continental tropical (cT) air masses are hot and dry; maritime tropical (mT) air masses are warm and moist; maritime polar (mP) air masses are cold and moist.

---

*The word *front* is used to denote the clashing of two air masses, probably because it resembles the fighting in Western Europe during World War I, when the term originated.

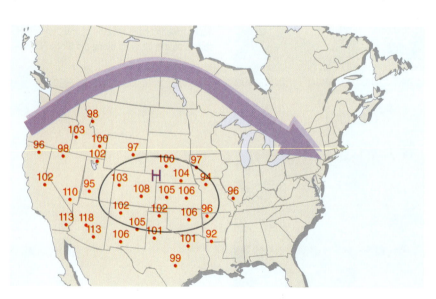

FIGURE 12.12
During June 29 and 30, 1990, cT air covered a large area of the central and western United States. Numbers on the map represent maximum temperatures (°F) during this period. The large H with the isobar shows the upper-level position of the subtropical high. Sinking air associated with the high contributed to the hot weather. Winds aloft were weak with the main flow shown by the heavy arrow.

FRONTS

Although we briefly looked at fronts in Chapter 1, we are now in a position to study them in depth, which will aid us in forecasting the weather. We will now learn about the general nature of fronts—how they move and what weather patterns are associated with them.

A **front** is the transition zone between two air masses of different densities. Since density differences are most often caused by temperature differences, fronts usually separate air masses with contrasting temperatures. Often, they separate air masses with different humidities as well. Remember that air masses have both horizontal and vertical extent; consequently, the upward extension of a front is referred to as a *frontal surface*, or a *frontal zone*.

Figure 12.13 shows a simplified weather map illustrating four different fronts. As we move from west to east across the map, the fronts appear in the following order: a stationary front between points A and B; a cold front between points B and C; a warm front between points C and D; and an occluded front between points C and L. Let's examine the properties of each of these fronts.

Stationary Fronts   A **stationary front** has essentially no movement. On a colored weather map, it is drawn as an alternating red and blue line. Semicircles face toward colder air on the red line and triangles point toward warmer air on the blue line. The stationary front between points A and B in Fig. 12.13 marks the boundary where cold dense cP air from Canada butts up against the north-south trending Rocky Mountains. Unable to cross the barrier, the cold air shows little or no westward movement. The stationary front is drawn along a line separating the cP from the milder mP air to the west. Notice that the surface winds tend to blow parallel to the front, but in opposite directions on either side of it. Upper-level winds often blow parallel to a stationary front.

The weather along the front is clear to partly cloudy, with much colder air lying on its eastern side. Because both air masses are dry, there is no precipitation. This is not, however, always the case. When warm moist air rides up and over the cold air, widespread cloudiness with light precipitation can cover a vast area. These are the conditions that prevail north of the east-west running stationary front depicted in Fig. 12.8, p. 309.

If the warmer air to the west begins to move and replace the colder air to the east, the front in Fig. 12.13 will no longer remain stationary; it will become a warm front. If, on the other hand, the colder air slides up over the mountain and replaces the warmer air on the other side, the front will become a cold front.

Cold Fronts   The **cold front** between points B and C on the surface weather map (Fig. 12.13) represents a zone where cold, dry stable polar air is replacing warm, moist unstable subtrop-

ical air. The front is drawn as a solid blue line with the triangles along the front showing its direction of movement. How did the meteorologist know to draw the front at that location? A closer look at the front will give us the answer.

The weather in the immediate vicinity of this cold front in the southeastern United States is shown in Fig. 12.14. The data plotted on the map represent the current weather at selected cities. The station model used to represent the data at each reporting station is a simplified one that shows temperature, dew point, present weather, cloud cover, sea level pressure, wind direction and speed. The little line in the lower right-hand corner of each station shows the *pressure tendency*—the pressure change, whether rising (/) or falling (\)—during the last three hours. With all of this information, the front can be properly located. (Appendix B explains the weather symbols and the station model more completely.)

The following criteria are used to locate a front on a surface weather map:

1. sharp temperature changes over a relatively short distance
2. changes in the air's moisture content (as shown by marked changes in the dew point)
3. shifts in wind direction
4. pressure and pressure changes
5. clouds and precipitation patterns

In Fig. 12.14, we can see a large contrast in air temperature and dew point on either side of the front. There is also a wind shift from southwesterly ahead of the front, to northwesterly behind it. Notice that each isobar kinks as it crosses the front, forming an elongated area of low pressure—a *trough*—which accounts for the wind shift. Since surface winds normally blow across the isobars toward lower pressure, we find winds with a southerly component ahead of the front and winds with a northerly component behind it.

Since the cold front is a trough of low pressure, sharp changes in pressure can be significant in locating the front's position. One important fact to remember is that the lowest pressure usually occurs just as the front passes a station. Notice that, as you move toward the front, the pressure drops, and, as you move away from it, the pressure rises. This is clearly shown by the pressure tendencies for each station on the map. Just

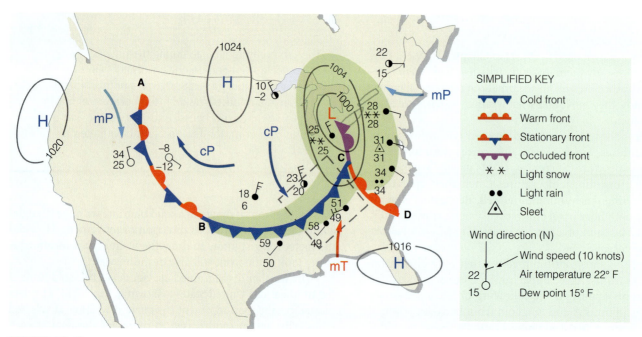

FIGURE 12.13

A simplified weather map showing surface pressure systems, air masses, and fronts. (Green-shaded area represents precipitation.)

FIGURE 12.14

A closer look at the surface weather associated with the cold front situated in the southeastern United States in Fig. 12.13.

FIGURE 12.15
A vertical view of the weather across the cold front in Fig. 12.14 along the line X–X'.

before the front passes, the pressure tendency shows a falling barometer (\), while just behind the front, the barometer is now beginning to rise (√), and farther behind the front, the barometer is rising steadily (/).

The cloud and precipitation patterns in Fig. 12.14 are better seen in a side view of the front along the line X–X' as illustrated in Fig. 12.15. We can see from Fig. 12.15 that, at the front, the cold, dense air wedges under the warm air, forcing the warm air upward, much like a snow shovel forces snow upward as the shovel glides through the snow. As the moist, unstable air rises, it condenses into a series of cumuliform clouds. Strong, upper-level westerly winds blow the delicate ice crystals (which form near the top of the cumulonimbus) into cirrostratus (Cs) and cirrus (Ci). These clouds usually appear far in advance of the approaching front. At the front itself, a relatively narrow band of thunderstorms (Cb) produces heavy showers with gusty winds. Behind the front, the air cools quickly. (Notice how the freezing level dips as it crosses the front.) The winds shift from southwesterly to northwesterly, pressure rises, and precipitation ends. As the air dries out, the skies clear, except for a few lingering fair weather cumulus clouds.

Observe that the leading edge of the front is steep. The steepness is due to friction, which slows the airflow near the ground. The air aloft pushes forward, blunting the frontal surface. If we could walk from where the front touches the surface back into the cold air, a distance of 50 km, the front would be about 1 km above us. Thus, the slope of the front—the ratio of vertical rise to horizontal distance—is 1:50. This is typical for a fast-moving cold front—those that move about 25 knots. In a slower-moving cold front—one that moves about 15 knots—the slope is much more gentle.

With slow-moving cold fronts, clouds and precipitation usually cover a broad area behind the front. When the ascending warm air is stable, stratiform clouds, such as nimbostratus, become the predominate cloud type and even fog may develop in the rainy area. Occasionally, out ahead of a fast-moving front, a line of active showers and thunderstorms, called a *squall line*, develops parallel to and often ahead of the advanc-

ing front, producing heavy precipitation and strong gusty winds.

As the temperature contrast across a front lessens, the front will often weaken and dissipate. Such a condition is known as **frontolysis**. On the other hand, an increase in the temperature contrast across a front can cause it to strengthen and regenerate into a more vigorous frontal system, a condition called **frontogenesis**.

An example of a regenerated front is shown in the infrared satellite pictures, Fig. 12.16. The cold front in Fig. 12.16a is weak, as indicated by the low clouds (gray tones) along the front. As the front moves offshore, over the warm Gulf Stream (Fig. 12.16b), it intensifies into a more vigorous frontal system as surface air becomes unstable and convective activity develops. Notice that extensive cloudiness and thunderstorms are forming along the frontal zone, as indicated by the bright areas in the infrared picture.

So far, we have considered the general weather patterns of "typical" cold fronts. There are, of course, exceptions. For example, if the rising warm air is dry and stable, scattered clouds are all that form, and there is no precipitation. In extremely dry weather, a marked change in the dew point, accompanied by a slight wind shift, may be the only clue to a passing front.[*] During the winter, a series of cold polar outbreaks may travel across the United States so quickly that warm air is unable to develop ahead of the front. In this case, frigid arctic air usually replaces cold polar air, and a drop in temperature is the only indication that a cold front has moved through your area. Along the West Coast, the Pacific Ocean modifies the air so much that cold fronts, such as those described in the previous section, are never seen. In fact, as a cold front moves inland from the Pacific Ocean, the surface temperature contrast across the front may be quite small.

---

[*]At the surface where the dew-point temperature changes rapidly, warm, humid air is separated from warm, dry air along a boundary called a *dry line*. Although not a cold front nor a warm front, convective clouds often form along a dry line and, as we will see in Chapter 15, some of these clouds may develop into severe thunderstorms.

(a)

(b)

FIGURE 12.16

(a) The weakening cold front over land on Tuesday morning, November 21, intensifies into (b) a vigorous front over water on Wednesday morning, November 22. (Infrared satellite images.)

Topographic features usually distort the wind pattern so much that locating the position of the front and the time of its passage are exceedingly difficult. In this case, the pressure tendency is the most reliable indication of a frontal passage.

In some instances along the West Coast, an approaching cold front (or upper-level trough) will cause cool marine air at the surface to surge into coastal and inland valleys. The cool air (which is often accompanied by a wind shift) may produce a sharp drop in air temperature. This may give the impression that a rather strong cold front has moved through, when in reality, the front may be many kilometers offshore.

Cold fronts usually move toward the south, southeast, or east. But sometimes they will move southwestward. In New England, this movement occurs when northeasterly surface winds, blowing clockwise around an anticyclone centered to the north over Canada, push a cold front southwestward often as far south as Boston. Because the cold front moves in from the east, or northeast, it is known as a **"back door" cold front**. As the front passes, westerly surface winds usually shift to easterly or northeasterly and temperatures drop as mP air flows in off the Atlantic Ocean.

An example of a "back door" cold front is shown in Fig. 12.17. This is a springtime situation where, behind the front, the weather is cold and damp with drizzle, as northeasterly winds sweep into the region from off the chilly Atlantic. To the south of the front, the weather is much warmer. Should the front move through this area, the more summerlike weather would change, in a matter of hours, to more winterlike. The cold, dense air behind the front is rather shallow. Consequently, the Appalachian Mountains act as a dam to the front's forward progress, halting its westward movement. This situation, where the cold, damp air is confined to the eastern side of the mountains, is called **cold air damming**. The stalled cold front now becomes a stationary front.

FIGURE 12.17

A "back door" cold front moving into New England during the spring. Notice that, behind the front, the weather is cold and damp with drizzle, while to the south, ahead of the front, the weather is partly cloudy and warm.

The cool air behind the front may linger for some time as warmer, less-dense air to the south rides up and over it. Forecasting how far south the "back door" cold front will move and when the entrenched cold air will leave can be a bit tricky.

Even though cold-front weather patterns have many exceptions, learning these patterns can be to your advantage if you live where well-defined cold fronts are experienced. Knowing them improves your own ability to make short-range weather forecasts. For your reference, Table 12.2 summarizes idealized cold-front weather.

TABLE 12.2 Typical Weather Conditions Associated with a Cold Front

WEATHER ELEMENT	BEFORE PASSING	WHILE PASSING	AFTER PASSING
Winds	South or southwest	Gusty, shifting	West or northwest
Temperature	Warm	Sudden drop	Steadily dropping
Pressure	Falling steadily	Minimum, then sharp rise	Rising steadily
Clouds	Increasing Ci, Cs, then either Tcu or Cb	Tcu or Cb*	Often Cu
Precipitation	Short period of showers	Heavy showers of rain or snow, sometimes with hail, thunder, and lightning	Decreasing intensity of showers, then clearing
Visibility	Fair to poor in haze	Poor, followed by improving	Good except in showers
Dew point	High; remains steady	Sharp drop	Lowering

*Tcu stands for towering cumulus, such as cumulus congestus; whereas Cb stands for cumulonimbus.

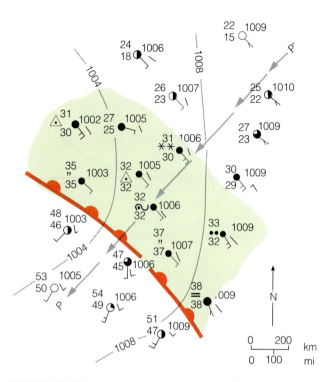

FIGURE 12.18

Surface weather associated with a typical warm front. (Green-shaded area represents precipitation.)

Warm Fronts

In Fig. 12.13, a **warm front** is drawn along the solid red line running from points C to D. Here, the leading edge of advancing warm, moist, subtropical air from the Gulf of Mexico replaces the retreating cold maritime polar air from the North Atlantic. The direction of frontal movement is given by the half circles, which point into the cold air; this front is heading toward the northeast. As the cold air recedes, the warm front slowly advances. The average speed of a warm front is about 10 knots, or about half that of an average cold front. During the day, as mixing occurs on both sides of the front, its movement may be much faster. Warm fronts often move in a series of rapid jumps, which show up on successive weather maps. At night, however, radiational cooling creates cool dense surface air behind the front. This inhibits both lifting and the front's forward progress. When the forward surface edge of the warm front passes a station, the wind shifts, the temperature rises, and the overall weather conditions improve. To see why, we will examine the weather commonly associated with the warm front both at the surface and aloft.

Look at Figs. 12.18 and 12.19 closely and observe that the warmer, less-dense air rides up and over the colder, more-dense surface air. This rising of warm air over cold, called **overrunning**, produces clouds and precipitation well in advance of the front's surface boundary. The warm front that separates the two air masses has an average slope of about 1:300*—a much more gentle or inclined shape than that of a typical cold front. Warm air overriding the cold air creates a stable atmosphere

---

*Typically, the slope of a warm front is on the order of 1:150 to 1:200.

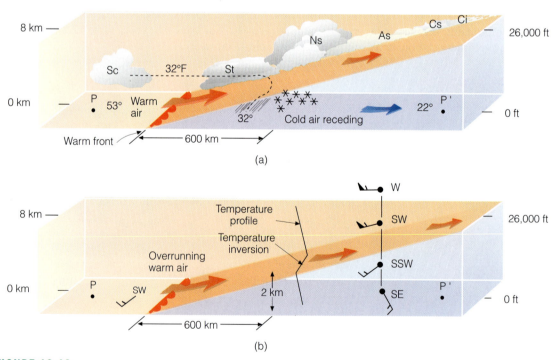

FIGURE 12.19

Vertical view of (a) clouds and precipitation and (b) the temperature and winds across the warm front in Fig. 12.18 along the line P–P'.

TABLE 12.3   Typical Weather Conditions Associated with a Warm Front

WEATHER ELEMENT	BEFORE PASSING	WHILE PASSING	AFTER PASSING
Winds	South or southeast	Variable	South or southwest
Temperature	Cool to cold, slow warming	Steady rise	Warmer, then steady
Pressure	Usually falling	Leveling off	Slight rise, followed by fall
Clouds	In this order: Ci, Cs, As, Ns, St, and fog; occasionally Cb in summer	Stratus-type	Clearing with scattered Sc; occasionally Cb in summer
Precipitation	Light-to-moderate rain, snow, sleet, or drizzle	Drizzle or none	Usually none; sometimes light rain or showers
Visibility	Poor	Poor, but improving	Fair in haze
Dew point	Steady rise	Steady	Rise, then steady

(see the vertical temperature measurements in Fig. 12.19b). Notice that a temperature inversion—called a **frontal inversion**—exists in the region of the upper-level front at the boundary where the warm air overrides the cold air. Another fact to notice is that the wind *veers* (shifts clockwise) with altitude, so that the southeasterly surface winds become southwesterly and westerly aloft.

Suppose we are standing at the position marked P' in Figs. 12.18 and 12.19. Note that we are over 1200 km (750 mi) ahead of the surface front. Here, the surface winds are light and variable. The air is cold and about the only indication of an approaching warm front is the high cirrus clouds overhead. We know the front is moving slowly toward us and that within a day or so it will pass our area. Suppose that, instead of waiting for the front to pass us, we drive toward it, observing the weather as we go.

Heading toward the front, we notice that the cirrus (Ci) clouds gradually thicken into a thin, white veil of cirrostratus (Cs) whose ice crystals cast a halo around the sun.* Almost imperceptibly, the clouds thicken and lower, becoming altocumulus (Ac) and altostratus (As) through which the sun shows only as a faint spot against an overcast gray sky. Snowflakes begin to fall, and we are still over 600 km (370 mi) from the surface front. The snow increases, and the clouds thicken into a sheetlike covering of nimbostratus (Ns). The winds become brisk and out of the southeast, while the barometer slowly falls. Within 400 km (250 mi) of the front, the cold surface air mass is now quite shallow. The surface air temperature moderates and, as we approach the front, the light snow changes first into sleet. It then becomes freezing rain and finally rain and drizzle as the air temperature climbs above freezing. Overall, the precipitation remains light or moderate but covers a broad area. Moving still closer to the front, warm, moist air mixes with cold, moist air producing ragged wind-blown stratus (St) and fog. (Thus, flying in the vicinity of a warm front is quite hazardous.)

Finally, after a trip of over 1200 km, we reach the warm front's surface boundary. As we cross the front, the weather changes are noticeable, but much less pronounced than those experienced with the cold front; they show up more as a gradual transition rather than a sharp change. On the warm side of the front, the air temperature and dew point rise, the wind shifts from southeast to south or southwest, and the barometer stops falling. The light rain ends and, except for a few stratocumulus (Sc), the fog and low clouds vanish.

This scenario of an approaching warm front represents average warm-front weather. In some instances, the weather can differ from this dramatically. For example, if the overrunning warm air is relatively dry and stable, only high and middle clouds will form, and no precipitation will occur. On the other hand, if the warm air is relatively moist and unstable (as is often the case during the summer), heavy showers can develop as thunderstorms become embedded in the cloud mass.

Along the West Coast, the Pacific Ocean significantly modifies the surface air so that warm fronts are difficult to locate on a surface weather map. Also, not all warm fronts move northward or northeastward. On rare occasions, a front will move into the eastern seaboard from the Atlantic Ocean as the front spins all the way around a deep storm positioned off the coast. Cold northeasterly winds ahead of the front usually become warm northeasterly winds behind it. Even with these exceptions, knowing the normal sequence of warm-front weather will be useful, especially if you live where warm fronts become well developed. You can look for certain cloud and weather patterns and make reasonably accurate short-range forecasts of your own. Table 12.3 summarizes typical warm-front weather.

*If the warm air is relatively unstable, ripples or waves of cirrocumulus clouds will appear as a "mackerel sky."

FIGURE 12.20

The formation of a cold-occluded front. The faster-moving cold front (a) catches up to the slower-moving warm front (b) and forces it to rise off the ground (c). (Green-shaded area represents precipitation.)

Occluded Fronts   When a cold front catches up to and overtakes a warm front, the frontal boundary created between the two air masses is called an **occluded front**, or, simply, an **occlusion**. On the surface weather map, it is represented as a purple line with alternating cold-front triangles and warm-front half circles; both symbols point in the direction toward which the front is moving. In Fig. 12.13 (p. 314), notice that the air behind the occluded front is colder than the air ahead of it. This is known as a *cold-type occluded front*, or **cold occlusion**. Let's see how this front develops.

The development of a cold occlusion is shown in Fig. 12.20. Along line *A–A'*, the cold front is rapidly approaching the slower-moving warm front. Along line *B–B'*, the cold front overtakes the warm front, and, as we can see in the vertical view

across *C–C'*, lifts both the warm front and the warm air mass off the ground. As a cold-occluded front approaches, the weather sequence is similar to that of a warm front, with high clouds lowering and thickening into middle and low clouds, with precipitation forming well in advance of the surface front. Since the front represents a trough of low pressure, southeasterly winds and falling barometers occur ahead of it. The frontal passage, however, brings weather similar to that of a cold front: heavy, often showery precipitation with winds shifting to west or northwest. After a period of wet weather, the sky begins to clear, barometers rise, and the air turns colder. The most violent weather usually occurs where the cold front is just overtaking the warm front, at the point of occlusion, where the greatest contrast in temperature occurs. Cold occlusions are the most prevalent type of front that moves into the Pacific coastal states. Occluded fronts frequently form over the North Pacific and North Atlantic, as well as in the vicinity of the Great Lakes.

Continental polar air over eastern Washington and Oregon may be much colder than milder maritime polar air moving inland from the Pacific Ocean. Figure 12.21 illustrates this sit-

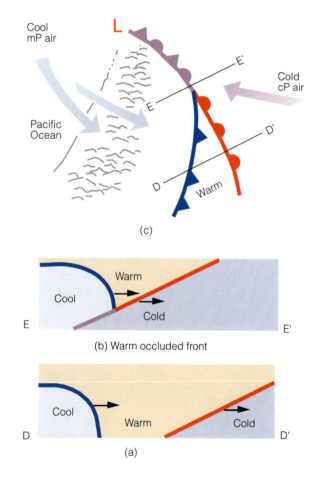

FIGURE 12.21

The formation of a warm-type occluded front. The faster-moving cold front in (a) overtakes the slower-moving warm front in (b). The lighter air behind the cold front rises up and over the denser air ahead of the warm front. Diagram (c) shows a surface map of the situation.

TABLE 12.4 Typical Weather Most Often Associated with Occluded Fronts

WEATHER ELEMENT	BEFORE PASSING	WHILE PASSING	AFTER PASSING
Winds	East, southeast, or south	Variable	West or northwest
Temperature			
Cold type	Cold or cool	Dropping	Colder
Warm type	Cold	Rising	Milder
Pressure	Usually falling	Low point	Usually rising
Clouds	In this order: Ci, Cs, As, Ns	Ns, sometimes Tcu and Cb	Ns, As, or scattered Cu
Precipitation	Light, moderate, or heavy precipitation	Light, moderate, or heavy continuous precipitation or showers	Light-to-moderate precipitation followed by general clearing
Visibility	Poor in precipitation	Poor in precipitation	Improving
Dew point	Steady	Usually slight drop, especially if cold-occluded	Slight drop, although may rise a bit if warm-occluded

uation. Observe that the air ahead of the warm front is colder than the air behind the cold front. Consequently, when the cold front catches up to and overtakes the warm front, the milder, lighter air behind the cold front is unable to lift the colder, heavier air off the ground. As a result, the cold front rides "piggyback" along the sloping warm front. This produces a *warm-type occluded front*, or a **warm occlusion**. The surface weather associated with a warm occlusion is similar to that of a warm front.

Contrast Fig. 12.20 and Fig. 12.21. Note that the primary difference between the warm- and cold-type occluded front is the location of the upper-level front. In a warm occlusion, the upper-level cold front *precedes* the surface occluded front, whereas in a cold occlusion the upper warm front follows the surface occluded front.

In the world of weather fronts, occluded fronts are the mavericks. In our discussion, we treated occluded fronts as forming when a cold front overtakes a warm front. Some may form in this manner, but others apparently form as new fronts, which develop when a surface storm intensifies in a region of cold air after its trailing cold and warm fronts have broken away and moved eastward. The new occluded front shows up on a surface chart as a trough of low pressure separating two cold air masses. Because of this, locating and defining occluded fronts at the surface is often difficult for the meteorologist. Similarly, you too may find it hard to recognize an occlusion. In spite of this, we will assume that the weather associated with occluded fronts behaves in a similar way to that shown in Table 12.4.

The frontal systems described in this chapter are actually part of a much larger storm system—the middle latitude cyclone. Chapter 13 details these storms, explaining where, why, and how they form.

SUMMARY

In this chapter, we considered the different types of air masses and the various weather each brings to a particular region. Continental arctic air masses are responsible for the extremely cold (arctic) outbreaks of winter, whereas continental polar air masses are responsible for cold, dry weather in winter and cool, pleasant weather in summer. Maritime polar air, having traveled over an ocean for a considerable distance, brings cool, moist weather to an area. The hot, dry weather of summer is associated with continental tropical air masses, while warm, humid conditions are due to maritime tropical air masses. Where air masses with sharply contrasting properties meet, we find weather fronts.

A front is a boundary between two air masses of different densities. Stationary fronts have essentially no movement, with cold air on one side and warm air on the other. Winds tend to blow parallel to the front, but in opposite directions on either side of it. Along the leading edge of a cold front, where colder air replaces warmer air, showers are prevalent, especially if the warmer air is moist and unstable. Along a warm front, warmer air rides up and over colder surface air, producing widespread cloudiness and precipitation that can cover thousands of square kilometers. Cold fronts typically move faster and are more steeply sloped than warm fronts. Occluded fronts, which are often difficult to locate and define on a surface weather map, may have characteristics of both cold and warm fronts.

Key Terms

The following terms are listed in the order they appear in the text. Define each. Doing so will aid you in reviewing the material covered in this chapter.

air mass
source regions (for air
    masses)
continental polar
    (air mass)
continental arctic
    (air mass)
lake-effect snows
maritime polar
    (air mass)
maritime tropical
    (air mass)
continental tropical
    (air mass)

air mass weather
front
stationary front
cold front
frontolysis
frontogenesis
"back door" cold front
cold air damming
warm front
overrunning
frontal inversion
occluded front (occlusion)
cold occlusion
warm occlusion

Questions for Review

1. If an area is described as a "good air mass source region," what information can you give about it?
2. It is summer. What type of afternoon weather would you expect from an air mass designated as mTk? Explain.
3. Why is cP air not welcome to the Central Plains in winter and yet very welcome in summer?
4. Explain why the central United States is not a good air mass source region.
5. Why do air temperatures tend to be a little higher on the eastern side of the Appalachian Mountains than on the western side, even though the same winter cP or cA air mass dominates both areas?
6. Explain how the airflow aloft regulates the movement of air masses.
7. List the temperature and moisture characteristics of each of the major air mass types.
8. What are lake-effect snows and how do they form?
9. Why are mP air masses along the East Coast of the United States usually colder than those along the nation's West Coast? Why are they also *less* prevalent?
10. Why are the boundaries between air masses more strongly developed in winter than in summer?
11. What type of air mass would be responsible for the weather conditions listed as follows?
    (a) heavy snow showers and low temperatures at Buffalo, New York
    (b) hot, muggy summer weather in the Midwest and the East
    (c) daily afternoon thunderstorms along the Gulf Coast
    (d) heavy snow showers along the western slope of the Rockies

    (e) refreshing, cool, dry breezes after a long summer hot spell on the Central Plains
    (f) heavy summer rainshowers in southern Arizona
    (g) drought with high temperatures over the Great Plains
    (h) persistent cold, damp weather with drizzle along the East Coast
    (i) summer afternoon thunderstorms forming along the eastern slopes of the Sierra Nevada
12. On a surface weather map, what do you know about a region where the word *frontogenesis* is marked?
13. Explain why barometric pressure usually falls with the approach of a cold front or occluded front.
14. Based on the following weather forecasts, what type of front will most likely pass the area?
    (a) Light rain and cold today, with temperatures just above freezing. Southeasterly winds shifting to westerly tonight. Turning colder with rain becoming heavy and possibly changing to snow.
    (b) Cool today with rain becoming heavy at times by this afternoon. Warmer tomorrow. Winds southeasterly becoming westerly by tomorrow morning.
    (c) Increasing cloudiness and warm today, with the possibility of showers by evening. Turning much colder tonight. Winds southwesterly, becoming gusty and shifting to northwesterly by tonight.
    (d) Increasing high cloudiness and cold this morning. Clouds increasing and lowering this afternoon, with a chance of snow or rain tonight. Precipitation ending tomorrow morning. Turning much warmer. Winds light easterly today, becoming southeasterly tonight and southwesterly tomorrow.
15. Describe the cloud types and patterns associated with a cold front, a warm front, and a cold-occluded front.
16. During the spring, on a warm, sunny day in Boston, Massachusetts, the wind shifts from southwesterly to northeasterly and the weather turns cold, damp, and overcast. What type of front moved through the Boston area? From what direction did the front apparently approach Boston?

Questions for Thought

1. Suppose an mP air mass moving eastward from the Pacific Ocean travels across the United States. Describe all of the modifications that could take place as this air mass moves eastward in winter. In summer.
2. Explain how an anticyclone during autumn can bring record-breaking low temperatures and cP air to the southeastern states, and only several days later very high temperatures and mT air.
3. In Fig. 12.4 (p. 306), there is a temperature inversion. How does this inversion differ from the frontal inversion illustrated in Fig. 12.19b (p. 318)?

4. When a very cold air mass covers half of the United States, a very warm air mass often covers the other half. Explain how this happens.

5. Explain why freezing rain more commonly occurs with warm fronts than with cold fronts.

6. In winter, cold-front weather is typically more violent than warm-front weather. Why? Explain why this is not necessarily true in summer.

7. When a cold front passes a station in the Northern Hemisphere, the wind shifts in a clockwise manner. How would the winds shift during the passage of a cold front in the Southern Hemisphere?

8. Why does the same cold front typically produce more rain over Kentucky than over western Kansas?

9. You are in Ithaca, New York and observe the wind shifting from East to South accompanied by a sudden rise in both the air temperature and dewpoint temperature. What type of front passed?

10. If Lake Erie freezes over in January, is it still possible to have lake-effect snows off Lake Erie in February? Why or why not?

Problems and Exercises

1. Make a sketch of North America and show the upper-air wind-flow pattern that would produce:
   (a) very cold cA air moving into the far western states in winter
   (b) cold cP air over the Central Plains in winter
   (c) warm mT air over the Midwest in winter
   (d) warm, moist mT air over southern California and Arizona during the summer

2. You are presently taking a weather observation. The sky is full of wispy cirrus clouds estimated to be about 6 km (20,000 ft) overhead. If a warm front is approaching from the south, about how far away is it (assuming a slope of 1:200)? If it is moving toward you at an average warm-front speed of about 10 knots, how long will it take before it passes your area?

Questions for Exploration

1. On the Blue Skies: College Edition CD-ROM go to the **Weather Analysis** section of the CD and click on "**Weather Maps.**"

2. On the Blue Skies: College Edition CD-ROM go to the **Weather Analysis** section of the CD and click on "**Find the Fronts.**"

A middle latitude cyclonic storm spins
counterclockwise over the eastern Atlantic.
(Photo:European Space Agency/
Science Photo Library.)

MIDDLE LATITUDE CYCLONES

CONTENTS

I t never for a moment occurred to me to regard this storm as a thing to prevent me from getting to New York. The Jackson Avenue station of the New Jersey Central was only five or six blocks from the house, and I anticipated not the slightest trouble in getting there. It was still snowing and blowing . . . and not a print had been made in the snow about the house. I got out to the gate and into the "street," and was then able to discover the real snow.

It was everywhere. Great piles of it rose up like gigantic arctic graves . . . in all directions. Every way that I turned I was confronted with these awful mounds. I took my bearings and steered for the Jackson Avenue station. Every step I took I went into my knees in snow and every other step I fell over on my face and tried to see how much of the stuff I could swallow. The wind was at my back and its accompanying snowflakes cut the back of my head and ears like a million icy lashes. I . . . plowed my way, jumping, falling, and crawling over the drifts, some of which were nine or ten feet high . . . and after an hour and ten minutes I got to the end of my six blocks. There were trains there, two of them, but they were stuck.

I gave up the idea of going to New York. My trip back to the house was simply awful. The wind was straight in my face and beat so in my eyes that I couldn't see a rod before me. My mustache was frozen stiff, and over my eyebrows were cakes of frozen snow. I stumbled along, falling down at almost every step, burying myself in the snow. Then, I began to feel like a crazy man. Every time I fell down, I shouted and cursed and beat the snow with my fists. Then it got dark, the wind howled and tore along, hurling the ice flakes in my face, and the very snow on the ground seemed to rise up and fling itself upon me.

In one of my crazy efforts to force ahead, I caught just a glimpse of the welcome gate posts, and then I laid down on my back and hollered. Somebody heard my cries, and just as I was going off comfortably to sleep my friend came plowing out through the snow, and he and this man dragged me into the house.

U.S. Department of Commerce, *American Weather Stories*

The storm described in our opening is now referred to as the "Blizzard of '88." This legendary storm of March, 1888, was accompanied by high winds of 60 miles per hour at Atlantic City, New Jersey, a severe cold wave, and unprecedented snowfall—up to 50 inches over portions of southeastern New York and southern New England, with drifts 30 to 40 feet high. In New York City, people died in the street, trapped in snowdrifts up to their hips. What atmospheric conditions are needed for such a monstrous storm to develop?

Early weather forecasters were aware that precipitation generally accompanied falling barometers and areas of low pressure. However, it was not until after the turn of this century that scientists began to piece together the information that yielded the ideas of modern meteorology and storm development.

Working largely from surface observations, a group of scientists in Bergen, Norway, developed a model explaining the life cycle of an extratropical storm; that is, a storm that forms at middle and high latitudes outside of the tropics. This extraordinary group of meteorologists included Vilhelm Bjerknes, his son Jakob, Halvor Solberg, and Tor Bergeron. They published their theory shortly after World War I. It was widely acclaimed and became known as the "polar front theory of a developing wave cyclone" or, simply, the **polar front theory**. What these meteorologists gave to the world was a working model of how a mid-latitude cyclone progresses through the stages of birth, growth, and decay. An important part of the model involved the development of weather along the polar front. As new information became available, the original work was modified, so that, today, it serves as a convenient way to describe the structure and weather associated with a migratory storm system.

POLAR FRONT THEORY

The development of a wave cyclone, according to the Norwegian model, begins along the polar front. You will remember (from our discussion of the general circulation in Chapter 11) that the polar front is a semicontinuous global boundary separating cold polar air from warm subtropical air. The stages of a developing wave cyclone are illustrated in the sequence of surface weather maps shown in Fig. 13.1.

Figure 13.1a shows a segment of the polar front as a stationary front. It represents a trough of lower pressure with higher pressure on both sides. Cold air to the north and warm air to the south flow parallel to the front, but in opposite directions. This type of flow sets up a cyclonic wind shear. You can conceptualize the shear more clearly if you place a pen between the palms of your hands and move your left hand toward your body; the pen turns counterclockwise, cyclonically.

Under the right conditions, a wavelike kink forms on the front, as shown in Fig. 13.1b. The wave that forms is known as a **frontal wave** or an *incipient cyclone*. Watching the formation of a frontal wave on a weather map is like watching a water wave from its side as it approaches a beach: It first builds, then breaks, and finally dissipates, which is why a cyclonic storm system is known as a **wave cyclone**. Figure 13.1b shows the newly formed wave with a cold front pushing southward and a warm front moving northward. The region of lowest pressure (called the *central pressure*) is at the junction of the two fronts. As the cold air displaces the warm air upward along the cold front, and as *overrunning* occurs ahead of the warm front, a narrow band of precipitation forms (shaded green area). Steered by the winds aloft, the system typically moves east or northeastward and gradually becomes a fully developed **open wave** in 12 to 24 hours (see Fig. 13.1c). The central pressure is now much lower, and several isobars encircle the wave's apex. These more tightly packed isobars create a stronger cyclonic flow, as the winds swirl counterclockwise and inward toward the low's center. Precipitation forms in a wide band *ahead* of the warm front and along a narrow band of the cold front. The region of warm air between the cold and warm fronts is known as the **warm sector**. Here, the weather tends to be partly cloudy, although scattered showers may develop if the air is unstable.

Energy for the storm is derived from several sources. As the air masses try to attain equilibrium, warm air rises and cold air sinks, transforming potential energy into kinetic energy—energy of motion. Condensation supplies energy to the system in the form of latent heat. And, as the surface air converges toward the low's center, wind speeds may increase, producing an increase in kinetic energy.

As the open wave moves eastward, its central pressure continues to decrease, and the winds blow more vigorously. The faster-moving cold front constantly inches closer to the warm front, squeezing the warm sector into a smaller area, as shown in Fig. 13.1d. Eventually, the cold front overtakes the warm front and the system becomes occluded. At this point, the storm is usually most intense, with clouds and precipitation covering a large area.

The point of occlusion where the cold front, warm front, and occluded front all come together in Fig. 13.1e is referred to as the *triple point*. Notice that in this region the cold and warm fronts appear similar to the open-wave cyclone in Fig. 13.1c. It is here where a new wave (called a **secondary low**) will occasionally form, move eastward, and intensify into a cyclonic storm. The center of the intense storm system shown in Fig. 13.1e gradually dissipates, because cold air now lies on both sides of the occluded front. The warm sector is still present, but is far removed from the center of the storm. Without the supply of energy provided by the rising warm, moist air, the old storm system dies out and gradually disappears (see Fig. 13.1f). We can think of the sequence of a developing wave cyclone as a whirling eddy in a stream of water that forms behind an obstacle, moves with the flow, and gradually vanishes downstream. The entire life cycle of a wave cyclone can last from a few days to over a week.

Figure 13.2 shows a series of wave cyclones at various stages of development along the polar front in winter. Such a succession of storms is known as a *"family" of cyclones*. Observe that to the north of the front are cold anticyclones; to the south over the Atlantic Ocean is the warm, semipermanent Bermuda

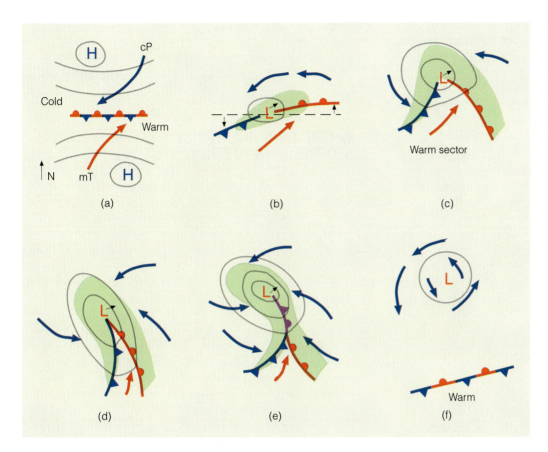

(a)   (b)   (c)

(d)   (e)   (f)

FIGURE 13.1
The idealized life cycle of a wave cyclone (a through f) in the Northern Hemisphere based on the polar front theory. As the life cycle progresses, the system moves eastward in a dynamic fashion. The small arrow next to each L shows the direction of storm movement.

FIGURE 13.2
A series of wave cyclones (a "family") of cyclones) forming along the polar front.

WEATHER WATCH

The winds of a mid-latitude cyclonic storm can be quite destructive. For example, a deep low-pressure area with winds gusting to 130 knots during January, 1921, destroyed 8 times as many Douglas fir trees in Washington and Oregon as did the great volcanic explosion of Mount St. Helens in May, 1980.

high. The polar front itself has developed into a series of loops, and at the apex of each loop is a cyclone. The cyclone over the northern plains (Low 1) is just forming; the one along the East Coast (Low 2) is an open wave; and the system near Iceland (Low 3) is dying out. If the average rate of movement of a wave cyclone from birth to decay is 25 knots, then it is entirely possible for a storm to develop over the central part of the United States, intensify into a large storm over New England, become occluded over the ocean, and reach the coast of England in its dissipating stage less than a week after it formed.

Up to now, we have considered the polar front model of a developing wave cyclone, which represents a rather simpli-

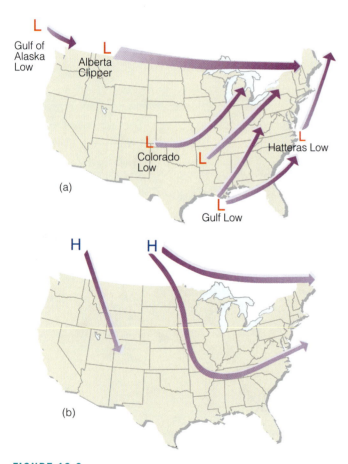

FIGURE 13.3
(a) Typical paths of winter mid-latitude cyclones. The lows are named after the region where they form. (b) Typical paths of winter anticyclones.

fied version of the stages that an extratropical storm system must go through. In fact, few (if any) storms adhere to the model exactly. Nevertheless, it serves as a good foundation for understanding the structure of storms. So keep the model in mind as you read the following sections.

WHERE DO MID-LATITUDE STORMS TEND TO FORM?

Any development or strengthening of a cyclone is called **cyclogenesis**. Within the United States, there are regions that show a propensity for cyclogenesis, including the eastern slopes of the Rockies, the Great Basin, the Gulf of Mexico, and the Atlantic Ocean east of the Carolinas. Near Cape Hatteras, North Carolina, for example, warm Gulf Stream water can supply moisture and warmth to the region south of a stationary front, thus increasing the contrast between air masses to a point where storms may suddenly spring up along the front. These cyclones, called *nor'easters,* normally move northeastward along the Atlantic Coast, bringing high winds and heavy snow or rain to coastal areas. Before the age of modern communications and weather prediction, such coastal storms would often go undetected during their formative stages; and sometimes an evening weather forecast of "fair and colder" along the eastern seaboard would have to be changed to "heavy snowfall" by morning. Fortunately, with today's weather information gathering and forecasting techniques, these storms rarely strike by surprise.

Figure 13.3 shows the typical paths taken in winter by mid-latitude cyclones and anticyclones. Notice in Fig. 13.3a that some of the lows are named after the region where they form, such as the *Hatteras Low* which develops off the coast near Cape Hatteras, North Carolina. The *Alberta Clipper* forms (or redevelops) on the eastern side of the Rockies in Alberta, Canada, then rapidly skirts across the northern tier states. The *Colorado Low*, in contrast, forms (or redevelops) on the leeward side of the Rockies, then usually moves eastward and northeastward toward the Great Lakes.

When extratropical cyclones deepen rapidly (in excess of 24 mb in 24 hours), the term *explosive cyclogenesis*, or "bomb," is sometimes used to describe them. As an example, explosive cyclogenesis occurred in a storm that developed over the warm Atlantic just east of New Jersey on September 10, 1978. As the central pressure of the storm dropped nearly 60 mb (1.8 in.) in 24 hours, hurricane force winds battered the ocean liner *Queen Elizabeth II* and sank the fishing vessel *Captain Cosmo.*

Frontal waves that develop into huge storms are called *unstable waves.* These waves form suddenly, grow in size, and then slowly dissipate with the entire process taking several days to a week. Other waves, called *stable waves*, remain small and never grow into giant weather producers. Why is it that some waves develop into huge storms, whereas others simply dissipate in a day or so?

This question poses one of the real challenges in weather forecasting. The answer is complex. Indeed, there are many surface conditions that do influence the formation of a wave, including mountain ranges and land-ocean temperature contrasts. However, the real key to the development of a wave cyclone is found in the *upper-wind flow*, in the region of the high-level westerlies. Therefore, before we can arrive at a reasonable answer to our question, we need to see how the winds aloft influence surface pressure systems.

VERTICAL STRUCTURE OF DEEP-PRESSURE SYSTEMS

In Chapter 10, we learned that thermal pressure systems are shallow and weaken with increasing elevation. On the other hand, developing surface storm systems are deep lows that usually intensify with height. This circumstance means that a surface low-pressure area will appear on an upper-level chart as either a closed low or a trough.

Suppose the upper-level low is directly above the surface low (see Fig. 13.4). Notice that only at the surface (because of friction) do the winds blow inward toward the low's center. As these winds converge (flow together), the air "piles up." This piling up of air, called **convergence**, causes air density to increase directly above the surface low. This increase in mass causes surface pressures to rise; gradually, the low fills and the surface low dissipates. The same reasoning can be applied to surface anticyclones. Winds blow outward, away from the center of a surface high. If a closed high or ridge lies directly over the surface anticyclone, **divergence** (the spreading out of air) at the surface will remove air from the column directly above the high. Surface pressures fall and the system weakens. Consequently, it appears that, if upper-level pressure systems were always located directly above those at the surface (such as shown in Fig. 13.4), cyclones and anticyclones would die out soon after they form (if they could form at all). What, then, is it that allows these systems to develop and intensify?

We know from Chapter 9 that lows aloft are generally associated with cold air. Note in Fig. 13.5 that behind the cold front there is cold air both at the surface and aloft. Consequently, the upper-level low is located *behind*, or to the *west* of, the surface low. Observe how the surface low tilts toward the northwest as we move up from the surface, showing up as a closed system at 500 mb and as a trough at 300 mb. Directly above the surface low, at 300 mb, the air spreads out and diverges (as indicated by the contour lines). This allows the converging surface air to rise and flow out of the top of the air column just below the tropopause, which acts as a constraint to vertical motions. We now have a mechanism for developing storms. *When upper-level divergence is stronger than surface convergence (more air is taken out at the top than is brought in at the bottom), surface pressures drop, and the low intensifies, or deepens. By the same token, when upper-level divergence is less than surface convergence (more air flows in at the bottom than is removed at the top), surface pressures rise, and the system weakens.*

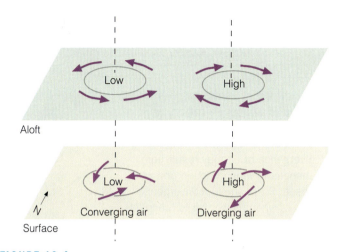

FIGURE 13.4
If lows and highs aloft were always directly above lows and highs at the surface, the surface systems would quickly dissipate.

We can also use Fig. 13.5 to explain the structure of the anticyclone. High pressures aloft are associated with warm air. At the surface and aloft, warm air lies to the southwest of the surface high. This causes the surface anticyclone to tilt toward the southwest at higher altitudes. Moving upward, we observe that the closed system becomes a ridge at the 300-mb level.

FIGURE 13.5
The vertical structure of cyclones and anticyclones.

A Closer Look at Convergence and Divergence

We know that *convergence* is the piling up of air above a region, while *divergence* is the spreading out of air above some region. Convergence and divergence of air may result from changes in wind direction and wind speed. For example, convergence occurs when moving air is funneled into an area, much in the way cars converge when they enter a crowded freeway. Divergence occurs when moving air spreads apart, much as cars spread out when a congested two-lane freeway becomes three lanes. On an upper-level chart, this type of convergence (also called *confluence*) occurs when contour lines move closer together, as a steady wind flows parallel to them (see Fig. 1). On the same chart, this type of divergence (also called *diffluence*) occurs when the

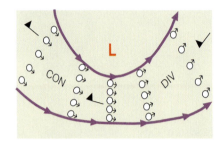

FIGURE 1

The formation of convergence (CON) and divergence (DIV) of air with a constant wind speed (indicated by flags). Circles represent air parcels that are moving parallel to the contour lines on a constant pressure chart.

contour lines move apart as a steady wind flows parallel to them. (In Fig. 13.5, observe the contour lines and the regions of convergence and divergence on the 300-mb chart.)

Convergence and divergence may also result from changes in wind speed. Speed convergence occurs when the wind slows down as it moves along, whereas speed divergence occurs when the wind speeds up. We can grasp these relationships more clearly if we imagine air molecules to be marching in a band. When the marchers in front slow down, the rest of the band members squeeze together, causing convergence; when the marchers in front start to run, the band members spread apart, or diverge. In summary, *speed convergence* takes place when the wind speed decreases downwind and *speed divergence* takes place when the wind speed increases downwind.

WEATHER WATCH

Mid-latitude storms have been known to change the course of history. One such storm, in March of 1776, forced the British to cancel a massive amphibious assault at Dorchester Heights, south of Boston, where General Washington held a well-fortified position. The storm kept Dorchester Heights from becoming a bloody, historic battlefield, such as Lexington, Concord, and Bunker Hill.

Also notice that directly above the surface high at 300 mb there is convergence of air (as indicated by the contour lines). Convergence causes an accumulation of air above the high, which allows the air to sink slowly and replace the diverging surface air. Hence, *when upper-level convergence exceeds low-level divergence (inflow at top is greater than outflow near the surface), surface pressures rise, and the anticyclone builds.* On the other hand, *when upper-level convergence is less than low-level divergence, the anticyclone weakens as surface pressures fall.* (Additional information on the subject of convergence and divergence is provided in the Focus section above.)

Look at the wind direction at the 500-mb level in Fig. 13.5. Winds at this altitude tend to steer surface systems in

the same direction that the winds are moving. Thus, the surface storm will move toward the northeast, while the surface anticyclone will move toward the southeast. As we can see in Fig. 13.3, these paths indicate the average movement of surface pressure systems in the eastern two-thirds of the United States. In general, the surface storm centers travel across the United States at about 16 knots in summer and about 27 knots in winter. The faster winter velocity reflects the stronger upper-level flow during this time of year.*

So far, we have seen that deep pressure systems exist at the surface and aloft throughout much of the troposphere. When the surface pressure system does not lie directly beneath the upper-level trough but tilts toward the west, the atmosphere is able to redistribute its mass. Regions of low-level convergence are compensated for by regions of upper-level divergence and vice versa. Cyclones and anticyclones can intensify and, steered by the winds aloft, move away from their region of formation.

Since regions of strong upper-level divergence and convergence typically occur when deep troughs and ridges—waves—exist in the flow aloft, the next section examines these waves and their influence on a developing storm.

---

*As a forecasting rule of thumb, surface pressure systems tend to move in the same direction as the wind at the 500-mb level. The speed at which the surface systems move is about half the speed of the 500-mb winds.

UPPER-LEVEL WAVES AND SURFACE STORMS

You may remember from the "dish-pan" experiment (found in the Focus section on p. 284) that, aloft, waves are a fundamental feature of an unevenly heated, rotating sphere, such as the earth. If we examine an upper-level chart that shows almost the entire Northern (or Southern) Hemisphere, such as Fig. 13.6, the waves appear as a series of troughs and ridges with significant amplitude that encircle the globe. The distance from trough to trough (or ridge to ridge), which is normally thousands of kilometers, is known as a **longwave**. At any one time there are usually between three and six of these waves looping around the earth. The fewer the number of waves, the longer their wavelengths. Since mountain ranges tend to disturb the upper-level wind flow, these waves are often found to the east of such topographic barriers as the Rockies and Tibetan Plateau. Sometimes the waves exhibit a relatively small amplitude (or north-to-south extent) and the flow is mainly zonal, or west to east. On other occasions, the waves exhibit considerable amplitude and the flow has a strong north-south (meridional) component to it.

Longwaves are also known as *planetary waves* and as **Rossby waves**, after C. G. Rossby, a famous meteorologist who carefully studied their motion. Imbedded in longwaves are **shortwaves**, which are small disturbances, or ripples (see Fig. 13.7a). Rossby found that the shorter the wavelength of a particular wave, the faster it moved downstream. Shortwaves tend to move eastward at a speed proportional to the average wind flow near the 700-mb level, about 3 km above sea level. Longwaves, on the other hand, often remain stationary, move

FIGURE 13.6

A 500-mb map of the Northern Hemisphere from a polar perspective shows five longwaves encircling the globe. Note that the wavelength of wave number 3 is greater than the width of the United States. Solid lines are contours. Dashed lines show the position of longwave troughs.

(a) DAY 1

(b) DAY 2 (24 hours later)

FIGURE 13.7

(a) Upper-air chart showing a longwave with three shortwaves (heavy dashed lines) embedded in the flow. (b) Twenty-four hours later the shortwaves have moved rapidly around the longwave. Notice that the shortwaves labeled 1 and 3 tend to deepen the longwave trough, while shortwave 2 has weakened as it moves into a ridge. Notice also that as the longwave deepens in diagram (b), its length actually shortens. Dashed lines are isotherms in °C. Solid lines are contours. Blue arrows indicate cold advection and red arrows warm advection.

eastward very slowly at less than 4° of longitude per day (about 8 knots), or even move westward (*retrograde*).* We can obtain a better idea of this wave movement if we think of longwaves as being huge meanders (loops) in a swiftly flowing stream of water. Water moves through the loops quickly, while the loops themselves move eastward very slowly, as the fast-flowing water cuts away at one bank and deposits material on the other. Suppose debris tumbles into the stream, disturbing the flow. The disturbed flow appears as a small wrinkle that travels downstream through the loops at a speed near the average stream flow. This wrinkle in the flow is analogous to a shortwave in the atmosphere.

Notice in Fig. 13.7b that, while the longwaves move eastward very slowly, the shortwaves move fairly quickly around the longwaves. Notice also that the shortwaves tend to deepen when they approach a longwave trough and weaken when they approach a ridge. Moreover, when a shortwave moves into a longwave trough, the trough tends to deepen. (Look at shortwave 3 in Fig. 13.7b.)

Where the contour lines in Fig. 13.7b are roughly parallel to the isotherms (dashed lines), the atmosphere is said to be **barotropic**. Since the winds at this level more or less parallel the contour lines, in a barotropic atmosphere the winds blow parallel to the isotherms. By comparison, where the isotherms *cross* the contour lines, temperature advection occurs and the atmosphere is said to be **baroclinic**.† Notice that the baroclinic region tends to be in a narrow zone in the vicinity of shortwaves 1 and 3. The shortwaves actually disturb the flow and accentuate the region of baroclinicity.

In the region of baroclinicity, winds cross the isotherms and produce *temperature advection*. **Cold advection** (or *cold air advection*) is the transport of cold air by the wind from a region of lower (colder) temperatures to a region of higher (warmer) temperatures. In the region of cold advection, the air temperature normally decreases. On the other hand, **warm advection** (or *warm air advection*) is the transport of warm air by the wind from a region of higher (warmer) temperatures to a region of lower (colder) temperatures. In the region of warm advection, the air temperature normally increases. For cold advection to occur, the wind must blow across the isotherms from colder to warmer regions, whereas for warm advection, the wind must blow across the isotherms from warmer to colder regions.

In the baroclinic region in Fig. 13.7b, observe that strong winds cross the isotherms, producing cold advection (blue arrows) on the trough's west side and warm advection (red arrows) on its east side. Below the baroclinic zone lies the polar front; above it flows the polar-front jet stream. The dis-

turbed flow created by the shortwaves is now capable of aiding in the development or intensification of a surface storm system. The theory explaining how this phenomenon occurs is known as the *baroclinic wave theory of developing cyclones.*

THE NECESSARY INGREDIENTS FOR A DEVELOPING WAVE CYCLONE

To better understand why wave cyclones develop and intensify into huge storms, we need to examine atmospheric conditions at the surface and aloft. Suppose that a portion of a longwave trough at the 500-mb level lies directly above a surface stationary front (see Fig. 13.8a). On the 500-mb chart, contour lines (solid lines) and isotherms (dashed lines) parallel each other and are crowded close together. Colder air is located in the northern half of the map, while warmer air is located to the south. Winds are blowing at fairly high velocities, which produce a sharp change in wind speed—a strong *wind speed shear*—from the surface up to this level. Suppose a shortwave moves through this region, disturbing the flow as illustrated in Fig. 13.8b. This sets up a kind of instability in the flow (as warmer air rises and colder air sinks) known as **baroclinic instability**.

Upper-Air Support   With the onset of baroclinic instability, horizontal and vertical air motions begin to enhance the formation of cyclonic storms. For example, as the flow aloft becomes disturbed, it begins to lend support for the intensification of surface pressure systems, as a region of converging air forms above position 1 in Fig. 13.8b and a region of diverging air forms above position 2. (Glance back at these two positions on the 300-mb chart in Fig. 13.5, p. 329.) The converging air aloft causes the surface air pressure to rise in the region marked *H* in Fig. 13.8b. Surface winds begin to blow out away from the region of higher pressure, and the air aloft gradually sinks to replace it. Meanwhile, diverging air aloft causes the surface air pressure to decrease beneath position 2, in the region marked *L* on the surface map. This initiates rising air, as the surface winds blow in toward the region of lower pressure.* As the converging surface air develops cyclonic spin, cold air flows southward and warm air northward. We can see in Fig. 13.8b that the western half of the stationary front is now a cold front and the eastern half a warm front. Cold air moves in behind the cold front, while warm air slides up along the warm front. These regions of cold and warm advection occur all the way up to the 500-mb level.

On the 500-mb chart in Fig. 13.8b, cold advection is occurring at position 1 as the wind crosses the isotherms,

---

*Retrograde wave motion means that the wave is actually moving in the opposite direction of the wind flow.

†Actually, on a constant pressure surface, baroclinic conditions exist where the air density varies, and barotropic conditions exist where the air density does not vary.

---

*It is interesting to note that even if initially there are no fronts on the surface map, they may begin to form where air masses having contrasting properties are brought together in the region where the surface air rises and the surrounding air flows inward.

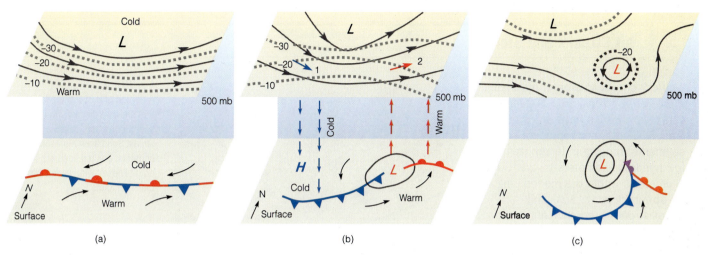

FIGURE 13.8
The formation of a wave cyclone during baroclinic instability. (a) A longwave trough at 500 mb lies parallel to and directly above the surface stationary front. (b) A shortwave (not shown) disturbs the flow aloft, which initiates temperature advection. The upper trough intensifies and provides the necessary vertical motions for the development of the surface cyclone. (c) The surface storm occludes, and a cold pool of air remains above it.

bringing cold air into the trough. The cold advection makes the air more dense and lowers the height of the air column from the surface up to the 500-mb level. (Recall that, on a 500-mb chart, lower heights mean the same as lower pressures; consequently, the pressure in the trough lowers and the trough deepens.) Meanwhile, at position 2, warm advection is taking place, which has the effect of raising the height of a column of air; here, the 500-mb heights increase and a ridge builds (strengthens). Therefore, *the overall effect of differential temperature advection is to intensify the wave.*

Regions of cold and warm advection are associated with vertical motions. Where there is cold advection, some of the cold, heavy air sinks; where there is warm advection, some of the warm, light air rises. Hence, due to advection, air must be sinking in the vicinity of position 1 and rising in the vicinity of position 2. The sinking of cold air and the rising of warm air provide energy for a developing cyclone, as potential energy is transformed into kinetic energy. Further, condensation in the ascending air releases latent heat, which strengthens the system even more. So, we now have a full-fledged middle latitude cyclone with all of the necessary ingredients for its development.

Eventually, the storm system occludes (see Fig. 13.8c).* Although some storms may continue to deepen (especially when the center of the upper low is not directly over the surface low), most storms begin to weaken at this point, as the supply of warm surface air is cut off and cold dry air behind the cold front is drawn in toward the center of the surface low.

_____
*If the occluded front should extend west of the low's center, it is sometimes referred to as a *bentback occlusion.*

Often, an upper-level pool of cold air (which has broken away from the main flow) lies almost directly above the surface low. Sometimes the upper low will break away entirely from the main flow, producing a **cut-off low**, which often appears as a single contour line on an upper-level chart. When the upper low lies directly above the surface low (as in Fig. 13.8c), the storm system is said to be *vertically stacked*. Usually the isotherms around the upper low parallel the contour lines, which indicates that no significant temperature advection is occurring. Without the necessary energy transformations, the surface system gradually dissipates. As its winds slacken and its central pressure gradually rises, the low is said to be *filling*. The upper-level low, however, may remain stationary for many days. If air is forced to ascend into this cold pocket, widespread clouds and precipitation may persist for some time, even though the surface storm system itself has moved east out of the picture.

The Role of the Jet Stream   Jet streams play an additional role in the development of surface cyclones (and anticyclones). In Chapter 11 we learned that the axis of the polar front jet stream pretty much coincides with the position of the polar front. We also learned that the region of strongest winds in the core of the jet is known as a *jet maximum* or *jet streak*. When the polar jet stream flows in a wavy west-to-east pattern (as illustrated in Fig. 13.9a), a jet streak tends to form in the trough of the jet, where pressure gradients are tight. The curving of the jet stream coupled with the changing wind speeds around the jet streak produces regions of strong convergence and divergence of air along the flanks of the jet. (More on this topic is given in the Focus section on p. 334.)

Jet Streaks and Storms

Figure 2 shows an area of maximum winds, a *jet streak*, on a 300-mb chart. Jet streaks have winds of at least 50 knots, and represent small segments (ranging in length from a few hundred kilometers to over 3000 kilometers) within the meandering jet stream flow.

Jet streaks are important in the development of surface storms because areas of convergence and divergence form at specific regions around them. To understand why, consider air moving through a straight jet streak (shaded area) in Fig. 3. As the air enters the front of the streak (known as the *entrance region*), it increases in speed; as it leaves the rear of the streak (known as the *exit region*), it decreases in speed. At this elevation in the atmosphere (about 10,000 m or 33,000 ft above the surface), the wind flow is nearly in geostrophic balance with the pressure gradient force (directed north) and the Coriolis force (directed south). As the air enters the jet streak, it increases in speed due to an increase in the pressure gradient force. The greater force temporarily exceeds the Coriolis force, and the air swings slightly to the north across the contour lines, which causes strong convergence of air at point 1 and weak divergence at point 2. Toward the middle of the jet streak, the increase in wind speed causes the Coriolis force to increase and the wind to become

nearly geostrophic again. However, as the air exits the jet streak, the pressure gradient force is reduced and the Coriolis force temporarily exceeds it, causing the air to swing slightly to the south. This process produces weak convergence at point 4 and strong divergence at point 3.

The conditions described so far exist in a straight jet streak that shows no curvature. When the jet stream becomes wavy, and the jet streak exhibits cyclonic curvature (as it does in Fig. 2), the areas of weak divergence (at point 2) and weak convergence (at point 4) all but disappear. What we are left with is a curving jet streak that exhibits strong divergence at point 3 (in the left exit region) and strong convergence at point 1 (in the left entrance region). Now look closely at Fig. 13.9a, p. 335. The curving nature of

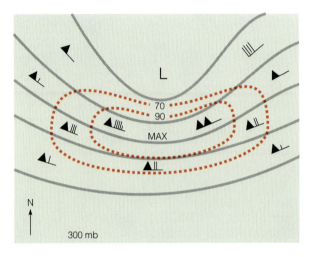

FIGURE 2
A portion of a 300-mb chart that shows a distinct wind speed maximum, called a jet streak. Solid lines represent contours. Dashed lines are equal lines of wind speed (isotachs) in knots.

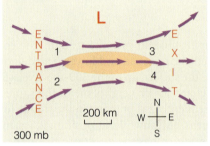

FIGURE 3
Changing air motions within a straight jet streak (shaded area) cause strong convergence of air at point 1 (left entrance region) and strong divergence at point 3 (left exit region).

the jet stream forms areas of divergence (marked D) and convergence (marked C) around its maximum winds. At the surface beneath the area of strong divergence is the deepening storm, while beneath the area of strong convergence is the building anticyclone.

Notice that the region of diverging air (marked *D* in Fig. 13.9a) draws warm surface air upward to the jet, which quickly sweeps the air downstream. Since the air above the storm is being removed more quickly than converging surface winds can supply air to the storm's center, the central pressure of the storm drops rapidly. As surface pressure gradients increase, the wind speed increases. Above the high-pressure area, a region of converging air (marked *C* in Fig. 13.9a) feeds air downward into the anticyclone. Hence, we find *the polar jet stream removing air above the surface cyclone and supplying air to the surface anticyclone.*

As the jet stream steers the storm along—toward the northeast in this case—the surface cold front catches up to the warm front and the storm occludes (see Fig. 13.9b). Since the surface low has moved out from under the pocket of diverging air aloft, the occluded storm gradually fills as surface air flows into the system.

Since the polar jet stream is strongest and moves farther south in winter, we can see why mid-latitude cyclonic storms are better developed and move more quickly during the colder months. During the summer when the polar jet shifts northward, developing mid-latitude storm activity shifts northward and occurs principally over the Canadian provinces of Alberta and the Northwest Territories.

In general, we now have a fairly good picture as to why some surface wave cyclones intensify into huge storms while others do not. For a storm to intensify, there must be an upper-level counterpart—a trough of low pressure—that lies to the *west* of the surface low. As shortwaves disturb the flow aloft, they cause regions of differential temperature advection to appear, leading to an intensification of the upper-level trough. At the same time, the polar jet forms into waves and swings slightly south of the developing storm. When these conditions exist, zones of converging and diverging air, along with rising and sinking air, provide energy conversions for the storm's growth. With this atmospheric situation, storms may form even where there are no pre-existing fronts. In regions where the upper-level flow is not disturbed by shortwaves or where no upper trough or jet stream exists, the necessary vertical and horizontal motions are insufficient to enhance cyclonic storm development and we say that the surface storm does not have the proper *upper-air support*. The horizontal and vertical motions, cloud patterns, and weather that typically occur with a developing open-wave cyclone are summarized in Fig. 13.10.

The Conveyor Belt Model of Rising and Descending Air

A three-dimensional model of a developing wave cyclone, especially for snowstorms that develop along the east coast of the United States, is illustrated in Fig. 13.11. The model describes rising and sinking air as traveling along three main "conveyor belts." According to the **conveyor belt model**, a warm air stream (known as a *warm conveyor belt*, red arrow in Fig. 13.11) originates at the surface in the warm sector, ahead of the cold front. As the warm airstream moves northward, it slowly rises along the sloping warm front. Water vapor in the rising air condenses, and clouds form. Aloft, the warm air flow gradually turns toward the northeast or east, parallel to the upper-level flow. The rising warm air causes cloudiness to appear well out ahead of the surface low and its surface warm front. From these clouds, rain or snow usually fall. Directly below the warm conveyor belt, a cold, moist airstream—the *cold conveyor belt*, blue arrow in Fig. 13.11—moves slowly westward from off the ocean, north of the surface low and the warm front. As the airstream moves into the vicinity of the surface low, rising air gradually forces the cold conveyor belt upward. The rising airstream usually turns as it ascends, producing a comma-shaped cloud similar to the one shown in Fig. 13.12. If the airstream rises high enough, it gets caught in the southwesterly flow aloft and swings northeastward. The last conveyor belt is a dry one that forms in the upper atmosphere. Called the *dry conveyor belt*, and shaded yellow in Fig. 13.11, this airstream slowly descends from the northwest behind the surface cold front, where it brings generally clearing weather. If a branch of this dry air sweeps into the storm, it produces a clear area that, in the comma cloud, appears to pinch off the comma's head from its tail. This

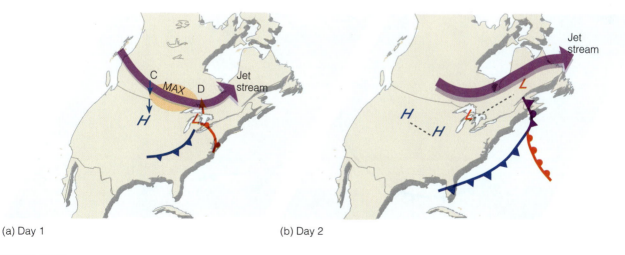

(a) Day 1 (b) Day 2

FIGURE 13.9

(a) As the polar jet stream and its area of maximum winds (the jet streak, or MAX), swings over a developing mid-latitude cyclone, an area of divergence (D) draws warm surface air upward and an area of convergence (C) allows cold air to sink. The jet stream, removing air above the surface storm and supplying air to the surface anticyclone, allows the two surface pressure systems to develop. When the surface storm moves northeastward and occludes (b), it no longer has the upper-level support of the jet stream, and the surface storm gradually dies out.

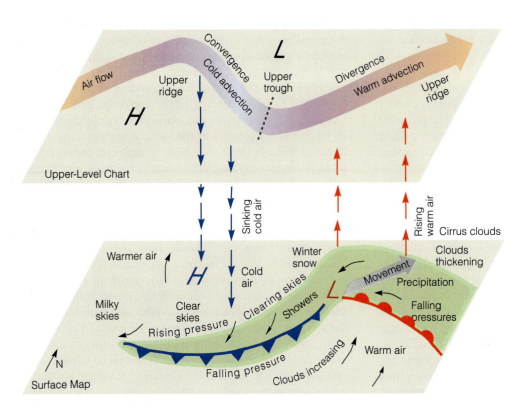

FIGURE 13.10
Summary of clouds, weather, and vertical motions associated with a developing wave cyclone.

FIGURE 13.11
The conveyor belt model of a developing wave cyclone. Cloud pattern for the storm would be in the shape of a comma. (After Kocin and Uccellini.)

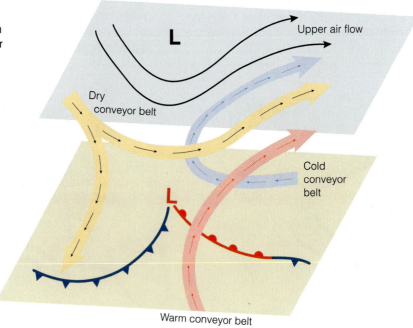

phenomenon tends to show up on satellite pictures as the storm becomes more fully developed.

We are now in a position to tie together many of the concepts we have learned about developing mid-latitude cyclones by examining a monstrous storm that formed during March, 1993.

A Developing Wave-Cyclone—The March Storm of 1993

A color-enhanced infrared satellite picture of a developing wave cyclone on the morning of March 13, 1993, is shown in Fig. 13.12. Notice that its cloud band is in the shape of a comma that covers the entire eastern seaboard. Such **comma clouds** indicate that the storm is still developing and intensifying. But this storm is not an ordinary wave cyclone—this storm intensified into a superstorm, which some forecasters dubbed "the storm of the century."

The surface weather map for the morning of March 13 (see Fig. 13.13) shows that the center of the open wave is over northern Florida. Observe in Fig. 13.12 that this position is in the head of the comma cloud. The central pressure of the storm is 975 mb (28.79 in.), which indicates an incredibly deep system, considering a typical open wave would have a central pressure closer to 996 mb (29.41 in.). A strong cold front stretches from the storm's center through western Florida. Behind the front, cold arctic air pours into the deep south. Ahead of the advancing front, a band of heavy thunderstorms (along a squall line) is pounding Florida with heavy rain, high winds, and tornadoes. Warm humid air in the warm sector is streaming northward, overrunning cold surface air ahead of the warm front, which is causing precipitation in the form of rain, snow, and sleet to fall over a vast area extending from Florida to New York.

The 500-mb chart for the morning of March 13 (see Fig. 13.14) shows that a deep trough extending southward out of Canada lies to the west of the surface low. A baroclinic atmosphere exists around the trough as isotherms intersect contour lines and strong winds produce differential temperature advection. Notice that warm advection is occurring on the trough's eastern side, ahead of the surface warm front shown in Fig. 13.13, while cold advection is occurring on the trough's western side, behind the position of the surface cold front. As temperature advection deepens the trough, rising and sinking air provide energy for the developing surface storm.

FIGURE 13.12
A color-enhanced infrared satellite picture that shows a developing wave cyclone at 2 A.M. (EST) on March 13, 1993. The darkest shades represent clouds with the coldest and highest tops. The dark cloud band moving through Florida represents a line of severe thunderstorms. Notice that the cloud pattern is in the shape of a comma.

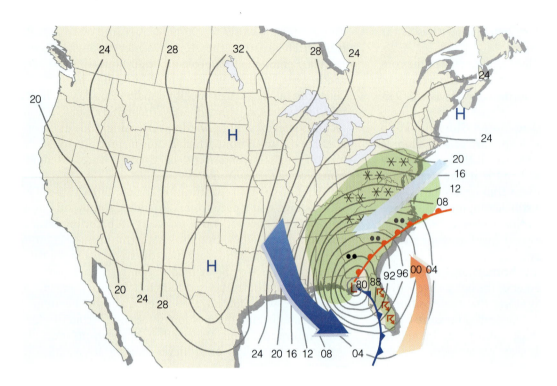

FIGURE 13.13
Surface weather map for 4 A.M. (EST) on March 13, 1993. Lines on the map are isobars. A reading of 96 is 996 mb and a reading of 00 is 1000 mb. (To obtain the proper pressure in millibars, place a 9 before those readings of 96 or lower, and place a 10 before those readings of 00 or higher.) Green shaded areas are receiving precipitation. (The large orange arrow represents warm humid air and the warm conveyor belt. The light blue arrow represents cold moist air and the cold conveyor belt; the dark blue arrow represents cold dry air and the dry conveyor belt.)

	Miles (statute) per hour	Knots
◎	Calm	Calm
—	1–2	1–2
⌐	3–8	3–7
└	9–14	8–12
⊢	15–20	13–17
⊦	21–25	18–22
⊨	26–31	23–27
⊫	32–37	28–32
⫴	38–43	33–37
⫵	44–49	38–42
⫶	50–54	43–47
◣	55–60	48–52
◤	61–66	53–57
◥	67–71	58–62
◸	72–77	63–67
◹	78–83	68–72
◺	84–89	73–77
◿	119–123	103–107

FIGURE 13.14
The 500-mb chart for 7 A.M. (EST) March 13, 1993. Solid lines are contours where 564 equals 5640 meters. Dashed lines are isotherms in °C. Wind entries in red show warm advection. Those in blue show cold advection. Those in gray indicate no appreciable temperature advection is occurring.

In Fig. 13.15, we can see that the storm began on March 12 as a frontal wave off the Texas coast. In the upper air, a short-wave, moving rapidly around a longwave, disturbed the flow, setting up the necessary ingredients for the surface storm's development. By the morning of March 13, the storm had intensified into a deep open-wave cyclone centered over Florida. In the upper air, a region of divergence positioned above the storm caused the storm's surface pressures to drop rapidly. Upper-level southwesterly winds (Fig. 13.14) directed the surface low northeastward, where it became occluded over Virginia during the afternoon of March 13. At this point, the storm's central pressure dropped to an incredibly low 960 mb (28.35 in.), a pressure comparable to a category 3 hurricane.* Although the surface winds were quite strong and gusty, they were not as strong as those in a category 3 hurricane because the isobars around the storm were spread farther apart than those in a hurricane and because surface friction slowed the winds. Higher up, away from the influence of the surface, the winds were much stronger as a wind of 125 knots was reported at the top of 1900-meter-high Mount Washington, New Hampshire.

The upper trough remained to the west of the surface low, and the storm continued its northeastward movement. In Fig. 13.15, we can see that by the morning of March 14 the storm (which was now a deep, bentback occluded system) had weakened slightly and was centered along the coast of Maine. Moving out from under its area of upper-level divergence, the storm weakened even more as it continued its northeastward journey, out over the North Atlantic. In all, the storm was one of the strongest ever. It blanketed deep snow from Alabama to Canada. Fierce winds piled the snow into huge drifts that closed roads, leaving motorists stranded. The storm shut down every major airport along the East Coast and more than 3 million people lost electric power at some point during the storm. "The storm of the century" damaged or destroyed hundreds of homes, produced 27 tornadoes, stranded hikers in North Carolina and Tennessee, caused an estimated $800 million in damage, and claimed the lives of at least 270 people.

BRIEF REVIEW   Up to this point, we have looked at the structure and development of mid-latitude storms. Before going on, here is a summary of a few of the important ideas presented so far:

- The polar front model of a developing wave cyclone represents a simplified but useful model of how an ideal storm progresses through the stages of birth, maturity, and dissipation.
- For a surface mid-latitude storm to develop or intensify (deepen), the upper-level low must be located to the west of (behind) the surface low.
- For a surface mid-latitude storm to intensify, upper-level divergence above the storm must be greater than surface

*As we will see in Chapter 16, a pressure of 960 mb is equivalent to the pressure in a category 3 hurricane on the Saffir-Simpson scale, which ranges from 1 to 5, with 5 being the strongest.

WEATHER WATCH

The great March storm of 1993 set record low barometric pressure readings in a dozen states, produced wind gusts exceeding 90 knots from New England to Florida, and deposited 50 billion tons of snow over the east coast of the United States. When melted, this is enough water to equal the flow over Niagara Falls for 100 days.

convergence (that is, more air must be removed above the storm than is brought in at the surface).

- When the upper-air flow develops into waves, winds often cross the isotherms, producing regions of cold advection and warm advection, which tend to intensify the wave. At the same time, vertical air motions begin to enhance the formation of the surface storm as the rising of warm air and the sinking of cold air provide the proper energy conversion for the storm's growth.
- When the polar-front jet stream develops into a looping wave, it provides an area of upper-level divergence for the surface mid-latitude storm and an area of upper-level convergence for the surface anticyclone.
- The curving nature of the polar-front jet stream tends to direct surface mid-latitude storms northeastward and surface anticyclones southeastward.

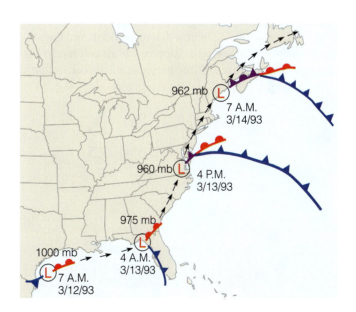

FIGURE 13.15
The development of a wave cyclone into the ferocious storm of March, 1993. A small wave in the western Gulf of Mexico intensifies into a deep open-wave cyclone over Florida. It moves northeastward and becomes occluded over Virginia where its central pressure drops to 960 mb (28.35 in.). As the occluded storm continues its northeastward movement, it gradually fills. The number next to the storm is its central pressure in millibars. Arrows show direction of movement. Time is Eastern Standard Time (EST).

The surface wave situated over the Gulf of Mexico in Fig. 13.15 could only develop into a deep low-pressure area if a region of strong upper-level divergence was situated above the developing wave cyclone. Meteorologists, then, are interested in locating regions of divergence on upper-level charts so that developing storms can be accurately predicted. We know from an earlier discussion that divergence and convergence of air are due to changes in either wind speed or wind direction. The problem is that it is a difficult task to measure divergence (or convergence) with any degree of accuracy using upper-level wind information. Therefore, meteorologists must look for something else that can be measured and, at the same time, can be related to regions of diverging (and converging) air. That something is called *vorticity*—the spinning of tiny air parcels.

VORTICITY, DIVERGENCE, AND DEVELOPING STORM SYSTEMS

When something spins, it has vorticity. The faster it spins, the greater its vorticity. In meteorology, **vorticity** is a measure of the spin of small air parcels. Although the spin can be in any direction, our concern will be with the spin of horizontally flowing air about a vertical axis, much like an ice skater spins about an imaginary vertical axis. Because our goal is to see how vorticity can be used to identify regions of divergence and convergence, we must give vorticity some quantitative value. When viewed from above, air that spins cyclonically (counterclockwise) has *positive vorticity* and air that spins anticyclonically (clockwise) has *negative vorticity*.

We can see in Fig. 13.16 how divergence aloft and the vorticity of surface air are related. Suppose the air column in Fig. 13.16 represents an area of low pressure with weak cyclonic

(positive) spin. Further, suppose that an invisible barrier separates the column (and the ice skater inside) from the surrounding air. Now suppose an area of divergence aloft (at the 300-mb level) moves directly over the air column. Divergence aloft means that within this region more air is leaving than is entering. This removal of air above the column lowers the atmospheric pressure at the surface. As the surface pressure lowers, the air surrounding the column converges on it, pushing inward on its sides. If we assume that the total mass of air in the column does not change, then squeezing the column causes it to shrink horizontally and stretch vertically, forcing the air in the column upward. Meanwhile, as the column stretches, the ice skater's arms are pulled in close, and the skater spins much faster. Hence, the rate at which air flows around the center of the column must also increase. Consequently, the vorticity of the column increases, becoming more positive. So we can see that divergence aloft causes an increase in the cyclonic (positive) vorticity of surface cyclones, which usually results in cyclogenesis and upward air motions.

Before we consider how vorticity on an upper-level chart ties in with developing weather systems, we need to examine two important types of this phenomenon: the earth's vorticity and relative vorticity.

Because the earth spins, it has vorticity. In the Northern Hemisphere, the **earth's vorticity** (also called *planetary vorticity*) is always positive because the earth spins counterclockwise about its vertical North Pole axis. The amount of earth vorticity imparted to any object—even to those that are not moving relative to the earth's surface—depends upon the latitude. In Fig. 13.17, an observer standing on the equator would not spin about his or her own *vertical axis*; further north, the observer would spin very slowly, while at the North Pole the observer would spin at a maximum rate of one revolution per day. It is now appar-

FIGURE 13.16

When upper-level divergence moves over an area of weak cyclonic circulation, the cyclonic circulation increases (that is, it becomes more positive), and air is forced upward.

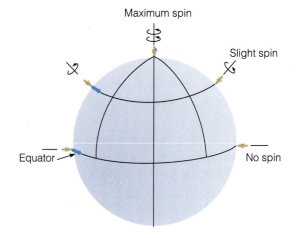

FIGURE 13.17

Due to the rotation of the earth, the rate of spin of observers about their vertical axes increases from zero at the equator to a maximum at the poles.

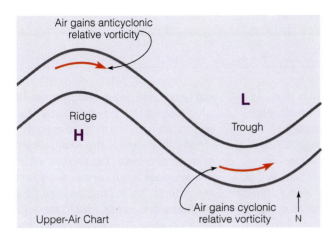

FIGURE 13.18

In a region where the contour lines curve, air moving through a ridge spins clockwise and gains anticyclonic relative vorticity. In the trough, the air spins counterclockwise and gains cyclonic relative vorticity.

FIGURE 13.19

Areas of cyclonic (positive) relative vorticity and anticyclonic (negative) relative vorticity can form in a region of strong wind-speed shear.

ent that any object on the earth has vorticity simply because the earth is spinning, and the amount of this earth vorticity increases from zero at the equator to a maximum at the poles.*

Moving air will generally have additional vorticity relative to the earth's surface. This type of vorticity, called **relative vorticity**, is the sum of two effects: the curving of the airflow (*curvature*) and the changing of the wind speed over a horizontal distance (*shear*). Figure 13.18 illustrates vorticity due to curvature. Air moving through a trough tends to spin cyclonically, increasing its relative vorticity. In the ridge, the spin tends to be anticyclonic, and the relative vorticity of the air increases, but in a negative direction. Whenever the wind blows faster on one side of an air parcel than on the other, a shear force is imparted on the parcel and it will spin and gain (or lose) relative vorticity (see Fig. 13.19).

The sum of the *earth's vorticity* and the *relative vorticity* is called the **absolute vorticity**. To further illustrate this concept, suppose you are watching an ice skater spin counterclockwise on a frozen lake. The ice skater possesses positive relative vorticity. If you could suddenly leave the earth and watch the same spinning skater from space, you would see the ice skater spinning on a rotating platform—the earth. The combination of the skater spinning about her vertical axis (her relative vorticity) plus the small spin imparted to her from the earth (earth's vorticity) yields the absolute vorticity of the skater. We are now in a position to see how vorticity aloft ties in with divergence and the development of middle latitude storms. (The concept of absolute vorticity can also be used to explain why the westerly flow aloft tends to form into waves. This topic is presented in the Focus section on p. 342.)

---

*The earth's vorticity at any latitude is equal to the product of twice the earth's angular rate of spin (2Ω) and the sine of the latitude (ϕ); that is, $2\Omega \sin(\phi)$. This expression is referred to as the *Coriolis parameter*.

Figure 13.20 shows an air parcel—a square blob of air in this case—moving through a ridge and a trough in the upper troposphere. (For simplicity, we will assume that the parcel has a constant speed and that there is no shear acting on it.) Notice that at every position the parcel has cyclonic spin. This situation occurs because the parcel's total, or absolute vorticity, is a combination of the earth's vorticity plus its relative vorticity. In middle and high latitudes, the earth's vorticity is great enough to make the parcel's spin cyclonic everywhere on the earth.

At position 1 (in the ridge), the air parcel has only a slight cyclonic spin because the relative vorticity, due to curvature, is anticyclonic and subtracts from the earth's vorticity. At position 2, the relative vorticity due to curvature is zero, which allows the earth's vorticity to spin the parcel faster. At position 3, the parcel spins even faster as the curvature is cyclonic, which adds to the earth's vorticity. At position 4, the parcel spins more slowly as the curvature is once again zero.

We can see in Fig. 13.20 that as the parcel moves from position 1 in the ridge to position 3 in the trough, its absolute vorticity increases with time. Within this region is typically found an area of upper-level convergence. As the parcel moves from position 3 to position 5, its absolute vorticity decreases with time. Within this region is typically found an area of upper-level divergence. (To refresh your memory, look back at Figs. 13.5 and 13.10.) We can now summarize the information in Fig. 13.20 by stating that as a parcel of air moves with the upper-level flow, an *increase in its absolute vorticity is related to upper-level convergence, and a decrease in its absolute vorticity is related to upper-level divergence.* And, in fact, as a parcel of air moves through a wave, the more negative the absolute vorticity becomes with time, the greater the divergence of air.

On upper-air charts that show absolute vorticity (such as on a 500-mb chart), we find that even though the clockwise flow around a high-pressure area produces negative relative

Vorticity and Longwaves

We know that aloft the atmosphere tends to form into a series of troughs and ridges known as longwaves (see Fig. 13.6). To understand how these waves form, we will assume that there is no divergence or convergence of air so that columns of air may not stretch or contract. Where these conditions prevail, the absolute vorticity of air will be conserved, meaning that the numerical value of the sum of the earth's vorticity and the relative vorticity will not change with time. Thus

Absolute vorticity
= earth's vorticity + relative vorticity
= constant.

If ζ_a is the absolute vorticity, ζ_r the relative vorticity, and f the earth's vorticity, then the expression becomes

$$\zeta_a = \zeta_r + f = \text{constant}$$

Hence, any decrease in the earth's vorticity must be compensated for by an increase in the relative vorticity and vice versa.

Consider, for example, that air in Fig. 4 is moving horizontally at a constant speed at the 500-mb level. At this level, divergence is usually near zero, so our initial assumption approaches a real situation. Since there is no wind speed shear, any change in the relative vorticity will be due to curvature. At position 1, the air is flowing southeastward. Heading equatorward, the air moves into a region of decreasing earth vorticity. To keep the absolute vorticity of the air constant, there must be a corresponding increase in the relative vorticity. Since increasing relative vorticity implies cyclonic curvature, the air turns counterclockwise at position 2 and heads northeastward. But now the air is moving into a region where the earth's vorticity steadily increases. To offset this increase, the relative vorticity must decrease. The relative vorticity will decrease if the curvature becomes anticyclonic, so at position 3 air turns clockwise and heads toward the equator once again. This again brings the air into a region where the earth's vorticity decreases. To compensate, the air must now turn cyclonically at position 4, and so on. In this manner, a series of upper-level longwaves may develop, encircling the entire earth.

FIGURE 4

The wavy path of air aloft due to the conservation of absolute vorticity. (See text for explanation.)

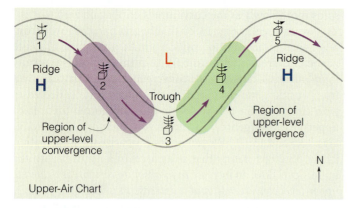

FIGURE 13.20

As an air parcel moves through the wave pattern from position 1 to position 3, its absolute vorticity increases with time. In this region (shaded purple), we normally experience an area of upper-level converging air. As an air parcel moves from position 3 to position 5, its absolute vorticity decreases with time. In this region (shaded green), we normally experience an area of upper-level diverging air.

vorticity, because of the earth's (positive) vorticity, the absolute vorticity is normally positive everywhere on the map. However, there are regions on the map (about the size of the state of Iowa) where the absolute vorticity is considerably greater. An area of high absolute vorticity is referred to as a *vorticity maximum* or *vort max*. A region of low absolute vorticity is known as a *vorticity minimum*.

Figure 13.21 illustrates how vorticity, divergence, vertical air motions, and surface storm development are linked together. The 500-mb chart (middle chart) in Fig. 13.21 shows a vorticity maximum. To the east of this maximum there is upper-level divergence and low-level convergence, which infers rising air and the possibility of clouds, precipitation, and cyclonic storm development. It follows that when a vorticity maximum aloft moves toward a stationary front, a wave will form along the front and a storm will develop. Even in the absence of fronts, a zone of organized clouds with precipitation may form in conjunction with a vorticity maximum. On the other hand, upwind from a vorticity maximum (to the west of the vorticity maximum in Fig. 13.21) or in a vorticity minimum, there is upper-level convergence, low-level divergence, slowly sinking air, and the generally fair weather we associate with surface anticyclones.

Because of the gaps in observational data, scientists often found it difficult to locate vorticity maximums on satellite pictures, especially in regions where there were no clouds. Now, however, geostationary satellites are equipped to observe atmospheric water vapor. The swirling patterns of moisture clearly identify the position of vorticity centers. In Fig. 13.22, observe the cyclonic swirl of water vapor associated with a region of maximum vorticity just off the coast of Oregon and Washington.

We are now in a position to place much of what we have learned into a real-life weather situation. Figure 13.23 illustrates the atmospheric conditions that turned an open-wave cyclone into a ferocious storm. The lower map shows a developing wave cyclone off the North Carolina coast on February 11, 1983. The counterclockwise circulation around the low is causing warm, moist air from subtropical waters to ride up and over very cold surface air that is entrenched over the eastern seaboard. Snow is falling in the cold air in advance of a warm front that extends from the low northeastward over the Atlantic Ocean.

Above the surface, the 500-mb chart (middle chart, Fig. 13.23) shows a broad longwave trough with a shortwave (heavy dashed line) moving through it. Notice that the shortwave is just to the west of the surface low. Apparently, the shortwave has disturbed the flow, as a strong area of baroclinicity exists to the west of the shortwave. Here, isotherms (dashed lines) are intersecting contour lines (solid lines), and cold advection is occurring.

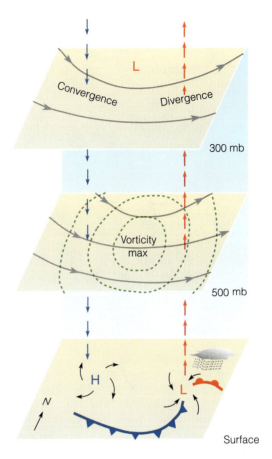

FIGURE 13.21

A region of high absolute vorticity—a vorticity maximum—on its downwind (eastern) side has diverging air aloft, converging surface air, and ascending air motions. On its upwind (western) side, there is converging air aloft, diverging surface air, and descending air motions.

FIGURE 13.22

Infrared water vapor image taken by GOES West. The cyclonic swirl of clouds off the coast of Oregon and Washington and over the Pacific define regions of maximum vorticity. The stretched-out band of clouds toward the bottom of the picture is the intertropical convergence zone.

FIGURE 13.23
The atmospheric conditions for February 11, 1983, at 7 A.M., EST. The bottom chart is the surface weather map. The middle chart is the 500-mb chart that shows contour lines (solid lines) in meters above sea level, and isotherms (dashed lines) in °C. The upper chart is the 200-mb chart that illustrates contours, winds, and the position of the jet stream (dark blue arrow). The dashed red line represents the jet streak. The letter D represents an area of strong divergence.

At the 200-mb level (upper chart, Fig. 13.23), the polar front jet stream swings just to the south of the surface low (marked by the letter D on the chart). *Recall from our earlier discussion that this location is most favorable for strong upper-level divergence.*

Figure 13.24 is a 500-mb chart that shows absolute vorticity for the same date and time as Fig. 13.23. Notice that the vorticity maximum of 14 (which is actually 14×10^{-5}/sec) is located in the same position as the shortwave in Fig. 13.23. Notice also that aloft, to the east of the vorticity maximum, lies the region of strong upper-level divergence associated with the jet stream. Directly below this region is the open-wave cyclone. Hence, the stage is set, and the necessary ingredients are in place for the surface low to develop into a major storm system.

The winds aloft steered the surface low northeastward, but the strong blocking high to the north over southern Canada (surface map, Fig. 13.23) slowed the storm's forward pace, allowing upper-level divergence to intensify the system. With the deep low just off the coast and the strong high to the north, pressure gradients increased, and strong, howling winds in excess of 30 knots battered New Jersey, eastern Pennsylvania, and southeastern New York. Meanwhile, huge quantities of moist ocean air produced copious snowfalls along the Atlantic coastal states. High above the surface, as cold air from the west moved over the region, the air became unstable; strong convective cells developed, and heavy "downpours" of snow were accompanied by lightning and thunder. At one point in the storm, Allentown, Pennsylvania, reported snow falling at the rate of five inches per hour.

The storm left a buried landscape from northern Virginia to Connecticut, where snow drifts in some areas rose to housetops. The storm shut down entire cities. It crippled

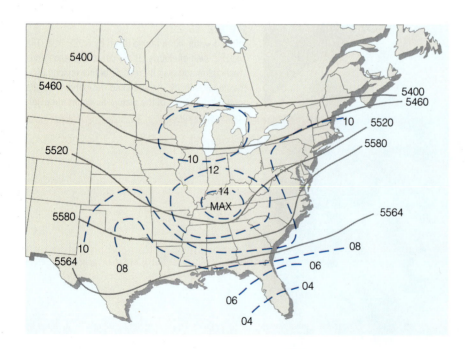

FIGURE 13.24
The 500-mb chart for February 11, 1983, at 7 A.M., EST. Solid lines are height contours in meters above sea level. Dashed lines are lines of constant absolute vorticity $\times 10^{-5}$/sec.

FOCUS ON A SPECIAL TOPIC

The Development of Lee-Side Lows

Vorticity helps to explain the development of storms on the leeward side of mountains, often called *lee cyclogenesis*. We can see why when we examine the flow of air over a mountain barrier (see Fig. 5). Assume that the wind is westerly and there is no shear, so that the relative vorticity (ζ_r) of the air column as it approaches the mountain will be zero. Theoretical studies show that the absolute vorticity (earth vorticity of $f + \zeta_r$) of the column divided by its depth (D) is equal to a constant, as long as the air is frictionless and no heating or cooling of the column takes place. (This expression is called the *potential vorticity*.) Thus

$$\text{Potential vorticity} = \frac{f + \zeta_r}{D} = \text{constant}.$$

As the column moves up the side of the mountain, the tropopause acts as a lid on the top of the column. This causes the column to shrink vertically and expand horizontally. To compensate for the column's decreasing depth, there must be a similar decrease in absolute vorticity. Since the relative vorticity (ζ_r) of the column is zero, the air must curve anticyclonically, or toward the southeast, into a region of decreasing earth vorticity (f must decrease).

Once the column begins to move downslope, it stretches vertically. Its depth increases, and so must its absolute vorticity. The relative vorticity of the southeastward-moving air is zero, but the air is moving into a region of decreasing earth vorticity. Consequently, there must be a bending of the airflow in the cyclonic sense (f must increase) to increase the column's absolute vorticity. The airflow now swings toward the north-east, which creates a trough of low pressure (a *lee-side low*) on the downwind side of the barrier. Now we can see why so many wave cyclones form along the flanks of the Rocky Mountains, especially in eastern Colorado, where the crest of the mountains reaches its maximum elevation.

Side view

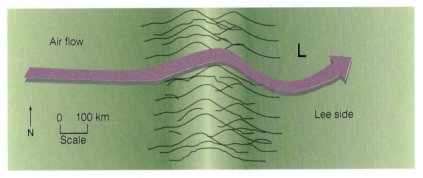

View from above

FIGURE 5

Due to changes in the air's vorticity, low-pressure areas (lee-side lows) tend to form on the leeward side of mountain ranges.

travel on the ground and in the air; closed businesses, government agencies, and schools; knocked out power to many thousands of homes; and caused several deaths. In addition, many cities reported their greatest 24-hour snowfall ever. The heaviest amount was 79 cm (31 in.) in Woodstock, Virginia; while Allentown, Pennsylvania, reported 64 cm (25.2 in.); Philadelphia, Pennsylvania, 54 cm (21.3 in.); and Baltimore, Maryland, 58 cm (22.8 in.). Because the heavy snow was accompanied by high winds and low temperatures, the storm has come to be known as "the Blizzard of '83."

Earlier in this chapter we saw that the eastern slopes of the Rockies were a favorable area for the development or strengthening of low-pressure areas (see Fig. 13.3a). Similarly, storms tend to intensify on the eastern side of most north-south trending mountain ranges. Such storms that form on the downwind (leeward) side of mountains are called **lee-side lows**. Changes in the air column's absolute vorticity as it crosses the mountain barrier help to explain how lee-side lows form. (More on this topic is given in the Focus section above.)

FIGURE 13.25
An enhanced infrared satellite image of an intense polar low situated over the Norwegian Sea, north of the Arctic Circle. Notice that convective clouds swirl counterclockwise about a clear area, or eye. Surprising similarities exist between polar lows and tropical hurricanes described in Chapter 16.

POLAR LOWS

Up to now, we have concentrated on middle latitude cyclones, especially those storms that form along the polar front. There are storms, however, that develop over water behind (or poleward) of the main polar front. Such storms are called **polar lows**. Although polar lows develop in both hemispheres, our discussion will center on those storms that form in the Northern Hemisphere in the cold polar air of the North Pacific, North Sea, and North Atlantic, especially in the region south of Iceland.

With diameters of 1000 to 500 km (600 to 300 mi) or less, polar lows are generally smaller in size than their mid-latitude cousin, the wave cyclone that tends to form along the polar front. Some polar lows have a comma-shaped cloud band. Others have a tight spiral of convective clouds that swirls counterclockwise about a clear area, or "eye," which resembles the eye of a tropical hurricane (see Fig. 13.25). In fact, like hurricanes, these smaller intense storms normally have a warmer central core, strong winds (often gale force or higher), and heavy showery precipitation that, unlike a hurricane, is in the form of snow.*

Polar lows typically form during the winter, from November through March. During this time, the sun is low on the horizon and absent for extended periods. This situation allows the air next to snow-and-ice-covered surfaces to cool rapidly and become incredibly cold. As this frigid air sweeps off the winter ice that covers much of the Arctic Ocean, it may come in contact with warmer air that is resting above a relatively warm ocean current. Where these two masses of contrasting air meet, the boundary separating them is called an *arctic front*.

Along the arctic front, the warmer (less dense) air rises, while the much colder (more dense) air slowly sinks beneath

---

*Tropical storms such as hurricanes are covered in Chapter 16. As we will see, the input of heat from the ocean surface into the hurricane increases as the wind speed increases, causing a positive feedback. Moreover, in hurricane environments, the ocean and air temperatures are about the same, whereas in the Arctic, the transfer of sensible heat from the surface is large because the air-sea temperature difference is large.

it. Recall from our discussion of developing middle latitude cyclones (p. 332) that the rising of warm air and the sinking of cold air establishes a condition known as *baroclinic instability*. As the warm air rises, some of its water vapor condenses, resulting in the formation of clouds and the release of latent heat, which warms the atmosphere. Meanwhile, at the ocean surface there is a transfer of sensible heat from the relatively warm water to the cold air above. This transfer drives convective updrafts directly from the surface. In addition, it tends to make the atmosphere more unstable, as heat is gained at the surface and lost to space at the top of the clouds as they radiate infrared energy upward.

The storm's development is enhanced if an upper trough lies to the west of the surface system and a shortwave disturbs the flow aloft. Similarly, the storm may intensify if a band of maximum winds—a jet streak—moves over the surface storm and a region of upper-level divergence forces the surface air upward. The developing cyclonic storm may attain a central pressure of 980 mb (28.94 in.) or lower. Generally, polar lows dissipate rapidly when they move over land.

SUMMARY

In this chapter, we discussed where, why, and how a wave cyclone forms. We began by examining the early polar front theory proposed by Norwegian scientists after World War I. We saw the important effect that the upper airflow has on the intensification and movement of surface high- and low-pressure areas.

We saw that when an upper-level trough lies to the west of a surface low-pressure area and when a shortwave disturbs the flow aloft, horizontal and vertical air motions begin to enhance the formation of the surface storm. Aloft, regions of divergence remove air from above surface mid-latitude cyclones, and regions of convergence supply air to surface anticyclones. The rising of warm air and the sinking of cold air provide the proper energy conversions for the storm's growth, as potential energy is transformed into kinetic energy. The polar jet provides additional support by removing air above the surface cyclone and supplying air to the surface anticyclone. The curving nature of the jet stream tends to direct mid latitude cyclones northeastward and anticyclones southeastward.

We looked at the concept of vorticity and how it relates to developing cyclones, longwaves, and lee-side lows. We found that on the downwind (eastern) side of a vorticity maximum, there is normally diverging air aloft, converging air at the surface, and cyclonic storm development. Finally, we examined polar lows—those storms that form over water in the cold air of the polar regions.

Key Terms

The following terms are listed in the order they appear in the text. Define each. Doing so will aid you in reviewing the material covered in this chapter.

polar front theory	baroclinic (atmosphere)
frontal wave	cold advection
wave cyclone	warm advection
open wave	baroclinic instability
warm sector	cut-off low
secondary low	conveyor belt model
cyclogenesis	comma cloud
convergence	vorticity
divergence	earth's vorticity
longwave	relative vorticity
Rossby waves	absolute vorticity
shortwave	lee-side lows
barotropic (atmosphere)	polar lows

Questions for Review

1. What are the sources of energy for a developing cyclone?
2. Why do wave cyclones "die out" after they become occluded?

3. List four regions in North America where wave cyclones tend to develop.

4. Explain this fact: Without upper-level divergence, a surface open wave would probably persist for less than a day.

5. Why do middle latitude surface low-pressure areas tilt westward with increasing height?

6. What is an Alberta Clipper? Where does it form and how does it move?

7. How are longwaves in the upper-level westerlies different from shortwaves?

8. What are the necessary ingredients for a wave cyclone to develop into a huge storm system?

9. Why do surface storms tend to fill when the upper-level low and the surface low become vertically stacked?

10. How does the polar front jet stream influence the formation of a wave cyclone?

11. Explain why, even though the polar front jet stream coincides with the polar front, some surface regions are more favorable for the development of wave cyclones than others.

12. Using a diagram, explain why a surface high-pressure area over North Dakota will typically move southeastward while, at the same time, a deep storm system over the Great Lakes will generally move northeastward.

13. What are the sources of energy for a developing cyclone?

14. As an air parcel moves through an upper-level longwave, absolute vorticity becomes more negative with time. Does the divergence or convergence of air increase?

15. Why does cyclogenesis often occur in the region just east of the Rockies?

16. What are polar lows? How and where do they form? What do some of them have in common with tropical hurricanes?

17. What are the roles of warm, cold, and dry conveyor belts in the development of a cyclonic storm system?

18. Explain why surface lows tend to deepen when a vorticity maximum aloft moves in from the west.

Questions for Thought

1. An English friend of yours says that last night's rain over northern Great Britain was caused by a storm that originally formed east of the Colorado Rockies. Explain how this could happen.

2. Would a wave cyclone intensify or dissipate if the upper trough were located to the *east* of the surface disturbance? Explain your answer with the aid of a diagram.

3. Explain why, at 500 millibars, when cold advection is occurring, the air temperature does not drop as fast as it should. (Hint: What type of vertical air motions are also occurring?)

4. Over the earth as a whole, would you expect the atmosphere to be mainly barotropic or baroclinic? Explain.

5. Baroclinic waves seldom form in the tropics. Why?

6. Suppose that the earth stops rotating. How would this affect the earth's vorticity? What would happen to the absolute vorticity of a moving air parcel? If the parcel were initially moving southwestward, how would its direction change, if at all?

7. Why do Pacific storms often redevelop on the eastern side of the Sierra Nevada?

8. If you only had isotherms on an upper-level chart, how would a cut-off low appear?

9. If polar lows form in cold polar air over water, how is the atmosphere made unstable so that towering convective cumulus clouds can form?

Problems and Exercises

1. As you refer to the 300-mb chart below, suppose the winds are blowing at a constant speed parallel to the contour lines.
   (a) On the chart, mark where regions of convergence and divergence are occurring.
   (b) Put an *L* on the chart where you might expect to observe a developing mid-latitude storm.
   (c) Put an *H* on the chart where you might expect to observe an anticyclone.
   (d) In which directions would the mid-latitude cyclone and anticyclone most likely move?
   (e) In terms of convergence and divergence, what are the necessary conditions for the intensification of the surface mid-latitude storm? For the building of the anticyclone?

2. Suppose there is a region of maximum absolute vorticity at position *M* on the chart.

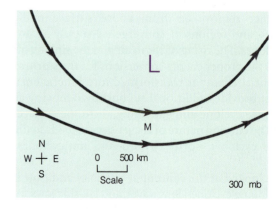

FIGURE 13.26
Chart for Exercises 1 and 2.

(a) How would the absolute vorticity be changing as an air parcel moves upstream with the flow?

(b) Based on absolute vorticity, circle on the map where you would expect to find upper-level divergence, surface convergence, and a developing storm.

 Questions for Exploration

1. On the Blue Skies: College Edition CD-ROM go to the **Weather Analysis** section of the CD and click on "**Weather Maps.**"
2. On the Blue Skies: College Edition CD-ROM go to the **Weather Forecasting** section of the CD and click on "**Cyclogenesis.**"

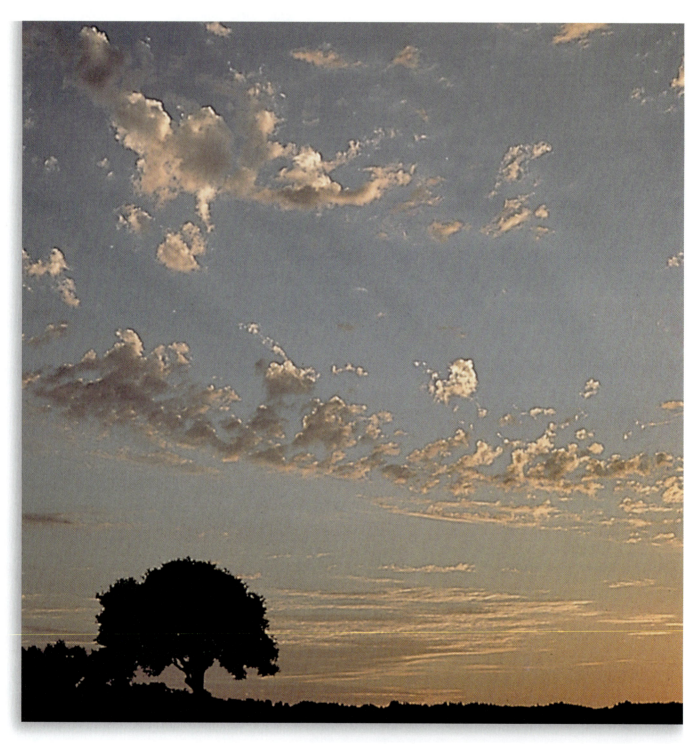

A watchful eye and a little weather information are important when making a local short-range weather forecast. These altocumulus castellanus clouds, northwesterly surface winds, a rising barometer with a sea level pressure of 1016 mb (30.00 in.), and the instant weather forecast chart (see Appendix E) all indicate that fair weather will remain into the next day. (Photo by author.)

WEATHER FORECASTING

CONTENTS

Weather forecasting can be a dangerous occupation. Weatherman Francisco Arias Olivera was a popular TV personality in the small Peruvian town of Sicuani with its 21,000 residents. The story goes that everyone trusted Francisco's weather predictions until one ill-fated day when a forecasted 2-inch rainfall turned into a 19-inch deluge that inundated the region. The Vilcanota River surged its banks and tore through the town, washing away 250 homes and drowning 17 unsuspecting residents. Outraged citizens stormed the town's only TV station, claiming that Francisco had not done his job. Several vigilantes dragged the bewildered weatherman from his office to a tree behind the station and lynched him. Six people were charged with murder but were released upon claiming justifiable homicide. Rumor has it that the town is almost back to normal, except that the forecasting position is still open at the TV station.

Weather forecasts are issued to save lives, to save property and crops, and to tell us what to expect in our atmospheric environment. In addition, knowing what the weather will be like in the future is vital to many human activities. For example, a summer forecast of extended heavy rain and cool weather would have construction supervisors planning work under protective cover, department stores advertising umbrellas instead of bathing suits, and ice cream vendors vacationing as their business dropped off. The forecast would alert farmers to harvest their crops before their fields became too soggy to support the heavy machinery needed for the job. And the commuter? Well, the commuter knows that prolonged rain could mean clogged gutters, flooded highways, stalled traffic, blocked railway lines, and late dinners.

On the other side of the coin, a forecast calling for extended high temperatures with low humidity has an entirely different effect. As ice cream vendors prepare for record sales, the dairy farmer anticipates a decrease in milk and egg production. The forest ranger prepares warnings of fire danger in parched timber and grasslands. The construction worker is on the job outside once again, but the workday begins in the early morning and ends by early afternoon to avoid the oppressive heat. And the commuter prepares for increased traffic stalls due to overheated car engines.

Put yourself in the shoes of a weather forecaster: It is your responsibility to predict the weather accurately so that thousands (possibly millions) of people will know whether to carry an umbrella, wear an overcoat, or prepare for a winter storm. Since weather forecasting is not an exact science, your predictions will occasionally be incorrect. If your erroneous forecast misleads many people, you may become the target of jokes, insults, and even anger. There are even people who expect you to be able to predict the unpredictable. For example, on Monday you may be asked whether two Mondays from now will be a nice day for a picnic. And, of course, what about next winter? Will it be bitterly cold?

Unfortunately, accurate answers to such questions are beyond meteorology's present technical capabilities. Will forecasters ever be able to answer such questions confidently? If so, what steps are being taken to improve the forecasting art? How are forecasts made, and why do they go wrong? These are just a few of the questions we will address in this chapter.

ACQUISITION OF WEATHER INFORMATION

Weather forecasting basically entails predicting how the present state of the atmosphere will change. Consequently, if we wish to make a weather forecast, present weather conditions over a large area must be known. To obtain this information, a network of observing stations is located throughout the world. Over 10,000 land-based stations and hundreds of ships provide surface weather information four times a day.* Most airports observe conditions hourly. Additional information, especially upper-air data, is supplied by radiosondes, aircraft, and satellites. Radiosonde data are usually available only at 0000 and 1200 UTC, while aircraft and satellite observations may occur throughout the day.

A United Nations agency—the World Meteorological Organization (WMO)—consists of over 130 nations. The WMO is responsible for the international exchange of weather data and certifies that the observation procedures do not vary among nations, an extremely important task, since the observations must be comparable.

After an observation is taken, it is immediately sent to a communication substation by electronic means, usually over dedicated phone lines or by satellite relay. From there, the data collected at many observation stations are sent to World Meteorological Centers (located in Melbourne, Australia; Moscow, Russia; and Washington, D.C.). Then, worldwide weather information is transmitted electronically to the National Center for Environmental Prediction (NCEP) (formerly called the National Meteorological Center) located in Camp Springs, Maryland, just outside Washington, D.C. Here, the massive job of analyzing the data, preparing weather maps and charts, and predicting the weather on a global and national basis begins.† From NCEP, this information is transmitted to public and private agencies worldwide.

The compiled charts, maps, and forecasts are sent electronically to *Weather Forecast Offices* (WFO). These stations use the data for preparing regional weather forecasts, as well as for advisories and warnings of impending severe weather. The region serviced by one of these offices is a state or a large portion of a state.

The public hears weather forecasts over radio or television. Many stations hire private meteorological companies or professional meteorologists to make their own forecasts aided by NCEP material or to modify a WSO forecast. Other stations hire meteorologically untrained announcers who usually read the forecasts of the National Weather Service word for word.

Today, the forecaster has access to hundreds of maps and charts, as well as vertical profiles (called *soundings*) of temperature, dew point, and winds. Also available are visible and infrared satellite images, as well as Doppler radar information that can detect and monitor the severity of precipitation and thunderstorms.

When hazardous weather is likely, the National Weather Service issues advisories in the form of weather watches and

---

*Observations are usually taken at 0000, 0600, 1200, and 1800 Coordinated Universal Time (UTC), which is also called Greenwich Mean Time (GMT)—local time at the Greenwich Observatory in England. To convert from UCT to your local time, see Appendix F.

†By international agreement, data are plotted using symbols illustrated in Appendix B.

warnings. A **watch** indicates that atmospheric conditions favor hazardous weather occurring over a particular region during a specified time period. A **warning**, on the other hand, indicates that hazardous weather is either imminent or actually occurring within the specified forecast area. *Advisories* are issued to inform the public of hazardous driving conditions caused by wind, dust, fog, snow, sleet, or freezing rain. (Additional information on watches, warnings, and advisories is given in the Focus section on p. 354.)

WEATHER FORECASTING METHODS

As late as the mid-1950s, all weather maps and charts were plotted by hand and analyzed by individuals. Meteorologists predicted the weather using certain rules that related to the particular weather system in question. For short-range forecasts of six hours or less, surface weather systems were moved along at a steady rate. Upper-air charts, especially the 500-mb chart, were used to predict where surface storms would develop and where pressure systems aloft would intensify or weaken. The predicted positions of these systems were extrapolated into the future using linear graphical techniques and current maps. Experience played a major role in making the forecast. In many cases, these forecasts turned out to be amazingly accurate. They were good but, with the advent of the modern computers, today's forecasts are even better.

The Computer and Weather Forecasting: Numerical Weather Prediction
Modern electronic computers can analyze large quantities of data extremely fast. Each day the many thousands of observations transmitted to NCEP are fed into a high-speed computer, which plots and draws lines on surface and upper-air charts. Meteorologists interpret the weather patterns and then correct any errors that may be present. The final chart is referred to as an **analysis.**

The computer not only plots and analyzes data, it also predicts the weather. The routine daily forecasting of weather by the computer has come to be known as **numerical weather prediction.**

Because the many weather variables are constantly changing, meteorologists have devised **atmospheric models** that describe the present state of the atmosphere. These are not physical models that paint a picture of a developing storm; they are, rather, mathematical models consisting of dozens of mathematical equations that describe how atmospheric temperature, pressure, and moisture will change with time. Actually, the models do not fully represent the real atmosphere but are approximations formulated to retain the most important aspects of the atmosphere's behavior.

The models are programmed into the computer, and surface and upper-air observations of temperature, pressure, moisture, winds, and air density are fed into the equations. To

WEATHER WATCH

Lewis F. Richardson was the first person to make a mathematical weather prediction without the use of a computer or even a calculator. In the early part of the twentieth century, Richardson trudged six long weeks through the myriad of computations that went into making a single forecast, for a single station, only six hours into the future. Unfortunately, due to an error (called *computational instability*), his forecast turned out to be way off.

determine how each of these variables will change, each equation is solved for a small increment of future time—say, five minutes—for a large number of locations called *grid points*, each situated a given distance apart.* In addition, each equation is solved for as many as 50 levels in the atmosphere. The results of these computations are then fed back into the original equations. The computer again solves the equations with the new "data," thus predicting weather over the following five minutes. This procedure is done repeatedly until it reaches some desired time in the future, usually 12, 24, 36, or 48 hours. The computer then prints this information, analyzes it, and draws the projected positions of pressure systems with their isobars or contour lines. The final forecast chart representing the atmosphere at a specified future time is called a **prognostic chart**, or, simply, a **prog.** Computer-drawn progs have come to be known as "machine-made" forecasts.

The computer solves the equations more quickly and efficiently than could be done by hand. For example, just to produce a 24-hour forecast chart for the Northern Hemisphere requires many hundreds of millions of mathematical calculations. It would, therefore, take a group of meteorologists working full time with hand calculators years to produce a single chart; by the time the forecast was available, the weather for that day would already be ancient history.

The forecaster uses the progs as a guide to predicting the weather. At present, there are a variety of models (and, hence, progs) from which to choose, each producing a slightly different interpretation of the weather for the same projected time and atmospheric level (see Fig. 14.1). The differences between progs may result from the way the models use the equations, or the distance between grid points. Some models predict some features better than others: One model works best in predicting the position of troughs on upper-level charts, while another forecasts the position of surface lows quite well. Some

---

*Some models have a grid spacing as small as 10 km, whereas the spacing in others exceeds 180 km. There are models that actually describe the atmosphere using a set of mathematical equations with wavelike characteristics rather than a set of discrete numbers associated with grid points.

FOCUS ON A SPECIAL TOPIC

Watches, Warnings, and Advisories

As we have seen, where severe or hazardous weather is either occurring or possible, the National Weather Service issues a forecast in the form of a watch or warning. The public, however, is not always certain as to what this forecast actually means. For example, a *high wind warning* indicates that there will be high winds—but how high and for how long? The following describes a few of the various watches, warnings, and advisories issued by the National Weather Service and the necessary precautions that should be taken during the event.

Wind advisory Issued when sustained winds reach 25 to 39 miles per hour or wind gusts are up to 57 miles per hour.

High wind warning Issued when sustained winds are at least 40 miles per hour, or wind gusts exceed 57 miles per hour. Caution should be taken when driving high-profile vehicles, such as trucks, trailers, and motor homes.

Wind-chill advisory Issued for wind-chill temperatures of –30° to –35°F or below.

Flash-flood watch Heavy rains may result in flash flooding in the specified area. Be alert and prepared for the possibility of a flood emergency that will require immediate action.

Flash-flood warning Flash flooding is occurring or is imminent in the specified area. Move to safe ground immediately.

Severe thunderstorm watch Thunderstorms (with winds exceeding 57 miles

Small craft advisory   Gale warning   Storm warning   Hurricane warning

FIGURE 1

Flags indicating advisories and warnings in maritime areas.

per hour and/or hail three-fourths of an inch or more in diameter) are possible.

Severe thunderstorm warning Severe thunderstorms have been visually sighted or indicated by radar. Be prepared for lightning, heavy rains, strong winds, and large hail. (Tornadoes can form with severe thunderstorms.)

Tornado watch Issued to alert people that tornadoes may develop within a specified area during a certain time period.

Tornado warning Issued to alert people that a tornado has been spotted either visually or by radar. Take shelter immediately.

Snow advisory In nonmountainous areas, expect a snowfall of 2 inches or more in 12 hours, or 3 inches or more in 24 hours.

Winter storm warning (formerly heavy snow warning) In nonmountainous areas, expect a snowfall of 4 inches or more in 12 hours or 6 inches or more in 24 hours. (Where heavy snow is

infrequent, a snow fall of several inches may justify a warning.)

Blizzard warning Issued when falling or blowing snow and winds of at least 35 miles per hour frequently restrict visibility to less than one-quarter mile for several hours.

Dense fog advisory Issued when fog limits visibility to less than one-quarter mile, or in some parts of the country to less than one-eighth mile.

WARNINGS OVER THE WATER

Small craft advisories Issued to alert mariners that weather or sea conditions might be hazardous to small boats. Expect winds of 18 to 34 knots (21 to 39 miles per hour). Figure 1 displays the posted advisory and warning flags.

Gale warning Winds will range between 34 and 47 knots (39 to 54 miles per hour) in the forecast area.

Storm warning Winds in excess of 47 knots (54 miles per hour) are to be expected in the forecast area.

Hurricane watch Issued when a tropical storm or hurricane becomes a threat to a coastal area. Be prepared to take precautionary action in case hurricane warnings are issued.

Hurricane warning Issued when it appears that the storm will strike an area within twenty-four hours. Expect wind speeds in excess of 64 knots (74 miles per hour).

---

models even forecast the state of the atmosphere 384 hours (16 days) into the future.

A good forecaster knows the idiosyncrasies of each model and carefully scrutinizes all the progs. The forecaster then makes a prediction based on the *guidance* of the computer, a personalized practical interpretation of the weather situation and any local geographic features that influence the weather within the specific forecast area. Table 14.1 gives a brief sum-

mary of a few "rules of thumb" that a forecaster might use when making a prediction using a prog.

Forecasting Tools   To help forecasters handle all the available charts and maps, high-speed data modeling systems using minicomputers are employed. The original communication system, known as *AFOS* (*Automation of Field Operations and Services*), provided weather information from any desired region on a television screen, as well as graphic overlays of satellite pictures and charts that enabled forecasters to view and analyze dynamic weather events over a broad area. However, presently the National Weather Service is undergoing a major modernization and restructuring program to replace the outdated AFOS system with a more advanced computerized system called **AWIPS** (*Advanced Weather Interactive Processing System*). The AWIPS system is shown in Fig. 14.2.

TABLE 14.1 A Few Forecasting "Rules of Thumb"*	
FORECAST QUESTION	**USE OF FORECAST CHART**
Cloudy or clear?	On the 700-mb forecast chart, the 70 percent relative humidity line usually encloses areas that are likely to have clouds.
Will it rain?	(a) On the 700-mb forecast chart, the 90 percent relative humidity line often encloses areas where precipitation is likely. If upward velocities are present, the chance of measurable precipitation is enhanced.
	(b) Along the west coast of North America, precipitation is much more likely north of the 5640-m height contour on the 500-mb forecast chart.
Will it rain or snow?	On the 850-mb forecast chart, snow is likely north of the –5°C (23°F) isotherm, whereas rain is likely south of this line.
Will the surface low intensify?	For the storm to intensify (deepen), an area of upper-level divergence must be over the cyclonic storm. On a 500-mb forecast chart that shows vorticity, look for a vorticity maximum (vort max) and remember from Chapter 13, p. 343: that to the east of an area of positive vorticity we usually find upper-level divergence, upward air motions, and cyclonic storm development.

*The forecast charts (progs) found in this table can be obtained from the World Wide Web.

(a)   (b)

FIGURE 14.1
Two 500-mb progs for 7 P.M. EST, May 4, 1999—48 hours into the future. Prog (a) is the MRF model, whereas Prog (b) is the ETA model. Solid lines on the map are height contours, where 552 equals 5520 m. Notice how the two progs (models) agree on most of the atmosphere's circulation.

FIGURE 14.2
The AWIPS computer work station provides various weather maps and overlays on different screens.

Composite Algorithm Output

STM ID	AZ	RAN	MESO	HAIL	SVRH	SIZE	MAXZ
4	303	111	NONE	100%	100%	>300	74
14	294	128	NONE	100%	100%	>300	75
6	299	115	NONE	100%	100%	>300	74
2	179	161	NONE	70%	50%	<100	62
5	336	166	NONE	70%	0%	<100	50

ORLANDO

KISSIMMEE

```
<0   3   9   14   20   25   31   36   42   47   53   58   64   >65   RF
```

Ctr az, ran: 298,107 k Mag: 6X Sel az, ran: 0, 0 k Hgt: 0.0 k Val: 0029.0
Vol scan: 22 Sweep: 0 Mode PPI Angle: 0.5 Nyquist: 31
Radar: KMLB   Site: KMLB   MMDDYY: 03/26/92 HHMMSS: 00:01:24

FIGURE 14.3
Doppler radar data from Melbourne, Florida, on March 25, 1992, during the time of a severe hailstorm that caused $60 million in damages in the Orlando area. In the table near the top of the display, the hail algorithm determined that there was 100 percent probability that the storm was producing hail and severe hail. The algorithm also estimated the maximum size of the hailstones to be greater than 3 inches. A forecaster can project the movement of the storm and adequately warn those areas in the immediate path of severe weather.

The AWIPS system has new data communications, storage, processing, and display capabilities (including graphical overlays) to better help the individual forecaster extract and assimilate information from the mass of available data. In addition, AWIPS is able to process information received from the Doppler radar system (the WSR-88D) and the new Automated Surface Observing Systems (*ASOS*) that are operational at selected airports and other sites throughout the United States. The *ASOS* system is designed to provide nearly continuous information about wind, temperature, pressure, cloud-base height, and runway visibility at various airports. Meteorologists believe that information from all of these sources will improve the accuracy of weather forecasts by providing previously unobtainable data for integration into numerical models. Moreover, much of the information from ASOS and Doppler radar is processed by software according to predetermined formulas, or *algorithms*, before it goes to the forecaster. Certain criteria or combinations of measurements can alert the forecaster to an impending weather situation, such as the severe weather illustrated in Fig. 14.3.

With so much information at the forecaster's disposal, it is essential that the data be easily accessible and in a format that

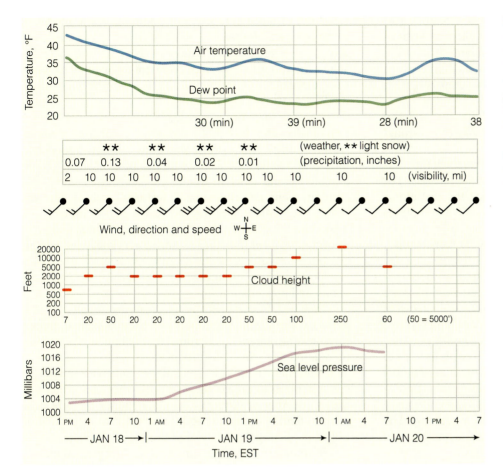

FIGURE 14.4
Meteogram illustrating predicted weather elements at Buffalo, New York, from 1 P.M. January 18, 1999, to 7 P.M. January 20, 1999. Notice that at 7 A.M. on January 19, the weather is projected to be: sea-level pressure 1007 mb; cloud height 2000 ft.; southwest winds at 15 knots; visibility 10 miles; light snow; air temperature 33°F; dew point 24°F; and the minimum temperature for the day should be 30°F. The forecast is derived from the model output statistics (MOS) of the Nested Grid Model (NGM).

allows several weather variables to be viewed at one time. The **meteogram** is a chart that shows how one or more weather variables has changed at a station over a given period of time. As an example, the chart may represent how air temperature, dew point, and sea-level pressure have changed over the past five days , or it may illustrate how these same variables are projected to change over the next five days (see Fig. 14.4).

Another aid in weather forecasting is the use of *soundings*—a two-dimensional vertical profile of temperature, dew point, and winds (see Fig. 14.5). The analysis of a sounding can be especially helpful when making a short-range forecast that covers a relatively small area, such as the mesoscale. The forecaster examines the sounding of the immediate area (or closest proximity), as well as the soundings of those sites upwind, to see how the atmosphere might be changing. Computer programs then automatically calculate from the sounding a number of meteorological *indexes* that can aid the forecaster in determining the likelihood of smaller-scale weather phenomena, such as thunderstorms, tornadoes, and hail. Soundings also provide information that can aid in the prediction of fog, air pollution alerts, and the downwind mixing of strong winds.

In the United States, a network of *wind profilers* (see Chapter 10) is providing forecasters with hourly wind speed and wind direction information at 72 different levels in a column

FIGURE 14.5
A sounding of air temperature, dew point, and winds at Pittsburgh, PA on January 14, 1999. Looking at this sounding, a forecaster would see that saturated air extends up to about 820 mb. The forecaster would also observe that below-freezing temperatures only exist in a shallow layer near the surface and that the freezing rain presently falling over the Pittsburgh area would continue or possibly change to rain, as cold easterly surface winds are swinging around to warmer southwesterly winds aloft. To the forecaster, these veering winds indicate warm advection is occurring above the Pittsburgh area.

of air 16 km thick. The almost continuous monitoring of winds is especially beneficial when briefing pilots on areas of strong headwinds and on regions of strong wind shear. Wind information from the profilers is also sent to the NCEP, where it is integrated into computer models used in preparing mid-range weather forecasts.

Since the forecaster has access to over 300 maps produced at NCEP each day, why is it that forecasts are sometimes wrong?

Why Forecasts Go Awry There are flaws inherent in the computer models that limit the accuracy of weather forecasts. For example, computer-forecast models idealize the real atmosphere, meaning that each model makes certain assumptions about the atmosphere. These assumptions may be on target for some weather situations and be way off for others. Consequently, the computer may produce a prog that on one day comes quite close to describing the actual state of the atmosphere and not so close on another. A forecaster who bases a prediction on an "off day" computer prog may find a forecast of "rain and windy" turning out to be a day of "clear and colder."

Another forecasting problem arises because the majority of models are not global in their coverage and errors are able to creep in along the model's boundaries. For example, a model that predicts the weather for North America may not accurately treat weather systems that move in along its boundary from the western Pacific. Obviously, a global model would be preferred. But a global model of similar sophistication would require an incredible number of computations.

Even though many thousands of weather observations are taken worldwide each day, there are still regions where observations are sparse, particularly over the oceans and at higher latitudes. Temperature data recorded by satellites have helped to alleviate this problem, as the newest *GOES** satellites, with advanced atmospheric sounders, are providing a more accurate profile of temperature for the computer models. Since the computer's forecast is only as good as the data fed into it, and since observing stations may be so far apart that they miss certain weather features, a denser network of observations is needed, especially in remote areas, to ensure better forecasts in the future.

Earlier, we saw that the computer solves the equations that represent the atmosphere at many locations called grid points, each about 10 to 200 km apart. As a consequence, weather systems larger than this distance, such as extensive mid-latitude cyclones and anticyclones, show up on computer progs, while systems much smaller than this distance, such as thunderstorms, do not. The computer models are, therefore, better at predicting the widespread precipitation associated with a large cyclonic storm than local showers and thunderstorms. In summer, when much of the precipitation falls as local showers, a computer prog may have indicated fair weather, while outside it is pouring rain. To capture the smaller-scale weather fea-

tures, the distance between grid points on some models is being reduced. However, as the horizontal spacing between grid points decreases, the number of computations increases. When the distance is halved, there are eight times as many computations to perform, and the time required to run the model goes up by a factor of 16.

Another forecasting problem is that computer models cannot adequately interpret many of the factors that influence surface weather, such as the interactions of water, ice, and local terrain on weather systems. Some models do take large geographic features (such as mountain chains and oceans) into account, while ignoring smaller features (such as hills and lakes). These smaller features can have a marked influence on the local weather. Given the effect of local terrain, as well as the impact of some of the other problems previously mentioned, computer forecasts presently do an inadequate job of predicting local weather conditions, such as surface temperatures, winds, and precipitation.

Even with a denser network of observing stations and near perfect computer models, there are countless small, unpredictable atmospheric fluctuations that fall under the heading of *chaos*. These small disturbances, as well as small errors (uncertainties) in the data, generally amplify with time as the computer tries to project the weather farther and farther into the future. After a number of days, these initial imperfections (errors) tend to dominate, and the forecast shows little or no skill in predicting the behavior of the real atmosphere.

In summary, imperfect numerical weather predictions may result from flaws in the computer models, from errors that creep in along the models' boundaries, from a sparseness of data, and/or from inadequate representation of the many pertinent processes, interactions, and inherently chaotic behavior that occurs within the atmosphere.

Up to this point, we have looked at weather forecasts made by high-speed computers using atmospheric models. There are, however, other forecasting methods, many of which have stood the test of time and are based mainly on the experience of the forecaster. Many of these techniques are of value, but often they give more of a general overview of what the weather should be like, rather than a specific forecast.

Other Forecasting Methods Probably the easiest weather forecast to make is a **persistence forecast**, which is simply a prediction that future weather will be the same as present weather. If it is snowing today, a persistence forecast would call for snow through tomorrow. Such forecasts are most accurate for time periods of several hours and become less and less accurate after that.

Another method of forecasting is the **steady-state**, or **trend method**. The principle involved here is that surface weather systems tend to move in the same direction and at approximately the same speed as they have been moving, providing no evidence exists to indicate otherwise. Suppose, for example, that a cold front is moving eastward at an average speed of 30 km per

---

*GOES stands for *Geostationary Operational Environmental Satellite.*

hour and it is 90 km west of your home. Using the steady-state method, we might extrapolate and predict that the front should pass through your area in three hours.

In recent years, the trend method has been employed in the making of forecasts from minutes for up to a few hours. Such short-term forecasting has come to be called **nowcasting**.

The **analogue method** is yet another form of weather forecasting. Basically, this method relies on the fact that existing features on a weather chart (or a series of charts) may strongly resemble features that produced certain weather conditions sometime in the past. To the forecaster, the weather map "looks familiar", and for this reason the analogue method is often referred to as **pattern recognition**. A forecaster might look at a prog and say "I've seen this weather situation before, and this happened." Prior weather events can then be utilized as a guide to the future. The problem here is that, even though weather situations may appear similar, they are never *exactly* the same. There are always sufficient differences in the variables to make applying this method a challenge.*

The analogue method can be used to predict a number of weather elements, such as maximum temperature. Suppose that in New York City the average maximum temperature on a particular date for the past 30 years is 10°C (50°F). By statistically relating the maximum temperatures on this date to other weather elements—such as the wind, cloud cover, and humidity—a relationship between these variables and maximum temperature can be drawn. By comparing these relationships with current weather information, the forecaster can predict the maximum temperature for the day.

Predicting the weather by **weather types** employs the analogue method. In general, weather patterns are categorized into similar groups or "types," using such criteria as the position of the subtropical highs, the upper-level flow, and the prevailing storm track. As an example, when the Pacific high is weak or depressed southward and the flow aloft is zonal (west-to-east), surface storms tend to travel rapidly eastward across the Pacific Ocean and into the United States without developing into deep systems. But when the Pacific high is to the north of its normal position and the upper airflow is meridional (north-south), looping waves form in the flow with surface lows usually developing into huge storms. As we saw in Chapter 13, these upper-level longwaves move slowly, usually remaining almost stationary for perhaps a few days to a week or more. Consequently, the particular surface weather at different positions around the wave is likely to persist for some time. Figure 14.6 presents an example of weather conditions most likely to prevail with a winter meridional weather type.

Weather types can be used as an approach to *long-range* (a month or more in advance) *weather forecasting*. Typically, the

*Presently, however, statistical predictions are made routinely of weather elements based on the past performance of computer models (the Model-Output Statistics, or MOS). These, in effect, are statistically weighted analogue forecast corrections to the computer model output.

WEATHER WATCH

Groundhog day, February 2, has its roots (so to speak) in religious ceremonies performed before the birth of Christ. Somehow, this date came to be considered the midpoint of winter, and people, in an attempt to forecast what the remaining half would be like, placed the burden of weather prognosticator on the backs (or rather, the shadows) of animals such as the groundhog.

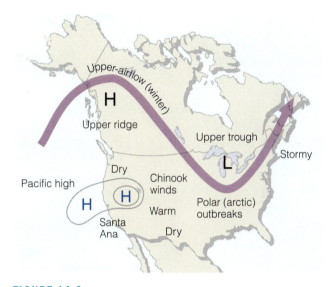

FIGURE 14.6
Winter weather type showing upper airflow (heavy arrow), surface position of Pacific high, and general weather conditions that should prevail.

upper-air circulation changes gradually from zonal to meridional over 4 to 6 weeks. As this slow change occurs in the upper air, the surface weather may repeat itself at specific intervals. For instance, winter cold fronts may sweep into New England every 4 days or so, bringing showers and below-normal temperatures. By projecting trends such as these, and assuming that the atmosphere's behavior will not change radically (an assumption not always valid), *extended weather forecasts* can be made. At best, these forecasts only show the broad-scale weather features. They do not adequately predict specific weather elements.

Currently, the Climate Prediction Center issues extended forecasts of 6 to 10 days, as well as a 30-day outlook for the coming month, and a 90-day seasonal outlook. These are not forecasts in the strict sense, but rather an overview of how average precipitation and temperature patterns may compare with normal conditions. The 30-day outlook is based on the relationship between the projected average flow at the 700-mb level and the surface weather conditions that this type of flow will create. Figure 14.7 gives a typical 90-day outlook. (The 90-day outlooks are based more on persistence statistics that

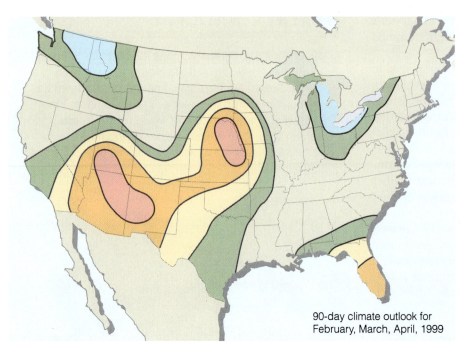

(a) Precipitation

FIGURE 14.7

The 90-day outlook for (a) precipitation and (b) temperature for February, March, and April, 1999. For precipitation (a), the darker the green color the greater the probability of precipitation being above normal, whereas the deeper the red color the greater the probability of precipitation being below normal. For temperature (b), the darker the orange/red color the greater the probability of temperature being above normal, whereas the darker the blue color, the greater the probability of temperature being below normal.

90-day climate outlook for
February, March, April, 1999

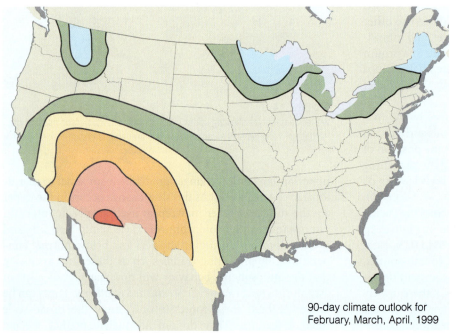

90-day climate outlook for
February, March, April, 1999

(b) Temperature

carry over the general weather patterns from immediately preceding months, seasons, and years.)

To improve medium- and extended-range forecasts, meteorologists are turning to a technique called **ensemble forecasting**. This approach is based on running several forecast models—or different versions (simulations) of a single model—each beginning with slightly different weather information to reflect the errors inherent in the measurements. If, at the end of a specified time, the models match each other fairly well, then the forecaster can issue a prediction with a high degree of confidence. If the models disagree, the forecaster, with little faith in the computer model prediction, issues a forecast with limited confidence, perhaps by giving a number ranging from 0 (no confidence) to 5 (great confidence). In essence, *the less agreement among the models, the less predictable the weather*. Consequently, it would not be wise to make outdoor plans for

Saturday when on Monday the weekend forecast calls for "sunny and warm" with a low degree of confidence.

A forecast based on the climatology (average weather) of a particular region is known as a **climatological forecast**. Anyone who has lived in Los Angeles for a while knows that July and August are practically rain-free. In fact, rainfall data for the summer months taken over many years reveal that rainfall amounts of more than a trace occur in Los Angeles about 1 day in every 90, or only about 1 percent of the time. Therefore, if we predict that it will not rain on some day next year during July or August in Los Angeles, our chances are nearly 99 percent that the forecast will be correct based on past records. Since it is unlikely that this pattern will significantly change in the near future, we can confidently make the same forecast for the year 2020.

When the Weather Service issues a forecast calling for rain, it is usually followed by a probability. For example: "The chance of rain is 60 percent." Does this mean (a) that it will rain on 60 percent of the forecast area or (b) that there is a 60 percent chance that it will rain within the forecast area? Neither one! The expression means that there is a 60 percent chance that any random place in the forecast area, such as your home, will receive measurable rainfall.* Looking at the forecast in another way, if the forecast for 10 days calls for a 60 percent chance of rain, it should rain where you live on 6 of those days. The verification of the forecast (as to whether it actually rained or not) is usually made at the Weather Service office. When the National Weather Service issues a forecast calling for a "slight chance of rain," what is the probability (percentage) that it will rain? Table 14.2 provides this information.

---

*The 60 percent chance of rain does not apply to a situation that involves rain showers. In the case of showers, the percentage refers to the expected area over which the showers will fall.

TABLE 14.2
Forecast wording used by the National Weather Service to describe the percentage probability of measurable precipitation (0.01 inch or greater) for steady precipitation and for convective, showery precipitation.

PERCENT PROBABILITY OF PRECIPITATION	FORECAST WORDING FOR STEADY PRECIPITATION	FORECAST WORDING FOR SHOWERY PRECIPITATION
20 percent	*Slight chance* of precipitation	*Widely scattered* showers
30 to 50 percent	*Chance* of precipitation	*Scattered* showers
60 to 70 percent	Precipitation *likely*	*Numerous* showers
≥ 80 percent	Precipitation,* rain, snow	*Showers**

*A forecast that calls for an 80 percent chance of rain in the afternoon might read like this: ". . . cloudy today with rain this afternoon. . . ." For an 80 percent chance of rain showers, the forecast might read ". . . cloudy today with rain showers this afternoon. . . ."

An example of a **probability forecast** using climatological data is given in Fig. 14.8. The map shows the probability of a "White Christmas"—1 inch or more of snow on the ground—across the United States. The map is based on the average of 30 years of data and gives the likelihood of snow in terms of a probability. For instance, the chances are 90 percent (9 Christmases out of 10) that portions of northern Minnesota, Michigan, and Maine will experience a White Christmas. In Chicago, it is 50 percent; and in Washington, D.C., about 20 percent. Many places in the far west and south have probabilities less

FIGURE 14.8
Probability of a "White Christmas"—one inch or more of snow on the ground—based on a 30-year average. The probabilities do not include the mountainous areas in the western United States.

than 5 percent, but nowhere is the probability exactly 0, for there is always some chance (no matter how small) that a mantle of white will cover the ground on Christmas day.

In most locations throughout North America, the weather is fair more often than rainy. Consequently, there is a forecasting bias toward fair weather, which means that, if you made a forecast of no-rain where you live for each day of the year, your forecast would be correct more than 50 percent of the time. But did you show any *skill* in making your correct forecast? What constitutes skill, anyway? And how accurate are the forecasts issued by the National Weather Service?

Accuracy and Skill in Forecasting In spite of the complexity and ever-changing nature of the atmosphere, forecasts made for between 12 and 24 hours are usually quite accurate. Those made for between 1 and 3 days are fairly good. Beyond about 7 days, however, forecast accuracy falls off rapidly. Although weather predictions made for up to 3 days are by no means perfect, they are far better than simply flipping a coin. But how accurate are they?

One problem with determining forecast accuracy is deciding what constitutes a right or wrong forecast. Suppose tomorrow's forecast calls for a minimum temperature of 5°C. If the official minimum turns out to be 6°C, is the forecast incorrect? Is it as incorrect as one 10 degrees off? By the same token, what about a forecast for snow over a large city, and the snow line cuts the city in half with the southern portion receiving heavy amounts and the northern portion none? Is the forecast right or wrong? At present, there is no clear-cut answer to the question of determining forecast accuracy.

How does forecast accuracy compare with forecast skill? Suppose you are forecasting the daily summertime weather in Los Angeles. It is not raining today and your forecast for tomorrow calls for "no rain." Suppose that tomorrow it doesn't rain. You made an accurate forecast, but did you show any skill in so doing? Earlier, we saw that the chance of measurable rain in Los Angeles on any summer day is very small indeed; chances are good that day after day it will not rain. For a forecast to show skill, it should be better than one based solely on the current weather (*persistence*) or on the "normal" weather (*climatology*) for a given region. Therefore, during the summer in Los Angeles, a forecaster will have many accurate forecasts calling for "no measurable rain," but will need skill to predict correctly on which summer days it will rain.

Meteorological forecasts, then, show skill when they are more accurate than a forecast utilizing only persistence or climatology. Persistence forecasts are usually difficult to improve upon for a period of time of several hours or less. Weather forecasts ranging from 12 hours to a few days generally show much more skill than those of persistence. However, as the range of the forecast period increases, the skill drops quickly. The 6– to 10–day mean outlooks both show some skill (which has been increasing over the last several decades) in predicting temperature and precipitation. However, the accuracy of precipitation forecasts is less than that for temperature. Presently, 7-day forecasts now show about as much skill as 5-day forecasts did a decade ago. Beyond 10 days, specific forecasts are only slightly better than climatology.

Forecasting large-scale weather events several days in advance (such as the blizzard of 1996 along the eastern seaboard of the United States) are far more accurate than forecasting the precise evolution and movement of small-scale, short-lived weather systems, such as tornadoes and severe thunderstorms. In fact, 3-day forecasts of the development and movement of a major low-pressure system show more skill today than 36-hour forecasts did 15 years ago.

Even though the *precise* location where a tornado will form is presently beyond modern forecasting techniques, the general area where the storm is *likely* to form can often be predicted up to 3 days in advance. With improved observing systems, such as Doppler radar and advanced satellite imagery, the lead time of watches and warnings for severe storms has increased. In fact, the lead time[*] for tornado warnings has more than doubled over the last decade.

Although scientists may never be able to skillfully predict the weather beyond about 10 days using available observations, the prediction of climatic trends appears to be more promising. Whereas individual weather systems vary greatly and are difficult to forecast very far in advance, global-scale patterns of winds and pressure frequently show a high degree of persistence and predictable change over periods of a few weeks to a month or more. With the latest generation of high-speed supercomputers, such as the CRAY at the National Center for Atmospheric Research, general circulation models (GCMs) are doing a far better job at predicting large-scale atmospheric behavior than did the earlier models. (The GCMs are numerical computer models that simulate global patterns of wind, pressure, and temperature, and how these phenomena change over time.) In fact, the new GCMs are able to simulate a number of global patterns quite well, such as *blocking highs* that can cause precipitation and temperature patterns to deviate considerably from average conditions. As new knowledge and methods of modeling are fed into the GCMs, it is hoped that they will become a reliable tool in the forecasting of weather and climate. (In Chapter 19, we will examine in more detail the climatic predictions based on numerical models.)

In Chapter 11, we saw how a vast warming of the equatorial tropical Pacific called *El Niño* can affect the weather in different regions of the world. These interactions, where a warmer tropical Pacific can influence rainfall in California, are called **teleconnections**. These types of interactions between widely separated regions are identified through statistical correlations. Look back at Fig. 11.23 (p. 295) and observe where temperature and precipitation patterns over North America

---

[*]Lead time is the interval of time between the issue of the warning and actual observance of the tornado.

tend to depart from normal during El Niño and La Niña events. Using this type of information, the Climate Prediction Center can issue a forecast of an impending wetter or drier winter, months in advance. Forecasts using teleconnections have shown promise. For example, as the tropical equatorial Pacific became much warmer than normal during the spring and early summer of 1997, forecasters predicted a wet rainfall season over central and southern California. Although the heavy rains didn't begin until December, the weather during the winter of 1997–1998 was wet and wild: Storm after storm pounded the region producing heavy rains, mud slides, road closures, and millions of dollars in damages.

BRIEF REVIEW   Up to this point, we have looked at the various methods of weather forecasting. Before going on, here is a review of some of the important ideas presented so far:

- The forecasting of weather by high-speed computers is known as *numerical weather prediction*. Mathematical models that describe how atmospheric temperature, pressure, and moisture will change with time are programmed into the computer. The computer then plots and draws surface and upper-air charts, and produces a variety of forecast charts called *progs*.

- After a number of days, flaws in the computer models and small errors in the data greatly limit the accuracy of weather forecasts.

- A *persistence forecast* is a prediction that future weather will be the same as the present weather, whereas a *climatological forecast* is based on the climatology of a particular region.

- For a forecast to show skill, it must be better than a persistence forecast or a climatological forecast.

- Ensemble forecasting is a technique based on running several forecast models (or different versions of a single model), each beginning with slightly different weather information to reflect errors in the measurements. If the different versions agree fairly well, a forecaster can place a high degree of confidence in the forecast. A low degree of confidence means that the models do not agree.

PREDICTING THE WEATHER FROM LOCAL SIGNS

Because the weather affects every aspect of our daily lives, attempts to predict it accurately have been made for centuries. One of the earliest attempts was undertaken by Theophrastus, a pupil of Aristotle, who in 300 B.C. compiled all sorts of weather indicators in his *Book of Signs*. A dominant influence in the field of weather forecasting for 2000 years, this work consists of ways to foretell the weather by examining natural signs, such as the color and shape of clouds, and the intensity

at which a fly bites. Some of these signs have validity and are a part of our own weather folklore—"a halo around the moon portends rain" is one of these. Today, we realize that the halo is caused by the bending of light as it passes through ice crystals and that ice crystal-type clouds (cirrostratus) are often the forerunners of an approaching storm.

Weather predictions can be made by observing the sky and using a little weather wisdom. If you keep your eyes open and your senses keenly tuned to your environment, you should, with a little practice, be able to make fairly good short-range local weather forecasts by interpreting the messages written in the weather elements. Table 14.3 is designed to help you with this endeavor.

The movement of clouds at different levels can assist you in predicting changes in the temperature of the air above you. Knowing how the temperature is changing aloft can help you predict the stability of the air (Chapter 7), as well as whether falling snow will change to rain, or vice versa. (This topic is explored more extensively in the Focus section beginning on p. 366.)

To further help you forecast the weather, the instant weather forecast chart (Appendix E) has been prepared by considering the relationship that the pressure and wind have to various weather systems. While the chart is applicable to much of the United States, local influences, such as mountains and large bodies of water, can affect the local weather to such an extent that the large-scale weather patterns on which the chart is based do not always show up clearly. (The chart works best during the fall, winter, and spring when the weather systems are active.)

The chart is simply a guide; in no way is it infallible. For example, rising pressures in most cases indicate improving weather conditions. On the Great Plains, however, a snowstorm may begin with a rising barometer as the storm system passes south or east of an area and upslope conditions prevail. Along the eastern seaboard, a slowly rising barometer combined with northwesterly winds could spell heavy precipitation as a storm drifts northeastward parallel to the coast.

By adding such information as cloud type, temperature, and the wind aloft (as indicated by cloud movement), each forecast on the chart can be improved upon. For example, suppose that the barometer reading is 1016 millibars, the pressure is falling, and the surface wind is blowing from the southeast. The instant weather forecast chart predicts that precipitation is on the way. This is a reasonable forecast, for Buys-Ballot's law (Chapter 9) suggests that the location of the surface storm is to the west. But how far west?

Suppose that the air temperature is near 10°C (50°F), and that overhead a layer of altostratus (at about 4 km, or 13,000 ft) is moving from southwest to northeast. Again, using Buys-Ballot's law (but this time for the winds aloft), we determine that there is an upper trough of low pressure to the west. As a general rule of thumb, the lower the stratiform cloud layer, the closer you are to the storm or frontal system. High cirrostratus

TABLE 14.3 Forecast at a Glance—Forecasting the Weather from Local Weather Signs.
Listed below are a few forecasting rules that may be applied when making a short-range local weather forecast.

OBSERVATION	INDICATION	LOCAL WEATHER FORECAST
Surface winds from the S or from the SW; clouds building to the west; warm (hot) and humid	Possible cool front and thunderstorms approaching from the west	Possible showers; possibly turning cooler; windy
Surface winds from the E or from the SE, cool or cold; high clouds thickening and lowering; halo around the sun or moon	Possible approach of a warm front	Possibility of precipitation within 12–24 hours; windy (rain with possible thunderstorms during the summer; snow changing to sleet or rain in winter)
Winter night		
(a) If clear, relatively calm with low humidity (low dew point temperature)	(a) Rapid radiational cooling will occur	(a) A very cold night
(b) If clear, relatively calm with low humidity and snow covering the ground	(b) Rapid radiational cooling will occur	(b) A very cold night with minimum temperatures lower than in (a)
(c) If cloudy, relatively calm with low humidity	(c) Clouds will absorb and radiate infrared (IR) energy back to surface	(c) Minimum temperature will not be as low as in (a) or (b)
Summer night		
(a) Clear, hot, humid (high dew points)	(a) Strong absorption and emission of IR energy back to surface by water vapor	(a) High minimum temperatures
(b) Clear and relatively dry	(b) More rapid radiational cooling	(b) Lower minimum temperatures
If surface winds are from the N and they become NE, then E, then SE (*veering winds*)	A surface high-pressure area may be moving to your E, and a surface low-pressure area may be approaching from the W	Increasing clouds, warmer with the possibility of precipitation within 24 hours
If surface winds are from the NE and they become N, then NW (*backing winds*)	A surface low-pressure area is moving to your E, and a surface high-pressure area may be approaching from the W	Clearing and colder (cooler in summer)
Scattered cumulus clouds that show extensive vertical growth by mid morning	Atmosphere is relatively unstable	Possible showers or thunderstorms by afternoon with gusty winds
Afternoon cumulus clouds with flat bases, and tops at just about the same level	Stable layer above clouds (region dominated by high pressure)	Continued partly cloudy with no precipitation; probably clearing by nightfall

clouds usually indicate that an approaching frontal system is 1 or 2 days away, while a layer of low stratus (especially with fog and drizzle) usually means that the front is close indeed. Altostratus clouds suggest that the approaching system is probably between 500 and 700 km to the west. The falling barometer, southeasterly surface winds, and southwesterly winds aloft suggest that the frontal system is moving toward us and that precipitation (probably rain) is likely within the next 6 to 8 hours. (If you have forgotten the typical weather associated with fronts, review Tables 12.2, p.317, 12.3, p. 319, and 12.4, p. 321, as well as the cyclone model, Fig. 13.10, p. 336.)

WEATHER FORECASTING USING SURFACE CHARTS

We are now in a position to forecast the weather, utilizing more sophisticated techniques. Suppose, for example, that we wish to make a short-range weather prediction and the only information available is a surface weather map. Can we make a forecast from such a chart? Most definitely. And our chances of that forecast being correct improve markedly if we have maps available from several days back. We can use these past maps to locate the previous position of surface features and predict their movement.

A simplified surface weather map is shown in Fig. 14.9. The map portrays early winter weather conditions on Tuesday

morning at 6:00 A.M. A single isobar is drawn around the pressure centers to show their positions without cluttering the map. Note that an open-wave cyclone is developing over the Central Plains. The weather conforms to the cyclone model (see Fig. 13.10, p. 336), with showers forming along the cold front and light rain and snow ahead of the warm front. The dashed lines on the map represent the position of the weather systems 6 hours ago. Our first question is: How will these systems move?

Determining the Movement of Weather Systems There are several methods we can use in forecasting the movement of

FIGURE 14.9
Surface weather map for 6:00 A.M. Tuesday. Dashed lines indicate positions of weather features six hours ago. Areas shaded green are receiving precipitation.

Forecasting Temperature Advection by Watching the Clouds

We know from Chapter 13 that when cold air is being brought into a region by the wind, we call this *cold advection*. When warm air is brought into a region, we call this *warm advection*.

A knowledge of temperature advection aloft is a valuable tool in forecasting the weather. In summer, when the surface is warm, cold advection aloft sets up instability and increases the likelihood of towering cumulus clouds and showers. On the other hand, warm advection aloft usually increases the temperature of the air, thus making it more stable. During the winter, this often leads to smoke and haze accumulating in the colder air near the surface.

By watching the movement of clouds, we get a good indication as to the wind direction at cloud level and also the type of advection. For instance, a cloud moving from the west indicates a west wind, a cloud from the south a south wind, and so on. Clouds at different levels frequently move in different directions, meaning that the wind direction is changing with height. Wind that changes direction in a clockwise sense (north to northeast to east, etc.) is a *veering wind*. Wind that changes direction in a counter-

FIGURE 2

(a) The wind veers with height, suggesting warm advection is occurring between the cloud layers. (b) The wind is backing with height, and cold advection is occurring between the cloud layers.

clockwise sense (north to northwest to west, etc.) is a *backing wind*. There are two general rules that will help us determine whether cold or warm advection is occurring in a layer of air above us:

1. Winds that *back* with height (change counterclockwise) indicate *cold advection*.

2. Winds that *veer* with height (change clockwise) indicate *warm advection*.

As an example, suppose we observe lower clouds moving from a southerly direction (south wind) and higher clouds moving from a westerly direction (west wind) (see Fig. 2a). The wind direction is veering with height; warm advection is occurring between

the cloud layers, and the air should be getting warmer. If, on some other day, we see lower clouds moving from a southerly direction and higher clouds moving from an easterly direction (Fig. 2b), the wind is backing with height and the atmosphere between the cloud layers is probably becoming colder.*

An example of the relationship between winds and advection is seen in the vertically shifting winds that accompany weather fronts. Figure 3b is a surface map showing a typical open-wave cyclone with its accompanying warm and cold fronts. Behind the cold front, swiftly moving cumulus clouds indicate a northwesterly wind exists about a kilometer or so above the surface.

*In both of these examples, we are assuming horizontal air motion only.

---

surface pressure systems and fronts. The following are a few of these forecasting rules of thumb:

1. For short-time intervals, storms and fronts tend to move in the same direction and at approximately the same speed as they did during the previous six hours (providing, of course, there is no evidence to indicate otherwise).
2. Lows tend to move in a direction that parallels the isobars in the warm air ahead of the cold front.
3. Lows tend to move toward the region of greatest pressure drop, while highs tend to move toward the region of greatest rise.
4. Surface pressure systems tend to move in the same direction as the wind at 5500 m (18,000 ft)—the 500-mb level. The

speed at which surface systems move is about half the speed of the winds at this level.

When the surface map (Fig. 14.9) is examined carefully and when rules of thumb 1 and 2 are applied, it appears that—based on present trends—the storm center over the Central Plains should move northeast. If *pressure tendencies** are plotted on our map, we can draw lines connecting points of equal pressure change. These lines, called **isallobars**, help us to visualize the regions of falling and rising pressure. The distribution of pressure change for our map might look like the one in Fig.

---

*The pressure tendency is the rate at which the pressure is changing during a given time, usually the past three hours.

Ahead of the advancing warm front, stratiform clouds indicate that here southeasterly winds prevail about a kilometer above the ground. We know from Chapter 13 that warm advection takes place *ahead* of the warm front and cold advection *behind* the cold front. In both cases, the advection usually occurs in a layer from the surface up to at least the 500-mb level.

At the 500-mb level (Fig. 3a), the position of the upper trough and the region of coldest air is to the west of the surface low. The direction of the wind and also the cloud movement is shown by arrows. Because of the upper trough's position, the winds aloft are westerly behind the cold front and southwesterly ahead of the warm front. Figure 3c shows how the wind direction changes from the surface to the 500-mb level. Behind the cold front, the winds back from northwesterly to westerly as we move upward. Cold advection is taking place as chilling air moves in from the west. Just ahead of the approaching warm front, the wind veers with height from southeasterly to southwesterly as warm air glides up and over the cool surface air.

We can use this information to improve upon a weather forecast. For instance, if you happen to be located

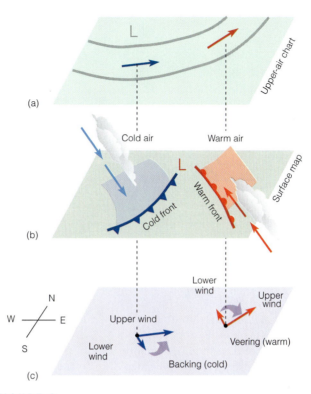

FIGURE 3

Clouds, winds, and advection associated with a cold and a warm front.

ahead of an advancing warm front and the winds above you are veering with height, the chances are that even if precipitation begins as snow it may change to rain as warm air moves in overhead. Behind a cold front where winds are backing with height, the influx of cold air may lower the temperature sufficiently so that rain first becomes mixed with snow, and then changes to snow before the storm moves eastward.

14.10. Drawn at 2-mb intervals, the isallobars show a broad region of falling pressure ahead of the warm front, with the largest drop occurring to the northeast of the storm. This pattern fits with the previous observations and strengthens the prediction that the low center will move toward the northeast. The area of rising pressure immediately behind the cold front suggests that the anticyclone near the United States–Canadian border will continue to move southeastward.

Pressure tendencies not only help predict the movement of highs and lows, they also indicate how the pressure systems are changing with time. The rapid fall in pressure in advance of the low indicates that the storm center is deepening as it moves. A deepening low means more closely spaced isobars, a greater pressure gradient, and stronger winds—something to take into account when we make our weather forecast. A drop in pressure, on the other hand, in the vicinity of an anticyclone suggests that it is weakening, while a rise in pressure means that its central pressure is increasing. Hence, the anticyclone moving out of Canada is strong (1034 mb) and will remain so, whereas the anticyclone centered off the Georgia coast is either moving eastward or weakening rapidly as indicated by the falling pressure in that area.

Before we complete our prediction about the movement of the pressure centers in Fig. 14.9, we need to look closely at the anticyclone off the Georgia coast. Strong highs, especially slow-moving ones, often retard the eastward progress of lows,

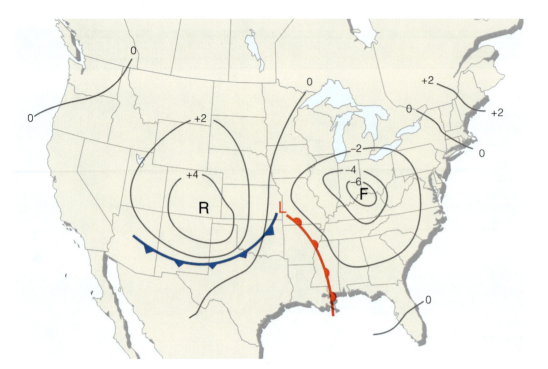

FIGURE 14.10
Isallobars—lines of equal 3-hour pressure change—for 6:00 A.M. Tuesday. The "F" represents the region of greatest pressure fall, while the "R" shows the region of greatest pressure rise. A plus 2 indicates a rise of 2 mb.

deflecting them either north or south. From all indications—falling pressures and past movement—this anticyclone is weakening and drifting slowly eastward. It should, therefore, pose no immediate problem to the northeastward movement of the storm center.

Even if we do not have access to pressure tendencies or previous weather maps, we can make an initial approximation of how pressure systems will move by using Fig. 13.3 (see p. 328), which shows the average tracks of lows and highs during the winter months. From this diagram, it appears that the cyclones and anticyclones in Fig. 14.9 are following rather typical trajectories.

If a 500-mb chart is available (such as Fig. 14.11), it would also indicate how the surface pressure systems should move, since the winds at this level tend to steer these systems along. From Fig. 14.11, it appears that, indeed, the surface low should move northeastward.

A Forecast for Six Cities

Our objective now is to make a weather forecast for six cities. To do this, we will project the pressure systems, fronts, and current weather into the future by assuming steady-state conditions. Figure 14.12 gives the 12- and 24-hour projected positions of these features.

A word of caution before we make our forecasts. We are assuming that the pressure systems and fronts are moving at a constant rate. This may or may not occur. Storm systems, for example, tend to accelerate until they occlude, after which their rate of movement slows. Furthermore, the direction of moving systems may change due to "blocking" highs and lows that exist in their path or because of shifting upper-level wind patterns. We will assume a constant rate of movement and forecast

accordingly, always keeping in mind that the longer our forecasts extend into the future, the more susceptible they are to error.

Using Fig. 14.12 to follow the storm center eastward, we can make a basic forecast. The cold front moving into north Texas on Tuesday morning is projected to pass Dallas by that evening, so a forecast for the Dallas area would be "warm with showers, then turning colder." But we can do much better than this. Knowing the weather conditions that accompany advancing pressure areas and fronts, we can make more detailed weather forecasts that will take into account changes in temperature, pressure, humidity, cloud cover, precipitation, and winds. Our forecast will include the 24-hour period from Tuesday morning to Wednesday morning for the cities of Augusta, Georgia; Washington, D.C.; Chicago, Illinois; Memphis, Tennessee; Dallas, Texas; and Denver, Colorado. We will begin with Augusta.

Weather Forecast for Augusta, Georgia On Tuesday morning, continental polar air associated with a high-pressure center brought freezing temperatures and fair weather to the Augusta area (see Fig. 14.9). Clear skies, light winds, and low humidities allowed rapid nighttime cooling so that, by morning, temperatures were in the low thirties. Now look closely at Fig. 14.12 and observe that the anticyclone is moving slowly eastward. Southerly winds on the western side of this system will bring warmer and more moist air to the region. Therefore, afternoon temperatures will be warmer than those of the day before. As the warm front approaches from the west, clouds will increase, appearing first as cirrus, then thickening and lowering into the normal sequence of warm-front clouds. Barometric pressure should fall. Clouds and high humidity should keep minimum temperatures well above freezing on Tuesday

	Miles (statute) per hour	Knots
◎	Calm	Calm
	1–2	1–2
	3–8	3–7
	9–14	8–12
	15–20	13–17
	21–25	18–22
	26–31	23–27
	32–37	28–32
	38–43	33–37
	44–49	38–42
	50–54	43–47
	55–60	48–52
	61–66	53–57
	67–71	58–62
	72–77	63–67
	78–83	68–72
	84–89	73–77
	119–123	103–107

FIGURE 14.11

A 500-mb chart for 6:00 A.M. Tuesday, showing wind flow. The light red L represents the position of the surface low. The winds aloft tend to steer surface pressure systems along and, therefore, indicate that the surface low should move northeastward at about half the speed of the winds at this level, or 25 knots.

night. Note that the projected area of precipitation (green-shaded region) does not quite reach Augusta. With all of this in mind, our forecast might sound something like this:

> Clear and cold this morning with moderating temperatures by afternoon. Increasing high clouds with skies becoming overcast by evening. Cloudy and not nearly as cold tonight and tomorrow morning. Winds will be light and out of the south or southeast. Barometric pressure will fall slowly.

Wednesday morning we discover that the weather in Augusta is foggy with temperatures in the upper 40s (°F). But fog was not in the forecast. What went wrong? We forgot to consider that the ground was still cold from the recent cold snap. The warm moist air moving over the cold surface was chilled below its dew point, resulting in fog. Above the fog were the low clouds we predicted. The minimum temperatures remained higher than anticipated because of the release of latent heat during fog formation and the absorption of infrared energy by the fog droplets. Not bad for a start. Now we will forecast the weather for Washington, D.C.

Rain or Snow for Washington, D.C.? Look at Fig. 14.12 and observe that the storm center is slowly approaching Washington, D.C., from the west. Hence, the clear weather, light south-westerly winds, and low temperatures on Tuesday morning (see Fig. 14.9) will gradually give way to increasing cloudiness, winds shifting to the southeast, and slightly higher temperatures. By Wednesday morning, the projected band of precipitation will be over the city. Will it be in the form of rain or snow? Without any data for temperatures aloft, this is difficult to determine. We can see in Fig. 14.9, however, that cities south of Washington, D.C.'s latitude are receiving snow. So a reasonable forecast would call for snow, possibly changing to rain as warm air moves in aloft in advance of the approaching fronts. A 24-hour forecast for Washington, D.C., might sound like this:

> Increasing clouds today and continued cold. Snow beginning by early Wednesday morning, possibly changing to rain. Winds will be out of the southeast. Pressures will fall.

Wednesday morning a friend in Washington, D.C., calls to tell us that the sleet began to fall but has since changed to rain. Sleet? Another fractured forecast! Well, almost. What we forgot to account for this time was the intensification of the storm. As the storm moved eastward, it deepened; central pressure lowered, pressure gradients tightened, and southeasterly winds blew stronger than anticipated. As air moved inland off the warmer Atlantic, it rode up and over the colder surface air. Snow falling into this warm layer at least partially melted; it then refroze as it entered the colder air near ground level. The

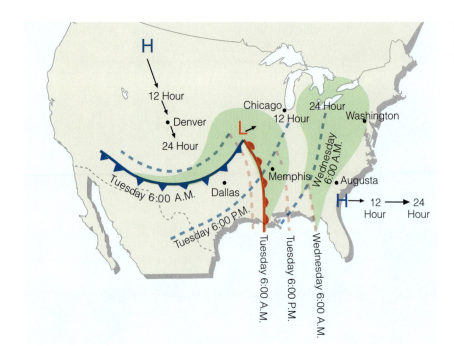

FIGURE 14.12
Projected 12- and 24-hour movement of fronts, pressure systems, and precipitation (shaded green area) from 6:00 A.M. Tuesday until 6:00 A.M. Wednesday.

advection of warmer air from the ocean slowly raised the surface temperatures, and the sleet soon became rain. Although we did not see this possibility when we made our forecast, a forecaster more familiar with local surroundings would have. Let's move on to Chicago.

Big Snowstorm for Chicago   From Figs. 14.9 and 14.12, it appears that Chicago is in for a major snowstorm. Overrunning of warm air has produced a wide area of snow which, from all indications, is heading directly for the Chicago area. Since cold air north of the low center will be over Chicago, precipitation reaching the ground should be frozen. On Tuesday morning the leading edge of precipitation is less than 6 hours away from Chicago. Based on the projected path of the storm, light snow should begin to fall around noon.

By evening, as the storm intensifies, snowfall should become heavy. It should taper off and finally end around midnight as the storm moves on east. If it snows for a total of 12 hours— 6 hours as light snow (around 1 inch every 3 hours) and 6 hours as heavy snow (around 1 inch per hour)—then the total expected accumulation will be between 6 and 10 inches. As the depression moves eastward, passing south of Chicago, winds on Tuesday will gradually shift from southeasterly to easterly, then northeasterly by evening. Since the system is intensifying, it should produce strong winds that will swirl the snow into huge drifts, which may bring traffic to a crawl.

The winds will continue to shift to the north and finally become northwesterly by Wednesday morning. By then the storm center will probably be far enough east so that skies should begin to clear. Cold air advected from the northwest behind the storm will cause temperatures to drop further. Barometer readings during the storm will fall as the low center approaches and reach a low value sometime Tuesday night, after which they will begin to rise. A weather forecast for Chicago might be:

> Cloudy and cold with light snow beginning by noon, becoming heavy by evening and ending by Wednesday morning. Total accumulations will range between 6 and 10 inches. Winds will be strong and gusty out of the east or northeast today becoming northerly tonight and northwesterly by Wednesday morning. Barometric pressure will fall sharply today and rise tomorrow.

A call Wednesday morning to a friend in Chicago reveals that our forecast was correct except that the total snow accumulation so far is 13 inches. We were off in our forecast because the storm system slowed as it became occluded. We did not consider this because we moved the system by the steady-state method. At this time of year (early winter), Lake Michigan is not quite frozen over and the added moisture picked up from the lake by the strong easterly winds also helped to produce a heavier-than-predicted snowfall. Again, a knowledge of the local surroundings would have helped make a more accurate forecast. The weather 700 km south of Chicago should be much different from this.

Mixed Bag of Weather for Memphis   Observe in Fig. 14.12 that, within 24 hours, both a warm and a cold front should move past Memphis. The light rain that began Tuesday morning should saturate the cool air, creating a blanket of low clouds and fog by midday. The warm front, as it moves through sometime Tuesday afternoon, should cause temperatures to rise slightly as winds shift to the south or southwest. At night, clear to partly cloudy skies should allow the ground and air above to cool, offsetting any tendency for a rapid rise in temperature. Falling pressures should level off in the warm sector, then fall once again as the cold front approaches. According to the projection in Fig. 14.12, the cold front should arrive sometime before midnight on Tuesday, bringing with it gusty northwesterly winds, showers, the possibility of thunderstorms, rising pressures, and colder air. Taking all of this into account, our weather forecast for Memphis will be:

> Cloudy and cool with light rain, low clouds, and fog early today, becoming partly cloudy and warmer by late this afternoon. Clouds increasing with possible showers and thunderstorms later tonight or early Wednesday morning and turning colder. Winds southeasterly this morning, becoming southerly or southwesterly this evening and shifting to northwesterly by Wednesday morning. Pressures falling this morning, leveling off this evening, then falling again tonight and rising by Wednesday morning.

A friend who lives near Memphis calls Wednesday to inform us that our forecast was correct except that the thunderstorms did not materialize and that Tuesday night dense fog formed in low-lying valleys, but by Wednesday morning it had dissipated. Apparently, in the warm sector, winds were not strong enough to mix the cold, moist air that had settled in the valleys with the warm air above. It's on to Dallas.

Cold Wave for Dallas   From Fig. 14.12, it appears that our weather forecast for Dallas should be straightforward, since a cold front is expected to pass the area around noon. Weather along the front is showery with a few thunderstorms developing; behind the front the air is clear but cold. By Wednesday morning it looks as if the cold front will be far to the east and south of Dallas and an anticyclone will be centered over Colorado. North or northwesterly winds on the east side of the high will bring cold continental polar air into Texas, dropping temperatures as much as 40°F within a 24-hour period. With minimum temperatures well below freezing, Dallas will be in the grip of a cold wave. Our weather forecast should therefore sound something like this:

> Increasing cloudiness and mild this morning with the possibility of showers and thunderstorms this afternoon. Clearing and turning much colder tonight and tomorrow. Winds will be southwesterly today, becoming gusty north or northwesterly this afternoon and tonight. Pressures falling this morning, then rising later today.

How did our forecast turn out? A quick call to Dallas on Wednesday morning reveals that the weather there is cold but not as cold as expected, and the sky is overcast. Cloudy weather? How can this be?

The cold front moved through on schedule Tuesday afternoon, bringing showers, gusty winds, and cold weather with it. Moving southward, the front gradually slowed and became stationary along a line stretching from the Gulf of Mexico westward through southern Texas and northern Mexico. (From the surface map alone we had no way of knowing this would happen.) Along the stationary front a wave formed. This disturbance caused warm, moist Gulf air to slide northward up and over the cold surface air. Clouds formed, minimum temperatures did not go as low as expected, and we are left with a fractured forecast. Let's give Denver a try.

Clear but Cold for Denver   In Fig. 14.12, we can see that, based on our projections, the cold anticyclone will be almost directly over Denver by Wednesday morning. Sinking air aloft should keep the sky relatively free of clouds. Weak pressure gradients will produce only weak winds and this, coupled with dry air, will allow for intense radiational cooling. Minimum temperatures will probably drop to well below 0°F. Our forecast should therefore read:

> Clear and cold through tomorrow. Northerly winds today becoming light and variable by tonight. Low temperatures tomorrow morning will be below zero. Barometric pressure will continue to rise.

Wednesday morning, we (almost reluctantly) inquire about the weather conditions at Denver. "Clear and very cold" is the reply. A successful forecast at last! We are told, however, that the minimum temperature did not go below zero; in fact, 13°F was as cold as it got. A downslope wind coming off the mountains to the west of Denver kept the air mixed and the minimum temperature higher than expected. Again, a forecaster familiar with the local topography of the Denver area would have foreseen the conditions that lead to such downslope winds and would have taken this into account when making the forecast.

A complete picture of the surface weather systems for 6:00 A.M. Wednesday morning is given in Fig. 14.13. By comparing this chart with Fig. 14.12, we can summarize why our forecasts did not turn out exactly as we had predicted. For one thing, the storm center near the Great Lakes moved slower than expected. This slow movement allowed a southeasterly flow of mild Atlantic air to overrun cooler surface air ahead of the storm while, behind the low, cities remained in the snow area for a longer time. The weak wave that developed along the trailing cold front brought cloudiness and precipitation to Texas and prevented the really cold air from penetrating deep into the south. Further west, the high originally over Montana moved more southerly than southeasterly, which set up a pressure gradient that brought westerly downslope winds to eastern Colorado.

The forecasting techniques discussed so far are those you can use when making a weather prediction. The following section describes how a meteorologist predicts the weather in a region where, to the west, surface weather features are extensively modified by a vast body of water and only scanty surface and upper-air data are available. Here, the forecaster must rely heavily on experience as well as more sophisticated tools, which include satellite data, upper-air charts, and computer progs.

FIGURE 14.13
Surface weather map for 6:00 A.M.
Wednesday.

A METEOROLOGIST MAKES A PREDICTION

It is late afternoon, and outside the weather forecast office near San Francisco the meteorologist mulls over what is going on in the sky. Overhead is a thin covering of cirrostratus; to the west, draped over the foothills, is the ever present stratus and fog. The air is cool and the winds are westerly. It is Sunday, March 25, and the forecaster's task is to predict the weather for the coastal area of central California.

What will tomorrow's weather be like? Will it be similar to today's or will it change markedly? A slowly falling barometer of 1016 mb (30 in.) and the high clouds moving in from the west point to an approaching storm system. A forecast of persistence might be good for the next several hours, but what about tomorrow morning or tomorrow afternoon?

The late afternoon surface analysis provides little assistance with these questions. The surface map for 4:00 P.M. (PST) Sunday (see Fig. 14.14) shows there are no weather fronts approaching the West Coast. In fact, the nearest front is a stationary one that has stalled over the Rockies. There is, however, a region of low pressure centered about 1100 km (700 mi) west of San Francisco, which (according to previous maps) has been there for several days. With a central pressure of only about 1012 mb (29.88 in.), the system is fairly weak. Could this weak depression be causing the increase in high cloudiness and the falling barometer? And will this pattern lead to rain tomorrow? A look at the 500-mb chart may help with these questions.

Help from the 500-mb Chart

Figure 14.15 shows the 500-mb analysis for 4:00 P.M. Sunday afternoon. While examining the chart, the meteorologist recognizes certain clues that will aid in making the forecast. For one thing, the 5640-m height line is over northern California. The forecaster knows that when this contour line is situated here or further south, the statistical probability of receiving measurable rainfall over central California increases greatly. There are, in fact, some forecasters who base their precipitation forecasts solely on the position of that line.

West of San Francisco the flow is meridional with a cut-off warm, upper high situated just south of Alaska. To the south both east and west of the high are troughs. Because the shape of this flow around the high resembles the Greek letter omega (Ω), the high and its accompanying ridge is known as an **omega high**. The forecaster recognizes the omega high as a *blocking high*, one that tends to persist in the same geographic location for many days. This blocking pattern also tends to keep the troughs in their respective positions, which has been the case for several days now. But the chart indicates that the cold upper trough located west of San Francisco may be changing somewhat.

Observe the spacing of the contour lines around this trough. Even with a limited number of actual wind observations, the close spacing of the contours to the west and northwest of the trough and the more widely spaced contours to the east of the trough hint that stronger winds exist to the west of the trough. The forecaster knows from past experience that this usually means the trough will deepen. Also note that west of the trough the winds are blowing across the isotherms (from colder regions), indicating that cold advection is occurring. The heavy dashed line on the west side of the trough represents the position of a shortwave trough, which is moving rapidly southward. The injection of cold air and the shortwave into the main trough should cause it to intensify. To the east of the main trough, the wind is crossing the isotherms and advecting warmer subtropical air northeastward. It is the lifting and condensing of this moist air that is producing the high clouds over San Francisco. All of these conditions—high wind speeds, regions of temperature advection, and a shortwave moving

FIGURE 14.14
Surface weather map for 4:00 P.M. Sunday, March 25.

FIGURE 14.15
The 500-mb chart for 4:00 P.M. Sunday, March 25. Arrows indicate wind flow. Red arrows indicate warm advection and blue arrows cold advection. Solid lines are height contours where 564 equals 5640 m. Dashed lines are isotherms in °C. Heavy dashed line shows position of shortwave trough.

into a longwave trough—manifest themselves as a deepening of the longwave trough. As the upper trough deepens, it should be capable of providing the necessary conditions favorable for the development of the surface low into a major storm system. (You may remember from Chapter 13 that the generation of this type of dynamic instability is called baroclinic instability.)

One of the main ingredients necessary for the development and intensification of the surface low is divergence of the airflow aloft. The forecaster knows that divergence aloft causes surface pressures to decrease beneath it. This in turn causes surface air to converge, rise, and condense into widespread cloudiness. But where will regions of divergence, convergence, and rising air be found on tomorrow's map? And how will tomorrow's map be different from today's? This is where the computer and the forecaster work together to come up with a prediction.

The Computer Provides Assistance The computer progs predict the future positions of weather systems. Some of the progs also predict where shortwave troughs will be located. It is important to know where the shortwaves will be found, be-

cause to the east of them there is usually upper-level divergence, lower-level convergence, rising air, clouds, and precipitation. Hence, predicting the position of a shortwave means predicting regions of inclement weather. (The position of the shortwaves also pretty much corresponds to the position of the vorticity maximums discussed in Chapter 13.)

Three models that predict the positions of the shortwaves, upper-level pressure systems, and flow aloft at the 500-mb level for 12, 24, and 36 hours into the future are shown in Fig. 14.16.* (Each prediction is made on Sunday afternoon.) Observe that there is good agreement among the models in that each model moves the upper trough slowly eastward and keeps it off the coast for the entire period. However, the actual positioning of

*Explaining the differences among the three models is beyond the scope of this book. Each model treats the atmosphere in a slightly different way. Some models have closer grid points. Some models have better resolution in the lower part of the atmosphere, whereas others have better resolution in the higher regions of the atmosphere. The idea in this forecasting example is *not* to illustrate the different models in use, but rather to show how a forecaster might use *any* numerical computer model as a forecasting tool.

FIGURE 14.16

Three computer-drawn progs (model A, model B, and model C) that show the 12-hour, 24-hour, and 36-hour projected 500-mb chart. Solid lines are contours. Dashed lines represent projected positions of shortwaves. (These predictions were made on Sunday, March 25, at 4:00 P.M., PST.)

the trough and the shortwaves (heavy dashed lines) differ for each model. For example, model A and model C move the upper trough eastward more quickly than does Model B. Also, the 12-hour prog for Model A and Model C both show several shortwaves moving around the upper trough with one shortwave (labeled 1), positioned west of San Francisco by Monday morning. Model B predicts that the same shortwave will be much farther to the west.

After examining each prog carefully, the forecaster must decide which model most accurately describes the future state of the atmosphere. Over the years, the forecaster knows that model A has performed well in predicting the positions of upper troughs that develop off the coast. Likewise, model C, because it uses more closely spaced grid points and a greater number of data points, has done an admirable job of forecasting the positions of upper troughs and shortwaves. Model B, on the other hand, tends to move the shortwaves along too slowly. Consequently, the forecaster puts more confidence into model A and model C.

Using experience and the progs, the meteorologist sets out to predict the weather. The 12-hour progs for model A and model C in Fig. 14.16 show a shortwave (labeled 1) approaching the California coast at 4 a.m. on Monday morning. As the shortwave approaches the coastline, clouds will increase and thicken and the likelihood of rain will increase. Because the main upper-level low is predicted to remain off the coast, southwesterly winds aloft will continue to pump moisture into the region and, as the 36-hour progs indicate, a series of shortwaves will move through the region, bringing a good chance of rain at least through Tuesday morning. Therefore, the precipitation forecast will sound like this:

Increasing cloudiness Sunday night with rain beginning Monday morning. Periods of rain likely through Tuesday morning.

A Valid Forecast   By early Monday morning, the maps begin to show the changes that the computer progs predicted. The surface map for 4:00 A.M. (PST) Monday morning (see Fig. 14.17) shows that the surface low in the Pacific has moved eastward and developed into a broad trough west of California. (Compare its position with Fig. 14.14.) The trough has deepened considerably as indicated by its central pressure of 1004 mb (29.65 in.). The approach of the storm is evidenced in San Francisco by thick middle clouds, southerly winds, and a falling barometer, nearly 4 mb lower than 12 hours ago. All these signs suggest that rain is on the way.

On the 500-mb chart for 4:00 a.m. Monday morning (Fig. 14.18), we can see that the injection of cold air and the shortwave into the main upper trough have caused it to deepen. Note that the height contours are now displaced farther south and that the contour in the middle of the trough is lower than on the previous 500-mb map (Fig. 14.15). Compare Fig. 14.18 with the 12-hour progs in model A and model C in Fig. 14.16 and notice how closely they match. The forecaster made a wise choice in showing confidence in these two computer models as they did a good job predicting the position of the upper-level low and shortwaves. Since the shortwaves are moving with the flow toward San Francisco, it should rain today. But at what time will the rain begin? Here is where satellite information assists the forecaster.

Assistance from the Satellite   The infrared satellite photograph taken at 6:45 A.M. Monday (see Fig. 14.19) shows that the middle clouds presently over California will soon give way to a *comma cloud*, a band of organized cumuliform clouds that look like a comma in the photograph. Observe that this cloud band lies almost beneath the shortwave approaching the coast, shown in Fig. 14.18. Also note that to the west a relatively unorganized mass of cumulus congestus clouds is beginning to form near the second shortwave in Fig. 14.18. Strong

FIGURE 14.17
Surface weather map for 4:00 A.M. (PST) Monday, March 26.

500 mb

FIGURE 14.18

The 500-mb analysis for 4:00 A.M. (PST) Monday, March 26. Heavy dashed lines show position of short-waves. Solid lines are height contours where 564 equals 5640 m. (Compare with Fig. 14.16, the 12-hour progs for model A and model C.)

FIGURE 14.19

Infrared satellite image taken at 6:45 A.M. (PST) Monday, March 26. The heavy dashed line shows the position of a comma cloud. Cumulus congestus clouds are developing beneath the second shortwave in Fig. 14.18.

southwesterly winds aloft should carry the large comma cloud and its weather into California.

By examining the movement of the cloud mass on successive satellite photographs, the forecaster can predict its arrival time and, hence, when rainfall will begin. According to satellite photographs, the leading edge of the comma cloud should be just offshore by Monday afternoon. Also, radar indicates that, just off the coast, light rain is now falling from the middle cloud layer. Consequently, rainfall should begin sometime in the morning, becoming heavier by the afternoon as the cumuliform clouds move in. Because the surface low is beneath the upper-level shortwave, the surface depression will probably continue to intensify, and pressure gradients around it will increase, creating strong and gusty winds from the south as the storm approaches. An amended forecast for San Francisco might read:

Rain beginning this morning becoming heavy by this afternoon. Strong and gusty southerly winds.

A Day of Rain and Wind   The first raindrops falling from altostratus clouds dampen city streets near the end of the morning rush hour. Quickly, the rain spreads inland, and by late Monday afternoon, Doppler radar shows that precipitation is falling throughout northern and central California as gusty southerly winds and moderate rain greet commuters on their way home.

The barometer has fallen sharply all day at San Francisco and by 4:00 P.M. the barometer reading is 1004 mb, a drop of 7 mb in just 6 hours. We can see the reason for this on the surface map for 4:00 P.M. Monday afternoon (Fig. 14.20). The low has not only moved closer to the coast, it has intensified considerably, as indicated by the drop of 11 mb in central pressure in just 12 hours. Spiraling around the low, a cold front marks the position of the comma cloud. At first, this may seem surprising, since no fronts were drawn on the previous map. However, remember that this is a baroclinic situa-

FIGURE 14.20
Surface weather map for 4:00 P.M. (PST) Monday, March 26.

Surface map

tion with cold advection behind the low, warm advection in front of it, and divergence in the flow aloft. At the surface, air masses with contrasting temperatures are being brought together in the region of the comma cloud. Since the cold air is on the western side of the comma cloud, the meteorologist saw fit to draw in a cold front. Notice that, to the north of the low, a stationary front marks the boundary between cold maritime polar air to the west and modified cool maritime polar air to the east.

As the surface low intensifies, it and the spiraling band of clouds move eastward more slowly. The front will, therefore, move through later than anticipated, sometime late Monday night or early Tuesday morning. The forecaster expects that the winds will remain strong and precipitation will be heavy as the front passes.

Early Tuesday morning, the front moves onshore, bringing with it heavy rain and winds with gusts exceeding 45 knots. Billowing cumulus clouds, and in some areas thunderstorms, drench the entire Pacific Coast with rain, and with snow at higher elevations. The storm center, constantly being drained of air by strong upper-level divergence, has deepened into a furious system with a central pressure of 988 mb (29.17 in.).

The satellite image for 9:00 A.M. Tuesday morning (see Fig. 14.21) provides us with a visual interpretation of the storm. We see superimposed on the photograph the positions of the surface low, fronts, and the winds aloft (heavy arrow). Note that a front with its heavy band of clouds stretches from Idaho southward into Nevada and southern California, while the surface low is still positioned off the northern California coast. Moving through central California is a band of clouds and showers associated with a shortwave, spinning counterclockwise around the low.

The upper flow indicates that the trough aloft is still off the coast just as the 36-hour computer progs on Sunday had predicted. If we follow the flow southward out of Canada, we see a patch of clear weather just offshore. Here, the air pushing south-

FIGURE 14.21
Visible satellite image for 9:00 A.M. (PST) Tuesday, March 27. Included in the picture are the positions of surface fronts, the upper-level flow (heavy arrow), and precipitation patterns.

ward off the land is cold and dry, so clouds do not form. However, as the air moves further south, it warms and picks up moisture from the water below. The rippled cloud pattern in the photograph is cumulus clouds, which form in the unstable air. Look closely at the photograph and notice that a comma cloud is developing to the southwest of the deep surface low. If you refer back to Fig. 14.16, you will see that the 36-hour prog for model A and model C had predicted that a shortwave would be located in this region on Tuesday morning. Is this cloud band organizing into another onslaught of high winds, heavy precipitation, and thunderstorms? If so, when will it arrive? And what about the storm off the coast? Will it deepen or fill, move inland or remain stationary? It's back to the drawing board—to the computer progs, the charts, and the satellite photographs. The challenge and anticipation of making another forecast are at hand.

SUMMARY

Forecasting tomorrow's weather entails a variety of techniques and methods. Persistence and steady-state forecasts are useful when making a short-range (0–6 hour) prediction. For a longer range forecast, the current analysis, satellite data, weather typing, intuition, experience, along with guidance from the many computer progs supplied by the National Weather Service, all go into making a prediction.

Different computer progs are based upon different atmospheric models that describe the state of the atmosphere and how it will change with time. Currently, flaws in the models—as well as tiny errors (uncertainties) in the data—generally amplify as the computer tries to project weather farther and farther into the future. At present, computer progs are better at forecasting the position of mid-latitude highs and lows and their development than local showers and thunderstorms. To skillfully forecast smaller features, the grid spacing on the models must be reduced. This results in many more computations and a longer running time.

As new information from atmospheric research programs is fed into the latest generation of computers, it is hoped that the progs will be able to show more skill in predicting the weather up to 10 days in the future. More promising at this time is the simulation of large-scale climatic trends by the most recent general circulation models.

The latter part of this chapter does not cover all the methods of weather prediction, but, rather, conveys an understanding of the problems confronting anyone who attempts to predict the behavior of this churning mass of air we call our atmosphere.

Most of the forecasting methods in this chapter apply mainly to skill in predicting events associated with large-scale weather systems, such as fronts and mid-latitude cyclones. The next chapter on severe weather deals with the formation and forecasting of smaller scale (mesoscale) systems, such as thunderstorms, squall lines, and tornadoes.

Key Terms

The following terms are listed in the order they appear in the text. Define each. Doing so will aid you in reviewing the material covered in this chapter.

weather watch
weather warning
analysis
numerical weather prediction
atmospheric models
prognostic chart (prog)
AWIPS
meteogram
persistence forecast
steady-state (trend) forecast
nowcasting

analogue forecasting
   method
pattern recognition
weather type forecasting
ensemble forecasting
climatological forecast
probability forecast
teleconnections
isallobars
omega high

Questions for Review

1. What is the function of the National Center for Environmental Prediction?
2. How does a *weather watch* differ from a *weather warning*?
3. In what ways has the computer assisted the meteorologist in making weather forecasts?
4. How does a prog differ from an analysis?
5. Briefly describe how the computer goes about making a weather forecast.
6. What are some of the problems associated with computer model forecasts?
7. Make a persistence forecast for your area for this same time tomorrow. Did you use any skill in making this prediction? Explain.
8. Describe four methods of forecasting the weather and give an example for each one.
9. If today's weather forecast calls for a "chance of snow," what is the probability that it will snow today? (Hint: See Table 14.2.)
10. Do all accurate forecasts show skill? Explain.
11. Would a forecast calling for a 20 percent chance of rain be high enough for you to cancel your plans for a picnic? Explain.
12. Do extended weather forecasts make specific predictions of rain or snow? Explain.
13. Use the instant weather forecast chart (Appendix E) as a guide to making a short-range weather forecast when the surface weather elements indicate those shown in Table 14.4.
14. If low clouds at an elevation of 1 km above you are moving from the southeast and clouds about 2 km higher are moving from the southwest, is cold or warm advection taking place between the cloud layers? Explain.

15. List four methods that you could use to predict the movement of a surface open-wave cyclone.
16. What is an omega high? What influence does it have on the movement of surface highs and lows?
17. Suppose that where you live, the middle of January is typically several degrees warmer than the rest of the month. If you forecast this "January thaw" for the middle of next January, what type of a weather forecast will you have made?
18. Given a map with isallobars, where will high and low pressure systems move?

Questions for Thought

1. From Fig. 14.8 determine the probability of a "White Christmas" for your area.
2. Suppose the chance for a "White Christmas" at your home is 10 percent. Last Christmas was a white one. If for next year you forecast a "nonwhite" Christmas, will you have shown any skill if your forecast turns out to be correct? Explain.
3. Suppose that it is presently warm and raining. A cold front will pass your area in 3 hours. Behind the front it is cold and snowing. Make a persistence forecast for your area 6 hours from now. Would you expect this forecast to be correct? Explain. Now, make a forecast for your area using the steady-state, or trend method.
4. Explain how the phrase "sensitive dependence on initial conditions" relates to the final outcome of a computer-based weather forecast.
5. You are in Calgary, AB, Canada, 100 km (62 mi) east of the Rockies. The current wind is from the north. Looking at a prog for tomorrow, you see that the wind will be from the west. Will tomorrow's temperature be warmer or cooler than today? Explain.

Problems and Exercises

1. From the worldwide web, the newspaper, or the National Weather Service, obtain a copy of next month's 30-day outlook. Based on the temperature and precipitation patterns of this forecast, draw several contour lines representing the upper-air pattern that would be necessary for such a forecast to come true.
2. When a persistent winter pattern at 500 mb appears similar to that shown in Fig. 14.22, show on the map where you would forecast the following: (a) good chance of precipitation; (b) above seasonal temperatures; (c) below seasonal temperatures; and (d) generally dry weather.

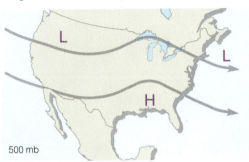

FIGURE 14.22
Diagram for Exercise 2.

3. In Fig. 14.9, p. 365, mark the position of the following cities: Cleveland, Ohio; Albuquerque, New Mexico; and New Orleans, Louisiana. Based on the projected movement of the surface weather systems in Fig. 14.12, p. 370, make a 24-hour weather forecast for each of these cities. In your forecast, include temperature, pressure, cloud cover, humidity, winds, and precipitation (if any). Compare your forecasts with the actual weather at the end of the period in Fig. 14.13, p. 372.

Questions for Exploration

1. On the Blue Skies: College Edition CD-ROM go to the **Weather Analysis** section of the CD and click on **"Isopleths."**
2. On the Blue Skies: College Edition CD-ROM go to the **Weather Forecasting** section of the CD and click on **"Forecasting."**

TABLE 14.4	Surface Weather Elements (to be used in answering Review Question 13)						
TIME	**AIR PRESSURE (MB)**	**ANY CHANGES**	**AIR TEMPERATURE (°F)**	**WIND DIRECTION**	**CLOUD TYPE**	**PRECIPITATION**	
Morning	1017	Falling slowly	29	SE	As	None	
Afternoon	1014	Rising rapidly	36	NW	Tcu	Rain showers	
Afternoon	1025	Steady	86	NW	Ci	None	
Afternoon	990	Falling rapidly	26	NE	Ns	Snow	
Early morning	1020	Rising	13	NW	Clear	None	
Nighttime	1012	Falling	31	SW	Sc	None	
Afternoon	1006	Falling	52	SE	Ns	Rain	
Nighttime	1005	Rising rapidly	45	W	Cb	Rain showers	

A supercell thunderstorm with
a tornado roars across the plains
of the central United States. (Photo
© Warren Faidley, Photo Network.)

THUNDERSTORMS AND TORNADOES

CONTENTS

Wednesday, March 18, 1925, was a day that began uneventfully, but within hours turned into a day that changed the lives of thousands of people and made meteorological history. Shortly after 1:00 P.M., the sky turned a dark greenish-black and the wind began whipping around the small town of Murphysboro, Illinois. Arthur and Ella Flatt lived on the outskirts of town with their only son, Art, who would be four years old in two weeks. Arthur was working in the garage when he heard the roar of the wind and saw the threatening dark clouds whirling overhead. Instantly concerned for the safety of his family, he ran toward the house as the tornado began its deadly pass over the area. With debris from the house flying in his path and the deafening thunder of destruction all around him, Arthur reached the front door. As he struggled in vain to get to his family, whose screams he could hear inside, the porch and its massive support pillars caved in on him. Inside the house, Ella had scooped up young Art in her arms and was making a panicked dash down the front hallway towards the door when the walls collapsed, knocking her to the floor, with Art cradled beneath her. Within seconds, the rest of the house fell down upon them. Both Arthur and Ella were killed instantly, but Art was spared, nestled safely under his mother's body.

As the dead and survivors were pulled from the devastation that remained, the death toll mounted. Few families escaped the grief of lost loved ones. The infamous tri-state tornado killed 234 people in Murphysboro and leveled 40 percent of the town.

The devastating tornado described in our opening cut a mile-wide path for a distance of more than 200 miles through the states of Missouri, Illinois, and Indiana. The tornado (which was probably a series of tornadoes) totally obliterated 4 towns, killed an estimated 695 persons, and left over 2000 injured. Tornadoes such as these, as well as much smaller ones, are associated with severe thunderstorms. Consequently, we will first examine the different types of thunderstorms. Later, we will focus on tornadoes, examining how and where they form, and why they are so destructive.

WHAT ARE THUNDERSTORMS?

It probably comes as no surprise that a *thunderstorm* is merely a storm containing lightning and thunder. Sometimes a thunderstorm produces gusty surface winds with heavy rain and hail. The storm itself may be a single cumulonimbus cloud, or several thunderstorms may form a cluster, or a line of thunderstorms may form that in some cases may extend for more than 100 km (62 mi).

The birth of a thunderstorm occurs when warm, humid air rises in a conditionally unstable environment. The trigger needed to start air moving upward may be the unequal heating of the surface, the effect of terrain, or the lifting of warm air along a frontal zone. Diverging upper-level winds, coupled with converging surface winds and rising air, also provide a favorable condition for thunderstorm development. Usually, several of these mechanisms work together to generate severe thunderstorms.

Scattered thunderstorms that form in summer are often referred to as **ordinary thunderstorms**, formerly *air-mass thunderstorms*, because they tend to develop in warm, humid air masses away from weather fronts. These storms are usually short-lived and rarely produce strong winds or large hail. On the other hand, *severe thunderstorms* may produce high winds, flash floods, damaging hail, and even tornadoes. Let's examine the ordinary thunderstorms first.

ORDINARY (AIR-MASS) THUNDERSTORMS

Extensive studies indicate that thunderstorms go through a cycle of development from birth, to maturity, to decay. The first stage is known as the **cumulus stage**. As humid air rises, it cools and condenses into a single cumulus cloud or a cluster of clouds (see Fig. 15.1). If you have ever watched a thunderstorm develop, you may have noticed that at first the cumulus clouds grow upward only a short distance, then they dissipate. This is because the cloud droplets evaporate as the drier air surrounding the cloud mixes with it. However, after the water drops evaporate, the air is more moist than before. So, the rising air is now able to condense at successively higher levels, and the cumulus cloud grows taller, often appearing as a rising dome or tower.

As the cloud builds, the transformation of water vapor into liquid or solid cloud particles releases large quantities of latent heat. This keeps the air inside the cloud warmer than the air surrounding it. The cloud continues to grow in the unstable atmosphere as long as it is constantly fed by rising air from below. In this manner, a cumulus cloud may show extensive vertical development in just a few minutes. During the cumulus stage, there is insufficient time for precipitation to form, and the updrafts keep water droplets and ice crystals suspended within the cloud. Also, there is no lightning or thunder during this stage.

As the cloud builds well above the freezing level, the cloud particles grow larger. They also become heavier. Eventually, the rising air is no longer able to keep them suspended, and they begin to fall. While this phenomenon is taking place, drier air from around the cloud is being drawn into it in a process called *entrainment*. The entrainment of drier air causes some of the raindrops to evaporate, which chills the air. The air, now being colder and heavier than the air around it, begins to descend as a *downdraft*. The downdraft may be enhanced as falling precipitation drags some of the air along with it.

The appearance of the downdraft marks the beginning of the **mature thunderstorm**. The downdraft and updraft within the mature thunderstorm constitute a *cell*.* In most storms, there are several cells, each of which may last for an hour or so.

During its mature stage, the thunderstorm is most intense. The top of the cloud, having reached a stable region of the atmosphere (which may be the stratosphere), begins to take on the familiar anvil shape, as strong upper-level winds spread the cloud's ice crystals horizontally (see Fig. 15.2). The cloud itself may extend upward to an altitude of over 12 km (40,000 ft) and be several kilometers in diameter near its base. Updrafts

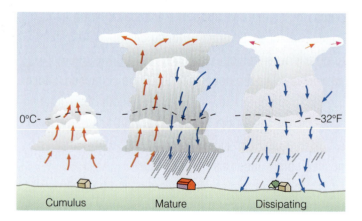

FIGURE 15.1
Simplified model depicting the life cycle of an ordinary thunderstorm that is nearly stationary. (Arrows show vertical air currents. Dashed line represents freezing level, 0°C isotherm.)

---

*In convection, the cell may be a single updraft or a single downdraft, or a couplet of the two, which defines the mature stage of the thunderstorm.

FIGURE 15.2
An ordinary thunderstorm in its mature stage. Note the distinctive anvil top.

and downdrafts reach their greatest strength in the middle of the cloud, creating severe turbulence. In some storms, the updrafts may intrude above the cloud top into the stable atmosphere, a condition known as *overshooting* (see Fig. 15.3). Lightning and thunder are also present in the mature stage. Heavy rain (and occasionally small hail) falls from the cloud. There is often a downrush of cold air with the onset of precipitation at the surface that may be felt as a strong wind gust. The rainfall, however, may or may not reach the surface, depending on the relative humidity beneath the storm. In the dry air of the desert Southwest, for example, a mature thunderstorm may look ominous and contain all of the ingredients of any other storm, except that the raindrops evaporate before reaching the ground.

After the storm enters the mature stage, it begins to dissipate in about 15 to 30 minutes. The **dissipating stage** occurs when the updrafts weaken and downdrafts tend to dominate throughout much of the cloud. Deprived of the rich supply of warm humid air, cloud droplets no longer form. Light precipitation now falls from the cloud, accompanied by only weak downdrafts. As the storm dies, the lower-level cloud particles evaporate rapidly, sometimes leaving only the cirrus anvil as the reminder of the once mighty presence (see Fig. 15.4).

A single ordinary thunderstorm may go through its three stages in an hour or less. The reason it does not last very long is that the storm's downdraft may cut off the storm's fuel supply by destroying the humid updrafts. But the storm lasts longer than one might expect. For example, a rising parcel of air constitutes a cell in a developing thunderstorm. Typically, it takes the parcel about 20 minutes to rise from the surface to its equilibrium level, high above the ground. Once the rising parcel reaches its destination, the thunderstorm should begin to

FIGURE 15.3
Rising air currents within this thunderstorm are causing these bulging clouds to extend into the stable atmosphere above the storm's anvil top. This storm, photographed at about 11 km (36,000 ft) above southern Michigan, produced gusty winds and heavy rain at the surface.

FIGURE 15.4
A dissipating thunderstorm. Most of the cloud particles in the lower half of the storm have evaporated.
Only the cirrus anvil stands out as a distinguishing feature.

dissipate, but the downdrafts of the storm force new cells upward, and, hence, the storm normally lasts considerably longer than 20 minutes.

Not only do thunderstorms produce summer rainfall for a large portion of the United States but they also bring with them momentary cooling after an oppressively hot day. The cooling comes during the mature stage, as the downdraft reaches the surface in the form of a blast of welcome relief. Sometimes, the air temperature may lower as much as 10°C (18°F) in just a few minutes. Unfortunately, the cooling effect is short-lived, as the downdraft diminishes or the thunderstorm moves on. In fact, after the storm has ended, the air temperature usually rises; and as the moisture from the rainfall evaporates into the air, the humidity increases, sometimes to a level where it actually feels more oppressive after the storm than it did before.

Upon reaching the surface, the cold downdraft has another effect. It may force warm, moist surface air upward. This rising air then condenses and gradually builds into a new thunderstorm. Thus, it is entirely possible for a series of thunderstorms to grow in a line, one next to the other, each in a different stage of development (see Fig. 15.5). Thunderstorms that form in this manner are termed **multicell storms**. Most ordinary thunderstorms are multicell storms, as are most severe thunderstorms.

As we saw earlier in this chapter, for a thunderstorm to develop there must be rising, moist air in a conditionally unstable atmosphere. The ingredient necessary to start the air

rising may be the unequal heating of the surface, a frontal boundary, a mountain range, or the leading edge of a sea breeze. Most of the thunderstorms that form in this manner are not severe, and their life cycle usually follows the pattern described for ordinary thunderstorms.

SEVERE THUNDERSTORMS

Severe thunderstorms are capable of producing large hail, strong, gusty surface winds, flash floods, and tornadoes.* Just as the ordinary thunderstorm, they form as moist air is forced to rise into a conditionally unstable atmosphere. But, severe thunderstorms also form in areas with a strong vertical wind shear.

Strong winds aloft may cause the updrafts in a severe thunderstorm to tilt in its mature stage, as depicted in Fig. 15.6. The storm is moving from left to right and the upper-level winds cause the system to tilt so that the updrafts move up and over the downdrafts. This situation allows the updraft to remain strong for an extended period of time. However, there are many severe thunderstorms that do not have a tilted updraft. In these storms, the updrafts may be so strong (sometimes up to 100 knots) that precipitation-size particles do not have enough time to form. Apparently, this type of thunder-

---

*The National Weather Service defines a severe thunderstorm as having ¾-inch hail and/or surface wind gusts of 50 knots (58 mi/hr).

FIGURE 15.5

A multicell storm. This storm is composed of a series of cells in successive stages of growth. The thunderstorm in the middle is in its mature stage, with a well-defined anvil. Heavy rain is falling from its base. To the right of this cell, a thunderstorm is in its cumulus stage. To the left, a well-developed cumulus congestus cloud is about ready to become a mature thunderstorm.

storm becomes severe when a strong vertical wind shear causes the storm to rotate. And, as we will discuss later in this chapter, it is the rotational aspect of thunderstorms that leads to the formation of tornadoes.

The updrafts in a severe thunderstorm may be so strong that the cloud top is able to intrude well into the stable stratosphere. In some cases, the top of the cloud may extend to more than 18 km (60,000 ft) above the surface. The violent updrafts keep hailstones suspended in the cloud long enough for them to grow

to considerable size. Once they are large enough, they either fall out the bottom of the cloud with the downdraft or a strong updraft may toss them out the side of the cloud, or even from the base of the anvil. Aircraft have actually encountered hail in clear air several kilometers from a storm. Also, downdrafts within the anvil may produce beautiful mammatus clouds.

As some of the falling precipitation evaporates, it cools the air and enhances the downdraft. The cool air that reaches the ground may act like a wedge, forcing warm, moist surface air up

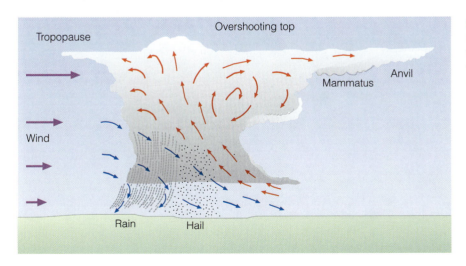

FIGURE 15.6

A simplified model describing air motions and other features associated with a severe thunderstorm that has a tilted updraft. The severity depends on the intensity of the storm's circulation pattern.

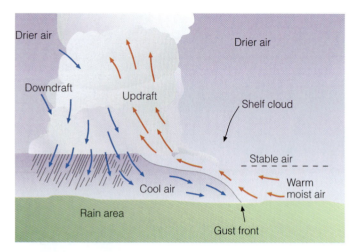

FIGURE 15.7
The lower half of a severe squall-line-type thunderstorm and some of the features associated with it.

into the system (see Fig. 15.7). Thus, the downdraft may help to maintain the updraft and vice versa, so that a severe thunderstorm with this type of updraft and downdraft configuration is able to maintain itself (for many hours in some cases).

The Gust Front and Microburst Look at Fig. 15.7 again and notice that the downdraft spreads laterally after striking the ground. The boundary separating this cold downdraft

from the warm surface air is known as a **gust front**. To an observer on the ground, the passage of the gust front resembles that of a cold front. During its passage, the wind shifts and becomes strong and gusty, with speeds occasionally exceeding 55 knots; temperatures drop sharply and, in the cold heavy air of the downdraft, the surface pressure rises. Sometimes it may jump several millibars, producing a small area of high pressure called a **mesohigh** (meaning mesoscale high). The cold air may linger close to the ground for several hours, well after the thunderstorm activity has ceased.

Along the leading edge of the gust front, the air is quite turbulent. Here, strong winds can pick up loose dust and soil and lift them into a huge tumbling cloud—the haboob that we described in Chapter 10. As warm, moist air rises along the forward edge of the gust front, a **shelf cloud** (also called an *arcus cloud*) may form, such as the one shown in Fig. 15.8. These clouds are especially prevalent when the atmosphere is very stable near the base of the thunderstorm. Look back at Fig. 15.7 and notice that the shelf cloud is attached to the base of the thunderstorm. Occasionally, an elongated ominous-looking cloud forms just behind the gust front. These clouds, which appear to slowly spin about a horizontal axis, are called **roll clouds** (see Fig. 15.9). Sometimes the leading edge of the gust front forces warm, moist air upward, producing new thunderstorms.

Beneath a severe thunderstorm, the downdraft may become localized so that it hits the ground and spreads horizontally in a radial burst of wind, much like water pouring from a

FIGURE 15.8
A dramatic example of a shelf cloud (or arcus cloud) associated with a severe thunderstorm.

FIGURE 15.9
A roll cloud forming behind a gust front. (Copyright Howard B. Bluestein.)

tap and striking the sink below. Such downdrafts are called **downbursts**. A downburst with winds extending only 4 km or less is termed a **microburst**. In spite of its small size, an intense microburst can induce damaging winds as high as 146 knots. (A larger downburst with winds extending more than 4 kilometers is termed a **macroburst**.) Figure 15.10 shows the dust clouds generated from a microburst north of Denver, Colorado. Since a microburst is an intense downdraft, its leading edge can evolve into a gust front.

Microbursts are capable of blowing down trees and inflicting heavy damage upon poorly built structures. In fact, microbursts may be responsible for some damage once attributed to tornadoes. Moreover, microbursts and their accompanying *wind shear* (that is, rapid changes in wind speed or wind direction) appear to be responsible for several airline crashes. When an aircraft flies through a microburst at a relatively low altitude, say 300 m (1000 ft) above the ground, it first encounters a headwind that generates extra lift. This is position (a) in Fig. 15.11. At this point, the aircraft tends to climb (it gains lift), and if the pilot noses the aircraft downward there could be grave consequences, for in a matter of seconds the aircraft encounters the powerful downdraft (position b), and the headwind is replaced by a tail wind (position c). This situation causes a sudden loss of lift and a subsequent decrease in the performance of the aircraft, which is now accelerating toward the ground.

One accident attributed to a microburst occurred north of Dallas–Fort Worth Regional Airport during August, 1985. Just as the aircraft was making its final approach, it encountered severe wind shear beneath a small, but intense thunderstorm. The aircraft then dropped to the ground and crashed, killing over 100 passengers.

To detect the hazardous wind shear associated with microbursts, a new and improved ground-based wind-shear detection system called LLWSAS (for *Low-Level Wind Shear Alert System*) has been installed at several airports. The system utilizes surface wind speed and direction sensors to measure the presence of wind shear. Unfortunately, LLWSAS can only detect a microburst after it hits the ground and only where the instruments are located. However, the installation of Doppler radars throughout the United States is providing a better system of detecting microbursts and other severe weather phenomena.

Studies conducted in the early 1980s at Denver's Stapleton Airport (JAWS, *Joint Airport Weather Studies*) and in the 1970s near Chicago, Illinois (NIMROD, *Northern Illinois Meteorological Research On Downbursts*), suggest that the leading edge of a microburst can contain an intense horizontally rotating vortex that is often filled with dust in a relatively dry region. At Denver, many microbursts emanate from virga—rain falling from a cloud but evaporating before reaching the ground. Apparently,

FIGURE 15.10
Dust clouds rising in response to the outburst winds of a microburst north of Denver, Colorado.

in these "dry" microbursts (Fig. 15.10), evaporating rain cools the air. The cooler heavy air then plunges downward through the warmer lighter air below. In humid regions, many microbursts are "wet" in that they are accompanied by blinding rain.

Microbursts can be associated with severe thunderstorms, producing strong, damaging winds. But studies show that they can also occur with clouds that produce only isolated showers—clouds that may or may not contain thunder and lightning.

The strong downburst winds associated with a cluster of severe thunderstorms can produce damaging *straight-line*

winds* that may exceed 90 knots. If the wind damage extends for at least 400 km (250 mi) along the storm's path, the winds are called a **derecho** (day-ray-sho), after the Spanish word for "straight ahead." Most derechoes are associated with severe thunderstorms that form in the early evening and last throughout the night. An especially powerful derecho roared through New York State during the early morning of July 15, 1995, where it blew down millions of trees in Adirondack State Park.

Supercell and Squall-Line Thunderstorms   The **supercell storm** is an enormous rotating thunderstorm whose updrafts and downdrafts are sufficiently structured so that it is able to maintain itself as a single entity for hours on end (see Fig. 15.12). Storms of this type are capable of producing updrafts that can exceed 90 knots, hail the size of grapefruit, damaging surface winds, and large, long-lasting tornadoes. The strong updrafts in the eastern side of the storm are usually sufficient to keep precipitation from forming. If precipitation does form, it may be swept laterally out of the region by the rapidly rotating updraft. The region in the thunderstorm where radar is unable to detect precipitation is known as the *bounded weak echo region (BWER)* or *echo free vault*.

Most supercell storms are *right-movers*. That is, they move to the right of the steering winds aloft.† Apparently, the rapidly

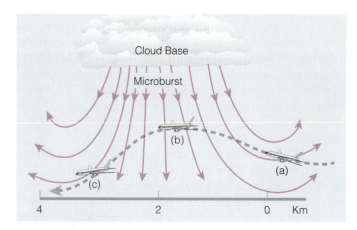

FIGURE 15.11
Flying into a microburst. At position (a), the pilot encounters a headwind; at position (b), a strong downdraft; and at position (c), a tailwind that reduces lift and causes the aircraft to lose altitude.

---

*Straight-line winds are thunderstorm-generated winds that are not associated with rotation.

†Some thunderstorms move to the left of the steering winds aloft. This movement may happen as a thunderstorm splits into two storms, with the northern half of the storm often being a left-mover and the southern half a right-mover.

FIGURE 15.12
A supercell storm
near Perryton, Texas.

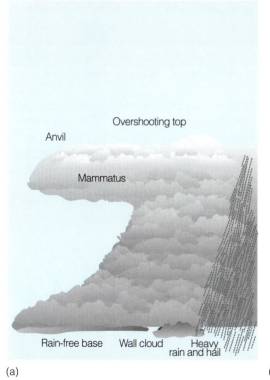

Anvil

Overshooting top

Mammatus

Rain-free base Wall cloud Heavy
rain and hail

(a)

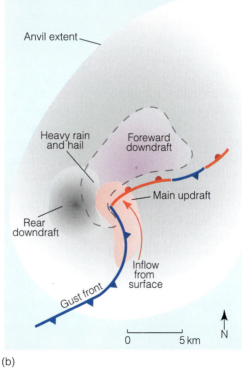

Anvil extent

Heavy rain
and hail

Forward
downdraft

Main updraft

Rear
downdraft

Inflow
from
surface

Gust front

0 5 km N

(b)

FIGURE 15.13
Diagram (a) illustrates some of the features of a classic supercell thunderstorm as viewed from the southeast.
Diagram (b) is a view of the thunderstorm from above, looking down on the
storm. The area enclosed by the dashed
line represents the radar image of the
storm. The area shaded blue represents
the rear downdraft, the area shaded
red, the main updraft. The gust front is
represented by a cold front symbol.

rising air of the updraft interacts with increasing horizontal winds that change direction with height (from more southerly to more westerly) in such a way that vertical pressure gradients are able to generate new updrafts on the right side of the storm. Hence, as the storm moves along, new cells form to the right of the winds aloft. An example of a *classic supercell* is shown in Fig. 15.13. Some supercells produce heavy precipita-

tion and large hail, which appears to fall near the center (or eastern) part of the storm. Such storms, called *HP supercells* (for *High Precipitation*), often produce extreme downbursts, flash flooding, and very large hail. If tornadoes are present, it is often difficult to see them, as they tend to form in the area of heavy precipitation. A supercell characterized by little precipi-tation is referred to as an *LP supercell* (for *Low Precipitation*).

FIGURE 15.14
An enhanced infrared satellite image showing a pre-frontal squall line (in dark red) that stretches from the Ohio Valley into the southern United States on November 15, 1989. Severe thunderstorms along the squall line spawned large hail and 25 tornadoes. The worst tornado struck Madison County, Alabama, killing 25 people and injuring 463 others.

These storms, which are capable of producing tornadoes and large hail, often have a vertical tower that, due to the storm's rotation, resembles a corkscrew.

The **squall line** forms as a line of thunderstorms. Sometimes they are right along a cold front but they also form in the warm air 100 to 300 km out ahead of it. These *pre-frontal squall-line thunderstorms* of the middle latitudes represent the largest and most severe type of squall line. The line of storms may extend for over 1000 km (600 mi), with huge thunderstorms causing severe weather over much of its length (see Fig. 15.14).

There is still debate as to exactly how pre-frontal squall-lines form. Models that simulate their formation suggest that,

initially, convection begins along the cold front then reforms further away. Moreover, the surging nature of the main cold front itself, or developing cumulus clouds along the front, may cause the air aloft to develop into waves (called *gravity waves*), much like the waves that form downwind of a mountain chain (see Fig. 15.15). Out ahead of the cold front, the rising motion of the wave may be the trigger that initiates the development of cumulus clouds and a pre-frontal squall line.

Squall lines that exhibit weaker updrafts and downdrafts tend to be more shallow than pre-frontal squall lines, and usually they have shorter life spans. These storms are referred to as *ordinary squall lines*. Severe weather may occur with them, but more typically they form as a line of thunderstorms that exhibit characteristics of ordinary thunderstorms. Ordinary squall lines may form with a stationary front, with a weak wave cyclone, or where no cyclonic storms are present. Many of the ordinary squall lines that form in the middle latitudes exhibit a structure similar to squall lines that form in the tropics.

The Dryline   Severe thunderstorms may form along or just east of a boundary called a dryline. The **dryline** represents a narrow zone where there is a sharp horizontal change in moisture. Because dew-point temperatures may drop along this boundary by as much as 9°C (16°F) per km, drylines have been referred to as *dew-point fronts*. Although drylines can occur as far north as the Dakotas, and as far east as the Texas-Louisiana border, they are most frequently observed in the western half of Texas, Oklahoma, and Kansas, especially during spring and early summer.

Figure 15.16 shows springtime weather conditions that can lead to the development of a dryline. The map shows a developing mid-latitude cyclone with a cold front, a warm front, and three distinct air masses. Behind the cold front, cold dry continental polar (cP) air or modified cool dry Pacific air pushes in from the northwest. In the warm sector, ahead of the cold front, warm dry continental tropical (cT) air moves in from the southwest. Further east, warm but very moist mar-

FIGURE 15.15
Pre-frontal squall-line thunderstorms may form ahead of an advancing cold front as the upper-air flow develops waves downwind from the cold front.

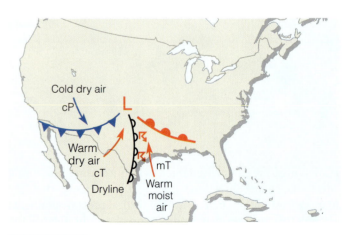

FIGURE 15.16
Surface conditions that can produce a dryline with severe thunderstorms.

itime tropical (mT) air sweeps northward from the Gulf of Mexico. The dryline is the north–south oriented boundary that separates the warm, dry air and the warm, moist air.

Along the cold front—where cold, dry air replaces warm, dry air—there is insufficient moisture for thunderstorm development. The moisture boundary lies along the dryline. Because the Central Plains of North America are elevated to the west, some of the hot, dry air from the southwest is able to ride over the slightly cooler, more humid air from the Gulf. This condition sets up a potentially unstable atmosphere just east of the dryline. Converging surface winds in the vicinity of the dryline, coupled with upper-level outflow, may result in rising air and the development of thunderstorms. As thunderstorms form, the cold air along the ground may initiate the uplift necessary for generating new (possibly more severe) thunderstorms. Many of the supercell thunderstorms that form along the dry line are of the low precipitation (LP supercell) variety. Hence, LP supercells are often referred to as *dryline storms*.

Floods and Flash Floods

Intense thunderstorms are often associated with **flash floods**—floods that rise rapidly with little or no advance warning. Such flooding often results when thunderstorms stall or move very slowly, causing heavy rainfall over a relatively small area. Or, they can occur when thunderstorms move quickly, but keep passing over the same area, a phenomenon called *training*. In recent years, flash floods in the United States have claimed an average of more than 100 lives a year, and have accounted for untold property and crop damage. (An example of a terrible flash flood that took the lives of more than 135 people is given in the Focus section on p. 392.)

In some areas, flooding occurs primarily in the spring when heavy rain and melting snow cause rivers to overflow their banks. During March, 1997, heavy downpours over the Ohio River Valley caused extensive flooding that forced thousands from their homes along rivers and smaller streams in Ohio, Kentucky, Tennessee, and West Virginia. One month later, heavy rain coupled with melting snow caused the Red

River to overflow its banks, inundating 75 percent of the town of Grand Forks, North Dakota. Flooding also occurs with tropical storms that deposit torrential rains over an extensive area.

During the summer of 1993, thunderstorm after thunderstorm rumbled across the upper midwest, causing the worst flood ever in that part of the United States. What began as a wetter than normal winter and spring for most of the upper midwest turned into "The Great Flood of 1993" by the end of July. In mid-June, thunderstorms began to form almost daily along a persistent frontal boundary that stretched across the upper midwest. The front (which remained nearly stationary for days on end) was positioned beneath the polar jet stream that was situated much farther south than usual for this time of year (see Fig. 15.17).* The jet stream provided pockets of upper-level divergence for the development of weak surface waves that rippled along the front. And converging surface air provided uplift for thunderstorm growth.

---

*As a note, the position of the jet stream caused the weather to be cooler than normal in the Pacific Northwest and warmer than normal in the East. While the Midwest was deluged with rain, the southeastern section of the United States was experiencing an extensive dry period.

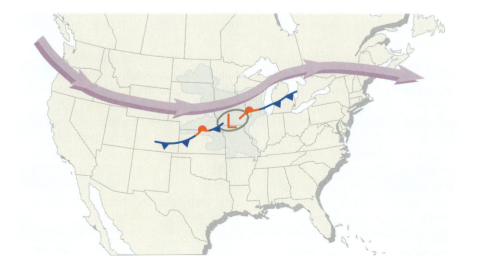

FIGURE 15.17

The heavy arrow represents the average position of the upper-level jet stream from mid-June through July, 1993. The jet stream helped fuel thunderstorms that developed in association with a stationary front that seemed to oscillate back and forth over the region as an alternating cold front and warm front. The "L" marks the center of a frontal wave that is moving along the front. Many of the thunderstorms that formed in conjunction with this pattern were severe, and over a period of weeks produced "The Great Flood of 1993." Most of the counties within the blue-shaded area were declared "disaster areas" due to flooding.

The Terrifying Flash Flood in the Big Thompson Canyon

July 31, 1976, was like any other summer day in the Colorado Rockies, as small cumulus clouds with flat bases and dome-shaped tops began to develop over the eastern slopes near the Big Thompson and Cache La Poudre rivers. At first glance, there was nothing unusual about these clouds, as almost every summer afternoon they form along the warm mountain slopes. Normally, strong upper-level winds push them over the plains, causing rainshowers of short duration. But the cumulus clouds on this day were different. For one thing, they were much lower than usual, indicating that the southeasterly surface winds were bringing in a great deal of moisture. Also, their tops were somewhat flattened, suggesting that an inversion aloft was stunting their growth. But these harmless-looking clouds gave no clue that later that evening in the Big Thompson Canyon more than 135 people would lose their lives in a terrible flash flood.

By late afternoon, a few of the cumulus clouds were able to puncture the inversion. Fed by moist southeasterly winds, these clouds soon developed into gigantic multicell thunderstorms with tops exceeding 18 km (60,000 ft). By early evening, these same clouds were producing incredible downpours in the mountains.

In the narrow canyon of the Big Thompson River, some places received as much as 30.5 cm (12 in.) of rain in the four hours between 6:30 P.M. and 10:30 P.M. local time. This is an incredible amount of precipitation, considering that the area normally receives about 40.5 cm (16 in.) for an entire year. The heavy downpours turned small creeks into raging torrents, and the Big Thompson River was quickly filled to capacity. Where the canyon narrowed, the river overflowed its banks and

FIGURE 1

Weather conditions that led to the development of severe thunderstorms that remained nearly stationary over the Big Thompson Canyon in the Colorado Rockies. The arrows within the thunderstorm represent air motions.

water covered the road. The relentless pounding of water caused the road to give way.

Soon cars, tents, mobile homes, resort homes, and campgrounds were being claimed by the river. Where the debris entered a narrow constriction, it became a dam. Water backed up behind it, then broke through, causing a wall of water to rush downstream. Of the approximately 2000 people in the canyon that evening, over 135 lost their lives. Property damage exceeded $35.5 million.

Figure 1 shows the weather conditions during the evening of July 31, 1976. A cool front moved through earlier in the day and is now south of Denver. The weak inversion layer associated with the front kept the cumulus clouds from building to great heights earlier in the afternoon. However, the strong southeasterly flow behind the cool front pushed unusually moist air upslope along the mountain range. Heated from below, the conditionally unstable air eventually punctured the

inversion and developed into a huge multicell thunderstorm complex that remained nearly stationary for several hours due to the weak southerly winds aloft. The deluge may have deposited 19 cm (7.5 in.) of rain on the main fork of the Big Thompson River in about one hour.

The flash floods associated with such intense thunderstorms occur with unpleasant frequency. On June 9, 1972, a destructive flood occurred at Rapid City, South Dakota. The weather conditions were very similar to those that caused the Big Thompson flood. A slowly moving cold front was situated to the south of Rapid City, with moist easterly winds behind it. Winds aloft were weak, so the severe thunderstorms moved slowly. Although there are other similarities between the two storms, we will not examine them here. However, a meteorologist, recognizing these conditions, would be able to alert a community to the probability of a flash flood and, hopefully, save many lives.

Fed by warm, humid air from the Gulf of Mexico, thunderstorms almost daily rolled through an area that stretched eastward from South Dakota into Minnesota and Wisconsin, and southward into Iowa, Illinois, and Missouri. Torrential rains from these storms quickly saturated the soil and soon runoff began to raise the water level in creeks and rivers. By the end of June, communities in the northern regions of the Mississippi River Valley were experiencing flooding.

As the thunderstorms continued into July, city after city was claimed by the rising waters (see Fig. 15.18). Between April and July, many areas had received twice their normal rainfall, and rivers continued to crest well above flood stage through July. By the time the water began to recede in August, more than 60 percent of the levees along the Mississippi River had been destroyed, and an area larger than the state of Texas had been covered with water (the blue-shaded area in Fig. 15.17). Estimates are that $6.5 billion in crops were lost as millions of acres of valuable farmland were inundated by flood waters. The worst flooding this area had ever seen took 45 human lives, damaged or destroyed 45,000 homes, and forced the evacuation of 74,000 people.

Mesoscale Convective Complexes   Where conditions are favorable for convection, a number of individual thunderstorms will occasionally grow in size and organize into a large convective weather system. These convectively driven systems, called **Mesoscale Convective Complexes (MCCs)**, are quite large—they can be as much as 1000 times larger than an individual ordinary thunderstorm. In fact, they are often large enough to cover an entire state, an area in excess of 100,000 square kilometers (see Fig. 15.19).

Within the MCC, the individual thunderstorms apparently work together to generate a long-lasting weather system that moves slowly (normally less than 20 knots) and often exists for periods exceeding 12 hours. The circulation of the MCC supports the growth of new thunderstorms as well as a region of widespread precipitation. These systems are beneficial, as they provide a significant portion of the growing season rainfall over much of the corn and wheat belts of the United States. However, MCCs can also produce a wide variety of severe weather, including hail, high winds, destructive flash floods, and tornadoes. In fact, about one-fourth of all MCCs produce damaging straight-line winds, or derechoes.

Mesoscale convective complexes tend to form during the summer in regions where the upper-level winds are weak, which is often beneath a ridge of high pressure. If a weak cold front should stall beneath the ridge, surface heating and moisture may be sufficient to generate thunderstorms on the cool side of the front. Often moisture from the south is brought into the system by a low-level jet stream. Most MCCs reach their maximum intensity in the early morning hours, which is partly due to the fact that the low-level jet reaches its maximum strength at night. Moreover, at night, the cloud tops cool rapidly by emitting infrared energy to space. Gradually, the atmosphere destabilizes as a vast amount of latent heat is released in the

FIGURE 15.18
Flooding during the summer of 1993 covered a vast area of the upper Midwest. Here, floodwaters near downtown Des Moines, Iowa, during July, 1993, inundate buildings of the Des Moines waterworks facility. Flood-contaminated water left 250,000 people without drinking water.

FIGURE 15.19
An enhanced infrared satellite image showing a Mesoscale Convective Complex extending from central Kansas across western Missouri. This organized mass of thunderstorms brought hail, heavy rain, and flooding to this area.

lower and middle part of the clouds. Within the system new thunderstorms form as older ones dissipate. With only weak upper-level winds, most MCCs move eastward very slowly.

Incidentally, there are a variety of Mesoscale Convective Systems (MCS), such as the squall line, that have many of the characteristics of an MCC, but do not last as long nor grow as large.

Thunderstorm Movement Most thunderstorms move roughly in the direction of the winds in the middle troposphere. However, as we saw earlier, supercell thunderstorms often move to the right of the steering winds aloft. And in some cases, a multicell storm over the central United States will take a peculiar path:

Individual cells will move with the wind, while the thunderstorm itself will move to the *right* of the wind. We will be able to understand this phenomenon if we take a closer look at the weather conditions necessary for the development of these storms.

Intense thunderstorms form over the central states when warm, moist surface air streaming northward from the Gulf of Mexico lies beneath drier southwesterly (or westerly) winds aloft. This unstable air generates huge thunderstorms, such as the one forming in Fig. 15.20. Warm, humid surface air fuels the storm, so that it is able to grow into a huge multicell storm. As the storm matures, it begins to deprive areas to its north of the warm, humid surface air that is necessary for new thun-

1 Developing cell
2 Mature thunderstorm
3 Dissipating thunderstorm

FIGURE 15.20
The formation of multicell thunderstorms that move to the right of the upper-level winds. The developing cell is steered to the northeast by the flow aloft. Downdrafts within the storm generate a new cell to the south. The new cell, which grows into a mature thunderstorm, cuts off the supply of moisture to the old cell, and so it dissipates. This entire sequence shows cells drifting to the northeast, while the whole thunderstorm moves slightly to the right of the upper-level winds.

derstorm growth. As the upper-level winds steer the storm toward the northeast, the storm's own downdraft pushes out its southern side, forcing moist air to rise and form new cells. Meanwhile, the old cell to the north, being deprived of its moisture, gradually dissipates. Therefore, as new cells form on the southern side of the thunderstorm, old cells to the north slowly die out, and the entire multicell thunderstorm moves to the right of the general wind flow.

DISTRIBUTION OF THUNDERSTORMS

It is estimated that more than 40,000 thunderstorms occur each day throughout the world. Hence, over 14 million occur annually. The combination of warmth and moisture make equatorial land masses especially conducive to thunderstorm formation. Here, thunderstorms occur on about one out of every three days. Thunderstorms are also prevalent over water along the intertropical convergence zone, where the low-level convergence of air helps to initiate uplift. The heat energy liberated in these storms helps the earth maintain its heat balance by distributing heat poleward (see Chapter 11). Thunderstorms rarely occur in dry climates, such as the polar regions and the desert areas of the subtropical highs.

Figure 15.21 shows the average annual number of days having thunderstorms in various parts of the United States. Notice that they occur most frequently in the southeastern states along the Gulf Coast with a maximum in Florida. A secondary maximum exists over the central Rockies. The region with the fewest thunderstorms is the Pacific coastal and interior valleys.

In many areas, thunderstorms form primarily in summer during the warmest part of the day when the surface air is most unstable. There are some exceptions, however. During the summer in the valleys of central and southern California, dry, sinking air produces an inversion that inhibits the development of towering cumulus clouds. In these regions

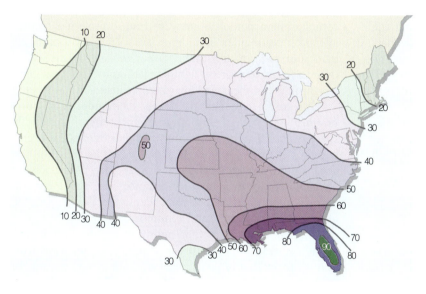

FIGURE 15.21
The average number of days each year on which thunderstorms are observed throughout the United States. (Due to the scarcity of data, the number of thunderstorms is underestimated in the mountainous west.)

FIGURE 15.22
The average number of days each year on which hail is observed throughout the United States.

thunderstorms are most frequent in winter and spring, particularly when cold, moist, unstable air aloft moves over moist, mild surface air. The surface air remains relatively warm because of its proximity to the ocean. Over the Central Plains, thunderstorms tend to form more frequently at night. These storms may be caused by a low-level southerly jet stream that forms at night, and not only carries humid air northward but also initiates areas of converging surface air, which helps to trigger uplift. As the thunderstorms build, their tops cool by radiating infrared energy to space. This cooling process tends to destabilize the atmosphere, making it more suitable for nighttime thunderstorm development.

At this point it is interesting to compare Fig. 15.21 and Fig. 15.22. Notice that, even though the greatest frequency of thunderstorms is near the Gulf Coast, the greatest frequency of hailstorms is over the western Great Plains. One reason for this situation is that conditions over the Great Plains are more favorable for the development of severe thunderstorms. We also find that, in summer along the Gulf Coast, a thick layer of warm moist air extends upward from the surface. Most hailstones falling into this layer will melt before reaching the ground. Over the plains, the warm surface layer is much shallower and drier. Falling hailstones do begin to melt, but the water around their periphery quickly evaporates in the dry air. This cools the hailstones and slows the melting rate so that many survive as ice all the way to the surface.

Now that we have looked at the development and distribution of thunderstorms, we are ready to examine an interesting, though yet not fully understood, aspect of all thunderstorms—lightning.

LIGHTNING AND THUNDER

Lightning is simply a discharge of electricity, a giant spark, which occurs in mature thunderstorms. Lightning may take place within a cloud, from one cloud to another, from a cloud to the surrounding air, or from a cloud to the ground. (The majority of lightning strikes occur within the cloud, while only about 20 percent or so occur between cloud and ground.) The lightning stroke can heat the air through which it travels to an incredible 30,000°C (54,000°F). This extreme heating causes the air to expand explosively, thus initiating a shock wave that becomes a booming sound wave—called **thunder**—that travels outward in all directions from the flash.

Light travels so fast that we see light instantly after a lightning flash. But the sound of thunder, traveling at only about 330 m/sec (1100 ft/sec), takes much longer to reach the ear. If we start counting seconds from the moment we see the lightning until we hear the thunder, we can determine how far away the stroke is. Because it takes sound about 3 seconds to travel 1 kilometer (5 seconds for each mile), if we see lightning and hear the thunder 15 seconds later, the lightning stroke occurred 5 km (3 mi) away.

When the lightning stroke is very close—100 m (330 ft) or less—thunder sounds like a clap or a crack followed immediately by a loud bang. When it is farther away, it often rumbles. The rumbling can be due to the sound emanating from different areas of the stroke. Moreover, the rumbling is accentuated when the sound wave reaches an observer after having bounced off obstructions, such as hills and buildings.

In some instances, lightning is seen but no thunder is heard. Does this mean that thunder was not produced by the lightning? Actually, there is thunder, but the atmosphere refracts (bends) and attenuates the sound waves, making the thunder inaudible. Sound travels faster in warm air than in cold atmosphere.* Because thunderstorms form in a conditionally unstable air, where the temperature normally drops rapidly with height, a sound wave moving outward away from a lightning stroke will often bend upward, away from an observer at the surface. Consequently, an observer closer than about 5 km to a lightning stroke will usually hear thunder, while an observer 15 km away will not.

However, even when a viewer is as close as several kilometers to a lightning flash, thunder may not be heard. For one thing, the complex interaction of sound waves and air molecules tends to attenuate the thunder. In addition, turbulent eddies of air less than 50 m in diameter scatter the sound waves. Hence, when thunder from a low-energy lightning flash travels several kilometers through turbulent air, it may become inaudible.

A sound occasionally mistaken for thunder is the **sonic boom**. Sonic booms are produced when an aircraft exceeds the speed of sound at the altitude at which it is flying. The aircraft compresses the air, forming a shock wave that trails out as a cone behind the aircraft. Along the shock wave, the air pressure changes rapidly over a short distance. The rapid pressure change causes the distinct boom. (Exploding fireworks generate a similar shock wave and a loud bang.)

Earlier, we learned that lightning occurs with mature thunderstorms. But lightning may also occur in snowstorms, in dust storms, in the gas cloud of an erupting volcano, and on very rare occasions in nimbostratus clouds. Lightning may also shoot from the top of thunderstorms into the upper atmosphere. More on this topic is given in the Focus section on p. 397.

What causes lightning? The normal fair weather electric field of the atmosphere is characterized by a negatively charged surface and a positively charged upper atmosphere. For lightning to occur, separate regions containing opposite electrical charges must exist within a cumulonimbus cloud. Exactly how this charge separation comes about is not totally comprehended; however, there are many theories to account for it.

Electrification of Clouds   One theory proposes that clouds become electrified as graupel and hail fall through a region of supercooled droplets and ice crystals. As liquid droplets collide with a hailstone, they freeze on contact and release latent heat.

---

*The speed of sound in calm air is equal to $20\sqrt{T}$, where T is the air temperature in Kelvins.

Red Sprites and Blue Jets

For many years airline pilots reported seeing strange bolts of light shooting upward, high above the tops of intense thunderstorms. These faint, mysterious flashes did not receive much attention, however, until they were first photographed in 1989. Photographs from sensitive, low light-level cameras on board jet aircraft revealed that the mysterious flashes were actually a colorful display called *red sprites* and *blue jets*, which seemed to dance above the clouds.

Sprites are massive, but dim, light flashes that appear directly above an intense thunderstorm system (see Fig. 2). Usually red, and lasting but a few millionths of a second, sprites tend to form simultaneously with lightning in the cloud below. However, they occur randomly and with only about one percent of all lightning strikes. Although it is not clear how they form, they are probably a type of lightning discharge that develops at high altitudes and

FIGURE 2
A red sprite flashes above an intense thunderstorm.

extends upward from the tops of clouds to as high as 95 km.

The other optical events that occur above active thunderstorms are blue jets. From the top of thunderstorms, blue jets usually dart upward at high speeds in narrow cones. Both blue

jets and red sprites are probably an integral feature of every major thunderstorm system on earth. Exactly how they form and what role they play in the earth's global electrical system has yet to be determined.

This keeps the surface of the hailstone warmer than that of the surrounding ice crystals. When the warmer hailstone comes in contact with a colder ice crystal, an important phenomenon occurs: *There is a net transfer of positive ions from the warmer object to the colder object.* Hence, the hailstone becomes negatively charged and the ice crystal positively charged, as the positive ions are incorporated into the ice. The same effect occurs when colder supercooled droplets freeze on contact with a warmer hailstone and tiny splinters of positively charged ice break off. These lighter, positively charged particles are then carried to the upper part of the cloud by updrafts. The larger hailstones, left with a negative charge, fall toward the bottom of the cloud. By this mechanism, the cold upper part of the cloud becomes positively charged, while the middle of the cloud becomes negatively charged. The lower part of the cloud is generally of negative and mixed charge except for an occasional positive region located in the falling precipitation near the melting level (see Fig. 15.23).

Another school of thought proposes that during the formation of precipitation, regions of separate charge exist within tiny cloud droplets and larger precipitation particles. In the upper part of these particles we find negative charge, while in the lower part we find positive charge. When falling precipitation collides with smaller particles, the larger precipitation particles become negatively charged and the smaller particles, positively charged. Updrafts within the cloud then sweep the smaller positively charged particles into the upper reaches of the cloud, while the smaller negatively charged particles either settle toward the lower part of the cloud or updrafts keep them suspended near the middle of the cloud.

The Lightning Stroke   Because unlike charges attract one another, the negative charge at the bottom of the cloud causes a region of the ground beneath it to become positively charged. As the thunderstorm moves along, this region of positive charge follows the cloud like a shadow. The positive charge is most dense on protruding objects, such as trees, poles, and buildings. The difference in charges causes an electric potential between the cloud and ground. In dry air, however, a flow of current does not occur because the air is a good electrical

FIGURE 15.23
The generalized charge distribution in a mature thunderstorm.

insulator. Gradually, the electrical potential gradient builds, and when it becomes sufficiently large (on the order of one million volts per meter), the insulating properties of the air break down, a current flows, and lightning occurs.

Cloud-to-ground lightning begins within the cloud when the localized electric potential gradient exceeds 3 million volts per meter along a path perhaps 50 m long. This situation causes a discharge of electrons to rush toward the cloud base and then toward the ground in a series of steps. Each discharge covers about 50 to 100 m, then stops for about 50-millionths of a second, then occurs again over another 50 m or so. This **stepped leader** is very faint and is usually invisible to the human eye. As the tip of the stepped leader approaches the ground, the potential gradient (the voltage per meter) increases, and a current of positive charge starts upward from the ground (usually along elevated objects) to meet it. After they meet, large numbers of electrons flow to the ground and a much larger, more luminous **return stroke** several centimeters in diameter surges upward to the cloud along the path followed by the stepped leader (Fig. 15.24). Hence, the downward flow of electrons establishes the bright channel of upward propagating current. Even though the bright return stroke travels from the ground up to the cloud, it happens so quickly—in one ten-thousandth of a second—that our eyes cannot resolve the motion, and we see what appears to be a continuous bright flash of light (see Fig. 15.25).

Sometimes there is only one lightning stroke, but more often the leader-and-stroke process is repeated in the same

(a)   (b)   (c)

FIGURE 15.24
The development of a lightning stroke. (a) When the negative charge near the bottom of the cloud becomes large enough to overcome the air's resistance, a flow of electrons—the stepped leader—rushes toward the earth. (b) As the electrons approach the ground, a region of positive charge moves up into the air through any conducting object, such as trees, buildings, and even humans. (c) When the downward flow of electrons meets the upward surge of positive charge, a strong electric current—a bright return stroke—carries positive charge upward into the cloud.

ionized channel at intervals of about four-hundredths of a second. The subsequent leader, called a **dart leader**, proceeds from the cloud along the same channel as the original stepped leader; however, it proceeds downward more quickly because the electrical resistance of the path is now lower. As the leader approaches the ground, normally a less energetic return stroke than the first one travels from the ground to the cloud. Typically, a lightning flash will have three or four leaders, each followed by a return stroke. A lightning flash consisting of many strokes (one photographed flash had 26 strokes) usually lasts less than a second. During this short period of time, our eyes may barely be able to perceive the individual strokes, and the flash appears to flicker.

Lightning may take on a variety of shapes and forms (see Fig. 15.26). When a dart leader moving toward the ground deviates from the original path taken by the stepped leader, the lightning appears crooked or forked, and it is called *forked lightning*. An interesting type of lightning is *ribbon lightning* that forms when the wind moves the ionized channel between each return stroke, causing the lightning to appear as a ribbon hanging from the cloud. If the lightning channel breaks up, or appears to break up, the lightning (called *bead lightning*) looks like a series of beads tied to a string. **Ball lightning** looks like a luminous sphere that appears to float in the air or slowly dart about for several seconds. Although many theories have been proposed, the actual cause of ball lightning remains an enigma. **Sheet lightning** forms when either the lightning flash occurs

WEATHER WATCH

The resultant electrical current from a single lightning bolt can provide enough energy to light a city of 200,000 people for one minute. It can also do some other interesting things. For example, the folks of Elgin, Manitoba, literally had their "goose cooked" during April, 1932, when a lightning bolt killed fifty-two geese that were flying overhead in formation. As the birds fell to the ground, they were reportedly gathered up and distributed to the townspeople for dinner.

inside a cloud or intervening clouds obscure the flash, such that a portion of the cloud (or clouds) appears as a luminous white sheet.

Distant lightning from thunderstorms that is seen but not heard is commonly called **heat lightning** because it frequently occurs on hot summer nights when the overhead sky is clear. As the light from distant electrical storms is refracted through the atmosphere, air molecules and fine dust scatter the shorter wavelengths of visible light, often causing heat lightning to appear orange (Fig. 15.26) to a distant observer.

As the electric potential near the ground increases, a current of positive charge moves up pointed objects, such as antennas and masts of ships. However, instead of a lightning stroke, a luminous greenish or bluish halo may appear above

FIGURE 15.25
Time exposure of lightning in Tulsa, Oklahoma. The bright flashes are return strokes. The lighter forked flashes are probably stepped leaders that did not make it to the ground.

them, as a continuous supply of sparks—a *corona discharge*—is sent into the air. This electric discharge, which can cause the top of a ship's mast to glow, is known as **St. Elmo's Fire**, named after the patron saint of sailors. St. Elmo's Fire is also seen around power lines and the wings of aircraft. When St. Elmo's Fire is visible and a thunderstorm is nearby, a lightning flash

FIGURE 15.28
A fulgurite that formed by lightning fusing sand particles.

FIGURE 15.27
The lightning rod extends above the building, increasing the likelihood that lightning will strike the rod rather than some other part of the structure. After lightning strikes the metal rod, it follows an insulated conducting wire harmlessly into the ground.

may occur in the near future, especially if the electric field of the atmosphere is increasing.

Lightning rods are placed on buildings to protect them from lightning damage. The rod is made of metal and has a pointed tip, which extends well above the structure (see Fig. 15.27). The positive charge concentration will be maximum on the tip of the rod, thus increasing the probability that the lightning will strike the tip and follow the metal rod harmlessly down into the ground, where the other end is deeply buried.

When lightning enters sandy soil, the extremely high temperature of the stroke may fuse sand particles together, producing a rootlike system of tubes called a **fulgurite**, after the Latin word for lightning (see Fig. 15.28). When lightning strikes an object such as a car, it normally leaves the passengers

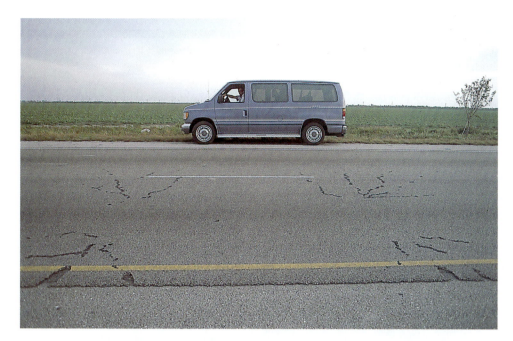

FIGURE 15.29
The four marks on the road surface represent areas where lightning, after striking a car traveling along south Florida's Sunshine State Parkway, entered the roadway through the tires. Lightning flattened three of the car's tires and slightly damaged the radio antenna. The driver and a six-year-old passenger were taken to a nearby hospital, treated for shock, and released.

unharmed because it usually takes the quickest path to the ground along the outside metal casing of the vehicle. The lightning then jumps to the road through the air, or it enters the roadway through the tires (see Fig. 15.29). If you should be caught in the open in a thunderstorm, what should you do? Of course, seek shelter immediately, but under a tree? If you are not sure, please read the Focus section on p. 402.

Lightning Detection and Suppression   For many years, lightning strokes were detected primarily by visual observation. Today, cloud-to-ground lightning is located by means of an instrument called a *lightning direction-finder*, which works by detecting the radio waves produced by lightning. Such waves are called **sferics**, a contraction from their earlier designation, *atmospherics*. A web of these magnetic devices is a valuable tool in pinpointing lightning strokes throughout the United States, Canada, and Alaska. Lightning detection devices allow scientists to examine in detail the lightning activity inside a storm as it intensifies and moves (see Fig. 15.30). This gives forecasters a better idea where intense lightning strokes might be expected.* In addition, when this information is correlated with satellite images, a more complete and precise structure of a thunderstorm is obtained.

Each year, approximately 10,000 fires are started by lightning in the United States alone and around $50 million worth of timber is destroyed. For this reason, tests have been conducted to see whether the number of cloud-to-ground lightning discharges can be reduced. One technique that has shown

some success in suppressing lightning involves seeding a cumulonimbus cloud with hair-thin pieces of aluminum about 10 centimeters long. The idea is that these pieces of metal will produce many tiny sparks, or *corona discharges*, and prevent the electrical potential in the cloud from building to a point where lightning occurs. While the results of this experiment

FIGURE 15.30
Cloud-to-ground lightning strikes in the vicinity of Chicago, Illinois as detected by the National Lightning Detection Network. (Global Atmospherics, Inc.)

*In fact, with the aid of these instruments and computer models of the atmosphere, the National Weather Service currently issues lightning probability forecasts for the western United States.

Don't Sit Under the Apple Tree

Because a single lightning stroke may involve a current as great as 100,000 amperes, animals and humans can be electrocuted when struck by lightning. The average yearly death toll in the United States attributed to lightning is nearly 100, with Florida accounting for the most fatalities. Many victims are struck in open places, riding on farm equipment, playing golf, or sailing in a small boat. Some live to tell about it, as did the champion golfer Lee Trevino. Others are less fortunate. When you see someone struck by lightning, immediately give CPR (cardiopulmonary resuscitation), as lightning normally leaves its victims unconscious without heartbeat and without breathing.

Most lightning fatalities occur in the vicinity of relatively isolated trees (see Fig. 3). Because a positive charge tends to concentrate in upward projecting objects, the upward return stroke that meets the stepped leader is most likely to originate from such objects. Clearly, sitting under a tree during an electrical

FIGURE 3

A cloud-to-ground lightning flash hitting a 65-foot sycamore tree. It should be apparent why one should *not* seek shelter under a tree during a thunderstorm.

storm is not wise. What *should* you do?

When caught in a thunderstorm, the best protection, of course, is to get inside a building. Automobiles and trucks (but not golf carts) may also provide protection. If no such shelter exists, be sure to avoid elevated places and isolated trees. If you are on level ground, try to keep your head as low as possible, but do not lie down. Because lightning channels usually emanate outward through the ground at the point of a lightning strike, a surface current may travel through your body and injure or kill you. Therefore, crouch down as low as possible and minimize the contact area you have with the ground. There are some warning signs to alert you to a strike. If your hair begins to stand on end or your skin begins to tingle and you hear clicking sounds, beware—lightning may be about to strike. And if you are standing upright, you may be acting as a lightning rod.

---

are inconclusive, many forestry specialists point out that nature itself may use a similar mechanism to prevent excessive lightning damage. The long, pointed needles of pine trees may act as tiny lightning rods, diffusing the concentration of electric charges and preventing massive lightning strokes.

BRIEF REVIEW In the last several sections, we examined thunderstorms with their associated lightning and thunder. Listed below for your review are several important concepts we considered:

- Thunderstorms tend to form when warm, humid air rises in a conditionally unstable atmosphere.

- While ordinary (air mass) thunderstorms are usually short-lived (typically less than an hour), and rarely produce strong

winds and hail, severe thunderstorms such as the super-cell and pre-frontal squall-line thunderstorms may exist for hours and are capable of producing large hail, strong surface winds (derechoes), flash floods, and tornadoes.

- Strong downdrafts of a thunderstorm, called downbursts (or if the downdrafts are smaller than 4 km, they are called microbursts) have been responsible for several airline crashes, because upon striking the surface, these winds produce extreme wind shear—rapid changes in wind speed and wind direction.

- While lightning is a visible electrical discharge (a giant spark that usually occurs in mature thunderstorms), thunder is the sound that results from the rapidly expanding heated air along the channel of the lightning stroke.

Now that we have looked at thunderstorms, we are now ready to explore a product of a thunderstorm that is one of nature's most awesome phenomena: the tornado, a rapidly spiraling column of air that can strike sporadically and violently.

TORNADOES

Tornadoes are rapidly rotating winds that blow around a small area of intense low pressure. A tornado's circulation is present on the ground either as a funnel-shaped cloud or as a swirling cloud of dust and debris. Sometimes called *twisters* or *cyclones*, tornadoes can assume a variety of shapes and forms that range from twisting ropelike funnels, to cylindrical-shaped funnels, to massive black funnels, to funnels that resemble an elephant's trunk hanging from a large cumulonimbus cloud. A **funnel cloud** is a tornado whose circulation has not reached the ground. When viewed from above, the majority of tornadoes rotate counterclockwise. A few have been seen rotating clockwise, but those are rare.

The diameter of most tornadoes is between 100 and 600 m (about 300 to 2000 ft), although some are just a few meters wide and others have diameters exceeding 1600 m (1 mi). Tornadoes that form ahead of an advancing cold front are often steered by southwesterly winds and, therefore, tend to move from the southwest toward the northeast at speeds usually between 20 and 40 knots. However, some have been clocked at speeds greater than 70 knots. Most tornadoes last only a few minutes and have an average path length of about 7 km (4 mi). There are cases where they have reportedly traveled for hundreds of kilometers and have existed for many hours, such as the one that lasted over 7 hours and cut a path 470 km (292 mi) long through portions of Illinois and Indiana on May 26, 1917.*

Major tornadoes usually evolve through a series of stages. The first stage is the *dust-whirl stage*, where dust swirling upward from the surface marks the tornado's circulation on the ground and a short funnel often extends downward from the thunderstorm's base. Damage during this stage is normally light. The next stage, called the *organizing stage*, finds the tornado increasing in intensity with an overall downward extent of the funnel. During the tornado's *mature stage*, damage normally is most severe as the funnel reaches its greatest width and is almost vertical (see Fig. 15.31). The *shrinking stage* is characterized by an overall decrease in the funnel's width, an increase in the funnel's tilt, and a narrowing of the damage swath at the surface, although the tornado may still be capable of intense and sometimes violent damage. The final stage, called the *decay stage*, usually finds the tornado stretched into the shape of a rope. Normally, the tornado becomes greatly contorted before it finally dissipates. Although these are the typical stages of a

major tornado, minor tornadoes may evolve only through the organizing stage. Some even skip the mature stage and go directly into the decay stage. However, when a tornado reaches its mature stage, its circulation usually stays in contact with the ground until it dissipates.

Each year, tornadoes take the lives of many people. The yearly average is about 100, although over 100 may die in a single day. In recent years, an alarming statistic is that 45 percent of all fatalities occurred in mobile homes. The deadliest tornadoes are those that occur in *families*; that is, different tornadoes spawned by the same thunderstorm. (Some thunderstorms produce a sequence of several tornadoes over 2 or more hours and over distances of 100 km or more.) Tornado families often are the result of a single, long-lived supercell thunderstorm. When a large number of tornadoes (typically 6 or more) form over a particular region, this constitutes what is termed a **tornado outbreak**. One of the most violent outbreaks ever recorded occurred on April 3 and 4, 1974. During a 16-hour period, 148 tornadoes cut through parts of 13 states, killing 307 people, injuring more than 6000, and causing an estimated $600 million in damage. Some of these tornadoes were among the most powerful ever witnessed. The combined path of all the tornadoes during this *super outbreak* amounted to 4181 km (2598 mi), well over half of the total path for an average year. The greatest loss of life attributed to tornadoes occurred during the tri-state outbreak of March 18, 1925, when an estimated 695 people died as 7 tornadoes traveled a total of 703 km (437 mi) across portions of Missouri, Illinois, and Indiana.

Tornado Occurrence   Tornadoes occur in many parts of the world, but no country experiences more tornadoes than the United States, which averages more than 900 annually and experienced a record 1293 tornadoes during 1992. Although tornadoes have occurred in every state, including Alaska and Hawaii, the greatest number occur in the tornado belt or *tornado alley* of the Central Plains, which stretches from central Texas to Nebraska* (see Fig. 15.32).

The Central Plains region is most susceptible to tornadoes because it provides the proper atmospheric setting for the development of the severe thunderstorms that spawn tornadoes. Here (especially in spring) warm, humid surface air is overlain by cooler, drier air aloft, producing a conditionally unstable atmosphere. When a strong vertical wind shear exists and the surface air is forced upward, large thunderstorms capable of spawning tornadoes may form. Therefore, tornado frequency is highest during the spring and lowest during the winter when the warm surface air is absent.

About three-fourths of all tornadoes in the United States develop from March to July. The month of May normally has the greatest number of tornadoes (the average is about 5 per

---

*Actually, this situation may have been several tornadoes (a family) that were generated by a single thunderstorm as it moved along.

*Many of the tornadoes that form along the Gulf Coast are generated by thunderstorms embedded within the circulation of hurricanes.

FIGURE 15.31
A mature tornado with winds exceeding 150 knots rips through southern Illinois.

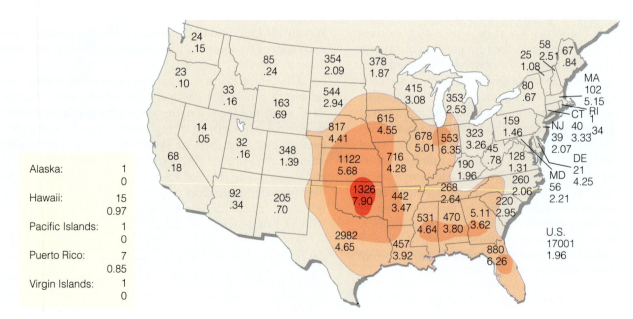

Alaska:	1	
	0	
Hawaii:	15	
	0.97	
Pacific Islands:	1	
	0	
Puerto Rico:	7	
	0.85	
Virgin Islands:	1	
	0	

FIGURE 15.32
Tornado incidence by state. The upper figure shows the number of tornadoes reported by each state during a 25-year period. The lower figure is the average annual number of tornadoes per 10,000 square miles. The darker the shading, the greater the frequency of tornadoes.

day) while the most violent tornadoes seem to occur in April when vertical wind shear tends to be present as well as when horizontal and vertical temperature and moisture contrasts are greatest. Although tornadoes have occurred at all times of the day and night, they are most frequent in the late afternoon (between 4:00 P.M. and 6:00 P.M.), when the surface air is most unstable; they are least frequent in the early morning before sunrise, when the atmosphere is most stable.

Although large, destructive tornadoes are most common in the Central Plains, they can develop anywhere if conditions are right. For example, a series of at least 36 tornadoes, more typical of those that form over the plains, marched through North and South Carolina on March 28, 1984, claiming 59 lives and causing hundreds of millions of dollars in damage. One tornado was enormous, with a diameter of at least 4000 m (2.5 mi) and winds that exceeded 200 knots. No place is totally immune to a tornado's destructive force. On March 1, 1983, a rare tornado cut a 5-km swath of destruction through downtown Los Angeles, California, damaging more than 100 homes and businesses and injuring 33 people.

Even in the central part of the United States, the statistical chance that a tornado will strike a particular place this year is quite small. However, tornadoes can provide many exceptions to statistics. Oklahoma City, for example, has been struck by tornadoes at least 32 times in the past 90 years. And the little town of Codell, Kansas, was hit by tornadoes in 3 consecutive years—1916, 1917, and 1918—and each time on the same date: May 20! Considering the many millions of tornadoes that must have formed during the geological past, it is likely that at least one actually moved across the land where your home is located, especially if it is in the Central Plains.

Tornado Winds   The strong winds of a tornado can destroy buildings, uproot trees, and hurl all sorts of lethal missiles into the air. People, animals, and home appliances all have been picked up, carried several kilometers, then deposited. Tornadoes have accomplished some astonishing feats, such as lifting a railroad coach with its 117 passengers and dumping it in a ditch 25 m away. Showers of toads and frogs have poured out of a cloud after tornadic winds sucked them up from a nearby pond. Other oddities include chickens losing all of their feathers and pieces of straw being driven into metal pipes. Miraculous events have occurred, too. In one instance, a schoolhouse was demolished and the 85 students inside were carried over 100 m without one of them being killed.

Our earlier knowledge of the furious winds of a tornado came mainly from observations of the damage done and the analysis of motion pictures. Today more accurate wind measurements are made with Doppler radar. Because of the destructive nature of the tornado, it was once thought that it packed winds greater than 500 knots. However, studies conducted after 1973 reveal that even the most powerful twisters seldom have winds exceeding 220 knots, and most tornadoes probably have winds of less than 125 knots. Nevertheless, being confronted with even a small tornado can be terrifying.

WEATHER WATCH

Tornadoes have been responsible for some rather unusual events. For example, in 1991, a Kansas tornado sucked a mother and her seven-year-old son out of their house in a bathtub. The tub with its occupants hit the ground hard, rose into the air, then hit the ground again, tossing its passengers into the neighbor's backyard. Battered and bruised with scratches and lumps, the mother and son survived. The bathtub, which disappeared in the tornado, was never found.

The high winds of the tornado cause the most damage as walls of buildings buckle and collapse when blasted by the extreme wind force. Also, as high winds blow over a roof, lower air pressure forms above the roof. The greater air pressure inside the building then lifts the roof just high enough for the strong winds to carry it away. A similar effect occurs when the tornado's intense low-pressure center passes overhead. Because the pressure in the center of a tornado may be more than 100 mb (3 in.) lower than that of its surrounding, there is a momentary drop in outside pressure when the tornado is above the structure. It was once thought that opening windows and allowing inside and outside pressures to equalize would minimize the chances of the building exploding. However, it now appears that opening windows during a tornado actually increases the pressure on the opposite wall and *increases* the chances that the building will collapse. (The windows are usually shattered by flying debris anyway.) Damage from tornadoes may also be inflicted on people and structures by flying debris. Hence, the wisest course to take when confronted with an approaching tornado is to *seek shelter immediately*.

At home, take shelter in a basement and stay away from windows. In a large building without a basement, the safest place is usually in a small room, such as a bathroom, closet, or interior hallway, preferably on the lowest floor and near the middle of the edifice. At school, move to the hallway and lie flat with your head covered. In a mobile home, leave immediately and seek substantial shelter. If none exists, lie flat on the ground in a depression or ravine. Don't try to outrun an oncoming tornado in a car or truck, as tornadoes often cover erratic paths with speeds sometimes exceeding 70 knots. If caught outdoors in an open field, look for a ditch, streambed, or ravine, and lie flat with your head covered.

When a tornado is approaching from the southwest, its strongest winds are on its southeast side. We can see why in Fig. 15.33. The tornado is heading northeast at 50 knots. If its rotational speed is 100 knots, then its forward speed will add 50 knots to its southeastern side (position D) and subtract 50 knots from its northwestern side (position A). Because the most destructive and extreme winds will be on the tornado's southeastern side, it is the southwest side of the building that will receive the maximum impact of the winds.

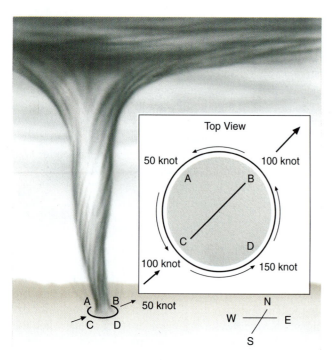

FIGURE 15.33
The total wind speed of a tornado is greater on one side than on the other. When facing an on-rushing tornado, the strongest winds will be on your left side.

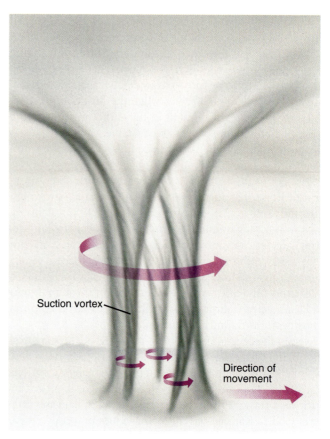

FIGURE 15.34
A powerful multi-vortex tornado with three suction vortices.

It now appears that many violent tornadoes (with winds exceeding 180 knots) contain smaller whirls that rotate within them. Such tornadoes are called *multi-vortex tornadoes* and the smaller whirls are called **suction vortices** (see Fig. 15.34). Suction vortices are only about 10 m (30 ft) in diameter, but they rotate very fast and apparently do a great deal of damage.

In the late 1960s, the late Dr. T. Theodore Fujita, a noted authority on tornadoes at the University of Chicago, proposed a scale (called the **Fujita scale**) for classifying tornadoes

according to their rotational wind speed and the damage done by the storm. Table 15.1 presents this scale.

Statistics reveal that the majority of tornadoes are F0 and F1 (weak tornadoes) and only a few percent each year are above the F3 classification (violent) with about one F5 tornado reported annually. However, it is the violent tornadoes that

TABLE 15.1	**Fujita Scale for Damaging Wind**			
SCALE	CATEGORY	MI/HR	KNOTS	EXPECTED DAMAGE
F0	Weak	40–72	35–62	Light: tree branches broken, sign boards damaged
F1		73–112	63–97	Moderate: trees snapped, windows broken
F2	Strong	113–157	98–136	Considerable: large trees uprooted, weak structures destroyed
F3		158–206	137–179	Severe: trees leveled, cars overturned, walls removed from buildings
F4	Violent	207–260	180–226	Devastating: frame houses destroyed
F5		261–318	227–276	Incredible: structures the size of autos moved over 100 meters, steel-reinforced structures highly damaged

FIGURE 15.35
A devastating F5 tornado about 200 m wide plows through Hesston, Kansas, on March 13, 1990, leaving almost 300 people homeless and 13 injured.

account for the majority of tornado-related deaths. As an example, a powerful F5 tornado marched through the southern part of Andover, Kansas, on the evening of April 26, 1991. The tornado, which stayed on the ground for nearly 110 km (68 mi), destroyed more than 100 homes and businesses, injured several hundred people, and out of the 39 tornado fatalities in 1991, this F5 tornado alone took the lives of 17. A powerful F5 tornado is shown in Fig. 15.35.

Tornado Formation Although everything is not known about the formation of a tornado, we do know that tornadoes tend to form with intense thunderstorms and that a conditionally unstable atmosphere is essential for their development. One atmospheric situation that frequently leads to severe thunderstorms with tornadoes in the spring is shown in Fig. 15.36.

At the surface, we find an open-wave middle latitude cyclone with cold, dry air moving in behind a cold front, and warm humid air pushing northward from the Gulf of Mexico behind a warm front. Above the warm surface air a wedge of warm, moist air is streaming northward. Directly above the moist layer is a wedge of colder, drier air moving in from the southwest. Higher up, at the 500-mb level, a trough of low pressure exists to the west of the surface low and, at the

300-mb level, the polar front jet stream swings over the region. At this level, the jet stream provides an area of divergence that initiates surface convergence and rising air. The stage is now set for the development of severe storms.

The boxed-off area on the surface map (Fig. 15.36) shows where tornadoes are most likely to form. They tend to form in this region because the position of cold air above warm air produces a conditionally unstable atmosphere. However, there is more to it than this. As warm air rises from the surface, many thunderstorms should form throughout the region of warm air ahead of the advancing cold front—the warm sector. But often this is not the case as the atmospheric conditions that generate severe thunderstorms tend to prevent many smaller ones from forming. To see why, we will examine the vertical profile of temperature and moisture—a *sounding*—in the warm sector ahead of the cold front.

Figure 15.37 shows a typical sounding of temperature and dew point in the warm sector before tornadoes occur. From the surface up to 800 mb, the conditionally unstable air is warm and very humid. At 800 mb, a shallow inversion (or simply, a very stable layer) acts like a cap on the moist layer below. Above the inversion, the air is cold and much drier. This air is also conditionally unstable, as the temperature drops at just

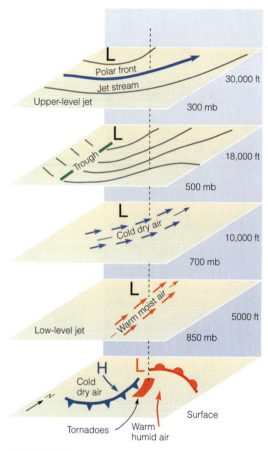

FIGURE 15.36

Conditions leading to the formation of severe thunderstorms that can spawn tornadoes.

about the dry adiabatic rate (10°C/1000 m). The cooling of this upper layer is due, mainly, to cold advection. Cold, dry, unstable air sitting above a warm, humid layer produces *convective instability*, which means that the atmosphere will become even

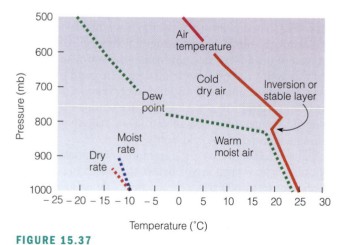

FIGURE 15.37

A typical sounding of air temperature and dew point that frequently precede the development of severe thunderstorms that spawn tornadoes.

more unstable if a layer of air is somehow forced upward (see Chapter 7).

The lifting of warm surface air can occur at the frontal zones, but the air may also begin to rise anywhere in the warm sector when the surface air heats up during the day. However, in the morning, the inversion acts as a lid on rising thermals and only small cumulus clouds form. As the day progresses (and the surface air heats even more), rising air breaks through the inversion at isolated places and clouds build rapidly, sometimes explosively, as the moist air is vented upward through the opening. Thus, we can see that the inversion prevents many small thunderstorms from forming. When the surface air is finally able to puncture the inversion, the upper-level jet stream rapidly draws the moist air up into the cold unstable air, and a large storm quickly develops. The severe thunderstorm soon builds to great heights, and an anvil forms. However, the strong updraft inside the storm causes its top to overshoot the anvil by 2 to 4 km.

While it is difficult to tell which thunderstorm will spawn a tornado, it is easier to predict where tornado-generating storms are most likely to form. Notice in Fig. 15.36 that the boxed-off area on the surface map is situated where the polar front jet stream and the cold, dry air cross the warm, humid surface air. Knowing this helps to explain why the region of greatest tornado activity shifts northward from winter to summer. During the winter, tornadoes are most likely to form over the southern Gulf states when the polar front jet is above this region, and the contrast between warm and cold air masses is greatest. In spring, humid Gulf air surges northward; contrasting air masses and the jet stream also move northward and tornadoes become more prevalent from the southern Atlantic states westward into the southern Great Plains. In summer, the contrast between air masses lessens, and the jet stream is normally near the Canadian border; hence, tornado activity tends to be concentrated from the northern plains eastward to New York State.

In order for the storm to spawn a tornado, the updraft must rotate. Remember from our earlier discussion that severe thunderstorms form in a region of strong vertical wind shear. In Fig. 15.36, the rapidly increasing wind speed with height (vertical wind speed shear) and the changing wind direction with height—from southerly at low levels to westerly at high levels (vertical wind direction shear)—may cause the updraft inside the storm to spin cyclonically. This rising, spinning column of air, perhaps 5 to 10 km across, is called a **mesocyclone**. Doppler radar observations have shown that, normally, rotation begins in the middle of the thunderstorm and gradually works downward. As air rushes in toward the low pressure of the mesocyclone, the rotational wind speeds increase much like a spinning skater increases in speed when the arms are brought in closer to the body (the conservation of angular momentum). At the same time, the mesocyclone stretches vertically and shrinks horizontally (see Fig. 15.38), and the spinning air is accelerated upward. Inside the mesocyclone, which is now between 2 and 4 km wide, a spinning vortex of increasing wind speed (a tornado) may—for reasons not fully under-

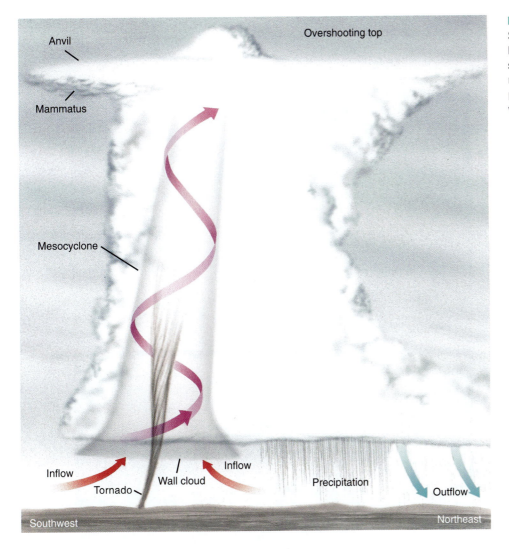

FIGURE 15.38
Some of the features associated with a tornado-breeding thunderstorm as viewed from the southeast. The thunderstorm is moving to the northeast. The tornado forms in the southwest part of the thunderstorm. Compare this Figure with Fig. 15.13 (p. 389), a classic supercell.

stood—appear near the midlevel of the cloud and gradually extend downward to the cloud base. At this point, the mesocyclone is called a *tornado cyclone*.

As air rushes into the low-pressure vortex from all directions, it expands, cools, and, if sufficiently moist, condenses into a visible cloud—the *funnel cloud*. As the air beneath the funnel is drawn into the core, it cools rapidly and condenses, and the funnel cloud descends toward the surface.* Upon reaching the ground, the tornado usually picks up dirt and debris, making it appear both dark and ominous. While the air along the outside of the funnel is spiraling upward, Doppler radar reveals that, within the core of violent tornadoes, the air is descending toward the extreme low pressure at the ground. As the air descends, it warms, causing the cloud droplets to evaporate. This process leaves the core free of clouds.

Observations reveal that the most strong and violent tornadoes develop near the right rear sector of a severe thunderstorm (on the southwestern side of an eastward moving storm, Fig. 15.38). However, weaker tornadoes may not only develop in the main updraft, but along the gust front, where the cool downdraft forces warm inflowing air upward. Tornadoes that form along the gust front are called **gustnadoes**. Although it appears that most strong and violent tornadoes form within the mesocyclone, not all mesocyclones produce tornadoes. In fact, weak tornadoes may form without a mesocyclone. Small tornadoes such as these (sometimes called *landspouts*) may form from thunderstorms or from cumulus congestus clouds. They tend to be similar to waterspouts described on p. 414.

The likelihood of a thunderstorm producing a tornado increases when the storm becomes a supercell, but not all supercells produce tornadoes. And not all tornadoes come from rotating thunderstorms (supercells). Certainly, all the processes that go into generating a tornado are not completely understood at this time. However, information on how a thunderstorm may develop rotation is given in the Focus section on p. 410.

Observing Tornadoes   As we have seen, the first sign that the thunderstorm is about to give birth to a tornado is the sight of

---

*Tornadoes may also begin as debris clouds on the ground in the absence of a condensation funnel. Some tornadoes actually begin at the ground and build upward.

Thunderstorm Rotation

Rotating clouds at the base of a severe thunderstorm often indicate that the storm is about to give birth to a tornado. But how do the clouds develop rotation?

Figure 4 illustrates how rotating vortices can develop near the surface. Notice that there is wind direction shear as surface winds are southeasterly; aloft they are westerly. There is also wind speed shear as the wind speed increases with increasing height. This wind shear causes the air near the surface to rotate about a horizontal axis, producing narrow tubes of spiraling air called *vortex tubes*. A strong updraft of a thunderstorm may then tilt a rotating tube and draw it into the storm as depicted in Fig. 5. This situation sets up two spinning vertical columns of air—one rotating clockwise and the other counterclockwise. As air is drawn more quickly into the storm, the spiraling columns spin faster.

If the thunderstorm has a more complicated structure (as most do) additional rotating air columns may form. This phenomenon normally induces the southern flank of the storm to rotate in one direction, usually counterclockwise (when viewed from above) and the northern flank in the other direction, usually clockwise. Hence, the thunderstorm rotates.

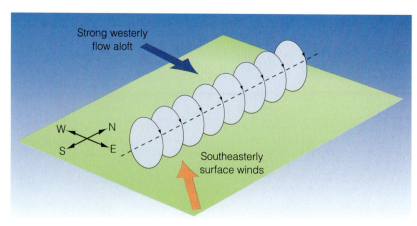

FIGURE 4

Spinning vortex tubes created by wind shear.

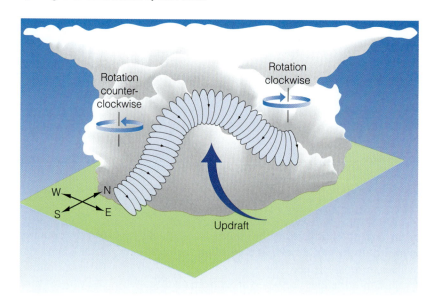

FIGURE 5

The strong updraft in the thunderstorm carries the vortex tube into the thunderstorm, producing two rotating air columns that are oriented in the vertical plane.

---

rotating clouds at the base of the storm.* If the area of rotating clouds lowers, it becomes a **wall cloud** (compare Fig. 15.38 with Fig. 15.39). Usually within the wall cloud, a smaller, rapidly rotating funnel extends toward the surface. Sometimes the air is so dry that the swirling wind remains invisible until it

reaches the ground and begins to pick up dust. Unfortunately, people have mistaken these "invisible tornadoes" for dust devils, only to find out (often too late) that they were not. Occasionally, the funnel cannot be seen due to falling rain, clouds of dust, or darkness. Even when not clearly visible, many tornadoes have a distinctive roar that can be heard for several kilometers. This sound, which has been described as "a roar like a thousand freight trains," appears to be loudest when the tornado is touching the surface. However, not all tornadoes make

---

*Occasionally, people will call a sky dotted with mammatus clouds "a tornado sky." Mammatus clouds may appear with both severe and nonsevere thunderstorms as well as with a variety of other cloud types (see Chapter 6). Mammatus clouds are not funnel clouds, do not rotate, and their appearance has no relationship to tornadoes.

this sound and, when these storms strike, they become silent killers.

When tornadoes are likely to form during the next few hours, a **tornado watch** is issued by the Storms Prediction Center in Norman, Oklahoma, to alert the public that tornadoes may develop within a specific area during a certain time period. Many communities have trained volunteer spotters, who look for tornadoes after the watch is issued. Once a tornado is spotted—either visually or on a radar screen—a **tornado warning** is issued by the local National Weather Service Office. In some communities, sirens are sounded to alert people of the approaching storm. Radio and television stations interrupt regular programming to broadcast the warning. Although not completely effective, this warning system is apparently saving many lives. Despite the large increase in population in the tornado belt during the past 30 years, tornado-related deaths have actually shown a decrease (see Table 15.2).

In an attempt to unravel some of the mysteries of the tornado, several studies are under way. In one called *VORTEX* (*V*erification of the *O*rigins of *R*otation in *T*ornadoes *Ex*periment) scientists using an armada of observational vehicles, aircraft, and state-of-the-art equipment, including Doppler radar, pursued tornado-generating thunderstorms over portions of the Central Plains during the summers of 1994 and 1995. From these data, observations are providing information on the inner workings of severe thunderstorms. At the same time, computer models are offering new insights into the formation and development of these same types of storms.

TABLE 15.2   Average Annual Number of Tornadoes and Tornado Deaths by Decade

DECADE	TORNADOES/YEAR	DEATHS/YEAR
1950–59	480	148
1960–69	681	94
1970–79	858	100
1980–89	819	52
1990–95	1,175*	44

*More tornadoes are being reported as populations increase and tornado-spotting technology improves.

SEVERE WEATHER AND DOPPLER RADAR

Most of our knowledge about what goes on inside a tornado-generating thunderstorm has been gathered through the use of Doppler radar. Remember from Chapter 8 that a radar transmitter sends out microwave pulses and that, when this energy strikes an object, a small fraction is scattered back to the antenna. Precipitation particles are large enough to bounce microwaves back to the antenna. As a consequence, the colorful area on the radar screen in Fig. 15.40 represents precipitation inside a severe thunderstorm, as viewed by the older

FIGURE 15.39
A tornado extends downward from a wall cloud over North Dakota.

FIGURE 15.40
A tornado-spawning thunderstorm shows a hook echo in its rainfall pattern on a conventional radar screen.

FIGURE 15.41
Doppler radar display of winds. A small supercell embedded in a squall line is moving toward Washington, D.C., on June 24, 1996. The dark circular area near the center of the image represents the hook echo of the storm. The orange and red colors indicate winds blowing away from the radar, whereas the blue and green colors indicate winds blowing toward the radar.

conventional-type radar. Notice that the pattern on the left side of the screen is in the shape of a hook. A **hook echo** such as this indicates the possible presence of a tornado. When tornadoes form, they do so near the tip of the hook. However, there is a problem here in that many severe thunderstorms (as well as smaller ones) do not show a hook echo, but still spawn tornadoes. Sometimes, when the hook echo does appear, the tornado is already touching the ground. Therefore, a better technique was needed in detecting tornado-producing storms. To address this need, Doppler radar was developed.

Doppler radar is like a conventional radar in that it can detect areas of precipitation and measure rainfall intensity. But a Doppler radar can do more—it can actually measure the speed at which precipitation is moving horizontally toward or away from the radar antenna. Because precipitation particles are carried by the wind, Doppler radar can peer into a severe storm and unveil its winds.

Doppler radar works on the principle that, as precipitation moves toward or away from the antenna, the returning radar pulse will change in frequency. A similar change occurs when the high-pitched sound (high frequency) of an approaching noise source, such as a siren or train whistle, becomes lower in pitch (lower frequency) after it passes by the person hearing it. This change in frequency is called the *Doppler shift* and this, of course, is where the Doppler radar gets its name.

A single Doppler radar cannot detect winds that blow parallel to the antenna. Consequently, two or more units probing the same thunderstorm are needed to give a complete three-dimensional picture of the winds within the storm. To help distinguish the storm's air motions, wind velocities can be displayed in color. Color contouring the wind field gives a good picture of the storm.

Even a single Doppler radar can uncover many of the features of a severe thunderstorm (see Fig. 15.41). For example, studies conducted in the 1970s revealed, for the first time, the existence of the swirling winds of the *mesocyclone* inside tornado-producing thunderstorms. Mesocyclones have a distinct image (signature) on the radar display. Studies show that about 30 percent of all mesocyclones produce tornadoes and about 95 percent produce severe weather. The time between mesocyclone identification and the tornado actually touching the ground is about 20 minutes.

Tornadoes also have a distinct signature, known as the *tornado vortex signature (TVS)*, which shows up as a region of rapidly changing wind directions within the mesocyclone (see Fig. 15.42). Unfortunately, the resolution of the Doppler radar is not high enough to measure actual wind speeds of most tornadoes, whose diameters are only a few hundred meters or less. However, a new and experimental Doppler system—called **Doppler lidar**—uses a light beam (instead of microwaves) to measure the change in frequency of falling precipitation, cloud particles, and dust. Because it uses a shorter wavelength of radiation, it has a narrower beam and a higher resolution than does Doppler radar. In an attempt to obtain tornado wind information at fairly close range (less than 10 km), smaller portable Doppler radar units (*Doppler on wheels*) are peering into tornado-generating storms (see Fig. 15.43).

The new network of 135 Doppler radar units deployed at selected weather stations within the continental United States is referred to as **NEXRAD** (an acronym for *NEX*t Generation *Weather RAD*ar). The NEXRAD system consists of the WSR-88D* Doppler radar and a set of computers that perform a variety of functions.

---

*The name WSR-88D stands for *Weather Surveillance Radar, 1988 Doppler.*

FIGURE 15.43
Graduate students from the University of Oklahoma use a portable Doppler radar to probe a tornado near Hodges, Oklahoma.

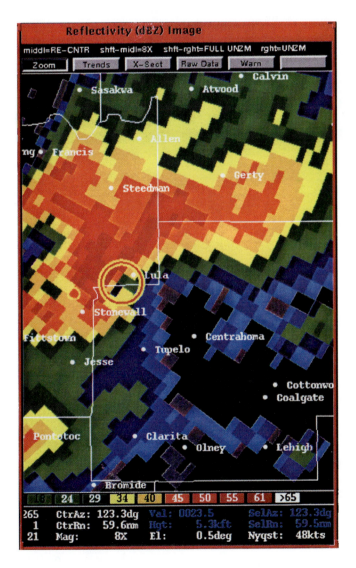

FIGURE 15.42
Doppler radar display showing a large supercell thunderstorm that is spawning an F4 tornado (circled area) near Lula, Oklahoma. The close packing of the winds indicate strong cyclonic rotation and the signature of a tornado. (Red and orange indicate winds blowing away from the radar. Green and blue indicate winds blowing toward the radar.)

The computers take in data, display it on a monitor, and run computer programs called *algorithms*, which, in conjunction with other meteorological data, detect severe weather phenomena, such as storm cells, hail, mesocyclones, and tornadoes. Algorithms provide a great deal of information to the forecasters that allows them to make better decisions as to which thunderstorms are most likely to produce severe weather and possible flash flooding. In addition, they give advanced and improved warning of an approaching tornado. More reliable warnings, of course, will cut down on the number of false alarms.

Because the Doppler radar shows air motions within a storm, it can help to identify the magnitude of other severe weather phenomena, such as gust fronts, microbursts, and wind shears that are dangerous to aircraft (see Fig. 15.44). Certainly, as more and more information from Doppler radar becomes available, our understanding of the processes that generate severe thunderstorms and tornadoes will be enhanced, and hopefully there will be an even better tornado and severe storm warning system, resulting in fewer deaths and injuries.

FIGURE 15.44
Doppler radar display showing a line of thunderstorms bent in the shape of a bow (colors red, orange, and yellow) as they move eastward across the San Joaquin Valley of California. Such *bow echos* often produce damaging surface winds near the center of the bow. Sometimes the left (usually northern) end of the bow will develop cyclonic rotation and produce a tornado.

FIGURE 15.45
A waterspout over the warm waters of the Florida Keys.

WATERSPOUTS

A **waterspout** is a rotating column of air over a large body of water. The waterspout may be a tornado that formed over land and then traveled over water. In such a case, the waterspout is sometimes referred to as a *tornadic waterspout*. Waterspouts that form over water, especially above the warm, shallow coastal waters of the Florida Keys, where almost 100 occur each month during the summer, are referred to as *"fair weather" waterspouts*.* These waterspouts are generally much smaller than an average tornado, as they have diameters usually between 3 and 100 m. Fair weather waterspouts are also less intense, as their rotating winds are typically less than 45 knots. In addition, they tend to move more slowly than tornadoes and they only last for about 10 to 15 minutes, although some have existed for up to one hour.

Fair weather waterspouts tend to form when the air is con-

ditionally unstable and clouds are developing. Unlike the tornado, they do not need a thunderstorm to generate them. Some form with small thunderstorms, but most form with developing cumulus congestus clouds whose tops are frequently no higher than 3600 m (12,000 ft) and do not extend to the freezing level. Apparently, the warm, humid air near the water helps to create atmospheric instability, and the updraft beneath the resulting cloud helps initiate uplift of the surface air. Studies even suggest that gust fronts and converging sea breezes may play a role in the formation of some of the waterspouts that form over the Florida Keys.

The waterspout funnel is similar to the tornado funnel in that both are clouds of condensed water vapor with converging winds that rise about a central core. Contrary to popular belief, the waterspout does not draw water up into its core; however, swirling spray may be lifted several meters when the waterspout funnel touches the water. Apparently, the most destructive waterspouts are those that begin as tornadoes over land, then move over water. A photograph of a particularly well-developed and intense fair weather waterspout near the Florida Keys is shown in Fig. 15.45.

---

*"Fair weather" waterspouts may form over any large body of warm water. Hence, they occur frequently over the Great Lakes in summer.

SUMMARY

In this chapter, we examined thunderstorms and the atmospheric conditions that produce them. The ingredients for an isolated ordinary (air-mass) thunderstorm are humid surface air, plenty of sunlight to heat the ground, and a conditionally unstable atmosphere. When these conditions prevail, small cumulus clouds may grow into towering clouds and thunderstorms within 20 minutes.

When conditions are ripe for thunderstorm development and a strong vertical wind shear exists, the stage is set for the generation of severe thunderstorms. Supercell thunderstorms may exist for many hours, as their updrafts and downdrafts are nearly in balance. Thunderstorms that form in a line, along or ahead of an advancing cold front, are called a squall line.

Lightning is a discharge of electricity that occurs in mature thunderstorms. The lightning stroke momentarily heats the air to an incredibly high temperature. The rapidly expanding air produces a sound called thunder. Along with lightning and thunder, severe thunderstorms produce violent weather, such as destructive hail, strong downdrafts, and the most feared of all atmospheric storms—the tornado.

Tornadoes are rapidly rotating columns of air that extend downward from the base of a thunderstorm. Most tornadoes are less than a few hundred meters wide with wind speeds less than 100 knots, although violent tornadoes may have wind speeds that exceed 250 knots. A violent tornado may actually have smaller whirls (suction vortices) rotating within it. With the aid of Doppler radar, scientists are probing tornado-spawning thunderstorms, hoping to better predict tornadoes and to better understand where, when, and how they form.

A normally small and less destructive cousin of the tornado is the "fair weather" waterspout that commonly forms above the warm waters of the Florida Keys and the Great Lakes in summer.

Key Terms

The following terms are listed in the order they appear in the text. Define each. Doing so will aid you in reviewing the material covered in this chapter.

ordinary thunderstorms	gust front
cumulus stage	mesohigh
mature thunderstorm	shelf cloud
dissipating stage	roll cloud
multicell storms	downburst
severe thunderstorms	microburst
macroburst	fulgurite
derecho	sferics
supercell storm	tornadoes
squall line	funnel cloud
dryline	tornado outbreak
flash floods	suction vortices
Mesoscale Convective	Fujita scale
Complexes (MCCs)	mesocyclone
lightning	gustnadoes
thunder	wall cloud
sonic boom	tornado watch
stepped leader	tornado warning
return stroke	hook echo
dart leader	Doppler radar
ball lightning	Doppler lidar
sheet lightning	NEXRAD
heat lightning	waterspout
St. Elmo's Fire	

Questions for Review

1. Describe the stages of development of an ordinary thunderstorm.
2. How do downdrafts form in thunderstorms?
3. What is necessary for a multicell thunderstorm to form?
4. Why do ordinary thunderstorms most frequently form in the afternoon?
5. Explain why ordinary thunderstorms tend to dissipate much sooner than severe thunderstorms.
6. (a) How do gust fronts form?
   (b) What type of weather does a gust front bring when it passes?
7. (a) Describe how a microburst forms.
   (b) Why is the term *wind shear* often used in conjunction with a microburst?
8. Why are severe thunderstorms not very common in polar latitudes?
9. Give a possible explanation for the generation of a prefrontal squall-line thunderstorm.
10. What is the difference between an HP supercell and LP supercell?
11. Where does the highest frequency of thunderstorms occur in the United States? Why there?
12. Why is large hail more common in Kansas than in Florida?

13. Describe one process by which thunderstorms become electrified.
14. Explain how a cloud-to-ground lightning stroke develops.
15. Why is it dangerous to seek shelter under a tree during a thunderstorm?
16. How is thunder produced?
17. Give some average statistics about tornado size, winds, and direction of movement.
18. Why is it suggested that one *not* open windows when a tornado is approaching?
19. What must the updraft do in order to produce a tornado?
20. Why is the central part of the United States more susceptible to tornadoes than any other region of the world?
21. (a) Describe how Doppler radar measures the winds inside a severe thunderstorm.
    (b) Why is Doppler radar superior to conventional radar for this purpose?
22. Describe the atmospheric conditions at the surface and aloft that are responsible for the development of the majority of tornado-spawning thunderstorms.
23. Explain both how and why there is a shift in tornado activity from winter to summer.
24. What conditions lead to the formation of "fair weather" waterspouts?

Questions for Thought

1. Why does the bottom half of a dissipating thunderstorm usually "disappear" before the top?
2. Sinking air warms, yet the downdrafts in a thunderstorm are cold. Why?
3. Explain why squall-line thunderstorms often form ahead of advancing cold fronts but seldom behind them.
4. A forecaster may say that he or she looks for "right moving" thunderstorms when predicting severe weather. What does this mean?
5. Why is the old adage "lightning never strikes twice in the same place" wrong?
6. Suppose while you are on a high mountain ridge a thundercloud passes overhead. What would be the wisest thing to do—stand upright? lie down? or crouch? Explain.
7. If you are confronted in an open field by a large tornado and there is no way that you could outrun it, probably the only thing that you could do would be to run and lie down in a depression. If given the choice, would you run toward your right or left as the tornado approaches? Explain your reasoning.

8. Tornadoes apparently form in the region of a strong updraft, yet they descend from the base of a cloud. Why?
9. In the Fujita scale for classifying tornadoes, into which category would the majority of waterspouts fall?

Problems and Exercises

1. On a map of the United States, place the surface weather conditions (air masses, fronts, and so on) that are necessary for the formation of most tornadoes.
2. On the surface weather map in Fig. 15.46, at which number would you most likely observe a line of thunderstorms forming? Explain why you chose that location.
3. A multi-vortex tornado with a rotational wind speed of 125 knots is moving from southwest to northeast at 30 knots. Assume the suction vortices within this tornado have rotational winds of 100 knots:
   (a) What is the maximum wind speed of this multi-vortex tornado?
   (b) If you are facing the approaching tornado, on which side (northeast, northwest, southwest, or southeast) would the strongest winds be found? the weakest winds? Explain both of your answers.
   (c) According to the Fujita scale (Table 15.1, p. 406), how would this tornado be classified?
4. If you see lightning and 10 seconds later you hear thunder, how far away is the lightning stroke?

FIGURE 15.46
Surface weather map for Exercise 2.

Questions for Exploration

1. On the Blue Skies: College Edition CD-ROM go to the **Severe Weather** section of the CD and click on **"Lightning."**

2. On the Blue Skies: College Edition CD-ROM go to the **Severe Weather** section of the CD and click on **"Microbursts."**

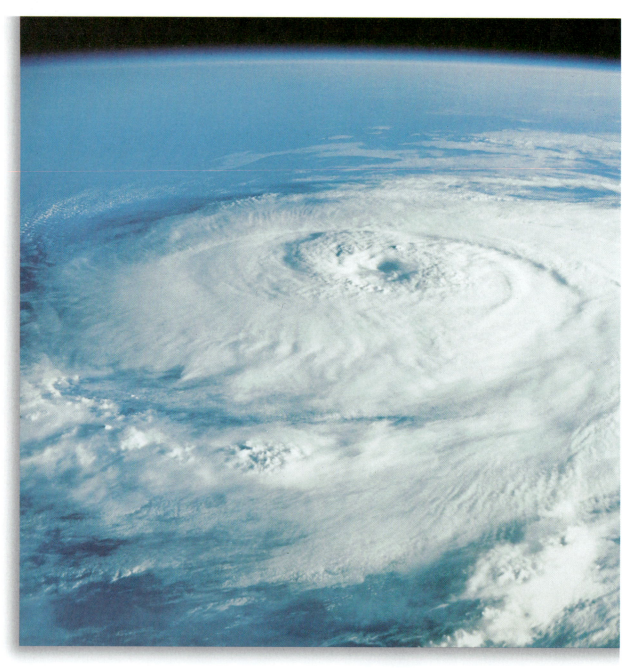

Hurricane Elena over the Gulf of Mexico, about 130 km southwest of Apalachicola, Florida, as photographed from the space shuttle Discovery during September, 1985. With surface winds of 105 knots and a central pressure near 955 mb, Elena rates as a category 3 on the Saffir-Simpson hurricane scale. (Photo: NASA.)

HURRICANES

CONTENTS

On September 18, 1926, as a hurricane approached Miami, Florida, everyone braced themselves for the devastating high winds and storm surge. Just before dawn the hurricane struck with full force—torrential rains, flooding, and easterly winds that gusted to over 100 miles per hour. Then, all of a sudden, it grew calm and a beautiful sunrise appeared. People wandered outside to inspect their property for damage. Some headed for work, and scores of adventurous young people crossed the long causeway to Miami Beach for the thrill of swimming in the huge surf. But the lull lasted for less than an hour. And from the south, ominous black clouds quickly moved overhead. In what seemed like an instant, hurricane force winds from the west were pounding the area and pushing water from Biscayne Bay over the causeway. Many astonished bathers, unable to swim against the great surge of water, were swept to their deaths. Hundreds more drowned as Miami Beach virtually disappeared under the rising wind-driven tide.

orn over warm tropical waters and nurtured by a rich supply of water vapor, the *hurricane* can indeed grow into a ferocious storm that generates enormous waves, heavy rains, and winds that may exceed 150 knots. What exactly are hurricanes? How do they form? And why do they strike the east coast of the United States more frequently than the west coast? These are some of the questions we will consider in this chapter.

TROPICAL WEATHER

In the broad belt around the earth known as the tropics—the region 23½° north and south of the equator—the weather is much different from that of the middle latitudes. In the tropics, the noon sun is always high in the sky, and so diurnal and seasonal changes in temperature are small. The daily heating of the surface and high humidity favor the development of cumulus clouds and afternoon thunderstorms. Most of these are individual thunderstorms that are not severe. Sometimes, however, the storms will align into a narrow band called a *non-squall cluster*. On other occasions, the thunderstorms will align into a row of vigorous convective cells known as a *tropical squall cluster*, or *squall line*. The passage of a squall line is usually noted by a sudden wind gust followed immediately by a heavy downpour that may produce 3 cm of rainfall in about 30 minutes. This deluge is then followed by several hours of relatively steady rainfall. Many of these tropical squall lines are similar to the middle latitude ordinary squall lines described in Chapter 15.

As it is warm all year long in the tropics, the weather is not characterized by four seasons which, for the most part, are determined by temperature variations. Rather, most of the tropics are marked by seasonal differences in precipitation. The greatest cloudiness and precipitation occur during the high-sun period, when the intertropical convergence zone moves into the region. Even during the dry season, precipitation can be irregular, as periods of heavy rain, lasting for several days, may follow an extremely dry spell.

The winds in the tropics generally blow from the east, northeast, or southeast. Because the variation of sea level pressure is normally quite small, drawing isobars on a weather map provides little useful information. Instead of isobars, **streamlines** that depict wind flow are drawn. Streamlines are useful because they show where surface air converges and diverges. Occasionally, the streamlines will be disturbed by a weak trough of low pressure called a **tropical wave**, or **easterly wave**, (see Fig. 16.1).

Tropical waves have wavelengths on the order of 2500 km (1550 mi) and travel from east to west at speeds between 10 and 20 knots. Look at Fig. 16.1 and observe that, on the western side of the trough (heavy dashed line), where easterly and northeasterly surface winds diverge, sinking air produces generally fair weather. On its eastern side, where the southeasterly winds converge, rising air generates showers and thunderstorms. Consequently, the main area of showers forms *behind* the trough. Occasionally, an easterly wave will intensify and grow into a hurricane.

ANATOMY OF A HURRICANE

A **hurricane** is an intense storm of tropical origin, with sustained winds exceeding 64 knots (74 mi/hr), which forms over the warm northern Atlantic and eastern North Pacific oceans. This same type of storm is given different names in different regions of the world. In the western North Pacific, it is called a **typhoon**, in the Philippines a *baguio* (or a typhoon), and in India and Australia a *cyclone*. By international agreement, *tropical cyclone* is the general term for all hurricane-type storms that originate over tropical waters. For simplicity, we will refer to all of these storms as hurricanes.

Figure 16.2 is a satellite image of Hurricane John situated over the eastern Pacific Ocean. The storm is approximately 550 km (340 mi) in diameter, which is about average for hurricanes. The area of broken clouds at the center is its **eye**. John's eye is almost 60 km (37 mi) wide, somewhat larger than the average 20- to 50-km eye diameter. Within the eye, winds are light and clouds are mainly broken. The surface air pressure is very low, nearly 965 mb (28.50 in.).* The dark blotches in the eye are regions where the sky is clear. Notice that the clouds align themselves into spiralling bands (called *spiral rain bands*) that swirl in toward the storm's center, where they wrap them-

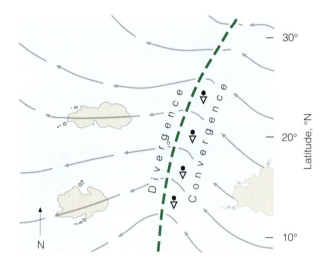

FIGURE 16.1

A tropical wave (also called an easterly wave) as shown by the bending of streamlines—lines that show wind flow patterns. (The heavy dashed line is the axis of the trough.) The wave moves slowly westward, bringing fair weather on its western side and showers on its eastern side.

---

*An extreme low pressure of 870 mb (25.70 in.) was recorded in Typhoon Tip during October, 1979, and Hurricane Gilbert had a pressure reading of 888 mb (26.22 in.) during September, 1988.

selves around the eye. Surface winds increase in speed as they blow counterclockwise and inward toward this center. Adjacent to the eye is the **eye wall**, a ring of intense thunderstorms that whirl around the storm's center and extend upward to almost 15 km above sea level. Within the eye wall, we find the heaviest precipitation and the strongest winds, which, in this storm, are 100 knots, with peak gusts of 115 knots.

If we were to venture from west to east (left to right) through the storm in Fig. 16.2, what might we experience? As we approach the hurricane, the sky becomes overcast with cirrostratus clouds; barometric pressure drops slowly at first, then more rapidly as we move closer to the center. Winds blow from the north and northwest with ever-increasing speed as we near the eye. The high winds, which generate huge waves over 10 m (33 ft) high, are accompanied by heavy rainshowers. As we

WEATHER WATCH

A Pacific typhoon, named Cobra, off the coast of Luzon in the Philippines, inflicted great damage on the United States Navy during the final year of World War II. In just a matter of hours during December, 1944, it claimed 3 destroyers, 146 aircraft, and the lives of 790 sailors.

move into the eye, the air temperature rises, winds slacken, rainfall ceases, and the sky brightens, as middle and high clouds appear overhead. The barometer is now at its lowest point (965 mb), some 50 mb lower than the pressure measured on the outskirts of the storm. The brief respite ends as we enter

FIGURE 16.2

Visible satellite image of Hurricane John. Because this storm is situated over the Pacific Ocean north of the equator, surface winds are blowing counterclockwise about its center (eye). The central pressure of the storm is 965 mb, with sustained winds of 100 knots near its eye.

FIGURE 16.3
A model that shows a vertical view of air motions, clouds, and precipitation in a typical hurricane.

the eastern region of the eye wall. Here, we are greeted by heavy rain and strong southerly winds. As we move away from the eye wall, the pressure rises, the winds diminish, the heavy rain lets up, and eventually the sky begins to clear.

This brief, imaginary venture raises many unanswered questions. Why, for example, is the surface pressure lowest at the center of the storm? And why is the weather clear almost immediately outside the storm area? To help us answer such questions, we need to look at a vertical view, a profile of the hurricane along a slice that runs directly through its center. A model that describes such a profile is given in Fig. 16.3.

The model shows that the hurricane is composed of an organized mass of thunderstorms that are an integral part of the storm's circulation. Near the surface, moist tropical air flows in toward the hurricane's center. Adjacent to the eye, this air rises and condenses into huge thunderstorms that produce heavy rainfall, as much as 25 cm (10 in.) per hour. Near the top of the thunder-

storms, the relatively dry air, having lost much of its moisture, begins to flow outward away from the center. This diverging air aloft actually produces a clockwise (anticyclonic) flow of air several hundred kilometers from the eye. As this outflow reaches the storm's periphery, it begins to sink and warm, inducing clear skies. In the vigorous thunderstorms of the eye wall, the air warms due to the release of large quantities of latent heat. This produces slightly higher pressures aloft, which initiate downward air motion within the eye. As the air subsides, it warms by compression. This process helps to account for the warm air and the absence of thunderstorms in the center of the storm.

As surface air rushes in toward the region of much lower surface pressure, it should expand and cool, and we might expect to observe cooler air around the eye, with warmer air further away. But, apparently, so much heat is added to the air from the warm ocean surface that the surface air temperature remains fairly uniform throughout the hurricane. Figure 16.4

FIGURE 16.4
A radar composite of Hurricane Danny showing several features associated with the storm.

is a 3-dimensional radar composite of Hurricane Danny as it sits near the mouth of the Mississippi River on July 18, 1997.

We are now left with an important question: Where and how do hurricanes form? Although not everything is known about their formation, it is known that certain necessary ingredients are required before a weak tropical disturbance will develop into a full-fledged hurricane.

HURRICANE FORMATION AND DISSIPATION

Hurricanes form over tropical waters where the winds are light, the humidity is high in a deep layer, and the surface water temperature is warm, typically 26°C (80°F) or greater, over a vast area* (see Fig. 16.5). Over the tropical and subtropical North Atlantic and North Pacific oceans these conditions prevail in summer and early fall; hence, the hurricane season normally runs from June through November.

For a mass of unorganized thunderstorms to develop into a hurricane, the surface winds must converge. In the Northern Hemisphere, converging air spins counterclockwise about an area of surface low pressure. Because this type of rotation will not develop on the equator where the Coriolis force is zero (see Chapter 9), hurricanes form in tropical regions, usually between 5° and 20° latitude. (In fact, about two-thirds of all tropical cyclones form between 10° and 20° of the equator.) Convergence may occur along a pre-existing atmospheric disturbance such as a front that has moved into the tropics from middle latitudes. Although the temperature contrast between the air on both sides of the front is gone, developing thunderstorms and converging surface winds may form, especially when the front is accompanied by a cold upper-level trough.

We know from Chapter 11 that the surface winds converge along the intertropical convergence zone (ITCZ). Occasionally, when a wave forms along the ITCZ, an area of low pressure develops, convection becomes organized, and the system grows into a hurricane. Weak convergence also occurs on the eastern side of a tropical wave, where hurricanes have been known to form. In fact, many if not most Atlantic hurricanes can be traced to tropical waves that form over Africa. However, only a small fraction of all of the tropical disturbances that form over the course of a year ever grow into hurricanes.

Even when all of the surface conditions appear near perfect for the formation of a hurricane (e.g., warm water, humid air, converging winds, and so forth), the storm may not develop if the weather conditions aloft are not just right. For example, in the region of the trade winds and especially near latitude 20°, the air is often sinking due to the subtropical high. The sinking air warms and creates an inversion known as the **trade wind inversion**. When the inversion is strong it can

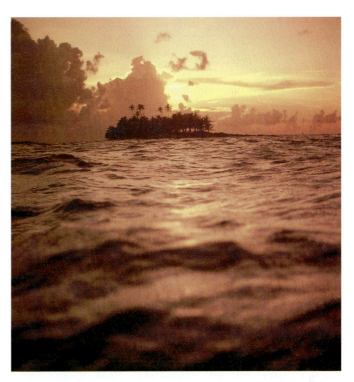

FIGURE 16.5
Hurricanes form over warm tropical waters where the winds are light and the humidity, in a deep layer, is high.

inhibit the formation of intense thunderstorms and hurricanes. Also, hurricanes do not form where the upper-level winds are strong. Strong winds tend to disrupt the organized pattern of convection and disperse the heat, which is necessary for the growth of the storm.

This situation of strong winds aloft typically occurs over the tropical Atlantic during a major El Niño event. As a consequence, during El Niño there are usually fewer Atlantic hurricanes than normal. However, the warmer water of El Niño in the northern tropical Pacific favors the development of hurricanes in that region. During the cold water episode in the tropical Pacific (known as La Niña), winds aloft over the tropical Atlantic usually weaken and become easterly—a condition that favors hurricane development.*

It is important to note that hurricanes tend to form where the upper-level winds are diverging and the air aloft is leaving a vertical column of air more quickly than the air at the surface is entering.

The energy for a hurricane comes from the direct transfer of sensible heat from the warm water into the atmosphere and from the transfer of latent heat from the ocean surface. One idea (known as the *organized convection theory*) proposes that for hurricanes to form, the thunderstorms must become orga-

---

*Not only must the ocean surface temperature be sufficiently warm, but it must be warm through a depth of about 200 m (600 ft) for hurricane formation to be possible.

---

*El Niño and La Niña are covered in Chapter 11 beginning on p. 292.

WEATHER WATCH

nized so that the latent heat that drives the system can be confined to a limited area. If thunderstorms start to organize along the ITCZ or along a tropical wave, and if the trade wind inversion is weak, the stage may be set for the birth of a hurricane. The likelihood of hurricane development is enhanced if the air aloft is unstable. Such instability can be brought on when a cold upper-level trough from middle latitudes moves over the storm area. When this situation occurs, the cumulonimbus clouds are able to build rapidly and grow into enormous thunderstorms (see Fig. 16.6).

Although the upper air is initially cold, it warms rapidly due to the huge amount of latent heat released during condensation. As this cold air is transformed into much warmer air, the air pressure in the upper troposphere above the developing

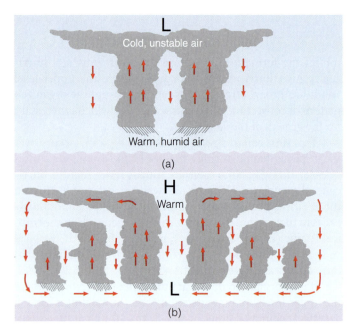

FIGURE 16.6

Development of a hurricane by the organized convection theory. (a) Cold air above an organized mass of tropical thunderstorms generates unstable air and large cumulonimbus clouds. (b) The release of latent heat warms the upper troposphere, creating an area of high pressure. Upper-level winds move outward away from the high. This movement, coupled with the warming of the air layer, causes surface pressures to drop. As air near the surface moves toward the lower pressure, it converges, rises, and fuels more thunderstorms. Soon a chain reaction develops, and a hurricane forms.

storm rises, producing an area of high pressure (see Chapter 9). Now the air aloft begins to move outward, away from the region of developing thunderstorms. This diverging air aloft, coupled with warming of the air layer, causes the surface pressure to drop, and a small area of surface low pressure forms. The surface air begins to spin counterclockwise and in toward the region of low pressure. As it moves inward, its speed increases, just as ice skaters spin faster as their arms are brought in close to their bodies (the conservation of angular momentum). The winds then generate rough seas, which increase the friction on the moving air. This increased friction causes the winds to converge and ascend about the center of the storm.

We now have a chain reaction in progress, or what meteorologists call a *feedback mechanism*. The rising air, having picked up added moisture and warmth from the choppy sea, fuels more thunderstorms and releases more heat, which causes the surface pressure to lower even more. The lower pressure near the center creates a greater friction, more convergence, more rising air, more thunderstorms, more heat, lower surface pressure, stronger winds, and so on until a full-blown hurricane is born.*

As long as the upper-level outflow of air is greater than the surface inflow, the storm will intensify and the surface pressure will drop. Because the air pressure within the system is controlled to a large extent by the warmth of the air, the storm will intensify only up to a point. The controlling factors are the temperature of the water and the release of latent heat. Consequently, when the storm is literally full of thunderstorms, it will use up just about all of the available energy, so that air temperature will no longer rise and pressure will level off. Because there is a limit to how intense the storm can become, peak wind gusts seldom exceed 200 knots. When the converging surface air near the center exceeds the outflow at the top, surface pressure begins to increase, and the storm dies out.

An alternative to the organized convection theory proposes that a hurricane is like a heat engine. In a heat engine, heat is taken in at a high temperature, converted into work, then ejected at a low temperature. In a hurricane, small swirling eddies transfer sensible and latent heat from the ocean surface into the overlying air. The warmer the water and the greater the wind speed, the greater the transfer of sensible and latent heat. As the air sweeps in toward the center of the storm, the rate of heat transfer increases because the wind speed increases toward the eye wall. Similarly, the higher wind speeds cause greater evaporation rates, and the overlying air becomes nearly saturated.

Near the eye wall, turbulent eddies transfer the warm moist air upward, where the water vapor condenses to form clouds.

---

*This feedback process, wherein cumulus convection becomes sufficiently organized for the growth and development of a hurricane, is called *conditional instability of the second kind*, or CISK. Conditional instability of the first kind refers to the type of instability described in Fig. 7.7, p. 164.

The release of latent heat inside the clouds causes the air temperature in the region of the eye wall to be much higher than the air temperature at the same altitude further out, away from the storm center. This situation causes a horizontal pressure gradient aloft that induces the air to move outward, away from the storm center in the anvils of the cumulonimbus clouds. At the top of the storm, heat is lost by clouds radiating infrared energy to space. Hence, in a hurricane, heat is taken in near the ocean surface, converted to kinetic energy (energy of motion) or wind, and lost at its top through radiational cooling.

In a heat engine, the amount of work done is proportional to the difference in temperature between its input and output region. The maximum strength a hurricane can achieve is determined by the difference in temperature between the ocean surface and the top of its clouds. As a consequence, the warmer the ocean surface, the lower the minimum pressure of the storm, and the higher its winds. Presently, there is much debate whether hurricanes are driven by the organized convection process, by the heat engine process, or by a combination of the two processes.

If the hurricane remains over warm water, it may survive for a long time. For example, Hurricane Tina (1992) traveled for thousands of kilometers over warm, tropical waters and maintained hurricane force winds for 24 days, making Tina one of the longest-lasting North Pacific hurricanes on record. However, most hurricanes last for less than a week; they weaken rapidly when they travel over colder water and lose their heat source. They also dissipate rapidly over land. Here, not only is their energy source removed, but their winds decrease in strength (due to the added friction) and blow more directly into the center, causing the central pressure to rise.

Hurricane Stages of Development   Hurricanes go through a set of stages from birth to death. Initially, the mass of thunderstorms with only a slight wind circulation is known as a **tropical disturbance**, or *tropical wave*. The tropical disturbance becomes a **tropical depression** when the winds increase to between 20 and 34 knots and several closed isobars appear about its center on a surface weather map. When the isobars are packed together and the winds are between 35 and 64 knots, the tropical depression becomes a **tropical storm**. The tropical storm is classified as a *hurricane* only when its winds exceed 64 knots (74 mi/hr).

Figure 16.7 shows four tropical systems in various stages of development. Moving from east to west, we see a weak tropical disturbance (a tropical wave) crossing over Panama. Further west, a tropical depression is organizing around a developing center with winds less than 25 knots. In a few days, this system will develop into Hurricane Gilma. Further west is a full-fledged hurricane with peak winds in excess of 110 knots. The swirling band of clouds to the northwest is Emilia;

FIGURE 16.7
Visible satellite image showing four tropical systems, each in a different stage of its life cycle.

once a hurricane (but now with winds less than 40 knots), it is rapidly weakening over colder water.

BRIEF REVIEW

Before reading the next several sections, here is a review of some of the important points about hurricanes.

- Hurricanes are tropical cyclones, comprised of an organized mass of thunderstorms.

- Hurricanes have peak winds about a central core (eye) that exceed 64 knots (74 mi/hr).

- Hurricanes form over warm tropical waters, where light surface winds converge, the humidity is high in a deep layer, and the winds aloft are weak.

- Hurricanes derive their energy from the warm tropical water and from the latent heat released as water vapor condenses into clouds.

- Hurricanes grow stronger as long as the air aloft moves outward, away from the storm center more quickly than the surface air moves in toward the center.

- Hurricanes dissipate rapidly when they move over colder water or over a large landmass.

Up to this point, it is probably apparent that tropical cyclones called hurricanes are similar to middle latitude cyclones in that, at the surface, both have central cores of low pressure and winds that spiral counterclockwise about their respective centers (Northern Hemisphere). However, there are many differences between the two systems, which are described in the Focus section on p. 427.

Hurricane Movement   Figure 16.8 shows where most hurricanes are born and the general direction in which they move. Notice that they form over tropical oceans, except in the South Atlantic and in the eastern South Pacific. The surface water temperatures are too cold in these areas for their development. It is also possible that the unfavorable location of the ITCZ during the Southern Hemisphere's warm season discourages their development.

Hurricanes that form over the North Pacific and North Atlantic are steered by easterly winds and move west or northwestward at about 10 knots for a week or so. Gradually, they swing poleward around the subtropical high, and when they move far enough north, they become caught in the westerly flow, which curves them to the north or northeast. In the middle latitudes, the hurricane's forward speed normally increases, sometimes to more than 50 knots. The actual path of a hurricane (which appears to be determined by the structure of the storm and the storm's interaction with the environment) may

FIGURE 16.8
Regions where tropical storms form, the names given to storms, and the typical paths they take.

How Do Hurricanes Compare with Middle Latitude Storms?

By now, it should be apparent that a hurricane is much different from the mid-latitude cyclone that we discussed in Chapter 13. A hurricane derives its energy from the warm water and the latent heat of condensation, whereas the mid-latitude storm derives its energy from horizontal temperature contrasts. The vertical structure of a hurricane is such that its central column of air is warm from the surface upward; consequently, hurricanes are called *warm-core lows*. A hurricane weakens with height, and the area of low pressure at the surface may actually become an area of high pressure above 12 km (40,000 ft). Mid-latitude cyclones, on the other hand, usually intensify with increasing height, and a cold upper-level low or trough exists to the west of the surface system. A hurricane usually contains an eye where the air is sinking, while mid-latitude cyclones are characterized by centers of rising air. Hurricane winds are strongest near the surface, whereas the strongest winds of the mid-latitude storm are found aloft in the jet stream.

Further contrasts can be seen on a surface weather map. Figure 1 shows Hurricane Allen over the Gulf of Mexico and a mid-latitude storm north of New England. Around the hurricane, the isobars are more circular, the pressure gradient is much steeper, and the winds are stronger. The hurricane has no fronts and is smaller (although Allen is larger than most hurricanes). There are similarities between the two systems: Both are areas of surface low pressure, with winds moving counterclockwise about their respective centers.

It is interesting to note that some northeasters (winter storms that

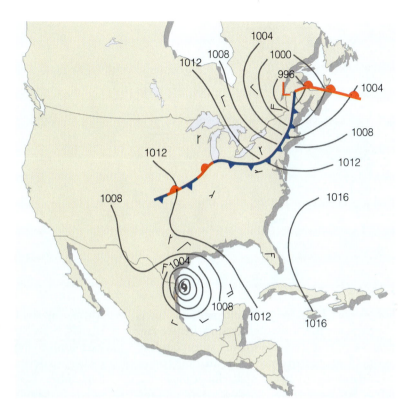

FIGURE 1

Surface weather map for the morning of August 9, 1980, showing Hurricane Allen over the Gulf of Mexico and a middle latitude storm system north of New England.

move northeastward along the coastline of North America, bringing with them heavy precipitation, high surf, and strong winds) may actually possess some of the characteristics of a hurricane. For example, a particularly powerful northeaster during January, 1989, was observed to have a cloud-free eye, with surface winds in excess of 85 knots spinning about a warm inner core. Moreover, some *polar lows*—lows that develop over polar waters during winter—may exhibit many of the observed characteristics of a hurricane, such as a symmetric band of thunderstorms spiraling inward around a cloud-free eye, a warm-core area of low pressure, and

strong winds near the storm's center. In fact, when surface winds within these polar storms reach 58 knots, they are sometimes referred to as *Arctic hurricanes*.

Even though hurricanes weaken rapidly as they move inland, their counterclockwise circulation may draw in air with contrasting properties. If the hurricane links with an upper-level trough, it may actually become a mid-latitude cyclone, which is what happened to Hurricane Hazel in 1954, to Agnes in 1972, and to Hugo in 1989.

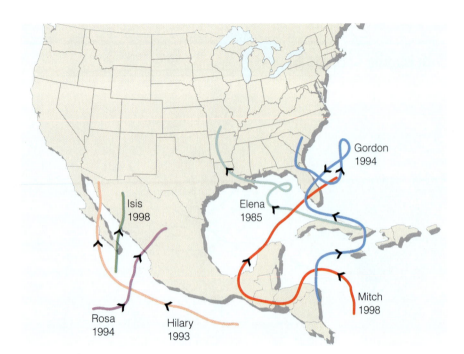

FIGURE 16.9
Some erratic paths taken by hurricanes.

vary considerably. Some take erratic paths and make odd turns that occasionally catch weather forecasters by surprise (see Fig. 16.9). There have been many instances where a storm heading directly for land suddenly veered away and spared the region from almost certain disaster. As an example, Hurricane Elena, with peak winds of 90 knots, moved northwestward into the Gulf of Mexico on August 29, 1985. It then veered eastward toward the west coast of Florida. After stalling offshore, it headed northwest. After weakening, it then moved onshore near Biloxi, Mississippi, on the morning of September 2.

As we saw in an earlier section, many hurricanes form off the coast of Mexico over the North Pacific. In fact, this area usually spawns about eight hurricanes each year, which is slightly more than the yearly average of six storms born over the tropical North Atlantic. Eastern North Pacific hurricanes normally move westward, away from the coast, and so little is heard about them. When one does move northwestward, it normally weakens rapidly over the cool water of the North Pacific. Occasionally, however, one will curve northward or even northeastward and slam into Mexico, causing destructive flooding. Hurricane Tico left 25,000 people homeless and caused an estimated $66 million in property damage after passing over Mazatlán, Mexico, in October, 1983. The remains of Tico even produced record rains and flooding in Texas and Oklahoma. Even less frequently, a hurricane will stray far enough north to bring summer rains to southern California and Arizona, as did the remains of Hurricane Boris during June, 1990, and Hurricane Lester during August, 1992.

The Hawaiian Islands, which are situated in the central North Pacific between about 20° and 23°N, appear to be in the direct path of many eastern Pacific hurricanes and tropical storms. By the time most of these storms have reached the islands, however, they have weakened considerably, and pass harmlessly to the south or northeast. The exceptions were Hurricane Iwa during November, 1982, and Hurricane Iniki during September, 1992. Iwa lashed part of Hawaii with 100-knot winds and huge surf, causing an estimated $312 million in damages. Iniki, the worst hurricane to hit Hawaii this century, battered the island of Kauai with torrential rain, sustained winds of 114 knots that gusted to 140 knots, and 20-foot waves that crashed over coastal highways. Major damage was sustained by most of the hotels and about 50 percent of the homes on the island. Iniki (the costliest hurricane in Hawaiian history with damage estimates of $1.8 billion) flattened sugar cane fields, destroyed the macadamia nut crop, injured about 100 people, and caused at least 7 deaths.

Hurricanes that form over the tropical North Atlantic also move westward or northwestward on a collision course with Central or North America. Most hurricanes, however, swing away from land and move northward, parallel to the coastline of the United States. A few storms, perhaps three per year, move inland, bringing with them high winds, huge waves, and torrential rain that may last for days. Figure 16.10 is a collection of infrared satellite images of Hurricane Georges, showing its path from September 18 to September 28, 1998. As Georges moved westward it ravaged the large Caribbean Islands, causing extensive damage and taking the lives of more than 350 people. After raking the Florida Keys with high winds and heavy rain, its path curved toward the northwest, where it eventually slammed into Mississippi with torrential rains and winds exceeding 100 knots. Four people in the United States died due to Hurricane Georges.

FIGURE 16.10
A composite of infrared satellite images of Hurricane Georges from September 18 to September 28, 1998, that shows its westward trek across the Caribbean, then northward into the United States.

A hurricane moving northward over the Atlantic will normally survive as a hurricane for a longer time than will its counterpart at the same latitude over the eastern Pacific. The reason is, of course, that the surface water of the Atlantic is much warmer (see Chapter 11).

Destruction and Warning When a hurricane is approaching from the east, its highest winds are usually on its north (poleward) side. The reason for this phenomenon is that the winds that push the storm along add to the winds on the north side and subtract from the winds on the south (equator) side. Hence, a hurricane with 110-knot winds moving westward at 10 knots will have 120-knot winds on its north side and 100-knot winds on its south side.

The same type of reasoning can be applied to a northward-moving hurricane. For example, as Hurricane Gloria moved northward along the coast of Virginia on the morning of September 27, 1985 (see Fig. 16.11), winds of 75 knots were swirling counterclockwise about its center. Because the storm was moving northward at about 25 knots, sustained winds on its eastern (right) side were about 100 knots, while on its western (left) side—on the coast—the winds were only about 50 knots. Even so, these winds were strong enough to cause significant beach erosion along the coasts of Maryland, Delaware, and New Jersey.

Even though Hurricane Gloria is moving northward in Fig. 16.11, there is a net transport of water directed eastward toward the coast. To understand this behavior, recall from Chapter 11 that as the wind blows over open water, the water beneath is set in motion. If we imagine the top layer of water to be broken into a series of layers, then we find each layer moving to the right of the layer above (Northern Hemisphere). This type of movement (bending) of water with depth (called the *Ekman Spiral*) causes a net transport of water (known as **Ekman transport**) to the right of the surface wind. Hence, the north wind on Hurricane Gloria's left (western) side causes a

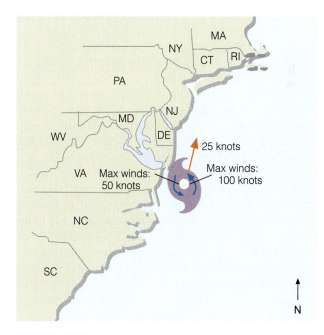

FIGURE 16.11
Hurricane Gloria on the morning of September 27, 1985. Moving northward at 25 knots, Gloria has sustained winds of 100 knots on its right side and 50 knots on its left side. (The central pressure of the storm is about 945 mb.)

FIGURE 16.12
Although the high winds of a hurricane can inflict a great deal of damage, the majority of destruction is usually brought about by huge waves and flooding. Here, Hurricane Kate slams into Key West, Florida, on November 20, 1985. The hurricane was responsible for seven deaths in the United States and over $1 billion in damages.

net transport of water toward the shore. Here, the water piles up and rapidly inundates the region.

The high winds of a hurricane also generate large waves, sometimes 10 to 15 m high. These waves move outward, away from the storm, in the form of *swells* that carry the storm's energy to distant beaches. Consequently, the effects of the storm may be felt days before the hurricane arrives.

Although the hurricane's high winds inflict a great deal of damage, it is the huge waves, high seas, and *flooding** that normally cause most of the destruction (see Fig. 16.12). The flooding is due, in part, to winds pushing water onto the shore and to the heavy rains, which may exceed 63 cm (25 in.) in 24 hours. Flooding is also aided by the low pressure of the storm. The region of low pressure allows the ocean level to rise (perhaps half a meter), much like a soft drink rises up a straw as air is withdrawn. (A drop of one millibar in air pressure produces a rise of one centimeter in ocean level.) The combined effect of high water (which

is usually well above the high-tide level), high winds, and the net Ekman transport toward the coast, produces the **storm surge**— an abnormal rise of several meters in the ocean level—which inundates low-lying areas and turns beachfront homes into piles of splinters (see Fig. 16.13). The storm surge is particularly damaging when it coincides with normal high tides.

Considerable damage may also occur from hurricane-spawned tornadoes. About one-fourth of the hurricanes that strike the United States produce tornadoes. The exact mechanism by which these tornadoes form is not yet known; however, studies suggest that surface topography may play a role by initiating the convergence (and, hence, rising) of surface air. Moreover, tornadoes tend to form in the right front quadrant of an advancing hurricane, where vertical wind speed shear is greatest. Studies also suggest that swathlike areas of extreme damage once attributed to tornadoes may actually be due to downbursts associated with the large thunderstorms around the eye wall. (Because of the potential destruction and loss of lives that hurricanes can inflict, attempts have been made to reduce their winds by seeding them. More on this topic is given in the Focus section on p. 431.)

---

*Hurricanes may sometimes have a beneficial aspect, in the sense that they can provide much needed rainfall in drought-stricken areas.

Modifying Hurricanes

Attempts have been made to reduce a hurricane's winds by seeding them with silver iodide. The idea is to seed the clouds just outside the eye wall with just enough artificial ice nuclei so that the latent heat given off will stimulate cloud growth in this area of the storm. These clouds, which grow at the expense of the eye wall thunderstorms, actually form a new eye wall farther away from the hurricane's center. As the storm center widens, its pressure gradient should weaken, which may cause its spiraling winds to decrease in speed. During project

STORMFURY, a joint effort of the National Oceanic and Atmospheric Administration (NOAA) and the U.S. Navy, several hurricanes were seeded by aircraft. In 1963, shortly after Hurricane Beulah was seeded with silver iodide, surface pressure in the eye began to rise and the region of maximum winds moved away from the storm's center. Even more encouraging results were obtained from the multiple seeding of Hurricane Debbie in 1969. After one day of seeding, Debbie showed a 30 percent reduction in maximum winds. However, the

question remains: Would the winds have lowered naturally had the storm not been seeded? One study even casts doubts upon the theoretical basis for this kind of hurricane modification because hurricanes appear to contain too little supercooled water and too much natural ice. Consequently, there are many uncertainties about the effectiveness of seeding hurricanes in an attempt to reduce their winds, and all endeavors to modify hurricanes have been discontinued since the 1970s.

In examining the extensive damage wrought by Hurricane Andrew during August, 1992, professor Theodore Fujita of the University of Chicago theorized that the areas of most severe damage might have been caused by **spin-up vortices (mini-swirls)**—small whirling eddies perhaps 30 to 100 m in diameter that occur in narrow bands. Lasting for about 10 seconds, the vortices appear to form in a region of strong wind speed shear in the hurricane's eye wall, where the air is rapidly rising. As intense updrafts stretch the vortices vertically, they shrink horizontally, which induces them to spin faster (perhaps as

fast as 70 knots), much like skaters spin faster as their arms are pulled inward. When the rotational winds of a vortice are added to the hurricane's steady wind, the total wind speed over a relatively small area may increase substantially. In the case of Hurricane Andrew, isolated wind speeds may have reached 174 knots (200 mi/hr) over narrow stretches of south Florida.

With the aid of ship reports, satellites, radar, buoys, and reconnaissance aircraft, the location and intensity of hurricanes are pinpointed and their movements carefully monitored. When a hurricane poses a direct threat to an area, a **hurricane**

FIGURE 16.13
When a storm surge moves in at high tide it can inundate and destroy a wide swath of coastal lowlands.

WEATHER WATCH

Even the aftermath of a hurricane can be deadly. In the Gulf states one of the greatest fears in the cleanup effort is encountering one of the poisonous snakes that seem to find their way into various nooks and crannies.

watch is issued, typically 24 to 48 hours before the storm arrives, by the National Hurricane Center in Miami, Florida, or by the Pacific Hurricane Center in Honolulu, Hawaii. When it appears that the storm will strike an area within 24 hours, a **hurricane warning** is issued. Along the east coast of North America, the warning is accompanied by a probability. The probability gives the percent chance of the hurricane's center passing within 105 km (65 mi) of a particular community. The warning is designed to give residents ample time to secure property and, if necessary, to evacuate the area.

A hurricane warning is issued for a rather large coastal area, usually about 550 km (342 mi) in length. Since the average swath of hurricane damage is normally about one-third this length, much of the area is "overwarned." As a consequence, many people in a warning area feel that they are needlessly forced to evacuate. The evacuation order is given by local authorities* and typically only for those low-lying coastal areas directly affected by the storm surge. People at higher elevations or further from the coast are not usually requested to leave, in part because of the added traffic problems this would create. This issue has engendered some controversy in the wake of Hurricane Andrew, since its winds were so devastating over inland south Florida during August, 1992. The time it takes to complete an evacuation puts a special emphasis on the timing and accuracy of the warning.

As new hurricane-prediction models are implemented, and as our understanding of the nature of hurricanes increases, improved forecasts of hurricane movement and intensification should become available.

Ample warning by the National Weather Service probably saved the lives of many people as Hurricane Allen moved onshore along the south Texas coast during the morning of August 10, 1980. The storm formed over the warm tropical Atlantic and moved westward on a rampage through the Caribbean, where it killed almost 300 people and caused extensive damage. After raking the Yucatán Peninsula with 150-knot winds, Allen howled into the warm Gulf of Mexico. It reintensified and its winds increased to 160 knots. Gale-force winds reached outward for 320 km (200 mi) north of its center. As it approached the south Texas coast, it was one of the greatest

storms to ever enter that area. The central pressure of the storm dropped to a low of 899 mb (26.55 in.). Up until this time, only the 1935 Labor Day storm that hit the Florida Keys with a pressure of 892 mb (26.35 in.) was stronger. But Allen's path became wobbly and it stalled offshore just long enough to lose much of its intensity. It moved sluggishly inland on the morning of August 10. Once it made *landfall,* * it quickly became a tropical storm with peak winds of less than 50 knots.

In recent years, the annual hurricane death toll in the United States has averaged between 50 and 100 persons, although over 200 people died in Mexico when Hurricane Gilbert slammed the Gulf Coast of Mexico during September, 1988. This relatively low total is partly due to the advance warning provided by the National Weather Service and to the fact that only a few really intense storms have reached land during the past 30 years. However, as the population density continues to increase in vulnerable coastal areas, the potential for a hurricane-caused disaster continues to increase also.

Hurricane Camille (1969) stands out as one of the most intense hurricanes to reach the coastline of the United States in recent decades. With a central pressure of 909 mb, tempestuous winds reaching 160 knots, and a storm surge more than 7 m (23 ft) above the normal high-tide level, Camille unleashed its fury on Mississippi, destroying thousands of buildings. During its rampage, it caused an estimated $1.5 billion in property damage and took more than 200 lives.

During September, 1989, Hurricane Hugo was born as a cluster of thunderstorms became a tropical depression off the coast of Africa, southeast of the Cape Verde Islands. The storm grew in intensity, tracked westward for several days, then turned northwestward, striking the island of St. Croix with sustained winds of 125 knots. After passing over the eastern tip of Puerto Rico, this large, powerful hurricane took aim at the coastline of South Carolina. With maximum winds estimated at about 120 knots (138 mi/hr), and a central pressure near 934 mb, Hugo made landfall near Charleston, South Carolina, about midnight on September 21st (see Figs. 16.14 and 16.15). The high winds and storm surge, which ranged between 2.5 and 6 m (8 and 20 ft), hurled a thundering wall of water against the shore. This knocked out power, flooded streets, and caused widespread destruction to coastal communities (see Fig. 16.16). The total damage in the United States attributed to Hugo was over $7 billion, with a death toll of 21 in the United States and 49 overall. But Hugo does not even come close to the costliest hurricane on record—that dubious distinction goes to Hurricane Andrew.

On August 21, 1992, as tropical storm Andrew churned westward across the Atlantic it began to weaken, prompting some forecasters to surmise that this tropical storm would never grow to hurricane strength. But Andrew moved into a region favorable for hurricane development. Even though it was out-

---

*In the state of New Jersey, the Board of Casinos and the Governor must be consulted before an evacuation can be ordered.

*Landfall is the position along a coast where the center of a hurricane passes from ocean to land.

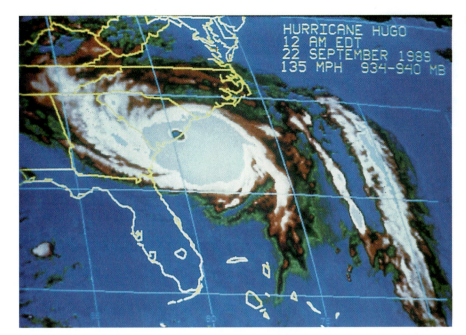

FIGURE 16.14
A color-enhanced infrared satellite image of Hurricane Hugo with its eye over the coast near Charleston, South Carolina.

side the tropics near latitude 25°N, warm surface water and weak winds aloft allowed Andrew to intensify rapidly. And in just two days Andrew's winds increased from 45 knots to 122 knots, turning an average tropical storm into one of the most intense hurricanes to strike Florida this century (see Table 16.1).

With steady winds of 126 knots (145 mi/hr) and a powerful storm surge, Andrew made landfall south of Miami on the morning of August 24 (see Fig. 16.17). The eye of the storm moved over Homestead, Florida. Andrew's fierce winds completely devastated the area (see Fig. 16.18), as 50,000 homes were destroyed, trees were leveled, and steel-reinforced tie beams weighing tons were torn free of townhouses and hurled as far as several blocks. Swaths of severe damage led scientists to postulate that peak winds may have approached 174 knots (200 mi/hr). Such winds may have occurred with *spin-up vortices* (swirling eddies of air) that added substantially to the storm's wind speed. In an instant, a wind gust of 142 knots (164 mi/hr) blew down a radar dome and inactivated several satellite dishes on the roof of the National Hurricane Center in Coral Gables. Observations reveal that some of Andrew's destruction may have been caused by microbursts in the severe thunderstorms of the eyewall. The hurricane roared westward across Southern Florida, weakened slightly, then regained strength over the warm Gulf of Mexico. Surging northwestward, Andrew slammed into Louisiana with 120-knot winds on the evening of August 25.

All told, Hurricane Andrew was the costliest natural disaster ever to hit the United States. It destroyed or damaged over 200,000 homes and businesses, left more than 160,000 people homeless, caused over $30 billion in damages, and took 53 lives, including 41 in Florida. Although Andrew may well be the most expensive hurricane on record, it is far from the deadliest.

Before the era of satellites and radar, catastrophic losses of life had occurred. In 1900, more than 6000 people lost their lives when a hurricane slammed into Galveston, Texas, with a huge storm surge. Most of the deaths occurred in the low-lying coastal regions as flood waters pushed inland. In October, 1893, nearly 2000 people perished on the Gulf Coast

FIGURE 16.15
Digitized radar display showing Hurricane Hugo's eye and eye wall with its heavy precipitation (dark red area), as the center of the storm moves onshore near Charleston (CHS), South Carolina.

(a)

(b)

FIGURE 16.16
Beach homes at Folly Beach, South Carolina, (a) before and (b) after Hurricane Hugo.

RANK	HURRICANE	YEAR	CENTRAL PRESSURE (millibars/inches)	CATEGORY	DEATH TOLL
1	Florida (Keys)	1935	892/26.35	5	408
2	Camille	1969	909/26.85	5	256
3	Andrew	1992	922/27.23	4	53
4	Florida (Keys)/South Texas	1919	927/27.37	4	>600*
5	Florida (Lake Okeechobee)	1928	929/27.43	4	1836
6	Donna	1960	930/27.46	4	50
7	Texas (Galveston)	1900	931/27.49	4	>6000
8	Louisiana (Grand Isle)	1909	931/27.49	4	350
9	Louisiana (New Orleans)	1915	931/27.49	4	275
10	Carla	1961	931/27.49	4	46
11	Hugo	1989	934/27.58	4	49
12	Florida (Miami)	1926	935/27.61	4	243

TABLE 16.1   The Twelve Most Intense Hurricanes (at Landfall) to Strike the United States from 1900 to 1998

*More than 500 of this total were lost at sea on ships. (The > symbol means "greater than.")

FIGURE 16.17
A color-enhanced infrared satellite image shows Hurricane Andrew moving across south Florida on the morning of August 24, 1992. The storm has a central pressure of 932 mb (27.52 in.) and sustained winds of 126 knots (145 mi/hr).

FIGURE 16.18
The remains of the Naranja Lakes Condominium near Homestead, Florida, after Hurricane Andrew devastated parts of south Florida on August 24, 1992.

of Louisiana as a giant storm surge swept that region. Spectacular losses are not confined to the Gulf Coast. Nearly 1000 people lost their lives in Charleston, South Carolina, during August of the same year. But these statistics are small compared to the more than 300,000 lives taken as a killer cyclone and storm surge ravaged the coast of Bangladesh with flood waters in 1970. Again in April, 1991, a similar cyclone devastated the area with reported winds of 127 knots and a storm surge of 7 m (23 ft). In all, the storm destroyed 1.4 million houses and killed 140,000 people and 1 million cattle. Unfortunately, the potential for a repeat of this type of disaster

remains high in Bangladesh, as many people live along the relatively low, wide flood plain that slopes outward to the bay. And, historically, this region is in a path frequently taken by tropical cyclones (look back at Fig. 16.8, p. 426).

Even with modern satellite observation techniques, hurricane disasters can reach epic proportions. For example, Hurricane Mitch during late October, 1998, became the most deadly hurricane to strike the Western Hemisphere since the Great Hurricane of 1780, which claimed approximately 22,000 lives in the eastern Caribbean. Mitch's high winds, huge waves (estimated maximum height 44 ft), and torrential rains destroyed

TABLE 16.2	Saffir-Simpson Hurricane Damage-Potential Scale							
SCALE NUMBER (CATEGORY)	CENTRAL PRESSURE		WINDS		STORM SURGE		DAMAGE
	mb	in.	mi/hr	knots	ft	m	
1	≥980*	≥28.94	74–95	64–82	4–5	~1.5	Damage mainly to trees, shrubbery, and unanchored mobile homes
2	965–979	28.50–28.91	96–110	83–95	6–8	~2.0–2.5	Some trees blown down; major damage to exposed mobile homes; some damage to roofs of buildings
3	945–964	27.91–28.47	111–130	96–113	9–12	~2.5–4.0	Foliage removed from trees; large trees blown down; mobile homes destroyed; some structural damage to small buildings
4	920–944	27.17–27.88	131–155	114–135	13–18	~4.0–5.5	All signs blown down; extensive damage to roofs, windows, and doors; complete destruction of mobile homes; flooding inland as far as 10 km (6 mi); major damage to lower floors of structures near shore
5	<920	<27.17	>155	>135	>18	>5.5	Severe damage to windows and doors; extensive damage to roofs of homes and industrial buildings; small buildings overturned and blown away; major damage to lower floors of all structures less than 4.5 m (15 ft) above sea level within 500 m of shore.

*Symbol > means "greater than"; < means "less than"; ≥ means "equal to or greater than"; ~ means "approximately equal to."

vast regions of coastal Central America (for Mitch's path, see Fig. 16.9, p. 428). In the mountainous regions of Honduras and Nicaragua, rainfall totals from the storm may have reached 190 cm (75 in.). The heavy rains produced floods and deep mudslides that swept away entire villages, including the inhabitants. Mitch caused over $5 billion in damages, destroyed hundreds of thousands of homes, and killed over 11,000 people. More than 3 million people were left homeless or were otherwise severely affected by this deadly storm.

In an effort to estimate the possible damage a hurricane's sustained winds and storm surge could do to a coastal area, the **Saffir-Simpson scale** was developed (see Table 16.2). The scale numbers (which range from 1 to 5) are based on actual conditions at some time during the life of the storm. As the hurricane intensifies or weakens, the category, or scale number, is reassessed accordingly. Major hurricanes are classified as Category 3 and above. In the western Pacific, a typhoon with sustained winds of at least 130 knots (150 mi/hr)—at the upper

end of the wind speed range in category 4 on the Saffir-Simpson scale—is called a **super-typhoon**.

Figure 16.19 shows the number of hurricanes that have made landfall along the coastline of the United States from 1900 to 1998. Out of a total of 162 hurricanes striking the American coastline, 65 (40 percent) were major hurricanes. Hence, along the Gulf and Atlantic coasts, on the average, about five hurricanes make landfall every three years, two of which are major hurricanes with winds in excess of 95 knots (110 mi/hr) and a storm surge exceeding 2.5 m.

NAMING HURRICANES

In reading the previous sections, you noticed that hurricanes were assigned names. (The name is assigned when the storm reaches tropical storm strength.) Before this practice was started, hurricanes were identified according to their latitude and longitude. This method was confusing, especially when two or more storms were present over the same ocean. To reduce the confusion, hurricanes were identified by letters of the alphabet. During World War II, names like Able and Baker were used. (These names correspond to the radio code words associated with each letter of the alphabet.) This method also seemed cumbersome so, beginning in 1953, the National Weather Service began using female names to identify hurricanes. The list of names for each year was in alphabetical order, so that the name of the season's first storm began with the letter *A*, the second with *B*, and so on.

From 1953 to 1977, only female names were used. However, beginning in 1978, hurricanes in the eastern Pacific were alternately assigned female and male names, but not just English names, as Spanish and French ones were used too. This practice was started for North Atlantic hurricanes in 1979. (Remember that a storm only gets a name when it reaches tropical storm strength.) Once a storm has caused great damage and it becomes infamous as a Category 3 or higher, its name is retired for at least ten years. Table 16.3 gives the proposed list of names for both North Atlantic and eastern Pacific hurricanes.

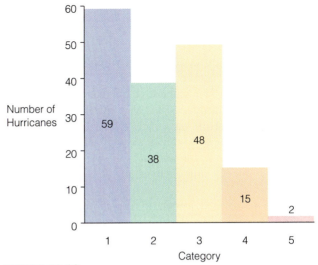

FIGURE 16.19

The number of hurricanes (by each category) that made landfall along the coastline of the United States from 1900 to 1998. All of the hurricanes struck the Gulf or Atlantic coasts. Categories 3, 4, and 5 are considered major hurricanes.

TABLE 16.3   Names for Hurricanes

NORTH ATLANTIC HURRICANE NAMES				EASTERN PACIFIC HURRICANE NAMES			
1999	2000	2001	2002	1999	2000	2001	2002
Arlene	Alberto	Allison	Arthur	Adrian	Aletta	Adolph	Alma
Bret	Beryl	Barry	Bertha	Beatriz	Bud	Barbara	Boris
Cindy	Chris	Chantal	Christobal	Calvin	Carlotta	Cosme	Cristina
Dennis	Debby	Dean	Dolly	Dora	Daniel	Dalilia	Douglas
Emily	Ernesto	Erin	Edouard	Eugene	Emilia	Erick	Elida
Floyd	Florence	Felix	Fay	Fernanda	Fabio	Flossie	Fausto
Gert	Gordon	Gabrielle	Gustav	Greg	Gilma	Gil	Genevieve
Harvey	Helene	Humberto	Hanna	Hilary	Hector	Henriette	Hernan
Irene	Isaac	Iris	Isidore	Irwin	Ileana	Israel	Iselle
Jose	Joyce	Jerry	Josephine	Jova	John	Juliette	Julio
Katrina	Keith	Karen	Kyle	Kenneth	Kristy	Kiko	Kenna
Lenny	Leslie	Lorenzo	Lili	Lidia	Lane	Lorena	Lowell
Maria	Michael	Michelle	Marco	Max	Miriam	Manual	Marie
Nate	Nadine	Noel	Nana	Norma	Norman	Narda	Norbert
Ophelia	Oscar	Olga	Omar	Otis	Olivia	Octave	Odile
Philippe	Patty	Pablo	Paloma	Pilar	Paul	Priscilla	Polo
Rita	Rafael	Rebekah	Rene	Ramon	Rosa	Raymond	Rachel
Stan	Sandy	Sebastien	Sally	Selma	Sergio	Sonia	Simon
Tammy	Tony	Tanya	Teddy	Todd	Tara	Tico	Trudy
Vince	Valerie	Van	Vicky	Veronica	Vicente	Velma	Vance
Wilma	William	Wendy	Wilfred	Wiley	Willa	Wallis	Winnie
				Xina	Xavier	Xina	Xavier
				York	Yolanda	York	Yolanda
				Zelda	Zeke	Zelda	Zeke

SUMMARY

Hurricanes are tropical cyclones with winds that exceed 64 knots (74 mi/hr) and blow counterclockwise about their centers in the Northern Hemisphere. A hurricane consists of a mass of organized thunderstorms that spiral in toward the extreme low pressure of the storm's eye. The most intense thunderstorms, the heaviest rain, and the highest winds occur outside the eye, in the region known as the eye wall. In the eye itself, the air is warm, winds are light, and skies may be broken or overcast.

Hurricanes are born over warm tropical waters where the air is humid, surface winds converge, and thunderstorms become organized in a region of weak upper-level winds. Surface convergence may occur along the ITCZ, on the eastern side of a tropical wave, or along a front that has moved into the tropics from higher latitudes. If the disturbance becomes more organized, it becomes a tropical depression. If central pressures drop and surface winds increase, the depression becomes a tropical storm. At this point, the storm is given a name. Some tropical storms continue to deepen into full-fledged hurricanes, as long as the outflow of air at the top of the storm is greater than the convergence of air at the surface.

In the organized convection theory, the energy source that drives the hurricane comes primarily from the release of latent heat. As vast quantities of latent heat are released in thunderstorms near the eye, surface pressures lower, inducing stronger winds, more rising air, and the development of even stronger thunderstorms. In the heat engine model, energy for the storm's growth is taken in at the surface in the form of sensible and latent heat, converted to kinetic energy in the form of winds, then lost at the cloud tops through radiational cooling.

The easterly winds in the tropics usually steer hurricanes westward. Most storms then gradually swing northwestward around the subtropical high to the north. If the storm moves into middle latitudes, the prevailing westerlies steer it northeastward. Because hurricanes derive their energy from the warm surface water and from the latent heat of condensation, they tend to dissipate rapidly when they move over cold water or over a large mass of land, where surface friction causes their winds to decrease and flow into their centers.

Although the high winds of a hurricane can inflict a great deal of damage, it is usually the huge waves and the flooding associated with the storm surge that cause the most destruction. The Saffir-Simpson hurricane scale was developed to estimate the potential destruction that a hurricane can cause.

Key Terms

The following terms are listed in the order they appear in the text. Define each. Doing so will aid you in reviewing the material covered in this chapter.

streamlines
tropical wave (easterly wave)
hurricane
typhoon
eye (of hurricane)
eye wall
trade wind inversion
tropical disturbance
tropical depression

tropical storm
Ekman transport
storm surge
spin-up vortices
    (mini-swirls)
hurricane watch
hurricane warning
Saffir-Simpson scale
super-typhoon

Questions for Review

1. What is a tropical (easterly) wave? How do these waves generally move in the Northern Hemisphere? Are showers found on the eastern or western side of the wave?
2. Why are streamlines, rather than isobars, used on surface weather maps in the tropics?
3. What is the name given to a hurricane-like storm over the western North Pacific Ocean?
4. Describe the horizontal and vertical structure of a hurricane.
5. What conditions at the surface and aloft are most conducive for the formation of a hurricane?
6. (a) Describe the formation of a hurricane using the organized convection theory. (b) In this theory, where do hurricanes derive their energy?
7. (a) Explain how the energy that drives a hurricane can be likened to a heat engine. (b) In this model, what determines the maximum strength (the highest winds) that the storm can achieve?
8. Would it be possible for a hurricane to form over land? Explain.
9. What factors tend to weaken hurricanes?
10. Distinguish among a tropical disturbance, a tropical depression, a tropical storm, and a hurricane.
11. In what ways is a hurricane different from a mid-latitude cyclone? In what ways are these two systems similar?
12. Why do most hurricanes move westward over tropical waters?

13. If the high winds of a hurricane are not responsible for inflicting the most damage, what is?
14. Explain how a storm surge forms.
15. (a) What are spin-up vortices? (b) How are they related to areas of extreme damage caused by Hurricane Andrew?
16. Hurricanes are given names when the storm is in what stage of development?
17. How does a hurricane watch differ from a hurricane warning?
18. Why have hurricanes been seeded with silver iodide?
19. Give two reasons why hurricanes are more likely to strike New Jersey than Oregon.

Questions for Thought

1. Why are hurricanes more apt to form in October than in May?
2. Would it be possible for a hurricane to form in the tropical North Atlantic or North Pacific during December? Explain.
3. Would the winds of a hurricane decrease more quickly as the storm moves over cooler water or over warmer land? Explain.
4. Explain why the surface water temperature of the ocean is usually cooler after the passage of a hurricane.
5. Suppose, in the North Atlantic, an eastward-moving ocean vessel is directly in the path of a westward-moving hurricane. What would be the ship's wisest course—to veer to the north of the storm or to the south of the storm? Explain.
6. Suppose this year five tropical storms develop into full-fledged hurricanes over the North Atlantic Ocean. Would the name of the third hurricane begin with the letter *C*? Explain.
7. You are in Darwin, Australia (on the north shore), and a hurricane approaches from the north. Where would the highest storm surge be, to the east or west? Explain.

8. Occasionally when a hurricane moves inland, it will encounter a mountain range. Describe what will happen when this occurs. What will happen to the hurricane's intensity? What will cause the most damage (winds, storm surge, flooding)? Why?

Problems and Exercises

1. A hurricane just off the coast of northern Florida is moving northeastward, parallel to the eastern seaboard. Suppose that you live in North Carolina along the coast:
   (a) How will the surface winds in your area change direction as the hurricane's center passes due east of you? Illustrate your answer by making a sketch of the hurricane's movement and the wind flow around it.
   (b) If the hurricane passes east of you, the strongest winds would most likely be blowing from which direction? Explain your answer. (Assume that the storm does not weaken as it moves northeastward.)
   (c) The lowest sea level pressure would most likely occur with which wind direction? Explain.
2. Use the Saffir-Simpson hurricane scale (see Table 16.2, p. 436) and the text material to determine the category of each of the following hurricanes:
   (a) Hurricane Gloria in Fig. 16.11, p. 429.
   (b) Hurricane Hugo in Fig. 16.14, p. 433.
   (c) Hurricane Andrew in Fig. 16.17, p. 435.

Questions for Exploration

1. On the Blue Skies: College Edition CD-ROM go to the **Hurricane** section of the CD and click on **"Virtual Hurricane."**
2. On the Blue Skies: College Edition CD-ROM go to the **Hurricane** section of the CD and click on **"Hurricane Forecasting."**

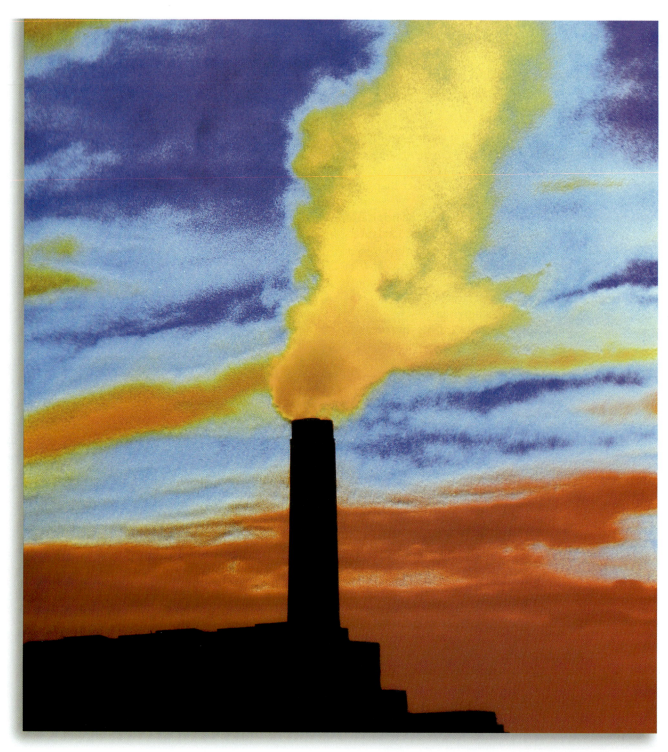

Smoke and sulfur oxides rise into the atmosphere from a chimney near Edinburgh, Scotland. (©Dr. Jeremy Burgess/Science Photo Library.)

AIR POLLUTION

CONTENTS

Air pollution makes the earth a less pleasant place to live. It reduces the beauty of nature. This blight is particularly noticed in mountain areas. Views that once made the pulse beat faster because of the spectacular panorama of mountains and valleys are more often becoming shrouded in smoke. When once you almost always could see giant boulders sharply etched in the sky and the tapered arrowheads of spired pines, you now often see a fuzzy picture of brown and green. The polluted air acts like a translucent screen pulled down by an unhappy God.

Louis J. Battan, *The Unclean Sky*

Every deep breath fills our lungs mostly with gaseous nitrogen and oxygen. Also inhaled, in minute quantities, may be other gases and particles, some of which could be considered pollutants. These contaminants come from car exhaust, chimneys, forest fires, factories, power plants, and other sources related to human activities.

Virtually every large city has to contend in some way with air pollution, which clouds the sky, injures plants, and damages property. Some pollutants merely have a noxious odor, whereas others can cause severe health problems. The cost is high. In the United States, for example, outdoor air pollution takes its toll in health care and lost work productivity at an annual expense that runs into *billions* of dollars. Estimates are that, worldwide, nearly one billion people in urban environments are continuously being exposed to health hazards from air pollutants.

This chapter takes a look at this serious contemporary concern. We begin by briefly examining the history of problems in this area, and then go on to explore the types and sources of air pollution, as well as the weather that can produce an unhealthful accumulation of pollutants. Finally, we investigate how air pollution influences the urban environment and also how it brings about unwanted acid precipitation.

A BRIEF HISTORY OF AIR POLLUTION

Strictly speaking, air pollution is not a new problem. More than likely it began when humans invented fire whose smoke choked the inhabitants of poorly ventilated caves. In fact, very early accounts of air pollution characterized the phenomenon as "smoke problems," the major cause being people burning wood and coal to keep warm.

To alleviate the smoke problem in old England, King Edward I issued a proclamation in 1273 forbidding the use of sea coal, an impure form of coal that produced a great deal of soot and sulfur dioxide when burned. One person was reputedly executed for violating this decree. In spite of such restrictions, the use of coal grew as a heating fuel during the fifteenth and sixteenth centuries.

As industrialization increased, the smoke problem worsened. In 1661, the prominent scientist John Evelyn wrote an essay deploring London's filthy air. And by the 1850s, London had become notorious for its "pea-soup" fog, a thick mixture of smoke and fog that hung over the city. These fogs could be dangerous. In 1873, one was responsible for as many as 700 deaths. Another in 1911 claimed the lives of 1150 Londoners. To describe this chronic atmospheric event, a physician, Harold Des Voeux, coined (around 1911) the word *smog*, meaning a combination of smoke and fog.

Little was done to control the burning of coal as time went by, primarily because it was extremely difficult to counter the basic attitude of the powerful industrialists: "Where there's muck, there's money." London's acute smog problem intensified. Then, during the first week of December, 1952, a major

disaster struck. The winds died down over London and the fog and smoke became so thick that people walking along the street literally could not see where they were going. People wore masks over their mouths and found their way along the sidewalks by feeling the walls of buildings. This particular disastrous smog lasted 5 days and took nearly 4000 lives, prompting Parliament to pass a Clean Air Act in 1956. Additional air pollution incidents occurred in England during 1956, 1957, and 1962, but due to the strong legislative measures taken against air pollution, London's air today is much cleaner, and "pea soup" fogs are a thing of the past.

Air pollution episodes were by no means limited to Great Britain. During the winter of 1930, for instance, Belgium's highly industrialized Meuse Valley experienced an air pollution tragedy when smoke and other contaminants accumulated in a narrow steep-sided valley. The tremendous buildup of pollutants caused about 600 people to become ill, and ultimately 63 died. Not only did humans suffer, but cattle, birds, and rats fell victim to the deplorable conditions.

The industrial revolution brought air pollution to the United States, as homes and coal-burning industries belched smoke, soot, and other undesirable emissions into the air. Soon, large industrial cities, such as St. Louis and Pittsburgh (which became known as the "Smoky City"), began to feel the effects of the ever-increasing use of coal. As early as 1911, studies documented the irritating effect of smoke particles on the human respiratory system and the "depressing and devitalizing" effects of the constant darkness brought on by giant, black clouds of smoke. By 1940, the air over some cities had become so polluted that automobile headlights had to be turned on during the day.

The first major documented air pollution disaster in the United States occurred at Donora, Pennsylvania, during October, 1948, when industrial pollution became trapped in the Monongahela River Valley. During the ordeal, which lasted 5 days, more than 20 people died and thousands became ill.* Several times during the 1960s, air pollution levels became dangerously high over New York City. Meanwhile, on the West Coast, in cities such as Los Angeles, the ever-rising automobile population, coupled with the large petroleum processing plants, were instrumental in generating a different type of pollutant—one that forms in sunny weather and irritates the eyes. Toward the end of World War II, Los Angeles had its first (of many) smog alerts.

Air pollution episodes in Los Angeles, New York, and other large American cities led to the establishment of much stronger emission standards for industry and automobiles. The Clean Air Act of 1970, for example, empowered the federal government to set emission standards that each state was required to enforce. The Clean Air Act was revised in 1977 and updated by Congress in 1990 to include even stricter emission

---

*Additional information about the Donora air pollution disaster is given in the Focus section on p. 462.

FIGURE 17.1
Easterly winds blow smoke from many large fires over a vast area of southern Africa.

requirements for autos and industry. The new version of the Act also includes incentives to encourage companies to lower emissions of those pollutants contributing to the current problem of acid rain. Moreover, amendments to the Act have identified 189 toxic air pollutants for regulation.

TYPES AND SOURCES OF AIR POLLUTANTS

Air pollutants are airborne substances (either solids, liquids, or gases) that occur in concentrations high enough to threaten the health of people and animals, to harm vegetation and structures, or to toxify a given environment. Air pollutants come from both natural sources and human activities. Examples of natural sources include wind picking up dust and soot from the earth's surface and carrying it aloft, volcanoes belching tons of ash and dust into our atmosphere, and forest fires producing vast quantities of drifting smoke (see Fig. 17.1).

Human-induced pollution enters the atmosphere from both *fixed sources* and *mobile sources*. Fixed sources encompass industrial complexes, power plants, homes, office buildings, and so forth; mobile sources include motor vehicles, ships, and jet aircraft. Certain pollutants are called **primary air pollutants** because they enter the atmosphere directly—from smoke stacks and tail pipes, for example. Other pollutants, known as **secondary air pollutants**, form only when a chemical reaction occurs between a primary pollutant and some other component of air, such as water vapor or another pollutant. Table 17.1 summarizes some of the sources of primary air pollutants.

We can see in Fig. 17.2 that transportation (motor vehicles, and so on) accounts for nearly 50 percent (by weight) of the pollution across the United States, with fuel combustion from stationary (fixed) sources coming in a distant second. Although hundreds of pollutants are found in our atmosphere, most fall into five groups, which are summarized in the following section.

Principal Air Pollutants   The term **particulate matter** represents a group of solid particles and liquid droplets that are small enough to remain suspended in the air. Collectively

(a) Primary Pollutants

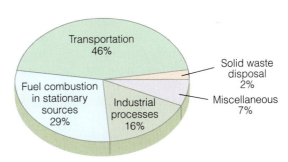

(b) Primary Sources

FIGURE 17.2
(a) Estimates of emissions of the primary air pollutants in the United States on a per weight basis; (b) the primary sources for the pollutants. (Data courtesy of United States Environmental Protection Agency.)

TABLE 17.1   Some of the Sources of Primary Air Pollutants

	SOURCES		POLLUTANTS
Natural			
	Volcanic eruptions		Particles (dust, ash), gases (SO_2, CO_2)
	Forest fires		Smoke, unburned hydrocarbons, CO_2, nitrogen oxides, ash
	Dust storms		Suspended particulate matter
	Ocean waves		Salt particles
	Vegetation		Hydrocarbons (VOCs),* pollens
	Hot springs		Sulfurous gases
Human caused			
Industrial	Paper mills		Particulate matter, sulfur oxides
	Power Plants	Coal	Ash, sulfur oxides, nitrogen oxides
		Oil	Sulfur oxides, nitrogen oxides, CO
	Refineries		Hydrocarbons, sulfur oxides, CO
	Manufacturing	Sulfuric acid	SO_2, SO_3, and H_2SO_4
		Phosphate fertilizer	Particulate matter, gaseous fluoride
		Iron and steel mills	Metal oxides, smoke, fumes, dusts, organic and inorganic gases
		Plastics	Gaseous resin
		Varnish/paint	Acrolein, sulfur compounds
Personal	Automobiles		CO, nitrogen oxides, hydrocarbons (VOCs), particulate matter
	Home furnaces/fireplaces		CO, particulate matter
	Open burning of refuse		CO, particulate matter

*VOCs are volatile organic compounds; they represent a class of organic compounds, most of which are hydrocarbons.

known as *aerosols*, this grouping includes solid particles that may irritate people but are usually not poisonous, such as soot (tiny solid carbon particles), dust, smoke, and pollen. Some of the more dangerous substances include asbestos fibers and arsenic. Tiny liquid droplets of sulfuric acid, PCBs, oil, and various pesticides are also placed into this category.

Because it often dramatically reduces visibility in urban environments, particulate matter pollution is the most noticeable (see Fig. 17.3). Some particulate matter collected in cities includes iron, copper, nickel, and lead. This type of pollution can immediately influence the human respiratory system. Once inside the lungs, it can make breathing difficult, particularly for those suffering from chronic respiratory disorders. Lead particles especially are dangerous as they tend to fall out of the atmosphere and become absorbed into the body through ingestion of contaminated food and water supplies. Lead accumulates in bone and soft tissues, and in high concentrations can cause brain damage, convulsions, and death. Even at low doses, lead can be particularly dangerous to fetuses, infants, and children who, when exposed, may suffer central nervous system damage.

Industrial processes account for nearly 40 percent of the estimated 6.6 million metric tons of particulate matter emitted over the United States during the course of a year, whereas highway vehicles account for about 17 percent. One main problem is that particulate pollution may remain in the atmosphere for some time depending on the size and the amount of precipitation that occurs. For example, larger, heavier particles with diameters greater than about 10 μm* (0.01 mm) tend to settle to the ground in about a day or so after being emitted; whereas fine, lighter particles with diameters less than 1 μm (0.001 mm) can remain suspended in the lower atmosphere for several weeks. Finer particles with diameters smaller than 10 μm are referred to as *PM-10*. These particles pose the greatest health risk, as they are small enough to penetrate the lung's natural defense mechanisms. Moreover, winds can carry these fine particles great distances before they finally reach the surface. In fact, suspended particles from sources in Europe and the former Soviet Union are believed responsible for the brownish cloud layer called *Arctic haze* that forms over the

---

*Recall that one micrometer (μm) is one-millionth of a meter.

(a)

(b)

Arctic each spring. Rain and snow remove many of these particles from the air; even the minute particles are removed by ice crystals and cloud droplets. In fact, numerical simulations of air pollution suggest that the predominant removal mechanism occurs when these particles act as nuclei for cloud droplets and ice crystals. As we will see in Chapter 19, a long-lasting accumulation of suspended particles (especially those rich in sulfur in the stratosphere) is not only aesthetically unappealing but has the potential for affecting the climate.

Many of the suspended particles are hygroscopic, as water vapor readily condenses onto them. As a thin film of water forms on the particles, they grow in size. When they reach a diameter between 0.1 and 1.0 μm these *wet haze* particles effec-

tively scatter incoming sunlight to give the sky a milky white appearance. The particles are usually sulfate or nitrate particular matter from combustion processes, such as those produced by diesel engines and power plants. The hazy air mass may become quite thick, and on humid summer days it often becomes well defined, as illustrated in Figure 17.4.

Carbon monoxide (CO), a major pollutant of city air, is a colorless, odorless, poisonous gas that forms during the incomplete combustion of carbon-containing fuels. From Fig. 17.2, we can see that carbon monoxide is the most plentiful of the primary pollutants.

The Environmental Protection Agency (EPA) estimates that over 60 million metric tons of carbon monoxide enter the

WEATHER WATCH

On any given day, estimates are that about 10 million tons of solid particulate matter are suspended in our atmosphere. And in a polluted environment, a volume of air about the size of a sugar cube can contain as many as 200,000 particles.

air annually over the United States alone—about half from highway vehicles. However, due to stricter air quality standards and the use of emission-control devices, carbon monoxide levels have decreased by about 40 percent since the early 1970s.

Fortunately, carbon monoxide is quickly removed from the atmosphere by microorganisms in the soil, for even in small amounts this gas is dangerous. Hence, it poses a serious problem in poorly ventilated areas, such as highway tunnels and underground parking garages. Because carbon monoxide cannot be seen or smelled, it can kill without warning. Here's how: Normally, your cells obtain oxygen through a blood pigment called *hemoglobin*, which picks up oxygen from the lungs, combines with it, and carries it throughout your body. Unfortunately, human hemoglobin prefers carbon monoxide to oxygen, so if there is too much carbon monoxide in the air you breathe, your brain will soon be starved of oxygen, and headache, fatigue, drowsiness, and even death may result.*

Sulfur dioxide (SO$_2$) is a colorless gas that comes primarily from the burning of sulfur-containing fossil fuels (such as coal and oil). Its primary source includes power plants, heating devices, smelters, petroleum refineries, and paper mills. How-

---

*Should you become trapped in your car during a snowstorm and you have your engine and heater running to keep warm, roll down the window just a little. This action will allow any carbon monoxide that may have entered the car through leaks in the exhaust system to escape.

ever, it can enter the atmosphere naturally during volcanic eruptions and as sulfate particles from ocean spray.

Sulfur dioxide readily oxidizes to form the secondary pollutants *sulfur trioxide* (SO$_3$) and, in moist air, highly corrosive *sulfuric acid* (H$_2$SO$_4$). Winds can carry these particles great distances before they reach the earth as undesirable contaminants. When inhaled into the lungs, high concentrations of sulfur dioxide aggravate respiratory problems, such as asthma, bronchitis, and emphysema. Sulfur dioxide in large quantities can cause injury to certain plants, such as lettuce and spinach, sometimes producing bleached marks on their leaves and reducing their yield.

Volatile organic compounds (VOCs) represent a class of organic compounds that are mainly **hydrocarbons**—individual organic compounds composed of hydrogen and carbon. At room temperature they occur as solids, liquids, and gases. Even though thousands of such compounds are known to exist, methane (which occurs naturally and poses no known dangers to health) is the most abundant. Other volatile organic compounds include benzene, formaldehyde, and some chlorofluorocarbons. The Environmental Protection Agency estimates that over 18 million metric tons of VOCs are emitted into the air over the United States each year, with about 34 percent of the total coming from vehicles used for transportation and about 50 percent from industrial processes.

Certain VOCs, such as benzene (an industrial solvent) and benzo-a-pyrene (a product of burning wood, tobacco and barbecuing), are known to be *carcinogens*—cancer-causing agents. Although many VOCs are not intrinsically harmful, some will react with nitrogen oxides in the presence of sunlight to produce secondary pollutants, which are harmful to human health.

Nitrogen oxides are gases that form when some of the nitrogen in the air reacts with oxygen during the high-temperature combustion of fuel. The two primary nitrogen pollutants are **nitrogen dioxide (NO$_2$)** and **nitric oxide (NO)**, which,

FIGURE 17.4
Cumulus clouds and a thunderstorm rise above the thick layer of haze that frequently covers the eastern half of the United States on humid summer days.

together, are commonly referred to as NO$_x$, or simply, *oxides of nitrogen*.

Although both nitric oxide and nitrogen dioxide are produced by natural bacterial action, their concentration in urban environments is between 10 and 100 times greater than in nonurban areas. In moist air, nitrogen dioxide reacts with water vapor to form corrosive nitric acid (HNO$_3$), a substance that adds to the problem of acid rain, which we will address later.

The primary sources of nitrogen oxides are motor vehicles, power plants, and waste disposal systems. High concentrations are believed to contribute to heart and lung problems, as well as lowering the body's resistance to respiratory infections. Studies on test animals suggest that nitrogen oxides may encourage the spread of cancer. Moreover, nitrogen oxides are highly reactive gases that play a key role in producing ozone and other ingredients of photochemical smog.

Ozone in the Troposphere

As mentioned earlier, the word **smog** originally indicated the combining of smoke and fog. Today, however, the word mainly refers to the type of smog that forms in large cities, such as Los Angeles. Because this type of smog forms when chemical reactions take place in the presence of sunlight (called *photochemical reactions*), it is termed **photochemical smog**, or *Los Angeles-type smog*. When the smog is composed of sulfurous smoke and foggy air, it is usually called *London-type smog*.

The main component of photochemical smog is the gas **ozone (O$_3$)**. Ozone is a noxious substance with an unpleasant odor that irritates eyes and the mucous membranes of the respiratory system, aggravating chronic diseases, such as asthma and bronchitis. Even in healthy people, exposure to relatively low concentrations of ozone for six or seven hours during periods of moderate exercise can significantly reduce lung function. This situation often is accompanied by symptoms such as chest pain, nausea, coughing, and pulmonary congestion. Ozone also attacks rubber, retards tree growth, and damages crops. Each year, in the United States alone, ozone is responsible for crop yield losses of several billion dollars.

We will see later that ozone forms naturally in the stratosphere through the combining of molecular oxygen and atomic oxygen. There, *stratospheric ozone* provides a protective shield against the sun's harmful ultraviolet rays. However, near the surface, in polluted air, ozone (often referred to as *tropospheric ozone*) is a secondary pollutant that is not emitted directly into the air. Rather, it forms from a complex series of chemical reactions involving other pollutants, such as nitrogen oxides and volatile organic compounds (hydrocarbons). Because sunlight is required to produce ozone, concentrations of tropospheric ozone are normally higher during the afternoons (see Fig. 17.5) and during the summer months, when sunlight is more intense.

In polluted air, ozone production occurs along the following lines. Sunlight (with wavelengths shorter than about

FIGURE 17.5

Average hourly concentrations of ozone measured at six major cities over a two-year period.

0.41 μm) dissociates nitrogen dioxide into nitric oxide and atomic oxygen, which may be expressed by

$$NO_2 + \text{solar radiation} \rightarrow NO + O.$$

The atomic oxygen then combines with molecular oxygen (in the presence of a third molecule, M), to form ozone, as

$$O_2 + O + M \rightarrow O_3 + M.$$

The ozone is then destroyed by combining with nitric oxide; thus

$$O_3 + NO \rightarrow NO_2 + O_2.$$

If sunlight is present, however, the newly formed nitrogen dioxide will break down into nitric oxide and atomic oxygen. The atomic oxygen then combines with molecular oxygen to form ozone again. Consequently, large concentrations of ozone can form in polluted air only if some of the nitric oxide reacts with other gases without *removing ozone in the process*. Under these conditions, certain hydrocarbons (emitted by autos and industrial sources) and the hydroxyl radical come into play.

The formation of the hydroxyl radical (OH) begins when ultraviolet radiation (at wavelengths of about 0.31 μm and below) dissociates some of the ozone into molecular oxygen and atomic oxygen; accordingly

$$O_3 + \text{UV radiation} \rightarrow O_2 + O.$$

The atomic oxygen formed is in an excited state, which means it can react with a variety of other molecules, including water vapor, to produce two hydroxyl radical molecules; thus

$$O + H_2O \rightarrow OH + OH.$$

The OH is called a "radical" because it contains an unpaired electron. This situation allows the OH molecule to react with many other atoms and molecules, including unburned or partially burned hydrocarbons (RH) released into the air by automobiles and industry:

$$OH + RH \rightarrow R\cdot + H_2O.$$

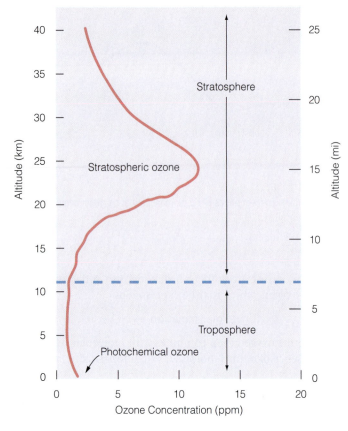

FIGURE 17.6

The average distribution of ozone above the earth's surface in the middle latitudes.

The product R· represents an organic hydrocarbon that can have a complex molecular structure. The R· is then able to react with molecular oxygen to form $RO_2·$, a reactive molecule that removes nitric oxide by combining with it to form nitrogen dioxide, shown by the expression:

$$RO_2· + NO \rightarrow NO_2 + \text{other products.}$$

In this manner, nitric oxide can react with hydrocarbons to form nitrogen dioxide *without removing ozone.* Hence, the reactive hydrocarbons in polluted air allow ozone concentrations to increase by preventing nitric oxide from destroying the ozone as rapidly as it is formed.

The hydrocarbons (VOCs) also react with oxygen and nitrogen dioxide to produce other undesirable contaminants, such as *PAN* (peroxyacetyl nitrate)—a pollutant that irritates eyes and is extremely harmful to vegetation—and organic compounds. Ozone, PAN, and small amounts of other oxidating pollutants are the ingredients of photochemical smog. Instead of being specified individually, these pollutants are sometimes grouped under a single heading called *photochemical oxidants.**

---

*An oxidant is a substance (such as ozone) whose oxygen combines chemically with another substance.

Hydrocarbons (VOCs) do occur naturally in the atmosphere, as they are given off by vegetation. Oxides of nitrogen drifting downwind from urban areas can react with these natural hydrocarbons and produce smog in relatively uninhabited areas. This phenomenon has been observed downwind of cities such as Los Angeles, London, and New York. Some regions have so much natural (background) hydrocarbon that it may be difficult to reduce ozone levels as much as desired.

In spite of vast efforts to control ozone levels in major metropolitan areas, results have been generally disappointing because ozone, as we have seen, is a secondary pollutant that forms from chemical reactions involving other pollutants. Ozone production should decrease in most areas when emissions of *both* nitrogen oxides and hydrocarbons (VOCs) are reduced. However, the reduction of only one of these pollutants will not necessarily diminish ozone production because the oxides of nitrogen act as a catalyst for producing ozone in the presence of hydrocarbons (VOCs). Air pollution models suggest that, in the Los Angeles basin, substantially reducing the emissions of hydrocarbons (VOCs) should also reduce ozone production.

Up to now, we have concentrated on ozone in the troposphere, primarily in a polluted environment. The next section examines the formation and destruction of ozone in the upper atmosphere—in the stratosphere.

Ozone in the Stratosphere   Recall from Chapter 1 that the stratosphere is a region of the atmosphere that lies above the troposphere between about 10 and 50 km (6 and 31 mi) above the earth's surface. The atmosphere is stable in the stratosphere, as the temperature in the lower part is isothermal (equal temperature), while above about 20 km there is a strong temperature inversion—the air temperature increases rapidly with height (look back at Fig. 1.9, p. 12). The inversion is due, in part, to the gas ozone that absorbs ultraviolet radiation at wavelengths below about 0.3 μm.

In the stratosphere, above middle latitudes, ozone is most dense at an altitude near 25 km (see Fig. 17.6). Even at this altitude, its concentration is quite small, as there are only about 12 ozone molecules for every million air molecules (12 ppm).* Although thin, this layer of ozone is significant, for it shields earth's inhabitants from harmful amounts of ultraviolet solar radiation. This protection is fortunate because ultraviolet radiation at wavelengths below 0.3 μm has enough energy to cause skin cancer in humans. Also, UV radiation at 0.26 μm can destroy acids in DNA (deoxyribonucleic acid), the substance that transmits the hereditary blueprint from one generation to the next.

If the concentration of stratospheric ozone decreases, the following might occur:

■ An increase in the number of cases of skin cancer.

■ A sharp increase in eye cataracts and sun burning.

---

*With a concentration of ozone of only 12 parts per million in the stratosphere, the composition of air here is about the same as it is near the earth's surface—mainly 78 percent nitrogen and 21 percent oxygen.

- Suppression of the human immune system.
- An adverse impact on crops and animals due to an increase in ultraviolet radiation.
- A reduction in the growth of ocean phytoplankton.
- A cooling of the stratosphere that could alter stratospheric wind patterns, possibly affecting the production (and destruction) of ozone.

Stratospheric Ozone: Production-Destruction Ozone (O_3) forms naturally in the stratosphere by the combining of atomic oxygen (O) with molecular oxygen (O_2)—in the presence of another molecule. Although it forms mainly above 25 km, ozone gradually drifts downward by mixing processes, producing a peak concentration in middle latitudes near 25 km. (In polar regions, its maximum concentration is found at lower levels.) Ozone is broken down into molecular and atomic oxygen by absorbing ultraviolet (UV) radiation with wavelengths between 0.2 and 0.3 μm (see Fig. 17.7). Thus

$$O_3 + UV \rightarrow O_2 + O.$$

Ozone is destroyed by colliding with other atoms and molecules. For example, ozone and atomic oxygen combine to form two oxygen molecules, as

$$O_3 + O \rightarrow 2\,O_2.$$

Likewise, the combination of two ozone molecules destroys ozone, as

$$O_3 + O_3 \rightarrow 3\,O_2.$$

These equations also represent the net result of a number of complex chemical reactions that include trace gases of nitrogen, hydrogen, and chlorine. For example, two natural destructive gases for ozone are nitric oxide (NO) and nitrogen dioxide (NO_2), which, as we have seen, are collectively known as *oxides of nitrogen*. The origin of these gases begins at the earth's surface as soil bacteria produce N_2O (nitrous oxide). This gas gradually finds its way into the stratosphere where, above about 25 km, solar energy converts some of it into ozone-destroying oxides of nitrogen. In the stratosphere just a small amount of nitric oxide can destroy a large amount of ozone. The following sequence of chemical reactions will illustrate why. In step 1, the nitric oxide quickly combines with ozone to form nitrogen dioxide and molecular oxygen. Then, in step 2, the nitrogen dioxide combines with atomic oxygen to form nitric oxide and molecular oxygen. Thus

Step 1 $NO + O_3 \rightarrow NO_2 + O_2$
Step 2 $NO_2 + O \rightarrow NO + O_2$

The nitric oxide (NO) released in step 2 is now ready to start destroying ozone again.

Stratospheric ozone is maintained by a delicate natural balance between production and destruction. Could this balance be upset?

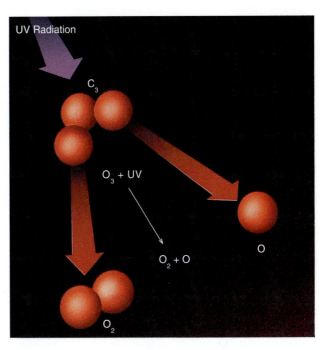

FIGURE 17.7
An ozone molecule absorbing ultraviolet radiation can become molecular oxygen (O_2) and atomic oxygen (O).

Stratospheric Ozone—Upsetting the Balance The concentration of ozone in the stratosphere may be changed by natural events. In the upper atmosphere, both cosmic rays* and solar particles can produce secondary electrons having sufficient energy to separate molecular nitrogen (N_2) into two nitrogen atoms (N). The nitrogen atoms combine with free atomic oxygen to form nitric oxide which, in turn, rapidly destroys ozone. Furthermore, large volcanic eruptions, such as the Philippine's Mt. Pinatubo in mid-June, 1991, can inject ozone-destroying chemicals into the stratosphere. And scientists, using measurements from satellites, discovered that even changes in ultraviolet radiation from the sun can cause small variations in the amount of stratospheric ozone.

It appears, too, that human activities are altering the amount of stratospheric ozone. This possibility was first brought to light in the early 1970s as Congress pondered over whether or not the United States should build a supersonic jet transport. One of the gases emitted from the engines of this aircraft is nitric oxide. Although the aircraft was designed to fly in the stratosphere below the level of maximum ozone, it was feared that the nitric oxide would eventually have an adverse effect on the ozone. This factor was one of many considered when Congress decided to halt the development of the United States' version of the supersonic transport.

---

*Cosmic rays are high-energy atomic nuclei and atomic particles that travel through space at extremely high speeds. Most cosmic rays are believed to come from supernovas, although some are produced in solar flares.

More recently, concerns involve emissions of chemicals at the earth's surface, such as nitrous oxide emitted from nitrogen fertilizers (which may drift into the stratosphere, where it could destroy ozone) and *chlorofluorocarbons* (CFCs). Until the late 1970s, when the United States banned all nonessential uses of chlorofluorocarbons, they were the most widely used propellants in spray cans, such as deodorants and hairsprays. In the troposphere, these gases are quite safe, being nonflammable, nontoxic, and unable to chemically combine with other substances.* Hence, these gases slowly diffuse upward without being destroyed. They apparently enter the stratosphere (1) near breaks in the tropopause, especially in the vicinity of jet streams; and (2) in building thunderstorms that penetrate the lower stratosphere.

Once chlorofluorocarbon molecules reach the middle stratosphere, ultraviolet energy that is normally absorbed by ozone breaks them up, releasing *chlorine* in the process. Note in the following sequence of reactions how chlorine destroys ozone rapidly. In step 1, chlorine (Cl) combines with ozone, forming chlorine monoxide (ClO)—a new substance—and molecular oxygen (O_2). Almost immediately, the chlorine monoxide combines with free atomic oxygen (step 2) to produce chlorine atoms and molecular oxygen. The free chlorine atoms are now ready to combine with and destroy more ozone molecules. Estimates are that a single chlorine atom removes as many as 100,000 ozone molecules before it is taken out of action by combining with other substances.

> *Step 1* $Cl + O_3 \rightarrow ClO + O_2$
> *Step 2* $ClO + O \rightarrow Cl + O_2$

Fortunately, chlorine atoms do not exist in the stratosphere forever. They are removed as chlorine monoxide combines with nitrogen dioxide to form chlorine nitrate, $ClONO_2$ (step 3). In step 4, free chlorine atoms combine with methane to form hydrogen chloride (HCl) and a new substance, CH_3.

> *Step 3* $ClO + NO_2 \rightarrow ClONO_2$
> *Step 4* $CH_4 + Cl \rightarrow HCl + CH_3$

Since the average lifetime of a CFC molecule is about 50 to 100 years, any increase in the concentration of CFCs is long-lasting and a genuine threat to the concentration of ozone. Given this fact and the additional knowledge that CFCs contribute to the earth's greenhouse effect, an international agreement signed in 1987—the *Montreal Protocol*—established a timetable for diminishing CFC emissions and the use of bromine compounds (halons), which destroy ozone at a rate 10 times greater than do chlorine compounds.

During November, 1992, representatives of more than half the world's nations met in Copenhagen to update and revise the treaty. Provisions of the meeting called for a quicker phase-out of the previously targeted ozone-destroying chemicals and the establishment of a permanent fund to help Third World nations find the technology to develop ozone-friendly chemicals.* The phase-out appears to be working, as the National Oceanic and Atmospheric Administration (NOAA) reported in 1996 that ground-based stations around the world had detected for the first time an overall slight decline (about 1 percent) in the total concentration of ozone-destroying chlorine and bromine compounds.

The question of just how much ozone is being depleted by chlorine is now under investigation. More than 5 billion kilograms of CFCs have already been released into the troposphere and will diffuse upward during the next few decades. In a 1991 study, an international panel of over 80 scientists concluded that the ozone layer thinned by about 3 percent during the summer from 1979 to 1991 over heavily populated areas of the Northern Hemisphere. Moreover, satellite measurements revealed that ozone concentrations during 1992 and 1993 dropped to record low levels over much of the globe. The decrease appears to stem from ozone-destroying chemicals and from the 1991 volcanic eruption of Mt. Pinatubo that sent tons of sulfur dioxide gas into the stratosphere, where it formed tiny droplets of sulfuric acid. These droplets not only enhance the ozone destructiveness of the chlorine chemicals but also alter the circulation of air in the stratosphere, making it more favorable for ozone depletion. In 1995, wintertime ozone levels dropped well below normal over much of the Northern Hemisphere. This decrease apparently was due to ozone-destroying pollution along with natural cold weather patterns that favored ozone reduction.

Presently, there are two major substitutes for CFCs, *hydrochlorofluorocarbons* (HCFCs) and *hydrofluorocarbons* (HFCs). The HCFCs contain fewer chlorine atoms per molecule than CFCs and, therefore, pose much less danger to the ozone layer, whereas HFCs contain no chlorine. These gases may have to be phased out, however, as both are greenhouse gases that can enhance global warming.

It appears that ozone is being destroyed more quickly above specific regions of the planet. Scientists point to the fact that ozone concentrations over springtime Antarctica have plummeted at an alarming rate. This sharp drop in ozone is known as the **ozone hole**. (More information on the ozone hole is provided in the Focus section on p. 451.)

Air Pollution: Trends and Patterns   Over the past decades, strides have been made in the United States to improve the quality of the air we breathe. Figure 17.8 shows the estimated emission trends over the United States for the primary pollutants. Notice that since the Clean Air Act of 1970, emissions of most pollutants have fallen off substantially, with lead showing the greatest reduction, primarily due to the gradual elimination of leaded gasoline.

Although the situation has improved, we can see from Fig. 17.8 that much more needs to be done, as large quantities of

---

*Recall that CFCs do act as strong greenhouse gases in the troposphere.

*The phase out of CFCs for domestic use has stimulated a thriving black market.

The Ozone Hole

In 1974, two chemists from the University of California at Irvine—F. Sherwood Rowland and Mario J. Molina—warned that increasing levels of CFCs would eventually deplete stratospheric ozone on a global scale. Their studies suggested that ozone depletion would occur gradually and would perhaps not be detectable for many years to come. It was surprising, then, when British researchers identified a year-to-year decline in stratospheric ozone over Antarctica. Their findings, corroborated later by satellites and balloon-borne instruments, showed that since the late 1970s ozone concentrations have diminished each year during the months of September and October. This decrease in stratospheric ozone over springtime Antarctica is known as the *ozone hole* (see Fig. 1). In years of severe depletion, such as in 1998, the ozone hole covers almost twice the area of the Antarctic continent.

To understand the causes behind the ozone hole, scientists in 1986 organized the first *National Ozone Expedition*, NOZE-1, which set up a fully instrumented observing station near McMurdo Sound, Antarctica. During 1987, with the aid of instrumented aircraft, NOZE-2 got under way. The findings from these research programs helped scientists put together the pieces of the ozone puzzle.

The stratosphere above Antarctica has one of the world's highest ozone concentrations. Most of this ozone forms over the tropics and is brought to the Antarctic by stratospheric winds. During September and October (spring in the Southern Hemisphere), a belt of stratospheric winds called the *polar vortex* encircles the Antarctic region near 66°S latitude, essentially isolating the cold Antarctic stratospheric air from the warmer air of the middle latitudes.

FIGURE 1

Ozone distribution over the Southern Hemisphere on October 17, 1994, as measured by the Total Ozone Mapping Spectrometer (TOMS) aboard the Russian *Meteor-3* satellite. Notice that the area of lowest ozone concentration or ozone hole (purple shades) is larger than Antarctica.

FIGURE 2

Color image of total ozone amounts over the Northern Hemisphere for March 24, 1997. Notice that minimum ozone values (purple shades) appear over a region near the North Pole. The color scale on the bottom of the image shows total ozone values in Dobson units (DU). A Dobson unit is the physical thickness of the ozone layer if it were brought to the earth's surface (500 DU equals 5 mm).

During the long dark Antarctic winter, temperatures inside the vortex can drop to –85°C (–121°F). This frigid air allows for the formation of *polar stratospheric clouds*. These ice clouds are critical in facilitating chemical interactions among nitrogen, hydrogen, and chlorine atoms, the end product of which is the destruction of ozone.

In 1986, the NOZE-1 study detected unusually high levels of chlorine compounds in the stratosphere, and, in 1987, the instrumented aircraft of NOZE-2 measured enormous increases in chlorine monoxide (ClO) when it entered the polar vortex. These findings, in conjunction with other chemical discoveries, allowed scientists to pinpoint *chlorine* from CFCs as the main cause of the ozone hole.

Chemistry alone, however, does not explain the entire ozone depletion problem in Antarctica. It does not, for example, explain the early decline in ozone readings just as the ozone hole begins to form. Perhaps dynamic events play a role in this phenomenon, such as the mixing of stratospheric ozone-rich air with tropospheric ozone-poor air from below.

In the polar Arctic, airborne instruments and satellites during the late 1980s and 1990s measured high levels of ozone-destroying chlorine compounds in the stratosphere. By 1997, springtime ozone levels in the Arctic were the lowest ever observed in the region—about 40 percent below average (see Fig. 2). But observations could not detect an ozone hole like the one that forms over the Antarctic.

Apparently, several factors inhibit massive ozone loss in the Arctic. For one thing, in the stratosphere, the circulation of air over the Arctic differs from that over the Antarctic. Then, too, the Arctic stratosphere is normally too warm for the widespread development of clouds that help activate chlorine molecules. How-

The Ozone Hole (continued)

ever, it appears that a very cold Arctic stratosphere, along with ozone-destroying chemicals were responsible for the low readings in 1997.

Ozone depletion is not just confined to the stratospheric Arctic and Antarctic. For example, in 1995, satellite measurements over Siberia revealed a 35 percent reduction in wintertime ozone levels compared to 1979 values. And over the United States in March, 1995, ozone levels fell between 15 and 20 percent below the values observed during March, 1979.

We still have much to learn about stratospheric ozone and the processes that both form and destroy it. Presently, atmospheric studies are providing more information so that a more complete assessment of the ozone problem will become available in the future.

---

pollutants still spew into our air. In fact, many areas of the country do not conform to the standards for air quality set by the Clean Air Act of 1990. A large part of the problem of pollution control lies in the fact that even with stricter emission laws, increasing numbers of autos (estimates are that more than 198 million are on the road today) and other sources can overwhelm control efforts.

Clean air standards are established by the Environmental Protection Agency. *Primary ambient air quality standards* are set to protect human health, whereas *secondary standards* protect human welfare, as measured by the effects of air pollution on visibility, crops, and buildings. The National Ambient Air Quality Standards (NAAQS) currently in effect for six pollu-

tants are given in Table 17.2. Those areas that do not meet air quality standards are called *nonattainment areas*. Even with stronger emission laws, estimates are that, more than 80 million Americans are breathing air that does not meet at least one of the standards.

To indicate the air quality in a particular region, the EPA developed the **pollutant standards index (PSI)**. The index includes the pollutants carbon monoxide, sulfur dioxide, nitrogen dioxide, particulate matter, and ozone. On any given day, the pollutant measuring the highest value is the one used in the index. The pollutant's measurement is then converted to a number that ranges from 0 to 500 (see Table 17.3). When the pollutant's value is the same as the primary ambient air quality

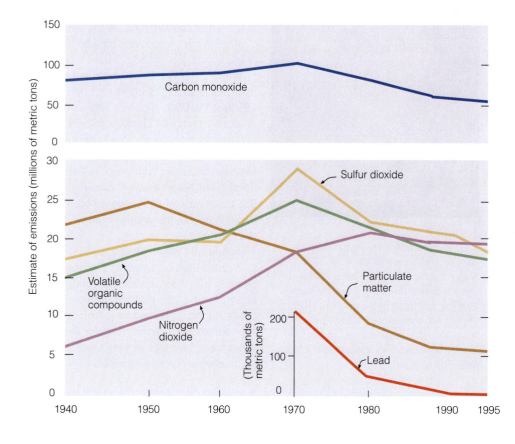

FIGURE 17.8

Emission estimates of six pollutants in the United States from 1940–1995. (Data courtesy of United States Environmental Protection Agency)

standard, the pollutant is assigned a PSI number of 100. A pollutant is considered unhealthful when its PSI value exceeds 100. Figure 17.9 shows the number of days when PSI values exceeded 100 at locations across the United States. When the PSI value is between 51 and 100, the air quality is described as "moderate." Although these levels may not be harmful to humans during a 24-hour period, they may exceed long-term standards. Table 17.3 also shows the health effects and the precautions that should be taken when the PSI value reaches a certain level.

Higher emission standards, along with cleaner fuels (such as natural gas), have made the air over our large cities cleaner today than it was years ago. In fact, total emissions of toxic chemicals spewed into the skies over the United States have been declining steadily since the EPA began its inventory of these chemicals in 1987. But the control of ozone in polluted

WEATHER WATCH

In Mexico City, Mexico, air pollution has become a severe problem. Clinics routinely fill up with patients suffering from pollution-related ailments. Residents are urged not to exercise outdoors. And during January, 1989, it became so intolerable that school children were given the entire month off.

air is still a pervasive problem. Because ozone is a secondary pollutant, its formation is controlled by the concentrations of other pollutants, namely nitrogen oxides and hydrocarbons (VOCs). Moreover, weather conditions play a vital role in ozone formation, as ozone reaches its highest concentrations

TABLE 17.2   The National Ambient Air Quality Standards

POLLUTANT	PRIMARY (HEALTH RELATED)		SECONDARY (WELFARE RELATED)	
	Type of Average	Standard Level Concentration[a]	Type of Average	Standard Level Concentrations
CO	8-hour[b]	9 ppm (10 $\mu g/m^3$)	No Secondary Standard	
	1-hour[b]	35 ppm (40 $\mu g/m^3$)	No Secondary Standard	
NO_2	Annual arithmetic mean	0.053 ppm (100 $\mu g/m^3$)	Same as Primary Standard	
O_3	Maximum daily 1-hour average[c]	0.12 ppm (235 $\mu g/m^3$)	Same as Primary Standard	
Pb	Maximum quarterly average	1.5 $\mu g/m^3$	Same as Primary Standard	
PM-10	Annual arithmetic mean[d]	50 $\mu g/m^3$	Same as Primary Standard	
	24-hour[d]	150 $\mu g/m^3$	Same as Primary Standard	
SO_2	Annual arithmetic mean	80 $\mu g/m^3$ (0.03 ppm)	3-hour[b]	1300 $\mu g/m^3$ (0.50 ppm)
	24-hour[b]	365 $\mu g/m^3$		

a Parenthetical value is an approximately equivalent concentration.

b Not to be exceeded more than once per year.

c The standard is attained when the expected number of days per calendar year with maximum hourly average concentrations above 0.12 ppm is equal to or less than 1.

d Particulate standards use PM-10 (particles less than 10 μm—or 10 micrometers—in diameter) as the indicator pollutant. The annual standard is attained when the expected annual arithmetic mean concentration is less than or equal to 50 $\mu g/m^3$; the 24-hour standard is attained when the expected number of days per calendar year above 150 $\mu g/m^3$ is equal to or less than 1.

TABLE 17.3   The Pollutant Standards Index

PSI VALUE	DESCRIPTION	GENERAL HEALTH EFFECTS	PSI EPISODE LEVEL	CAUTIONARY STATEMENTS
0–50	Good	None		
51–100	Moderate	None		
101–199	Unhealthful	Mild aggravation of symptoms in susceptible persons, with irritation symptoms in the healthy population.		Persons with existing heart or respiratory ailments should reduce physical exertion and outdoor activity.
200–299	Very unhealthful	Significant aggravation of symptoms and decreased exercise tolerance in persons with heart or lung disease, with widespread symptoms in the healthy population.	Stage 1 Health advisory alert	Elderly and persons with existing heart or lung disease should stay indoors and reduce physical activity.
300–399	Hazardous	Premature onset of certain diseases in addition to significant aggravation of symptoms and decreased exercise tolerance in healthy persons.	Stage 2 Health advisory warning	Elderly and persons with existing heart or lung disease should stay indoors and avoid physical exertion. General population should avoid outdoor activity.
400–500	Hazardous	Premature death of ill and elderly. Healthy people will experience adverse symptoms that affect their normal activity.	Stage 3 Emergency	All persons should remain indoors, keeping windows and doors closed. All persons should minimize physical exertion and avoid traffic.

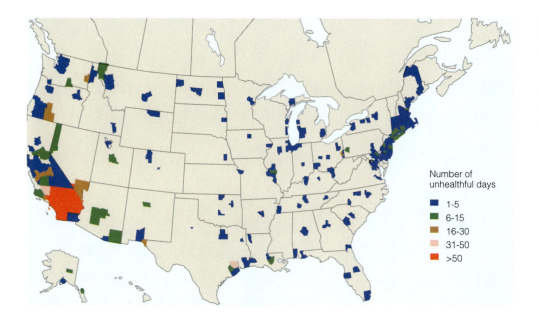

FIGURE 17.9

The number of unhealthful days (by county) across the United States when the PSI value for any one of the five pollutants (CO, SO_2, NO_2, O_3, and particulate matter) exceeded 100 during 1990. (Data courtesy of United States Environmental Protection Agency)

Number of unhealthful days

- 1-5
- 6-15
- 16-30
- 31-50
- >50

in hot sunny weather when surface winds are light and a stagnant high-pressure area covers the region. As a result of these factors, year-to-year ozone trends are quite variable (see Fig. 17.10). In 1991, the EPA designated 98 nonattainment areas for ozone across the United States. Of these areas, 43 were designated as marginal, 31 as moderate, 14 as serious, 9 as severe, and 1 (Los Angeles) as extreme. However, Los Angeles has shown a decline in the number of unhealthful days due to ozone. In 1980, Los Angeles had 100 stage 1 (very unhealthful) smog alert days, whereas in 1991, it had 47.

BRIEF REVIEW   Before going on to the next several sections, here is a brief review of some of the important points presented so far.

- Near the surface, air pollutants (such as particulate matter, CO, SO_2, NO, NO_2, and VOCs) enter the atmosphere directly, whereas secondary air pollutants (such as O_3) form when a chemical reaction takes place between a primary pollutant and some other component of air.

- The word "smog" (coined in London in the early 1900s) originally meant the combining of smoke and fog. Today, the word mainly refers to photochemical smog—pollutants that form in the presence of sunlight.

- Stratospheric ozone forms naturally in the stratosphere and provides a protective shield against the sun's harmful ultraviolet rays. Tropospheric ozone that forms in polluted air is a health hazard and is the primary ingredient of photochemical smog.

- Human-induced chemicals, such as chlorofluorocarbons (CFCs), appear to be altering the amount of ozone in the stratosphere by releasing chlorine, which rapidly destroys ozone.

Up to now we have concentrated on outdoor air pollution. The Focus section on p. 456 looks at pollutants inside the home.

FACTORS THAT AFFECT AIR POLLUTION

If you live in a region that periodically experiences smog, you may have noticed that these episodes often occur with clear skies, light winds, and generally warm sunny weather. Although this may be "typical" air pollution weather, it by no means represents the only weather conditions necessary to produce high concentrations of pollutants, as we will see in the following sections.

The Role of the Wind   The wind speed plays a role in diluting pollution. When vast quantities of pollutants are spewed

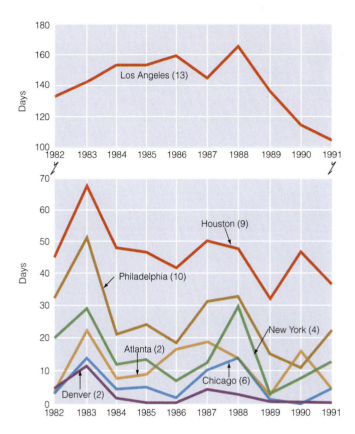

FIGURE 17.10
The number of days at selected cities when the PSI for ozone exceeded a value of 100 (unhealthful or worse) from 1982 to 1992. A hot, sunny summer created favorable weather conditions for ozone formation over the eastern United States during 1983 and again in 1988. The number next to each city represents the number of ozone monitoring sites. (Data courtesy of the United States Environmental Protection Agency.)

into the air, the wind speed determines how quickly the pollutants mix with the surrounding air and, of course, how fast they move away from their source. Strong winds tend to lower the concentration of pollutants by spreading them apart as they move downstream. Moreover, the stronger the wind, the more turbulent the air. Turbulent air produces swirling eddies that dilute the pollutants by mixing them with the cleaner surrounding air. Hence, when the wind dies down, pollutants are not readily dispersed, and they tend to become more concentrated (see Fig. 17.11).

The Role of Stability and Inversions   Recall from Chapter 7 that atmospheric stability determines the extent to which air will rise. Remember also that an unstable atmosphere favors vertical air currents, whereas a stable atmosphere strongly resists upward vertical motions. Consequently, smoke emitted into a stable atmosphere tends to spread horizontally, rather than mix vertically.

FOCUS ON A SPECIAL TOPIC

Indoor Air Pollution

When people think of air pollution, most think of outside air where automobiles, factories, and power plants spew countless tons of contaminants into the air. But, surprisingly, the air we breathe inside our homes and other structures can be between 5 and 100 times more polluted than the air we breathe outdoors, even in the largest, most industrialized, cities! Since we spend so much time indoors, the quality of air we breathe there can have a far greater impact on our health and well-being than the air outdoors.

The Environmental Protection Agency (EPA) has identified many sources of indoor air pollution, ranging from building materials, pressed wood products, furnishings, and home cleaning products, to pesticides, adhesives, and personal care products. In addition, heating sources (such as unvented kerosene heaters, wood stoves, and fireplaces) can release a variety of pollutants into a home. The pollution impact of any heating source depends upon several factors, such as how old the source is, the level of maintenance it receives, as well as its location and access to ventilation. For

example, a gas cooking stove or heating stove with improper fittings and adjustments can cause a significant emission of carbon monoxide. New carpets and padding (as well as the adhesives used in their installation) emit volatile organic compounds.

Some pollution sources, such as building materials and foam insulation, produce a constant stream of pollutants, whereas activities like tobacco smoking emit pollutants into the air on an intermittent basis. In certain instances, outside pollution is brought indoors. Some pollutants enter homes and other structures through cracks and holes in foundations and basements, as is the case with radon.

Radon is a colorless, odorless gas— a natural radioactive compound—that forms as the uranium in soil and rock breaks down. Radon is found everywhere on earth and only becomes a problem when it leaks out of the soil and becomes trapped inside homes and buildings. The radon gas that seeps in through cracks and other openings in a building can accumulate to levels that create a serious health threat. Studies

by the EPA have shown that as many as 10 percent of the homes (about 8 million) in the United States may have elevated levels of radon. The level of radon varies greatly from structure to structure and can only be measured by devices known as *radon detectors*. Inside a home, the radon decays into *polonium*, a solid substance that attaches itself to dust in the air. The tiny dust particles can be inhaled deep into the lungs, where they attach to lung tissue. As the polonium decays, it damages the lung tissue, sometimes producing mutated cells that may develop into lung cancer.

Another major chemical pollutant found inside our homes is *formaldehyde*. It is a colorless, pungent-smelling gas, used widely to manufacture building materials, insulation, and other household products. In most homes, the significant sources of formaldehyde are urethane-formaldehyde foam insulation and pressed wood products, such as particle board and plywood paneling. Emissions of formaldehyde from new products can be greatly affected by indoor temperatures and ventilation.

The stability of the atmosphere is determined by the way the air temperature changes with height (the lapse rate). When the measured air temperature decreases rapidly as we move up into the atmosphere, the atmosphere is unstable. If, however, the measured air temperature either decreases quite slowly as we ascend, or actually increases with height (remember that this is called an inversion), the atmosphere is stable. An inversion represents an extremely stable atmosphere where warm

WIND SPEED
5 meters/sec

(a)

WIND SPEED
10 meters/sec

(b)

FIGURE 17.11

If each chimney emits a puff of smoke every second, then where the wind speed is low (a), the smoke puffs are closer together and more concentrated. Where the wind speed is greater (b), the smoke puffs are farther apart and more diluted as turbulent eddies mix the smoke with the surrounding air.

Exposure to formaldehyde can cause watery eyes, burning sensations in the nose and throat, breathing difficulties, and nausea. It can also trigger attacks in individuals suffering from asthma.

Another polluter of our indoor air is *asbestos*, a mineral fiber once used in insulation and as a fire-retardant. In recent years, manufacturers have voluntarily reduced the use of asbestos, but much asbestos still remains in furnace and pipe insulation, texturing materials, and floor tiles of older buildings. The most lethal fibers of asbestos are invisible. When inhaled, these tiny particles can accumulate and remain deep in the lungs for extended periods of time, where they damage tissue and potentially cause cancer or *asbestosis*, a permanent scarring of the lung that can be fatal.

Smoking tobacco indoors can also create an extremely dangerous health situation. Environmental tobacco smoke is a complex mixture of more than 4700 different compounds. These pollutants enter the body as particles and as gases, such as carbon monoxide and hydrogen cyanide. Exposure to tobacco smoke greatly increases the risk of developing lung cancer in both smokers and nonsmokers (studies show that the nonsmoking spouses of smokers experience a 30 percent increase in the occurrence of lung cancer). The small children of smokers are also more likely to fall victim to such illnesses as bronchitis and pneumonia. Heart disease is closely linked with exposure to tobacco smoke, as is premature aging of the skin.

The amount of ventilation an indoor area receives plays a large part in how highly concentrated pollutants can become. Outside air enters the home in several ways. Air flows in through cracks around windows and doors, from crawl spaces and basements. Mechanical heating and cooling systems and outdoor-vented fans help to mix indoor and outdoor air. Simply opening doors and windows provides some exchange of air. But in our attempts to conserve energy, many of these possibilities for ventilation are cut off. Cracks are caulked, weather-stripping is applied to doors and windows, and modern heating and cooling systems recirculate indoor air, thus trapping pollutants in our indoor environment. Even homes and buildings that have not undergone these measures can have pollutants build to high levels.

In recent years, a phenomenon known as *"sick building syndrome"* has become the focus of public concern. Several illnesses, such as *Legionnaire's disease* (which is spread through contaminated air conditioning systems), have been attributed to specific building problems, but other, less definite, illnesses have been traced to the indoor office environment as well. These complaints range from dry mucous membranes, sneezing, fatigue, and irritability to forgetfulness and nausea. Unfortunately, the nature of these symptoms and their random occurrence makes tracing the source difficult, if not impossible.

Indoor pollutants are responsible for a wide variety of health problems. Irritation of the eyes, nose, and throat, headaches, fatigue, and dizziness are but a few of the maladies attributed to indoor air pollutants. While these symptoms can be annoying, other life-threatening diseases can occur after prolonged exposure to many of the substances mentioned earlier, such as radon, asbestos, and other toxic compounds.

air lies above cool air. Any air parcel that attempts to rise into the inversion will, at some point, be cooler and heavier (more dense) than the warmer air surrounding it. Hence, the inversion acts like a lid on vertical air motions.

One type of very stable atmosphere usually exists during the night and early morning hours. If the sky is clear and the winds are light, the air near the ground can become much cooler than the air higher up. Recall that this situation is called a *radiation* (or *surface*) *inversion* (see Chapter 3).

Figure 17.12 shows a strong radiation inversion on a clear calm winter night. Notice that within the stable inversion, the smoke from the shorter stack does not rise very high, but spreads out, contaminating the area around it. In the relatively unstable air above the inversion, smoke from the taller stack is able to rise and become dispersed. Since radiation inversions are often rather shallow, it should be apparent why taller chimneys have replaced many of the shorter ones. In fact, taller chimneys disperse pollutants better than shorter ones even in the absence of a surface inversion because the taller chimneys are able to mix pollutants throughout a greater volume of air. Although these taller stacks do improve the air quality in their immediate area, they may also contribute to the acid rain problem by allowing the pollutants to be swept great distances downwind.

As the sun rises and the surface warms, the radiation inversion normally weakens and disappears before noon. By afternoon, the atmosphere is sufficiently unstable so that, with adequate winds, pollutants are able to disperse vertically. The changing atmospheric stability, from stable in the early morning to conditionally unstable in the afternoon, can have a profound effect on the daily concentrations of pollution in certain regions. For example, on a busy city street corner, carbon monoxide levels can be considerably higher in the early morning than in the early afternoon (with the same flow of traffic). Changes in atmospheric stability can also cause smoke plumes

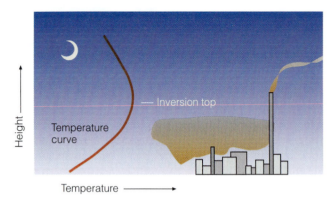

FIGURE 17.12
The smoke from the shorter stack is trapped within the inversion, while the smoke from the taller stack, above the inversion, rises, mixes, and disperses downwind.

from chimneys to change during the course of a day. (Some of these changes are described in the Focus section on p. 459.)

Radiation inversions normally last just a few hours, while *subsidence inversions* may persist for several days or longer. Subsidence inversions, therefore, are the ones commonly associated with major air pollution episodes. They form as the air above a deep anticyclone slowly sinks (subsides) and warms.*

A typical temperature profile of a subsidence inversion is shown in Fig. 17.13. Notice that in the relatively unstable air beneath the inversion, the pollutants are able to mix vertically up

---
*Remember from Chapter 2 that sinking air always warms because it is being compressed by the surrounding air.

to the inversion base. The stable air of the inversion, however, inhibits vertical mixing and acts like a lid on the pollution below, preventing it from entering into the inversion (see Fig. 17.14).

In Fig. 17.13, the region of relatively unstable (well mixed) air that extends from the surface to the base of the inversion is referred to as the **mixing layer**. The vertical extent of the mixing layer is called the **mixing depth**. Observe that if the inversion rises, the mixing depth increases and the pollutants would be dispersed throughout a greater volume of air; if the inversion lowers, the mixing depth would decrease and the pollutants would become more concentrated, sometimes reaching unhealthy levels. Since the atmosphere tends to be most unstable in the afternoon and most stable in the early morning, we typically find the greatest mixing depth in the afternoon and the most shallow one (if one exists at all) in the early morning.

The position of the semipermanent Pacific high off the coast of California contributes greatly to the air pollution in that region. The Pacific high promotes subsiding air, which warms the air aloft. Surface winds around the high promote upwelling of ocean water (see Chapter 11). Upwelling makes the surface water cool, which, in turn, cools the air above. Warm air aloft coupled with cool, surface (marine) air, together produce a strong and persistent subsidence inversion—one that exists 80 to 90 percent of the time over the city of Los Angeles between June and October, the smoggy months. The pollutants trapped within the cool marine air are occasionally swept eastward by a sea breeze. This action carries smog from the coastal regions into the interior valleys (see Fig. 17.15).

The Role of Topography   The shape of the landscape (topography) plays an important part in trapping pollutants. We

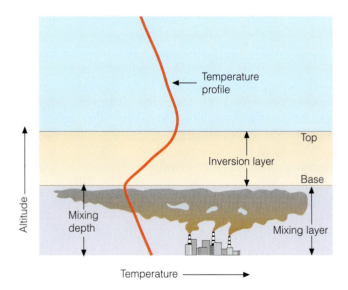

FIGURE 17.13
The inversion layer prevents pollutants from escaping into the air above it. If the inversion lowers, the mixing depth decreases and the pollutants are concentrated within a smaller volume.

FIGURE 17.14
A thick layer of polluted air is trapped in the valley. The top of the polluted air marks the base of a subsidence inversion.

Smoke Stack Plumes

We know that the stability of the air (especially near the surface) changes during the course of a day. These changes can influence the pollution near the ground as well as the behavior of smoke leaving a chimney. Figure 3 illustrates different smoke plumes that can develop with adequate wind, but different types of stability.

In Fig. 3, diagram a, it is early morning, the winds are light, and a radiation inversion extends from the surface to well above the height of the smoke stack. In this stable environment, there is little up and down motion, so the smoke spreads horizontally rather than vertically. When viewed from above, the smoke plume resembles the shape of a fan. For this reason, it is referred to as a *fanning smoke plume.*

Later in the morning, the surface air warms quickly and destabilizes as the radiation inversion gradually disappears from the surface upward (Fig. 3, diagram b). However, the air above the chimney is still stable, as indicated by the presence of the inversion. Consequently, vertical motions are confined to the region near the surface. Hence, the smoke mixes downwind, increasing the concentration of pollution at the surface—sometimes to dangerously high levels. This effect is called *fumigation.* Here again, we can see why a taller smokestack is preferred. A taller stack extends upward into the stable layer, producing a fanning plume that does not mix downward toward the ground.

If daytime heating of the ground continues, the depth of atmospheric instability increases. Notice in Fig.

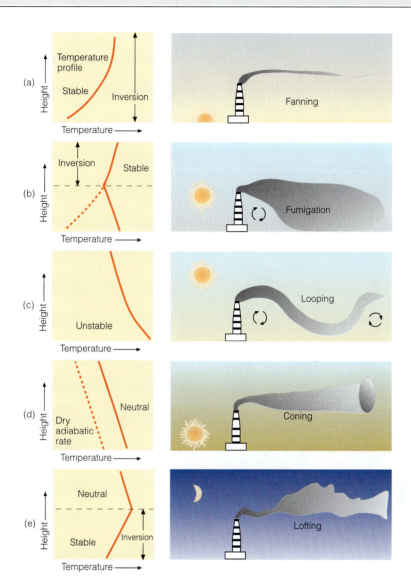

FIGURE 3

As the vertical temperature profile changes during the course of a day (a through e), the pattern of smoke emitted from the stack changes as well.

3, diagram c, that the inversion has completely disappeared. Light-to-moderate winds combine with rising and sinking air to cause the smoke to move up and down in a wavy pattern, producing a *looping smoke plume.*

The continued rising of warm air and sinking of cool air can cause the temperature profile to equal that of the dry adiabatic rate (Fig. 3, diagram d). In this neutral atmosphere, vertical and horizontal motions are

(continued on p. 460)

Smoke Stack Plumes (continued)

about equal, and the smoke from the stack tends to take on the shape of a cone, forming a *coning smoke plume.*

After sunset, the ground cools rapidly and the radiation inversion reappears. When the top of the inversion extends upward to slightly above the stack, stable air is near the ground with neutral air above (Fig. 3, diagram e). Because the stable air in the inversion prevents the smoke from mixing downward, the smoke is carried upward, producing a *lofting smoke plume.* Thus, smoke plumes provide a clue to the stability of the atmosphere, and knowing the stability yields important information about the dispersion of pollutants.

Of course, other factors influence the dispersion of pollutants from a chimney, including the pollutants' temperature and exit velocity, wind speed and direction, and, of course, the chimney's height. Overall, taller chimneys, greater wind speeds, and higher exit velocities result in a lower concentration of pollutants.

know from Chapter 3 that, at night, cold air tends to drain downhill, where it settles into low-lying basins and valleys. The cold air can have several effects: It can strengthen a pre-existing surface inversion, and it can carry pollutants downhill from the surrounding hillsides (see Fig. 17.16).

Valleys prone to pollution are those completely encased by mountains and hills. The surrounding mountains tend to block the prevailing wind. With light winds, and a shallow mixing layer, the poorly ventilated cold valley air can only slosh back and forth like a murky bowl of soup.

Air pollution concentrations in mountain valleys tend to be greatest during the colder months. During the warmer months, daytime heating can warm the sides of the valley to the point that upslope valley winds vent the pollutants upward,

FIGURE 17.15
The leading edge of cool, marine air carries pollutants into Riverside, California.

like a chimney. Valleys susceptible to stagnant air exist in just about all mountainous regions.

The pollution problem in several large cities is, at least, partly due to topography. For example, the city of Los Angeles is surrounded on three sides by hills and mountains. Cool marine air from off the ocean moves inland and pushes against the hills, which tend to block the air's eastward progress. Unable to rise, the cool air settles in the basin, trapping pollutants from industry and millions of autos. Baked by sunlight, the pollutants become the infamous photochemical smog. By the same token, the "mile high" city of Denver, Colorado, sits in a broad shallow basin that frequently traps both cold air and pollutants.

Severe Air Pollution Potential

The greatest potential for an episode of severe air pollution occurs when all of the factors mentioned in the previous sections come together simultaneously. Ingredients for a major buildup of atmospheric pollution are:

- many sources of air pollution (preferably clustered close together)
- a deep high-pressure area that becomes stationary over a region
- light surface winds that are unable to disperse the pollutants
- a strong subsidence inversion produced by the sinking of air aloft
- a shallow mixing layer with poor ventilation
- a valley where the pollutants can accumulate
- clear skies so that radiational cooling at night will produce a surface inversion, which can cause an even greater buildup of pollutants near the ground
- and, for photochemical smog, adequate sunlight to produce secondary pollutants, such as ozone

Light winds and poor vertical mixing can produce a condition known as **atmospheric stagnation**. When this condition prevails for several days to a week or more, the buildup of pollutants can lead to some of the worst air pollution disasters on record, such as the one in the valley city of Donora, Pennsylvania, where in 1948 seventeen people died within fourteen hours. (Additional information on the Donora disaster is found in the Focus section on p. 462.)

AIR POLLUTION AND THE URBAN ENVIRONMENT

For more than 100 years, it has been known that cities are generally warmer than surrounding rural areas. This region of city warmth, known as the **urban heat island**, can influence the concentration of air pollution. However, before we look at its influence, let's see how the heat island actually forms.

The urban heat island is due to industrial and urban development. In rural areas, a large part of the incoming solar

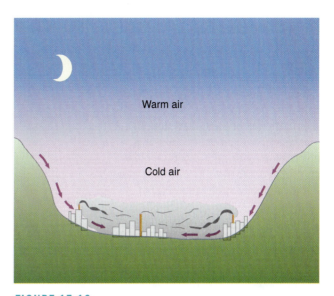

FIGURE 17.16
At night, cold air and pollutants drain downhill and settle in low-lying valleys.

energy is used to evaporate water from vegetation and soil. In cities, where less vegetation and exposed soil exists, the majority of the sun's energy is absorbed by urban structures and asphalt. Hence, during warm daylight hours, less evaporative cooling in cities allows surface temperatures to rise higher than in rural areas.*

At night, the solar energy (stored as vast quantities of heat in city buildings and roads) is slowly released into the city air. Additional city heat is given off at night (and during the day) by vehicles and factories, as well as by industrial and domestic heating and cooling units. The release of heat energy is retarded by the tall vertical city walls that do not allow infrared radiation to escape as readily as do the relatively level surfaces of the surrounding countryside. The slow release of heat tends to keep nighttime city temperatures higher than those of the faster cooling rural areas. Overall, the heat island is strongest (1) at night when compensating sunlight is absent, (2) during the winter when nights are longer and there is more heat generated in the city, and (3) when the region is dominated by a high-pressure area with light winds, clear skies, and less humid air. Over time, increasing urban heat islands affect climatological temperature records, producing artificial warming in climatic records taken in cities. As we will see in Chapter 19, this warming must be accounted for in interpreting climate change over the past century.

The constant outpouring of pollutants into the environment may influence the climate of a city. Certain particles

---

*The cause of the urban heat island is quite involved. Depending on the location, time of year, and time of day, any or all of the following differences between cities and their surroundings can be important: albedo (reflectivity of the surface), surface roughness, emissions of heat, emissions of moisture, and emissions of particles that affect net radiation and the growth of cloud droplets.

Five Days in Donora—An Air Pollution Episode

On Tuesday morning, October 26, 1948, a cold surface anticyclone moved over the eastern half of the United States. There was nothing unusual about this high-pressure area; with a central pressure of only 1025 mb (30.27 in.), it was not exceptionally strong (see Fig. 4). Aloft, however, a large blocking-type ridge formed over the region, and the jet stream, which moves the surface pressure features along, was far to the west. Consequently, the surface anticyclone became entrenched over Pennsylvania and remained nearly stationary for five days.

The widely spaced isobars around the high-pressure system produced a weak pressure gradient and generally light winds throughout the area. These light winds, coupled with the gradual sinking of air from aloft, set the stage for a disastrous air pollution episode.

On Tuesday morning, radiation fog gradually settled over the moist ground in Donora, a small town nestled in the Monongahela Valley of western Pennsylvania. Because Donora rests on bottom land, surrounded by rolling hills, its residents were accustomed to fog, but not to what was to follow.

The strong radiational cooling that formed the fog, along with the sinking air of the anticyclone, combined to produce a strong temperature inversion. Light, downslope winds spread cool air and contaminants over Donora from the community's steel mill, zinc smelter, and sulfuric acid plant.

The fog with its burden of pollutants lingered into Wednesday. Cool drainage winds during the night strengthened the inversion and added more effluents to the already filthy air. The dense fog layer blocked sunlight from reaching the ground. With essentially no surface heating, the mixing depth lowered and the pollution

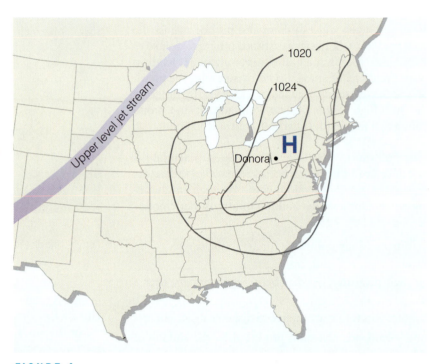

FIGURE 4

Surface weather map that shows a stagnant anticyclone over the eastern United States on October 26, 1948. The heavy arrow represents the position of the jet stream.

became more concentrated. Unable to mix and disperse both horizontally and vertically, the dirty air became confined to a shallow, stagnant layer.

Meanwhile, the factories continued to belch impurities into the air (primarily sulfur dioxide and particulate matter) from stacks no higher than 40 m tall. The fog gradually thickened into a moist clot of smoke and water droplets. By Thursday, the visibility had decreased to the point where one could barely see across the street. At the same time, the air had a penetrating, almost sickening, smell of sulfur dioxide. At this point, a large percentage of the population became ill.

The episode reached a climax on Saturday, as seventeen deaths were reported. As the death rate mounted, alarm swept through the town. An

emergency meeting was called between city officials and factory representatives to see what could be done to cut down on the emission of pollutants.

The light winds and unbreathable air persisted until, on Sunday, an approaching storm generated enough wind to vertically mix the air and disperse the pollutants. A welcome rain then cleaned the air further. All told, the episode had claimed the lives of 22 people. During the five-day period, about half of the area's 14,000 inhabitants experienced some ill effects from the pollution. Most of those affected were older people with a history of cardiac or respiratory disorders.

reflect solar radiation, thereby reducing the sunlight that reaches the surface. Some particles serve as nuclei upon which water and ice form. Water vapor condenses onto these particles when the relative humidity is as low as 70 percent, forming haze that greatly reduces visibility. Moreover, the added nuclei increase the frequency of city fog.

Studies suggest that precipitation may be greater in cities than in the surrounding countryside. This may be due in part to the increased roughness of city terrain, brought on by large structures that cause surface air to slow and gradually converge. This piling-up of air over the city then slowly rises, much like toothpaste does when its tube is squeezed. At the same time, city heat warms the surface air, making it more unstable. This enhances rising air motions, which aid in forming clouds and thunderstorms. This helps explain why both tend to be more frequent over cities. Table 17.4 summarizes the environmental influence of cities by contrasting the urban environment with the rural.*

On clear still nights when the heat island is pronounced, a small thermal low-pressure area forms over the city. Sometimes a light breeze—called a **country breeze**—blows from the countryside into the city. If there are major industrial areas along the city's outskirts, pollutants are carried into the heart of town, where they tend to concentrate. This is especially true if an inversion inhibits vertical mixing and dispersion (see Fig. 17.17).

Pollutants from urban areas may even affect the weather downwind from them. In a controversial study conducted at La Porte, Indiana—a city located about 50 km downwind of the industries of south Chicago—scientists suggested that La Porte had experienced a notable increase in annual precipitation since 1925. Because this rise closely followed the increase in steel production, it was suggested that the phenomenon was due to the additional emission of particles or moisture (or both) by industries to the west of La Porte.

A study conducted in St. Louis, Missouri (the Metropolitan Meteorological Experiment, or *METROMEX*), indicated that the average annual precipitation downwind from this city increased by about 10 percent. These increases closely followed industrial development upwind. This study also demonstrated that precipitation amounts were significantly greater on weekdays (when pollution emissions were higher) than on weekends (when pollution emissions were lower). Corroborative findings have been reported for Paris, France, and for other cities as well. However, in areas with marginal humidity to support the formation of clouds and precipitation, studies suggest that the rate of precipitation may actually decrease as excess pollutant particles (nuclei) compete for the available moisture, similar to the effect of overseeding a cloud, discussed in Chapter 8.

TABLE 17.4 Contrast of the Urban and Rural Environment (Average Conditions)*	
CONSTITUENTS	URBAN AREA (CONTRASTED TO RURAL AREA)
Mean pollution level	higher
Mean sunshine reaching the surface	lower
Mean temperature	higher
Mean relative humidity	lower
Mean visibility	lower
Mean wind speed	lower
Mean precipitation	higher
Mean amount of cloudiness	higher
Mean thunderstorm (frequency)	higher

*Values are omitted because they vary greatly depending upon city, size, type of industry, and season of the year.

ACID DEPOSITION

Air pollution emitted from industrial areas, especially products of combustion, such as oxides of sulfur and nitrogen, can be carried many kilometers downwind. Either these particles and gases slowly settle to the ground in dry form (*dry deposition*) or they are removed from the air during the formation of cloud particles and then carried to the ground in rain and snow (*wet deposition*). **Acid rain** and *acid precipitation* are common terms used to describe wet deposition, while **acid deposition** encompasses both dry and wet acidic substances. How, then, do these substances become acidic?

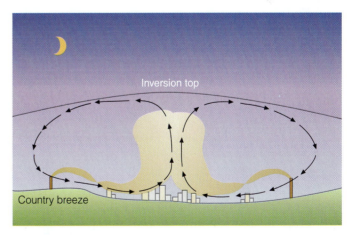

FIGURE 17.17
On a clear, relatively calm night, a weak country breeze carries pollutants from the outskirts into the city, where they concentrate and rise due to the warmth of the city's urban heat island. This effect may produce a pollution (or dust) dome from the suburb to the center of town.

---

*The impact that tiny liquid and solid particles, aerosols, may have on a larger scale is complex and depends upon a number of factors, which are addressed in Chapter 19.

WEATHER WATCH

Cubatao, Brazil, just may be the most polluted city in the world. Located south of São Paulo, this heavily industrialized area of 100,000 people lies in a coastal valley—known by local residents as "the valley of death." Temperature inversions and stagnant air combine to trap the many pollutants that spew daily into the environment. Recently, nearly one-third of the downtown residents suffered from respiratory disease, and more babies are born deformed there than anywhere else in South America.

Emissions of sulfur dioxide (SO_2) and oxides of nitrogen may settle on the local landscape, where they transform into acids as they interact with water, especially during the formation of dew or frost. The remaining airborne particles may transform into tiny dilute drops of sulfuric acid (H_2SO_4) and nitric acid (HNO_3) during a complex series of chemical reactions involving sunlight, water vapor, and other gases. These acid particles may then fall slowly to earth, or they may adhere to cloud droplets or to fog droplets, producing **acid fog**. They may even act as nuclei on which the cloud droplets begin to grow. When precipitation occurs in the cloud, it carries the acids to the ground. Because of this, precipitation is becoming increasingly acidic in many parts of the world, especially downwind of major industrial areas.

Airborne studies conducted during the middle 1980s revealed that high concentrations of pollutants that produce acid rain can be carried great distances from their sources. For example, scientists in one study discovered high concentrations of pollutants about 600 km off the east coast of North America. It is suspected that they came from industrial East Coast cities. Although most pollutants are washed from the atmosphere during storms, some may be swept over the Atlantic, reaching places like Bermuda and Ireland. Acid rain knows no national boundaries.

Although studies suggest that acid precipitation may be nearly worldwide in distribution, regions noticeably affected are eastern North America, central Europe, and Scandinavia. Sweden contends that most of the sulfur emissions responsible for its acid precipitation are coming from factories in England. In some places, acid precipitation occurs naturally, such as in northern Canada, where natural fires in exposed coal beds produce tremendous quantities of sulfur dioxide. By the same token, acid fog can form by natural means (see Fig. 17.18).

Precipitation is naturally somewhat acidic. The carbon dioxide occurring naturally in the air dissolves in precipitation, making it slightly acidic with a pH between 5.0 and 5.6. Consequently, precipitation is considered acidic when its pH is below about 5.0 (see Fig. 17.19). In the northeastern United States, where emissions of sulfur dioxide are primarily responsible for the acid precipitation, typical pH values range between 4.0 and 4.5 (see Fig. 17.20). But acid precipitation is not confined to the Northeast; the acidity of precipitation has increased rapidly during the past 20 years in the southeastern states, too. Further west, rainfall acidity also appears to be on

FIGURE 17.18

Acid fog produced by natural means. Here, acid fog, with an acidity comparable to orange juice (pH 3.9), drifts over Kruzof Island in southeast Alaska. The source of the acidity is thought to be the oxidation of natural hydrocarbons (VOCs) and sulfide gases emitted by forests, bogs, and marine ecosystems.

the increase. Along the West Coast, the main cause of acid deposition appears to be the oxides of nitrogen released in automobile exhaust. In Los Angeles, acid fog is a more serious problem than acid rain, especially along the coast, where fog is most prevalent. The fog's pH is usually between 4.4 and 4.8, although pH values of 3.0 and below have been measured.

High concentrations of acid deposition can damage plants and water resources (freshwater ecosystems seem to be particularly sensitive to changes in acidity). Concern centers chiefly on areas where interactions with alkaline soil are unable to neutralize the acidic inputs. Studies indicate that thousands of lakes in the United States and Canada are so acidified that entire fish populations may have been adversely affected. In an attempt to reduce acidity, lime(calcium carbonate, $CaCO_3$) is being poured into some lakes. Natural alkaline soil particles can be swept into the air where they neutralize the acid. If it were not for airborne alkaline dust, China and the western United States would have a more severe acid rain problem than they do.

About a third of the trees in Germany show signs of a blight that is due, in part, to acid deposition. Apparently, acidic particles raining down on the forest floor for decades have caused a chemical imbalance in the soil that, in turn, causes serious deficiencies in certain elements necessary for the trees' growth. The trees are thus weakened and become susceptible to insects and drought. The same type of processes may be affecting North American forests, but at a much slower pace, as many forests at higher elevations from New England to South Carolina appear to be in serious decline (see Fig. 17.21). Moreover, acid precipitation is a problem in the mountainous West

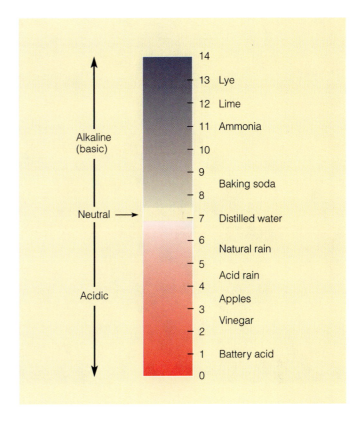

FIGURE 17.19

The pH scale ranges from 0 to 14, with a value of 7 considered neutral. Values greater than 7 are alkaline and below 7 are acidic. The scale is logarithmic, which means that rain with pH 3 is 10 times more acidic than rain with pH 4 and 100 times more acidic than rain with pH 5.

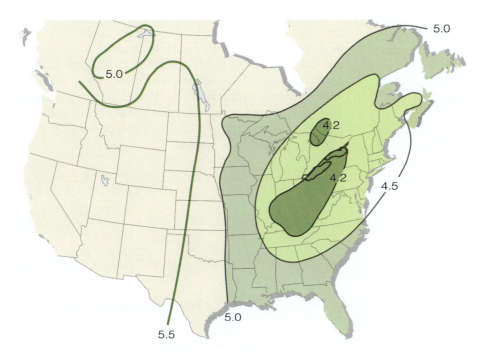

FIGURE 17.20

Annual average value of pH in precipitation weighted by the amount of precipitation in the United States and Canada for 1980.

FIGURE 17.21
The effect of acid rain on trees at Mt. Mitchell, North Carolina.

where high mountain lakes and forests seem to be most affected.

Also, acid deposition is eroding the foundations of structures in many cities throughout the world. In Rome, the acidity of rainfall is beginning to disfigure priceless outdoor fountain sculptures and statues. The estimated annual cost of this damage to building surfaces, monuments, and other structures is more than $2 billion.

Control of acid deposition is a difficult political problem because those affected by acid rain can be quite distant from those who cause it. Technology can control sulfur emissions (for example, stack scrubbers and fluidized bed combustion) and nitrogen emissions (catalytic converters on cars), but some people argue the cost is too high. If the United States turns

more to coal-fired power plants, which are among the leading sources of sulfur oxide emissions, many scientists believe that the acid deposition problem will become more acute.

In an attempt to better understand acid deposition, the National Center for Atmospheric Research (NCAR) and the Environmental Protection Agency have been working to develop computer models that better describe the many physical and chemical processes contributing to acid deposition. To deal with the acid deposition problem, the Clean Air Act of 1990 imposed a reduction in the United States' emissions of sulfur dioxide and nitrogen dioxide. Canada has imposed new pollution control standards and set a goal of reducing industrial air pollution by 50 percent.

SUMMARY

In this chapter, we found that air pollution has plagued humanity for centuries. Air pollution problems began when people tried to keep warm by burning wood and coal. These problems worsened during the industrial revolution as coal became the primary fuel for both homes and industry. Even though many American cities do not meet all of the air quality standards set by the federal Clean Air Act of 1990, the air over our large cities is cleaner today than it was 50 years ago due to stricter emission standards and cleaner fuels.

We examined the types and sources of air pollution and found that primary air pollutants enter the atmosphere directly, whereas secondary pollutants form by chemical reactions that involve other pollutants. The secondary pollutant ozone is the main ingredient of photochemical smog—a smog that irritates the eyes and forms in the presence of sunlight. In polluted air, ozone forms during a series of chemical reactions involving nitrogen oxides and hydrocarbons (VOCs). In the stratosphere, ozone is a naturally occurring gas that protects us from the sun's harmful ultraviolet rays. We examined how ozone forms in the stratosphere and how it may be altered by natural means. We also learned that human-induced gases, such as chlorofluorocarbons, work their way into the stratosphere where they release chlorine that rapidly destroys ozone, especially in polar regions.

We looked at the pollutant standards index and found that a number of areas across the United States still have days considered unhealthy by the standards set by the United States Environmental Protection Agency. We also looked at the main factors affecting air pollution and found that most air pollution episodes occur when the winds are light, skies are clear, the mixing layer is shallow, the atmosphere is stable, and a strong inversion exists. These conditions usually prevail when a high-pressure area stalls over a region.

We observed that, on the average, urban environments tend to be warmer and more polluted than the rural areas that surround them. We saw that pollution from industrial areas can modify environments downwind of them, as oxides of sulfur and nitrogen are swept into the air, where they may transform into acids that fall to the surface. Acid deposition, a serious problem in many regions of the world, knows no national boundaries—the pollution of one country becomes the acid rain of another.

Key Terms

The following terms are listed in the order they appear in the text. Define each. Doing so will aid you in reviewing the material covered in this chapter.

air pollutants
primary air pollutants
secondary air pollutants
particulate matter
carbon monoxide (CO)
sulfur dioxide (SO_2)
volatile organic
    compounds (VOCs)
hydrocarbons
nitrogen dioxide (NO_2)
nitric oxide (NO)
smog
photochemical smog

ozone (O_3)
ozone hole
pollutant standards
    index (PSI)
mixing layer
mixing depth
atmospheric stagnation
urban heat island
country breeze
acid rain
acid deposition
acid fog

Questions for Review

1. What are some of the main sources of air pollution?
2. List a few of the substances that fall under the category of particulate matter.
3. Why does the particulate matter referred to as PM-10 pose the greatest risk to human health?
4. List two ways particulate matter is removed from the atmosphere.
5. Describe the primary sources and some of the health problems associated with each of the following pollutants:
    (a) carbon monoxide (CO)
    (b) sulfur dioxide (SO_2)
    (c) volatile organic compounds (VOCs)
    (d) nitrogen oxides
6. How does London-type smog differ from Los Angeles-type smog?
7. What is the main component of photochemical smog? What are some of the adverse health effects of photochemical smog? Why is the production of photochemical smog more prevalent during the summer and early fall than during the middle of winter?
8. In polluted air, (a) describe the role that NO_2 plays in the production of tropospheric ozone, and (b) the role that NO plays in the destruction of tropospheric ozone. What role do hydrocarbons (VOCs) play in the production of ozone and other photochemical oxidants?
9. Describe the main processes that account for stratospheric ozone production and destruction. What natural and human-produced substances could alter the concentration of ozone in the stratosphere?

10. (a) How are CFCs related to the destruction of stratospheric ozone?
    (b) If all of the ozone in the stratosphere were destroyed, what possible effects might this have on the earth's inhabitants?
11. Explain how scientists believe the Antarctic "ozone hole" forms.
12. Describe some of the sources of pollution found inside a home or building.
13. Why are high levels of radon inside a home so dangerous?
14. What are primary ambient air quality standards and secondary standards intended to do? What are nonattainment areas?
15. Describe the role that the wind plays in diluting pollutants.
16. How is an accumulation of pollutants related to the stability of the atmosphere?
17. Why is it that polluted air and inversions seem to go hand in hand?
18. Give several reasons why taller smoke stacks are better than shorter ones at improving the air quality in their immediate area.
19. Explain how the mixing depth normally changes during the course of a day. As the mixing depth changes, how does it affect the concentration of pollution near the surface?
20. How does topography influence the concentration of pollutants in cities such as Los Angeles and Denver? In mountainous terrain?
21. List the factors that can lead to a major buildup of atmospheric pollution.
22. Discuss the main causes of the urban heat island.
23. What causes the "country breeze"? Why is it usually more developed at night than during the day? Would it be more easily developed in summer or winter? Explain.
24. How can pollution play a role in influencing the precipitation downwind of certain large industrial complexes?
25. Why is acid deposition considered a serious problem in many regions of the world? How does precipitation become acidic?

Questions for Thought

1. (a) Suppose clouds of nitrogen dioxide drift slowly from a major industrial complex over a relatively unpopulated area. If the area is essentially "free" of hydrocarbons, would you expect high levels of tropospheric ozone to form? Explain.
   (b) Now suppose that the clouds of nitrogen dioxide drift slowly over an area that has a high concentration of hydrocarbons (VOCs), from both natural and industrial sources.

Would you expect high levels of tropospheric ozone to form under these conditions? Explain your reasoning.
2. For least-polluted conditions, what would be the best time of day for a farmer to burn agricultural debris? Explain why you chose that time of day.
3. Why are most severe air pollution episodes associated with subsidence inversions rather than radiation inversions?
4. What surface and upper-air conditions lead to atmospheric stagnation?
5. Table 17.4, p. 463 shows that cloudiness is generally greater in urban areas than in rural areas. Since clouds reflect a great deal of incoming sunlight, they tend to keep daytime temperatures lower. Why then, during the day, are urban areas generally warmer than surrounding rural areas?
6. Acid snow can be a major problem. In the high mountains of the western United States, especially downwind of a major metropolitan or industrial area, explain why, for a high-mountain lake, acid snow can be a greater problem than acid rain, even when both have the same pH.
7. What atmospheric conditions would produce a fumigation-type plume downwind of a smoke stack during the *middle* of the afternoon?
8. Why do we want to reduce high ozone concentrations at the Earth's surface while also wishing to increase ozone concentration in the stratosphere?

Problems and Exercises

1. Keep a log of the daily PSI readings in your area and note the pollutants listed in the index. Also, keep a record of the daily weather conditions, such as cloud cover, high temperature for the day, average wind direction and speed, etc. See if there is any relationship between these weather conditions and high PSI readings for certain pollutants.
2. Suppose the PSI reading for ozone is 350.
   (a) How would the air be described on this day?
   (b) What PSI episode level would exist?
   (c) What would be the general health effects, and what precautions should a person take under these conditions?
3. Suppose the air temperature at the surface is 30°C. Further suppose that the air temperature decreases at the dry adiabatic rate (10°C/km) up to the base of a strong subsidence inversion, situated about 2000 m above the surface. If the surface air temperature increases to 40°C, and the air temperature continues to decrease at the dry adiabatic rate up to the base of the inversion, will the mixing depth increase or decrease? Explain your answer with a diagram. Will pollutants found within the mixing layer be more concentrated or less concentrated? Explain.
4. Rain with a pH of 2 is how much more acidic than rain with a pH of 5?

5. Keep a log of daily PSI readings for 1 urban and 1 rural location for days on which both areas have similar weather conditions; note the weather conditions. Compare them. Do you notice less or more difference between the PSI readings for the two locations based on weather conditions?

 Questions for Exploration

1. On the Blue Skies: College Edition CD-ROM go to the **Atmospheric Chemistry** section of the CD and click on "**Urban Ozone**."

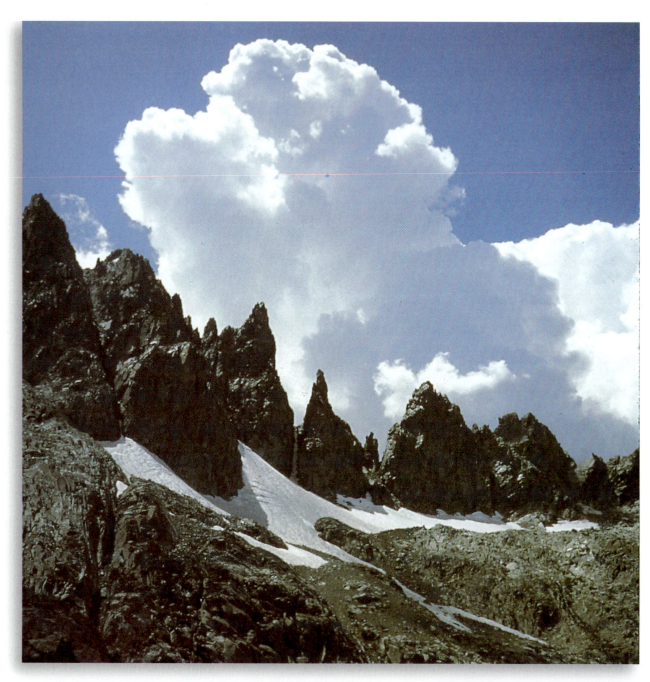

CHAPTER 18

The cool summer climate near the top of the Minarets in the Sierra Nevada allows winter snow to survive in sheltered coves and valleys. Afternoon thunderstorms that build along the windward slopes produce ample summer rains. However, just a few hundred kilometers to the east and a few thousand meters lower in elevation, the climate is hot (with afternoon summer temperatures hovering near 40°C or 104°F) and dry (as the region is in the rain shadow of this mountain range). (Photo: T. Ansel Toney.)

GLOBAL CLIMATE

CONTENTS

The climate is unbearable . . . At noon today the highest temperature measured was −33°C. We really feel that it is late in the season. The days are growing shorter, the sun is low and gives no warmth, katabatic winds blow continuously from the south with gales and drifting snow. The inner walls of the tent are like glazed parchment with several millimeters thick ice-armour . . . Every night several centimeters of frost accumulate on the walls, and each time you inadvertently touch the tent cloth a shower of ice crystals falls down on your face and melt. In the night huge patches of frost from my breath spread around the opening of my sleeping bag and melt in the morning. The shoulder part of the sleeping bag facing the tent-side is permeated with frost and ice, and crackles when I roll up the bag . . . For several weeks now my fingers have been permanently tender with numb fingertips and blistering at the nails after repeated frostbites. All food is frozen to ice and it takes ages to thaw out everything before being able to eat. At the depot we could not cut the ham, but had to chop it in pieces with a spade. Then we threw ourselves hungrily at the chunks and chewed with the ice crackling between our teeth. You have to be careful with what you put in your mouth. The other day I put a piece of chocolate from an outer pocket directly in my mouth and promptly got frostbite with blistering of the palate.

Ove Wilson (Quoted in David M. Gates, *Man and His Environment*)

Our opening comes from a report by Norwegian scientists on their encounter with one of nature's cruelest climates—that of Antarctica. Their experience illustrates the profound effect that climate can have on even ordinary events, such as eating a piece of chocolate. Though we may not always think about it, climate profoundly affects nearly everything in the middle latitudes, too. For instance, it influences our housing, clothing, the shape of landscapes, agriculture, how we feel and live, and even where we reside, as most people will choose to live on a sunny hillside rather than in a cold, dark, and foggy river basin. Entire civilizations have flourished in favorable climates and have moved away from, or perished in, unfavorable ones. We learned early in this text that *climate* is the average of the day-to-day weather over a long duration. But the concept of climate is much larger than this, for it encompasses, among other things, the daily and seasonal extremes of weather within specified areas.

When we speak of climate, then, we must be careful to specify the spatial location we are talking about. For example, the Chamber of Commerce of a rural town may boast that its community has mild winters with air temperatures seldom below freezing. This may be true several meters above the ground in an instrument shelter, but near the ground the temperature may drop below freezing on many winter nights. This small climatic region near or on the ground is referred to as a **microclimate**. Because a much greater extreme in daily air temperatures exists near the ground than several meters above, the microclimate for small plants is far more harsh than the thermometer in an instrument shelter would indicate.

When we examine the climate of a small area of the earth's surface, we are looking at the **mesoclimate**. The size of the area may range from a few acres to several square kilometers. Mesoclimate includes regions such as forests, valleys, beaches, and towns. The climate of a much larger area, such as a state or a country, is called **macroclimate**. The climate extending over the entire earth is often referred to as **global climate**.

In this chapter, we will concentrate on the larger scales of climate. We will begin with the factors that regulate global climate, then we will discuss how climates are classified. Finally, we will examine the different types of climate.

A WORLD WITH MANY CLIMATES

The world is rich in climatic types. From the teeming tropical jungles to the frigid polar "wastelands," there seems to be an almost endless variety of climatic regions. The factors that produce the climate in any given place—the **climatic controls**—are the same that produce our day-to-day weather. Briefly, the controls are the:

1. intensity of sunshine and its variation with latitude
2. distribution of land and water
3. ocean currents
4. prevailing winds
5. positions of high- and low-pressure areas
6. mountain barriers
7. altitude

We can ascertain the effect these controls have on climate by observing the global patterns of two weather elements—temperature and precipitation.

Global Temperatures   Figure 18.1 shows mean annual temperatures for the world. To eliminate the distorting effect of topography, the temperatures are corrected to sea level.* Notice that in both hemispheres the isotherms are oriented east-west, reflecting the fact that locations at the same latitude receive nearly the same amount of solar energy. In addition, the annual solar heat that each latitude receives decreases from low to high latitude; hence, annual temperatures tend to decrease from equatorial toward polar regions.†

The bending of the isotherms along the coastal margins is due in part to the unequal heating and cooling properties of land and water, and to ocean currents and upwelling. For example, along the west coast of North and South America, ocean currents transport cool water equatorward. In addition to this, the wind in both regions blows toward the equator, parallel to the coast. This situation favors upwelling of cold water (see Chapter 11), which cools the coastal margins. In the area of the eastern North Atlantic Ocean (north of 40°N), the poleward bending of the isotherms is due to the Gulf Stream and the North Atlantic Drift, which carry warm water northward.

The fact that land masses heat up and cool off more quickly than do large bodies of water means that variation in temperature between summer and winter will be far greater over continental interiors than along the west coastal margins of continents. By the same token, the climates of interior continental regions will be more extreme, as they have (on the average) higher summer temperatures and lower winter temperatures than their west-coast counterparts. In fact, west-coast climates are typically quite mild for their latitude.

The highest mean temperatures do not occur in the tropics, but rather in the subtropical deserts of the Northern Hemisphere. Here, the subsiding air associated with the subtropical anticyclones produces generally clear skies and low humidity. In summer, the high sun beating down upon a relatively barren landscape produces scorching heat.

The lowest mean temperatures occur over large land masses at high latitudes. The coldest area of the world is the Antarctic.

---

*This correction is made by adding to each station above sea level an amount of temperature that would correspond to the normal temperature lapse rate of 6.5°C per 1000 m (3.6°F per 1000 ft).

†Average global temperatures for January and July are given in Figs. 3.21 and 3.22, respectively, on p. 69.

During part of the year, the sun is below the horizon; when it is above the horizon, it is low in the sky and its rays do not effectively warm the surface. Consequently, the land remains snow- and ice-covered year-round. The snow and ice reflect perhaps 80 percent of the sunlight that reaches the surface. Much of the unreflected solar energy is used to transform the ice and snow into water vapor. The relatively dry air and the Antarctic's high elevation permit rapid radiational cooling during the dark winter months, producing extremely cold surface air. The extremely cold Antarctic helps to explain why, overall, the Southern Hemisphere is cooler than the Northern Hemisphere. Other contributing factors for a cooler Southern Hemisphere include the fact that polar regions of the Southern Hemisphere reflect more incoming sunlight, and the fact that less land area is found in tropical and subtropical areas of the Southern Hemisphere.

The Hottest and Coldest Places on Earth Most people are aware of the extreme heat that exists during the summer in the desert southwest of the United States. But how hot does it get there? On July 10, 1913, Greenland Ranch in Death Valley, California, reported the highest temperature ever observed in North America: 57°C, or 134°F. Here, air temperatures are persistently hot throughout the summer, with the average maximum for July being 47°C (116°F). During the summer of 1917,

there was an incredible period of 43 consecutive days when the maximum temperature reached 49°C (120°F) or higher.

Probably the hottest urban area in the United States is Yuma, Arizona. Located along the California–Arizona border, Yuma's high temperature during July averages 42°C (108°F). In 1937, the high reached 38°C (100°F) or more for 101 consecutive days.

In a more humid climate, the maximum temperature rarely climbs above 41°C (106°F). However, during the record heat wave of 1936, the air temperature reached 49°C (121°F) near Alton, Kansas. And during the heat wave of 1983, which destroyed about $7 billion in crops and increased the nation's air-conditioning bill by an estimated $1 billion, Fayetteville reported North Carolina's all-time record high temperature when the mercury hit 43°C (110°F).

FIGURE 18.1
Average annual sea level temperatures throughout the world (°F).

TABLE 18.1   Some Record High Temperatures Throughout the World

LOCATION (LATITUDE)	RECORD HIGH TEMPERATURE		RECORD FOR:	DATE
	(°C)	(°F)		
El Azizia, Libya (32°N)	58	136	The world	September 13, 1922
Death Valley, CA (36°N)	57	134	Western Hemisphere	July 10, 1913
Tirat Tsvi, Israel (32°N)	54	129	Middle East	June 21, 1942
Cloncurry, Queensland (21°S)	53	128	Australia	January 16, 1889
Seville, Spain (37°N)	50	122	Europe	August 4, 1881
Rivadavia, Argentina (35°S)	49	120	South America	December 11, 1905
Midale, Saskatchewan (49°N)	45	113	Canada	July 5, 1937
Fort Yukon, Alaska (66°N)	38	100	Alaska	June 27, 1915
Pahala, Hawaii (19°N)	38	100	Hawaii	April 27, 1931
Esparanza, Antarctica (63°S)	14	58	Antarctica	October 20, 1956

These readings, however, do not hold a candle to the hottest place in the world. That distinction probably belongs to Dallol, Ethiopia. Dallol is located south of the Red Sea, near latitude 12°N, in the hot, dry Danakil Depression. A prospecting company kept weather records at Dallol from 1960 to 1966. During this time, the average daily maximum temperature exceeded 38°C (100°F) every month of the year, except during December and January, when the average maximum lowered to 37°C (98°F) and 36°C (97°F), respectively. On many days, the air temperature exceeded 49°C (120°F). The average annual temperature for the six years at Dallol was 34°C (94°F). In comparison, the average annual temperature in Yuma is 23°C (74°F) and at Death Valley, 24°C (76°F). The highest temperature reading on earth (under standard conditions) occurred about 3500 km northeast of Dallol at El Azizia, Libya (32°N), when, on September 13, 1922, the temperature reached a scorching 58°C (136°F). Table 18.1 gives record high temperatures throughout the world.

One of the coldest spots in the United States is International Falls, Minnesota, where the average temperature for January is −16°C (3°F). Located about 400 km to the south, Minneapolis–St. Paul, with an average temperature of −9°C (16°F) for the three winter months, is the coldest major urban area in the nation. For duration of extreme cold, Minneapolis reported 186 consecutive hours of temperatures below −18°C (0°F) during the winter of 1911–1912. Within the 48 adjacent states, however, the record for the longest duration of severe cold belongs to Langdon, North Dakota, where the thermometer remained below −18°C (0°F) for 41 consecutive days during the winter of 1936. The official record for the lowest temperature in the 48 adjacent states belongs to Rogers Pass, Montana, where on the morning of January 20, 1954, the mercury dropped to −57°C (−70°F). The lowest official temperature for Alaska, −62°C (−80°F), occurred at Prospect Creek on January 23, 1971.

The coldest areas in North America are found in the Yukon and Northwest Territories of Canada. Resolute, Canada (latitude 75°N), has an average temperature of −32°C (−26°F) for the month of January.

The coldest winters in the Northern Hemisphere are found in the interior of Siberia and Greenland. For example, the average January temperature in Yakutsk, Siberia (latitude 62°N), is −43°C (−46°F). There, the mean temperature for the entire year is a bitter cold −11°C (12°F). At Eismitte, Greenland, the average temperature for February (the coldest month) is −47°C (−53°F), with the mean annual temperature being a frigid −30°C (−22°F). Even though these temperatures are extremely low, they do not come close to the coldest area of the world: the Antarctic.

At the geographical South Pole, nearly 2800 m (9200 ft) above sea level, where the Amundsen-Scott scientific station has been keeping records for more than 25 years, the average temperature for the month of July (winter) is −59°C (−74°F) and the mean annual temperature is −49°C (−57°F). The lowest temperature ever recorded there (−83°C, or −117°F) occurred under clear skies with a light wind on the morning of June 23, 1983. Cold as it was, it was not the record low for the world. That belongs to the Russian station at Vostok, Antarctica (latitude 78°S), where the temperature plummeted to −89°C (−129°F) on July 21, 1983. See Table 18.2 for record low temperatures throughout the world.

Global Precipitation   Figure 18.2 (pages 476–477) shows the worldwide general pattern of annual precipitation, which varies from place to place. There are, however, certain regions that stand out as being wet or dry. For example, equatorial regions are typically wet, while the subtropics and the polar regions are relatively dry. The global distribution of precipitation is closely

TABLE 18.2   Some Record Low Temperatures Throughout the World

LOCATION (LATITUDE)	RECORD LOW TEMPERATURE (°C)	(°F)	RECORD FOR:	DATE
Vostok, Antarctica (78°S)	−89	−129	The world	July 21, 1983
Verkhoyansk, Russia (67°N)	−68	−90	Northern Hemisphere	February 7, 1892
Northice, Greenland (72°N)	−66	−87	Greenland	January 9, 1954
Snag, Yukon (62°N)	−63	−81	North America	February 3, 1947
Prospect Creek, Alaska (66°N)	−62	−80	Alaska	January 23, 1971
Rogers Pass, Montana (47°N)	−57	−70	United States (excluding Alaska)	January 20, 1954
Sarmiento, Argentina (34°S)	−33	−27	South America	June 1, 1907
Ifrane, Morocco (33°N)	−24	−11	Africa	February 11, 1935
Charlotte Pass, Australia (36°S)	−22	−8	Australia	July 22, 1949
Mt. Haleakala, Hawaii (20°N)	−10	14	Hawaii	January 2, 1961

tied to the general circulation of the atmosphere (Chapter 11) and to the distribution of mountain ranges and high plateaus.

Figure 18.3 shows in simplified form how the general circulation influences the north-to-south distribution of precipitation to be expected on a uniformly water-covered earth. Precipitation is most abundant where the air rises; least abundant where it sinks. Hence, one expects a great deal of precipitation in the tropics and along the polar front, and little near subtropical highs and at the poles. Let's look at this in more detail.

In tropical regions, the trade winds converge along the Intertropical Convergence Zone (ITCZ), producing rising air,

towering clouds, and heavy precipitation all year long. Poleward of the equator, near latitude 30°, the sinking air of the subtropical highs produces a "dry belt" around the globe. The Sahara Desert of North Africa is in this region. Here, annual rainfall is exceedingly light and varies considerably from year to year. Because the major wind belts and pressure systems shift with the season—northward in July and southward in January—the area between the rainy tropics and the dry subtropics is influenced by both the ITCZ and the subtropical highs.

In the cold air of the polar regions there is little moisture, so there is little precipitation. Winter storms drop light,

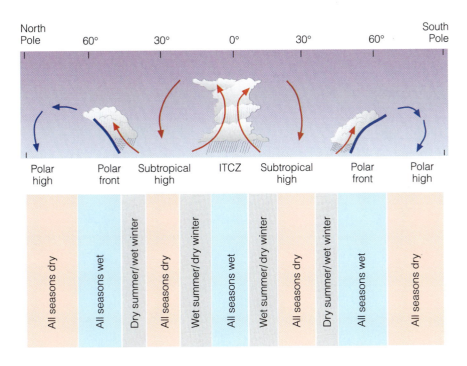

FIGURE 18.3

A vertical cross section along a line running north to south illustrates the main global regions of rising and sinking air and how each region influences precipitation.

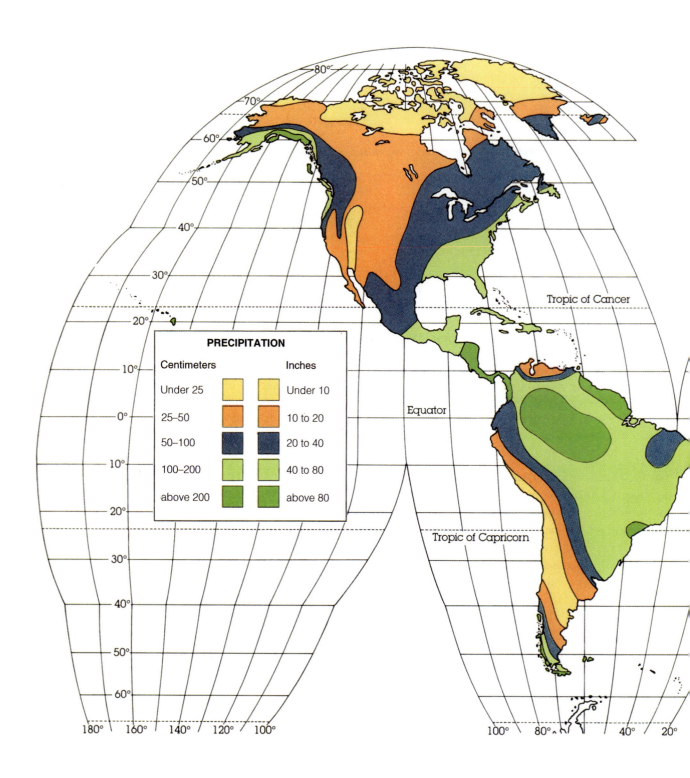

FIGURE 18.2
Annual global pattern of precipitation.

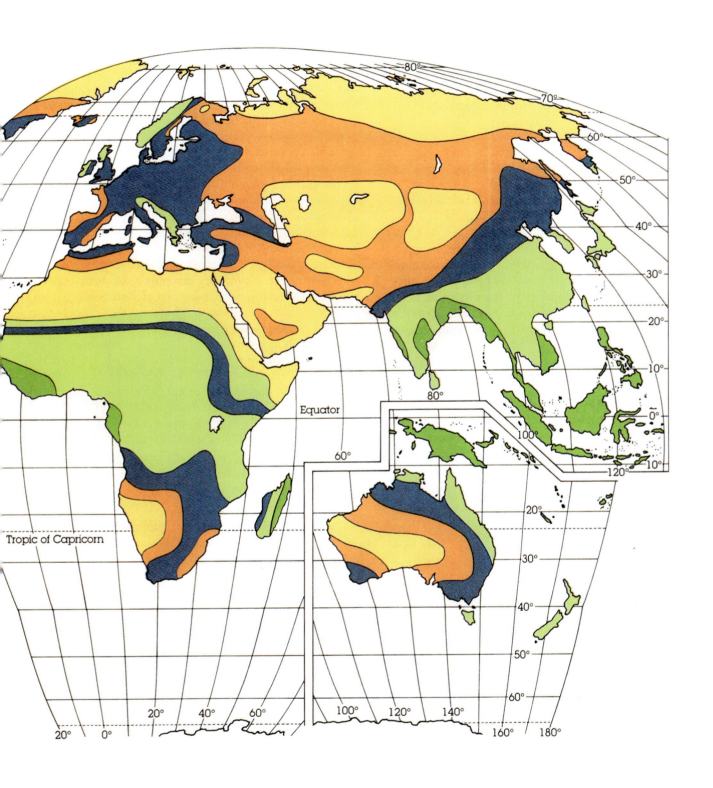

Equator

Tropic of Capricorn

WEATHER WATCH

Wynoochee Oxbow, Washington, on the Olympic Peninsula, is considered the wettest weather station in the continental United States, with an average rainfall of 366 cm (144 in.)—a total 86 times greater than the average 4.3 cm (1.7 in.) for Death Valley, California.

powdery snow that remains on the ground for a long time because of the low evaporation rates. In summer, a ridge of high pressure tends to block storm systems that would otherwise travel into the area; hence, precipitation in polar regions is meager in all seasons.

There are exceptions to this idealized pattern. For example, in middle latitudes the migrating position of the subtropical anticyclones also has an effect on the west-to-east distribution of precipitation. The sinking air associated with these systems is more strongly developed on their eastern side. Hence, the air along the eastern side of an anticyclone tends to be more stable; it is also drier, as cooler air moves equatorward because of the circulating winds around these systems. In addition, along coastlines, cold upwelling water cools the surface air even more, adding to the air's stability. Consequently, in summer, when the Pacific high moves to a position centered off the California coast, a strong, stable subsidence inversion forms above coastal regions. With the

strong inversion and the fact that the anticyclone tends to steer storms to the north, central and southern California areas experience little, if any, rainfall during the summer months.

On the western side of subtropical highs, the air is less stable and more moist, as warmer air moves poleward. In summer, over the North Atlantic, the Bermuda high pumps moist tropical air northward from the Gulf of Mexico into the eastern two-thirds of the United States. The humid air is conditionally unstable to begin with, and by the time it moves over the heated ground, it becomes even more unstable. If conditions are right, the moist air will rise and condense into cumulus clouds, which may build into towering thunderstorms.

In winter, the subtropical North Pacific high moves south, allowing storms traveling across the ocean to penetrate the western states, bringing much needed rainfall to California after a long, dry summer. The Bermuda high also moves south in winter. Across much of the United States, intense winter storms develop and travel eastward, frequently dumping heavy precipitation as they go. Usually, however, the heaviest precipitation is concentrated in the eastern states, as moisture from the Gulf of Mexico moves northward ahead of these systems. Therefore, cities on the plains typically receive more rainfall in summer, those on the west coast have maximum precipitation in winter, while cities in the midwest and east usually have abundant precipitation all year long. The contrast in seasonal precipitation among a west coast city (San Francisco), a central plains city (Kansas City), and an eastern city (Baltimore) is clearly shown in Fig. 18.4.

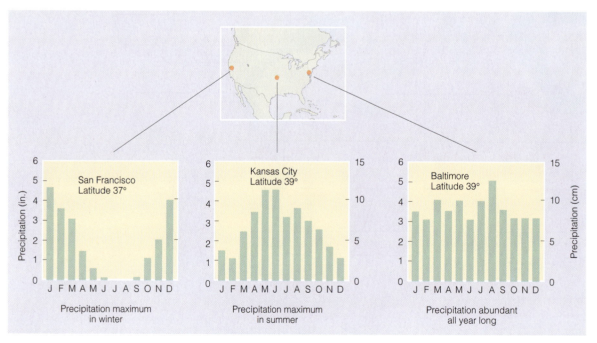

FIGURE 18.4
Variation in annual precipitation for three Northern Hemisphere cities.

FIGURE 18.5
The effect of topography on average annual precipitation along a line running from the Pacific Ocean through central California into western Nevada.

Mountain ranges disrupt the idealized pattern of global precipitation (1) by promoting convection (because their slopes are warmer than the surrounding air) and (2) by forcing air to rise along their windward slopes (*orographic uplift*). Consequently, the windward side of mountains tends to be "wet." As air descends and warms along the leeward side, there is less likelihood of clouds and precipitation. Thus, the leeward side of mountains tends to be "dry." As Chapter 7 points out, a region on the leeward side of a mountain where precipitation is noticeably less is called a *rain shadow*.

A good example of the rain shadow effect occurs in the northwestern part of Washington State. Situated on the western side at the base of the Olympic Mountains, the Hoh River Valley annually receives an average 380 cm (150 in.) of precipitation. On the eastern (leeward) side of this range, only about 100 km (62 mi) from the Hoh rain forest, the mean annual precipitation is less than 43 cm (17 in.), and irrigation is necessary to grow certain crops. Figure 18.5 shows a classic example of how topography produces several rain shadow effects. (Additional information on precipitation extremes is given in the Focus section on p. 480.)

BRIEF REVIEW   Before going on to the section on climate classification, here is a brief review of some of the facts we covered so far:

- The climate controls are the factors that govern the climate of any given region.
- The hottest places on earth tend to occur in the subtropical deserts of the Northern Hemisphere, where clear skies and sinking air, coupled with low humidity and a high summer sun beating down upon a relatively barren landscape, produce extreme heat.
- The coldest places on earth tend to occur in the interior of high-latitude land masses. The coldest areas of the Northern Hemisphere are found in the interior of Siberia and Greenland, whereas the Antarctic holds the world record for extreme cold.
- The wettest places in the world tend to be located on the windward side of mountains where warm, humid air rises upslope.

CLIMATIC CLASSIFICATION

The climatic controls interact to produce such a wide array of different climates that no two places experience exactly the same climate. However, the similarity of climates within a given area allows us to divide the earth into climatic regions.

The Ancient Greeks   By considering temperature and worldwide sunshine distribution, the ancient Greeks categorized the world into three climatic regions:

1. A low-latitude *torrid zone;* bounded by the northern and southern limit of the sun's vertical rays ($23\frac{1}{2}°$N and $23\frac{1}{2}°$S); here, the noon sun is always high, day and night are of nearly equal length, and it is warm year-round.
2. A high-latitude *polar* (or *frigid*) *zone;* bounded by the Arctic or Antarctic Circle; cold all year long due to long periods of winter darkness and a low summer sun.

Precipitation Extremes

Most of the "rainiest" places in the world are located on the windward side of mountains. For example, Mount Waialeale on the island of Kauai, Hawaii, has the greatest annual average rainfall on record: 1168 cm (460 in.). Cherrapunji, on the crest of the southern slopes of the Khasi Hills in northeastern India, receives an average of 1080 cm (425 in.) of rainfall each year, the majority of which falls during the summer monsoon, between April and October. Cherrapunji, which holds the greatest twelve-month rainfall total of 2647

cm (1042 in.), once received 380 cm (150 in.) of rain in just five days.

Record rainfall amounts are often associated with tropical storms. On the island of La Réunion (about 650 km east of Madagascar in the Indian Ocean), a tropical cyclone dumped 135 cm (53 in.) of rain on Belouve in twelve hours. Heavy rains of short duration often occur with severe thunderstorms that move slowly or stall over a region. On July 4, 1956, 3 cm (1.2 in.) of rain fell from a thunderstorm on Unionville, Maryland, in one minute.

Snowfalls tend to be heavier where cool, moist air rises along the windward slopes of mountains. One of the snowiest places in North America is located at the Paradise Ranger Station in Mt. Rainier National Park, Washington. Situated at an elevation of 1646 m (5400 ft) above sea level, this station receives an average 1575 cm (620 in.) of snow annually. However, a record 2850 cm (1122 in.) was received during the winter of 1971–1972.

As we noted earlier, the driest regions of the world lie in the frigid

KEY TO MAP

❶	World's greatest annual average rainfall	1168 cm (460 in.)	Mt. Waialeale, Hawaii
❷	Greatest 1-month rainfall total	930 cm (366 in.)	Cherrapunji, India, July, 1861
❸	Greatest 12-hour rainfall total	135 cm (53 in.)	Belouve, La Réunion Island, February 28, 1964
❹	Greatest 24-hour rainfall total in United States	109 cm (43 in.)	Alvin, Texas, July 25, 1979
❺	Greatest 42-minute rainfall total	30 cm (12 in.)	Holt, Missouri, June 22, 1947
❻	Greatest 1-minute rainfall total in United States	3 cm (1.2 in.)	Unionville, MD, July 4, 1956
❼	Lowest annual average rainfall in Northern Hemisphere	3 cm (1.2 in.)	Bataques, Mexico
❽	Lowest annual average rainfall in the world	0.08 cm (0.03 in.)	Arica, Chile
❾	Greatest annual snowfall in United States	2850 cm (1122 in.)	Paradise Ranger Station, Mt. Rainier, WA, 1971–1972
❿	Greatest snowfall in 1 month	991 cm (390 in.)	Tamarack, CA, January, 1911
⓫	Greatest snowfall in 24 hours	193 cm (76 in.)	Silverlake, Boulder, CO April 14–15, 1921
⓬	Longest period without measurable precipitation in U. S. (993 days)	0.0 cm (0.0 in.)	Bagdad, CA August 1909 to May 1912

3. A *temperate zone;* sandwiched between the other two zones; has distinct summer and winter, so exhibits characteristics of both extremes.

Such a sunlight, or temperature-based, climatic scheme is, of course, far too simplistic. It excludes precipitation, so there is no way to differentiate between wet and dry regions. The best classification of climates would take into account as many meteorological factors as can possibly be obtained.

The Köppen System   A widely used classification of world climates based on the annual and monthly averages of temperature and precipitation was devised by the famous German scientist Waldimir Köppen (1846–1940). Initially published in 1918, the original **Köppen classification system** (Table 18.3) has since been modified and refined. Faced with the lack of adequate observing stations throughout the world, Köppen related the distribution and type of native vegetation to the

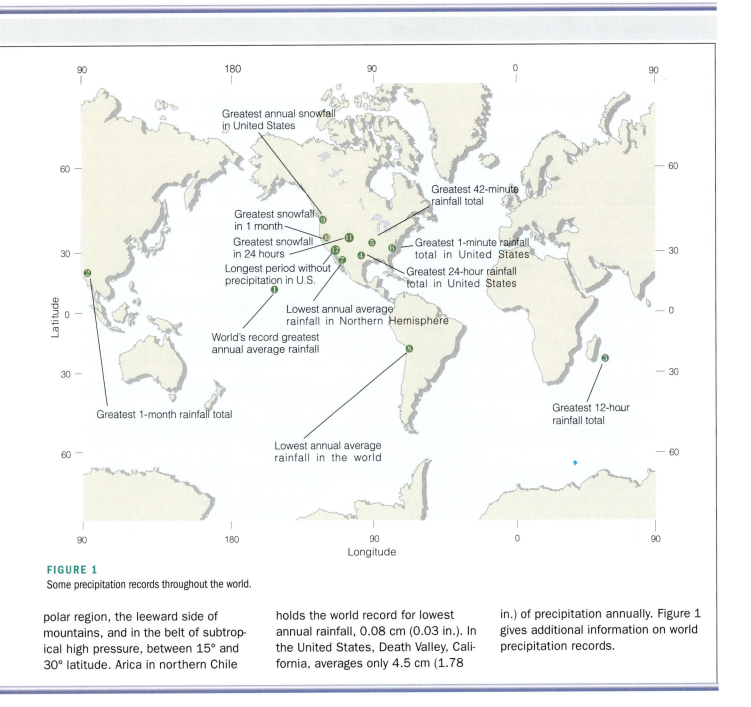

FIGURE 1

Some precipitation records throughout the world.

polar region, the leeward side of mountains, and in the belt of subtropical high pressure, between 15° and 30° latitude. Arica in northern Chile holds the world record for lowest annual rainfall, 0.08 cm (0.03 in.). In the United States, Death Valley, California, averages only 4.5 cm (1.78 in.) of precipitation annually. Figure 1 gives additional information on world precipitation records.

various climates. In this way, climatic boundaries could be approximated where no climatological data were available.

Köppen's scheme employs five major climatic types; each type is designated by a capital letter:

A *Tropical moist climates*: All months have an average temperature above 18°C (64°F). Since all months are warm, there is no real winter season.

B *Dry climates*: Deficient precipitation most of the year. Potential evaporation and transpiration exceed precipitation.

C *Moist mid-latitude climates with mild winters*: Warm-to-hot summers with mild winters. The average temperature of the coldest month is below 18°C (64°F) and above −3°C (27°F).

D *Moist mid-latitude climates with severe winters*: Warm summers and cold winters. The average temperature of the warmest month exceeds 10°C (50°F), and the coldest monthly average drops below –3°C (27°F).

E *Polar climates*: Extremely cold winters and summers. The average temperature of the warmest month is below 10°C (50°F). Since all months are cold, there is no real summer season.

Each group contains subregions that describe special regional characteristics, such as seasonal changes in temperature and precipitation. In mountainous country, where rapid changes in elevation bring about sharp changes in climatic type, delineating the climatic regions is impossible. These regions are designated by the letter H, for highland climates.

Köppen's system has been criticized primarily because his boundaries (which relate vegetation to monthly temperature and precipitation values) do not correspond to the natural boundaries of each climatic zone. In addition, the Köppen system implies that there is a sharp boundary between climatic zones, when in reality there is a gradual transition.

The Köppen system has been revised several times, most notably by the German climatologist Rudolf Geiger, who worked with Köppen on amending the climatic boundaries of certain regions. A popular modification of the Köppen system was developed by the American climatologist Glenn T. Trewartha, who redefined some of the climatic types and altered the climatic world map by putting more emphasis on the lengths of growing seasons and average summer temperatures.

Thornthwaite's System   To correct some of the Köppen deficiencies, the American climatologist C. Warren Thornthwaite (1899–1963) devised a new classification system in the early 1930s. Both systems utilized temperature and precipitation measurements and both related natural vegetation to climate. However, to emphasize the importance of precipitation (P) and evaporation (E) on plant growth, Thornthwaite developed a *P/E ratio*, which is essentially monthly precipitation divided by monthly evaporation. The annual sum of the P/E ratios gives the **P/E index**. Using this index, the Thornthwaite system defines five major humidity provinces and their characteristic vegetations: rain forest, forest, grassland, steppe, and desert.

To better describe the moisture available for plant growth, Thornthwaite proposed a new classification system in 1948 and slightly revised it in 1955. His new scheme emphasized the concept of *potential evapotranspiration*[*] (PE), which is the amount of moisture that would be lost from the soil and vegetation if the moisture were available.

Thornthwaite incorporated potential evapotranspiration into a moisture index that depends essentially on the differences between precipitation and PE. The index is high in moist climates and negative in arid climates. An index of 0 marks the boundary between wet and dry climates.

THE GLOBAL PATTERN OF CLIMATE

Figure 18.6 (pages 484–485) displays how the major climatic regions of the world are distributed, based mainly on the work of Köppen. (The major climatic types along with their subdivisions are given in Table 18.3.) We will first examine humid tropical climates in low latitudes and then we'll look at middle latitude and polar climates. Bear in mind that each climatic region has many subregions of local climatic differences wrought by such factors as topography, elevation, and large bodies of water. Remember, too, that boundaries of climatic regions represent gradual transitions. Thus, the major climatic characteristics of a given region are best observed away from its periphery.

Tropical Moist Climates (Group A)

General characteristics: year-round warm temperatures (all months have a mean temperature above 18°C, or 64°F); abundant rainfall (typical annual average exceeds 150 cm, or 59 in.).

Extent: northward and southward from the equator to about latitude 15° to 25°.

Major types (based on seasonal distribution of rainfall): *tropical wet* (Af), *tropical monsoon* (Am), and *tropical wet and dry* (Aw).

At low elevations near the equator, in particular the Amazon lowland of South America, the Congo River Basin of Africa, and the East Indies from Sumatra to New Guinea, high temperatures and abundant yearly rainfall combine to produce a dense, broadleaf, evergreen forest called a **tropical rain forest**. Here, many different plant species, each adapted to differing light intensity, present a crudely layered appearance of diverse vegetation. In the forest, little sunlight is able to penetrate to the ground through the thick crown cover. As a result, little plant growth is found on the forest floor. However, at the edge of the forest, or where a clearing has been made, abundant sunlight allows for the growth of tangled shrubs and vines, producing an almost impenetrable *jungle* (see Fig. 18.7).

Within the **tropical wet climate**[*] (Af), seasonal temperature variations are small (normally less than 3°C) because the noon sun is always high and the number of daylight hours is relatively constant. However, there is a greater variation in temperature between day (average high about 32°C) and night (average low about 22°C) than there is between the warmest and coolest months. This is why people remark that winter comes to the tropics at night. The weather here is monotonous

---

[*]Evapotranspiration refers to the evaporation from soil and transpiration of plants.

[*]The tropical wet climate is also known as the *tropical rain forest climate*.

TABLE 18.3 **Köppen's Climatic Classification System**

LETTER SYMBOL 1ST	2ND	3RD	CLIMATIC CHARACTERISTICS	CRITERIA
A			Humid tropical	All months have an average temperature of 18°C (64°F) or higher
	f		Tropical wet (rain forest)	Wet all seasons; all months have at least 6 cm (2.4 in.) of rainfall
	w		Tropical wet and dry (savanna)	Winter dry season; rainfall in driest month is less than 6 cm (2.4 in.) and less than $10 - P/25$ (P is mean annual rainfall in cm)
	m		Tropical monsoon	Short dry season; rainfall in driest month is less than 6 cm (2.4 in.) but equal to or greater than $10 - P/25$
B			Dry	Potential evaporation and transpiration exceed precipitation. The dry/humid boundary is defined by the following formulas: $p = 2t + 28$ when 70% or more of rain falls in warmer 6 months (dry winter) $p = 2t$ when 70% or more of rain falls in cooler 6 months (dry summer) $p = 2t + 14$ when neither half year has 70% or more of rain (p is the mean annual precipitation in cm and t is the mean annual temperature in °C)*
	S		Semi-arid (steppe)	The BS/BW boundary is exactly one-half the dry/humid boundary
	W		Arid (desert)	
		h	Hot and dry	Mean annual temperature is 18°C (64°F) or higher
		k	Cool and dry	Mean annual temperature is below 18°C (64°F)
C			Moist with mild winters	Average temperature of coolest month is below 18°C (64°F) and above –3°C (27°F)
	w		Dry winters	Average rainfall of wettest summer month at least 10 times as much as in driest winter month
	s		Dry summers	Average rainfall of driest summer month less than 4 cm (1.6 in.); average rainfall of wettest winter month at least 3 times as much as in driest summer month
	f		Wet all seasons	Criteria for w and s cannot be met
		a	Summers long and hot	Average temperature of warmest month above 22°C (72°F); at least 4 months with average above 10°C (50°F)
		b	Summers long and cool	Average temperature of all months below 22°C (72°F); at least 4 months with average above 10°C (50°F)
		c	Summers short and cool	Average temperature of all months below 22°C (72°F); 1 to 3 months with average above 10°C (50°F)
D			Moist with cold winters	Average temperature of coldest month is –3°C (27°F) or below; average temperature of warmest month is greater than 10°C (50°F)
	w		Dry winters	Same as under Cw
	s		Dry summers	Same as under Cs
	f		Wet all seasons	Same as under Cf
		a	Summers long and hot	Same as under Cfa
		b	Summers long and cool	Same as under Cfb
		c	Summers short and cool	Same as under Cfc
		d	Summers short and cool; winters severe	Average temperature of coldest month is –38°C (–36°F) or below
E			Polar climates	Average temperature of warmest month is below 10°C (50°F)
	T		Tundra	Average temperature of warmest month is greater than 0°C (32°F) but less than 10°C (50°F)
	F		Ice cap	Average temperature of warmest month is 0°C (32°F) or below

*The dry/humid boundary is defined in English units as: $p = 0.44t - 3$ (dry winter); $p = 0.44t - 14$ (dry summer); and $p = 0.44t - 8.6$ (rainfall evenly distributed). Where p is mean annual rainfall in inches and t is mean annual temperature in °F.

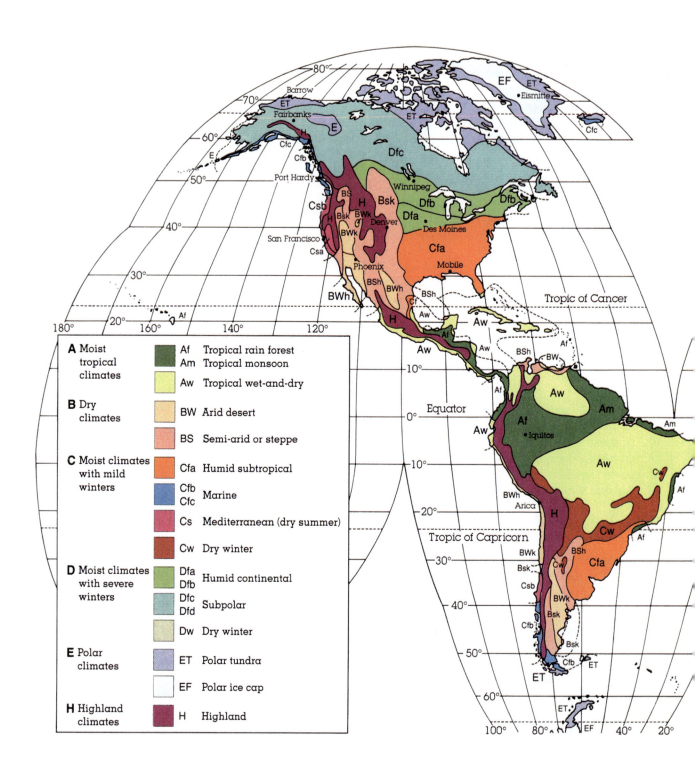

FIGURE 18.6
Worldwide distribution of climatic regions (after Köppen).

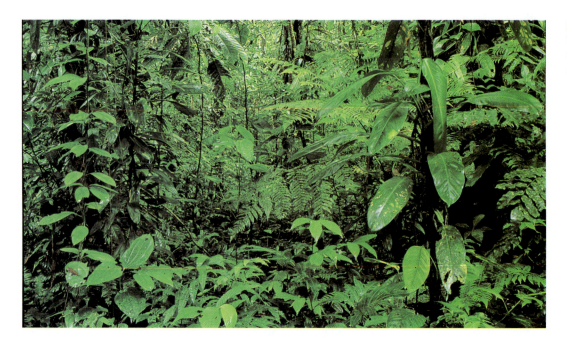

FIGURE 18.7
Tropical rain forest near Iquitos, Peru. (Climatic information for this region is presented in Fig. 18.8.)

FIGURE 18.8
Temperature and precipitation data for Iquitos, Peru, latitude 4°S. A station with a tropical wet climate (Af).

and sultry. There is little change in temperature from one day to the next. Furthermore, almost every day, towering cumulus clouds form and produce heavy, localized showers by early afternoon. As evening approaches, the showers usually end and skies clear. Typical annual rainfall totals are greater than 150 cm (59 in.) and, in some cases, especially along the windward side of hills and mountains, the total may exceed 400 cm (157 in.).

The high humidity and cloud cover tend to keep maximum temperatures from reaching extremely high values. In fact, summer afternoon temperatures are normally higher in middle latitudes than here. Nighttime cooling can produce saturation and, hence, a blanket of dew and—occasionally—fog covers the ground.

An example of a station with a tropical wet climate (Af) is Iquitos, Peru (see Fig. 18.8). Located near the equator (latitude 4°S), in the low basin of the upper Amazon River, Iquitos has an average annual temperature of 25°C (77°F), with an annual temperature range of only 2.2°C (4°F). Notice also that the monthly rainfall totals vary more than do the monthly temperatures. This is due primarily to the migrating position of the Intertropical Convergence Zone (ITCZ) and its associated wind-flow patterns. Although monthly precipitation totals vary considerably, the average for each month exceeds 6 cm, and consequently no month is considered deficient of rainfall.

Take a minute and look back at Fig. 18.7. From the photo, one might think that the soil beneath the forest's canopy would be excellent for agriculture. Actually, this is not true. As heavy rain falls on the soil, the water works its way downward, removing nutrients in a process called *leaching*. Strangely enough, many of the nutrients needed to sustain the lush forest actually come from dead trees that decompose. The roots of the living trees absorb this matter before the rains leach it away.

When the forests are cleared for agricultural purposes, or for the timber, what is left is a thick red soil called **laterite**. When exposed to the intense sunlight of the tropics, the soil may harden into a bricklike consistency, making cultivation almost impossible.

Köppen classified tropical wet regions, where the monthly precipitation totals drop below 6 cm for perhaps one or two months, as **tropical monsoon climates** (Am). Here, yearly rainfall totals are similar to those of the tropical wet climate, usually exceeding 150 cm a year. Because the dry season is brief and copious rains fall throughout the rest of the year, there is sufficient soil moisture to maintain the tropical rain forest through the short dry period. Tropical monsoon climates can be seen in Fig. 18.6 along the coasts of Southeast Asia, India, and in northeastern South America.

Poleward of the tropical wet region, total annual rainfall diminishes, and there is a gradual transition from the tropical wet climate to the **tropical wet-and-dry climate** (Aw), where a distinct dry season prevails. Even though the annual precipitation usually exceeds 100 cm, the dry season, where the monthly rainfall is less than 6 cm (2.4 in.), lasts for more than two months. Because tropical rain forests cannot survive this "drought," the jungle gradually gives way to tall, coarse **savanna grass**, scattered with low, drought-resistant deciduous trees (see Fig. 18.9). The dry season occurs during the winter (low sun period), when the region is under the influence of the subtropical highs. In summer, the ITCZ moves poleward, bringing with it heavy precipitation, usually in the form of showers. Rainfall is enhanced by slow moving shallow lows that move through the region.

Tropical wet-and-dry climates not only receive less total rainfall than the tropical wet climates, but the rain that does occur is much less reliable, as the total rainfall often fluctuates widely from one year to the next. In the course of a single year, for example, destructive floods may be followed by serious droughts. As with tropical wet regions, the daily range of temperature usually exceeds the annual range, but the climate here is much less monotonous. There is a cool season in winter when the maximum temperature averages 30°C to 32°C (86°F to 90°F). At night, the low humidity and clear skies allow for rapid radiational cooling and, by early morning, minimum temperatures drop to 20°C (68°F) or below.

From Fig. 18.6, pp. 484–485, we can see that the principal areas having a tropical wet-and-dry climate (Aw) are those located in western Central America, in the region both north and south of the Amazon Basin (South America), in southcentral and eastern Africa, in parts of India and Southeast Asia, and in northern Australia. In many areas (especially within India and Southeast Asia), the marked variation in precipitation is associated with the *monsoon* —the seasonal reversal of winds.

As we saw in Chapter 10, the monsoon circulation is due in part to differential heating between land masses and oceans.

FIGURE 18.9
Acacia trees illustrate typical trees of the East African grassland savanna, a region with a tropical wet-and-dry climate (Aw).

FIGURE 18.10
Climatic data for Timbo, Guinea, latitude 11°N. A station with a tropical wet-and-dry climate (Aw).

During winter in the Northern Hemisphere, winds blow outward, away from a cold, shallow high-pressure area centered over continental Siberia. These downslope, relatively dry northeasterly winds from the interior provide India and Southeast Asia with generally fair weather and the dry season. In summer, the wind-flow pattern reverses as air flows into a developing thermal low over the continental interior. The humid air from the water rises and condenses, resulting in heavy rain and the wet season.

An example of a station with a tropical wet-and-dry climate (Aw) is given in Fig. 18.10. Located at latitude 11°N in west Africa, Timbo, Guinea, receives an annual average 163 cm (64 in.) of rainfall. Notice that the rainy season is during the summer when the ITCZ has migrated to its most northern position. Note also that practically no rain falls during the months of December, January, and February, when the region comes under the domination of the subtropical high-pressure area and its sinking air.

The monthly temperature patterns at Timbo are characteristic of most tropical wet-and-dry climates. As spring approaches, the noon sun is slightly higher, and the more intense sunshine produces greater surface heating and higher afternoon temperatures—usually above 32°C (90°F) and occasionally above 38°C (100°F)—creating hot, dry desertlike conditions. After this brief hot season, a persistent cloud cover and the evaporation of rain tends to lower the temperature during the summer. The warm, muggy weather of summer often resembles that of the tropical wet climate (Af). The rainy summer is followed by a warm, relatively dry period, with afternoon temperatures usually climbing above 30°C (86°F).

Poleward of the tropical wet-and-dry climate, the dry season becomes more severe. Clumps of trees are more isolated and the grasses dominate the landscape. When the potential annual water loss through evaporation and transpiration exceeds the annual water gain from precipitation, the climate is described as dry.

Dry Climates (Group B)

General characteristics: deficient precipitation most of the year; potential evaporation and transpiration exceed precipitation.

Extent: the subtropical deserts extend from roughly 20° to 30° latitude in large continental regions of the middle latitudes, often surrounded by mountains.

Major types: arid (BW)—the "true desert"—and semi-arid (BS).

A quick glance at Fig. 18.6, pp. 484–485, reveals that, according to Köppen, the dry regions of the world occupy more land area (about 26 percent) than any other major climatic type. Within these dry regions, a deficiency of water exists. Here, the potential annual loss of water through evaporation is greater than the annual water gained through precipitation. Thus, classifying a climate as dry depends not only on precipitation totals but also on temperature, which greatly influences evaporation. For example, 35 cm (14 in.) of precipitation in a hot climate will support only sparse vegetation, while the same amount of precipitation in northcentral Canada will support a conifer forest. In addition, a region with a low annual rainfall total is more likely to be classified as dry if the majority of precipitation is concentrated during the warm summer months, when evaporation rates are greater.

Precipitation in a dry climate is both meager and irregular. Typically, the lower the average annual rainfall, the greater its variability. For example, a station that reports an annual rainfall of 5 cm may actually measure no rainfall for two years; then, in a single downpour, it may receive 10 cm.

The major dry regions of the world can be divided into two primary categories. The first includes the area of the subtropics (between latitude 15° and 30°), where the sinking air of the subtropical anticyclones produces generally clear skies. The second is found in the continental areas of the middle lati-

FIGURE 18.11
Rain streamers (virga) are common in dry climates, as falling rain evaporates into the drier air before ever reaching the ground.

tudes. Here, far removed from a source of moisture, areas are deprived of precipitation. Dryness here is often accentuated by mountain ranges that produce a rain shadow effect.

Köppen divided dry climates into two types based on their degree of dryness: the *arid* (BW)* and the *semi-arid*, or steppe (BS). These two climatic types can be divided even further. For example, if the climate is hot and dry with a mean annual temperature above 18°C (64°F), it is either BWh or BSh (the *h* is for heiss, meaning hot in German). On the other hand, if the climate is cold (in winter, that is) and dry with a mean annual temperature below 18°C, then it is either BWk or BSk (where the *k* is for *kalt*, meaning cold in German).

The **arid climates** (BW) occupy about 12 percent of the world's land area. From Fig. 18.6, pp. 484–485, we can see that this climatic type is found along the west coast of South America and Africa and over much of the interior of Australia. Notice, also, that a swath of arid climate extends from northwest Africa all the way into central Asia. In North America, the arid climate extends from northern Mexico into the southern interior of the United States and northward along the leeward

slopes of the Sierra Nevada. This region includes both the Sonoran and Mojave deserts and the Great Basin.

The southern desert region of North America is dry because it is dominated by the subtropical high most of the year, and winter storm systems tend to weaken before they move into the area. The northern region is in the rain shadow of the Sierra Nevada. These regions are deficient in precipitation all year long, with many stations receiving less than 13 cm (5 in.) annually. As noted earlier, the rain that does fall is spotty, often in the form of scattered summer afternoon showers. Some of these showers can be downpours that change a gentle gully into a raging torrent of water. More often than not, however, the rain evaporates into the dry air before ever reaching the ground, and the result is rainstreamers (virga) dangling beneath the clouds (see Fig. 18.11).

Contrary to popular belief, few deserts are completely without vegetation. Although meager, the vegetation that does exist must depend on the infrequent rains. Thus, most of the native plants are **xerophytes**—those capable of surviving prolonged periods of drought (see Fig. 18.12). Such vegetation includes various forms of cacti and short-lived plants that spring up during the rainy periods.

---

*The letter *W* is for *Wüste*, the German word for desert.

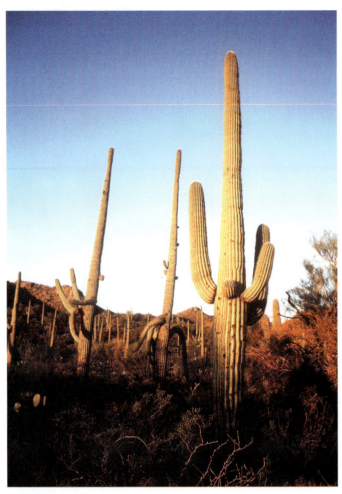

FIGURE 18.12
Creosote bushes and cactus are typical of the vegetation found in the arid southwestern American deserts (BWh).

In low-latitude deserts (BWh), intense sunlight produces scorching heat on the parched landscape. Here, air temperatures are as high as anywhere in the world. Maximum daytime readings during the summer can exceed 50°C (122°F), although 40°C to 45°C (104°F to 113°F) are more common. In the middle of the day, the relative humidity is usually between 5 and 25 percent. At night, the air's relatively low water vapor content allows for rapid radiational cooling. Minimum temperatures often drop below 25°C (77°F). Thus, arid climates have large daily temperature ranges, often between 15°C and 25°C (27°F and 45°F) and occasionally higher.

During the winter, temperatures are more moderate, and minimums may, on occasion, drop below freezing. The variation in temperature from summer to winter produces large annual temperature ranges. We can see this in the climate record for Phoenix, Arizona (see Fig. 18.13), a city in the southwestern United States with a BWh climate. Notice that the average annual temperature in Phoenix is 22°C (72°F), and that the average temperature of the warmest month (July)

reaches a sizzling 32°C (90°F). As we would expect, rainfall is meager in all months. There is, however, a slight maximum in July and August. This is due to the summer monsoon, when more humid, southerly winds are likely to sweep over the region and develop into afternoon showers and thunderstorms.

In middle latitude deserts (BWk), average annual temperatures are lower. Summers are typically warm to hot, with afternoon temperatures frequently reaching 40°C (104°F). Winters are usually extremely cold, with minimum temperatures sometimes dropping below −35°C (−31°F). Many of these deserts lie in the rainshadow of an extensive mountain chain, such as the Sierra Nevada and the Cascade mountains in North America, the Himalayan Mountains in Asia, and the Andes in South America. The meager precipitation that falls comes from an occasional summer shower or a passing mid-latitude cyclone in winter.

Again, refer to Fig. 18.6 and notice that around the margins of the arid regions, where rainfall amounts are greater, the climate gradually changes into **semi-arid** (BS). This region is called **steppe** and typically has short bunch grass, scattered low bushes, trees, or sagebrush (see Fig. 18.14). In North America, this climatic region includes most of the Great Plains, the southern coastal sections of California, and the northern

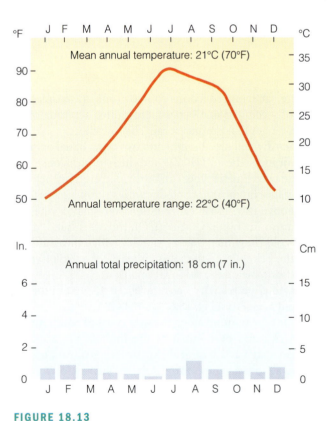

FIGURE 18.13
Climatic data for Phoenix, Arizona, latitude 33.5°N. A station with an arid climate (BWh).

FIGURE 18.14
Cumulus clouds forming over the steppe grasslands of western North America, a region with a semi-arid climate (BS).

valleys of the Great Basin. As in the arid region, northern areas experience lower winter temperatures and more frequent snowfalls. Annual precipitation is generally between 20 and 40 cm (8 and 16 in.). The climatic record for Denver, Colorado (see Fig. 18.15), exemplifies the semi-arid (BSk) climate.

As average rainfall amounts increase, the climate gradually changes to one that is more humid. Hence, the semi-arid (steppe) climate marks the transition between the arid and the humid climatic regions. (Before reading about moist climates, you may wish to read the Focus section on p. 492 about deserts that experience drizzle but little rainfall.)

Moist Subtropical Mid-latitude Climates (Group C)

General characteristics: humid with mild winters (i.e., average temperature of the coldest month below 18°C, or 64°F, and above −3°C, or 27°F).

Extent: on the eastern and western regions of most continents, from about 25° to 40° latitude.

Major types: humid subtropical (Cfa), marine (Cfb), and dry-summer subtropical, or Mediterranean (Cs).

The Group C climates of the middle latitudes have distinct summer and winter seasons. Additionally, they have ample precipitation to keep them from being classified as dry. Al-

though winters can be cold, and air temperatures can change appreciably from one day to the next, no month has a mean temperature below −3°C (27°F), for if it did, it would be classified as a D climate—one with severe winters.

The first C climate we will consider is the **humid subtropical climate** (Cfa). Notice in Fig. 18.6, that Cfa climates are found principally along the east coasts of continents, roughly between 25° and 40° latitude. They dominate the southeastern section of the United States, as well as eastern China and southern Japan. In the Southern Hemisphere, they are found in southeastern South America and along the southeastern coasts of Africa and Australia.

A trademark of the humid subtropical climate is its hot, muggy summers. This sultry summer weather occurs because Cfa climates are located on the western side of subtropical highs, where maritime tropical air from lower latitudes is swept poleward into these regions. Generally, summer dew-point temperatures are high (often exceeding 23°C, or 73°F) and so is the relative humidity, even during the middle of the day. The high humidity combines with the high air temperature (usually above 32°C, or 90°F) to produce more oppressive conditions than are found in equatorial regions. Summer morning low temperatures often range between 21°C and 27°C (70°F and 81°F). Occasionally, a weak summer cool front will bring temporary relief from the sweltering conditions.

A Desert with Clouds and Drizzle

We already know that not all deserts are hot. By the same token, not all deserts are sunny. In fact, some coastal deserts experience considerable cloudiness, especially low stratus and fog.

Amazingly, these coastal deserts are some of the driest places on earth. They include the Atacama Desert of Chile and Peru, the coastal Sahara Desert of northwest Africa, the Namib Desert of southwestern Africa, and a portion of the Sonoran Desert in Baja, California (see Fig. 2). On the Atacama Desert, for example, some regions go without measurable rainfall for decades. And Arica, in northern Chile, has an annual rainfall of only 0.08 cm (0.03 in.).

The cause of this aridity is, in part, due to the fact that each region is adjacent to a large body of relatively cool water. Notice in Fig. 2 that these deserts are located along the western coastal margins of continents, where a subtropical high-pressure area causes prevailing winds to move cool water from higher latitudes along the coast. In addition, these winds help to accentuate the water's coldness by initiating *upwelling*—the rising of cold water from lower levels. The combination of these conditions tends to produce coastal water temperatures between 10°C and 15°C (50°F and 59°F), which is quite cool for such low latitudes. As surface air sweeps across the cold water, it is

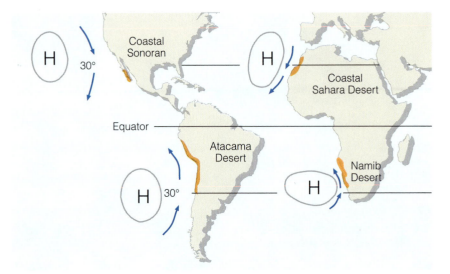

FIGURE 2
Location of coastal deserts that experience frequent fog, drizzle, and low clouds.

chilled to its dew point, often producing a blanket of fog and low clouds, from which drizzle falls. The drizzle, however, accounts for very little rainfall. In most regions, it is only enough to dampen the streets with a mere trace of precipitation.

As the cool stable air moves inland, it warms, and the water droplets evaporate. Hence, most of the cloudiness and drizzle is found along the immediate coast. Although the relative humidity of this air is high, the dew-point temperature is comparatively low (often near that of the coastal surface water). Inland, further warming causes the air to rise. However, a stable subsidence

inversion, associated with the subtropical highs, inhibits vertical motions by capping the rising air, causing it to drift back toward the ocean, where it sinks, completing a rather strong sea breeze circulation. The position of the subtropical highs, which tend to remain almost stationary, plays an additional role by preventing the Intertropical Convergence Zone with its rising, unstable air from entering the region.

And so we have a desert with clouds and drizzle—a desert that owes its existence, in part, to its proximity to rather cold ocean water and, in part, to the position and air motions of the subtropical highs.

However, devastating heat waves, sometimes lasting many weeks, can occur when an upper-level ridge moves over the area.

Winters tend to be relatively mild, especially in the lower latitudes, where air temperatures rarely dip much below freezing. Poleward regions experience winters that are colder and harsher. Here, frost, snow, and ice storms are more common,

but heavy snowfalls are rare. Winter weather can be quite changeable, as almost summerlike conditions can give way to cold rain and wind in a matter of hours when a middle latitude storm and its accompanying fronts pass through the region.

Humid subtropical climates experience adequate and fairly well-distributed precipitation throughout the year, with

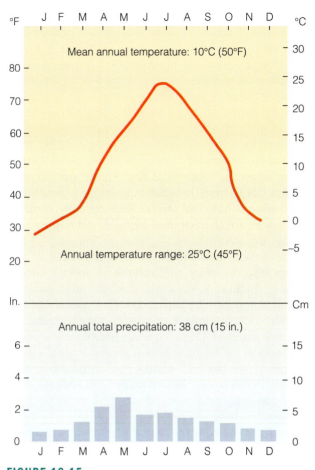

FIGURE 18.15
Climatic data for Denver, Colorado, latitude 40°N. A station with a semi-arid climate (BSk).

where summers are longer (but still cool), the climate is classified as *west coast marine*, or simply **marine**, Cfb.

Where mountains parallel the coastline, such as along the west coasts of North and South America, the marine influence is restricted to narrow belts. Unobstructed by high mountains, prevailing westerly winds pump ocean air over much of western Europe and thus provide this region with a marine climate (Cfb).

During much of the year, marine climates are characterized by low clouds, fog, and drizzle. The ocean's influence produces adequate precipitation in all months, with much of it falling as light or moderate rain associated with maritime polar air masses. Snow does fall, but frequently it turns to slush after only a day or so. In some locations, topography greatly enhances precipitation totals. For example, along the west coast of North America, coastal mountains not only force air upward enhancing precipitation, they also slow the storm's

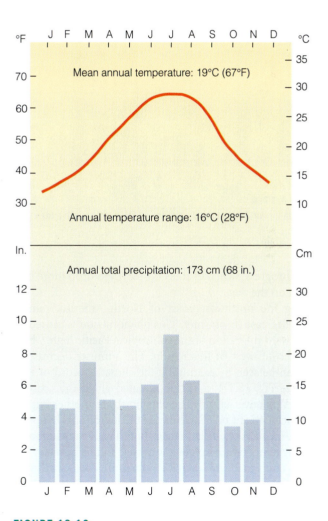

FIGURE 18.16
Climatic data for Mobile, Alabama, latitude 30°N. A station with a humid subtropical climate (Cfa).

typical annual averages between 80 and 165 cm (31 and 65 in.). In summer, when thunderstorms are common, much of the precipitation falls as afternoon showers. Tropical storms entering the United States and China can substantially add to their summer and autumn rainfall totals. Winter precipitation most often occurs with eastward-trekking middle latitude storms. In the southeastern United States, the abundant rainfall supports a thick pine forest that becomes mixed with oak at higher latitudes. The climate data for Mobile, Alabama, a city with a Cfa climate, is given in Fig. 18.16.

Glance back at Fig. 18.6, pp. 484–485, and observe that C climates extend poleward along the western side of most continents from about latitude 40° to 60°. These regions are dominated by prevailing winds from the ocean that moderate the climate, keeping winters considerably milder than stations located at the same latitude farther inland. In addition to this, summers are quite cool. When the summer season is both short and cool, the climate is designated as Cfc.* Equatorward,

---

*Table 18.3, p. 483, details the necessary criteria for each climatic type.

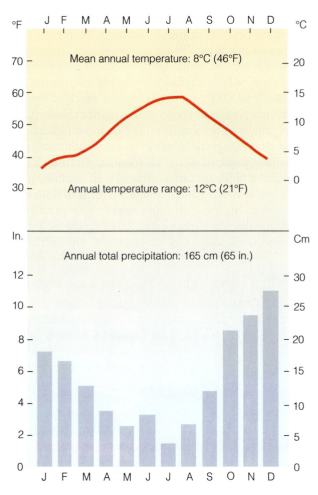

FIGURE 18.17

Climatic data for Port Hardy, Canada, latitude 51°N. A station with a marine climate (Cfb).

eastward progress, which enables the storm to drop more precipitation on the area.

Along the northwest coast of North America, rainfall amounts decrease in summer. This phenomenon is caused by the northward migration of the subtropical Pacific high, which is located southwest of this region. The summer decrease in rainfall can be seen by examining the climatic record of Port Hardy (see Fig. 18.17), a station situated along the coast of Canada's Vancouver Island. The data illustrate another important characteristic of marine climates: the low annual temperature range for such a high-latitude station. The ocean's influence keeps daily temperature ranges low as well. In this climate type, it rains on many days and when it is not raining, skies are usually overcast. The heavy rains produce a dense forest of Douglas fir.

Moving equatorward of marine climates, the influence of the subtropical highs becomes greater, and the summer dry period more pronounced. Gradually, the climate changes from

marine to one of **dry-summer subtropical** (Cs), or **Mediterranean**, because it also borders the coastal areas of the Mediterranean Sea. Along the west coast of North America, Portland, Oregon, because it has rather dry summers, marks the transition between the marine climate and the dry-summer subtropical climate to the south.

The extreme summer aridity of the Mediterranean climate, which in California may exist for five months, is caused by the sinking air of the subtropical highs. In addition, these anticyclones divert summer storm systems poleward. During the winter, when the subtropical highs move equatorward, mid-latitude storms from the ocean frequent the region, bringing with them much needed rainfall. Consequently, Mediterranean climates are characterized by mild, wet winters, and mild-to-hot, dry summers.

Where surface winds parallel the coast, upwelling of cold water helps keep the water itself and the air above it cool all summer long. In these coastal areas, which are often shrouded in low clouds and fog, the climate is called *coastal Mediterranean* (Csb). Here, summer daytime maximum temperatures usually reach about 21°C (70°F), while overnight lows often drop below 15°C (59°F). Inland, away from the ocean's influence, summers are hot and winters are a little cooler than coastal areas. In this *interior Mediterranean climate* (Csa), summer afternoon temperatures usually climb above 34°C (93°F) and occasionally above 40°C (104°F).

Figure 18.18 contrasts the coastal Mediterranean climate of San Francisco, California, with the interior Mediterranean climate of Sacramento, California. While Sacramento is only 130 km (80 mi) inland from San Francisco, Sacramento's average July temperature is 9°C (16°F) higher. As we would expect, Sacramento's annual temperature range is considerably higher, too. Although Sacramento and San Francisco both experience an occasional frost, snow in these areas is a rarity.

In Mediterranean climates, yearly precipitation amounts range between 30 and 90 cm (14 and 35 in.). However, much more precipitation falls on surrounding hillsides and mountains. Because of the summer dryness, the land supports only a scrubby type of low-growing woody plants and trees called *chaparral* (see Fig. 18.19).

At this point, we should note that summers are not as dry along the Mediterranean Sea as they are along the west coast of North America. Moreover, coastal Mediterranean areas are also warmer, due to the lack of upwelling in the Mediterranean Sea.

Before leaving our discussion of C climates, note that when the dry season is in winter, the climate is classified as Cw. Over northern India and portions of China, the relatively dry winters are the result of northerly winds from continental regions circulating southward around the cold Siberian high. Many lower-latitude regions with a Cw climate would be tropical if it were not for the fact they are too high in elevation and, consequently, too cool to be designated as tropical.

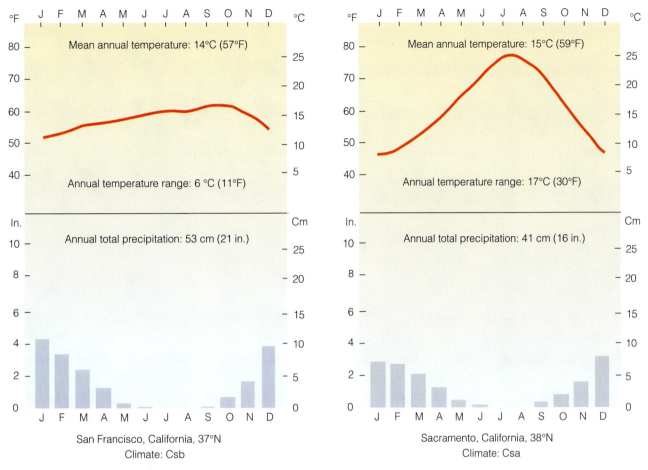

FIGURE 18.18
Comparison of a coastal Mediterranean climate, Csb (San Francisco, at left), with an interior Mediterranean climate, Csa (Sacramento, at right).

Moist Continental Climates (Group D)

General characteristics: warm-to-cool summers and cold winters (i.e., average temperature of warmest month exceeds 10°C, or 50°F, and the coldest monthly average drops below −3°C, or 27°F); winters are severe with snowstorms, blustery winds, bitter cold; climate controlled by large continent.

Extent: north of moist subtropical mid-latitude climates.

Major types: humid continental with hot summers (Dfa), humid continental with cool summers (Dfb), and subpolar (Dfc).

The D climates are controlled by large land masses. Therefore, they are found only in the Northern Hemisphere. Look at the climate map, Fig. 18.6 on pp. 484–485, and notice that D climates extend across North America and Eurasia, from about latitude 40°N to almost 70°N. In general, they are characterized by cold winters and warm-to-cool summers.

As we know, for a station to have a D climate, the average temperature of its coldest month must dip below −3°C (27°F). This is not an arbitrary number. Köppen found that, in Europe, this temperature marked the southern limit of persistent snow cover in winter.* Hence, D climates experience a great deal of winter snow that stays on the ground for extended periods. When the temperature drops to a point such that no month has an average temperature of 10°C (50°F), the climate is classified as polar (E). Köppen found that the average monthly temperature of 10°C tended to represent the minimum temperature required for tree growth. So no matter how cold it gets in a D climate (and winters can get extremely cold), there is enough summer warmth to support the growth of trees.

---

*In North America, studies suggest that an average monthly temperature of 0°C (32°F) or below for the coldest month seems to correspond better to persistent winter snow cover.

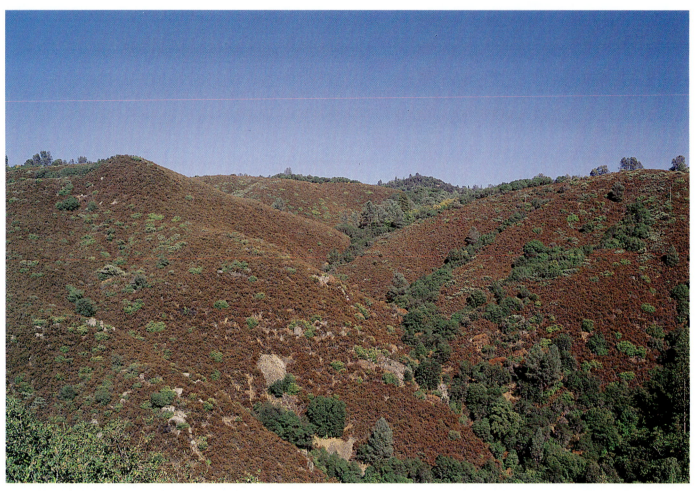

FIGURE 18.19
In the Mediterranean-type climates of North America, typical chaparral vegetation includes chamise, manzanita, and foothill pine.

There are two basic types of D climates: the **humid continental** (Dfa and Dfb) and the **subpolar** (Dfc). Humid continental climates are observed from about latitude 40°N to 50°N (60°N in Europe). Here, precipitation is adequate and fairly evenly distributed throughout the year, although interior stations experience maximum precipitation in summer. Annual precipitation totals usually range from 50 to 100 cm (20 to 40 in.). Native vegetation in the wetter regions includes forests of spruce, fir, pine, and oak. In autumn, nature's pageantry unveils itself as the leaves of deciduous trees turn brilliant shades of red, orange, and yellow (see Fig. 18.20).

Humid continental climates are subdivided on the basis of summer temperatures. Where summers are long and hot,* the climate is described as *humid continental with hot summers* (Dfa). Here summers are often hot and humid, especially in the southern regions. Midday temperatures often exceed 32°C (90°F) and occasionally 40°C (104°F). Summer nights are usually warm and humid, as well. The frost-free season normally lasts from five to six months, long enough to grow a wide variety of crops. Winters tend to be windy, cold, and snowy. Farther north, where summers are shorter and not as hot,* the climate is described as *humid continental with long cool summers* (Dfb). In Dfb climates, summers are not only cooler but much less humid. Temperatures may exceed 35°C (95°F) for a time, but extended hot spells lasting many weeks are rare. The frost-free season is shorter than in the Dfa climate, and normally lasts between three and five months. Winters are long, cold, and windy. It is not uncommon for temperatures to drop below –30°C (–22°F) and

---

*Hot means that the average temperature of the warmest month is above 22°C (72°F) and at least four months have a monthly mean temperature above 10°C (50°F).

*"Not as hot" means that the average temperature of the warmest month is below 22°C (72°F) and at least four months have a monthly mean temperature above 10°C (50°F).

stay below −18°C (0°F) for days and sometimes weeks. Autumn is short, with winter often arriving right on the heels of summer. Spring, too, is short, as late spring snowstorms are common, especially in the more northern latitudes.

Figure 18.21 compares the Dfa climate of Des Moines, Iowa, with the Dfb climate of Winnipeg, Canada. Notice that both cities experience a large annual temperature range. This is characteristic of climates located in the northern interior of continents. In fact, as we move poleward, the annual temperature range increases. In Des Moines, it is 31°C (56°F), while 950 km (590 mi) to the north in Winnipeg, it is 38°C (68°F). The summer precipitation maximum expected for these interior

WEATHER WATCH

Does it actually get warmer in polar climates than in tropical climates? Well, the highest temperature ever measured in Prospect Creek, Alaska (100°F), a city situated on the Arctic Circle with a subpolar climate, is exactly 2°F higher than the warmest temperature ever measured in Belem, Brazil, a city situated near the equator with a tropical wet climate.

FIGURE 18.20
The leaves of deciduous trees burst into brilliant color during autumn over the countryside of Adirondack Park, a region with a humid continental climate.

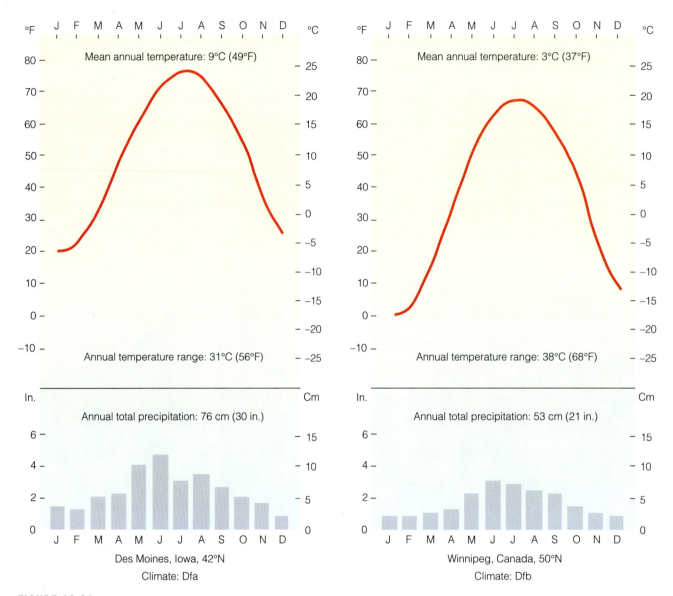

FIGURE 18.21
Comparison of a humid continental hot summer climate, Dfa (Des Moines, at left), with a humid continental cool summer climate, Dfb (Winnipeg, at right).

continental locations shows up well in Fig. 18.21. Most of the summer rain is in the form of air-mass convective showers, although an occasional weak frontal system can produce more widespread precipitation, as can a cluster of thunderstorms—the Mesoscale Convective Complex described in Chapter 15. The weather in both climatic types can be quite changeable, especially in winter, when a brief warm spell is replaced by blustery winds and temperatures plummeting well below −30°C (−22°F).

When winters are severe and summers short and cool, with only one to three months having a mean temperature exceeding 10°C (50°F), the climate is described as *subpolar*

(Dfc). From Fig. 18.6, we can see that, in North America, this climate occurs in a broad belt across Canada and Alaska; in Eurasia, it stretches from Norway over much of Siberia. The exceedingly low temperatures of winter account for these areas being the primary source regions for continental polar and arctic air masses. Extremely cold winters coupled with cool summers produce large annual temperature ranges, as exemplified by the climate data in Fig. 18.22 for Fairbanks, Alaska.

Precipitation is comparatively light in the subpolar climates, especially in the interior regions, with most places receiving less than 50 cm (20 in.) annually. A good percentage

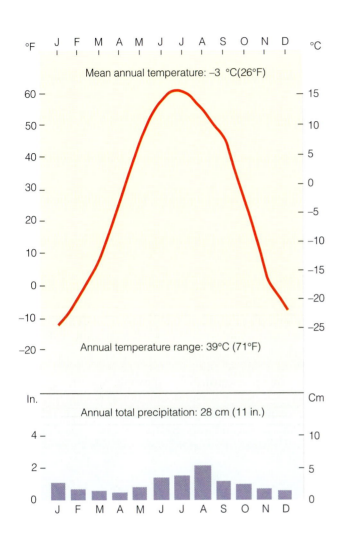

FIGURE 18.22
Climatic data for Fairbanks, Alaska, latitude 65°N. A station with a subpolar climate (Dfc).

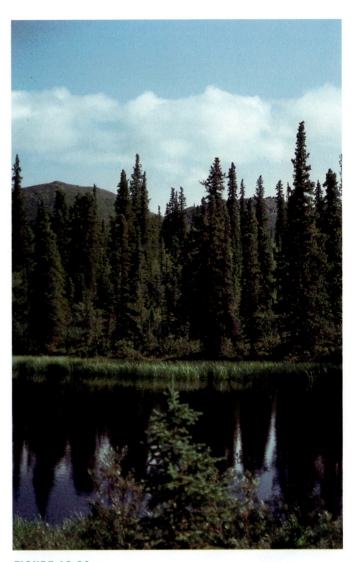

FIGURE 18.23
Coniferous forests (taiga) such as this occur where winter temperatures are low and precipitation is abundant.

of the precipitation falls when weak cyclonic storms move through the region in summer. The total snowfall is usually not large but the cold air prevents melting, so snow stays on the ground for months at a time. Because of the low temperatures, there is a low annual rate of evaporation that ensures adequate moisture to support the boreal forests of conifers and birches known as **taiga** (see Fig. 18.23). Hence, the subpolar climate is known also as a *boreal climate* and as a *taiga climate*.

In the taiga region of northern Siberia and Asia, where the average temperature of the coldest month drops to a frigid −38°C (−36°F) or below, the climate is designated Dfd. Where the winters are considered dry, the climate is designated Dwd.

Polar Climates (Group E)

General characteristics: year-round low temperatures (i.e., average temperature of the warmest month is below 10°C, or 50°F).

Extent: northern coastal areas of North America and Eurasia; Greenland and Antarctica.

Major types: polar tundra (ET) and polar ice caps (EF).

In the **polar tundra** (ET), the average temperature of the warmest month is below 10°C (50°F), but above freezing. (See Fig. 18.24, the climate data for Barrow, Alaska.) Here, the ground is permanently frozen to depths of hundreds of meters,

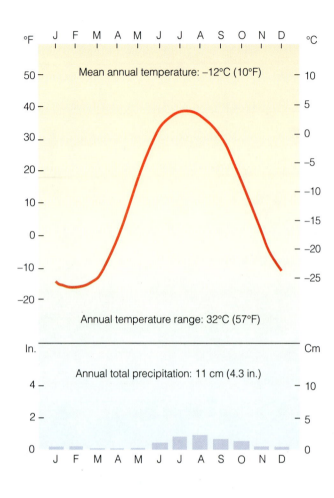

a condition known as **permafrost**. Summer weather, however, is just warm enough to thaw out the upper meter or so of soil. Hence, during the summer, the tundra turns swampy and muddy. Annual precipitation on the tundra is meager, with most stations receiving less than 20 cm (8 in.). In lower latitudes, this would constitute a desert, but in the cold polar regions evaporation rates are very low and moisture remains adequate. Because of the extremely short growing season, *tundra vegetation* consists of mosses, lichens, dwarf trees, and scattered woody vegetation, fully grown and only several centimeters tall (see Fig. 18.25).

Even though summer days are long, the sun is never very high above the horizon. Additionally, some of the sunlight that reaches the surface is reflected by snow and ice, while some is used to melt the frozen soil. Consequently, in spite of the long hours of daylight, summers are quite cool. The cool summers and the extremely cold winters produce large annual temperature ranges.

When the average temperature for every month drops below freezing, plant growth is impossible, and the region is perpetually covered with snow and ice. This climatic type is known as **polar ice cap** (EF). It occupies the interior ice sheets of Greenland and Antarctica, where the depth of ice in some places measures thousands of meters. In this region, tem-

FIGURE 18.25
Tundra vegetation in Alaska. This type of tundra is composed mostly of sedges and dwarfed wildflowers that bloom during the brief growing season.

FIGURE 18.26

Climatic data for Eismitte, Greenland, latitude 71°N. Located in the interior of Greenland at an elevation of 3030 m (9941 ft), Eismitte has a polar ice cap climate (EF).

peratures are never much above freezing, even during the middle of "summer." The coldest places in the world are located here. Precipitation is extremely meager with many places receiving less than 10 cm (4 in.) annually. Most precipitation falls as snow during the "warmer" summer. Strong downslope katabatic winds frequently whip the snow about, adding to the climate's harshness. The data in Fig. 18.26 for Eismitte, Greenland, illustrate the severity of an EF climate.

Highland Climates (Group H)

It is not necessary to visit the polar regions to experience a polar climate. Because temperature decreases with altitude, climatic changes experienced when climbing 300 m (1000 ft) in elevation are about equivalent in high latitudes to horizontal changes experienced when traveling 300 km (186 mi) northward. (This distance is equal to about 3° latitude.) Therefore, when ascending a high mountain, one can travel through many climatic regions in a relatively short distance.

Figure 18.27 shows how the climate and vegetation change along the western slopes of the central Sierra Nevada. (See Fig. 18.5, p. 479, for the precipitation patterns for this region.) Notice that, at the base of the mountains, the climate and vegetation represent semi-arid conditions, while in the foothills the climate becomes Mediterranean and the vegetation changes to chaparral. Higher up, thick fir and pine forests prevail. At still higher elevations, the climate is subpolar and the taiga gives way to dwarf trees and tundra vegetation. Near the summit there are permanent patches of ice and snow, with some small glaciers nestled in protected areas. Hence, in less than 13,000 vertical feet, the climate has changed from semi-arid to polar.

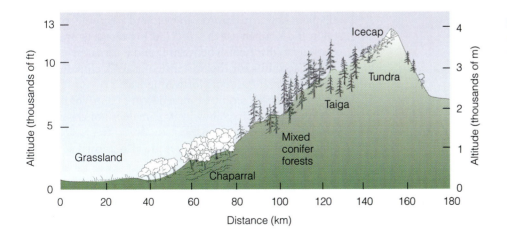

FIGURE 18.27

Vertical view of changing vegetation and climate due to elevation in the central Sierra Nevada.

SUMMARY

In this chapter, we examined global temperature and precipitation patterns, as well as the various climatic regions throughout the world. Tropical climates are found in low latitudes, where the noon sun is always high, day and night are of nearly equal length, every month is warm, and no real winter season exists. Some of the rainiest places in the world exist in the tropics, especially where warm, humid air rises upslope along mountain ranges.

Dry climates prevail where potential evaporation and transpiration exceed precipitation. Some deserts, such as the Sahara, are mainly the result of sinking air associated with the subtropical highs, while others, due to the rain shadow effect, are found on the leeward side of mountains. Many deserts form in response to both of these effects.

Middle latitudes are characterized by a distinct winter and summer season. Winters tend to be milder in lower latitudes and more severe in higher latitudes. Along the east coast of some continents, summers tend to be hot and humid as moist air sweeps poleward around the subtropical highs. The air often rises and condenses into afternoon thunderstorms in this humid subtropical climate. The west coasts of many continents tend to be drier, especially in summer, as the combination of cool ocean water and sinking air of the subtropical highs, to a large degree, inhibit the formation of cumuliform clouds.

In the middle of large continents, such as North America and Eurasia, summers are usually wetter than winters. Winter temperatures are generally lower than those experienced in coastal regions. As one moves northward, summers become shorter and winters longer and colder. Polar climates prevail at high latitudes, where winters are severe and there is no real summer. When ascending a high mountain, one can travel through many climatic zones in a relatively short distance.

Key Terms

The following terms are listed in the order they appear in the text. Define each. Doing so will aid you in reviewing the material covered in this chapter.

microclimate	tropical monsoon climate
mesoclimate	tropical wet-and-dry
macroclimate	climate
global climate	savanna grass
climatic controls	arid climate
Köppen classification system	xerophytes
P/E index	semi-arid climate
tropical rain forest	steppe
tropical wet climate	humid subtropical climate
laterite	marine climate

dry-summer subtropical	polar tundra climate
(Mediterranean)	permafrost
humid continental climate	polar ice cap climate
subpolar climate	
taiga	

Questions for Review

1. Describe as many factors as you can that influence the global pattern of precipitation.
2. Explain why, in North America, precipitation typically is a maximum along the west coast in winter, a maximum on the central plains in summer, and fairly evenly distributed between summer and winter along the east coast.
3. What constitutes a dry climate?
4. What climatic information did Köppen use in classifying climates?
5. In which climatic region would each of the following be observed: tropical rain forest, xerophytes, steppe, taiga, tundra, and savanna?
6. Why do marine climates (Cs) usually lie along the west coast of continents?
7. Why are large annual temperature ranges characteristic of D-type climates?
8. Why are D climates only found in the Northern Hemisphere?
9. Explain why a tropical rain forest climate will support a tropical rain forest, while a tropical wet-and-dry climate will not.
10. What is the primary distinction between a Cfa and a Dfa climate?
11. Explain how arid deserts can be found adjacent to oceans.
12. Why did Köppen use the 10°C (50°F) average temperature for July to distinguish between D and E climates?
13. What accounts for the existence of a BWk climate in the western Great Basin of North America?
14. Barrow, Alaska, receives a mere 11 cm (4.3 in.) of precipitation annually. Explain why its climate is not classified as arid or semi-arid.
15. Explain why subpolar climates are also known as boreal climates and taiga climates.

Questions for Thought

1. Why do cities directly east of the Rockies (such as Denver, Colorado) receive much more precipitation than cities east of the Sierra Nevada (such as Reno and Lovelock, Nevada)?

2. What climatic controls affect the climate in your area?
3. Why do Atlantic coastal cities, such as Boston, Massachusetts, have continental climates instead of marine climates?
4. Many structures in a polar tundra climate are built on pilings. Explain why.
5. Why are summer afternoon temperatures in a humid subtropical climate (Cfa) often higher than in a tropical wet climate (Af)?

Problems and Exercises

1. According to Köppen's climatic system (see Fig. 18.6, pp. 484–485), what major climatic type is most abundant (a) in North America, (b) in South America, (c) throughout the world?

2. Suppose a city has the mean annual precipitation and temperature given in Table 18.4. Based on Köppen's climatic types, how would this climate be classified? On a map of North America, approximately where would this city be located? What type of vegetation would you expect to see there? Answer these same questions for the data in Table 18.5.

3. Compare the following climate classifications for your area:
   (a) Ancient Greeks
   (b) Köppen system
   (c) Thornthwaite's system
   Which classification system is best for your area's mesoclimate? Macroclimate?

TABLE 18.4

	JAN.	FEB.	MAR.	APR.	MAY	JUNE	JULY	AUG.	SEPT.	OCT.	NOV.	DEC.	YEAR
Temperature (°F)	40	42	50	60	68	77	80	79	73	62	49	42	60
Precipitation (in.)	4.9	4.2	5.3	3.7	3.8	3.2	4.0	3.3	2.7	2.5	3.4	4.1	45

TABLE 18.5

	JAN.	FEB.	MAR.	APR.	MAY	JUNE	JULY	AUG.	SEPT.	OCT.	NOV.	DEC.	YEAR
Temperature (°F)	18	18	29	42	55	65	70	68	60	48	36	23	44
Precipitation (in.)	1.9	1.5	2.2	2.6	2.9	3.6	3.8	3.0	3.1	2.9	2.8	1.9	32

Ten thousand years ago, additional water from precipitation and melting glaciers formed giant Lake Bonneville—a lake that covered over 50,000 square kilometers of northwestern Utah. Since then, the climate has changed. Evaporation in this semi-arid region has gradually lowered the lake until, today, most of the landscape is a flat, salt-covered bed that occasionally becomes dotted with puddles after a summer rain shower. (Photo by author.)

CLIMATE CHANGE

CONTENTS

A change in our climate however is taking place very sensibly. Both heats and colds are becoming much more moderate within the memory even of the middle-aged. Snows are less frequent and less deep. They do not often lie, below the mountains, more than one, two, or three days, and very rarely a week. They are remembered to have been formerly frequent, deep, and of long continuance. The elderly inform me the earth used to be covered with snow about three months in every year. The rivers, which then seldom failed to freeze over in the course of the winter, scarcely ever do now. This change has produced an unfortunate fluctuation between heat and cold, in the spring of the year, which is very fatal to fruits. In an interval of twenty-eight years, there was no instance of fruit killed by the frost in the neighborhood of Monticello. The accumulated snows of the winter remaining to be dissolved all together in the spring, produced those overflowings of our rivers, so frequent then, and so rare now.

Thomas Jefferson, *Notes on the State of Virginia*, 1781

The climate is always changing. Evidence shows that climate has changed in the past, and nothing suggests that it will not continue to change. In Chapter 17, we saw that as the urban environment grows, its climate differs from that of the region around it. Sometimes the difference is striking, as when city nights are warmer than the nights of the outlying rural areas. Other times, the difference is subtle, as when a layer of smoke and haze covers the city or when the climate of a relatively small area—the microclimate—becomes modified by the light and warmth of a city street lamp (see Fig. 19.1). Climate change, in the form of a persistent drought or a delay in the annual monsoon rains, can adversely affect the lives of millions. Even small changes can have an adverse effect when averaged over many years, as when grasslands once used for grazing gradually become uninhabited deserts. In this chapter, first we will investigate how the global climate has changed, then we will examine some theories on why it has changed.

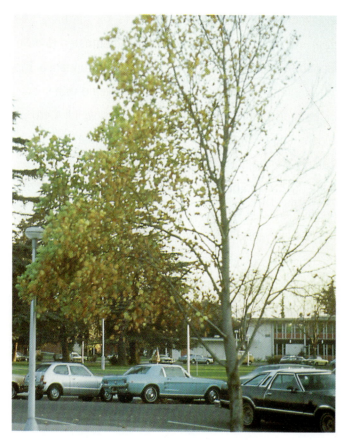

FIGURE 19.1

Altering the microclimate. Notice in the picture that the leaves are still on the tree near the streetlight. Apparently, this sodium vapor lamp emits enough warmth and light during the night to trick the leaves into behaving as if it were September rather than the middle of November.

THE EARTH'S CHANGING CLIMATE

Not only is the earth's climate always changing, but a mere 18,000 years ago the earth was in the grip of a cold spell, with *alpine glaciers* extending their ice fingers down river valleys and huge ice sheets (*continental glaciers*) covering vast areas of North America and Europe. The ice measured several kilometers thick and extended as far south as New York and the Ohio River Valley. Perhaps the glaciers advanced 10 times during the last 2 million years, only to retreat. In the warmer periods, between glacier advances, average global temperatures were slightly higher than at present. Hence, some scientists feel that we are still in an ice age, but in the comparatively warmer part of it.

Presently, glaciers cover only about 10 percent of the earth's land surface. The total volume of ice over the face of the earth amounts to a little more than 25 million cubic kilometers. Most of this ice is in the Greenland and Antarctic ice sheets, and its accumulation over time has allowed scientists to measure past climatic changes. If global temperatures were to rise enough so that all of this ice melted, the level of the ocean would rise about 65 m (213 ft) (see Fig. 19.2). Imagine the catastrophic results: Many major cities (such as New York, Tokyo, and London) would be inundated. Even a rise in global temperature of several degrees Celsius might be enough to raise sea level by about half a meter or so, flooding coastal lowlands.

Determining Past Climates   The study of the geological evidence left behind by advancing and retreating glaciers is one factor suggesting that global climate has undergone slow but continuous changes. To reconstruct past climates, scientists must examine and then carefully piece together all the available evidence. Unfortunately, the evidence only gives a general understanding of what the climate was like. For example, fossil pollen of a tundra plant collected in a layer of sediment in New England and dated to be 12,000 years old suggests that the climate of that region was much colder than it is today.

Other evidence of global climatic change comes from core samples taken from ocean floor sediments and ice from Greenland. A multiuniversity research project known as CLIMAP (*Climate: long-range investigation mapping and prediction*) studied the past million years of global climate. Thousands of meters of ocean sediment obtained with a hollow-centered drill were analyzed. The sediment contains the remains of calcium carbonate shells of organisms that once lived near the surface. Because certain organisms live within a narrow range of temperature, the distribution and type of organisms within the sediment indicate the surface water temperature.

In addition, the oxygen-isotope* ratio of these shells provides information about the sequence of glacier advances. For

---

*Isotopes are atoms whose nuclei have the same number of protons but different numbers of neutrons.

FIGURE 19.2
If all the ice locked up in glaciers and ice sheets were to melt, estimates are that this coastal area of south Florida would be under 65 m (213 ft) of water. Even a relatively small 1-m rise in sea level would threaten half of the world's population with rising seas.

example, most of the oxygen in sea water is composed of 8 protons and 8 neutrons in its nucleus, giving it an atomic weight of 16. However, about one out of every thousand oxygen atoms contains an extra 2 neutrons, giving it an atomic weight of 18. When ocean water evaporates, the heavy oxygen 18 tends to be left behind. Consequently, during periods of glacier advance, the oceans, which contain less water, have a higher concentration of oxygen 18. Since the shells of marine organisms are constructed from the oxygen atoms existing in ocean water, determining the ratio of oxygen 18 to oxygen 16 within these shells yields information about how the climate may have varied in the past. A higher ratio of oxygen 18 to oxygen 16 in the sediment record suggests a colder climate, while a lower ratio suggests a warmer climate. Using data such as these, the CLIMAP project was able to reconstruct the earth's surface ocean temperature for various times during the past (see Fig. 19.3).

Vertical ice cores extracted from glaciers in Antarctica and Greenland provide additional information on past temperature patterns. Glaciers form over land where temperatures are sufficiently low so that, during the course of a year, more snow falls than will melt. Successive snow accumulations over many years compact the snow, which slowly recrystallizes into ice. Under the influence of gravity, the ice begins to move, and a glacier is born. Since ice is composed of hydrogen and oxygen, examining the oxygen-isotope ratio in ancient cores provides a past record of temperature trends. Generally, the colder the air when the snow fell, the richer the concentration of oxygen 16 in the core. Moreover, bubbles of ancient air trapped in the ice can be analyzed to determine the past composition of the atmosphere (see Fig. 19.11, p. 515).

Scientists are even using the calcium carbonate material that forms into tiny stones (*otoliths*) in the inner ears of fish to reconstruct past temperatures of the Great Lakes region. As the otoliths grow, they extract oxygen from the lake water. The oxygen-isotope ratio then provides scientists with information on changes in water temperature over the life of the fish, whether it died last week or 10,000 years ago.

Still other evidence of climatic change comes from the study of annual growth rings of trees, called **dendrochronology**. As a tree grows, it produces a layer of wood cells under its bark. Each year's growth appears as a ring. The changes in thickness of the rings indicate climatic changes that may have taken place from one year to the next. Tree rings are only useful in regions that experience an annual cycle and in trees that are stressed by temperature or moisture during their growing season. The growth of tree rings has been correlated

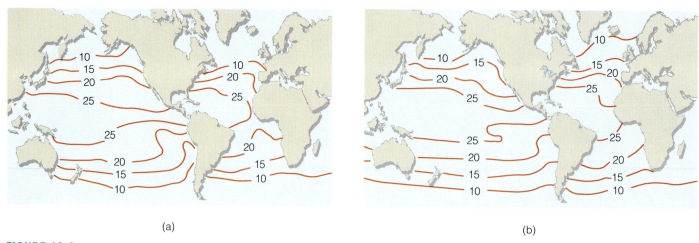

(a)   (b)

FIGURE 19.3

(a) Sea surface isotherms (°C) during August 18,000 years ago and (b) during August today.

with precipitation and temperature patterns for hundreds of years into the past in various regions of the world.

Other data have been used to reconstruct past climates, such as:

1. records of natural lake-bottom sediment and soil deposits
2. the study of pollen in deep ice caves, soil deposits, and sea sediments
3. certain geologic evidence (ancient coal beds, sand dunes, and fossils) and the change in the water level of closed basin lakes
4. documents concerning droughts, floods, and crop yields
5. the study of oxygen-isotope ratios of corals
6. dating calcium carbonate layers of stalactites in caves

Despite all of these data, our knowledge about past climates is still incomplete. Now that we have reviewed *how* the climatologist gains information about the past, let's look at *what* this information reveals.

Climate Through the Ages   Throughout much of the earth's history, the global climate was probably between 8°C and 15°C warmer than it is today. During most of this time, the polar regions were free of ice. These comparatively warm conditions, however, were interrupted by several periods of glaciation. Geologic evidence suggests that one glacial period occurred about 700 million years ago (m.y.a.) and another about 300 m.y.a. The most recent one—the *Pleistocene epoch* or, simply, the **Ice Age**—began about 2 m.y.a. Let's summarize the climatic conditions that led up to the Pleistocene.

About 65 m.y.a., the earth was warmer than it is now; polar ice caps did not exist. Beginning about 55 m.y.a., the earth entered a long cooling trend. After millions of years, polar ice appeared. As average temperatures continued to lower, the ice grew thicker, and by about 10 m.y.a. a deep blan-

ket of ice covered the Antarctic. Meanwhile, snow and ice began to accumulate in high mountain valleys of the Northern Hemisphere, and alpine, or valley, glaciers soon appeared.

About 2 m.y.a., continental glaciers appeared in the Northern Hemisphere, marking the beginning of the Pleistocene epoch. The Pleistocene, however, was not a period of continuous glaciation but a time when glaciers alternately advanced and retreated (melted back) over large portions of North America and Europe. Between the glacial advances were warmer periods called **interglacial periods**, which lasted for 10,000 years or more.

It was once thought that interglacial periods represented a more stable type of climate, unlike the large climate variations experienced during the colder part of the Ice Age. But analysis of Greenland ice cores suggests that the warm *Eemian interglacial period* (which lasted 19,000 years from 133,000 to 114,000 years ago) may have consisted of two major cold spells, lasting 2000 years and 6000 years. Some scientists speculate, however, that the cooling may represent a shift in the location of Greenland's moisture supply (which would change the oxygen-isotope ratio) rather than actual coolings, as these cold spells do not show up in the Antarctic ice core record.

The most recent North American glaciers reached their maximum thickness and extent about 18,000–22,000 years ago (y.a.). At that time, the sea level was perhaps 125 m (395 ft) lower than it is now. The lower sea level exposed vast areas of land, such as the *Bering land bridge* (a strip of land that connected Siberia to Alaska), which allowed human and animal migration from Asia to North America.

The ice began to retreat about 14,000 y.a. as surface temperatures slowly rose (see Fig. 19.4). Then, about 11,000 y.a., the average temperature suddenly dropped and northeastern North America and northern Europe reverted back to glacial conditions. About 1000 years later, the cold spell (known as the

FIGURE 19.4
The average air temperature variations for the past 18,000 years. (Modified from J. T. Houghton, et al., *Climate Change: The IPCC Assessment*, Cambridge University Press, Cambridge, England, 1990.)

Younger-Dryas*) ended abruptly, and by 8000 y.a. the continental ice sheets over North America were gone. From about 6000–5000 y.a., the climate was probably 1°C warmer than at present. This time frame represents the warmest of the current interglacial period, or *Holocene epoch*. For this reason, the warm spell is referred to as the **mid-Holocene maximum** and, because this warm period favored the development of plants, it is also known as the *climatic optimum*. About 5000 y.a., a cooling trend set in, during which extensive alpine glaciers returned, but not continental ice sheets.

It is interesting to note that ice core data from Greenland reveal that rapid shifts in climate (from ice age conditions to a much warmer state) took place in as little as 3 years over central Greenland around the end of the Younger-Dryas. The data also reveal that similar rapid shifts in climate occurred several times toward the end of the Ice Age.

Climate During the Last 1000 Years

About 1000 y.a., the Northern Hemisphere was relatively warm and dry. During this time, vineyards flourished and wine was produced in England, indicating warm, dry summers and the absence of cold springs. It was during this warm, tranquil period of several hundred years (known as the **Medieval Climatic Optimum**) that the Vikings colonized Iceland and Greenland.

Sometime around A.D. 1200, the mild climate of western Europe began to show extreme variations. For several hundred years the climate grew stormy. Both great floods and great droughts occurred. Extremely cold winters were followed by relatively warm ones. During the cold spells, the English vineyards and the Viking settlements suffered. Europe experienced several famines during the 1300s.

Around 1400 to 1550, the climate moderated (see Fig. 19.5). However, starting in the middle 1550s, the average temperature began to drop. This cooling trend (which continued for almost 300 years) is known as the **Little Ice Age**. During this time, the global mean temperature dropped by about 0.5°C, which allowed alpine glaciers to increase in size and advance down river canyons. Winters were long and severe, summers short and wet. The vineyards in England vanished and farming became impossible in the more northerly latitudes. Cut off from the rest of the world by an advancing ice pack, the Viking colony in Greenland perished.

During the Little Ice Age, one particular year stands out: 1816. In Europe that year, bad weather contributed to a poor wheat crop, and famine spread across the land. In North America, unusual blasts of cold arctic air moved through Canada and the northeastern United States between May and September. The cold spells brought heavy snow in June and killing frosts in July and August. In the warmer days that followed each cold snap, farmers replanted, only to have another cold outbreak damage the planting. The year 1816 has come to be known as "the year without a summer" or "eighteen hundred and froze-to-death." The unusually cold summer was followed by a bitterly cold winter.

In the late 1800s, the average global temperature began to rise (see Fig. 19.6). From about 1900 until about 1940, the average temperature of the lower atmosphere rose nearly 0.5°C.

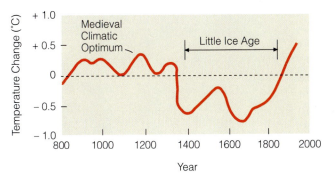

FIGURE 19.5
The average temperature variations of eastern Europe for about the last 1200 years.

---

*This exceptionally cold spell is named after the *Dryas*, an arctic flower.

WEATHER WATCH

Winters were much colder over North America during the Little Ice Age. In fact, the cold winters allowed soldiers during the Revolutionary War to drag cannons from Staten Island to Manhattan across the frozen Upper New York Bay.

Following the warmer period, the earth began to cool slightly over the next 25 years or so. In the late 1960s and 1970s, the cooling trend ended over most of the Northern Hemisphere. During the 1970s and into the 1980s, the average yearly temperature showed considerable fluctuation from year to year and from region to region, with the overall trend pointing to warming. The warming trend continued into the 1990s, with recent years being among the warmest of the twentieth century.

We can see in Fig. 19.6 that over the last hundred years the earth's surface has warmed by about 0.6°C (about 1°F). The warming, however, is not uniform, as the greatest warming has occurred over the mid-latitude continents in winter and spring, while a few areas (such as the North Atlantic Ocean) have actually cooled in recent decades. The United States has experienced little warming as compared to the rest of the world. Moreover, most of the warming has occurred at night.

The changes in air temperature shown in Fig. 19.6 are derived from three main sources: air temperatures over land, air temperatures over ocean, and sea surface temperatures. There are, however, uncertainties in the temperature record. For example, during this time period recording stations have moved, and techniques for measuring temperature have varied. Also, marine observing stations are scarce. Moreover, urbanization (especially in developed nations) tends to artificially raise average temperatures as cities grow (the urban heat island effect). When urban warming is taken into account and improved sea surface temperature information is incorporated

into the data, the warming over the past hundred years measures between 0.3°C and 0.6°C.

A global increase in temperature between 0.3° and 0.6°C may seem very small, but in Fig. 19.4 we can see that global temperatures have varied no more than 1.5°C during the past 10,000 years. Consequently, an increase of 0.6°C becomes significant when compared with temperature changes over thousands of years.

Up to this point we have examined the temperature record of the earth's surface and observed that during the past century the earth has been in a warming trend. Most climate scientists believe that at least part of the warming is due to an enhanced greenhouse effect caused by increasing levels of greenhouse gases, such as CO_2.* If increasing levels of CO_2 are at least partly responsible for the warming, why did the climate begin to cool after 1940? And what caused the Little Ice Age from about 1550 to 1850? These are a few of the questions we will address in the following sections.

POSSIBLE CAUSES OF CLIMATIC CHANGE

Why the earth's climate changes naturally is not totally understood. Many theories attempt to explain the changing climate, but no single theory alone can satisfactorily account for *all* the climatic variations of the past.

Why hasn't the riddle of a fluctuating climate been completely solved? One major problem facing any comprehensive theory is the intricate interrelationship of the elements involved. For example, if temperature changes, many other elements may be altered as well. The interactions among the atmosphere, the oceans, and the ice are extremely complex and the number of

---

*The earth's atmospheric greenhouse effect is due mainly to the absorption and emission of infrared radiation by gases, such as water vapor, CO_2, methane, nitrous oxide, and chlorofluorocarbons. Refer back to Chapter 2 for additional information on this topic.

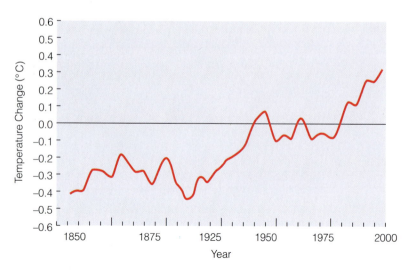

FIGURE 19.6

Changes in the average global (land and sea) surface air temperature from 1850 to 1997. The zero line represents the average surface air temperature from 1961 to 1990.

possible interactions among these systems is enormous. No climatic element within the system is isolated from the others. With this in mind, we will first investigate how feedback systems work; then we will consider some of the current theories of climatic change.

Climate Change and Feedback Mechanisms

In Chapter 2, we learned that the earth-atmosphere system is in a delicate balance between incoming and outgoing energy. If this balance is upset, even slightly, global climate can undergo a series of complicated changes.

Let's assume that the earth-atmosphere system has been disturbed to the point that the earth has entered a slow warming trend. Over the years the temperature slowly rises, and water from the oceans rapidly evaporates into the warmer air. The increased quantity of water vapor absorbs more of the earth's infrared energy, thus strengthening the atmospheric greenhouse effect. This raises the air temperature even more, which, in turn, allows more water vapor to evaporate into the atmosphere. The greenhouse effect becomes even stronger and the air temperature rises even more. This situation is known as the **water vapor-temperature rise feedback**. It represents a **positive feedback mechanism** because the initial increase in temperature is reinforced by the other processes. If this feedback were left unchecked, the earth's temperature would increase until the oceans evaporated away. Such a chain reaction is called a *runaway greenhouse effect*. The earth-atmosphere system has a number of checks and balances that help it to counteract tendencies of climate change. For example, as the earth's temperature increases, its outgoing infrared energy increases proportional to the fourth power of its absolute temperature (Stefan-Boltzmann law, see Chapter 2, p. 34). This increase in radiant energy from the surface would greatly slow the temperature change and help to stabilize the climate. Hence, there is no evidence that a runaway greenhouse effect ever occurred on earth, and it is not very likely that it will occur in the future.

Another positive feedback mechanism is the **snow-albedo feedback**, where an increase in global surface air temperature might cause snow and ice to melt in polar latitudes. This melting would reduce the albedo (reflectivity) of the surface, allowing more solar energy to reach the surface, which would further raise the temperature. But helping to counteract the positive feedback mechanisms are **negative feedback mechanisms**—those that tend to weaken the interactions among the variables rather than reinforce them. Suppose, for example, that as the surface warms, more water evaporates from the oceans and global low cloudiness increases. Low clouds tend to reflect a large percentage of incoming sunlight, and with less solar energy to heat the surface, the warming slows.

All feedback mechanisms work simultaneously and in both directions. We just saw that the snow-albedo feedback produces a positive feedback on a warming planet, but it can produce a positive feedback on a cooling planet as well. For example, suppose the earth were in a slow global cooling trend that lasted for hundreds or even thousands of years. Lower temperatures might allow for a greater snow cover in middle and high latitudes, which would increase the albedo of the surface so that much of the incident sunlight would be reflected back to space. Less sunlight absorbed at the surface might cause a further drop in temperature. This action might further increase the snow cover, lowering the temperature even more. If left unchecked, the snow-albedo feedback would produce a *runaway ice age* which, of course, is not likely on earth because other feedback mechanisms in the atmospheric system are constantly working to moderate the magnitude of the cooling.

Climate Change, Plate Tectonics, and Mountain Building

During the geologic past, the earth's surface has undergone extensive modifications. One involves the slow shifting of the continents and the ocean floors. This motion is explained in the widely accepted **theory of plate tectonics** (formerly called the *theory of continental drift*). According to this theory, the earth's outer shell is composed of huge plates that fit together like pieces of a jigsaw puzzle. The plates, which slide over a partially molten zone below them, move in relation to one another. Continents are embedded in the plates and move along like luggage riding piggyback on a conveyor belt. The rate of motion is extremely slow, only a few centimeters per year.

Besides providing insights into many geological processes, plate tectonics also helps to explain past climates. For example, we find glacial features near sea level in Africa today, suggesting that the area underwent a period of glaciation hundreds of millions of years ago. Were temperatures at low elevations near the equator ever cold enough to produce ice sheets? Probably not. The ice sheets formed when this land mass was located at a much higher latitude. Over the many millions of years since then, the land has slowly moved to its present position. Along the same line, we can see how the fossil remains of tropical vegetation can be found under layers of ice in polar regions today.

According to plate tectonics, the now existing continents were at one time joined together in a single huge continent, which broke apart. Its pieces slowly moved across the face of the earth, thus changing the distribution of continents and ocean basins (see Fig. 19.7). Some scientists feel that, when land masses are concentrated in middle and high latitudes, ice sheets are more likely to form. During these times, there is a greater likelihood that more sunlight will be reflected back into space and that the snow-albedo feedback mechanism mentioned earlier will amplify the cooling.

The various arrangements of the continents may also influence the path of ocean currents. This would alter the transport of heat from low to high latitudes and change both the global wind system and the climate in middle and high latitudes. As an example, suppose that plate movement "pinches off" a rather large body of high latitude ocean water such that the transport of warm water into the region is cut off. In winter, the surface water would eventually freeze over with ice. This freezing would, in turn, reduce the amount of sensible

FIGURE 19.7
Geographical distribution of (a) land masses about 180 million years ago, and (b) today. Arrows show the relative direction of continental movement.

and latent heat given up to the atmosphere. Furthermore, the ice allows snow to accumulate on top of it, thereby setting up conditions that could lead to even lower temperatures.

There are other mechanisms by which tectonic processes* may influence climate. In Fig. 19.8, notice that the formation of oceanic plates (plates that lie beneath the ocean) begins at a *ridge*, where dense, molten material from inside the earth wells up to the surface, forming new sea floor material as it hardens. Spreading (on the order of several centimeters a year) takes place at the ridge center, where two oceanic plates move away

---

*Tectonic processes are large-scale processes that deform the earth's crust.

from one another. When an oceanic plate encounters a lighter continental plate, it responds by diving under it, in a process called *subduction*. Heat and pressure then melt a portion of the subducting rock, which usually consists of volcanic rock and calcium-rich ocean sediment. The molten rock may then gradually work its way to the surface, producing volcanic eruptions that spew water vapor, carbon dioxide, and minor amounts of other gases into the atmosphere. The release of these gases (called *degassing*) usually takes place at other locations as well, for instance, at ridges where new crustal rock is forming.

Some scientists speculate that climatic change, taking place over millions of years, might be related to the rate at which the

FIGURE 19.8
The earth is composed of a series of moving plates. The rate at which plates move (spread) may influence global climate. During times of rapid spreading, increased volcanic activity may promote global warming by enriching the CO_2 content of the atmosphere.

plates move and, hence, related to the amount of CO_2 in the air. For example, during times of rapid spreading, a relatively wide ridge forms, causing sea level to rise relative to the continents. At the same time, an increase in volcanic activity vents large quantities of CO_2 into the atmosphere, which enhances the atmospheric greenhouse effect, causing global temperatures to rise. Moreover, a higher sea level means that there is less exposed landmass and, presumably, less chemical weathering of rocks*—a process that removes CO_2 from the atmosphere. However, as global temperatures climb, increasing temperatures promote chemical weathering that removes atmospheric CO_2 at a faster rate.

Millions of years later, when spreading rates decrease, less volcanic activity means less degassing. The changing shape of the underwater ridge causes the sea level to drop relative to the continents, exposing more rocks for chemical attack and the removal of CO_2 from the air. A reduction in CO_2 levels weakens the greenhouse effect, which causes global temperatures to drop. The accumulation of ice and snow over portions of the continents may promote additional cooling by reflecting more sunlight back to space. The cooling, however, will not go unchecked, as lower temperatures retard both the chemical weathering of rocks and the depletion of atmospheric CO_2.

A chain of volcanic mountains forming above a subduction zone may disrupt the airflow over them. By the same token, mountain building that occurs when two continental plates collide (like that which presumably formed the Himalayan mountains and Tibetan highlands) can have a marked influence on global circulation patterns and, hence, on the climate of an entire hemisphere.

Up to now, we have examined how climatic variations can take place over millions of years due to the movement of continents and the associated restructuring of landmasses, mountains, and oceans. We will now turn our attention to variations in the earth's orbit that may account for climatic fluctuations that take place on a time scale of tens of thousands of years.

Climate Change and Variations in the Earth's Orbit

A theory ascribing climatic changes to variations in the earth's orbit is the **Milankovitch theory**, named for the astronomer Milutin Milankovitch, who first proposed the idea in the 1930s. The basic premise of this theory is that, as the earth travels through space, three separate cyclic movements combine to produce variations in the amount of solar energy that falls on the earth.

The first cycle deals with changes in the shape (**eccentricity**) of the earth's orbit as the earth revolves about the sun. Notice in Fig. 19.9 that the earth's orbit changes from being elliptical to being nearly circular. To go from less elliptical to more elliptical and back again takes about 100,000 years. The greater the eccentricity of the orbit (that is, the more eccentric

FIGURE 19.9
For the earth's orbit to stretch from nearly a circle (dashed line) to an elliptical orbit (solid line) and back again takes nearly 100,000 years. (Diagram is highly exaggerated and is not to scale.)

the orbit), the greater the variation in solar energy received by the earth between its closest and farthest approach to the sun.

Presently, we are in a period of low eccentricity. The earth is closer to the sun in January and farther away in July (see Chapter 3). The difference in distance (which only amounts to about 3 percent) is responsible for a nearly 7 percent increase in the solar energy received at the top of the atmosphere from July to January. When the difference in distance is 9 percent (a highly eccentric orbit), the difference in solar energy received will be on the order of 20 percent. In addition, the more eccentric orbit will change the length of seasons in each hemisphere by changing the length of time between the vernal and autumnal equinoxes.*

The second cycle takes into account the fact that, as the earth rotates on its axis, it wobbles like a spinning top. This wobble, known as the **precession** of the earth's axis, occurs in a cycle of about 23,000 years. Presently, the earth is closer to the sun in January and farther away in July. Due to precession, the reverse will be true in about 11,000 years (see Fig. 19.10). In about 23,000 years we will be back to where we are today. This means, of course, that if everything else remains the same, 11,000 years from now seasonal variations in the Northern Hemisphere should be greater than at present. The opposite would be true for the Southern Hemisphere.

The third cycle takes about 41,000 years to complete and relates to the changes in tilt (**obliquity**) of the earth as it orbits the sun. Presently, the earth's orbital tilt is $23\frac{1}{2}°$, but during the 41,000-year cycle the tilt varies from about 22° to $24\frac{1}{2}°$. The smaller the tilt, the less seasonal variation there is between summer and winter in middle and high latitudes. Thus, winters tend to be milder and summers cooler. During the warmer winters, more snow would probably fall in polar regions due to the air's increased capacity for water vapor. And during the cooler summers, less snow would melt. As a consequence, the periods of smaller tilt would tend to promote the formation of

---

*Chemical weathering is the process by which rocks decompose.

*Although rather large percentage changes in solar energy can occur between summer and winter, the globally and annually averaged change in solar energy received by the earth (due to orbital changes) hardly varies at all. It is the distribution of incoming solar energy that changes, not the totals.

WEATHER WATCH

If all the snow that normally falls over central Canada during the course of one year were to stay on the ground and not melt (even during the summer), it would take nearly 3000 years to build an ice sheet comparable to the one that existed there 18,000 years ago.

glaciers in high latitudes. In fact, when all of the cycles are taken into account, the present trend should be toward a cooler climate over the Northern Hemisphere.

In summary, the Milankovitch cycles that combine to produce variations in solar radiation received at the earth's surface include:

1. changes in the shape (*eccentricity*) of the earth's orbit about the sun
2. *precession* of the earth's axis of rotation, or wobbling
3. changes in the tilt (*obliquity*) of the earth's axis

In the 1970s, scientists of the CLIMAP project found strong evidence in deep-ocean sediments that variations in climate during the past several hundred thousand years were closely associated with the Milankovitch cycles. More recent studies have strengthened this premise. For example, studies conclude that during the past 800,000 years, ice sheets have peaked about every 100,000 years. This conclusion corresponds naturally to variations in the earth's eccentricity. Superimposed on this situation are smaller ice advances that show up at intervals of about 41,000 years and 23,000 years. It appears, then, that eccentricity is the *forcing factor*—the external cause—for the frequency of glaciation, as it appears to control the severity of the climatic variation.

But orbital changes alone are probably not totally responsible for ice buildup and retreat. Evidence (from trapped air bubbles in the ice sheets of Greenland and Antarctica representing thousands of years of snow accumulation) reveals that CO_2 levels were about 30 percent lower during colder glacial periods than during warmer interglacial periods (see Fig. 19.11). Analysis of air bubbles in Antarctic ice cores reveals that methane follows a similar pattern to that of CO_2. This knowledge suggests that lower atmospheric CO_2 levels may have had the effect of amplifying the cooling initiated by the orbital changes. Likewise, increasing CO_2 levels at the end of the glacial period may have accounted for the rapid melting of the ice sheets. Just why atmospheric CO_2 levels have varied as glaciers expanded and contracted is not clear, but it appears to be due to changes in biological activity taking place in the oceans.

Perhaps, also, changing levels of CO_2 indicate a shift in ocean circulation patterns. Such shifts, brought on by changes in precipitation and evaporation rates, may alter the distribution of heat energy around the world. Alteration wrought in this manner could, in turn, affect the global circulation of winds, which may explain why alpine glaciers in the Southern Hemisphere expanded and contracted in tune with Northern Hemisphere glaciers during the last ice age, even though the Southern Hemisphere (according to the Milankovitch cycles) was not in an orbital position for glaciation.

Still other factors may work in conjunction with the earth's orbital changes to explain the temperature variations between glacial and interglacial periods. Some of these are:

1. the amount of dust and other aerosols in the atmosphere
2. the reflectivity of the ice sheets
3. the concentration of other trace gases, such as methane
4. the changing characteristics of clouds
5. the rebounding of land, having been depressed by ice

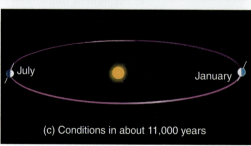

FIGURE 19.10

(a) Like a spinning top, the earth's axis of rotation slowly moves and traces out the path of a cone in space. (b) Presently the earth is closer to the sun in January, when the Northern Hemisphere experiences winter. (c) In about 11,000 years, due to precession, the earth will be closer to the sun in July, when the Northern Hemisphere experiences summer.

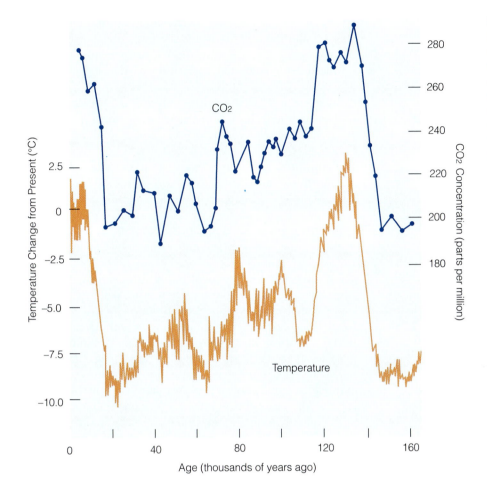

FIGURE 19.11
Analysis of trapped bubbles of ancient air in the polar ice sheet at the Vostok station in Antarctica reveals that over the past 160,000 years, CO_2 levels (upper curve) correlate well with air temperature changes (bottom curve). Temperatures are derived from the analysis of oxygen isotopes. Note that CO_2 levels were about 30 percent lower and Antarctic temperatures about 10°C (18°F) lower during the colder glacial periods.

Hence, the Milankovitch cycles, in association with other natural factors, may explain the advance and retreat of ice over periods of 10,000 to 100,000 years. But what caused the Ice Age to begin in the first place? And why have periods of glaciation been so infrequent during geologic time? The Milankovitch theory does not attempt to answer these questions.

Climate Change and Atmospheric Particles

Microscopic liquid and solid particles (*aerosols*) that enter the atmosphere from both human-induced (anthropogenic) and natural sources can have an effect on climate. The effect, however, is exceedingly complex and depends upon a number of factors, such as the particle's size, shape, color, chemical composition, and vertical distribution above the surface. In this section, we will first examine aerosols in the lower atmosphere. Then we will examine the effect that volcanic aerosols in the stratosphere have on climate.

Aerosols in the Troposphere   Aerosols enter the lower atmosphere in a variety of ways—from factory and auto emissions, agricultural burning, wildland fires, and dust storms. Some particles (such as soil dust and sulfate particles) mainly reflect and scatter incoming sunlight, while others (such as smoky soot) readily absorb sunlight, which warms the air around them. Aerosols that reduce the amount of sunlight reaching the earth's surface tend to cause net cooling of the surface air during the day. Certain aerosols also selectively absorb and emit infrared energy back to the surface, producing a net warming of the surface air at night. However, the overall net effect of human-induced aerosols on climate is to *cool the surface.*

In recent years, the effect of highly reflective **sulfate aerosols** on climate has been extensively researched. In the lower atmosphere, the majority of these particles come from the combustion of sulfur-containing fossil fuels but emissions from smoldering volcanoes can also be a significant source of tropospheric sulfate aerosols. Sulfur pollution, which has more than doubled globally since preindustrial times, enters the atmosphere mainly as sulfur dioxide gas. There, it transforms into tiny sulfate droplets or particles. Since these aerosols usually remain in the atmosphere for only a few days, they do not have time to spread around the globe. Hence, they are not well mixed and their effect is felt mostly over the Northern Hemisphere, especially over polluted regions. Over the oceans, a major source of sulfate aerosols comes from tiny drifting aquatic plants—phytoplankton—that produce *dimethylsulphide* (DMS). The DMS slowly diffuses into the atmosphere

FIGURE 19.12

Large volcanic eruptions rich in sulfur can affect climate. As sulfur gases in the stratosphere transform into tiny reflective sulfuric acid particles, they prevent a portion of the sun's energy from reaching the surface. Here, the Philippine volcano Mount Pinatubo erupts during June, 1991.

where it oxidizes to form sulfur dioxide, which in turn converts to sulfate aerosols.

Sulfate aerosols not only scatter incoming sunlight back to space, but they also serve as cloud condensation nuclei. Consequently, they have the potential for altering the physical characteristics of clouds. For example, if the number of sulfate aerosols and, hence, condensation nuclei inside a cloud should increase, the cloud would have to share its available moisture with the added nuclei, a situation that should produce many more (but smaller) cloud droplets. The greater number of droplets would reflect more sunlight and have the effect of brightening the cloud and reducing the amount of sunlight that reaches the surface.

In summary, sulfate aerosols reflect incoming sunlight, which tends to lower the earth's surface temperature during the day. Sulfate aerosols may also modify clouds by increasing their reflectivity. Because sulfate pollution has increased significantly

over industrialized areas of eastern Europe, northeastern North America and China, the cooling effect brought on by these particles may explain: (1) why the Northern Hemisphere has warmed less than the Southern Hemisphere during the past several decades, (2) why the United States has experienced little warming compared to the rest of the world, and (3) why most of the global warming has occurred at night and not during the day, especially over polluted areas. Research is still being done, and the overall effect of tropospheric aerosols on the climate system is not totally understood. (Information regarding the possible effect on climate from particles injected into the atmosphere during nuclear war is given in the Focus section on p. 517.)

Volcanic Eruptions and Aerosols in the Stratosphere Volcanic eruptions can have a definitive impact on climate. During volcanic eruptions, fine particles of ash and dust (as well as gases) can be ejected into the stratosphere (see Fig. 19.12). Scientists agree that the volcanic eruptions having the greatest impact on climate are those rich in sulfur gases. These gases, over a period of about 2 months, combine with water vapor in the presence of sunlight to produce tiny, reflective sulfuric acid particles that grow in size, forming a dense layer of haze. The haze may reside in the stratosphere for several years, absorbing and reflecting back to space a portion of the sun's incoming energy. The absorption of the sun's energy along with the absorption of infrared energy from the earth, warms the stratosphere. The reflection of incoming sunlight by the haze tends to cool the air at the earth's surface, especially in the hemisphere where the eruption occurs.

The two largest volcanic eruptions so far this century in terms of their sulfur-rich veil, were that of El Chichón in Mexico during April, 1982, and Mount Pinatubo in the Philippines during June, 1991.* Mount Pinatubo ejected an estimated 20 million tons of sulfur dioxide (more than twice that of El Chichón) that gradually worked its way around the globe (see Fig. 19.13). For major eruptions such as this one, mathematical models predict that average hemispheric temperatures can drop by about 0.2° to 0.5°C or more for one to three years after the eruption. Model predictions agreed with temperature changes brought on by the Pinatubo eruption, as in early 1993 the mean global surface temperature had decreased by about 0.5°C (see Fig. 19.14). The cooling might even have been greater had the eruption not coincided with a major El Niño event that began in 1990 and lasted until early 1995 (see Chapter 11 for information on El Niño).

As previously noted, volcanic eruptions rich in sulfur warm the lower stratosphere. During the winter, the tropical

---

*The eruption of Mount Pinatubo in 1991 was many times greater than that of Mount St. Helens in the Pacific Northwest in 1980. In fact, the largest eruption of Mount St. Helens was a lateral explosion that pulverized a portion of the volcano's north slope. The ensuing dust and ash (and very little sulfur) had virtually no effect on global climate as the volcanic material was confined mostly to the lower atmosphere and fell out quite rapidly over a large area of the northwestern United States.

FOCUS ON A SPECIAL TOPIC

Nuclear Winter—Climate Change Induced by Nuclear War

A number of studies indicate that a nuclear war involving hundreds or thousands of nuclear detonations would drastically modify the earth's climate.

Researchers assume that a nuclear war would raise an enormous pall of thick, sooty smoke from massive fires that would burn for days, even weeks, following an attack. The smoke would drift higher into the atmosphere, where it would be caught in the upper-level westerlies and circle the middle latitudes of the Northern Hemisphere. Unlike soil dust, which mainly scatters and reflects incoming solar radiation, soot particles readily absorb sunlight. Hence, for several weeks after the war, sunlight would virtually be unable to penetrate the smoke layer, bringing darkness or, at best, twilight at midday.

Such reduction in solar energy would cause surface air temperatures over land masses to drop below freezing, even during the summer, resulting in extensive damage to plants and crops and the death of millions (or perhaps billions) of people. The dark, cold, and gloomy conditions that would be brought on by nuclear war is often referred to as **nuclear winter**.

As the lower troposphere cools, the solar energy absorbed by the smoke particles in the upper troposphere would cause this region to warm. The end result would be a strong, stable temperature inversion extending from the surface up into the higher atmosphere. A strong inversion would lead to a number of adverse effects, such as suppressing convection, altering precipitation processes, and causing major changes in the general wind patterns.

The heating of the upper part of the smoke cloud would cause it to rise upward into the stratosphere, where it would then drift southward. Thus, about one-third of the smoke would remain in the atmosphere for a year or longer. The other two-thirds would be washed out in a month or so by precipitation. This smoke lofting, combined with persisting sea ice formed by the initial cooling, would produce climatic change that would remain for several years.

Virtually all research on nuclear winter, including models and analog studies, confirms this gloomy scenario. Observations of forest fires show lower temperatures under the smoke, confirming part of the theory. A three-year study involving more than 300 scientists from more than 30 countries conducted by the Scientific Committee On Problems of the Environment (SCOPE) of the International Council of Scientific Unions has detailed the climatic, environmental, and agricultural effects of nuclear winter. The implications of nuclear winter are clear: A nuclear war would drastically alter global climate and would devastate our living environment.

Even with improved global superpower relations, and the end of the Cold War, the danger of nuclear winter remains a possibility. Presently, the current global nuclear arsenal is more than that needed to produce the effects of a nuclear winter. As other nations develop nuclear capability, the potential for nuclear winter remains with us. It will not disappear until the global nuclear weapons arsenal numbers in the hundreds, not in the thousands.

stratosphere can become much warmer than the polar stratosphere. This situation produces a strong horizontal pressure gradient and strong west-to-east (zonal) stratospheric winds. These winds apparently work their way down into the upper troposphere, where they direct milder maritime surface air from off the ocean onto the continents. The milder ocean air produces warmer winters over Northern Hemisphere continents during the first or second winter after the eruption occurs.

An infamous cold spell often linked to volcanic activity occurred during the year 1816, "the year without a summer" mentioned earlier. Apparently, a rather stable longwave pattern in the atmosphere produced unseasonably cold summer weather over eastern North America and western Europe. The cold weather followed the massive eruption in 1815 of Mount Tambora in Indonesia. In addition to this, major volcanic eruptions occurred in the 4 years preceding Tambora. If, indeed, the cold weather pattern was brought on by volcanic eruptions, it was probably an accumulation of several volca-

FIGURE 19.13
Sulfur dioxide plume (dark red and green areas) from the eruption of Mount Pinatubo as measured by the Upper Atmosphere Research Satellite on September 21, 1991. Only three months after the eruption, the plume girdles the equator in the stratosphere at an altitude near 25 km.

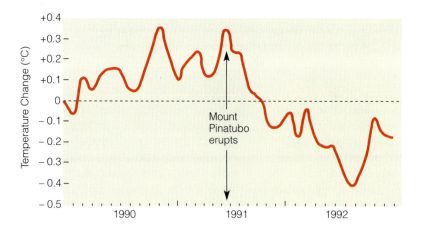

FIGURE 19.14
Changes in average global air temperature from 1990–1992. After the eruption of Mount Pinatubo in June, 1991, the average global temperature by July, 1992, decreased by almost 0.5°C (0.9°F) from the 1981–1990 average (dashed line).

noes loading the stratosphere with particles—particles that probably remained there for several years.

In an attempt to correlate sulfur-rich volcanic eruptions with long-term trends in global climate, scientists are measuring the acidity of annual ice layers in Greenland and Antarctica. Generally, the greater the concentration of sulfuric acid particles in the atmosphere, the greater the acidity of the ice layer. Relatively acidic ice has been uncovered from about A.D. 1350 to about 1700, a time that corresponds to the Little Ice Age. Such findings suggest that sulfur-rich volcanic eruptions may have played an important role in triggering this comparatively cool period and, perhaps, other cool periods during the geologic past. Moreover, recent core samples taken from the northern Pacific Ocean reveal that volcanic eruptions in the northern Pacific were at least 10 times larger 2.6 million years ago (a time when Northern Hemisphere glaciation began) than previous volcanic events recorded elsewhere in the sediment.

Climate Change and Variations in Solar Output   In the past, it was thought that solar energy does not vary by more than a fraction of a percent over many years. However, measurements made by sophisticated radiometers aboard satellites suggest that the sun's energy output may vary considerably more than was thought. Moreover, the sun's energy output appears to change slightly with sunspot activity.

Sunspots are huge magnetic storms on the sun that show up as cooler (darker) regions on the sun's surface. They occur in cycles, with the number and size reaching a maximum approximately every 11 years. During periods of maximum sunspots, the sun emits more energy (about 0.1 percent more)* than during periods of sunspot minimums (see Fig. 19.15). Evidently, the greater number of bright areas (*faculae*) around the sunspots radiate more energy, which offsets the effect of the dark spots.

---

*As a reference, an increase of 1 percent in solar energy output would be comparable to the total effect that increasing levels of CO_2 and other greenhouse gases have had on warming the atmosphere to date.

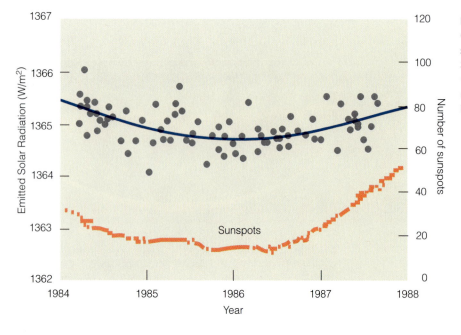

FIGURE 19.15
Changes in solar energy output (upper curve) in watts per square meter as measured by the Earth Radiation Budget Satellite. Bottom curve represents the yearly average number of sunspots. As sunspot activity increases from minimum to maximum, the sun's energy output increases by about 0.1 percent.

FIGURE 19.16
The sun's energy output (brightness) changes slightly over a period of years. Such changes may cause variations in the earth's climate system.

It appears that the 11-year sunspot cycle has not always prevailed. Apparently, between 1645 and 1715, during the period known as the **Maunder minimum**,* there were few, if any, sunspots. It is interesting to note that the minimum occurred during the coldest stage of the Little Ice Age—a time when global mean temperature decreased by about 0.5°C over the long-term average. Some scientists suggest that a reduction in solar brightness was, in part, responsible for this cold spell.

In an attempt to better understand the sun's behavior, solar researchers are examining stars that are similar in age and mass to our sun. Recent observations suggest that, in some of these stars, energy output may vary by as much as 0.4 percent, leading some scientists to speculate that changes in the sun's brightness might account for part of the global warming during the last century.

The sun's magnetic field varies with sunspot activity and actually reverses every 11 years. Because it takes 22 years to return to its original state, the sun's *magnetic cycle* is 22 years, rather than 11. Some researchers point to the fact that periodic 20-year droughts on the Great Plains of the United States seem to correlate with this 22-year solar cycle. More recently, scientists have found a relationship between the 11-year sunspot cycle and weather patterns across the Northern Hemisphere. It appears that winter warmings might be related to variations in sunspots and to a pattern of reversing stratospheric winds over the tropics, known as the *quasi-biennial oscillation* (because it takes about 2 years for the winds to complete the cycle).

Earlier, we learned that the total energy output of the sun varies over an 11-year sunspot cycle by only about 0.1 percent.

However, most of this variation is in the ultraviolet wavelength, which is absorbed by stratospheric ozone. But ozone varies with the sunspot cycle, reaching a maximum during sunspot maximum. Consequently, during sunspot maximum, more ozone is able to absorb the excess ultraviolet radiation. This absorption warms the stratosphere, producing temperature gradients and winds that appear to influence the circulation in the troposphere, especially the tropical Hadley cell (see Chapter 11) and the mid-latitude storm track.

To sum up, fluctuations in solar output may account for climatic changes over time scales of decades and centuries. To date, many theories have been proposed linking solar variations to climate change, but none have been proven. However, instruments aboard satellites and solar telescopes on the earth are monitoring the sun to observe how its energy output may vary (see Fig. 19.16). Because many years of data are needed, it may be some time before we fully understand the relationship between solar activity and climate change on earth.

BRIEF REVIEW Before going on to the next section, here is a brief review of some of the facts and concepts we covered so far:

- The earth's climate is constantly undergoing change. Evidence suggests that throughout much of the earth's history, the earth's climate was much warmer than it is today.

- The most recent glacial period (or Ice Age) began about 2 million years ago. During this time, glacial advances were interrupted by warmer periods called *interglacial periods*. In North America, glaciers reached their maximum thickness and extent about 18,000 to 22,000 years ago and disappeared completely from North America by about 8000 years ago.

---

*This period is named after E. W. Maunder, the British solar astronomer who first discovered the low sunspot period sometime in the late 1880s.

■ The Younger-Dryas event represents a time about 11,000 years ago when northeastern North America and northern Europe reverted back to glacier conditions.

■ Over the last hundred years or so, the earth's surface temperature has increased by about 0.6°C (about 1°F).

■ The shifting of continents, along with volcanic activity and mountain-building, proposes how climate variations can take place over millions of years.

■ The Milankovitch theory (in association with other natural forces) proposes that altering glacial and interglacial episodes during the past 2 million years are the result of small variations in the tilt of the earth's axis and in the geometry of the earth's orbit around the sun.

■ Trapped air bubbles in the ice sheets of Greenland and Antarctica reveal that CO_2 levels (and methane levels) were lower during colder glacial periods and higher during warmer interglacial periods.

■ Sulfate aerosols in the troposphere reflect incoming sunlight, which tends to lower the earth's surface temperature during the day. Sulfate aerosols may also modify clouds by increasing the cloud's reflectivity.

■ Volcanic eruptions, rich in sulfur, may be responsible for cooler periods in the geologic past.

■ Fluctuation in solar output (brightness) may account for climatic changes over time scales of decades and centuries.

In previous sections, we saw how increasing levels of CO_2 may have contributed to changes in global climate spanning thousands and even millions of years. Today, we may be undertaking a global scientific experiment by injecting vast quantities of CO_2 into our atmosphere without really knowing the long-term consequences. The next section describes how CO_2 and other trace gases may be enhancing the earth's greenhouse effect, producing global warming.

CARBON DIOXIDE, THE GREENHOUSE EFFECT, AND RECENT GLOBAL WARMING

We know from Chapter 2 that CO_2 is a greenhouse gas that strongly absorbs infrared radiation and plays a major role in warming the lower atmosphere. We also know that CO_2 has been increasing steadily in the atmosphere, primarily due to the burning of fossil fuel (see Fig. 1.4, p. 4). However, deforestation may also be adding to this increase as tropical rain forests are cut down and replaced with plants less efficient in removing CO_2 from the atmosphere. In 1998, the annual average of CO_2 was about 365 parts per million, and present estimates are that if CO_2 levels continue to increase at the same rate that they have been (about 1.5 parts per million per year), atmospheric concentrations will rise to about 500 parts per million by the year

2100. To complicate the picture, trace gases such as methane (CH_4), nitrous oxide (N_2O), and chlorofluorocarbons (CFCs), all of which readily absorb infrared radiation, have been increasing in concentration over the past century.* Collectively, the increase in these gases is about equal to CO_2 in their ability to enhance the atmospheric greenhouse effect.

Most numerical climate models (mathematical models that simulate climate) predict that by the year 2100 increasing concentrations of greenhouse gases will result in a mean global warming of surface air between 1° and 3.5°C (between about 2° and 6°F). The newest, most sophisticated models take into account a number of important relationships, including the interactions between the oceans and the atmosphere, the processes by which CO_2 is removed from the atmosphere, and the cooling effect produced by sulfate aerosols in the lower atmosphere. The models also predict that as the air warms, additional water vapor will evaporate from the oceans into the air. The added water vapor (which is the most abundant greenhouse gas) will produce a positive feedback by enhancing the atmospheric greenhouse effect and accelerating the temperature rise. (This is the *water vapor-temperature rise feedback* described on p. 511.) Without this feedback produced by the added water vapor, the models predict that by the year 2100 the warming will be on the order of about 1.2°C (2°F).

Oceans, Clouds, and Global Warming   There are some uncertainties in the climate picture. The oceans, for example, play a major role in the climate system, yet the exact effect they will have on rising levels of CO_2 and global warming is uncertain (see Fig. 19.17). The oceans are huge storehouses for CO_2. Microscopic plants (phytoplankton) extract CO_2 from the atmosphere during photosynthesis and store some of it below the ocean surface when they die. Would a warmer earth trigger a larger blooming of these tiny plants, in effect reducing CO_2 in the atmosphere? Or, would a gradual rise in ocean temperature increase the amount of CO_2 in the air due to the fact that warmer oceans can't hold as much CO_2 as colder ones? Furthermore, the oceans have a large capacity for storing heat energy. Thus, as they slowly warm, they should retard the rate at which the atmosphere warms. Overall, the response of ocean temperatures, ocean circulations, and sea ice to global warming will probably determine the global pattern and speed of climate change.

Some studies suggest that during the Ice Age worldwide ocean circulations were probably different from those of today. In fact, a breakdown in the deep ocean circulation in the North Atlantic, and the subsequent cutting off of warm water from low latitudes, may have been responsible for the Younger-Dryas event—the cold spell that took place several thousand years after the continental glaciers began to retreat. (More on this topic is given in the Focus section on pp. 522–523.)

---

*Refer back to Chapter 1 and to Table 1.1, p. 3 for additional information on the concentration of these gases.

Although most climate scientists believe that over the twenty-first century the earth will warm at an unprecedented rate (a process that might cause many problems), they also recognize that increasing levels of CO_2 in the atmosphere may have some positive consequences. For example, some scientists contend that the higher level of CO_2 will act as a "fertilizer" for some plants, accelerating their growth. Increased plant growth consumes more CO_2, which might retard the increasing rate of CO_2 in the environment.

Other scientists feel that the increased plant growth might force some insects to eat more, resulting in a net loss of vegetation. There is concern also that a major increase in CO_2 might upset the balance of nature, with some plant species becoming so dominant that others are eliminated. In cold climates, where crops are now grown only marginally, the warming effect may actually increase crop yield, while in tropical areas, where many developing nations are located, the warming may decrease crop yield.

If the atmosphere's water vapor content increases, so might global cloudiness. How, then, would clouds—which come in a variety of shapes and sizes and form at different altitudes—affect the climate system? Clouds reflect incoming sunlight back to space, a process that tends to cool the climate, but they also absorb infrared radiation from the earth, which tends to warm it. Just how the climate will respond to changes in cloudiness will probably depend on the type of clouds that form and their physical properties, such as liquid water (or ice) content and droplet size distribution. For example, high, thin cirriform clouds (composed mostly of ice) tend to promote a net warming effect: They allow a good deal of sunlight to pass through (which warms the earth's surface), yet because they are cold, they warm the atmosphere by absorbing more infrared radiation from the earth than they emit upward. Low stratified clouds, on the other hand, tend to promote a net cooling effect. Composed mostly of water droplets, they reflect much of the sun's incoming energy, and, because their tops are relatively warm, they radiate away much of the infrared energy they receive from the earth. Satellite data from the Earth Radiation Budget Experiment confirm that, overall, clouds presently have a *net cooling effect* on our planet, which means that, without clouds, our atmosphere would be warmer.

Additional clouds in a warmer world would not necessarily have a net cooling effect, however. Their influence on the average surface air temperature would depend on their extent and on whether low or high clouds dominate the climate scene. Consequently, the feedback from clouds could potentially enhance or reduce the warming produced by increasing greenhouse gases. Most models show that as the surface air warms, there will be more convection, more convective-type clouds, and an increase in cirrus clouds. This situation would tend to provide a positive feedback on the climate system, and the effect of clouds on cooling the earth would be diminished.

Some scientists speculate that an increase in towering cumuliform clouds, brought on by enhanced convection, will

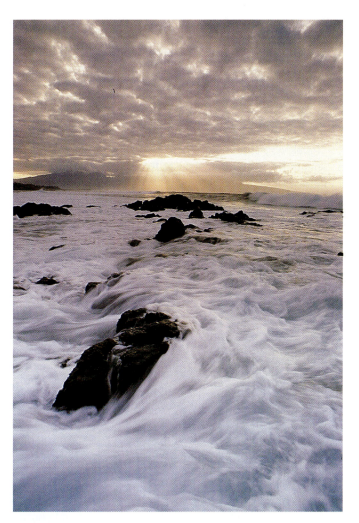

FIGURE 19.17

Oceans and clouds play an important part in the earth's climate system. How they will respond to increasing global temperatures is not clear. Oceans may well add water vapor to the atmosphere, which might promote warming by enhancing the greenhouse effect. An increase in global cloudiness could potentially enhance or reduce the warming produced by increasing greenhouse gases.

promote another negative feedback on global warming. They contend that as cumulus clouds develop, much of their water vapor will condense and fall to the surface as rain, leaving the upper part of the clouds relatively dry. Additionally, sinking air filling the space around the clouds produces warmer and dryer air aloft. Less water vapor, they feel, will diminish the effect of greenhouse warming. All modeling and observational studies, however, do not support these ideas, as convection generally moistens rather than dries the middle and upper troposphere.

In addition to the amount and distribution of clouds, the way in which climate models calculate the cloud's optical properties (such as albedo) can have a large influence on the model's calculations. For example, to illustrate the effect that cloud properties might have on climate models, researchers at

The Ocean Conveyor Belt and Climate Change

Earlier in this chapter we learned that, during the last glacier period, the climate around Greenland (and probably other areas of the world) underwent shifts, from ice-age temperatures to much warmer conditions in a matter of years. What could bring about such large fluctuations in temperature over such a short period of time? It now appears that a vast circulation of ocean water, known as the **conveyor belt**, plays a major role in the climate picture.

Figure 1 illustrates the movement of the ocean conveyor belt, or *thermohaline circulation*.* The conveyor-like circulation begins in the north Atlantic near Greenland and Iceland, where salty surface water is cooled through contact with cold Arctic air masses. The cold, dense water sinks and flows southward through the deep Atlantic Ocean, around Africa, and into the Indian and Pacific Oceans. In the North Atlantic, the sinking of cold water draws warm water northward from lower latitudes.

*Thermohaline circulations are ocean circulations produced by differences in temperature and/or salinity. Changes in ocean water temperature or salinity create changes in water density.

As this water flows northward, evaporation increases the water's salinity (dissolved salt content) and density. When this salty, dense water reaches the far regions of the North Atlantic, it gradually sinks to great depths. This warm part of the conveyor (or Gulf Stream) delivers an incredible amount of tropical heat to the northern Atlantic. During the winter, this heat is transferred to the overlying atmosphere, and evaporation moistens the air. Strong westerly winds then carry this warmth and moisture into northern and western Europe, where it causes winters to be much warmer and wetter than one would normally expect for this latitude.

Ocean sediment records along with ice-core records from Greenland suggest that the giant conveyor belt has switched on and off during the last glacial period. Such events have apparently coincided with rapid changes in climate. For example, when the conveyor belt is strong, winters in northern Europe tend to be wet and relatively mild. However, when the conveyor belt is weak or stops altogether, winters in northern Europe appear to turn much colder. This switching from a period of milder winters to one of severe cold shows

up many times in the climate record. One such event—The Younger-Dryas —illustrates how quickly climate can change and how western and northern Europe's climate can cool within a matter of decades, then quickly return back to milder conditions.

Apparently, the mechanism that switches the conveyor belt off is a massive influx of freshwater. For example, about 11,000 years ago during the Younger-Dryas event, freshwater from a huge glacial lake began to flow down the St. Lawrence River and into the North Atlantic. This massive inflow of freshwater reduced the salinity (and, hence, density) of the surface water to the point that it stopped sinking. The conveyor shut down for about 1000 years during which time severe cold engulfed much of northern Europe. The conveyor belt started up again when the influx of fresh water began to drain down the Mississippi rather than into the North Atlantic. It was during this time that milder conditions returned to northern Europe.

Will increasing levels of CO_2 have an effect on the conveyor belt? Some climate models predict that as CO_2 levels increase, more precipitation will fall over the North Atlantic. This

---

the British Meteorological Office altered the representation of clouds in their model. Initially, the model projected a global temperature rise of about 5°C, accompanying a doubling of atmospheric CO_2. However, when water clouds replaced ice clouds in the simulation, the projected temperature rise was less than 2°C. It is no wonder, then, that a study conducted with 14 climate models showed good agreement on how the global climate would respond if current values of CO_2 were doubled under clear skies. But when clouds were incorporated into the models, the models did not agree, and, in fact, varied greatly over a wide range.

Possible Consequences of Global Warming   Most climate experts feel that increasing levels of greenhouse gases will, by the year 2100, cause the earth to warm by between 1° and 3.5°C. Look back to Fig. 19.4, p. 509, and observe that even if the warming turns out to be 1°C, the earth will be warmer than at any other time during the past 5000 years. A warming of 3.5°C would be equivalent to the rise in temperature since the Younger-Dryas event. Climate models predict that the warming should be greatest over the high northern latitudes in winter with little warming over the Arctic in summer. Overall, in winter, surface warming should be greater over land areas than over the oceans.

situation reduces the density of the sea water and slows down the conveyor belt. In fact, if CO_2 levels double, computer models predict that the conveyor belt will slow by about 30 percent. If CO_2 levels quadruple, models predict that the conveyor belt will stop and severe cold will return to northern Europe, even though global temperatures will likely increase dramatically.

FIGURE 1

The ocean conveyor belt. In the North Atlantic cold, salty water sinks, drawing warm water northward from lower latitudes. The warm water provides warmth and moisture for the air above, which is then swept into northern Europe by westerly winds that keep the climate of that region milder than one would normally expect. When the conveyor belt stops, winters apparently turn much colder over northern Europe.

Some climatic models predict that, if average global temperatures increase by about 3°C, the jet stream will weaken and global winds will shift from their "normal" position. The added surface warmth will enhance evaporation, which will lead to a greater worldwide average precipitation. However, the shifting upper-level winds might reduce precipitation over certain areas, which, in turn, would put added stress on certain agricultural areas, especially when the models predict that more precipitation will fall in winter over higher latitudes. Several models indicate that precipitation intensity should increase, suggesting a possibility for more extreme rainfall events, such as floods and severe droughts. If the planet warms, total rainfall must increase to balance the increase in evaporation. But at this point, climate models are unable to determine *exactly* how global precipitation patterns will change.

In a warmer world, most of the precipitation might fall as rain, even in the mountainous regions of eastern and western North America, which might allow much of the winter runoff to end up in the sea, rather than in reservoirs that capture melting snow during the spring.

Other consequences of global warming might be a rise in the sea level as alpine glaciers recede, polar ice melts, and the

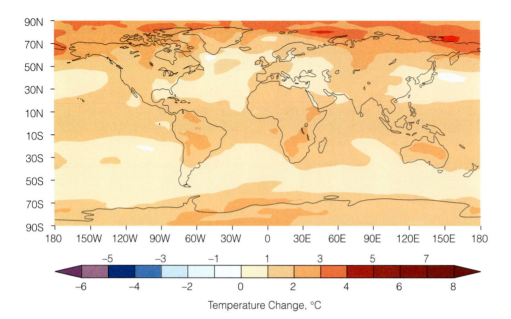

FIGURE 19.18

Projected changes in surface air temperature due to a doubling of CO_2 and human-induced sulfide emissions with an Atmospheric Ocean General Circulation Model (AOGCM). Notice that the greatest warming is projected for the northern polar latitudes. [After F. B. Mitchell, et al., "Transient climate response to increasing sulphate aerosols and greenhouse gases," *Nature* (1995) 376: 501-504.]

oceans expand as they slowly warm. Presently estimates are that by the year 2100 sea level will rise about 50 cm (20 in.) from its present level. Taking into account both lower and higher temperature projections, the rise may be as low as 15 cm or as high as 95 cm. Rising ocean levels might have a damaging influence on coastal ecosystems. In addition, coastal groundwater supplies might become contaminated with salt water.

Climate models predict that the warming should be greater in northern regions (see Fig. 19.18). In the high latitudes of the Northern Hemisphere, the dark green boreal forests absorb up to three times as much solar energy as the snow-covered tundra. Consequently, the winter temperatures in subarctic regions are, on the average, about 11.5°C (21°F) higher than they would be without trees. If warming allows the boreal forests to expand into the tundra, the forests may accelerate the warming in that region. As the temperature rises, organic matter in the soil should decompose at a faster rate, adding more CO_2 to the air, which might accelerate the warming even more. Moreover, trees that grow in a climate zone defined by temperature may become especially hard hit as rising temperatures place them in an inhospitable environment. In a weakened state, they may become more susceptible to insects and disease.

The effect that increasing levels of CO_2 might have on the upper atmosphere is not totally clear. However, climate models suggest that while the lower atmosphere (troposphere) steadily warms, the upper atmosphere (stratosphere, mesosphere, and thermosphere) will cool. The cooling is brought on by the additional molecules of CO_2 (and other trace gases) emitting more infrared radiation both upward and downward.

Is the Warming Real?   Earlier in this chapter we saw that, since the beginning of the twentieth century, the average global surface air temperature has risen between 0.3° and 0.6°C. Is this warming real? Certainly, the greenhouse effect is real. We know from Chapter 2 that our world without water vapor, CO_2, and other greenhouse gases would be about 33°C (59°F) colder than at present. With an average surface temperature of about –18°C (0°F), much of the planet would be uninhabitable. In Chapter 2, we also learned that when the rate of incoming solar energy balances the rate of outgoing infrared energy from the earth's surface and atmosphere, the earth-atmosphere system is in a state of *radiative equilibrium*. Increasing concentrations of greenhouse gases can disturb this equilibrium and are, therefore, referred to as **radiative forcing agents**. The *radiative forcing** provided by extra CO_2 and other greenhouse gases increased by about 1 watt per square meter (1 W/m²) from the middle 1700s to 1950. From 1950 to 1990, the radiative forcing of these gases had increased to about 2.5 W/m². Consequently, most climate scientists contend that *some* of the warming during this century is due to increasing greenhouse gases, but the exact amount of warming is uncertain.

In an attempt to find a signal that suggests that greenhouse gases are altering earth's climate, scientists initially turned their attention to polar regions, where the warming should be greater. Here, they examined ice sheets (especially the west Antarctic ice sheet) to see if shrinkage of the ice might be occurring. But in polar regions, as elsewhere around the globe, rising temperatures produce complex interactions among temperature, precipitation, and wind patterns. Hence, it is now believed that as temperatures rise in south polar regions, more snow will fall in the warmer (but still cold) air, causing snow and ice to build up over the continent of Antarctica. Perhaps, this idea explains why high mountain glaciers in the tropics

---

*Radiative forcing is interpreted as an increase (positive) or a decrease (negative) in net radiant energy observed over an area at the tropopause. All factors being equal, an increase in radiative forcing may induce surface warming, whereas a decrease may induce surface cooling.

and middle latitudes of the Northern Hemisphere are shrinking at record rates, while those in polar regions are not.

A few scientists even contend that the problem of global warming is overstated. They believe that there are too many uncertainties in the climate models to adequately represent the highly complex atmosphere, especially with respect to clouds and oceans. Some point to the fact that many climate models predict that, due to increasing levels of greenhouse gases, average global temperatures should have already risen by at least 1°C, instead of less than 1°C, as observed. Others feel that the warming is statistical in nature and falls within the earth's *natural variability* of climate change.

With levels of CO_2 increasing by more than 28 percent over the past 100 years, why has the observed increase in global temperature been so small? Some researchers suggest that the large heat capacity of the ocean is delaying the warming of the atmosphere. In addition, other natural factors (such as changes in the radiation output of the sun and volcanic eruptions that inject large quantities of sulfur dioxide into the stratosphere) may be countering the warming. Recent findings, however, indicate that human-related emissions of sulfur, leading to sulfuric aerosols (plus human impact on other aerosols), appear to be responsible for the climate response over recent decades being smaller than one would have expected from increasing levels of CO_2 and other greenhouse gases.

Indeed, because the interactions between the earth and its atmosphere are so complex, it is difficult to unequivocally *prove* that the recent warming trend has been due primarily to increasing concentrations of greenhouse gases. The problem is that any human-induced signal of climate change is superimposed on a background of natural climatic variations ("noise") such as the El Niño–Southern Oscillation (ENSO) phenomenon (discussed in Chapter 11). Moreover, in the temperature observations it is difficult to separate a signal from the noise of natural climate variability.

The Intergovernmental Panel on Climate Change (IPCC), a committee of over 2000 leading earth scientists, considered the issues of climate change in a report published in 1990. Updated in 1992, and again in 1995, the report concluded that:

- Emissions resulting from human activities are substantially increasing the atmospheric concentrations of the greenhouse gases carbon dioxide, methane, and nitrous oxide.

- Many greenhouse gases remain in the atmosphere for a long time (decades to centuries in the case of carbon dioxide and nitrous oxide); hence, such gases enhance the greenhouse effect up to hundreds of years.

- Global mean surface air temperature has increased by between 0.3° to 0.6°C since the late nineteenth century.

- The cooling effect of sulfate aerosols resulting from sulfur emissions may have offset a significant part of the greenhouse warming during the past several decades.

- Carbon dioxide concentrations are likely to reach 500 parts per million by the year 2100.

WEATHER WATCH

Global warming is expected to lead to an increase in the occurrence of extremely hot days and a decrease in the occurrence of extremely cold days. One study indicates that if the world warmed by 4°C (7°F), the number of days each year when the air temperature exceeded 32°C (90°F) in Washington, D.C., would increase from an average of 35 to 85.

- Increasing levels of greenhouse gases are likely to cause the global mean surface air temperature to increase by between 1° and 3.5°C (with the best estimate being 2°C) by the year 2100.

- Global sea level has risen by between 10 and 25 cm over the past 100 years, and much of the rise may be related to the increase in global mean temperature.

- Average sea level is expected to rise between 15 and 95 cm (with the best estimate being 50 cm) by the year 2100.

- Warmer global temperatures will lead to a more vigorous hydrological (water) cycle. This prediction translates into prospects for more severe droughts and/or floods in some places and less severe droughts and/or floods in other places.

- There are many uncertainties in climate predictions, particularly with regard to the timing, magnitude, and regional patterns of climate change.

- There are uncertainties in the magnitude and patterns of long-term natural climatic variability. Nevertheless, *the balance of evidence* suggests that there is a discernible human influence on global climate.

But how much influence does humanity have on global climate? Earlier, we learned that over the past 100 years or so, the earth's average surface temperature has risen by as much as 0.6°C. Will the climate slowly continue to warm, or will it warm at an accelerated rate due to increasing concentrations of greenhouse gases? Are we in a natural cycle where the temperature will soon level off, then slowly drop? In centuries to come, will the Northern Hemisphere enter a cooler period as predicted by the Milankovitch cycles? Certainly, climate has changed in the past, but how much and how quickly will it change in the future? Unfortunately, at present, we have no answers to these important questions. However, as we learn more about our complicated and imperfectly understood atmosphere, we will gain better insight into what the future may have in store for the climate of our globe.

Even though there is uncertainty about the rate and patterns of future global warming, cutting down on the emissions of greenhouse gases and pollutants could have potentially positive benefits, such as reducing acid rain, diminishing haze, and

The Sahel—An Example of Climatic Variability and Human Existence

The Sahel is in North Africa, located between about 14° and 18°N latitude (see Fig. 2). Bounded on the north by the dry Sahara and on the south by the grasslands of the Sudan, the Sahel is a semi-arid region of variable rainfall. Precipitation totals may exceed 50 cm (20 in.) in the southern portion while in the north, rainfall is scanty. Yearly rainfall amounts are also variable as a year with adequate rainfall can be followed by a dry one.

During the winter, the Sahel is dry but, as summer approaches, the Intertropical Convergence Zone (ITCZ) with its rain usually moves into the region. The inhabitants of the Sahel are mostly nomadic people who migrate to find grazing land for their cattle and goats. In the early and middle 1960s, adequate rainfall led to improved pasture lands; herds grew larger and so did the population. However, in 1968, the annual rains did not reach as far north as usual, marking the beginning of a series of dry years and a severe drought.

Rain fell in 1969, but the totals were far below those of the favorable years in the mid-1960s. The decrease in rainfall, along with overgrazing, turned thousands of square kilometers of pasture into barren wasteland. By 1973, when the severe drought reached its climax, rainfall totals were 50 percent of the long-term average, and perhaps 50 percent of the cattle and goats had died. The Sahara Desert had migrated southward into the northern fringes of the region, and a great famine had taken the lives of more than 100,000 people. Many more of the 2 million or so inhabitants would have perished had it not been for massive outside aid.

Although low rainfall years have been followed by wetter ones, relatively dry conditions have persisted over the region for the past 30 years or so. The wetter years of the 1950s and 1960s appear to be due to the northward displacement of the ITCZ. The drier years, however, appear to be more related to the intensity of rain that falls during the so-called rainy season. But what causes the lack of intense rain? Some scientists feel that this situation is due to a *biogeophysical feedback mechanism* wherein less rainfall and reduced vegetation cover modify the surface and promote a positive feedback relationship: Surface changes act to reduce convective activity, which in turn promotes or reinforces the dry conditions. As an example, when the vegetation is removed from the surface (perhaps through overgrazing or excessive cultivation), the surface albedo (reflectivity) increases, and the surface temperature drops. But studies show that less vegetation cover does not always result in a higher albedo.

Since the mid-1970s the Sahara Desert has not progressively migrated southward into the Sahel. In fact, during dry years, the desert does migrate southward, but in wet years, it retreats. By the same token, vegetation cover throughout the Sahel is more extensive during the wetter years. Consequently, desertification is not presently overtaking the Sahel, nor is the albedo of the region showing much year-to-year change.

So the question remains: Why did the Sahel go from a period of abundant rainfall in the 1950s and early 1960s to relatively dry conditions since then? Was there a large change in the surface albedo brought on by reduced vegetation? Without adequate satellite imagery during those years, it is impossible to tell. Does this relatively dry spell indicate a long-term fluctuation in climate, or will the wetter years of the 1950s return? And if global temperatures rise into the next century, how will precipitation patterns change? At present, we have no answers.

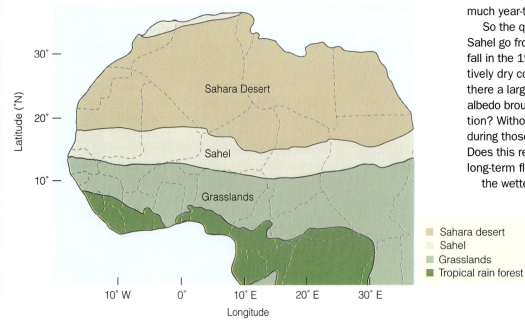

FIGURE 2
The semi-arid Sahel of North Africa is bounded by the Sahara Desert to the north and grasslands to the south.

slowing stratospheric ozone depletion. Even if the greenhouse warming proves to be less than today's consensus, these measures would certainly benefit humanity.

In Perspective As scientists debate the causes and effects of global warming, modification of the earth's surface, taking place right now, could potentially influence the immediate climate of certain regions. For example, studies show that about half the rainfall in the Amazon River Basin is returned to the atmosphere through evaporation and through transpiration from the leaves of trees. Consequently, clearing large areas of tropical rain forests in South America to create open areas for farms and cattle ranges will most likely cause a decrease in evaporative cooling. This decrease, in turn, could lead to a warming in that area of at least several degrees Celsius. In turn,

the reflectivity of the deforested area will change. Similar changes in albedo result from the overgrazing and excessive cultivation of grasslands in semi-arid regions, causing an increase in desert conditions (a process known as **desertification**).

Currently, billions of acres of the world's range and cropland, along with the welfare of millions of people, are affected by desertification. Annually, millions of acres are reduced to a state of near or complete uselessness. The main cause is overgrazing, although overcultivation, poor irrigation practices, and deforestation also play a role. The effect this will have on climate, as surface albedos increase and more dust is swept into the air, is uncertain. (For a look at how a modified surface influences the inhabitants of a region in Africa, read the Focus section on p. 526.)

SUMMARY

In this chapter, we considered some of the many ways the earth's climate can be changed. First, we saw that the earth's climate has undergone considerable change during the geologic past. Some of the evidence for a changing climate comes from tree rings (dendrochronology), chemical analysis of oxygen isotopes in ice cores and fossil shells, and geologic evidence left behind by advancing and retreating glaciers. The evidence from these suggest that, throughout much of the geologic past (before humanity arrived on the scene), the earth was much warmer than it is today. There were cooler periods, however, during which glaciers advanced over large sections of North America and Europe.

We examined some of the possible causes of climate change, noting that the problem is extremely complex, as a change in one variable in the climate system almost immediately changes other variables. One theory suggests that the shifting of the continents, along with volcanic activity and mountain building, may account for variations in climate that take place over millions of years.

The Milankovitch theory proposes that alternating glacial and interglacial episodes during the past 2 million years are the result of small variations in the tilt of the earth's axis and in the geometry of the earth's orbit around the sun. Another theory suggests that certain cooler periods in the geologic past may have been caused by volcanic eruptions rich in sulfur. Still another theory postulates that climatic variations on earth might be due to variations in the sun's energy output.

We examined how sophisticated climate models project that the earth's surface will continue to warm by between 1°C and 3.5°C (with the best estimate being 2°C) by the year 2100, as increasing levels of CO_2 and other trace gases enhance the atmospheric greenhouse effect. We learned that more research is needed to improve these climate models, especially in the

area of the representation of clouds, in order to better refine these estimates.

Key Terms

The following terms are listed in the order they appear in the text. Define each. Doing so will aid you in reviewing the material covered in this chapter.

dendrochronology	negative feedback
Ice Age	mechanism
interglacial period	theory of plate tectonics
Younger-Dryas (event)	Milankovitch theory
mid-Holocene maximum	eccentricity
Medieval Climatic	precession
Optimum	obliquity
Little Ice Age	sulfate aerosols
water vapor-temperature	nuclear winter
rise feedback	Maunder minimum
positive feedback	conveyor belt (ocean)
mechanism	radiative forcing agents
snow-albedo feedback	desertification

Questions for Review

1. Describe some of the methods scientists use to determine past climatic conditions.
2. How does the overall climate of the world today compare with the so-called "normal" climate throughout earth's history?

3. Explain how the changing climate influenced the formation of the Bering land bridge.

4. What are some of the uncertainties in the temperature record during the past 100 years?

5. How does the temperature during the mid-Holocene maximum compare with average global temperatures today?

6. When did the Younger-Dryas episode occur? Was it a warm or a cold spell?

7. When did the Little Ice Age occur? Did it occur during a period of high sunspot activity? Explain.

8. How does a positive feedback mechanism differ from a negative feedback mechanism? Is the water vapor-temperature rise feedback considered positive or negative? Explain.

9. How does the theory of plate tectonics explain climate change over periods of millions of years?

10. Describe the Milankovitch theory of climatic change by explaining how each of the three cycles alters the amount of solar energy reaching the earth.

11. Given the analysis of air bubbles trapped in polar ice during the past 160,000 years, were CO_2 levels generally higher or lower during colder glacial periods?

12. How do sulfate aerosols in the lower atmosphere affect surface air temperatures during the day?

13. Describe the scenario of nuclear winter.

14. Volcanic eruptions rich in sulfur warm the stratosphere. Do they warm or cool the earth's surface? Explain.

15. Explain how variations in the sun's energy might influence global climate.

16. Most climate models predict that increasing levels of CO_2 will cause the mean global surface temperature to rise by as much as 3.5°C by the year 2100. What other greenhouse gas must also increase in concentration in order for this condition to occur?

17. (a) Describe how clouds influence the climate system.
    (b) Which clouds would tend to promote surface cooling: high clouds or low clouds?

18. Even though CO_2 concentrations have risen dramatically over the past 100 years, how do scientists explain the fact that global temperatures have risen only slightly?

19. List some of the consequences that increasing levels of CO_2 and other greenhouse gases might have on the atmosphere and its inhabitants.

20. Is CO_2 the only greenhouse gas we should be concerned with for anthropogenic climate change? If not, what are the other gases?

21 Why are glacial features often found in coastal Africa?

Questions for Thought

1. About 18,000 years ago, when glaciation was at a maximum, was global precipitation greater or less than at present? Explain your reasoning.

2. Explain why periods of glacial advance in the higher latitudes tend to occur with warmer winters and cooler summers.

3. Explain two different ways that an increase in sulfate particles might lower surface air temperatures.

4. Are ice ages in the Northern Hemisphere more likely when: (a) the tilt of the earth is at a maximum or a minimum? (b) the sun is closest to the earth during summer in the Northern Hemisphere, or during winter? Explain your reasoning for both (a) and (b).

5. If the climate continues to warm, explain how an increase in global cloudiness might either enhance or diminish the rate of warming.

6. Most global circulation models (GCMs) show that the poles will warm faster than the tropics. What effect will this have on winter storms in mid-latitudes?

7. The oceans are a major sink of CO_2. One hypothesis states that as warming increases, less CO_2 will be dissolved in the oceans. Would you expect the earth to warm or cool further? Why?

Problems and Exercises

1. If the annual precipitation near Hudson Bay (latitude 55°N) is 38 cm (15 in.) per year, calculate how long it would take snow falling on this region to reach a thickness of 3000 m (about 10,000 ft). (Assume that all the precipitation falls as snow, that there is no melting during the summer, and that the annual precipitation remains constant. To account for compaction of the snow, use a water equivalent of 5 to 1.)

2. On a warming planet, the snow-albedo feedback produces a positive feedback. Make a diagram (or several diagrams) to illustrate this phenomenon. Now, with another diagram, show that the snow-albedo feedback produces a positive feedback on a cooling planet.

Questions for Exploration

1. On the Blue Skies: College Edition CD-ROM go to the **Atmospheric Chemistry** section of the CD and click on "**Temperature Trends**."

UNITS, CONVERSIONS, ABBREVIATIONS, AND EQUATIONS

LENGTH

1 kilometer (km)	=	1000 m
	=	3281 ft
	=	0.62 mi
1 mile (mi)	=	5280 ft
	=	1609 mi
	=	1.61 km
1 meter (m)	=	100 cm
	=	3.28 ft
	=	39.37 in.
1 foot (ft)	=	12 in.
	=	30.48 cm
	=	0.305 m
1 centimeter (cm)	=	0.39 in.
	=	0.01 m
	=	10 mm
1 inch (in.)	=	2.54 cm
	=	0.08 ft
1 millimeter (mm)	=	0.1 cm
	=	0.001 m
	=	0.039 in.
1 micrometer (μm)	=	0.0001 cm
	=	0.000001 m
1 degree latitude	=	111 km
	=	60 nautical mi
	=	69 statute mi

AREA

1 square centimeter (cm^2)	=	0.15 in.^{2}
1 square inch (in.^{2})	=	6.45 cm^2
1 square meter (m^2)	=	10.76 ft^2
1 square foot (ft^2)	=	0.09 m^2

VOLUME

1 cubic centimeter (cm^3)	=	0.06 in.^{3}
1 cubic inch (in.^{3})	=	16.39 cm^3
1 liter (l)	=	1000 cm^3
	=	0.264 gallon (gal) U.S.

SPEED

1 knot	=	1 nautical mi/hr
	=	1.15 statute mi/hr
	=	0.51 m/sec
	=	1.85 km/hr
1 mile per hour (mi/hr)	=	0.87 knot
	=	0.45 m/sec
	=	1.61 km/hr
1 kilometer per hour (km/hr)	=	0.54 knot
	=	0.62 mi/hr
	=	0.28 m/sec
1 meter per second (m/sec)	=	1.94 knots
	=	2.24 mi/hr
	=	3.60 km/hr

FORCE

1 dyne	=	1 gram centimeter per second per second
	=	2.2481×10^{-6} pound (lb)
1 newton (N)	=	1 kilogram meter per second per second
	=	10^5 dynes
	=	0.2248 lb

MASS

1 gram (g)	=	0.035 ounce
	=	0.002 lb
1 kilogram (kg)	=	1000 g
	=	2.2 lb

ENERGY

1 erg	=	1 dyne per cm
	=	2.388×10^{-8} cal
1 joule (J)	=	1 newton meter
	=	0.239 cal
	=	10^7 erg
1 calorie (cal)	=	4.186 J
	=	4.186×10^7 erg

PRESSURE

1 millibar (mb)	=	1000 dynes/cm²
	=	0.75 millimeter of mercury (mm Hg)
	=	0.02953 inch of mercury (in. Hg)
	=	0.01450 pound per square inch (lb/in.²)
	=	100 pascals (Pa)
1 standard atmosphere	=	1013.25 mb
	=	760 mm Hg
	=	29.92 in. Hg
	=	14.7 lb/in.²
1 inch of mercury	=	33.865 mb
1 millimeter of mercury	=	1.3332 mb
1 pascal	=	0.01 mb
	=	1 N/m²
1 hectopascal (hPa)	=	1 mb
1 kilopascal (kPa)	=	10 mb

POWER

1 watt (W)	=	1 J/sec
	=	14.3353 cal/min
1 cal/min	=	0.06973 W
1 horse power (hp)	=	746 W

POWERS OF TEN

Prefix

nano	one-billionth	=	10^{-9}	=	0.000000001
micro	one-millionth	=	10^{-6}	=	0.000001
milli	one-thousandth	=	10^{-3}	=	0.001
centi	one-hundredth	=	10^{-2}	=	0.01
deci	one-tenth	=	10^{-1}	=	0.1
hecto	one hundred	=	10^2	=	100
kilo	one thousand	=	10^3	=	1000
mega	one million	=	10^6	=	1,000,000
giga	one billion	=	10^9	=	1,000,000,000

TEMPERATURE

$$°C = \tfrac{5}{9}\,(°F - 32)$$

To convert degrees Fahrenheit (°F) to degrees Celsius (°C): Subtract 32 degrees from °F, then divide by 1.8.

To convert degrees Celsius (°C) to degrees Fahrenheit (°F): Multiply °C by 1.8, then add 32 degrees.

To convert degrees Celsius (°C) to Kelvins (K): Add 273 to Celsius temperature, as

$$K = °C + 273.$$

TABLE A.1 Temperature Conversions

°F	°C	°F	°C	°F	°C	°F	°C	°F	°C	°F	°C	°F	°C	°F	°C
−40	−40	−20	−28.9	0	−17.8	20	−6.7	40	4.4	60	15.6	80	26.7	100	37.8
−39	−39.4	−19	−28.3	1	−17.2	21	−6.1	41	5.0	61	16.1	81	27.2	101	38.3
−38	−38.9	−18	−27.8	2	−16.7	22	−5.6	42	5.6	62	16.8	82	27.8	102	38.9
−37	−38.3	−17	−27.2	3	−16.1	23	−5.0	43	6.1	63	17.2	83	28.3	103	39.4
−36	−37.8	−16	−26.7	4	−15.6	24	−4.4	44	6.7	64	17.8	84	28.9	104	40.0
−35	−37.2	−15	−26.1	5	−15.0	25	−3.9	45	7.2	65	18.3	85	29.4	105	40.6
−34	−36.7	−14	−25.6	6	−14.4	26	−3.3	46	7.8	66	18.9	86	30.0	106	41.1
−33	−36.1	−13	−25.0	7	−13.9	27	−2.8	47	8.3	67	19.4	87	30.6	107	41.7
−32	−35.6	−12	−24.4	8	−13.3	28	−2.2	48	8.9	68	20.0	88	31.1	108	42.2
−31	−35.0	−11	−23.9	9	−12.8	29	−1.7	49	9.4	69	20.6	89	31.7	109	42.8
−30	−34.4	−10	−23.3	10	−12.2	30	−1.1	50	10.0	70	21.1	90	32.2	110	43.3
−29	−33.9	−9	−22.8	11	−11.7	31	−0.6	51	10.6	71	21.7	91	32.8	111	43.9
−28	−33.3	−8	−22.2	12	−11.1	32	0.0	52	11.1	72	22.2	92	33.3	112	44.4
−27	−32.8	−7	−21.7	13	−10.6	33	0.6	53	11.7	73	22.8	93	33.9	113	45.0
−26	−32.2	−6	−21.1	14	−10.0	34	1.1	54	12.2	74	23.3	94	34.4	114	45.6
−25	−31.7	−5	−20.6	15	−9.4	35	1.7	55	12.8	75	23.9	95	35.0	115	46.1
−24	−31.1	−4	−20.0	16	−8.9	36	2.2	56	13.3	76	24.4	96	35.6	116	46.7
−23	−30.6	−3	−19.4	17	−8.3	37	2.8	57	13.9	77	25.0	97	36.1	117	47.2
−22	−30.0	−2	−18.9	18	−7.8	38	3.3	58	14.4	78	25.6	98	36.7	118	47.8
−21	−29.4	−1	−18.3	19	−7.2	39	3.9	59	15.0	79	26.1	99	37.2	119	48.3

TABLE A.2 SI Units* and Their Symbols

QUANTITY	NAME	UNITS	SYMBOL
length	meter	m	m
mass	kilogram	kg	kg
time	second	sec	sec
temperature	Kelvin	K	K
density	kilogram per cubic meter	kg/m³	kg/m³
speed	meter per second	m/sec	m/sec
force	newton	m · kg/sec²	N
pressure	pascal	N/m²	Pa
energy	joule	N · m	J
power	watt	J/sec	W

*SI stands for Systēme International, which is the international system of units and symbols.

TABLE A.3 Some Useful Equations and Constants

NAME	EQUATION	CONSTANTS AND ABBREVIATIONS
Gas law (equation of state)	$p = \rho RT$	$R = 287\ \text{J/kg} \cdot \text{K (SI) or}$ $R = 2.87 \times 10^6\ \text{erg/g} \cdot \text{K}$ $p = $ pressure is N/m² (SI) $\rho = $ density (kg/m³) $T = $ temperature (K)
Stefan-Boltzmann law	$E = \sigma T^4$	$\sigma = 5.67 \times 10^{-8}$ W/m² · K⁴ (SI) or $\sigma = 5.67 \times 10^{-5}$ erg/cm² · K⁴ · sec $E = $ radiation emitted in W/m² (SI)
Wien's law	$\lambda_{max} = \dfrac{w}{T}$	$w = 0.2897\ \mu\text{m K}$ $\lambda_{max} = $ wavelength (μm)
Solar constant		1376 W/m² (SI) 1.97 cal/cm²/min
Geostrophic wind equation	$V_g = \dfrac{1}{2\Omega\sin\phi\rho}\ \dfrac{\Delta p}{d}$	$V_g = $ geostrophic wind (m/sec) $\Omega = 7.29 \times 10^{-5}$ radian*/sec $\phi = $ latitude $d = $ distance (m) $\Delta p = $ pressure difference (N/m²)
Coriolis parameter	$f = 2\Omega\sin\phi$	
Hydrostatic equation	$\dfrac{\Delta p}{\Delta z} = -\rho g$	$g = $ force of gravity (9.8 m/sec²) $\Delta z = $ change in height (m)

*2π radians equal 360°.

WEATHER SYMBOLS AND THE STATION MODEL

SIMPLIFIED SURFACE-STATION MODEL

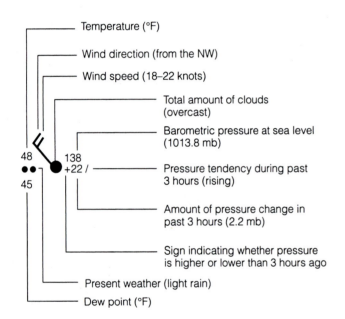

Temperature (°F)

Wind direction (from the NW)

Wind speed (18–22 knots)

Total amount of clouds (overcast)

Barometric pressure at sea level (1013.8 mb)

Pressure tendency during past 3 hours (rising)

Amount of pressure change in past 3 hours (2.2 mb)

Sign indicating whether pressure is higher or lower than 3 hours ago

Present weather (light rain)

Dew point (°F)

CLOUD COVERAGE

Clear

1/8

Scattered

3/8

4/8

5/8

Broken

7/8

Overcast

Obscured

Missing

UPPER-AIR MODEL (500 mb)

Temperature (°C)

Height of pressure surface in meters with first 3 digits given (5640 m)

12 hour height change in meters (04 equals 40 m)

Sign indicating whether height is rising or falling

Dew point depression (difference between air temperature and dew point, °C)

Wind speed (58–62 knots)

Wind direction (from the southwest)

COMMON WEATHER SYMBOLS

Light rain

Moderate rain

Heavy rain

Light snow

Moderate snow

Heavy snow

Light drizzle

Ice pellets (sleet)

Freezing rain

Freezing drizzle

Rain shower

Snow shower

Showers of hail

Drifting or blowing snow

Dust storm

Fog

Haze

Smoke

Thunderstorm

Hurricane

WIND ENTRIES

	Miles (statute) per hour	Knots	Kilometers per Hour
◎	Calm	Calm	Calm
	1–2	1–2	1–3
	3–8	3–7	4–13
	9–14	8–12	14–19
	15–20	13–17	20–32
	21–25	18–22	33–40
	26–31	23–27	41–50
	32–37	28–32	51–60
	38–43	33–37	61–69
	44–49	38–42	70–79
	50–54	43–47	80–87
	55–60	48–52	88–96
	61–66	53–57	97–106
	67–71	58–62	107–114
	72–77	63–67	115–124
	78–83	68–72	125–134
	84–89	73–77	135–143
	119–123	103–107	144–198

PRESSURE TENDENCY

∧ Rising, then falling

⟋ Rising, then steady; or rising, then rising more slowly

╱ Rising steadily or unsteadily

✓ Falling or steady, then rising; or rising, then rising more quickly

— Barometer now higher than 3 hours ago

— Steady, same as 3 hours ago

∨ Falling, then rising, same or lower than 3 hours ago

╲ Falling, then steady; or falling, then falling more slowly

╲ Falling steadily, or unsteadily

∧ Steady or rising, then falling; or falling, then falling more quickly

— Barometer now lower than 3 hours ago

FRONT SYMBOLS

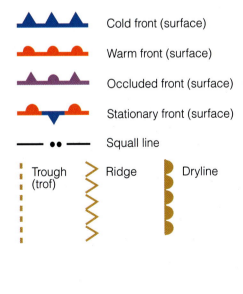

Cold front (surface)

Warm front (surface)

Occluded front (surface)

Stationary front (surface)

Squall line

Trough (trof)   Ridge   Dryline

BEAUFORT WIND SCALE (Over Land)

TABLE C.1   Estimating Wind Speed from Surface Observations

BEAUFORT NUMBER	DESCRIPTION	WIND SPEED			OBSERVATIONS
		mi/hr	knots	km/hr	
0	Calm	0–1	0–1	0–2	Smoke rises vertically
1	Light air	1–3	1–3	2–6	Direction of wind shown by drifting smoke, but not by wind vanes
2	Slight breeze	4–7	4–6	7–11	Wind felt on face; leaves rustle; wind vanes moved by wind; flags stir
3	Gentle breeze	8–12	7–10	12–19	Leaves and small twigs move; wind will extend light flag
4	Moderate breeze	13–18	11–16	20–29	Wind raises dust and loose paper; small branches move; flags flap
5	Fresh breeze	19–24	17–21	30–39	Small trees with leaves begin to sway; flags ripple
6	Strong breeze	25–31	22–27	40–50	Large tree branches in motion; whistling heard in telegraph wires; umbrellas used with difficulty
7	High wind	32–38	28–33	51–61	Whole trees in motion; inconvenience felt walking against wind; flags extend
8	Gale	39–46	34–40	62–74	Wind breaks twigs off trees; walking is difficult
9	Strong gale	47–54	41–47	75–87	Slight structural damage occurs (signs and antennas blown down)
10	Whole gale	55–63	48–55	88–101	Trees uprooted; considerable damage occurs
11	Storm	64–74	56–64	102–119	Winds produce widespread damage
12	Hurricane	≥ 75	≥ 65	≥ 120	Winds produce extensive damage

HUMIDITY AND DEW-POINT TABLES (Psychromatic Tables)

To obtain the dew point (or relative humidity), simply read down the temperature column and then over to the wet-bulb depression. For example, in Table D.1, a temperature of 10°C with a wet-bulb depression of 3°C produces a dew-point temperature of 4°C. (Dew-point temperature and relative humidity readings are appropriate for pressures near 1000 mb.)

TABLE D.1   Dew-Point Temperature (°C)

AIR (DRY-BULB) TEMPERATURE (°C)	WET-BULB DEPRESSION (DRY-BULB TEMPERATURE MINUS WET-BULB TEMPERATURE) (°C)																
	0.5	1.0	1.5	2.0	2.5	3.0	3.5	4.0	4.5	5.0	7.5	10.0	12.5	15.0	17.5	20.0
−20	−25	−33														
−17.5	−21	−27	−38													
−15	−19	−23	−28													
−12.5	−15	−18	−22	−29												
−10	−12	−14	−18	−21	−27	−36										
−7.5	−9	−11	−14	−17	−20	−26	−34									
−5	−7	−8	−10	−13	−16	−19	−24	−31								
−2.5	−4	−6	−7	−9	−11	−14	−17	−22	−28	−41						
0	−1	−3	−4	−6	−8	−10	−12	−15	−19	−24						
2.5	1	0	−1	−3	−4	−6	−8	−10	−13	−16						
5	4	3	2	0	−1	−3	−4	−6	−8	−10	−48					
7.5	6	6	4	3	2	1	−1	−2	−4	−6	−22					
10	9	8	7	6	5	4	2	1	0	−2	−13					
12.5	12	11	10	9	8	7	6	4	3	2	−7	−28				
15	14	13	12	12	11	10	9	8	7	5	−2	−14				
17.5	17	16	15	14	13	12	12	11	10	8	2	−7	−35			
20	19	18	18	17	16	15	14	14	13	12	6	−1	−15			
22.5	22	21	20	20	19	18	17	16	16	15	10	3	−6	−38		
25	24	24	23	22	21	21	20	19	18	18	13	7	0	−14		
27.5	27	26	26	25	24	23	23	22	21	20	16	11	5	−5	−32	
30	29	29	28	27	27	26	25	25	24	23	19	14	9	2	−11	
32.5	32	31	31	30	29	29	28	27	26	26	22	18	13	7	−2	
35	34	34	33	32	32	31	31	30	29	28	25	21	16	11	4	
37.5	37	36	36	35	34	34	33	32	32	31	28	24	20	15	9	0
40	39	39	38	38	37	36	36	35	34	34	30	27	23	18	13	6
42.5	42	41	41	40	40	39	38	38	37	36	33	30	26	22	17	11
45	44	44	43	43	42	42	41	40	40	39	36	33	29	25	21	15
47.5	47	46	46	45	45	44	44	43	42	42	39	35	32	28	24	19
50	49	49	48	48	47	47	46	45	45	44	41	38	35	31	28	23

TABLE D.2   Relative Humidity (Percent)

AIR (DRY-BULB) TEMPERATURE (°C)	WET-BULB DEPRESSION (DRY-BULB TEMPERATURE MINUS WET-BULB TEMPERATURE) (°C)																	
	0.5	1.0	1.5	2.0	2.5	3.0	3.5	4.0	4.5	5.0	7.5	10.0	12.5	15.0	17.5	20.0	22.5	25.0
−20	70	41	11															
−17.5	75	51	26	2														
−15	79	58	38	18														
−12.5	82	65	47	30	13													
−10	85	69	54	39	24	10												
−7.5	87	73	60	48	35	22	10											
−5	88	77	66	54	43	32	21	11	1									
−2.5	90	80	70	60	50	42	37	22	12	3								
0	91	82	73	65	56	47	39	31	23	15								
2.5	92	84	76	68	61	53	46	38	31	24								
5	93	86	78	71	65	58	51	45	38	32	1							
7.5	93	87	80	74	68	62	56	50	44	38	11							
10	94	88	82	76	71	65	60	54	49	44	19							
12.5	94	89	84	78	73	68	63	58	53	48	25	4						
15	95	90	85	80	75	70	66	61	57	52	31	12						
17.5	95	90	86	81	77	72	68	64	60	55	36	18	2					
20	95	91	87	82	78	74	70	66	62	58	40	24	8					
22.5	96	92	87	83	80	76	72	68	64	61	44	28	14	1				
25	96	92	88	84	81	77	73	70	66	63	47	32	19	7				
27.5	96	92	89	85	82	78	75	71	68	65	50	36	23	12	1			
30	96	93	89	86	82	79	76	73	70	67	52	39	27	16	6			
32.5	97	93	90	86	83	80	77	74	71	68	54	42	30	20	11	1		
35	97	93	90	87	84	81	78	75	72	69	56	44	33	23	14	6		
37.5	97	94	91	87	85	82	79	76	73	70	58	46	36	26	18	10	3	
40	97	94	91	88	85	82	79	77	74	72	59	48	38	29	21	13	6	
42.5	97	94	91	88	86	83	80	78	75	72	61	50	40	31	23	16	9	2
45	97	94	91	89	86	83	81	78	76	73	62	51	42	33	26	18	12	6
47.5	97	94	92	89	86	84	81	79	76	74	63	53	44	35	28	21	15	9
50	97	95	92	89	87	84	82	79	77	75	64	54	45	37	30	23	17	11

TABLE D.3 Dew-Point Temperature (°F)

AIR (DRY-BULB) TEMPERATURE (°F)	WET-BULB DEPRESSION (DRY-BULB TEMPERATURE MINUS WET-BULB TEMPERATURE) (°F)																							
	1	2	3	4	5	6	7	8	9	10	11	12	13	14	15	16	17	18	19	20	25	30	35	40
0	-7	-20																						
5	-1	-9	-24																					
10	5	-2	-10	-27																				
15	11	6	0	-9	-26																			
20	16	12	8	2	-7	-21																		
25	22	19	15	10	5	-3	-15	-51																
30	27	25	21	18	14	8	2	-7	-25															
35	33	30	28	25	21	17	13	7	0	-11														
40	38	35	33	30	28	25	21	18	13	7	-1	-14												
45	43	41	38	36	34	31	28	25	22	18	13	7	-1	-14										
50	48	46	44	42	40	37	34	32	29	26	22	18	13	8	0	-13								
55	53	51	50	48	45	43	41	38	36	33	30	27	24	20	15	9	1	-12						
60	58	57	55	53	51	49	47	45	43	40	38	35	32	29	25	21	17	11	4	-8				
65	63	62	60	59	57	55	53	51	49	47	45	42	40	37	34	31	27	24	19	14				
70	69	67	65	64	62	61	59	57	55	53	51	49	47	44	42	39	36	33	30	26	-11			
75	74	72	71	69	68	66	64	63	61	59	57	55	54	51	49	47	44	42	39	36	15			
80	79	77	76	74	73	72	70	68	67	65	63	62	60	58	56	54	52	50	47	44	28	-7		
85	84	82	81	80	78	77	75	74	72	71	69	68	66	64	62	61	59	57	54	52	39	19		
90	89	87	86	85	83	82	81	79	78	76	75	73	72	70	69	67	65	63	61	59	48	32		
95	94	93	91	90	89	87	86	85	83	81	80	79	78	76	74	73	71	70	68	66	56	43	24	
100	99	98	96	95	94	93	91	90	89	87	86	85	83	82	80	79	77	76	74	72	63	52	37	12
105	104	103	101	100	99	98	96	95	94	93	91	90	89	87	86	84	83	82	80	78	70	61	48	30
110	109	108	106	105	104	103	102	100	99	98	97	95	94	93	91	90	89	87	86	84	77	68	57	43
115	114	113	112	110	109	108	107	106	104	103	102	101	99	98	97	96	94	93	92	90	83	75	65	54
120	119	118	117	115	114	113	112	111	110	108	107	106	105	104	102	101	100	98	97	96	89	81	73	63

TABLE D.4   Relative Humidity (Percent)

WET-BULB DEPRESSION (DRY-BULB TEMPERATURE MINUS WET-BULB TEMPERATURE) (°F)

AIR (DRY-BULB) TEMPERATURE (°F)

	1	2	3	4	5	6	7	8	9	10	11	12	13	14	15	16	17	18	19	20	25	30	35	40
0	67	31	1																					
5	73	46	20																					
10	78	56	34	13																				
15	82	64	46	29	11																			
20	85	70	55	40	26	12																		
25	87	74	62	49	37	25	13	1																
30	89	78	67	56	46	36	26	16	6															
35	91	81	72	63	54	45	36	27	19	10	2													
40	92	83	75	68	60	52	45	37	29	22	15	7												
45	93	86	78	71	64	57	51	44	38	31	25	18	12	6										
50	93	87	80	74	67	61	55	49	43	38	32	27	21	16	10	5								
55	94	88	82	76	70	65	59	54	49	43	38	33	28	23	19	14	9							
60	94	89	83	78	73	68	63	58	53	48	43	39	34	30	26	21	17	13	9	5				
65	95	90	85	80	75	70	66	61	56	52	48	44	39	35	31	27	24	20	16	12				
70	95	90	86	81	77	72	68	64	59	55	51	48	44	40	36	33	29	25	22	19	3			
75	96	91	86	82	78	74	70	66	62	58	54	51	47	44	40	37	34	30	27	24	9			
80	96	91	87	83	79	75	72	68	64	61	57	54	50	47	44	41	38	35	32	29	15	3		
85	96	92	88	84	80	76	73	69	66	62	59	56	52	49	46	43	41	38	35	32	20	8		
90	96	92	89	85	81	78	74	71	68	65	61	58	55	52	49	47	44	41	39	36	24	13	3	
95	96	93	89	85	82	79	75	72	69	66	63	60	57	54	51	49	46	43	41	38	27	17	7	
100	96	93	89	86	83	80	77	73	70	68	65	62	59	56	54	51	49	46	44	41	30	21	12	4
105	97	93	90	87	83	80	77	74	71	69	66	63	60	58	55	53	50	48	46	43	33	23	15	7
110	97	93	90	87	84	81	78	75	73	70	67	65	62	60	57	55	52	50	48	46	36	26	18	11
115	97	94	91	88	85	82	79	76	74	71	68	66	63	61	58	56	54	52	49	47	37	28	21	13
120	97	94	91	88	85	82	80	77	74	72	69	67	65	62	60	58	55	53	51	49	40	31	23	17

INSTANT WEATHER FORECAST CHART

This chart is a guide to forecasting the weather. It is applicable to most of the United States, especially the eastern two-thirds. It works best during the fall, winter, and spring, when the weather systems are most active.

TABLE E.1   Instant Weather Forecast Chart

SEA LEVEL PRESSURE (mb)	PRESSURE TENDENCY	SURFACE WIND DIRECTION	SKY CONDITION	24-HR WEATHER FORECAST (SEE WEATHER FORECAST CODE)
1023 or higher (30.21 in.)	rising, steady, or falling	any direction	clear, high clouds, Cu	1, 18 (in winter, 14)
1022 to 1016 (30.20 in. to 30.00 in.)	rising or steady	SW, W, NW, N	clear, high clouds or Cu	1, 18
	falling or steady	SW, S, SE	clear, high clouds	1, 3, 17, 5
	falling	SW, S, SE	middle or low clouds	6, 17
	falling	E, NE	middle or low clouds	6, 14
	falling or steady	E, NE	clear or high clouds	3, 5, 14
1015 to 1009 (29.99 in. to 29.80 in.)	rising	SW, W, NW, N	clear	1, 14
			overcast	2, 16
			precipitation	11, 2, 16
	falling	any direction	clear	3, 17 (dry climate summer, 1, 15)
	falling or steady	SW, S, SE	high clouds	3, 17, 5
	falling	SW, S, SE	middle or low clouds	7
	falling	E, NE	middle or low clouds	7, 12, 14
	falling	SE, E, NE	overcast, precipitation	9
	falling	S, SW	overcast, precipitation	10, 13
1008 or below (29.79 in.)	rising	SW, W, NW, N	clear	1, 12
	rising	SW, W, NW, N	overcast	2, 12, 16
	rising	SW, W, NW, N	overcast with precipitation	11, 12, 16
	rising	NE	overcast	4, 12, 13, 14
	rising	NE	overcast with precipitation	11, 12, 13, 14
	rising or steady	SW, S, SE	clear	3, 6, 8, 12, 15
	falling	SW, S, SE	overcast	7, 8, 12, 13
	falling	SW, S, SE	overcast with precipitation	8, 10, 12, 13, 16
	falling	N	overcast	4, 14
	falling or steady	E, NE	overcast	7, 12, 14
	falling	E, NE	overcast with precipitation	8, 9, 12, 13

WEATHER FORECAST CODE

1 = clear or scattered clouds
2 = clearing
3 = increasing clouds
4 = continued overcast
5 = precipitation possible within 24 hours
6 = precipitation possible within 12 hours

7 = precipitation possible within 8 hours
8 = possible period of heavy precipitation
9 = precipitation continuing
10 = precipitation ending within 12 hours
11 = precipitation ending within 6 hours
12 = windy

13 = possible wind shift to W, NW, or N
14 = continued cool or cold
15 = continued mild or warm
16 = turning colder
17 = slowly rising temperatures
18 = little temperature change

CHANGING GMT AND UTC TO LOCAL TIME

The system of time used in meteorology is Greenwich Mean Time (GMT), which is also known as Coordinated Universal Time (UTC), and as Zulu (Z) Time. This is the time measured on the prime meridian (0° longitude) in Greenwich, England. Because Eastern Standard Time (EST) is 5 hours slower than GMT, to convert from GMT to EST simply requires subtracting 5 hours from GMT. Conversely, to change EST to GMT entails adding 5 hours to EST. Figure F.1 shows how to convert to GMT in various time zones of North America. Since in meteorology the time is given on a 24-hour clock, Table F.1 shows the relationship between the familiar two 12-hour periods of A.M. and P.M. and the 24-hour system.

TABLE F.1   Conversion of A.M. and P.M. Time System to 24-Hour System

TIME	24-HR SYSTEM TIME	TIME	24-HR SYSTEM TIME	TIME	24-HR SYSTEM TIME	TIME	24-HR SYSTEM TIME
A.M.		A.M.		P.M.		P.M.	
12:00 (midnight)	0000	6:00	0600	12:00 (noon)	1200	6:00	1800
1:00	0100	7:00	0700	1:00	1300	7:00	1900
2:00	0200	8:00	0800	2:00	1400	8:00	2000
3:00	0300	9:00	0900	3:00	1500	9:00	2100
4:00	0400	10:00	1000	4:00	1600	10:00	2200
5:00	0500	11:00	1100	5:00	1700	11:00	2300

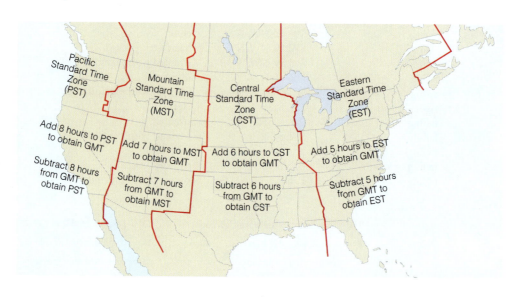

FIGURE F.1
The time zones of North America.

HEAT INDEX (HI) TABLE

TABLE G.1   Air Temperature (°F) and Relative Humidity Are Combined to Determine an Apparent Temperature or Heat Index (HI). An Air Temperature of 95°F with a Relative Humidity of 55 Percent Produces an Apparent Temperature (HI) of 110°F.

Relative Humidity (%)

Air Temp (°F)	0	5	10	15	20	25	30	35	40	45	50	55	60	65	70	75	80	85	90	95	100
140	125																				
135	120	128																			
130	117	122	131																		
125	111	116	123	131	141																
120	107	111	116	123	130	139	148														
115	103	107	111	115	120	127	135	143	151												
110	99	102	105	108	112	117	123	130	137	143	150										
105	95	97	100	102	105	109	113	118	123	129	135	142	149								
100	91	93	95	97	99	101	104	107	110	115	120	126	132	138	144						
95	87	88	90	91	93	94	96	98	101	104	107	110	114	119	124	130	136				
90	83	84	85	86	87	88	90	91	93	95	96	98	100	102	106	109	113	117	122		
85	78	79	80	81	82	83	84	85	86	87	88	89	90	91	93	95	97	99	102	105	108
80	73	74	75	76	77	77	78	79	79	80	81	81	82	83	85	86	86	87	88	89	91
75	69	69	70	71	72	72	73	73	74	74	75	75	76	76	77	77	78	78	79	79	80
70	64	64	65	65	66	66	67	67	68	68	69	69	70	70	70	70	71	71	71	71	72

Heat Index (or apparent temperature)

STANDARD ATMOSPHERE

ALTITUDE				PRESSURE	TEMPERATURE		DENSITY
meters	feet	kilometers	miles	millibars	°C	°F	kg/m³
0	0	0.0	0.0	1013.25	15.0	(59.0)	1.225
500	1,640	0.5	0.3	954.61	11.8	(53.2)	1.167
1,000	3,280	1.0	0.6	898.76	8.5	(47.3)	1.112
1,500	4,921	1.5	0.9	845.59	5.3	(41.5)	1.058
2,000	6,562	2.0	1.2	795.01	2.0	(35.6)	1.007
2,500	8,202	2.5	1.5	746.91	−1.2	(29.8)	0.957
3,000	9,842	3.0	1.9	701.21	−4.5	(23.9)	0.909
3,500	11,483	3.5	2.2	657.80	−7.7	(18.1)	0.863
4,000	13,123	4.0	2.5	616.60	−11.0	(12.2)	0.819
4,500	14,764	4.5	2.8	577.52	−14.2	(6.4)	0.777
5,000	16,404	5.0	3.1	540.48	−17.5	(0.5)	0.736
5,500	18,045	5.5	3.4	505.39	−20.7	(−5.3)	0.697
6,000	19,685	6.0	3.7	472.17	−24.0	(−11.2)	0.660
6,500	21,325	6.5	4.0	440.75	−27.2	(−17.0)	0.624
7,000	22,965	7.0	4.3	411.05	−30.4	(−22.7)	0.590
7,500	24,606	7.5	4.7	382.99	−33.7	(−28.7)	0.557
8,000	26,247	8.0	5.0	356.51	−36.9	(−34.4)	0.526
8,500	27,887	8.5	5.3	331.54	−40.2	(−40.4)	0.496
9,000	29,528	9.0	5.6	308.00	−43.4	(−46.1)	0.467
9,500	31,168	9.5	5.9	285.84	−46.6	(−51.9)	0.440
10,000	32,808	10.0	6.2	264.99	−49.9	(−57.8)	0.413
11,000	36,089	11.0	6.8	226.99	−56.4	(−69.5)	0.365
12,000	39,370	12.0	7.5	193.99	−56.5	(−69.7)	0.312
13,000	42,651	13.0	8.1	165.79	−56.5	(−69.7)	0.267
14,000	45,932	14.0	8.7	141.70	−56.5	(−69.7)	0.228
15,000	49,213	15.0	9.3	121.11	−56.5	(−69.7)	0.195
16,000	52,493	16.0	9.9	103.52	−56.5	(−69.7)	0.166
17,000	55,774	17.0	10.6	88.497	−56.5	(−69.7)	0.142
18,000	59,055	18.0	11.2	75.652	−56.5	(−69.7)	0.122
19,000	62,336	19.0	11.8	64.674	−56.5	(−69.7)	0.104
20,000	65,617	20.0	12.4	55.293	−56.5	(−69.7)	0.089
25,000	82,021	25	15.5	25.492	−51.6	(−60.9)	0.040
30,000	98,425	30	18.6	11.970	−46.6	(−51.9)	0.018
35,000	114,829	35	21.7	5.746	−36.6	(−33.9)	0.008
40,000	131,234	40	24.9	2.871	−22.8	(−9.0)	0.004
45,000	147,638	45	28.0	1.491	−9.0	(15.8)	0.002
50,000	164,042	50	31.1	0.798	−2.5	(27.5)	0.001
60,000	196,850	60	37.3	0.220	−26.1	(−15.0)	0.0003
70,000	229,659	70	43.5	0.052	−53.6	(−64.5)	0.00008
80,000	262,467	80	49.7	0.010	−74.5	(−102.1)	0.00002

HURRICANE TRACKING CHART

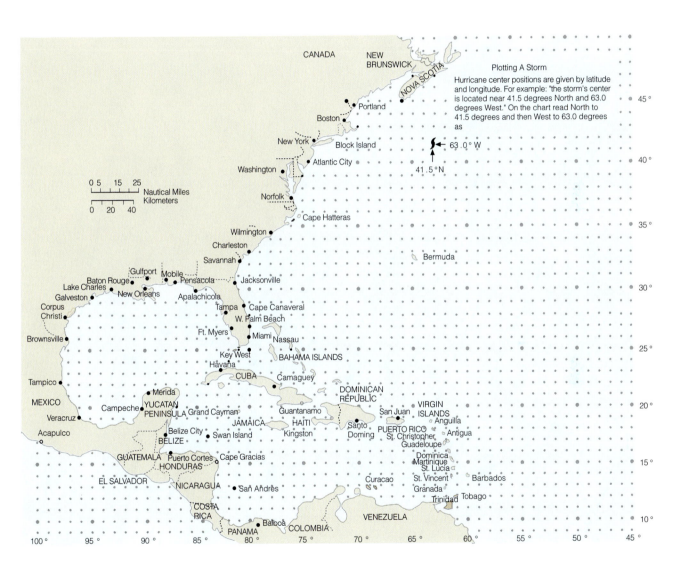

Plotting A Storm

Hurricane center positions are given by latitude and longitude. For example: "the storm's center is located near 41.5 degrees North and 63.0 degrees West." On the chart read North to 41.5 degrees and then West to 63.0 degrees as

63.0 ° W

41.5 °N

ADIABATIC CHART

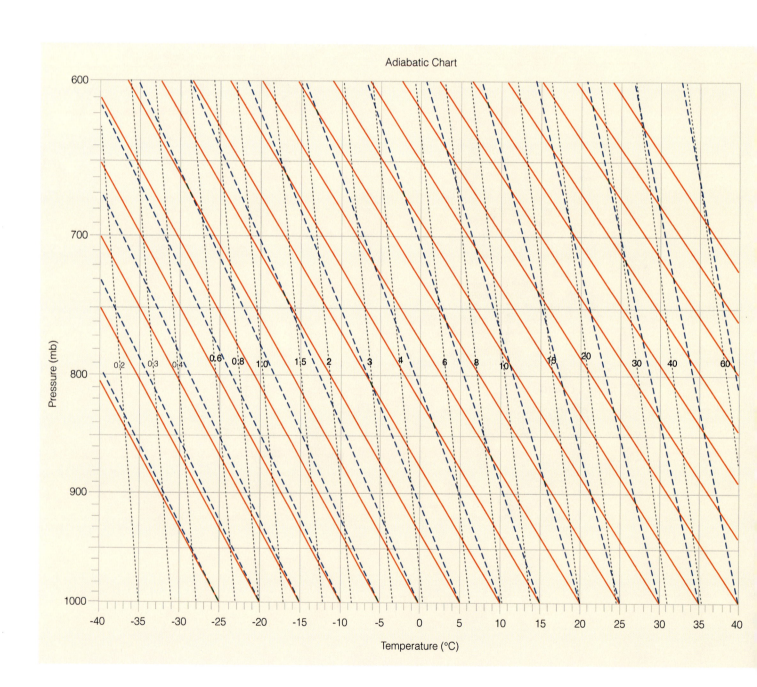

CLIMATIC DATA FOR CITIES THROUGHOUT THE WORLD

The top numbers for each city represent average temperatures in degrees Fahrenheit. The bottom numbers represent average precipitation in inches. The location of each city is shown on the map on page A-26.

NORTH AMERICA

1. ALBUQUERQUE, NM   ELEVATION 1619 m (5311 Ft)   CLIMATE Bsk   LAT. 35°N   LONG. 106.5°W

	JAN.	FEB.	MAR.	APRIL	MAY	JUNE	JULY	AUG.	SEPT.	OCT.	NOV.	DEC.	YEAR
t (°F)	35	40	46	56	65	75	78	76	70	58	44	37	57
p (in.)	0.4	0.4	0.5	0.5	0.8	0.6	1.2	1.3	0.9	0.8	0.4	0.5	8.1

2. ANCHORAGE, AK   ELEVATION 35 m (114 Ft)   CLIMATE Dfc   LAT. 61°N   LONG. 150°W

	JAN.	FEB.	MAR.	APRIL	MAY	JUNE	JULY	AUG.	SEPT.	OCT.	NOV.	DEC.	YEAR
t (°F)	12	19	25	37	47	56	58	56	48	36	22	14	36
p (in.)	0.8	0.7	0.5	0.4	0.5	1.0	1.9	2.6	2.5	1.9	1.0	0.9	14.7

3. ATLANTA, GA   ELEVATION 308 m (1010 Ft)   CLIMATE Cfa   LAT. 33.5°N   LONG. 84.5°W

	JAN.	FEB.	MAR.	APRIL	MAY	JUNE	JULY	AUG.	SEPT.	OCT.	NOV.	DEC.	YEAR
t (°F)	45	46	51	60	69	77	79	78	73	62	51	45	61
p (in.)	4.4	4.5	5.4	4.5	3.2	3.8	4.7	3.6	3.3	2.4	3.0	4.4	47.1

4. BARROW, AK   ELEVATION 9.4 m (31 Ft)   CLIMATE ET   LAT. 71°N   LONG. 157°W

	JAN.	FEB.	MAR.	APRIL	MAY	JUNE	JULY	AUG.	SEPT.	OCT.	NOV.	DEC.	YEAR
t (°F)	−16	−18	−15	0	18	33	39	38	30	17	−1	−11	10
p (in.)	0.2	0.2	0.1	0.1	0.4	0.8	0.9	0.6	0.5	0.2	0.2	0.2	4.3

5. CALGARY, CANADA   ELEVATION 1079 m (3504 Ft)   CLIMATE Dfb   LAT. 51°N   LONG. 114°W

	JAN.	FEB.	MAR.	APRIL	MAY	JUNE	JULY	AUG.	SEPT.	OCT.	NOV.	DEC.	YEAR
t (°F)	13	17	26	40	50	56	62	60	51	42	28	19	39
p (in.)	0.5	0.5	0.8	1.0	2.3	3.1	2.5	2.3	1.5	0.7	0.7	0.6	16.7

6. CHESTERFIELD, CANADA   ELEVATION 9.8 m (32 Ft)   CLIMATE ET   LAT. 63.5°N   LONG. 92°W

	JAN.	FEB.	MAR.	APRIL	MAY	JUNE	JULY	AUG.	SEPT.	OCT.	NOV.	DEC.	YEAR
t (°F)	−26	−26	−13	3	21	37	46	46	37	21	0	−15	10
p (in.)	0.3	0.4	0.4	0.7	0.5	1.0	1.8	1.7	1.5	1.2	0.8	0.6	10.9

7.   CHICAGO, IL   ELEVATION 185 m (607 Ft)   CLIMATE Dfa   LAT. 42°N   LONG. 88°W

	JAN.	FEB.	MAR.	APRIL	MAY	JUNE	JULY	AUG.	SEPT.	OCT.	NOV.	DEC.	YEAR
t (°F)	26	28	36	49	60	70	76	74	66	55	40	29	51
p (in.)	1.9	1.6	2.7	3.0	3.7	4.1	3.4	3.2	2.7	2.8	2.2	1.9	33.2

8.   CHURCHILL, CANADA   ELEVATION 35 m (115 Ft)   CLIMATE Dfc   LAT. 59°N   LONG. 94°W

	JAN.	FEB.	MAR.	APRIL	MAY	JUNE	JULY	AUG.	SEPT.	OCT.	NOV.	DEC.	YEAR
t (°F)	−19	−16	−6	14	30	43	54	52	43	27	6	−11	16
p (in.)	0.5	0.6	0.9	0.9	0.9	1.9	2.2	2.7	2.3	1.4	1.0	0.7	16.0

9.   COPPERMINE, CANADA   ELEVATION 9.1 m (30 Ft)   CLIMATE Dfc   LAT. 68°N   LONG. 115°W

	JAN.	FEB.	MAR.	APRIL	MAY	JUNE	JULY	AUG.	SEPT.	OCT.	NOV.	DEC.	YEAR
t (°F)	−18	−15	−4	14	28	43	54	54	43	30	10	−8	18
p (in.)	0.7	0.7	0.8	1.4	1.6	1.9	1.5	2.4	2.0	1.6	1.8	1.0	17.4

10.   DALLAS, TX   ELEVATION 147 m (481 Ft)   CLIMATE Cfa   LAT. 33°N   LONG. 97°W

	JAN.	FEB.	MAR.	APRIL	MAY	JUNE	JULY	AUG.	SEPT.	OCT.	NOV.	DEC.	YEAR
t (°F)	46	50	56	65	73	81	85	85	78	68	55	48	66
p (in.)	2.3	2.6	2.8	4.0	4.8	3.2	1.9	1.9	2.8	2.7	2.7	2.7	34.6

11.   DENVER, CO   ELEVATION 1609 m (5280 Ft)   CLIMATE BSk   LAT. 40°N   LONG. 105°W

	JAN.	FEB.	MAR.	APRIL	MAY	JUNE	JULY	AUG.	SEPT.	OCT.	NOV.	DEC.	YEAR
t (°F)	28	32	36	46	56	66	73	72	63	51	38	32	50
p (in.)	0.6	0.7	1.2	2.1	2.7	1.4	1.5	1.3	1.1	1.0	0.7	0.5	14.8

12.   FAIRBANKS, AK   ELEVATION 133 m (436 Ft)   CLIMATE Dfc   LAT. 65°N   LONG. 148°W

	JAN.	FEB.	MAR.	APRIL	MAY	JUNE	JULY	AUG.	SEPT.	OCT.	NOV.	DEC.	YEAR
t (°F)	−11	−2	9	30	48	59	60	55	43	27	5	−8	26
p (in.)	0.9	0.5	0.4	0.3	0.7	1.4	1.8	2.2	1.1	0.9	0.6	0.5	11.3

13.   FARGO, ND   ELEVATION 273 m (895 Ft)   CLIMATE Dfb   LAT. 47°N   LONG. 97°W

	JAN.	FEB.	MAR.	APRIL	MAY	JUNE	JULY	AUG.	SEPT.	OCT.	NOV.	DEC.	YEAR
t (°F)	7	11	25	42	55	65	71	69	59	46	28	13	41
p (in.)	0.6	0.7	0.9	1.9	2.2	3.0	2.3	2.7	1.7	1.3	0.9	0.6	18.7

14.   FORT SIMPSON, CANADA   ELEVATION 169 m (554 Ft)   CLIMATE Dfc   LAT. 62°N   LONG. 121°W

	JAN.	FEB.	MAR.	APRIL	MAY	JUNE	JULY	AUG.	SEPT.	OCT.	NOV.	DEC.	YEAR
t (°F)	−17	−9	−5	27	46	57	63	57	46	30	7	−11	25
p (in.)	0.7	0.7	0.5	0.7	1.4	1.5	2.0	1.5	1.3	1.1	0.9	0.8	13.1

15.   FROBISHER BAY, CANADA   ELEVATION 33.5 m (110 Ft)   CLIMATE ET   LAT. 64°N   LONG. 68.5°W

	JAN.	FEB.	MAR.	APRIL	MAY	JUNE	JULY	AUG.	SEPT.	OCT.	NOV.	DEC.	YEAR
t (°F)	−15	−13	−6	9	28	39	46	45	37	23	7	−8	16
p (in.)	0.7	0.9	0.8	0.8	0.7	0.9	1.5	2.0	1.8	1.1	1.1	1.0	13.3

16.   LOS ANGELES, CA   ELEVATION 29.6 m (97 Ft)   CLIMATE BSk   LAT. 34°N   LONG. 118.5°W

	JAN.	FEB.	MAR.	APRIL	MAY	JUNE	JULY	AUG.	SEPT.	OCT.	NOV.	DEC.	YEAR
t (°F)	54	55	57	59	62	65	69	69	68	65	61	57	62
p (in.)	2.7	2.9	1.8	1.0	0.1	0.1	0.0	0.0	0.2	0.4	1.1	2.4	12.6

17. MEXICO CITY, MEXICO   ELEVATION 2237 m (7340 Ft)   CLIMATE Cwb   LAT. 19.5°N   LONG. 99°W

	JAN.	FEB.	MAR.	APRIL	MAY	JUNE	JULY	AUG.	SEPT.	OCT.	NOV.	DEC.	YEAR
t (°F)	54	57	61	64	65	64	62	62	61	59	56	54	60
p (in.)	0.2	0.3	0.5	0.7	1.9	4.1	4.5	4.3	4.1	1.6	0.5	0.3	23.0

18. MIAMI, FL   ELEVATION 2 m (7 Ft)   CLIMATE Am   LAT. 26°N   LONG. 80°W

	JAN.	FEB.	MAR.	APRIL	MAY	JUNE	JULY	AUG.	SEPT.	OCT.	NOV.	DEC.	YEAR
t (°F)	67	68	70	74	78	81	82	82	81	78	72	68	75
p (in.)	2.0	1.9	2.3	3.9	6.4	7.4	6.8	7.0	9.5	8.2	2.8	1.7	59.8

19. MONTREAL, CANADA   ELEVATION 57 m (187 Ft)   CLIMATE Dfa   LAT. 45.5°N   LONG. 73.5°W

	JAN.	FEB.	MAR.	APRIL	MAY	JUNE	JULY	AUG.	SEPT.	OCT.	NOV.	DEC.	YEAR
t (°F)	16	18	28	43	57	66	72	70	61	48	36	22	45
p (in.)	3.8	3.0	3.5	2.6	3.1	3.4	3.7	3.5	3.7	3.4	3.5	3.6	40.8

20. NEW ORLEANS, LA   ELEVATION 2.7 m (9 Ft)   CLIMATE Cfa   LAT. 30°N   LONG. 90°W

	JAN.	FEB.	MAR.	APRIL	MAY	JUNE	JULY	AUG.	SEPT.	OCT.	NOV.	DEC.	YEAR
t (°F)	56	58	63	70	76	82	83	83	80	73	62	57	70
p (in.)	4.8	4.2	6.6	5.4	5.4	5.6	7.1	6.4	5.8	3.7	4.0	4.6	63.5

21. NEW YORK, NY   ELEVATION 4 m (13 Ft)   CLIMATE Cfa   LAT. 41°N   LONG. 74°W

	JAN.	FEB.	MAR.	APRIL	MAY	JUNE	JULY	AUG.	SEPT.	OCT.	NOV.	DEC.	YEAR
t (°F)	32	32	39	49	60	70	76	74	68	58	46	35	53
p (in.)	3.2	2.9	4.2	3.5	3.7	3.4	4.0	5.0	4.2	3.2	3.5	3.2	43.9

22. PORTLAND, ME   ELEVATION 13 m (43 Ft)   CLIMATE Dfb   LAT. 44°N   LONG. 70°W

	JAN.	FEB.	MAR.	APRIL	MAY	JUNE	JULY	AUG.	SEPT.	OCT.	NOV.	DEC.	YEAR
t (°F)	22	23	31	42	53	62	68	67	59	49	38	26	45
p (in.)	4.4	3.8	4.3	3.7	3.4	3.2	2.9	2.4	3.5	3.2	4.2	3.8	42.8

23. PORTLAND, OR   ELEVATION 9.1 m (30 Ft)   CLIMATE Csb   LAT. 45.5°N   LONG. 124°W

	JAN.	FEB.	MAR.	APRIL	MAY	JUNE	JULY	AUG.	SEPT.	OCT.	NOV.	DEC.	YEAR
t (°F)	40	44	48	54	59	64	68	68	64	56	47	42	55
p (in.)	5.4	4.9	4.2	2.4	1.9	1.6	0.4	0.6	1.8	3.5	6.0	7.1	39.9

24. RENO, NV   ELEVATION 1340 m (4397 Ft)   CLIMATE BSk   LAT. 39.5°N   LONG. 120°W

	JAN.	FEB.	MAR.	APRIL	MAY	JUNE	JULY	AUG.	SEPT.	OCT.	NOV.	DEC.	YEAR
t (°F)	31	36	41	48	55	62	70	67	60	51	40	33	50
p (in.)	1.0	1.0	0.7	0.5	0.5	0.4	0.2	0.2	0.2	0.6	0.6	0.9	7.0

25. ST JOHN'S, CANADA   ELEVATION 74 m (243 Ft)   CLIMATE Dfb   LAT. 47.5°N   LONG. 53°W

	JAN.	FEB.	MAR.	APRIL	MAY	JUNE	JULY	AUG.	SEPT.	OCT.	NOV.	DEC.	YEAR
t (°F)	24	22	28	35	42	52	60	61	54	46	37	29	44
p (in.)	5.3	4.9	4.6	4.2	3.6	3.5	3.5	3.7	3.8	5.3	5.9	5.5	53.8

26. ST. LOUIS, MO   ELEVATION 163 m (535 Ft)   CLIMATE Cfa   LAT. 38.5°N   LONG. 90°W

	JAN.	FEB.	MAR.	APRIL	MAY	JUNE	JULY	AUG.	SEPT.	OCT.	NOV.	DEC.	YEAR
t (°F)	32	35	43	55	64	74	78	77	70	58	44	35	55
p (in.)	2.0	2.0	3.1	3.7	3.7	4.3	3.3	3.0	2.8	2.9	2.6	2.0	35.3

27. | SAN DIEGO, CA ELEVATION 4 m (13 Ft) CLIMATE BSk LAT. 33°N LONG. 117°W

	JAN.	FEB.	MAR.	APRIL	MAY	JUNE	JULY	AUG.	SEPT.	OCT.	NOV.	DEC.	YEAR
t (°F)	55	56	59	62	64	66	70	72	70	66	62	57	63
p (in.)	2.0	2.2	1.6	0.8	0.2	0.1	0.0	0.1	0.2	0.5	0.9	2.0	10.4

28. | SEATTLE, WA ELEVATION 118 m (387 Ft) CLIMATE Csb LAT. 47.5°N LONG. 122.5°W

	JAN.	FEB.	MAR.	APRIL	MAY	JUNE	JULY	AUG.	SEPT.	OCT.	NOV.	DEC.	YEAR
t (°F)	38	41	44	49	56	60	65	64	60	52	44	41	51
p (in.)	5.7	4.2	3.8	2.4	1.7	1.6	0.8	1.0	2.0	4.0	5.4	6.3	38.9

CENTRAL AND SOUTH AMERICA

29. | ASUNCIÓN, PARAGUAY ELEVATION 139 m (456 Ft) CLIMATE Aw LAT. 25°S LONG. 57.5°W

	JAN.	FEB.	MAR.	APRIL	MAY	JUNE	JULY	AUG.	SEPT.	OCT.	NOV.	DEC.	YEAR
t (°F)	81	80	78	72	67	64	64	66	70	73	76	80	72
p (in.)	5.5	5.1	4.3	5.2	4.6	2.7	2.2	1.5	3.1	5.5	5.9	6.2	51.8

30. | BELEM, BRAZIL ELEVATION 12.8 m (42 Ft) CLIMATE Af LAT. 1.5°S LONG. 48.5°W

	JAN.	FEB.	MAR.	APRIL	MAY	JUNE	JULY	AUG.	SEPT.	OCT.	NOV.	DEC.	YEAR
t (°F)	80	79	80	80	80	80	80	80	80	80	81	80	80
p (in.)	12.5	14.1	14.1	12.6	10.2	6.7	5.9	4.4	3.5	3.3	2.6	6.1	96

31. | BUENOS AIRES, ARGENTINA ELEVATION 27 m (89 Ft) CLIMATE Cfa LAT. 34°S LONG. 58.5°W

	JAN.	FEB.	MAR.	APRIL	MAY	JUNE	JULY	AUG.	SEPT.	OCT.	NOV.	DEC.	YEAR
t (°F)	74	73	69	61	55	50	49	51	55	60	66	71	61
p (in.)	3.1	2.6	4.4	3.5	3.0	2.4	2.4	2.4	3.1	3.4	3.2	3.9	37.4

32. | GOIAS, BRAZIL ELEVATION 520 m (1706 Ft) CLIMATE Aw LAT. 16°S LONG. 50°W

	JAN.	FEB.	MAR.	APRIL	MAY	JUNE	JULY	AUG.	SEPT.	OCT.	NOV.	DEC.	YEAR
t (°F)	74	76	76	77	76	72	72	76	79	78	76	74	76
p (in.)	12.5	9.9	10.2	4.6	0.4	0.3	0.0	0.3	2.3	5.3	9.4	9.5	64.7

33. | LIMA, PERU ELEVATION 120 m (394 Ft) CLIMATE BWh LAT. 12°S LONG. 77°W

	JAN.	FEB.	MAR.	APRIL	MAY	JUNE	JULY	AUG.	SEPT.	OCT.	NOV.	DEC.	YEAR
t (°F)	70	72	72	68	64	61	60	59	59	61	63	67	65
p (in.)	0.1	0.0	0.0	0.0	0.2	0.2	0.3	0.3	0.3	0.1	0.1	0.0	1.60

34. | MANAGUA, NICARAGUA ELEVATION 56 m (184 FT) CLIMATE Aw LAT. 12°N LONG. 86°W

	JAN.	FEB.	MAR.	APRIL	MAY	JUNE	JULY	AUG.	SEPT.	OCT.	NOV.	DEC.	YEAR
t (°F)	78	80	82	83	83	81	80	80	80	80	79	78	80
p (in.)	0.2	0.0	0.2	0.2	3.0	11.7	5.3	5.1	7.2	9.6	2.3	0.2	45.0

35. | MANAUS, BRAZIL ELEVATION 44 m (144 Ft) CLIMATE Am LAT. 3°S LONG. 60°W

	JAN.	FEB.	MAR.	APRIL	MAY	JUNE	JULY	AUG.	SEPT.	OCT.	NOV.	DEC.	YEAR
t (°F)	82	82	82	81	82	82	82	83	84	84	84	83	82
p (in.)	14.4	12.8	15.0	15.8	7.3	5.2	2.3	1.5	3.8	4.2	5.6	8.0	95.9

36. | **SANTIAGO, CHILE** | **ELEVATION 520 m (1706 Ft)** | **CLIMATE Csb** | **LAT. 33.5°S** | **LONG. 71°W**

	JAN.	FEB.	MAR.	APRIL	MAY	JUNE	JULY	AUG.	SEPT.	OCT.	NOV.	DEC.	YEAR
t (°F)	67	66	62	56	51	46	46	48	52	56	61	66	56
p (in.)	0.1	0.1	0.2	0.5	2.5	3.3	3.0	2.2	1.2	0.6	0.3	0.2	14.2

37. | **SARMIENTO, ARGENTINA** | **ELEVATION 268 m (879 Ft)** | **CLIMATE BWk** | **LAT. 45.5°S** | **LONG. 69°W**

	JAN.	FEB.	MAR.	APRIL	MAY	JUNE	JULY	AUG.	SEPT.	OCT.	NOV.	DEC.	YEAR
t (°F)	65	64	58	52	45	38	37	42	46	53	58	62	52
p (in.)	0.2	0.3	0.3	0.4	0.8	0.8	0.6	0.5	0.4	0.3	0.2	0.3	5.10

AFRICA

38. | **BENGHAZI, LIBYA** | **ELEVATION 25 m (82 Ft)** | **CLIMATE BSh** | **LAT. 32°N** | **LONG. 20°E**

	JAN.	FEB.	MAR.	APRIL	MAY	JUNE	JULY	AUG.	SEPT.	OCT.	NOV.	DEC.	YEAR
t (°F)	55	57	63	66	72	75	78	79	78	75	66	59	68
p (in.)	2.6	1.6	0.8	0.2	0.1	0.0	0.0	0.0	0.1	0.7	1.8	2.6	10.5

39. | **BULAWAYO, ZIMBABWE** | **ELEVATION 1343 m (4405 Ft)** | **CLIMATE BSh** | **LAT. 20°S** | **LONG. 28.5°E**

	JAN.	FEB.	MAR.	APRIL	MAY	JUNE	JULY	AUG.	SEPT.	OCT.	NOV.	DEC.	YEAR
t (°F)	71	70	69	68	62	57	57	61	68	72	72	72	67
p (in.)	5.6	4.3	3.3	0.7	0.4	0.1	0.0	0.0	0.2	0.8	3.2	4.8	23.4

40. | **CAIRO, EGYPT** | **ELEVATION 116 m (381 Ft)** | **CLIMATE BWh** | **LAT. 30°N** | **LONG. 31.5°E**

	JAN.	FEB.	MAR.	APRIL	MAY	JUNE	JULY	AUG.	SEPT.	OCT.	NOV.	DEC.	YEAR
t (°F)	55	57	63	70	76	80	82	82	78	74	65	58	70
p (in.)	0.2	0.2	0.2	0.1	0.1	0.0	0.0	0.0	0.0	0.0	0.1	0.2	1.1

41. | **CAPE TOWN, SOUTH AFRICA** | **ELEVATION 17 m (56 Ft)** | **CLIMATE Csb** | **LAT. 34°S** | **LONG. 18.5°E**

	JAN.	FEB.	MAR.	APRIL	MAY	JUNE	JULY	AUG.	SEPT.	OCT.	NOV.	DEC.	YEAR
t (°F)	69	70	68	62	58	56	54	55	57	61	64	67	62
p (in.)	0.6	0.3	0.7	1.9	3.1	3.3	3.5	2.6	1.7	1.2	0.7	0.4	20.0

42. | **DAKAR, SENEGAL** | **ELEVATION 40 m (131 Ft)** | **CLIMATE BSh** | **LAT. 14.5°N** | **LONG. 17.5°W**

	JAN.	FEB.	MAR.	APRIL	MAY	JUNE	JULY	AUG.	SEPT.	OCT.	NOV.	DEC.	YEAR
t (°F)	72	72	72	73	76	80	82	82	82	82	80	74	77
p (in.)	0.0	0.0	0.0	0.0	0.0	0.7	3.5	10.0	5.2	1.5	0.1	0.3	21.0

43. | **DOUALA, CAMEROON** | **ELEVATION 8 m (26 Ft)** | **CLIMATE Aw** | **LAT. 4°N** | **LONG. 9.5°E**

	JAN.	FEB.	MAR.	APRIL	MAY	JUNE	JULY	AUG.	SEPT.	OCT.	NOV.	DEC.	YEAR
t (°F)	80	80	80	80	80	78	76	76	76	76	78	79	78
p (in.)	1.8	3.7	8.0	9.1	11.8	21.2	29.2	27.3	20.9	16.9	6.1	2.5	158.5

44. | **IN SALAH, ALGERIA** | **ELEVATION 280 m (919 Ft)** | **CLIMATE BWh** | **LAT. 27°N** | **LONG. 2.5°E**

	JAN.	FEB.	MAR.	APRIL	MAY	JUNE	JULY	AUG.	SEPT.	OCT.	NOV.	DEC.	YEAR
t (°F)	56	61	68	77	84	95	98	96	91	80	66	58	78
p (in.)	0.1	0.1	0.0	0.0	0.0	0.0	0.0	0.1	0.0	0.0	0.2	0.1	0.6

45.

KHARTOUM, SUDAN		ELEVATION 390 m (1279 Ft)		CLIMATE BWh		LAT. 15.5°N		LONG. 32.5°E				
	JAN.	FEB.	MAR.	APRIL	MAY	JUNE	JULY	AUG.	SEPT.	OCT.	NOV.	DEC.	YEAR
t (°F)	74	73	76	80	85	90	94	94	90	84	81	77	83
p (in.)	0.0	0.0	0.0	0.0	0.1	0.3	2.1	2.8	0.7	0.2	0.0	0.0	6.2

46.

LUANDA, ANGOLA		ELEVATION 59 m (194 Ft)		CLIMATE Bsh		LAT. 9°S		LONG. 13°E				
	JAN.	FEB.	MAR.	APRIL	MAY	JUNE	JULY	AUG.	SEPT.	OCT.	NOV.	DEC.	YEAR
t (°F)	78	80	81	81	78	72	70	69	72	75	77	78	76
p (in.)	1.0	1.4	3.0	4.6	0.5	0.0	0.0	0.0	0.1	0.2	1.1	0.8	12.7

47.

MARRAKECH, MOROCCO		ELEVATION 460 m (1509 Ft)		CLIMATE Bsh		LAT. 31.5°N		LONG. 8°W				
	JAN.	FEB.	MAR.	APRIL	MAY	JUNE	JULY	AUG.	SEPT.	OCT.	NOV.	DEC.	YEAR
t (°F)	52	55	59	67	69	77	82	85	76	70	62	54	67
p (in.)	1.0	1.1	1.3	1.2	0.6	0.3	0.1	0.1	0.4	0.9	1.2	1.2	9.4

48.

MOMBASA, KENYA		ELEVATION 16 m (52 Ft)		CLIMATE Aw		LAT. 4°S		LONG. 39.5°E				
	JAN.	FEB.	MAR.	APRIL	MAY	JUNE	JULY	AUG.	SEPT.	OCT.	NOV.	DEC.	YEAR
t (°F)	81	82	82	81	78	78	76	76	77	79	80	80	79
p (in.)	1.0	0.7	2.5	7.7	12.6	4.7	3.5	2.5	2.5	3.4	3.8	2.4	47.3

MADAGASCAR

49.

TAMATAVE, MADAGASCAR		ELEVATION 6 m (20 Ft)		CLIMATE Af		LAT. 18°S		LONG. 49.5°E				
	JAN.	FEB.	MAR.	APRIL	MAY	JUNE	JULY	AUG.	SEPT.	OCT.	NOV.	DEC.	YEAR
t (°F)	80	80	79	78	74	72	70	70	72	74	77	79	75
p (in.)	14.4	14.8	17.8	15.7	10.4	11.1	11.9	8.0	5.2	3.9	4.6	10.3	128.1

EUROPE AND ASIA

50.

ASTRAKHAN, RUSSIA		ELEVATION 18 m (59 Ft)		CLIMATE BSk		LAT. 46°N		LONG. 48°E				
	JAN.	FEB.	MAR.	APRIL	MAY	JUNE	JULY	AUG.	SEPT.	OCT.	NOV.	DEC.	YEAR
t (°F)	22	23	35	51	66	72	76	74	64	50	39	28	50
p (in.)	0.6	0.4	0.5	0.5	0.6	0.8	0.4	1.0	0.9	0.6	0.6	0.7	7.6

51.

BAGHDAD, IRAQ		ELEVATION 34 m (111 Ft)		CLIMATE BWh		LAT. 33.5°N		LONG. 44.5°E				
	JAN.	FEB.	MAR.	APRIL	MAY	JUNE	JULY	AUG.	SEPT.	OCT.	NOV.	DEC.	YEAR
t (°F)	50	53	60	71	82	89	93	93	87	76	64	53	73
p (in.)	0.9	1.0	1.1	0.5	0.1	0.0	0.0	0.0	0.0	0.1	0.8	1.0	5.5

52.

BANGKOK, THAILAND		ELEVATION 2 m (7 Ft)		CLIMATE Aw		LAT. 14°N		LONG. 100.5°E				
	JAN.	FEB.	MAR.	APRIL	MAY	JUNE	JULY	AUG.	SEPT.	OCT.	NOV.	DEC.	YEAR
t (°F)	78	82	84	86	85	84	83	83	82	82	80	78	82
p (in.)	0.3	0.8	1.4	2.3	7.8	6.3	6.3	6.9	12.0	8.1	2.6	0.2	55.0

53.   BEIJING (PEKING), CHINA   ELEVATION 52 m (171 Ft)   CLIMATE Dwa   LAT. 40°N   LONG. 116°E

	JAN.	FEB.	MAR.	APRIL	MAY	JUNE	JULY	AUG.	SEPT.	OCT.	NOV.	DEC.	YEAR
t (°F)	24	28	41	58	68	76	79	77	68	55	38	28	53
p (in.)	0.2	0.2	0.3	0.7	1.4	3.1	9.6	5.6	2.3	0.6	0.4	0.1	24.5

54.   BELGRADE, SERBIA   ELEVATION 132 m (433 Ft)   CLIMATE Cfb   LAT. 45°N   LONG. 20.5°E

	JAN.	FEB.	MAR.	APRIL	MAY	JUNE	JULY	AUG.	SEPT.	OCT.	NOV.	DEC.	YEAR
t (°F)	32	36	44	54	64	69	72	72	66	56	45	37	54
p (in.)	1.9	1.8	1.8	2.1	2.9	3.8	2.4	2.2	2.0	2.2	2.4	2.2	27.7

55.   BERGEN, NORWAY   ELEVATION 44 m (144 Ft)   CLIMATE Cfb   LAT. 60.5°N   LONG. 5.5°E

	JAN.	FEB.	MAR.	APRIL	MAY	JUNE	JULY	AUG.	SEPT.	OCT.	NOV.	DEC.	YEAR
t (°F)	34	34	36	42	49	55	58	57	52	45	39	36	45
p (in.)	7.9	6.0	5.3	4.4	3.9	4.3	5.2	7.3	9.4	9.2	8.1	8.2	79.2

56.   BERLIN, GERMANY   ELEVATION 50 m (164 Ft)   CLIMATE Cfb   LAT. 52.5°N   LONG. 13.5°E

	JAN.	FEB.	MAR.	APRIL	MAY	JUNE	JULY	AUG.	SEPT.	OCT.	NOV.	DEC.	YEAR
t (°F)	31	32	39	48	58	64	67	66	59	49	40	34	49
p (in.)	1.6	1.5	1.2	1.5	1.7	2.4	2.6	2.6	1.8	1.8	1.8	1.5	22.0

57.   BOMBAY, INDIA   ELEVATION 11.3 m (37 Ft)   CLIMATE Aw   LAT. 19°N   LONG. 73°E

	JAN.	FEB.	MAR.	APRIL	MAY	JUNE	JULY	AUG.	SEPT.	OCT.	NOV.	DEC.	YEAR
t (°F)	76	75	79	82	86	84	81	80	80	82	81	78	80
p (in.)	0.1	0.1	0.1	0.0	0.7	19.1	24.3	13.4	10.4	2.5	0.5	0.1	71.3

58.   CHERRAPUNJI, INDIA   ELEVATION 1313 m (4309 Ft)   CLIMATE Cwb   LAT. 25°N   LONG. 91.5°E

	JAN.	FEB.	MAR.	APRIL	MAY	JUNE	JULY	AUG.	SEPT.	OCT.	NOV.	DEC.	YEAR
t (°F)	53	56	62	65	67	68	68	69	69	66	60	55	63
p (in.)	0.7	2.1	7.3	26.2	50.4	106.1	96.3	70.1	43.3	19.4	2.7	0.5	425.0

59.   HELSINKI, FINLAND   ELEVATION 46 m (151 Ft)   CLIMATE Dfb   LAT. 60°N   LONG. 25°E

	JAN.	FEB.	MAR.	APRIL	MAY	JUNE	JULY	AUG.	SEPT.	OCT.	NOV.	DEC.	YEAR
t (°F)	22	20	26	37	48	58	63	60	52	42	34	27	41
p (in.)	2.2	1.7	1.4	1.7	1.6	2.0	2.7	2.8	2.8	2.9	2.7	2.6	27.1

60.   HONG KONG   ELEVATION 33 m (109 Ft)   CLIMATE Cwa   LAT. 22°N   LONG. 114°E

	JAN.	FEB.	MAR.	APRIL	MAY	JUNE	JULY	AUG.	SEPT.	OCT.	NOV.	DEC.	YEAR
t (°F)	60	59	64	71	78	82	82	82	81	77	70	64	72
p (in.)	1.3	1.8	2.9	5.4	11.5	15.8	15.0	14.2	10.1	4.5	1.7	1.2	85.4

61.   IRKUTSK, RUSSIA   ELEVATION 467 m (1532 Ft)   CLIMATE Dfc   LAT. 52°N   LONG. 104°E

	JAN.	FEB.	MAR.	APRIL	MAY	JUNE	JULY	AUG.	SEPT.	OCT.	NOV.	DEC.	YEAR
t (°F)	−6	−2	14	31	44	56	60	58	46	31	11	−4	28
p (in.)	0.5	0.4	0.3	0.6	1.3	2.2	3.1	2.8	1.7	0.7	0.6	0.6	14.8

62.   JAKARTA, INDONESIA   ELEVATION 8 m (26 Ft)   CLIMATE Am   LAT. 6°S   LONG. 107°E

	JAN.	FEB.	MAR.	APRIL	MAY	JUNE	JULY	AUG.	SEPT.	OCT.	NOV.	DEC.	YEAR
t (°F)	79	79	81	81	81	81	80	81	81	81	80	80	80
p (in.)	11.8	11.8	8.3	5.8	4.5	3.8	2.5	1.7	2.6	4.4	5.6	8.0	70.8

63. | LISBON, PORTUGAL   ELEVATION 77 m (253 Ft)   CLIMATE Csa   LAT. 38.5°N   LONG. 9°W

	JAN.	FEB.	MAR.	APRIL	MAY	JUNE	JULY	AUG.	SEPT.	OCT.	NOV.	DEC.	YEAR
t (°F)	52	53	56	60	63	68	72	72	70	65	58	52	62
p (in.)	4.3	3.0	4.2	2.1	1.7	0.6	0.1	0.2	1.3	2.4	3.7	4.1	27.7

64. | LONDON, ENGLAND   ELEVATION 4.9 m (16 Ft)   CLIMATE Cfb   LAT. 51.5°N   LONG. 0.5°W

	JAN.	FEB.	MAR.	APRIL	MAY	JUNE	JULY	AUG.	SEPT.	OCT.	NOV.	DEC.	YEAR
t (°F)	40	40	44	49	54	61	64	64	58	52	46	42	52
p (in.)	2.1	1.6	1.5	1.5	1.8	1.8	2.2	2.3	1.9	2.2	2.5	1.9	23.3

65. | MANILA, PHILIPPINES   ELEVATION 14 m (47 Ft)   CLIMATE Aw   LAT. 14.5°N   LONG. 121°E

	JAN.	FEB.	MAR.	APRIL	MAY	JUNE	JULY	AUG.	SEPT.	OCT.	NOV.	DEC.	YEAR
t (°F)	78	78	81	83	84	83	82	81	82	82	80	78	81
p (in.)	0.9	0.5	0.7	1.3	5.1	10.0	17.0	16.6	14.0	7.6	5.7	2.6	82.0

66. | MOSCOW, RUSSIA   ELEVATION 154 m (505 Ft)   CLIMATE Dfb   LAT. 56°N   LONG. 37.5°E

	JAN.	FEB.	MAR.	APRIL	MAY	JUNE	JULY	AUG.	SEPT.	OCT.	NOV.	DEC.	YEAR
t (°F)	14	15	24	40	53	62	66	63	52	40	28	20	40
p (in.)	1.5	1.4	1.1	1.9	2.2	2.9	3.0	2.9	1.9	2.7	1.7	1.6	24.8

67. | MUSCAT, OMAN   ELEVATION 4.6 m (15 Ft)   CLIMATE BWh   LAT. 23.5°N   LONG. 58.5°E

	JAN.	FEB.	MAR.	APRIL	MAY	JUNE	JULY	AUG.	SEPT.	OCT.	NOV.	DEC.	YEAR
t (°F)	72	72	78	84	92	94	92	88	88	86	80	74	83
p (in.)	1.1	0.7	0.4	0.4	0.0	0.1	0.0	0.0	0.0	0.1	0.4	0.7	3.9

68. | NARVIK, NORWAY   ELEVATION 40 m (131 Ft)   CLIMATE Dfc   LAT. 68.5°N   LONG. 17.5°E

	JAN.	FEB.	MAR.	APRIL	MAY	JUNE	JULY	AUG.	SEPT.	OCT.	NOV.	DEC.	YEAR
t (°F)	24	24	28	35	43	50	58	56	48	39	32	28	39
p (in.)	2.2	1.9	2.4	1.8	1.7	2.6	2.3	3.3	3.8	3.4	2.3	2.2	29.9

69. | PARIS, FRANCE   ELEVATION 75 m (246 Ft)   CLIMATE Cfb   LAT. 49°N   LONG. 2.5°E

	JAN.	FEB.	MAR.	APRIL	MAY	JUNE	JULY	AUG.	SEPT.	OCT.	NOV.	DEC.	YEAR
t (°F)	38	40	46	52	58	64	67	66	62	53	45	40	53
p (in.)	2.2	1.8	1.4	1.7	2.2	2.1	2.3	2.5	2.2	2.0	2.0	2.0	24.4

70. | SVERDLOVSK, UKRAINE   ELEVATION 272 m (894 Ft)   CLIMATE Dfc   LAT. 57°N   LONG. 60.5°E

	JAN.	FEB.	MAR.	APRIL	MAY	JUNE	JULY	AUG.	SEPT.	OCT.	NOV.	DEC.	YEAR
t (°F)	0	8	18	34	48	57	62	58	48	32	16	5	32
p (in.)	0.5	0.4	0.5	0.7	1.9	2.7	2.6	2.7	1.6	1.2	1.1	0.8	16.7

71. | TOKYO, JAPAN   ELEVATION 6 m (19 Ft)   CLIMATE Cfa   LAT. 35.5°N   LONG. 140°E

	JAN.	FEB.	MAR.	APRIL	MAY	JUNE	JULY	AUG.	SEPT.	OCT.	NOV.	DEC.	YEAR
t (°F)	38	40	45	55	62	70	76	79	72	62	52	42	58
p (in.)	1.9	2.9	4.2	5.3	5.8	6.5	5.6	6.0	9.2	8.2	3.8	2.2	61.6

72. | URUMCHI, CHINA   ELEVATION 906 m (2972 Ft)   CLIMATE BSk   LAT. 44°N   LONG. 87.5°E

	JAN.	FEB.	MAR.	APRIL	MAY	JUNE	JULY	AUG.	SEPT.	OCT.	NOV.	DEC.	YEAR
t (°F)	3	7	22	48	60	66	70	68	58	40	22	12	40
p (in.)	0.6	0.3	0.5	1.5	1.1	1.5	0.7	1.0	0.6	1.7	1.6	0.4	11.5

73.	VERKHOYANSK, SIBERIA			ELEVATION 100 m (328 Ft)			CLIMATE Dfd		LAT. 67.5°N		LONG. 134°E	
	JAN.	FEB.	MAR.	APRIL	MAY	JUNE	JULY	AUG.	SEPT.	OCT.	NOV.	DEC.	YEAR
t (°F)	−58	−48	26	4	32	54	56	48	36	4	−35	−54	5
p (in.)	0.2	0.2	0.1	0.2	0.3	0.9	1.1	1.0	0.5	0.3	0.3	0.2	5.3

74.	VLADIVOSTOK, RUSSIA			ELEVATION 29 m (94 Ft)			CLIMATE Dwb		LAT. 43°N		LONG. 132°E	
	JAN.	FEB.	MAR.	APRIL	MAY	JUNE	JULY	AUG.	SEPT.	OCT.	NOV.	DEC.	YEAR
t (°F)	6	14	26	40	49	58	66	70	62	48	30	14	40
p (in.)	0.3	0.4	0.7	1.2	2.1	2.9	3.3	4.7	4.3	1.9	1.2	0.6	23.6

AUSTRALIA AND NEW ZEALAND

75.	ALICE SPRINGS, AUSTRALIA			ELEVATION 579 m (1901 Ft)			CLIMATE BWh		LAT. 23.5°S		LONG. 134°E	
	JAN.	FEB.	MAR.	APRIL	MAY	JUNE	JULY	AUG.	SEPT.	OCT.	NOV.	DEC.	YEAR
t (°F)	84	82	76	68	60	54	53	58	65	73	78	82	69
p (in.)	1.7	1.3	1.1	0.4	0.6	0.5	0.3	0.3	0.3	0.7	1.2	1.5	9.9

76.	AUCKLAND, NEW ZEALAND			ELEVATION 49 m (161 Ft)			CLIMATE Cfb		LAT. 37.5°S		LONG. 175°E	
	JAN.	FEB.	MAR.	APRIL	MAY	JUNE	JULY	AUG.	SEPT.	OCT.	NOV.	DEC.	YEAR
t (°F)	67	67	65	61	57	53	51	52	55	61	60	64	59
p (in.)	3.3	4.1	2.8	4.3	4.8	5.5	5.5	4.3	3.8	4.2	3.2	3.1	48.9

77.	DARWIN, AUSTRALIA			ELEVATION 32 m (104 Ft)			CLIMATE Aw		LAT. 12.5°S		LONG. 131°E	
	JAN.	FEB.	MAR.	APRIL	MAY	JUNE	JULY	AUG.	SEPT.	OCT.	NOV.	DEC.	YEAR
t (°F)	84	83	84	84	82	79	77	79	83	85	86	85	83
p (in.)	15.2	12.3	10.0	3.9	0.6	0.1	0.0	0.1	0.5	2.0	4.7	9.4	58.7

78.	PERTH, AUSTRALIA			ELEVATION 60 m (197 Ft)			CLIMATE Csa		LAT. 32°S		LONG. 116°E	
	JAN.	FEB.	MAR.	APRIL	MAY	JUNE	JULY	AUG.	SEPT.	OCT.	NOV.	DEC.	YEAR
t (°F)	74	74	71	66	61	57	56	56	58	62	66	71	64
p (in.)	0.3	0.4	0.8	1.7	5.1	7.1	6.7	5.7	3.4	2.2	0.8	0.5	34.7

79.	SYDNEY, AUSTRALIA			ELEVATION 42 m (138 Ft)			CLIMATE Cfa		LAT. 34°S		LONG. 151°E	
	JAN.	FEB.	MAR.	APRIL	MAY	JUNE	JULY	AUG.	SEPT.	OCT.	NOV.	DEC.	YEAR
t (°F)	72	72	70	64	59	55	53	56	59	64	67	70	63
p (in.)	3.5	4.0	5.0	5.3	5.0	4.6	4.6	3.0	2.9	2.8	2.9	2.9	46.5

GREENLAND AND ICELAND

80.	ANGMAGSSALIK, GREENLAND			ELEVATION 29 m (95 Ft)			CLIMATE ET		LAT. 65.5°N		LONG. 37.5°W	
	JAN.	FEB.	MAR.	APRIL	MAY	JUNE	JULY	AUG.	SEPT.	OCT.	NOV.	DEC.	YEAR
t (°F)	19	19	21	27	36	43	45	45	39	32	27	23	31
p (in.)	2.9	2.4	2.6	2.1	2.0	1.8	1.5	2.1	3.3	4.7	3.0	2.7	31.1

81.		IVIGTUT, GREENLAND		ELEVATION 30 m (98 Ft)		CLIMATE ET		LAT. 61°N		LONG. 48°W		
	JAN.	FEB.	MAR.	APRIL	MAY	JUNE	JULY	AUG.	SEPT.	OCT.	NOV.	DEC.	YEAR
t (°F)	19	19	24	31	40	47	49	47	41	34	26	21	33
p (in.)	3.3	2.6	3.4	2.5	3.5	3.2	3.1	3.7	5.9	5.7	4.6	3.1	44.6

82.		REYKJAVIK, ICELAND		ELEVATION 28 m (92 Ft)		CLIMATE Cfc		LAT. 64°N		LONG. 22°W		
	JAN.	FEB.	MAR.	APRIL	MAY	JUNE	JULY	AUG.	SEPT.	OCT.	NOV.	DEC.	YEAR
t (°F)	31	32	34	38	44	50	52	51	48	42	35	32	41
p (in.)	4.0	3.1	3.0	2.1	1.6	1.7	2.0	2.6	3.1	3.4	3.6	3.7	33.9

The numbers on the map show the location of the cities represented in Appendix K.

PERIODICALS

Selected nontechnical periodicals that contain articles on weather and climate.

Bulletin of the American Meteorological Society. Monthly. The Amercan Meteorological Society, 45 Beacon St., Boston, MA 02108.

Earth. Bimonthly. Kalmbach Publishing Co., 21027 Crossroads Circle, Waukesha, WI 53187.

Meteorological Magazine. Monthly. British Meteorological Office, British Information Services, 845 Third Avenue, New York, NY.

National Weather Digest. Quarterly. National Weather Association, 4400 Stamp Road, Room 404, Marlow Heights, MD 20031. (Deals mainly with weather forecasting.)

Weather. Monthly. Royal Meteorological Society, James Glaisher House, Grenville Place, Bracknell, Berkshire, England.

Weatherwise. Bimonthly. Heldref Publications, 4000 Albermarle St., N.W., Washington, DC 20016.

SELECTED TECHNICAL PERIODICALS

EOS—Transaction of the American Geophysical Union. American Geophysical Union (AGS), Washington, DC.

Journal of Applied Meteorology. American Meteorological Society (AMS), Boston, MA.

Journal of Atmospheric and Oceanic Technology. AMS, Boston, MA.

Journal of Atmospheric Science. AMS, Boston, MA.

Journal of Climate. AMS, Boston, MA.

Journal of Geophysical Research. American Geophysical Union, Washington, DC.

Monthly Weather Review. AMS, Boston, MA.

Weather and Forecasting. AMS, Boston, MA.

Additional periodicals that frequently contain articles of meteorological interest.

American Scientist. Bimonthly. Sigma Xi, the Scientific Research Society, Inc., New Haven, CT.

Science. Weekly. American Association for the Advancement of Science, Washington, DC.

Scientific American. Monthly. Scientific American, Inc., New York, NY.

Smithsonian. Monthly. The Smithsonian Association, Washington, DC.

BOOKS

The titles listed below may be drawn upon for additional information. Many are written at the introductory level. Those that are more advanced are marked with an asterisk.

Anderson, Bette R. *Weather in the West*, American West Publishing Co., Palo Alto, CA, 1975.

Anthes, R. A. *Tropical Cyclones: Their Evolution, Structure, and Effect*, American Meteorological Society, Boston, MA, 1982.

Arya, Pal S. *Air Pollution Meteorology*, Oxford University Press, New York, 1998.

Battan, Louis J. *Harvesting the Clouds*, Anchor Books, Doubleday and Co., Garden City, NY, 1961.

——. *The Nature of Violent Storms*, Anchor Books, Doubleday and Co., Garden City, NY, 1961.

——. *The Unclean Sky*, Anchor Books, Doubleday and Co., Garden City, NY, 1966.

Bigg, Grant R. *The Oceans and Climate*, Cambridge University Press, New York, 1996.

*Bluestein, Howard B. *Synoptic-Dynamic Meteorology in Midlatitudes. Vol. 1: Principles of Kinematics and Dynamics*, Oxford University Press, New York, 1992.

——. *Synoptic-Dynamic Meteorology in Midlatitudes. Vol. II: Observations and Theory of Weather Systems*, Oxford University Press, New York, 1993.

——. *Tornado Alley: Monster Storms of the Great Plains*, Oxford University Press, New York, 1999.

Bohren, Craig F. *Clouds in a Glass of Beer: Simple Experiments in Atmospheric Physics*, Wiley, New York, 1987.

——. *What Light Through Yonder Window Breaks?*, Wiley, New York, 1991.

Boubel, Richard W., et al., *Fundamentals of Air Pollution* (3rd ed.), Academic Press, New York, 1994.

Burgess, Eric, and Douglass Torr. *Into the Thermosphere: The Atmosphere Explorers*, National Aeronautics and Space Administration, Washington, DC, 1987.

Burroughs, William J. *Watching the World's Weather*, Cambridge University Press, New York, 1991.

*Byers, H. R. *General Meteorology*, McGraw-Hill, New York, 1974.

*Cotton, W. R., and R. A. Anthes. *Storm and Cloud Dynamics*, Academic Press, New York, 1989.

Cotton, William R. *Storms*, ASTeR Press, Fort Collins, CO, 1990.

Cotton, William R., and Roger A. Pielke. *Human Impacts on Weather and Climate*, Cambridge University Press, New York, 1995.

De Blij, H. J. *Nature on the Rampage*, Smithsonian Books, Washington, DC, 1994.

Eagleman, Joe R. *Air Pollution Meteorology*, Trimedia Publishing Co., Lenexa, KS, 1991.

Edinger, James G. *Watching for the Wind*, Anchor Books, Doubleday and Co., Garden City, NY, 1967.

Elsner, James B., and A. Biral Kara. *Hurricanes of the North Atlantic*, Oxford University Press, New York, 1999.

Elsom, Derek M. *Atmospheric Pollution: A Global Problem* (2nd ed.), Blackwell Publishers, Oxford, England, 1992.

Encyclopedia of Climate and Weather, Vol. 1 and Vol. 2, Stephen H. Schneider, Ed., Oxford University Press, New York, 1996.

Energy and Climate Change. Report of the DOE Multi-Laboratory Climate Change Committee, Lewis Publishers, Chelsea, MI, 1991.

England, Gary A. *Weathering the Storm*, University of Oklahoma Press, Norman, OK, 1996.

Firor, John. *The Changing Atmosphere: A Global Challenge*, Yale University Press, New Haven, CT, 1990.

*Fleagle, Robert G., and Joost A. Businger. *An Introduction to Atmospheric Physics* (2nd ed.), Academic Press, New York, 1980.

Fujita, T. T. *The Downburst—Microburst and Macroburst*, University of Chicago Press, Chicago, 1985.

Geiger, R. *The Climate Near the Ground*, Harvard University Press, Cambridge, MA, 1965.

Glossary of Weather and Climate. Ira W. Geer, Ed., American Meteorological Society, Boston, MA, 1996.

Goody, Richard M., and C. G. Walker. *Atmospheres*, Prentice-Hall, Englewood Cliffs, NJ, 1972.

*Graedel, T. E. and Paul J. Crutzen. *Atmospheric Change: An Earth System Perspective*, W. H. Freeman, New York, 1993.

Graedel, Thomas E., and Paul J. Crutzen. *Atmosphere, Climate, and Change*, W. H. Freeman, New York, 1995.

Greenler, Robert. *Rainbows, Halos and Glories*, Cambridge University Press, New York, 1980.

*Grotjahn, Richard. *Global Atmospheric Circulations: Observations and Theories*, Oxford University Press, Oxford, England, 1993.

Hewitt, Paul G. *Conceptual Physics: A New Introduction to Your Environment* (6th ed.), Little, Brown, Boston, MA, 1988.

*Hobbs, Peter V. *Basic Physical Chemistry for Atmospheric Sciences*, Cambridge University Press, New York, 1995.

*Holton, James R. *An Introduction to Dynamic Meteorology* (2nd ed.), Academic Press, New York, 1979.

Houghton, David D. *Handbook of Applied Meteorology*, Wiley (through AMS), Boston, MA, 1985.

Houghton, J. T., G. J. Jenkins, and J. J. Ephraums. *Climate Change: The IPCC Scientific Assessment*, Cambridge University Press, Cambridge, England, 1990.

——. *Climate Change 1992: The Supplementary Report to the IPCC Scientific Assessment*, Cambridge University Press, Cambridge, England, 1992.

Houghton, J. T., et al., Eds. *Climate Change 1994. Radiative Forcing of Climate Change*, Cambridge University Press, Cambridge, England, 1995.

—— *Climate Change 1995. The Science of Climate Change*, Cambridge University Press, Cambridge, England, 1996.

Hoyt, Douglas V. and Kenneth H. Schatten. *The Role of the Sun in Climate Change*, Oxford University Press, New York, 1997.

Hughes, Patrick. *American Weather Stories*, U.S. Department of Commerce, NOAA, Washington, DC, 1976.

Imbrie, John, and K. P. Imbrie. *Ice Ages: Solving the Mystery*, Enslow Publishers, Short Hills, NJ, 1979.

International Cloud Atlas. World Meteorological Organization, Geneva, Switzerland, 1987.

James, Bruce P., et al., Eds. *Climate Change 1995. Economic and Social Dimensions of Climate Change*, Cambridge University Press, Cambridge, England, 1996.

Keen, Richard A. *Skywatch: The Western Weather Guide*. Fulcrum Incorporated, Golden, CO, 1987.

——. *Skywatch East: A Weather Guide*, Fulcrum Incorporated, Golden, CO, 1992.

Kessler, Edwin. *Thunderstorm Morphology and Dynamics* (2nd ed.), University of Oklahoma Press, Norman, OK, 1986.

Kocin, Paul J., and L. W. Uccellini. *Snowstorms along the Northeastern Coast of the United States: 1955 to 1985*, American Meteorological Society, Boston, MA, 1990.

Ladurie, E. L. R. *Times of Feast, Times of Famine: A History of Climate since the Year 1000*, Doubleday and Co., Garden City, NY, 1971.

Landsberg, H. E. *Weather and Health*, Anchor Books, Doubleday and Co., Garden City, NY, 1969.

*Ludlam, F. H. *Clouds and Storms: The Behavior and Effects of Water in the Atmosphere*, Pennsylvania State University Press (through AMS), Boston, MA, 1980.

Ludlum, D. M. *The Audubon Society Field Guide to North American Weather*, Alfred A. Knopf, New York, 1991.

——. *The Country Journal New England Weather Book*, Houghton Mifflin, Boston, MA, 1976.

——. *Early American Winters: 1604–1870*, American Meteorological Society, Boston, MA, 1967.

——. *Weather Record Book*, Weatherwise, Inc., Princeton, NJ, 1971.

Lynch, David K., and William Livingston. *Color and Light in Nature*, Cambridge University Press, New York, 1995.

Mason, B. J. *Acid Rain: Its Causes and Its Effects on Inland Waters*, Oxford University Press, New York, 1992.

——. *Clouds, Rain and Rainmaking* (2nd ed.), Cambridge University Press, New York, 1975.

Meinel, Aden, and Marjorie Meinel. *Sunsets, Twilights and Evening Skies.* Cambridge University Press, New York, 1983.

Middleton, W. E. K. *The Invention of the Meteorological Instruments*, Johns Hopkins University Press, Baltimore, MD, 1969.

National Academy Press. *Acid Deposition—Atmospheric Processes in Eastern North America*, Washington, DC, 1983.

——. *Causes and Effects of Stratospheric Ozone Reduction: An Update*, Washington, DC, 1982.

——. *Changing Climate: A Report of the Carbon Dioxide Assessment Committee*, Washington, DC, 1983.

——. *The Effects on the Atmosphere of a Major Nuclear Exchange*, Washington, DC, 1985.

——. *Ozone Depletion, Greenhouse Gases, and Climate Change*, Washington, DC, 1989.

——. *Solar Variability, Weather, and Climate*, Washington, DC, 1982.

*Palmen, E., and C. W. Newton. *Atmospheric Circulation Systems*, Academic Press, New York, 1969.

Prospects for Future Climate. Special US/USSR Report on Climate and Climate Change, Lewis Publishers, Chelsea, MI, 1990.

*Ray, Peter S., Ed. *Mesoscale Meteorology and Forecasting*, American Meteorological Society, Boston, MA, 1986.

Righter, Robert W. *Wind Energy in America*, University of Oklahoma Press, Norman, OK, 1996.

Roberts, Walter Orr and Henry Landsford. *The Climate Mandate*, W. H. Freeman, San Francisco, 1979.

*Rogers, R. R. *A Short Course in Cloud Physics* (3rd ed.), Pergamon Press, Oxford, England, 1989.

Schaeter, Vincent J., and John A. Day. *A Field Guide to the Atmosphere*, Houghton Mifflin (through AMS), Boston, MA, 1981.

——. *Peterson First Guide to Clouds and Weather*, Houghton Mifflin, Boston, MA, 1991.

Scorer, Richard S., and Arjen Verkaik. *Spacious Skies*, David and Charles Publishers, London, 1989.

Simpson, Robert H. and Herbert Riehl. *The Hurricane and Its Impact*, Louisiana State University Press, Baton Rouge, LA, 1981.

Stanford, John L. *Tornado: Accounts of Tornadoes in Iowa*, Iowa State University Press, Ames, IA, 1977.

Stull, Roland B. *Meteorology Today for Scientists and Engineers*, West Publishing Co., Minneapolis, MN, 1995.

*Wallace, J. M., and P. V. Hobbs. *Atmospheric Science: An Introductory Survey*, Academic Press, New York, 1977.

*Watson, Robert T., et al., Eds. *Climate Change 1995. Impacts, Adaptations and Mitigation of Climate Change: Scientific-Technical Analyses*, Cambridge University Press, Cambridge, England, 1996.

Williams, Jack. *The USA Today Weather Book*, Vintage Books, Random House, New York, 1992.

GLOSSARY

Absolute humidity   The mass of water vapor in a given volume of air. It represents the density of water vapor in the air.

Absolute vorticity   *See* Vorticity.

Absolute zero   A temperature reading of –273°C, –460°F, or 0K. Theoretically, there is no molecular motion at this temperature.

Absolutely stable atmosphere   An atmospheric condition that exists when the environmental lapse rate is less than the moist adiabatic rate. This results in a lifted parcel of air being colder than the air around it.

Absolutely unstable atmosphere   An atmospheric condition that exists when the environmental lapse rate is greater than the dry adiabatic rate. This results in a lifted parcel of air being warmer than the air around it.

Accretion   The growth of a precipitation particle by the collision of an ice crystal or snowflake with a supercooled liquid droplet that freezes upon impact.

Acid deposition   The depositing of acidic particles (usually sulfuric acid and nitric acid) at the earth's surface. Acid deposition occurs in dry form (*dry deposition*) or wet form (*wet deposition*). Acid rain and acid precipitation often denote wet deposition. (*See* Acid rain.)

Acid fog   *See* Acid rain.

Acid rain   Cloud droplets or raindrops combining with gaseous pollutants, such as oxides of sulfur and nitrogen, to make falling rain (or snow) acidic—pH less than 5.0. If fog droplets combine with such pollutants it becomes *acid fog*.

Actual Vapor Pressure   *See* Vapor pressure.

Adiabatic process   A process that takes place without a transfer of heat between the system (such as an air parcel) and its surroundings. In an adiabatic process, compression always results in warming, and expansion results in cooling.

Advection   The horizontal transfer of any atmospheric property by the wind.

Advection fog   Occurs when warm, moist air moves over a cold surface and the air cools to below its dew point.

Advection-radiation fog   Fog that forms as relatively warm moist air moves over a colder surface that cooled mainly by radiational cooling.

Aerosols   Tiny suspended solid particles (dust, smoke, etc.) or liquid droplets that enter the atmosphere from either natural or human (anthropogenic) sources, such as the burning of fossil fuels. Sulfur-containing fossil fuels, such as coal, produce *sulfate aerosols*.

Aerovane   A wind instrument that indicates or records both wind speed and wind direction. Also called a *skyvane*.

AFOS   Acronym for *Automation of Field Operations and Services*. Electronic-computerized system that displays weather information on TV-type consoles.

Aggregation   The clustering together of ice crystals to form snowflakes.

Air density   *See* Density.

Air glow   A faint glow of light emitted by excited gases in the upper atmosphere. Air glow is much fainter than the aurora.

Air mass   A large body of air that has similar horizontal temperature and moisture characteristics.

Air-mass thunderstorm   *See* Ordinary thunderstorm.

Air mass weather   A persistent type of weather that may last for several days (up to a week or more). It occurs when an area comes under the influence of a particular air mass.

Air parcel   *See* Parcel of air.

Air pollutants   Solid, liquid, or gaseous airborne substances that occur in concentrations high enough to threaten the health of people and animals, to harm vegetation and structures, or to toxify a given environment.

Air pressure (atmospheric pressure)   The pressure exerted by the mass of air above a given point, usually expressed in millibars (mb), inches of mercury (Hg) or in hectopascals (hPa).

Aitken nuclei   *See* Condensation nuclei.

Albedo   The percent of radiation returning from a surface compared to that which strikes it.

Aleutian low   The subpolar low-pressure area that is centered near the Aleutian Islands on charts that show mean sea level pressure.

Altimeter   An instrument that indicates the altitude of an object above a fixed level. Pressure altimeters use an aneroid barometer with a scale graduated in altitude instead of pressure.

Altocumulus   A middle cloud, usually white or gray. Often occurs in layers or patches with wavy, rounded masses or rolls.

Altocumulus castellanus   An altocumulus cloud showing vertical development. Individual cloud elements have towerlike tops, often in the shape of tiny castles.

Altostratus   A middle cloud composed of gray or bluish sheets or layers of uniform appearance. In the thinner regions, the sun or moon usually appears dimly visible.

Analogue method of forecasting   A forecast made by comparison of past large-scale synoptic weather patterns that resemble a given (usually current) situation in its essential characteristics.

Analysis   The drawing and interpretation of the patterns of various weather elements on a surface or upper-air chart.

Anemometer   An instrument designed to measure wind speed.

Aneroid barometer   An instrument designed to measure atmospheric pressure. It contains no liquid.

Angular momentum   The product of an object's mass, speed, and radial distance of rotation.

Annual range of temperature   The difference between the warmest and coldest months at any given location.

Anticyclone   An area of high pressure around which the wind blows clockwise in the Northern Hemisphere and counterclockwise in the Southern Hemisphere. Also called a *high*.

Apparent temperature   What the air temperature "feels like" for various combinations of air temperature and relative humidity.

Arcus cloud   *See* Shelf cloud.

Arid climate   An extremely dry climate—drier than the semi-arid climate. Often referred to as a "true desert" climate.

Asbestos   A general name for the fibrous variety of silicate minerals that are incombustible and resist chemicals and heat. Once used for fireproofing, electrical insulation, building materials, chemical filters, and brake linings.

ASOS   Acronym for *Automated Surface Observing Systems*. A system designed to provide continuous information of wind, temperature, pressure, cloud base height, and runway visibility at selected airports.

Atmosphere   The envelope of gases that surround a planet and are held to it by the planet's gravitational attraction. The earth's atmosphere is mainly nitrogen and oxygen.

Atmospheric boundary layer   The layer of air from the earth's surface usually up to about 1 km (3300 ft) where the wind is influenced by friction of the earth's surface and objects on it. Also called the *planetary boundary layer* and the *friction layer*.

Atmospheric greenhouse effect   The warming of an atmosphere by its absorbing and emitting infrared radiation while allowing shortwave radiation to pass on through. The gases mainly responsible for the earth's atmospheric greenhouse effect are water vapor and carbon dioxide. Also called the *greenhouse effect*.

Atmospheric models   Simulation of the atmosphere's behavior by mathematical equations or by physical models.

Atmospheric stagnation   A condition of light winds and poor vertical mixing that can lead to a high concentration of pollutants. Air stagnations are most often associated with fair weather, an inversion, and the sinking air of a high-pressure area.

Atmospheric window   The wavelength range between 8 and 11 μm in which little absorption of infrared radiation takes place.

Attenuation   Any process in which the rate of flow of a beam of energy decreases (mainly due to absorption or scattering) with increasing distance from the energy source.

Aurora   Glowing light display in the nighttime sky caused by excited gases in the upper atmosphere giving off light. In the Northern Hemisphere it is called the *aurora borealis* (northern lights); in the Southern Hemisphere, the *aurora australis* (southern lights).

Autumnal equinox   The equinox at which the sun approaches the Southern Hemisphere and passes directly over the equator. Occurs around September 23.

AWIPS   Acronym for *Advanced Weather Interactive Processing System*. New computerized system that integrates and processes data received at a Weather Forecasting Office from NEXRAD, ASOS, and analysis and guidance products prepared by NMC.

Back-door cold front   A cold front moving south or southwest along the Atlantic seaboard of the United States.

Backing wind   A wind that changes direction in a counterclockwise sense (e.g., north to northwest to west).

Ball lightning   A rare form of lightning that may consist of a reddish, luminous ball of electricity or charged air.

Banner cloud   A cloud extending downwind from an isolated mountain peak, often on an otherwise cloud-free day.

Baroclinic (atmosphere)   The state of the atmosphere where surfaces of constant pressure intersect surfaces of constant density. On an isobaric chart, isotherms cross the contour lines, and temperature advection exists.

Baroclinic instability   A type of instability arising from a meridional (north to south) temperature gradient, a strong vertical wind speed shear, temperature advection, and divergence in the flow aloft. Many mid-latitude cyclones develop as a result of this instability.

Barograph   A recording barometer.

Barometer   An instrument that measures atmospheric pressure. The two most common barometers are the *mercury barometer* and the *aneroid barometer*.

Barotropic (atmosphere)   A condition in the atmosphere where surfaces of constant density parallel surfaces of constant pressure.

Bead lightning   Lightning that appears as a series of beads tied to a string.

Bergeron process   *See* Ice crystal process.

Bermuda high   *See* Subtropical high.

Billow clouds   Broad, nearly parallel lines of clouds oriented at right angles to the wind.

Bimetallic thermometer   A temperature-measuring device usually consisting of two dissimilar metals that expand and contract differentially as the temperature changes.

Blackbody   A hypothetical object that absorbs all of the radiation that strikes it. It also emits radiation at a maximum rate for its given temperature.

Blizzard   A severe weather condition characterized by low temperatures and strong winds (greater than 35 mi/hr) bearing a great amount of snow either falling or blowing. When these conditions continue after the falling snow has ended, it is termed a *ground blizzard*.

Boulder winds   Fast-flowing, local downslope winds that may attain speeds of 100 knots or more. They are especially strong along the eastern foothills of the Rocky Mountains near Boulder, Colorado.

Boundary layer   *See* Atmospheric boundary layer.

Brocken bow   A bright ring of light seen around the shadow of an observer's head as the observer peers into a cloud or fog bank. Formed by *diffraction* of light.

Buys-Ballot's law   A law describing the relationship between the wind direction and the pressure distribution. In the Northern Hemisphere, if you stand with your back to the surface wind, then turn clockwise about 30°, lower pressure will be to your left. In the Southern Hemisphere, stand with your back to the surface wind, then turn counterclockwise about 30°, lower pressure will be to your right.

California current   The ocean current that flows southward along the west coast of the United States from about Washington to Baja, California.

California norther   A strong, dry, northerly wind that blows in late spring, summer, and early fall in northern and central California. Its warmth and dryness are due to downslope compressional heating.

Cap cloud   *See* Pileus cloud.

Carbon dioxide (CO_2)   A colorless, odorless gas whose concentration is about 0.036 percent (365 ppm) in a volume of air near sea level. It is a selective absorber of infrared radiation and, consequently, it is important in the earth's atmospheric greenhouse effect. Solid CO_2 is called *dry ice*.

Carbon monoxide (CO)   A colorless, odorless, toxic gas that forms during the incomplete combustion of carbon-containing fuels.

Ceiling   The height of the lowest layer of clouds when the weather reports describe the sky as broken or overcast.

Ceiling balloon   A small balloon used to determine the height of the cloud base. The height is computed from the balloon's ascent rate and the time required for its disappearance into the cloud.

Ceilometer   An instrument that automatically records cloud height.

Celsius scale   A temperature scale where zero is assigned to the temperature where water freezes and 100 to the temperature where water boils (at sea level).

Centripetal acceleration   The inward-directed acceleration on a particle moving in a curved path.

Centripetal force   The radial force required to keep an object moving in a circular path. It is directed toward the center of that curved path.

Chinook wall cloud   A bank of clouds over the Rocky Mountains that signifies the approach of a chinook.

Chinook wind   A warm, dry wind on the eastern side of the Rocky Mountains. In the Alps, the wind is called a *Foehn*.

Chlorofluorocarbons (CFCs)   Compounds consisting of methane (CH_4) or ethane (C_2H_6) with some or all of the hydrogen replaced by chlorine or fluorine. Used in fire extinguishers, as refrigerants, as solvents for cleaning electronic microcircuits, and as propellants. CFCs contribute to the atmospheric greenhouse effect and destroy ozone in the stratosphere.

Cirrocumulus   A high cloud that appears as a white patch of clouds without shadows. It consists of very small elements in the form of grains or ripples.

Cirrostratus   High, thin, sheetlike clouds, composed of ice crystals. They frequently cover the entire sky and often produce a halo.

Cirrus   A high cloud composed of ice crystals in the form of thin, white, featherlike clouds in patches, filaments, or narrow bands.

Clear air turbulence (CAT)   Turbulence encountered by aircraft flying through cloudless skies. Thermals, wind shear, and jet streams can each be a factor in producing CAT.

Clear ice   A layer of ice that appears transparent because of its homogeneous structure and small number and size of air pockets.

Climate   The accumulation of daily and seasonal weather events over a long period of time.

Climatic controls   The relatively permanent factors that govern the general nature of the climate of a region.

Climatic optimum   *See* Mid-Holocene maximum.

Climatological forecast   A weather forecast, usually a month or more in the future, which is based upon the climate of a region rather than upon current weather conditions.

Cloud   A visible aggregate of tiny water droplets and/or ice crystals in the atmosphere above the earth's surface.

Cloudburst   Any sudden and heavy rain shower.

Cloud seeding   The introduction of artificial substances (usually silver iodide or dry ice) into a cloud for the purpose of either modifying its development or increasing its precipitation.

Cloud streets   Lines or rows of cumuliform clouds.

Coalescence   The merging of cloud droplets into a single larger droplet.

Cold advection (or *cold air advection*)   The transport of cold air by the wind from a region of lower temperatures to a region of higher temperatures.

Cold air damming   A shallow layer of cold air that is trapped between the Atlantic coast and the Appalachian Mountains.

Cold fog   *See* Supercooled cloud.

Cold front   A transition zone where a cold air mass advances and replaces a warm air mass.

Cold occlusion   *See* Occluded front.

Cold wave   A rapid fall in temperature within 24 hours that often requires increased protection for agriculture, industry, commerce, and human activities.

Comma cloud   A band of organized cumuliform clouds that looks like a comma on a satellite photograph.

Computer enhancement   A process where the temperatures of radiating surfaces are assigned different shades of gray (or different colors) on an infrared picture. This allows specific features to be more clearly delineated.

Condensation   The process by which water vapor becomes a liquid.

Condensation level   The level above the surface marking the base of a cumuliform cloud.

Condensation nuclei   Also called *cloud condensation nuclei*. Tiny particles upon whose surfaces condensation of water vapor begins in the atmosphere. Small nuclei less than 0.2 µm in radius are called *Aitken nuclei*; those with radii between 0.2 and 1 µm are *large nuclei*, while *giant nuclei* have radii larger than 1 µm.

Conditionally unstable atmosphere   An atmospheric condition that exists when the environmental lapse rate is less than the dry adiabatic rate but greater than the moist adiabatic rate. Also called *conditional instability*.

Conduction   The transfer of heat by molecular activity from one substance to another, or through a substance. Transfer is always from warmer to colder regions.

Constant height chart (constant-level chart)   A chart showing variables, such as pressure, temperature, and wind, at a specific altitude above sea level. Variation in horizontal pressure is depicted by isobars. The most common constant-height chart is the surface chart, which is also called the *sea level chart* or *surface weather map*.

Constant pressure chart (isobaric chart)   A chart showing variables, such as temperature and wind, on a constant pressure surface. Variations in height are usually shown by lines of equal height (contour lines).

Contact freezing   The process by which contact with a nucleus such as an ice crystal causes supercooled liquid droplets to change into ice.

Continental arctic air mass   An air mass characterized by extremely low temperatures and very dry air.

Continental polar air mass   An air mass characterized by low temperatures and dry air. Not as cold as arctic air masses.

Continental tropical air mass   An air mass characterized by high temperatures and low humidity.

Contour line   A line that connects points of equal elevation above a reference level, most often sea level.

Contrail (condensation trail)   A cloudlike streamer frequently seen forming behind aircraft flying in clear, cold, humid air.

Controls of temperature   The main factors that cause variations in temperature from one place to another.

Convection   Motions in a fluid that result in the transport and mixing of the fluid's properties. In meteorology, convection usually refers to atmospheric motions that are predominantly vertical, such as rising air currents due to surface heating. The rising of heated surface air and the sinking of cooler air aloft is often called *free convection*. (Compare with *forced convection*.)

Convective instability   Instability arising in the atmosphere when a column of air exhibits warm, moist, nearly saturated air near the surface and cold, dry air aloft. When the lower part of the layer is lifted and saturation occurs, it becomes unstable.

Convergence   An atmospheric condition that exists when the winds cause a horizontal net inflow of air into a specified region.

Conveyor belt model (for mid-latitude storms)   A three-dimensional picture of a mid-latitude cyclone and the various air streams (called conveyor belts) that interact to produce the weather associated with the storm.

Cooling degree-day A form of degree-day used in estimating the amount of energy necessary to reduce the effective temperature of warm air. A cooling degree-day is a day on which the average temperature is one degree above a desired base temperature.

Coriolis force An apparent force observed on any free-moving object in a rotating system. On the earth, this deflective force results from the earth's rotation and causes moving particles (including the wind) to deflect to the right in the Northern Hemisphere and to the left in the Southern Hemisphere.

Corona (optic) A series of colored rings concentrically surrounding the disk of the sun or moon. Smaller than the halo, the corona is often caused by the diffraction of light around small water droplets of uniform size.

Country breeze A light breeze that blows into a city from the surrounding countryside. It is best observed on clear nights when the urban heat island is most pronounced.

Crepuscular rays Alternating light and dark bands of light that appear to fan out from the sun's position, usually at twilight.

Cumulonimbus An exceptionally dense and vertically developed cloud, often with a top in the shape of an anvil. The cloud is frequently accompanied by heavy showers, lightning, thunder, and sometimes hail. It is also known as a *thunderstorm cloud*.

Cumulus A cloud in the form of individual, detached domes or towers that are usually dense and well defined. It has a flat base with a bulging upper part that often resembles cauliflower. Cumulus clouds of fair weather are called *cumulus humilis*. Those that exhibit much vertical growth are called *cumulus congestus* or *towering cumulus*.

Cumulus stage The initial stage in the development of an air mass thunderstorm in which rising, warm, humid air develops into a cumulus cloud.

Curvature effect In cloud physics, as cloud droplets decrease in size, they exhibit a greater surface curvature that causes a more rapid rate of evaporation.

Cut-off low A cold upper-level low that has become displaced out of the basic westerly flow and lies to the south of this flow.

Cyclogenesis The development or strengthening of middle latitude (extratropical) cyclones.

Cyclone An area of low pressure around which the winds blow counterclockwise in the Northern Hemisphere and clockwise in the Southern Hemisphere.

Daily range of temperature The difference between the maximum and minimun temperatures for any given day.

Dart leader The discharge of electrons that proceeds intermittently toward the ground along the same ionized channel taken by the initial lightning stroke.

Dendrochronology The analysis of the annual growth rings of trees as a means of interpreting past climatic conditions.

Density The ratio of the mass of a substance to the volume occupied by it. Air density is usually expressed as g/cm³ or kg/m³.

Deposition A process that occurs in subfreezing air when water vapor changes directly to ice without becoming a liquid first.

Deposition nuclei Tiny particles (ice nuclei) upon which an ice crystal may grow by the process of deposition.

Derecho Strong, damaging, straight-line winds associated with a cluster of severe thunderstorms that most often form in the evening or at night.

Desertification A general increase in the desert conditions of a region.

Desert pavement An arrangement of pebbles and large stones that remains behind as finer dust and sand particles are blown away by the wind.

Dew Water that has condensed onto objects near the ground when their temperatures have fallen below the dew point of the surface air.

Dew cell An instrument used to determine the dew-point temperature.

Dew point (dew-point temperature) The temperature to which air must be cooled (at constant pressure and constant water vapor content) for saturation to occur.

Dew point hygrometer An instrument that determines the dew-point temperature of the air.

Diffraction The bending of light around objects, such as cloud and fog droplets, producing fringes of light and dark or colored bands.

Dispersion The separation of white light into its different component wavelengths.

Dissipating stage The final stage in the development of an air mass thunderstorm when downdrafts exist throughout the cumulonimbus cloud.

Divergence An atmospheric condition that exists when the winds cause a horizontal net outflow of air from a specific region.

Doldrums The region near the equator that is characterized by low pressure and light, shifting winds.

Doppler lidar The use of light beams to determine the velocity of objects such as dust and falling rain by taking into account the *Doppler shift*.

Doppler radar A radar that determines the velocity of falling precipitation either toward or away from the radar unit by taking into account the *Doppler shift*.

Doppler shift (effect) The change in the frequency of waves that occurs when the emitter or the observer is moving toward or away from the other.

Downburst A severe localized downdraft that can be experienced beneath a severe thunderstorm. (Compare *Microburst* and *Macroburst*.)

Drizzle Small water drops between 0.2 and 0.5 mm in diameter that fall slowly and reduce visibility more than light rain.

Drought A period of abnormally dry weather sufficiently long enough to cause serious effects on agriculture and other activities in the affected area.

Dry adiabatic rate The rate of change of temperature in a rising or descending unsaturated air parcel. The rate of adiabatic cooling or warming is about 10°C per 1000 m (5.5°F per 1000 ft).

Dry adiabats Lines on an adiabatic chart that show the dry adiabatic rate for rising or descending air. They represent lines of constant potential temperature.

Dry bulb temperature The air temperature measured by the dry-bulb thermometer of a psychrometer.

Dry climate A climate deficient in precipitation where annual potential evaporation and transpiration exceed precipitation.

Dry haze *See* Haze.

Dry line A boundary that separates warm, dry air from warm, moist air. It usually represents a zone of instability along which thunderstorms form.

Dry-summer subtropical climate A climate characterized by mild, wet winters and warm to hot, dry summers. Typically located between 30 and 45 degrees latitude on the western side of continents. Also called *Mediterranean climate*.

Dust devil (or whirlwind) A small but rapidly rotating wind made visible by the dust, sand, and debris it picks up from the surface. It develops best on clear, dry, hot afternoons.

Easterly wave   A migratory wavelike disturbance in the tropical easterlies. Easterly waves occasionally intensify into tropical cyclones. They are also called *tropical waves.*

Eccentricity (of the earth's orbit)   The deviation of the earth's orbit from elliptical to nearly circular.

Eddy   A small volume of air (or any fluid) that behaves differently from the larger flow in which it exists.

Eddy viscosity   The internal friction produced by turbulent flow.

Ekman spiral   An idealized description of the way the wind-driven ocean currents vary with depth. In the atmosphere it represents the way the winds vary from the surface up through the friction layer or planetary boundary layer.

Ekman transport   Net surface water transport due to the Ekman spiral. In the Northern Hemisphere the transport is 90° to the right of the surface wind direction.

Electrical hygrometer   *See* Hygrometer.

Electrical thermometers   Thermometers that use elements that convert energy from one form to another (transducers). Common electrical thermometers include the electrical resistance thermometer, thermocouple, and thermistor.

Electromagnetic waves   *See* Radiant energy.

El Niño   An extensive ocean warming that begins along the coast of Peru and Ecuador. Major El Niño events occur once every 2 to 7 years as a current of nutrient-poor tropical water moves southward along the west coast of South America.

Embryo   In cloud physics, a tiny ice crystal that grows in size and becomes an ice nucleus.

Energy   The property of a system that generally enables it to do work. Some forms of energy are kinetic, radiant, potential, chemical, electric, and magnetic.

Ensemble forecasting   A forecasting technique that entails running several forecast models, each beginning with slightly different weather information. The forecaster's level of confidence is based on how well the models agree (or disagree) at the end of some specified time.

ENSO (El Niño/southern oscillation)   A condition in the Tropical Pacific whereby the reversal of surface air pressure at opposite ends of the Pacific Ocean induces westerly winds, a strengthening of the equatorial countercurrent, and extensive ocean warming.

Entrainment   The mixing of environmental air into a pre-existing air current or cloud so that the environmental air becomes part of the current or cloud.

Environmental lapse rate   The rate of decrease of air temperature with elevation. It is most often measured with a radiosonde.

Equilibrium vapor pressure   The necessary vapor pressure around liquid water that allows the water to remain in equilibrium with its environment. Also called *saturation vapor pressure.*

Evaporation   The process by which a liquid changes into a gas.

Evaporation (mixing) fog   Fog produced when sufficient water vapor is added to the air by evaporation, and the moist air mixes with relatively drier air. The two common types are *steam fog,* which forms when cold air moves over warm water, and *frontal fog,* which forms as warm raindrops evaporate in a cool air mass.

Exosphere   The outermost portion of the atmosphere.

Extratropical cyclone   A cyclonic storm that most often forms along a front in middle and high latitudes. Also called a *middle latitude storm,* a *depression,* and a *low.* It is not a tropical storm or hurricane.

Eye   A region in the center of a hurricane (tropical storm) where the winds are light and skies are clear to partly cloudy.

Eye wall   A wall of dense thunderstorms that surrounds the eye of a hurricane.

Fahrenheit scale   A temperature scale where 32 is assigned to the temperature where water freezes and 212 to the temperature where water boils (at sea level).

Fall streaks   Falling ice crystals that evaporate before reaching the ground.

Fall wind   A strong, cold katabatic wind that blows downslope off snow-covered plateaus.

Fata Morgana   A complex mirage that is characterized by objects being distorted in such a way as to appear as castlelike features.

Feedback mechanism   A process whereby an initial change in an atmospheric process will tend to either reinforce the process (*positive feedback*) or weaken the process (*negative feedback*).

Ferrel cell   The name given to the middle latitude cell in the 3-cell model of the general circulation.

Fetch   The distance that the wind travels over open water.

Flash flood   A flood that rises and falls quite rapidly with little or no advance warning, usually as the result of intense rainfall over a relatively small area.

Foehn   *See* Chinook wind.

Fog   A cloud with its base at the earth's surface.

Forced convection   On a small scale, a form of mechanical stirring taking place when twisting eddies of air are able to mix hot surface air with the cooler air above. On a larger scale, it can be induced by the lifting of warm air along a front (*frontal uplift*) or along a topographic barrier (*orographic uplift*).

Forked lightning   Cloud-to-ground lightning that exhibits downward-directed crooked branches.

Formaldehyde   A colorless gaseous compound (HCHO) used in the manufacture of resins, fertilizers, and dyes. Also used as an embalming fluid, a preservative, and a disinfectant.

Free convection   *See* Convection.

Freeze   A condition occurring over a widespread area when the surface air temperature remains below freezing for a sufficient time to damage certain agricultural crops. A freeze most often occurs as cold air is advected into a region, causing freezing conditions to exist in a deep layer of surface air. Also called *advection frost.*

Freezing nuclei   Particles that promote the freezing of supercooled liquid droplets.

Freezing rain and freezing drizzle   Rain or drizzle that falls in liquid form and then freezes upon striking a cold object or ground. Both can produce a coating of ice on objects which is called *glaze.*

Friction layer   The atmospheric layer near the surface usually extending up to about 1 km (3300 ft) where the wind is influenced by friction of the earth's surface and objects on it. Also called the *atmospheric boundary layer* and *planetary boundary layer.*

Front   The transition zone between two distinct air masses.

Frontal fog   *See* Evaporation fog.

Frontal inversion   A temperature inversion encountered upon ascending through a sloping front, usually a warm front.

Frontal thunderstorms   Thunderstorms that form in response to forced convection (forced lifting) along a front. Most go through a cycle similar to those of ordinary thunderstorms.

Frontal wave   A wavelike deformation along a front in the lower levels of the atmosphere. Those that develop into storms are termed *unstable waves,* while those that do not are called *stable waves.*

Frontogenesis   The formation, strengthening, or regeneration of a front.

Frontolysis   The weakening or dissipation of a front.

Frost (also called hoarfrost)   A covering of ice produced by deposition on exposed surfaces when the air temperature falls below the frost point.

Frostbite   The partial freezing of exposed parts of the body, causing injury to the skin and sometimes to deeper tissues.

Frost point   The temperature at which the air becomes saturated with respect to ice when cooled at constant pressure and constant water vapor content.

Frozen dew   The transformation of liquid dew into tiny beads of ice when the air temperature drops below freezing.

Fujita scale   A scale developed by T. Theodore Fujita for classifying tornadoes according to the damage they cause and their rotational wind speed.

Fulgurite   A rootlike tube (or several tubes) that form when a lightning stroke fuses sand particles together.

Funnel cloud   A tornado whose circulation has not reached the ground. Often appears as a rotating conelike cloud that extends downward from the base of a thunderstorm.

Galaxy   A huge assembly of stars (between millions and hundreds of millions) held together by gravity.

Gas law   The thermodynamic law applied to a perfect gas that relates the pressure of the gas to its density and absolute temperature.

General circulation of the atmosphere   Large-scale atmospheric motions over the entire earth.

Geostationary satellite   A satellite that orbits the earth at the same rate that the earth rotates and thus remains over a fixed place above the equator.

Geostrophic wind   A theoretical horizontal wind blowing in a straight path, parallel to the isobars or contours, at a constant speed. The geostrophic wind results when the Coriolis force exactly balances the horizontal pressure gradient force.

Giant nuclei   *See* Condensation nuclei.

Glaciated cloud   A cloud or portion of a cloud where only ice crystals exist.

Global scale   The largest scale of atmospheric motion. Also called the *planetary scale*.

Glory   Colored rings that appear around the shadow of an object.

Gradient wind   A theoretical wind that blows parallel to curved isobars or contours.

Graupel   Ice particles between 2 and 5 mm in diameter that form in a cloud often by the process of accretion. Snowflakes that become rounded pellets due to riming are called *graupel* or *snow pellets*.

Green flash   A small green color that occasionally appears on the upper part of the sun as it rises or sets.

Greenhouse effect   *See* Atmospheric greenhouse effect.

Ground fog   *See* Radiation fog.

Growing degree-day   A form of the degree-day used as a guide for crop planting and for estimating crop maturity dates.

Gulf stream   A warm, swift, narrow ocean current flowing along the east coast of the United States.

Gust front   A boundary that separates a cold downdraft of a thunderstorm from warm, humid surface air. On the surface its passage resembles that of a cold front.

Gustnado   A relatively weak tornado associated with a thunderstorm's outflow. It most often forms along the gust front.

Gyre   A large circular, surface ocean current pattern.

Haboob   A dust or sandstorm that forms as cold downdrafts from a thunderstorm turbulently lift dust and sand into the air.

Hadley cell   A thermal circulation proposed by George Hadley to explain the movement of the trade winds. It consists of rising air near the equator and sinking air near 30° latitude.

Hailstones   Transparent or partially opaque particles of ice that range in size from that of a pea to that of golf balls.

Hailstreak   The accumulation of hail at the earth's surface along a relatively long (10 km), narrow (2 km) band.

Hair hygrometer   *See* Hygrometer.

Halos   Rings or arcs that encircle the sun or moon when seen through an ice crystal cloud or a sky filled with falling ice crystals. Halos are produced by refraction of light.

Haze   Fine dry or wet dust or salt particles dispersed through a portion of the atmosphere. Individually these are not visible but cumulatively they will diminish visibility. *Dry haze* particles are very small, on the order of 0.1 μm. *Wet haze* particles are larger.

Heat   A form of energy transferred between systems by virtue of their temperature differences.

Heat capacity   The ratio of the heat absorbed (or released) by a system to the corresponding temperature rise (or fall).

Heat index (HI)   An index that combines air temperature and relative humidity to determine an apparent temperature—how hot it actually feels.

Heating degree-day   A form of the degree-day used as an index for fuel consumption.

Heat lightning   Distant lightning that illuminates the sky but is too far away for its thunder to be heard.

Heatstroke   A physical condition induced by a person's overexposure to high air temperatures, especially when accompanied by high humidity.

Heiligenschein   A faint white ring surrounding the shadow of an observer's head on a dew-covered lawn.

Heterosphere   The region of the atmosphere above about 85 km where the composition of the air varies with height.

High   *See* Anticyclone.

High inversion fog   A fog that lifts above the surface but does not completely dissipate because of a strong inversion (usually subsidence) that exists above the fog layer.

Homogenous freezing   The freezing of pure water. For tiny cloud droplets, homogeneous freezing does not occur until the air temperature reaches about −40°C.

Homosphere   The region of the atmosphere below about 85 km where the composition of the air remains fairly constant.

Hook echo   The shape of an echo on a radar screen that indicates the possible presence of a tornado.

Horse latitudes   The belt of latitude at about 30° to 35° where winds are predominantly light and the weather is hot and dry.

Humid continental climate   A climate characterized by severe winters and mild to warm summers with adequate annual precipitation. Typically located over large continental areas in the Northern Hemisphere between about 40° and 70° latitude.

Humidity   A general term that refers to the air's water vapor content. (*See* Relative humidity.)

Humid subtropical climate   A climate characterized by hot muggy summers, cool to cold winters, and abundant precipitation throughout the year.

Hurricane   A severe tropical cyclone having winds in excess of 64 knots (74 mi/hr).

Hurricane warning A warning given when it is likely that a hurricane will strike an area within 24 hours.

Hurricane watch A hurricane watch indicates that a hurricane poses a threat to an area (often within several days) and residents of the watch area should be prepared.

Hydrocarbons Chemical compounds composed of only hydrogen and carbon—they are included under the general term volatile organic compounds (VOCs).

Hydrologic cycle A model that illustrates the movement and exchange of water among the earth, atmosphere, and oceans.

Hydrophobic The ability to resist the condensation of water vapor. Usually used to describe "water-repelling" condensation nuclei.

Hydrostatic equation An equation that states that the rate at which the air pressure decreases with height is equal to the air density times the acceleration of gravity. The equation relates to how quickly the air pressure decreases in a column of air.

Hydrostatic equilibrium The state of the atmosphere when there is a balance between the vertical pressure gradient force and the downward pull of gravity.

Hygrometer An instrument designed to measure the air's water vapor content. The sensing part of the instrument can be hair (*hair hygrometer*), a plate coated with carbon (*electrical hygrometer*), or an infrared sensor (*infrared hygrometer*).

Hygroscopic The ability to accelerate the condensation of water vapor. Usually used to describe condensation nuclei that have an affinity for water vapor.

Hypothermia The deterioration in one's mental and physical condition brought on by a rapid lowering of human body temperature.

Hypoxia A condition experienced by humans when the brain does not receive sufficient oxygen.

Ice Age *See* Pleistocene epoch.

Ice crystal process A process that produces precipitation. The process involves tiny ice crystals in a supercooled cloud growing larger at the expense of the surrounding liquid droplets. Also called the *Bergeron process.*

Ice fog A type of fog that forms at very low temperatures, composed of tiny suspended ice particles.

Icelandic low The subpolar low-pressure area that is centered near Iceland on charts that show mean sea level pressure.

Ice nuclei Particles that act as nuclei for the formation of ice crystals in the atmosphere.

Ice pellets *See* Sleet.

Indian summer An unseasonably warm spell with clear skies near the middle of autumn. Usually follows a substantial period of cool weather.

Inferior mirage *See* Mirage.

Infrared radiation Electromagnetic radiation with wavelengths between about 0.7 and 1000 μm. This radiation is longer than visible radiation but shorter than microwave radiation.

Infrared radiometer An instrument designed to measure the intensity of infrared radiation emitted by an object. Also called *infrared sensor.*

Insolation The *in*coming *sol*ar rad*iation* that reaches the earth and the atmosphere.

Instrument shelter A boxlike wooden structure designed to protect weather instruments from direct sunshine and precipitation.

Interglacial period A time interval of relatively mild climate during the Ice Age when continental ice sheets were absent or limited in extent to Greenland and the Antarctic.

Intertropical convergence zone (ITCZ) The boundary zone separating the northeast trade winds of the Northern Hemisphere from the southeast trade winds of the Southern Hemisphere.

Inversion An increase in air temperature with height.

Ion An electrically charged atom, molecule, or particle.

Ionosphere An electrified region of the upper atmosphere where fairly large concentrations of ions and free electrons exist.

Iridescence Brilliant spots or borders of colors, most often red and green, observed in clouds up to about 30° from the sun.

Isallobar A line of equal change in atmospheric pressure during a specified time interval.

Isobar A line connecting points of equal pressure.

Isobaric map *See* Constant pressure chart.

Isobaric surface A surface along which the atmospheric pressure is everywhere equal.

Isotach A line connecting points of equal wind speed.

Isotherm A line connecting points of equal temperature.

Isothermal layer A layer where the air temperature is constant with increasing altitude. In an isothermal layer, the air temperature lapse rate is zero.

Jet maximum *See* Jet streak.

Jet streak A region of high wind speed that moves through the axis of a jet stream. Also called *jet maximum.*

Jet stream Relatively strong winds concentrated within a narrow band in the atmosphere.

Katabatic (fall) wind Any wind blowing downslope. It is usually cold.

Kelvin A unit of temperature. A Kelvin is denoted by K and 1 K equals 1°C. Zero Kelvin is absolute zero, or −273.15°C.

Kelvin scale A temperature scale with zero degrees equal to the theoretical temperature at which all molecular motion ceases. Also called the *absolute scale.* The units are sometimes called "degrees Kelvin"; however, the correct SI terminology is "Kelvins," abbreviated K.

Kinetic energy The energy within a body that is a result of its motion.

Kirchhoff's law A law that states: good absorbers of a given wavelength of radiation are also good emitters of that wavelength.

Knot A unit of speed equal to 1 nautical mile per hour. 1 knot equals 1.15 mi/hr.

Köppen classification system A system for classifying climates developed by W. Köppen that is based mainly on annual and monthly averages of temperature and precipitation.

Lake breeze A wind blowing onshore from the surface of a lake.

Lake-effect snows Localized snowstorms that form on the downwind side of a lake. Such storms are common in late fall and early winter near the Great Lakes as cold, dry air picks up moisture and warmth from the unfrozen bodies of water.

Laminar flow A nonturbulent flow in which the fluid moves smoothly in parallel layers or sheets.

Land breeze A coastal breeze that blows from land to sea, usually at night.

La Niña A condition where the central and eastern tropical Pacific Ocean turns cooler than normal.

Lapse rate The rate at which an atmospheric variable (usually temperature) decreases with height. (*See* Environmental lapse rate.)

Latent heat The heat that is either released or absorbed by a unit mass of a substance when it undergoes a change of state, such as during evaporation, condensation, or sublimation.

Laterite A soil formed under tropical conditions where heavy rainfall leaches soluble minerals from the soil. This leaching leaves the soil hard and poor for growing crops.

Lee-side low Storm systems (extratropical cyclones) that form on the downwind (lee) side of a mountain chain. In the United States lee-side lows frequently form on the eastern side of the Rockies and Sierra Nevada mountains.

Lenticular cloud A cloud in the shape of a lens.

Level of free convection The level in the atmosphere at which a lifted air parcel becomes warmer than its surroundings in a conditionally unstable atmosphere.

Lidar An instrument that uses a laser to generate intense pulses that are reflected from atmospheric particles of dust and smoke. Lidars have been used to determine the amount of particles in the atmosphere as well as particle movement that has been converted into wind speed. Lidar means *light detection and ranging*.

Lifting condensation level (LCL) The level at which a parcel of air, when lifted dry adiabatically, would become saturated.

Lightning A visible electrical discharge produced by thunderstorms.

Liquid-in-glass thermometer *See* Thermometer.

Little Ice Age The period from about 1550 to 1850 when average global temperatures were lower, and alpine glaciers increased in size and advanced down mountain canyons.

Local winds Winds that tend to blow over a relatively small area; often due to regional effects, such as mountain barriers, large bodies of water, local pressure differences, and other influences.

Longwave radiation A term most often used to describe the infrared energy emitted by the earth and the atmosphere.

Longwaves in the westerlies A wave in the major belt of westerlies characterized by a long length (thousands of kilometers) and significant amplitude. Also called *Rossby waves*.

Low *See* Extratropical cyclone.

Low-level jet streams Jet streams that typically form near the earth's surface below an altitude of about 2 km and usually attain speeds of less than 60 knots.

Macroburst A strong downdraft (*downburst*) greater than 4 km wide that can occur beneath thunderstorms. A downburst less than 4 km across is called a *microburst*.

Macroclimate The general climate of a large area, such as a country.

Macroscale The normal meteorological synoptic scale for obtaining weather information. It can cover an area ranging from the size of a continent to the entire globe.

Magnetic storm A worldwide disturbance of the earth's magnetic field caused by solar disturbances.

Magnetosphere The region around the earth in which the earth's magnetic field plays a dominant part in controlling the physical processes that take place.

Mammatus clouds Clouds that look like pouches hanging from the underside of a cloud.

Marine climate A climate controlled largely by the ocean. The ocean's influence keeps winters relatively mild and summers cool.

Maritime air Moist air whose characteristics were developed over an extensive body of water.

Maritime polar air mass An air mass characterized by low temperatures and high humidity.

Maritime tropical air mass An air mass characterized by high temperatures and high humidity.

Mature thunderstorm The second stage in the three-stage cycle of an air-mass thunderstorm. This stage is characterized by heavy showers, lightning, thunder, and violent vertical motions inside cumulonimbus clouds.

Maunder minimum A period from about 1645 to 1715 when few, if any, sunspots were observed.

Maximum thermometer A thermometer with a small constriction just above the bulb. It is designed to measure the maximum air temperature.

Mean annual temperature The average temperature at any given location for the entire year.

Mean daily temperature The average of the highest and lowest temperature for a 24-hour period.

Mechanical turbulence Turbulent eddy motions caused by obstructions, such as trees, buildings, mountains, and so on.

Medieval climatic optimum A relatively warm period that lasted from about A.D. 950–1250, when air temperatures were especially warm, perhaps 1°C higher than at present.

Mediterranean climate *See* Dry-summer subtropical climate.

Mercury barometer A type of barometer that uses mercury to measure atmospheric pressure. The height of the mercury column is a measure of atmospheric pressure.

Meridional flow A type of atmospheric circulation pattern in which the north-south component of the wind is pronounced.

Mesoclimate The climate of an area ranging in size from a few acres to several square kilometers.

Mesocyclone A vertical column of cyclonically rotating air within a severe thunderstorm.

Mesohigh A relatively small area of high atmospheric pressure that forms beneath a thunderstorm.

Mesopause The top of the mesosphere. The boundary between the mesosphere and the thermosphere, usually near 85 km.

Mesoscale The scale of meteorological phenomena that range in size from a few km to about 100 km. It includes local winds, thunderstorms, and tornadoes.

Mesoscale convective complex (MCC) A large organized convective weather system comprised of a number of individual thunderstorms. The size of an MCC can be 1000 times larger than an individual air-mass thunderstorm.

Mesosphere The atmospheric layer between the stratosphere and the thermosphere. Located at an average elevation between 50 and 80 km above the earth's surface.

Meteogram A chart that shows how one or more weather variables has changed at a station over a given period of time or how the variables are likely to change with time.

Meteorology The study of the atmosphere and atmospheric phenomena as well as the atmosphere's interaction with the earth's surface, oceans, and life in general.

Microburst A strong localized downdraft (downburst) less than 4 km wide that occurs beneath thunderstorms. A strong downburst greater than 4 km across is called a *macroburst*.

Microclimate The climate structure of the air space near the surface of the earth.

Micrometer (μm) A unit of length equal to one-millionth of a meter.

Microscale The smallest scale of atmospheric motions.

Middle latitudes The region of the world typically described as being between 30° and 50° latitude.

Middle latitude cyclones *See* Extratropical cyclone.

Mid-Holocene maximum A period in geologic history (about 5000 to 6000 years ago) when global temperatures were perhaps 1°C higher than at present.

Milankovitch theory   A theory proposed by Milutin Milankovitch in the 1930s suggesting that changes in the earth's orbit were responsible for variations in solar energy reaching the earth's surface and climatic changes.

Millibar (mb)   A unit for expressing atmospheric pressure. Sea level pressure is normally close to 1013 mb.

Minimum thermometer   A thermometer designed to measure the minimum air temperature during a desired time period.

Mini-swirls   Small whirling eddies perhaps 30 to 100 m in diameter that form in a region of strong wind shear of a hurricane's eye wall. Same as *spin-up vortices*.

Mirage   A refraction phenomenon that makes an object appear to be displaced from its true position. When an object appears higher than it actually is, it is called a *superior mirage*. When an object appears lower than it actually is, it is an *inferior mirage*.

Mixed cloud   A cloud containing both water drops and ice crystals.

Mixing depth   The vertical extent of the mixing layer.

Mixing layer   The unstable atmospheric layer that extends from the surface up to the base of an inversion. Within this layer, the air is well stirred.

Mixing ratio   The ratio of the mass of water vapor in a given volume of air to the mass of dry air.

Moist adiabatic rate   The rate of change of temperature in a rising or descending saturated air parcel. The rate of cooling or warming varies but a common value of 6°C per 1000 m (3.3°F per 1000 ft) is used.

Moist adiabats   Lines on an adiabatic chart that show the moist adiabatic rate for rising and descending air.

Molecular viscosity   The small-scale internal fluid friction that is due to the random motion of the molecules within a smooth-flowing fluid, such as air.

Molecule   A collection of atoms held together by chemical forces.

Monsoon depressions   Weak low-pressure areas that tend to form in response to divergence in an upper-level jet stream. The circulation around the low strengthens the monsoon wind system and enhances precipitation during the summer.

Monsoon wind system   A wind system that reverses direction between winter and summer. Usually the wind blows from land to sea in winter and from sea to land in summer.

Mountain and valley breeze   A local wind system of a mountain valley that blows downhill (*mountain breeze*) at night and uphill (*valley breeze*) during the day.

Multicell storms   Thunderstorms in a line, each of which may be in a different stage of development.

Nacreous clouds   Clouds of unknown composition that have a soft, pearly luster and that form at altitudes about 25 to 30 km above the earth's surface. They are also called *mother-of-pearl clouds*.

Negative feedback mechanism   *See* Feedback mechanism.

Neutral stability (neutrally stable atmosphere)   An atmospheric condition that exists in dry air when the environmental lapse rate equals the dry adiabatic rate. In saturated air the environmental lapse rate equals the moist adiabatic rate.

NEXRAD   An acronym for *Next* Generation Weather *Rad*ar. The main component of NEXRAD is the WSR 88-D, Doppler radar.

Nimbostratus   A dark, gray cloud characterized by more or less continuously falling precipitation. It is not accompanied by lightning, thunder, or hail.

Nitric oxide (NO)   A colorless gas produced by natural bacterial action in soil and by combustion processes at high temperatures. In polluted air, nitric oxide can react with ozone and hydrocarbons to form other substances. In this manner, it acts as an agent in the production of photochemical smog.

Nitrogen (N_2)   A colorless and odorless gas that occupies about 78 percent of dry air in the lower atmosphere.

Nitrogen dioxide (NO_2)   A reddish-brown gas, produced by natural bacterial action in soil and by combustion processes at high temperatures. In the presence of sunlight, it breaks down into nitric oxide and atomic oxygen. In polluted air, nitrogen dioxide acts as an agent in the production of photochemical smog.

Nitrogen oxides (NO_X)   Gases produced by natural processes and by combustion processes at high temperatures. In polluted air, nitric oxide (NO) and nitrogen dioxide (NO_2) are the most abundant oxides of nitrogen, and both act as agents for the production of photochemical smog.

Noctilucent clouds   Wavy, thin, bluish-white clouds that are best seen at twilight in polar latitudes. They form at altitudes about 80 to 90 km above the surface.

Nocturnal inversion   *See* Radiation inversion.

Northeaster   A name given to a strong, steady wind from the northeast that is accompanied by rain and inclement weather. It often develops when a storm system moves northeastward along the coast of North America. Also called *Nor'easter*.

Northern lights   *See* Aurora.

Nowcasting   Short-term weather forecasts varying from minutes up to a few hours.

Nuclear winter   The dark, cold, and gloomy conditions that presumably would be brought on by nuclear war.

Nucleation   Any process in which the phase change of a substance to a more condensed state (such as condensation, deposition, and freezing) is initiated about a particle (nucleus) or at a certain locus.

Numerical weather prediction (NWP)   Forecasting the weather based upon the solutions of mathematical equations by high-speed computers.

Obliquity (of the earth's axis)   The tilt of the earth's axis. It represents the angle from the perpendicular to the plane of the earth's orbit.

Occluded front (occlusion)   A complex frontal system that ideally forms when a cold front overtakes a warm front. When the air behind the front is colder than the air ahead of it, the front is called a *cold occlusion*. When the air behind the front is milder than the air ahead of it, it is called a *warm occlusion*.

Oceanic front   A boundary that separates masses of water with different temperatures and densities.

Offshore wind   A breeze that blows from the land out over the water. Opposite of an onshore wind.

Omega high   A ridge in the middle or upper troposphere that has the shape of a Greek letter omega (Ω).

Onshore wind   A breeze that blows from the water onto the land. Opposite of an offshore wind.

Open wave   The stage of development of a wave cyclone (mid-latitude cyclonic storm) where a cold front and a warm front exist, but no occluded front. The center of lowest pressure in the wave is located at the junction of the two fronts.

Orchard heaters   Oil heaters placed in orchards that generate heat and promote convective circulations to protect fruit trees from damaging low temperatures. Also called *smudge pots*.

Ordinary thunderstorm   (also called *air-mass thunderstorm*) A thunderstorm produced by local convection within a conditionally unstable air mass. It often forms in the afternoon and does not reach the intensity of a severe thunderstorm.

Orographic uplift   The lifting of air over a topographic barrier. Clouds that form in this lifting process are called *orographic clouds*.

Outgassing   The release of gases dissolved in hot, molten rock.

Overrunning   A condition that occurs when air moves up and over another layer of air.

Oxygen (O_2)   A colorless and odorless gas that occupies about 21 percent of dry air in the lower atmosphere.

Ozone (O_3)   An almost colorless gaseous form of oxygen with an odor similar to weak chlorine. The highest natural concentration is found in the stratosphere where it is known as *stratospheric ozone*. It also forms in polluted air near the surface where it is the main ingredient of photochemical smog. Here, it is called *tropospheric ozone*.

Ozone hole   A sharp drop in stratospheric ozone concentration observed over the Antarctic during the spring.

Pacific high   *See* Subtropical high.

Parcel of air   An imaginary small body of air a few meters wide that is used to explain the behavior of air.

Parhelia   *See* Sundog.

Particulate matter   Solid particles or liquid droplets that are small enough to remain suspended in the air. Also called *aerosols*.

Pattern recognition   An analogue method of forecasting where the forecaster uses prior weather events (or similar weather map conditions) to make a forecast.

Permafrost   A layer of soil beneath the earth's surface that remains frozen throughout the year.

Persistence forecast   A forecast that the future weather condition will be the same as the present condition.

Photochemical smog   *See* Smog.

Photodissociation   The splitting of a molecule by a photon.

Photon   A discrete quantity of energy that can be thought of as a packet of electromagnetic radiation traveling at the speed of light.

Photosphere   The visible surface of the sun from which most of its energy is emitted.

Pileus cloud   A smooth cloud in the form of a cap. Occurs above, or is attached to, the top of a cumuliform cloud. Also called a *cap cloud*.

Pilot balloon   A small balloon that rises at a constant rate and is tracked by a theodolite in order to obtain wind speed and wind direction at various levels above the earth's surface.

Planetary boundary layer   *See* Atmospheric boundary layer.

Planetary scale   The largest scale of atmospheric motion. Sometimes called the *global scale*.

Plasma   *See* Solar wind.

Plate tectonics   The theory that the earth's surface down to about 100 km is divided into a number of plates that move relative to one another across the surface of the earth. Once referred to as continental drift.

Pleistocene Epoch (or Ice Age)   The most recent period of extensive continental glaciation that saw large portions of North America and Europe covered with ice. It began about 2 million years ago and ended about 10,000 years ago.

Polar easterlies   A shallow body of easterly winds located at high latitudes poleward of the subpolar low.

Polar front   A semipermanent, semicontinuous front that separates tropical air masses from polar air masses.

Polar front jet stream (Polar jet)   The jet stream that is associated with the polar front in middle and high latitudes. It is usually located at altitudes between 9 and 12 km.

Polar front theory   A theory developed by a group of Scandinavian meteorologists that explains the formation, development, and overall life history of cyclonic storms that form along the polar front.

Polar ice cap climate   A climate characterized by extreme cold, as every month has an average temperature below freezing.

Polar low   An area of low pressure that forms over polar water behind (poleward of) the main polar front.

Polar orbiting satellite   A satellite whose orbit closely parallels the earth's meridian lines and thus crosses the polar regions on each orbit.

Polar tundra climate   A climate characterized by extremely cold winters and cool summers, as the average temperature of the warmest month climbs above freezing but remains below 10°C (50°F).

Pollutants   Any gaseous, chemical, or organic matter that contaminates the atmosphere, soil, or water.

Pollutant standards index (PSI)   An index of air quality that provides daily air pollution concentrations. Intervals on the scale relate to potential health effects.

Positive feedback mechanism   *See* Feedback mechanism.

Potential energy   The energy that a body possesses by virtue of its position with respect to other bodies in the field of gravity.

Potential evapotranspiration (PE)   The amount of moisture that, if it were available, would be removed from a given land area by evaporation and transpiration.

Potential temperature   The temperature that a parcel of dry air would have if it were brought dry adiabatically from its original position to a pressure of 1000 mb.

Precession (of the earth's axis of rotation)   The wobble of the earth's axis of rotation that traces out the path of a cone over a period of about 23,000 years.

Precipitation   Any form of water particles—liquid or solid—that falls from the atmosphere and reaches the ground.

P/E index (precipitation-evaporation index)   An index that gives the long-range effectiveness of precipitation in promoting plant growth.

P/E ratio (precipitation-evaporation ratio)   An expression devised for the purpose of classifying climates; based on monthly totals of precipitation and evaporation.

Pressure   The force per unit area. *See also* Air pressure.

Pressure gradient   The rate of decrease of pressure per unit of horizontal distance. On the same chart, when the isobars are close together, the pressure gradient is steep. When the isobars are far apart, the pressure gradient is weak.

Pressure gradient force (PGF)   The force due to differences in pressure within the atmosphere that causes air to move and, hence, the wind to blow. It is directly proportional to the pressure gradient.

Pressure tendency   The rate of change of atmospheric pressure within a specified period of time, most often three hours. Same as *barometric tendency*.

Prevailing westerlies   The dominant westerly winds that blow in middle latitudes on the poleward side of the subtropical high-pressure areas. Also called *westerlies*.

Prevailing wind   The wind direction most frequently observed during a given period.

Primary air pollutants   Air pollutants that enter the atmospheric directly.

Probability forecast   A forecast of the probability of occurrence of one or more of a mutually exclusive set of weather conditions.

Prognostic chart (prog)   A chart showing expected or forecasted conditions, such as pressure patterns, frontal positions, contour height patterns, and so on.

Prominence   *See* Solar flare.

Psychrometer   An instrument used to measure the water vapor content of the air. It consists of two thermometers (dry bulb and wet bulb). After whirling the instrument, the dew point and relative humidity can be obtained with the aid of tables.

Radar   An electronic instrument used to detect objects (such as falling precipitation) by their ability to reflect and scatter microwaves back to a receiver.

Radiant energy (radiation)   Energy propagated in the form of electromagnetic waves. These waves do not need molecules to propagate them, and in a vacuum they travel at nearly 300,000 km per sec (186,000 mi per sec).

Radiational cooling   The process by which the earth's surface and adjacent air cool by emitting infrared radiation.

Radiation fog   Fog produced over land when radiational cooling reduces the air temperature to or below its dew point. It is also known as *ground fog* and *valley fog*.

Radiation inversion   An increase in temperature with height due to radiational cooling of the earth's surface. Also called a *nocturnal inversion*.

Radiative equilibrium temperature   The temperature achieved when an object, behaving as a black body, is absorbing and emitting radiation at equal rates.

Radiative forcing agent   Any factor (such as increasing greenhouse gases and variations in solar output) that can change the balance between incoming energy from the sun and outgoing energy from the earth and the atmosphere.

Radiometer   *See* Infrared radiometer.

Radiosonde   A balloon-borne instrument that measures and transmits pressure, temperature, and humidity to a ground-based receiving station.

Radon   A colorless, odorless, radioactive gas that forms naturally as uranium in soil and rock breaks down.

Rain   Precipitation in the form of liquid water drops that have diameters greater than that of drizzle.

Rainbow   An arc of concentric colored bands that spans a section of the sky when rain is present and the sun is positioned at the observer's back.

Rain gauge   An instrument designed to measure the amount of rain that falls during a given time interval.

Rain shadow   The region on the leeside of a mountain where the precipitation is noticeably less than on the windward side.

Rawinsonde observation   A radiosonde observation that includes wind data.

Reflection   The process whereby a surface turns back a portion of the radiation that strikes it.

Refraction   The bending of light as it passes from one medium to another.

Relative humidity   The ratio of the amount of water vapor in the air compared to the amount required for saturation (at a particular temperature and pressure). The ratio of the air's actual vapor pressure to its saturation vapor pressure.

Relative vorticity   *See* Vorticity.

Return stroke   The luminous lightning stroke that propagates upward from the earth to the base of a cloud.

Ribbon lightning   Lightning that appears to spread horizontally into a ribbon of parallel luminous streaks when strong winds are blowing parallel to the observer's line of sight.

Ridge   An elongated area of high atmospheric pressure.

Rime   A white or milky granular deposit of ice formed by the rapid freezing of supercooled water drops as they come in contact with an object in below-freezing air.

Riming   *See* Accretion.

Roll cloud   A dense, roll-shaped, elongated cloud that appears to slowly spin about a horizontal axis behind the leading edge of a thunderstorm's gust front.

Rossby waves   *See* Longwaves in the westerlies.

Rotor cloud   A turbulent cumuliform type of cloud that forms on the leeward side of large mountain ranges. The air in the cloud rotates about an axis parallel to the range.

Rotors   Turbulent eddies that form downwind of a mountain chain, creating hazardous flying conditions.

Saffir-Simpson scale   A scale relating a hurricane's central pressure and winds to the possible damage it is capable of inflicting.

St. Elmo's fire   A bright electric discharge that is projected from objects (usually pointed) when they are in a strong electric field, such as during a thunderstorm.

Sand dune   A hill or ridge of loose sand shaped by the winds.

Santa Ana wind   A warm, dry wind that blows into southern California from the east off the elevated desert plateau. Its warmth is derived from compressional heating.

Saturation (of air)   An atmospheric condition whereby the level of water vapor is the maximum possible at the existing temperature and pressure.

Saturation vapor pressure   The maximum amount of water vapor necessary to keep moist air in equilibrium with a surface of pure water or ice. It represents the maximum amount of water vapor that the air can hold at any given temperature and pressure. (*See* Equilibrium vapor pressure.)

Savanna   A tropical or subtropical region of grassland and drought-resistant vegetation. Typically found in tropical wet-and-dry climates.

Scales of motion   The hierarchy of atmospheric circulations from tiny gusts to giant storms.

Scattering   The process by which small particles in the atmosphere deflect radiation from its path into different directions.

Scintillation   The apparent twinkling of a star due to its light passing through regions of differing air densities in the atmosphere.

Sea breeze   A coastal local wind that blows from the ocean onto the land. The leading edge of the breeze is termed a *sea breeze front*.

Sea breeze convergence zone   A region where sea breezes, having started in different regions, flow together and converge.

Sea level pressure   The atmospheric pressure at mean sea level.

Secondary air pollutants   Pollutants that form when a chemical reaction occurs between a primary air pollutant and some other component of air. Tropospheric ozone is a secondary air pollutant.

Secondary low   A low pressure area (often an open wave) that forms near, or in association with, a main low-pressure area.

Seiches   Standing waves that oscillate back and forth over an open body of water.

Selective absorbers   Substances such as water vapor, carbon dioxide, clouds, and snow that absorb radiation only at particular wavelengths.

Semi-arid climate   A dry climate where potential evaporation and transpiration exceed precipitation. Not as dry as the arid climate. Typical vegetation is short grass.

Semipermanent highs and lows   Areas of high pressure (anticyclones) and low pressure (extratropical cyclones) that tend to persist at a particular latitude belt throughout the year. In the Northern Hemisphere, typically they shift slightly northward in summer and slightly southward in winter.

Sensible heat   The heat we can feel and measure with a thermometer.

Sensible temperature   The sensation of temperature that the human body feels in contrast to the actual temperature of the environment as measured with a thermometer.

Severe thunderstorms   Intense thunderstorms capable of producing heavy showers, flash floods, hail, strong and gusty surface winds, and tornadoes.

Sferics   Radio waves produced by lightning. A contraction of *atmospherics.*

Shear   *See* Wind shear.

Sheet lightning   Occurs when the lightning flash is not seen but the flash causes the cloud (or clouds) to appear as a diffuse luminous white sheet.

Shelf cloud   A dense, arch-shaped, ominous-looking cloud that often forms along the leading edge of a thunderstorm's gust front, especially when stable air rises up and over cooler air at the surface. Also called an *arcus cloud.*

Shelterbelt   A belt of trees or shrubs arranged as a protection against strong winds.

Shortwave (in the atmosphere)   A small wave that moves around longwaves in the same direction as the air flow in the middle and upper troposphere. Shortwaves are also called *shortwave troughs.*

Shortwave radiation   A term most often used to describe the radiant energy emitted from the sun, in the visible and near ultraviolet wavelengths.

Shower   Intermittent precipitation from a cumuliform cloud, usually of short duration but often heavy.

Siberian high   A strong, shallow area of high pressure that forms over Siberia in winter.

Sleet   A type of precipitation consisting of transparent pellets of ice 5 mm or less in diameter. Same as *ice pellets.*

Smog   Originally smog meant a mixture of smoke and fog. Today, smog means air that has restricted visibility due to pollution, or pollution formed in the presence of sunlight—*photochemical smog.*

Smog front (also smoke front)   The leading edge of a sea breeze that is contaminated with smoke or pollutants.

Smudge pots   *See* Orchard heaters.

Snow   A solid form of precipitation composed of ice crystals in complex hexagonal form.

Snow-albedo feedback   A positive feedback whereby increasing surface air temperatures enhance the melting of snow and ice in polar latitudes. This reduces the earth's albedo and allows more sunlight to reach the surface, which causes the air temperature to rise even more.

Snowflake   An aggregate of ice crystals that falls from a cloud.

Snow flurries   Light showers of snow that fall intermittently.

Snow grains   Precipitation in the form of very small, opaque grains of ice. The solid equivalent of drizzle.

Snow pellets   White, opaque, approximately round ice particles between 2 and 5 mm in diameter that form in a cloud either from the sticking together of ice crystals or from the process of accretion. Also called *graupel.*

Snow rollers   A cylindrical spiral of snow shaped somewhat like a child's muff and produced by the wind.

Snow squall (shower)   An intermittent heavy shower of snow that greatly reduces visibility.

Solar constant   The rate at which solar energy is received on a surface at the outer edge of the atmosphere perpendicular to the sun's rays when the earth is at a mean distance from the sun. The value of the solar constant is about two calories per square centimeter per minute or about 1376 W/m^2 in the SI system of measurement.

Solar flare   A rapid eruption from the sun's surface that emits high energy radiation and energized charged particles.

Solar wind   An outflow of charged particles from the sun that escapes the sun's outer atmosphere at high speed.

Solute effect   The dissolving of hygroscopic particles, such as salt, in pure water, thus reducing the relative humidity required for the onset of condensation.

Sonic boom   A loud explosive-like sound caused by a shock wave eminating from an aircraft (or any object) traveling at or above the speed of sound.

Sounding   An upper-air observation, such as a radiosonde observation. A vertical profile of an atmospheric variable such as temperature or winds.

Source regions   Regions where air masses originate and acquire their properties of temperature and moisture.

Southern oscillation   The reversal of surface air pressure at opposite ends of the tropical Pacific Ocean that occur during major El Niño events.

Specific heat   The ratio of the heat absorbed (or released) by the unit mass of the system to the corresponding temperature rise (or fall).

Specific humidity   The ratio of the mass of water vapor in a given parcel to the total mass of air in the parcel.

Spin-up vortices   Small whirling eddies perhaps 30 to 100 m in diameter that form in a region of strong wind shear in a hurricane's eye wall. Same as *mini-swirls.*

Squall line   A line of thunderstorms that form along a cold front or out ahead of it.

Stable air   *See* Absolutely stable atmosphere.

Standard atmosphere   A hypothetical vertical distribution of atmospheric temperature, pressure, and density in which the air is assumed to obey the gas law and the hydrostatic equation. The lapse rate of temperature in the troposphere is taken as 6.5°C/1000 m or 3.6°F/1000 ft.

Standard atmospheric pressure   A pressure of 1013.25 millibars (mb), 29.92 inches of mercury (Hg), 760 millimeters (mm) of mercury, 14.7 pounds per square inch (lb/in.²), or 1013.25 hectopascals (hPa).

Standard rain gauge   A nonrecording rain gauge with an 8-inch diameter collector funnel and a tube that amplifies rainfall by tenfold.

Stationary front   A front that is nearly stationary with winds blowing almost parallel and from opposite directions on each side of the front.

Station pressure   The actual air pressure computed at the observing station.

Steady-state forecast   A weather prediction based on the past movement of surface weather systems. It assumes that the systems will move in the same direction and at approximately the same speed as they have been moving. Also called *trend forecasting.*

Steam fog   *See* Evaporation (mixing) fog.

Stefan-Boltzman law   A law of radiation which states that the amount of radiant energy emitted from a unit surface area of an object (ideally a black body) is proportional to the fourth power of the object's absolute temperature.

Steppe   An area of grass-covered, treeless plains that has a semi-arid climate.

Stepped leader   An initial discharge of electrons that proceeds intermittently toward the ground in a series of steps in a cloud-to-ground lightning stroke.

Storm surge   An abnormal rise of the sea along a shore; primarily due to the winds of a storm, especially a hurricane.

Stratocumulus A low cloud, predominantly stratiform, with low, lumpy, rounded masses, often with blue sky between them.

Stratosphere The layer of the atmosphere above the troposphere and below the mesosphere (between 10 km and 50 km), generally characterized by an increase in temperature with height.

Stratospheric polar night jet A jet stream that forms near the top of the stratosphere over polar latitudes during the winter months.

Stratus A low, gray cloud layer with a rather uniform base whose precipitation is most commonly drizzle.

Streamline A line that shows the wind flow pattern.

Sublimation The process whereby ice changes directly into water vapor without melting.

Subpolar climate A climate observed in the Northern Hemisphere that borders the polar climate. It is characterized by severely cold winters and short, cool summers. Also known as *taiga climate* and *boreal climate*.

Subpolar low A belt of low pressure located between 50° and 70° latitude. In the Northern Hemisphere, this "belt" consists of the *Aleutian low* in the North Pacific and the *Icelandic low* in the North Atlantic. In the Southern Hemisphere, it exists around the periphery of the Antarctic continent.

Subsidence The slow sinking of air, usually associated with high-pressure areas.

Subsidence inversion A temperature inversion produced by compressional warming—the adiabatic warming of a layer of sinking air.

Subtropical front A zone of temperature transition in the upper troposphere over subtropical latitudes, where warm air carried poleward by the Hadley cell meets the cooler air of the middle latitudes.

Subtropical high A semipermanent high in the subtropical high-pressure belt centered near 30° latitude. The *Bermuda high* is located over the Atlantic Ocean off the east coast of North America. The *Pacific high* is located off the west coast of North America.

Subtropical jet stream The jet stream typically found between 20° and 30° latitude at altitudes between 12 and 14 km.

Suction vortices Small, rapidly rotating whirls perhaps 10 m in diameter that are found within large tornadoes.

Sulfate aerosols *See* Aerosols.

Sulfur dioxide (SO_2) A colorless gas that forms primarily in the burning of sulfur-containing fossil fuels.

Summer solstice Approximately June 21 in the Northern Hemisphere when the sun is highest in the sky and directly overhead at latitude $23\frac{1}{2}°N$, the Tropic of Cancer.

Sundog A colored luminous spot produced by refraction of light through ice crystals that appears on either side of the sun. Also called *parhelia*.

Sun pillar A vertical streak of light extending above (or below) the sun. It is produced by the reflection of sunlight off ice crystals.

Sunspots Relatively cooler areas on the sun's surface. They represent regions of an extremely high magnetic field.

Supercell storm An enormous severe thunderstorm whose updrafts and downdrafts are nearly in balance, allowing it to maintain itself for several hours. It can produce large hail and tornadoes.

Supercooled cloud (or cloud droplets) A cloud composed of liquid droplets at temperatures below 0°C (32°F). When the cloud is on the ground it is called *supercooled fog* or *cold fog*.

Superior mirage *See* Mirage.

Supernova A tremendous explosion of a massive star.

Supersaturation A condition whereby the atmosphere contains more water vapor than is needed to produce saturation with respect to a flat surface of pure water or ice, and the relative humidity is greater than 100 percent.

Super typhoon A tropical cyclone (typhoon) in the western Pacific that has sustained winds of 130 knots or greater.

Surface inversion *See* Radiation inversion.

Synoptic scale The typical weather map scale that shows features such as high- and low-pressure areas and fronts over a distance spanning a continent. Also called the *cyclonic scale*.

Taiga (boreal forest) The open northern part of the coniferous forest. Taiga also refers to subpolar climate.

Tangent arc An arc of light tangent to a halo. It forms by refraction of light through ice crystals.

Tcu An abbreviation sometimes used to denote a towering cumulus cloud (cumulus congestus).

Teleconnections A linkage between weather changes occurring in widely separated regions of the world.

Temperature The degree of hotness or coldness of a substance as measured by a thermometer. It is also a measure of the average speed or kinetic energy of the atoms and molecules in a substance.

Temperature inversion An increase in air temperature with height, often simply called an *inversion*.

Terminal velocity The constant speed obtained by a falling object when the upward drag on the object balances the downward force of gravity.

Texas norther A strong, cold wind from between the northeast and northwest associated with a cold outbreak of polar air that brings a sudden drop in temperature. Sometimes called a *blue norther*.

Theodolite An instrument used to track the movements of a pilot balloon.

Thermal A small, rising parcel of warm air produced when the earth's surface is heated unevenly.

Thermal belts Horizontal zones of vegetation found along hillsides that are primarily the result of vertical temperature variations.

Thermal circulations Air flow resulting primarily from the heating and cooling of air.

Thermal lows and thermal highs Areas of low and high pressure that are shallow in vertical extent and are produced primarily by surface temperatures.

Thermal tides Atmospheric pressure variations due to the uneven heating of the atmosphere by the sun.

Thermal turbulence Turbulent vertical motions that result from surface heating and the subsequent rising and sinking of air.

Thermograph An instrument that measures and records air temperature.

Thermometer An instrument for measuring temperature. The most common are liquid-in-glass, which have a sealed glass tube attached to a glass bulb filled with liquid.

Thermosphere The atmospheric layer above the mesosphere (above about 85 km) where the temperature increases rapidly with height.

Thunder The sound due to rapidly expanding gases along the channel of a lightning discharge.

Thunderstorm A local storm produced by cumulonimbus clouds. Always accompanied by lightning and thunder.

Tornado An intense, rotating column of air that protrudes from a cumulonimbus cloud in the shape of a funnel or a rope whose circulation is present on the ground. (*See* Funnel cloud.)

Tornado outbreak A series of tornadoes that forms within a particular region—a region that may include several states. Often associated with widespread damage and destruction.

Tornado vortex signature (TVS) An image of a tornado on the Doppler radar screen that shows up as a small region of rapidly changing wind directions inside a mesocyclone.

Tornado warning A warning issued when a tornado has actually been observed either visually or on a radar screen. It is also issued when the formation of tornadoes is imminent.

Tornado watch A forecast issued to alert the public that tornadoes may develop within a specified area.

Trace (of precipitation) An amount of precipitation less than 0.01 in. (0.025 cm).

Trade wind inversion A temperature inversion frequently found in the subtropics over the eastern portions of the tropical oceans.

Trade winds The winds that occupy most of the tropics and blow from the subtropical highs to the equatorial low.

Transpiration The process by which water in plants is transferred as water vapor to the atmosphere.

Tropical depression A mass of thunderstorms and clouds generally with a cyclonic wind circulation of between 20 and 34 knots.

Tropical disturbance An organized mass of thunderstorms with a slight cyclonic wind circulation of less than 20 knots.

Tropical easterly jet A jet stream that forms on the equatorward side of the subtropical highs near 15 km.

Tropical monsoon climate A tropical climate with a brief dry period of perhaps one or two months.

Tropical rain forest A type of forest consisting mainly of lofty trees and a dense undergrowth near the ground.

Tropical storm Organized thunderstorms with a cyclonic wind circulation between 35 and 64 knots.

Tropical wave A migratory wavelike disturbance in the tropical easterlies. Tropical waves occasionally intensify into tropical cyclones. They are also called *easterly waves*.

Tropical wet-and-dry climate A tropical climate poleward of the tropical wet climate where a distinct dry season occurs, often lasting for two months or more.

Tropical wet climate A tropical climate with sufficient rainfall to produce a dense tropical rain forest.

Tropopause The boundary between the troposphere and the stratosphere.

Tropopause jets Jet streams found near the tropopause, such as the polar front and subtropical jet streams.

Troposphere The layer of the atmosphere extending from the earth's surface up to the tropopause (about 10 km above the ground).

Trough An elongated area of low atmospheric pressure.

Turbulence Any irregular or disturbed flow in the atmosphere that produces gusts and eddies.

Twilight The time at the beginning of the day immediately before sunrise and at the end of the day after sunset when the sky remains illuminated.

Typhoon A hurricane that forms in the western Pacific Ocean.

Ultraviolet radiation Electromagnetic radiation with wavelengths longer than X-rays but shorter than visible light.

Unstable air *See* Absolutely unstable atmosphere.

Upslope fog Fog formed as moist, stable air flows upward over a topographic barrier.

Upslope precipitation Precipitation that forms due to moist, stable air gradually rising along an elevated plain. Upslope precipitation is common over the western Great Plains, especially east of the Rocky Mountains.

Upwelling The rising of water (usually cold) toward the surface from the deeper regions of a body of water.

Urban heat island The increased air temperatures in urban areas as contrasted to the cooler surrounding rural areas.

Valley breeze *See* Mountain breeze.

Valley fog *See* Radiation fog.

Vapor pressure The pressure exerted by the water vapor molecules in a given volume of air.

Veering wind The wind that changes direction in a clockwise sense—north to northeast to east, and so on.

Vernal equinox The equinox at which the sun approaches the Northern Hemisphere and passes directly over the equator. Occurs around March 20.

Virga Precipitation that falls from a cloud but evaporates before reaching the ground. (*See* Fall streaks.)

Viscosity The resistance of fluid flow. *See* Molecular viscosity *and* Eddy viscosity.

Visible radiation (light) Radiation with a wavelength between 0.4 and 0.7 μm.

Visible region *See* Visible radiation.

Visibility The greatest distance an observer can see and identify prominent objects.

Volatile organic compounds (VOCs) A class of organic compounds that are released into the atmosphere from sources such as motor vehicles, paints, and solvents. VOCs (which include hydrocarbons) contribute to the production of secondary pollutants, such as ozone.

Vorticity A measure of the spin of a fluid, usually small air parcels. *Absolute vorticity* is the combined vorticity due to the earth's rotation and the vorticity due to the air's circulation relative to the earth. *Relative vorticity* is due to the curving of the air flow and wind shear.

Wall cloud An area of rotating clouds that extends beneath a severe thunderstorm and from which a funnel cloud may appear. Also called a *collar cloud*.

Warm advection (or *warm air advection*) The transport of warm air by the wind from a region of higher temperatures to a region of lower temperatures.

Warm-core low A low-pressure area that is warmer at its center than at its periphery. Tropical cyclones exhibit this temperature pattern.

Warm front A front that moves in such a way that warm air replaces cold air.

Warm occlusion *See* Occluded front.

Warm sector The region of warm air within a wave cyclone that lies between a retreating warm front and an advancing cold front.

Water equivalent The depth of water that would result from the melting of a snow sample. Typically about 10 inches of snow will melt to 1 inch of water, producing a water equivalent of 10 to 1.

Waterspout A column of rotating wind over water that has characteristics of a dust devil and tornado.

Water vapor Water in a vapor (gaseous) form. Also called *moisture*.

Water vapor-temperature rise feedback A positive feedback whereby increasing surface air temperatures cause an increase in the evaporation of water from the oceans. Increasing concentrations of atmospheric water vapor enhance the greenhouse effect, which causes the surface air temperature to rise even more.

Watt (W) The unit of power in SI units where 1 watt is equivalent to 1 joule per second.

Wave cyclone An extratropical cyclone that forms and moves along a front. The circulation of winds about the cyclone tends to produce a wavelike deformation on the front.

Wavelength   The distance between successive crests, troughs, or identical parts of a wave.

Weather   The condition of the atmosphere at any particular time and place.

Weather elements   The elements of *air temperature, air pressure, humidity, clouds, precipitation, visibility,* and *wind* that determine the present state of the atmosphere, the weather.

Weather type forecasting   A forecasting method where weather patterns are categorized into similar groups or types.

Weather types   Certain weather patterns categorized into similar groups. Used as an aid in weather prediction.

Weather warning   A forecast indicating that hazardous weather is either imminent or actually occurring within the specified forecast area.

Weather watch   A forecast indicating that atmospheric conditions are favorable for hazardous weather to occur over a particular region during a specified time period.

Westerlies   The dominant westerly winds that blow in the middle latitudes on the poleward side of the subtropical high-pressure areas.

Wet-bulb depression   The difference in degrees between the air temperature (dry-bulb temperature) and the wet-bulb temperature.

Wet-bulb temperature   The lowest temperature that can be obtained by evaporating water into the air.

Wet haze   *See* Haze.

Whirlwinds   *See* Dust devils.

Wien's law   A law of radiation which states that the wavelength of maximum emitted radiation by an object (ideally a blackbody) is inversely proportional to the object's absolute temperature.

Wind   Air in motion relative to the earth's surface.

Wind-chill factor   The cooling effect of any combination of temperature and wind, expressed as the loss of body heat. Also called *wind-chill index.*

Wind direction   The direction *from which* the wind is blowing.

Wind machines   Fans placed in orchards for the purpose of mixing cold surface air with warmer air above.

Wind profiler   A Doppler radar capable of measuring the turbulent eddies that move with the wind. Because of this, it is able to provide a vertical picture of wind speed and wind direction.

Wind rose   A diagram that shows the percent of time that the wind blows from different directions at a given location over a given time.

Wind-sculptured trees   Trees whose branches are bent, twisted, and broken off on one side by strong prevailing winds. Also called *flag trees.*

Wind shear   The rate of change of wind speed or wind direction over a given distance.

Wind sock   A tapered fabric shaped like a cone that indicates wind direction by pointing away from the wind. Also called a *wind cone.*

Wind vane   An instrument used to indicate wind direction.

Windward side   The side of an object facing into the wind.

Wind waves   Water waves that form due to the flow of air over the water's surface.

Winter chilling   The amount of time the air temperature during the winter must remain below a certain value so that fruit and nut trees will grow properly during the spring and summer.

Winter solstice   Approximately December 21 in the Northern Hemisphere when the sun is lowest in the sky and directly overhead at latitude $23\frac{1}{2}°$S, the Tropic of Capricorn.

Xerophytes   Drought-resistant vegetation.

Younger-Dryas event   A cold episode that took place about 11,000 years ago, when average temperatures dropped suddenly and portions of the Northern Hemisphere reverted back to glacial conditions.

Zonal wind flow   A wind that has a predominate west-to-east component.

INDEX

PHOTO CREDITS

Chapter 13

Fig. 13.12 and 13.22 NOAA photos.
Fig. 13.25 Photo courtesy Steven Businger, University of Hawaii, Manoa.

Chapter 14

Fig. 14.1 Courtesy of NOAA/NWS.
Fig. 14.2 Photo courtesy Jan Null.
Fig. 14.3 Figure courtesy J. T. Johnson, National Severe Storms Laboratory, Norman, OK.
Fig. 14.7 National Weather Service/NOAA.
Fig. 14.8 Data taken from "Statistical Probabilities for a White Christmas," U.S. Department of Commerce.
Fig. 14.16 Courtesy NWS.
Figs. 14.19 and 14.21 NOAA photos.

Chapter 15

Figs. 15.2, 15.3, 15.4, 15.5, 15.10, 15.28, and 15.29 Photos by author.
Fig. 15.8 Photo by Ron Smith.
Fig. 15.9 Photo copyright Howard B. Bluestein.
Fig. 15.12 © Charles Doswell.
Fig. 15.13 Adapted from several sources including the National Weather Association's Meteorological monographs 1–93.
Figs. 15.14 and 15.19 NOAA photos.
Fig. 15.18 Wide World Photos.
Fig. 15.21 After *Thunderstorms, Vol. 2*, U.S. Government Printing Office, Washington, D.C., 1982.
Fig. 15.22 Data courtesy of NOAA.
Fig. 2 Geophysical Institute, University of Alaska, Fairbanks.
Fig. 15.25 Photo by Mike Spurgat.
Fig. 15.26 National Center for Atmospheric Research/University Corporation for Atmospheric Research/National Science Foundation. Photographer: D. Baumhefner.
Fig. 15.30 Global Atmospherics, Inc.
Fig. 3 Photo by Johnny Autery.
Fig. 15.31 Photo by Mary K. Hurley.

Fig. 15.32 Data courtesy of U.S. Department of Commerce.
Fig. 15.35 Photo copyright © Wade Balzer, Weatherstock.
Fig. 15.39 Photo © by Edi Ann Otto.
Fig. 15.40 National Center for Atmospheric Research/University Corporation for Atmospheric Research/National Science Foundation. Photographer: Jim Wilson.
Figs. 15.41 and 15.42 National Severe Storms Laboratory and the NEXRAD Operational Support Facility, courtesy Andy White.
Fig. 15.43 Photo copyright © Howard B. Bluestein.
Fig. 15.44 Doppler radar, National Weather Service.
Fig. 15.45 Courtesy Joseph H. Golden, NOAA.

Chapter 16

Fig. 16.2 NOAA photo.
Fig. 16.4 Image courtesy of Intellicast.com. © 1998 WSI Corporation. Reprinted with permission.
Fig. 16.5 Photo © Comstock.
Figs. 16.7 and 16.10 NOAA/National Weather Service.
Fig. 16.12 Photo © M. Laca, Weatherstock.
Figs. 16.14, 16.15, 16.16, and 16.17 NOAA photos.
Fig. 16.18 Photo by author.
Tables 16.1, 16.2, and 16.3 NOAA/National Hurricane Center.

Chapter 17

Fig. 17.1 NASA photo.
Fig. 17.3a and 17.3b Courtesy John Day and the University of Colorado Health Sciences Center.
Fig. 17.4 Photo by author.
Figs. 1 and 2 NASA photos.
Fig. 17.14 Photo by J. L. Medeiros.
Fig. 17.15 Photo by Jim Edwards, Riverside Press Enterprise.
Fig. 17.18 National Center for Atmospheric Research/University Corporation for Atmospheric Research/National Science Foundation. Photographer: Lee Klinger.
Fig. 17.20 After United States and Canada Work Group No. 2, 1982.
Fig. 17.21 © Don Hinrichsen.

Chapter 18

Fig. 18.7 © Frans Lanting, Minden Pictures.
Figs. 18.9, 18.12, and 18.23 Photos by J. L. Medeiros.
Figs. 18.11, 18.14, and 18.19 Photos by author.
Fig. 18.20 © Carr Clifton, Minden Pictures.
Fig. 18.25 © Michio Hoshino, Minden Pictures.

Chapter 19

Figs. 19.1, 19.2, and 19.16 Photos by author.
Fig. 19.3 Modified from *CLIMAP*, 1976.
Fig. 19.5 Modified from *Understanding Climate Change*, National Academy of Sciences, 1975.
Fig. 19.6 Modified from Robert T. Watson, et al., *Climate Change 1995: Impacts, Adaptations and Mitigation of Climate Change*, Cambridge University Press, Cambridge, England, 1996.
Fig.19.11 From J. M. Barnola, et al., "Vostok ice core provides 160,000-year record of atmospheric CO_2, *Nature* (1 Oct., 1987), 329: 460.
Fig. 19.12 Gamma Liaison.
Fig. 19.13 NASA photo.
Fig. 19.14 Data courtesy of John Christy, University of Alabama, Huntsville, and R. Spencer, NASA Marshall Space Flight Center.
Fig. 19.15 From V. Ramanathan, B. R. Barstrom, and E. F. Harrison, "Climate and the earth's radiation budget," Physics Today (May, 1989), Fig. 5, p. 27.
Fig. 19.17 © Frans Lanting, Minden Pictures.
Fig. 19.18 After F. B. Mitchell, et al., "Transient climate response to increasing sulphate aerosols and greenhouse gases." *Nature* (1995) 376: 501–504.

Cumulus Small, puffy clouds with relatively flat bases and limited vertical growth.

Cumulus congestus Cumulus clouds that show extensive vertical development. They may form in rows, such as these, or as individual clouds that appear as a head of cauliflower. Showery precipitation may fall from cumulus congestus.

Cumulus The sprouting clouds with pronounced vertical growth are *cumulus congestus*. The clouds with much smaller vertical extent are called *cumulus humilis*. The very small ragged-looking clouds are called *cumulus fractus*.

CLOUD CHART

Ahrens
Meteorology
Today

Nimbostratus A dark gray-looking cloud that often covers the entire sky. Steady rain or snow falls from its base. The fragmented cloud beneath the nimbostratus is *Stratus Fractus*, or *scud*.

Stratus This low, unifrom, grayish-looking cloud typically has a base below 2000 m (6500 ft). Drizzle may fall from its base. It is distinguished from altostratus in that the sun is not usually visible through stratus.

Stratocumulus A low, lumpy-looking wide spread cloud with dark and light shading. (If you hold your hand at arm's length, the individual cloud elements will appear to be about the size of your fist.)

Cirrus High ice crystal clouds blown by the wind. Typically, they are observed at altitudes above about 6000 m (20,000 ft).

Cirrostratus A wide-spread high cloud composed of ice crystals, that is normally white and usually covers a large portion of the sky. Sometimes halos around the sun or moon are the only indication of its presence.

Cirrocumulus These high clouds are small rounded white puffs composed of ice crystals. (The hole in the center may have been produced by a jet aircraft descending through the cloud deck.)

Altocumulus Middle clouds composed of water droplets and ice crystals at elevations usually between 2000 m and 7000 m (6500 ft and 23,000 ft) above the surface. (If you hold your hand at arm's length, the individual puffs will be about the size of your thumbnail.)

Altostratus This gray-looking water droplet and ice crystal middle cloud often blurs the sun, making it appear watery or "dimly visible".

Cumulonimbus This cumuliform cloud may develop vertically to great heights, often with a sprouting anvil-shaped top. It produces heavy showers of rain or snow, lightning and thunder, and strong, gusty surface winds.

Mammatus Downward moving air causes these clouds to hang from the base of another cloud—most often a cumulonimbus or an altostratus.

Lenticular These lens-shaped clouds often form in waves that develop downwind of a mountain. They normally remain in one place as the air rushes through them.